METHODS IN PHARMACOLOGY AND TOXICOLOGY

Series Editor
Y. James Kang
Department of Pharmacology and Toxicology
University of Louisville
Louisville, KY, USA

For further volumes:
http://www.springer.com/series/7653

Methods in Pharmacology and Toxicology publishes cutting-edge techniques, including methods, protocols, and other hands-on guidance and context, in all areas of pharmacological and toxicological research. Each book in the series offers time-tested laboratory protocols and expert navigation necessary to aid toxicologists and pharmaceutical scientists in laboratory testing and beyond. With an emphasis on details and practicality, *Methods in Pharmacology and Toxicology* focuses on topics with wide-ranging implications on human health in order to provide investigators with highly useful compendiums of key strategies and approaches to successful research in their respective areas of study and practice.

Green Chemistry in Drug Discovery

From Academia to Industry

Edited by

Paul F. Richardson

World Wide Medicinal Chemistry, Pfizer Inc., San Diego, CA, USA

Humana Press

Editor
Paul F. Richardson
World Wide Medicinal Chemistry
Pfizer Inc.
San Diego, CA, USA

ISSN 1557-2153 ISSN 1940-6053 (electronic)
Methods in Pharmacology and Toxicology
ISBN 978-1-0716-1577-5 ISBN 978-1-0716-1579-9 (eBook)
https://doi.org/10.1007/978-1-0716-1579-9

This Humana imprint is published by the registered company Springer Science+Business Media, LLC, part of Springer Nature.
The registered company address is: 1 New York Plaza, New York, NY 10004, U.S.A.

Preface

Since the landmark publication from Paul Anastas and John Warner detailing the 12 Principles of Green Chemistry over 20 years ago (Fig. 1), there has been an exponential growth in the implementation of sustainable strategies within the pharmaceutical industry [1]. A recent publication highlights Amgen's transformation within this space saving both time and money by making drugs in a more sustainable manner and represents a call to arms for industrial research stating that "Drug Companies must adopt Green Chemistry" [2]. The rationale provided herein links the Principles of Green Chemistry to the "Triple Bottom Line" model showing how each of these can have beneficial impacts for profit, people, and the planet [3]. Surveys have shown that while many pharmaceutical companies implement green chemistry initiatives to both manufacturing and process workflows, their subsequent uptake within discovery groups is often more flexible based on a desire not to limit the

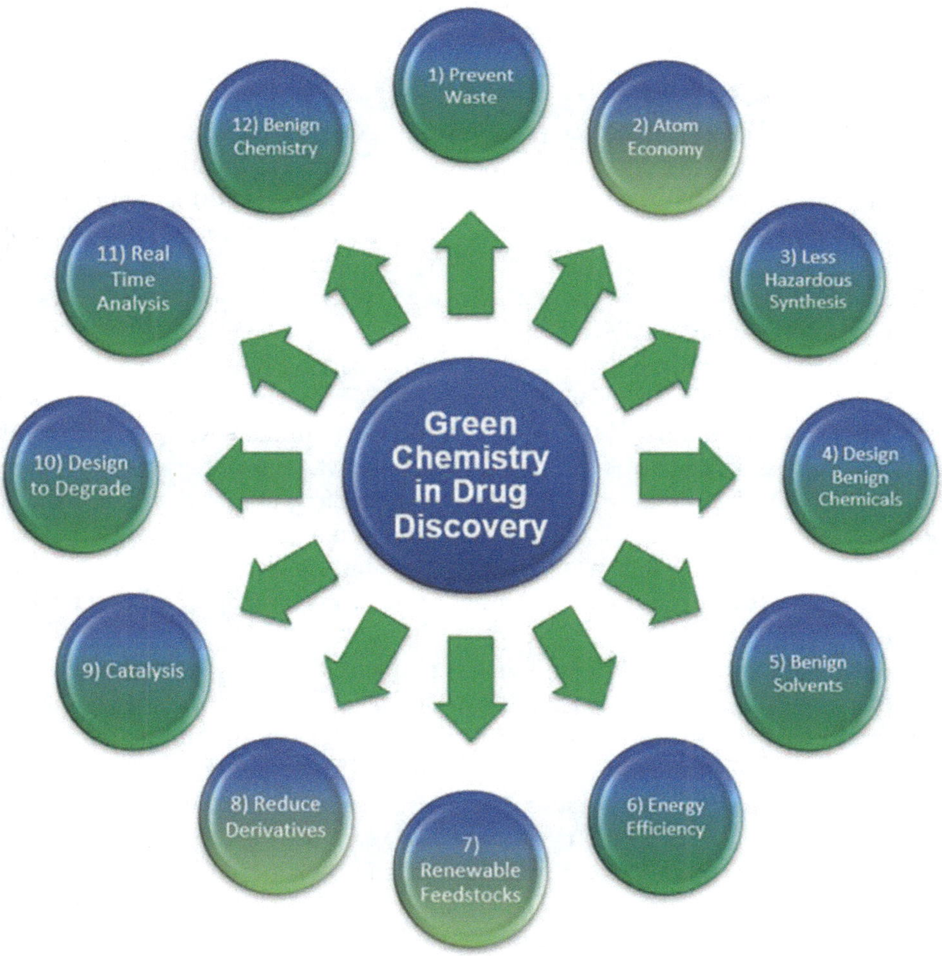

Fig. 1 Twelve principles of green chemistry

scientific freedom and innovation of chemists during this phase of medicinal chemistry programs [4].

There are some unique challenges presented to a medicinal chemist, specifically often the time pressure to deliver a compound for testing as well as the diverse scope of chemical substrate that one might be working on at any given time [5]. In addition, the relatively small scale of the analogue synthesis at this stage is often cited as a rationale for perhaps not making judicious choices in terms of sustainable reaction conditions. In addition, developing synthetic sequences that allow the variation of different vectors within a lead molecular series as well as the established poor probability of a compound's success of becoming a drug also often pushes the onus on addressing the challenges associated with the sustainability of a synthesis onto chemical development. However, in discovery, it should be noted that significant gains in this space are made in terms of making macroscopic changes with synthetic practices, for example, in solvent selections associated with chromatography.

When one considers "developing a sustainable synthetic route" to a compound, there are certain strategies one can implement [6]. In drug discovery, typically when trying to accelerate the development of a compound, the discovery route is adopted initially with modifications to crucially address any safety concerns while looking to make other changes, for example switching out undesirable solvents (and reagents) as well as eliminating chromatography. At this time, brainstorming sessions will be undertaken to evaluate and identify long-term synthetic approaches to the molecule suitable for manufacturing, which will place a critical focus on the availability of starting materials as well as overall cost of goods.

It should be stated right from the outset that this book is not intended as a comprehensive overview on implementation of green chemistry in medicinal chemistry drug discovery. There are numerous guides out there providing detailed guidance on, among other topics, reagent choice, solvent selection, catalysis, greener alternatives to chromatography, and comparing the sustainability of synthetic routes [7–9]. This collection though is intended to highlight several emerging areas with a specific focus on their application to the expeditious discovery of new biologically active entities.

The chapters herein are loosely divided into three parts. Initially, thought was given to delineating content through which of the Principles of Green Chemistry the topic was most closely aligned, though this idea was quickly abandoned as in many cases a topic can have a broader impact across several areas. To take an example, whereas Biocatalysis is well aligned with the ninth Principle to utilize catalysis, the reactions often minimize waste (Principle 1), are atom economical (Principle 2), are inherently benign and safe running under aqueous conditions (Principles 3, 5, and 12), and are typically conducted at room temperature (Principle 6) while adherence even to a number of the other Principles of Green Chemistry here can also be proposed.

Alternatively, the chapters of this collection are divided in a more arbitrary manner. The first part considers "Greener Approaches to Classical Transformations," focusing on several chemical transformations that are both prevalent and have been highlighted as challenging within the pharmaceutical industry [10, 11]. This is illustrated by chapters on both the synthesis of heterocycles and the synthesis of amides. Heterocyclic chemistry has represented the cornerstone of medicinal chemistry for over 100 years with numerous studies analyzing the impact of their incorporation on the physicochemical properties and potency of drugs [12, 13]. However, their syntheses often rely on classical methods that employ harsh reaction conditions and are not compatible with a wide scope of substrates. In a similar manner, a number of studies have demonstrated that the formation of amide bonds is the most common reaction carried out in medicinal chemistry programs with a reliance also on employing this transformation in the synthesis of combinatorial libraries owing to its robust nature [14, 15]. While reliable, the classical reaction to form an amide between a carboxylic acid and an amine typically involves use of a stoichiometric reagent to activate one of the substrates leading to subsequent generation of waste. The chapter herein not only describes a catalytic approach to amide bond formation but also enables the use of an alternative pool of monomers (alcohols as opposed to carboxylic acids) to allow the medicinal chemist to create this biologically relevant linkage in a different way.

In principle, the direct substitution of alcohols represents an ideal process as displacement of a hydroxyl group would be both atom-economical and lead to water as the sole by-product. The substitution of alcohols is often carried out in organic synthesis, but it is rarely a direct displacement, and developing improved sustainable approaches to achieve this goal is the topic of the next two chapters. The Mitsunobu reaction was first reported in 1967 and represents a robust method for the clean displacement of alcohols with subsequent inversion of configuration [16]. The reaction is mediated by stoichiometric amounts of an azodicarboxylate and a phosphine reagent proceeding through a relatively complex mechanism though being applicable to a range of nucleophiles in addition to the initially disclosed carboxylic acids. While this reaction has been widely used within the pharmaceutical industry, development efforts typically focus on facilitating the work-up to remove the by-products resulting from the reagents without resorting to chromatography. Development of a catalytic version of the Mitsunobu reaction would be a significant advance for this methodology. The other approach, which is typically used for the substitution of alcohols, involves stoichiometric activation through conversion to potentially genotoxic alkyl halides or sulfonates followed by subsequent displacement [17]. While there are both thermodynamic and kinetic barriers to affecting a direct displacement, progress has been made on methodologies based upon classical S_N1 and S_N2 reactions as well as redox hydrogen transfer approaches that lead to the desired overall net substitution reaction.

In a similar manner, the Friedel-Crafts set of reaction for introducing either alkyl or acyl substituents to aromatic nuclei through an electrophilic substitution process has been known since 1877 [18]. The reaction is typically mediated by a strong Lewis-Acid and is carried out under relatively forcing conditions (depending to a large extent on the electronic nature of the aromatic substrate). Again, while both a useful and robust transformation, there is considerable interest in developing new systems to mediate this transformation to allow both an increase in scope to more densely functionalized substrates and the reaction to proceed under milder conditions as well as generating less waste.

Through analysis of metrics, the use of solvents represents by far the most significant contributor to the process mass intensity (PMI) within the pharmaceutical industry [19]. With this, the optimization of solvent recycling and the quest to identify more reaction media with more environmentally benign properties represent important research initiatives. Though the earliest example of an ionic liquid is believed to have been reported in 1888, it is only with their emergence as potential electrolytes in batteries in the 1970s that has stimulated a growth in interest in their potential utility [20]. While their favorable properties such as low combustibility, thermal stability, and excellent solvating properties for a range of both nonpolar and polar compounds have led to use of ionic liquids for a range of applications, the ability to both easily tune the properties and functionalize the ionic liquid being used for a specific purpose is particularly attractive in a research setting.

The second part investigates topics with an impact on overall synthetic strategy with the first two considering approaches to primarily both maximize atom economy and minimize the generation of waste [21]. Whereas closely related, cascade reactions and multicomponent reactions are subtly different. The former (sometimes referred to as a domino or tandem reaction) involves at least two consecutive reactions in which each subsequent reaction occurs due to chemical functionality formed in the previous step without the need to isolate intermediates [22]. These reactions are often cited to mimic nature in terms of assembling complex molecules from relatively simple building blocks. It should be emphasized that often the telescoping of organic reactions is utilized in medicinal chemistry and this involves taking a compound into a discrete synthetic step (different conditions, new reagents) without intermediate purification, and is very different to a cascade process. A multicomponent reaction refers to a process where three or more components react to form a single product [23]. Mechanistically these consist usually of reactions occurring in a sequential manner with a high degree of selectivity with the products retaining a majority of atoms from the starting materials. Multicomponent reactions have been heavily exploited in medicinal chemistry in the synthesis of combinatorial libraries owing to the ability to generate a large collection of relatively complex drug-like molecules from simple starting materials in an expeditious manner.

The functionalization of C–H bonds represents an attractive option within medicinal chemistry as it can avoid numerous functional group manipulations as well as protection/deprotection sequences to access the molecules of interest in an atom-economical and waste-sparing fashion [24, 25]. However, there are significant hurdles to realize this transformation not only with the high activation energy required for C–H bond cleavage but also with regard to realizing the desired selectivity given the plethora of C–H moieties likely to be present in the substrate of interest. This subject has been an area of exponential growth within the chemistry community over the past 15–20 years with both transition-metal-mediated and more recently photocatalytic approaches to achieve C–H bond activation being described [26]. Typical strategies involve either the utilization of proximal metal-directing groups or incorporation of removable, modifiable, or traceless directing groups within the substrate to achieve selectivity. While challenges still exist in this area in addition to those noted such as the requirement to often add additional oxidants as well as control of the enantioselectivity of such processes, significant progress has been made with both approaches to achieving controlled C–H functionalization in complex systems.

While the topics so far within developing synthetic strategy have presented a general theme with regard to atom economy/waste minimization with a focus to some degree on avoiding unnecessary synthetic manipulations/isolations, the final topic further develops this idea by enabling regioselective transformations to be achieved through the in situ protection of functional groups that are likely to react preferentially or at best compete with the desired conversion. While masked during the reaction of interest, deprotection is generally achieved in a seamless fashion during reaction workup. One particularly elegant example of this approach is through masking an amine simply through protonation enabling reactions to take place selectively at alternative sites in the molecule of interest with the free amine liberated through adjustment of the pH after reaction.

The final part considers the implementation and impact of a range of enabling technologies within medicinal chemistry. While in this environment there are pressures both in terms of time and resource constraints, the utilization of technology can serve to bridge these gaps and realize efficiency gains. Catalysis has long been recognized as a mainstay within Green Chemistry programs with specifically the numerous successes realized through the implementation of enzymatic catalysis highlighted as a major achievement in reducing the environmental burden of the synthesis of complex active pharmaceuticals. Developments continue within this field through the ability to bioengineer enzymes to mediate specific transformations with exquisite levels of selectivity while also increasing the utility of biocatalysis through production of screening kits for use in discovery chemistry [27]. Without doubt, Pd-catalyzed cross-coupling for the formation of C-C bonds has revolutionized drug discovery programs over the past 30 years though, despite the low loadings of Pd that can be achieved in these processes, there are concerns regarding the long-term sustainability of this approach based not only on long-term Pd supply but also on the resource drain in accessing it [28]. New directions in coupling chemistry evaluate emerging methods based on Cu-, Fe-, and Ni-based catalysis and the potential reactivity and sustainability advantages that they possess [29].

One of the main challenges with running synthetic reactions is identifying the optimal conditions for a specific transformation, and this problem is exacerbated with emerging catalytic manifolds in which numerous options exist for catalysts, ancillary ligands, various additives, and choice of reaction solvents. Whereas systematic screening of such reaction variables has been routine within chemical development for some time, implementation of such workflows has been challenging within medicinal chemistry owing to both the scarcity of materials and the lack of appropriate markers for the desired products. These issues have largely been resolved through miniaturization enabling reactions to be run on sub-milligram scale coupled with high resolution analysis allowing 96-well plate high-throughput experimentation (HTE) to be realized [30, 31]. While such experimentation allows a broad overview of reaction parameters to be evaluated in a rapid manner, the speculative incorporation of more environmentally benign solvents/reagents allows greener options to be identified. In a similar manner, the emergence of flow chemistry presents many advantages within the pharmaceutical community [32, 33]. While from a manufacturing perspective this offers the opportunity for more precise reaction control with a reduced footprint for operations, within medicinal chemistry, the ability to access "forbidden" hazardous chemistries or promote transformations at high temperatures and pressures to explore novel chemical space are particularly enticing. The combined benefits from a reaction safety perspective and the ability to scale chemistry as a function of time as opposed to vessel size also make this technology attractive within the pharmaceutical industry.

PET imaging technology represents an invaluable tool utilized for disease diagnosis and progression as well as to the in vivo evaluation of drug therapies [34]. While the longer half-life of fluorine-18 (*ca.* 110 min) presents advantages for synthesis and imaging time, carbon-11 ($t_{1/2}$ *ca.* 20 min) based on the omnipresent nature of carbon can more frequently be substituted into biological molecules without causing chemical alterations that may influence imaging or interactions with the target of interest. While reagents derived from $[^{11}C]$-CO_2 have often been utilized in this manner, given the relatively short half-life, there are advantages to directly using $[^{11}C]$-CO_2 (given that this is the cyclotron-generated feedstock reagent) with robust synthetic methods emerging to enable its incorporation into relatively complex substrates under mild conditions.

The nature of the topics covered in this book is all currently of significant active interest to both the academic and medicinal chemistry research communities, and as such new developments are constantly being reported. However, it is hoped that the current collection will provide both a fundamental insight to the progress that has been made and some of the challenges that still exist for these to be effectively implemented in the drug discovery process in a routine manner.

San Diego, CA, USA ***Paul F. Richardson***

References

1. Anastas PT, Warner JC (1998) Green chemistry: theory and practice. Oxford University Press, New York, NY
2. Tucker JL, Faul MM (2016) Industrial research: drug companies must adopt green chemistry. Nature 534(7605):27–29
3. Elkington J (1998) Cannibals with forks: the triple bottom line of 21st century business. New Society Publishers, Gabriola Island, BC; Stony Creek, CT
4. Watson WJW (2012) How do the fine chemical, pharmaceutical, and related industries approach green chemistry and sustainability? Green Chem 14:251–259
5. Lombardino JG, Lowe JA III (2004) The role of the medicinal chemist in drug discovery - then and now. Nat Rev Drug Disc 3:853–862
6. Li J, Eastgate MD (2019) Making better decisions during synthetic route design: leveraging prediction to achieve greenness-by-design. React Chem Eng 4:1595–1607
7. Bryan MC, Dillon B, Hamann LG, Hughes GJ, Kopach ME, Peterson EA, Pourashraf M, Raheem I, Richardson P, Richter D, Sneddon HF (2013) Sustainable practices in medicinal chemistry: current state and future directions. J Med Chem 56(15):6007–6021
8. Aliagas I, Berger R, Goldberg K, Nishimura RT, Reilly J, Richardson P, Richter D, Sherer EC, Sparling BA, Bryan MC (2017) Sustainable practices in medicinal chemistry Part 2: Green by design. J Med Chem 60(14):5955–5968
9. Peterson EA, Dillon B, Raheem I, Richardson P, Richter D, Schmidt R, Sneddon HF (2014) Sustainable chromatography (an oxymoron?). Green Chem 16:4060–4075
10. Constable DJC, Dunn PJ, Hayler JD, Humphrey GR, Leazer JL Jr, Linderman RJ, Lorenz K, Manley J, Pearlman BA, Wells A, Zaks A, Zhang TY (2007) Key green chemistry research areas – a perspective from pharmaceutical manufacturers. Green Chem 9:411–420
11. Bryan MC, Dunn PJ, Entwistle D, Gallou F, Koenig SG, Hayler JD, Hickey MR, Hughes S, Kopach ME, Moine G, Richardson P, Roschangar F, Steven A, Weiberth FJ (2018) Key green chemistry research areas from a pharmaceutical manufacturers' perspective revisited. Green Chem 20:5082–5103
12. Vitaku E, Smith DT, Njardarson JT (2014) Analysis of the structural diversity, substitution patterns, and frequency of nitrogen heterocycles among U.S. FDA approved pharmaceuticals. J Med Chem 57 (24):10257–10274
13. Taylor AP, Robinson RP, Fobian YM, Blakemore DC, Jones LH, Fadeyi O (2016) Modern advances in heterocyclic chemistry in drug discovery. Org Biomol Chem 14:6611–6637
14. Brown DG, Boström J (2016) Analysis of past and present synthetic methodologies on medicinal chemistry: where have all the new reactions gone? J Med Chem 59:4443–4458
15. Boström J, Brown DG, Young RJ, Keserü GM (2018) Expanding the medicinal chemistry synthetic toolbox. Nat Rev Drug Disc 17:709–727
16. Fletcher S (2015) The Mitsunobu reaction in the 21st century. Org Chem Front 2:739–752
17. Szekely G, Amores de Sousa MC, Gil M, Ferreira FC, Heggie W (2015) Genotoxic impurities in pharmaceutical manufacturing: sources, regulations, and mitigation. Chem Rev 115(16):8182–8229
18. (2018) Charles Friedel (1832-1899) and James Mason Crafts (1839-1917): the Friedel-Crafts alkylation and acylation reactions. Synform 4:A49–A52
19. Jiminez-Gonzalez C, Ponder CS, Broxterman Q, Manley JB (2011) Using the right green yardstick: why process mass intensity is used in the pharmaceutical industry to drive more sustainable processes. Org Process Res Dev 15(4):912–917
20. Welton T (2018) Ionic liquids: a brief history. Biophys Rev 10:691–706
21. Hayashi Y (2016) Pot economy and one-pot synthesis. Chem Sci 7:866–880
22. Nicolaou KC, Chen JS (2009) The art of total synthesis through cascade reactions. Chem Soc Rev 38:2993–3009
23. Zhi S, Ma Z, Zhang W (2019) Consecutive multicomponent reactions for the synthesis of complex molecules. Org Biomol Chem 17:7632–7650

24. Abrams DJ, Provencher PA, Sorensen EJ (2018) Recent applications of C–H functionalization in complex natural product synthesis. Chem Soc Rev 47:8925–8967

25. Davies HML, Morton D (2016) Recent advances in C–H functionalization. J Org Chem 81(2):343–350

26. Revathi L, Ravindar L, Fang W-Y, Rakesh KP, Qin H-L (2018) Visible light-induced C–H bond functionalization: a critical review. Adv Synth Catal 360(24):4652–4698

27. Devine PN, Howard RM, Kumar R, Thompson MP, Truppo MD, Turner NJ (2018) Extending the application of biocatalysis to meet the challenges of drug development. Nat Chem 2:409–421

28. Johansson Seechurn CCC, Kitching MO, Colacot TJ, Snieckus V (2012) Palladium-catalyzed cross-coupling: a historical contextual perspective to the 2010 nobel prize. Angew Chem Int Ed 51 (21):5062–5085

29. Boit TB, Bulger AS, Dander JE, Garg NK (2020) Activation of C-O and C-N bonds using non-precious-metal catalysis. ACS Catal 10(20):12109–12126. See for example

30. Krska SW, DiRocco DA, Dreher SD, Shevlin M (2017) The evolution of chemical high-throughput experimentation to address challenging problems in pharmaceutical synthesis. Acc Chem Res 50 (12):2976–2985

31. Shevlin M (2017) Practical high-throughput experimentation for chemists. ACS Med Chem Lett 8 (6):601–607

32. Gioiello A, Piccinno A, Lozza AM, Cerra B (2020) Medicinal chemistry in the era of machines and automation: recent advances in continuous flow technology. J Med Chem 63:6624–6647

33. Guidi M, Seeberger PH, Gilmore K (2020) How to approach flow chemistry. Chem Soc Rev 49:8910–8932

34. Vaquero JJ, Kinahan P (2015) Positron emission tomography: current challenges and opportunities for technological advances in clinical and preclinical imaging systems. Annu Rev Biomed Eng 17:385–414

Contents

Contributors

JOEL W. BEATTY • *Department of Chemistry, University of Michigan, Ann Arbor, MI, USA*
SIMON BERRITT • *WRDM Pfizer Inc., Groton, CT, USA*
HYEJIN CHO • *Department of Chemistry, University of Massachusetts Boston, Boston, MA, USA*
RAZVAN CIOC • *Department of Chemistry & Pharmaceutical Sciences, Amsterdam Institute of Molecules, Medicines & Systems, VU University Amsterdam, Amsterdam, The Netherlands*
PIER GIORGIO COZZI • *Dipartimento di Chimica "G. Ciamician", ALMA MATER STUDIORUM, Università di Bologna, Bologna, Italy*
VY M. DONG • *Department of Chemistry, University of California, Irvine, CA, USA*
GARY M. GALLEGO • *Oncology Medicinal Chemistry, Pfizer, La Jolla, CA, USA*
REBECCA A. GALLEGO • *Oncology Medicinal Chemistry, Pfizer, La Jolla, CA, USA*
ANDREA GUALANDI • *Dipartimento di Chimica "G. Ciamician", ALMA MATER STUDIORUM, Università di Bologna, Bologna, Italy*
HE HUANG • *Department of Chemistry & Chemical Biology, University of New Mexico, Albuquerque, NM, USA*
EUSEBIO JUARISTI • *Departamento de Química, Centro de Investigación y de Estudios Avanzados, Instituto Politécnico Nacional, México D.F., México*
RAJESH KUMAR • *Pfizer Worldwide Research and Development, Chemical Research and Development, Groton, CT, USA*
NICHOLAS E. LEADBEATER • *Department of Chemistry, University of Connecticut, Storrs, CT, USA*
ELISABETTA MANONI • *Dipartimento di Chimica "G. Ciamician", ALMA MATER STUDIORUM, Università di Bologna, Bologna, Italy*
CARLOS A. MARTINEZ • *Pfizer Worldwide Research and Development, Chemical Research and Development, Groton, CT, USA*
LUCA MENGOZZI • *Dipartimento di Chimica "G. Ciamician", ALMA MATER STUDIORUM, Università di Bologna, Bologna, Italy*
JAMES J. MOUSSEAU • *Halda Therapeutics, New Haven, CT, USA*
ARTURO OBREGÓN-ZÚÑIGA • *Facultad de Ciencias Químicas, Universidad Autónoma de Nuevo León, San Nicolás de los Garza, Nuevo León, México*
NATHAN J. OLDENHUIS • *Department of Chemistry, University of New Hampshire, Durham, NH, USA*
ROMANO V. A. ORRU • *Department of Chemistry & Pharmaceutical Sciences, Amsterdam Institute of Molecules, Medicines & Systems, VU University Amsterdam, Amsterdam, The Netherlands; Aachen-Maastricht Institute for Biobased Materials (AMIBM), Maastricht University, Brightlands Chemelot Campus, Center Court, Urmonderbaan, Geleen, Netherlands*
ANA INÉS BELLOMO PERAZA • *Centro de Investigaciones en Bionanociencias (CIBION), Buenos Aires, Argentina*
PAUL F. RICHARDSON • *Oncology Medicinal Chemistry, Pfizer, La Jolla, CA, USA*

Benjamin H. Rotstein • *Department of Biochemistry, Microbiology and Immunology, University of Ottawa, Ottawa, ON, Canada; University of Ottawa Heart Institute, Ottawa, ON, Canada*
Eelco Ruijter • *Department of Chemistry & Pharmaceutical Sciences, Amsterdam Institute of Molecules, Medicines & Systems, VU University Amsterdam, Amsterdam, The Netherlands*
Christian Schäfer • *Department of Chemistry, University of Massachusetts Boston, Boston, MA, USA*
Jason R. Schmink • *Mount Hope Estate & Winery, Manheim, PA, USA*
Grigoriy Sereda • *Department of Chemistry, University of South Dakota, Vermillion, SD, USA*
Chunquan Sheng • *Department of Chemistry & Chemical Biology, University of New Mexico, Albuquerque, NM, USA; Department of Medicinal Chemistry, School of Pharmacy, Second Military Medical University, Shanghai, P. R. China*
Antonia F. Stepan • *pRED, Pharma Research & Early Development, Roche Innovation Center Basel, F. Hoffmann-La Roche Ltd., Basel, Switzerland*
Corey R. J. Stephenson • *Department of Chemistry, University of Michigan, Ann Arbor, MI, USA*
Alan Steven • *Pharmaceutical Technology and Development, AstraZeneca, Macclesfield, UK*
Tsuyoshi Taniguchi • *School of Pharmaceutical Sciences, Institute of Medical, Pharmaceutical and Health Sciences, Kanazawa University, Kanazawa, Japan; Graduate School of Natural Science and Technology, Kanazawa University, Kanazawa, Japan*
Béla Török • *Department of Chemistry, University of Massachusetts Boston, Boston, MA, USA*
Neil Vasdev • *Azrieli Centre for Neuro-Radiochemistry, Brain Health Imaging Centre, Centre for Addiction and Mental Health & Department of Psychiatry, University of Toronto, Toronto, ON, Canada*
Wei Wang • *Department of Chemistry & Chemical Biology, University of New Mexico, Albuquerque, NM, USA; College of Pharmacy, University of Arizona, Tucson, AZ, USA*
Aaron M. Whittaker • *Department of Process Research and Development, MRL, Merck & Co. Inc., Rahway, NJ, USA*
Claire Margaret Wilson • *School of Chemistry, Faculty of Science, Agriculture and Engineering, Newcastle University, Newcastle upon Tyne, UK*
John W. Wong • *Pfizer Worldwide Research and Development, Chemical Research and Development, Groton, CT, USA*
Chenguang Yu • *Department of Chemistry & Chemical Biology, University of New Mexico, Albuquerque, NM, USA*

Part I

Greener Approaches to Classical Transformations

Chapter 1

Green Synthesis of Common Heterocycles

Christian Schäfer, Hyejin Cho, and Béla Török

Abstract

Recent advances in the environmentally benign synthesis of common heterocycles are described. This account features three main parts: the preparation of non-aromatic heterocycles, one-ring aromatic heterocycles, and their condensed analogs. Due to the great variety of and high interest in these compounds, the work focuses on providing representative examples of the preparation of the target compounds.

Key words Sustainable synthesis, Non-aromatic heterocycles, Aromatic heterocycles, Condensed heterocycles

1 Introduction

Heterocyclic compounds possess a broad range of applications in a variety of industries from the pharmaceutical to agrochemical, dyestuff, or polymer industry. Their widespread use is well illustrated by the extremely high ratio (>70%) of heterocyclic compound among drugs. There is a vast array of methods available for the synthesis of heterocycles. Many of these syntheses, including most classical methods, however, do not comply with the ever strengthening rules of environmental safety and sustainable production. Due to this pressure, the syntheses of heterocycles, among other compounds, have been revamped, and a number of environmentally sustainable methods have been developed over the past two decades. Although due to the large body of work regarding the synthesis of heterocycles it is not possible to cover every development, an effort will be made to provide, as comprehensive as possible, an account that will focus on the environmentally benign synthesis of these compounds. In addition, emphasis will be placed on the green synthesis of the most common heterocycles [1].

Paul F. Richardson (ed.), *Green Chemistry in Drug Discovery: From Academia to Industry*, Methods in Pharmacology and Toxicology, https://doi.org/10.1007/978-1-0716-1579-9_1,

2 Non-aromatic Heterocycles

Non-aromatic heterocyclic compounds are versatile organic compounds that are used as building blocks in a multitude of reactions. Besides their synthetic utility, they are found in biologically active natural products and synthetic derivatives [2].

2.1 3-Membered Rings

The most common examples of the 3-membered heteroatom containing rings are the epoxides and aziridines that are important starting materials in synthetic applications and also structural motifs in several bioactive compounds. Accordingly, the synthesis and chemical properties of such compounds are regularly reviewed [3, 4].

Due to the importance of epoxides as synthetic building blocks as well as reagents for multiple purposes, their synthesis attracted extensive attention. The most typical preparation of epoxides is the oxidation of a C=C double bond by oxidants (Scheme 1). The oxidation is commonly carried out by hydrogen peroxide or molecular oxygen as environmentally benign oxidants.

The key aspects of the process are using hydrogen peroxide or oxygen as an oxidant and the catalyst that will enable the alkene and the oxidants to react and form the epoxide. Thus, finding the best performing catalyst has been a top priority [5]. Several different catalysts have been recently introduced for the synthesis of epoxides such as gold nanoparticles/O_2 [6], Ti-Beta zeolites/H_2O_2 [7], supported gold catalysts [8], ionic liquids [9, 10], Fe(III)/H_2O_2 [11], or Pd + Pt/titanium silicalite [12].

The enantioselective catalytic oxidation of alkenes had been developed by Sharpless et al. [13]. Finding environmentally better catalysts for this reaction is in the forefront of enantioselective catalysis [14]. Shibata et al. have developed a method that is representative of the progress being made in enantioselective catalysis. The epoxidation of α,β-unsaturated carbonyl compounds was carried out using organocatalysis, namely, methylhydrazine as an initiator and a cinchona derivative as a chiral modifier (Scheme 2) [15]. The reaction provided the products with excellent yields (91–99%) and very high ee's (>97%).

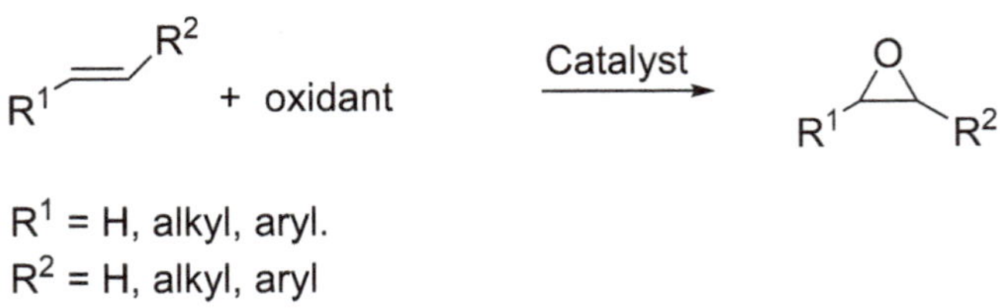

Scheme 1 General prepation of epoxides

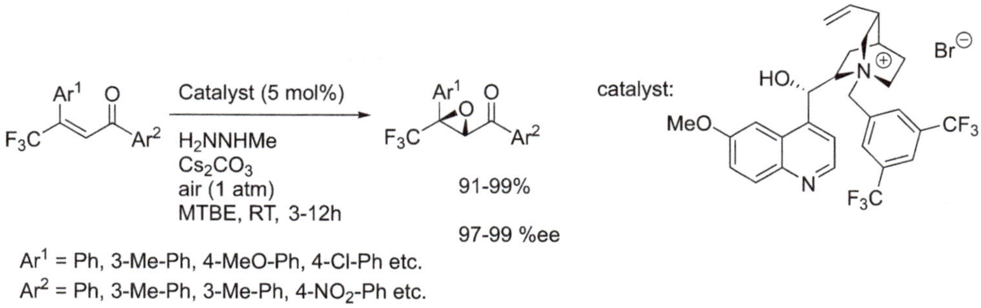

Scheme 2 Enantioselective epoxidation by organocatalysis

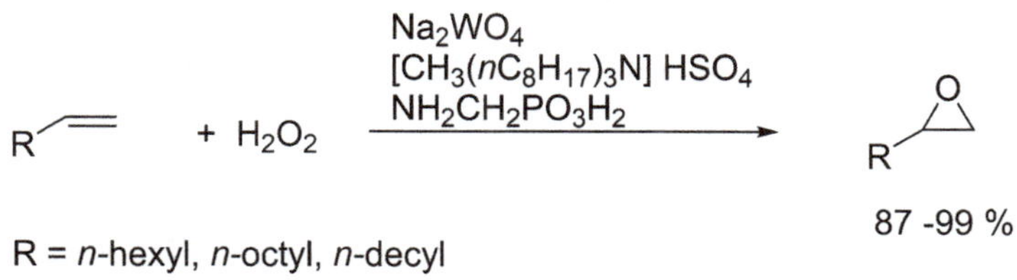

Scheme 3 A halide-free epoxidation with hydrogen peroxide

The epoxidation has also been carried out by enzymatic catalysis. In a recent review. Aouf et al. summarized lipase-catalyzed epoxidation of fatty acids and phenolic compounds [16]. Other enzymes used for epoxidation include peroxygenase [17]. The epoxidation of fatty acids from soybean oil appeared to benefit from the application of ultrasonic irradiation [18]. Noyori and co-workers developed an environmentally benign epoxidation process based on the application of hydrogen peroxide as an oxidant and sodium tungstate dihydrate, (aminomethyl)phosphonic acid, and methyltrioctylammonium hydrogen sulfate [19, 20] as the catalytic system (Scheme 3). While the substrates were limited to terminal olefins, the process can be carried out without a solvent and also is applicable at industrial scale.

The three major pathways applied to prepare aziridines are the addition of a carbene equivalent to a C=N double bond, the addition of an azide or other suitable N-source to a C=C double bond, or the intramolecular cyclization of amine derivatives (Scheme 4). Advances in the stereoselective aziridination have been recently reviewed [21].

An example of the first method is well illustrated by the azide addition to styrenes in the presence of a Co-complex catalyst (Scheme 5) [22]. The reaction provided the products in good to excellent yields and with excellent enantioselectivities. While most of these methods apply reagents that are generally considered to be of high reactivity, they represent a high atom economy approach for the synthesis of aziridines [23]. A similar methodology is followed by Cramer and Jenkins; the synthesis of aziridines is carried out

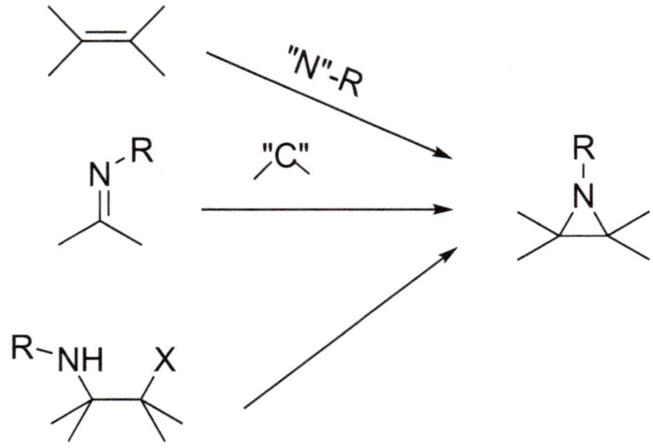

Scheme 4 Three major pathways for stereoselective aziridination

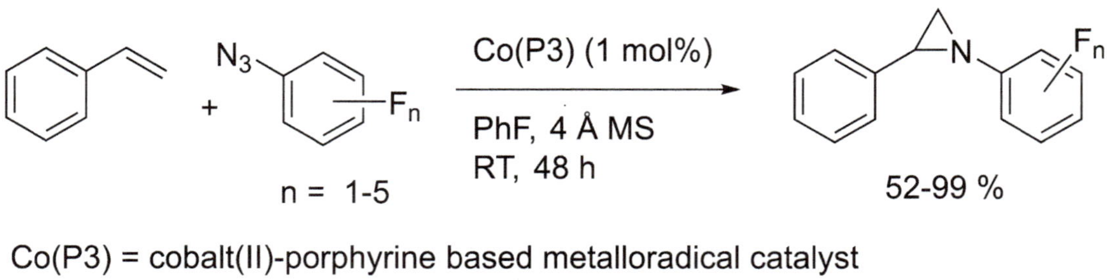

Scheme 5 Enantioselective aziridination of alkenes with fluoroaryl azides

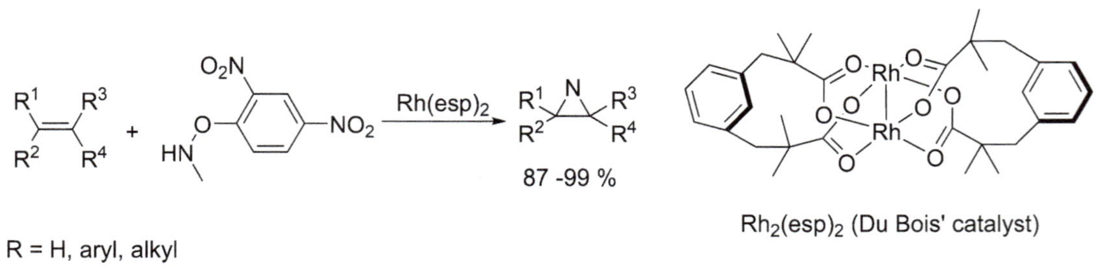

Scheme 6 Stereospecific aziridination by the Du Bois's catalyst from olefins

from alkenes and aryl azides with a reusable macrocyclic tetracarbene iron catalyst in moderate to excellent yields [24]. There are several other available transition metal catalysts for this reaction as summarized in a recent account [25].

A very recent article by Jat et al. [26] describes the application of the Du Bois' catalyst in the aziridination of a broad variety of alkenes (Scheme 6). The method appears generally applicable to alkenes of various structures, including styrenes, terminal and non-terminal aliphatic alkenes, stilbenes, and cyclic alkenes as well.

The second method, the addition of a suitable C-source, appears significantly more applicable for green synthesis. The

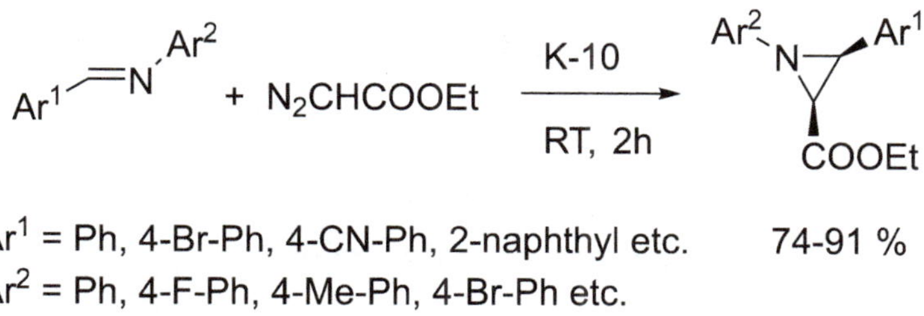

Scheme 7 K-10 catalyzed diastereoselective synthesis of aziridines

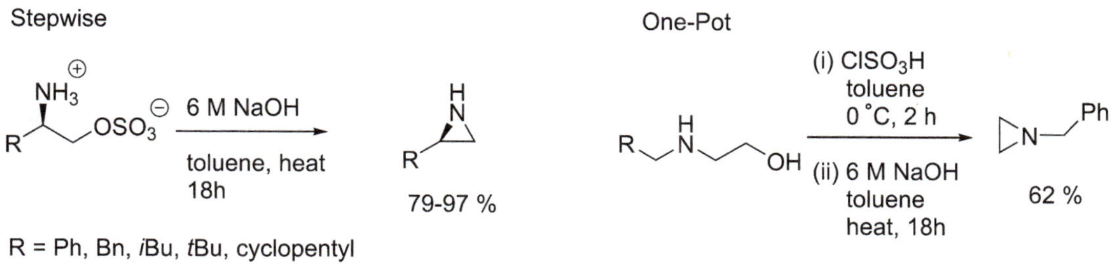

Scheme 8 Modified Wenker synthesis both in a stepwise and one-pot conditions

stereospecific synthesis of *cis*-aziridines from imines and ethyl diazoacetate has been described by Borkin et al. (Scheme 7). The reaction was catalyzed by an environmentally friendly solid acid, K-10 montmorillonite. The reaction readily occurred at room temperature and was completed in a relatively short time (2 h). In addition to the high yields (74–91%), the *cis* product is formed with exclusive selectivity (>99%) [27].

The third approach, an intramolecular cyclization, is commonly a nucleophilic substitution by an amino group on α-carbon atoms bearing a good leaving group. Buckley et al. applied the modified Wenker synthesis both in a stepwise manner and under one-pot conditions by using chlorosulfonic acid to transform the OH group of the aminoalcohol to a sulfate ester which will readily undergo cyclization under mild conditions (Scheme 8) [28].

2.2 4-Membered Rings

There has been significantly less emphasis placed on the synthesis of oxetanes and azetidines. Oxetanes are commonly synthesized by the photochemical Paterno-Buchi reaction [29, 30] or the Mitsonobu type cyclization [31] that is strongly similar to the approach depicted in Scheme 6.

Azetidines have been synthesized by a similar intramolecular cyclization of unsaturated amines via iodocyclization (Scheme 9). While the reaction provided the products in good to excellent yields, it appears that the regioselectivity of the amine addition to the C=C double bond is strongly dependent on the reaction temperature. At room temperature, the azetidine forms with exclusive selectivity; however, elevated temperatures result in the formation

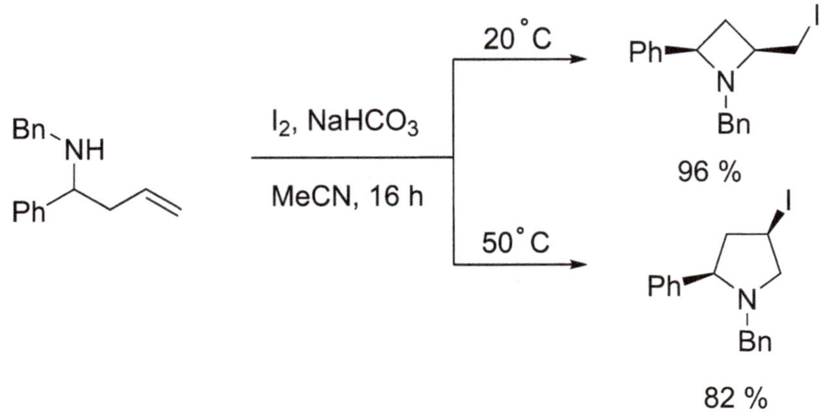

Scheme 9 Synthesis of azetidines and pyrrolidines via iodocyclization at different temperatures

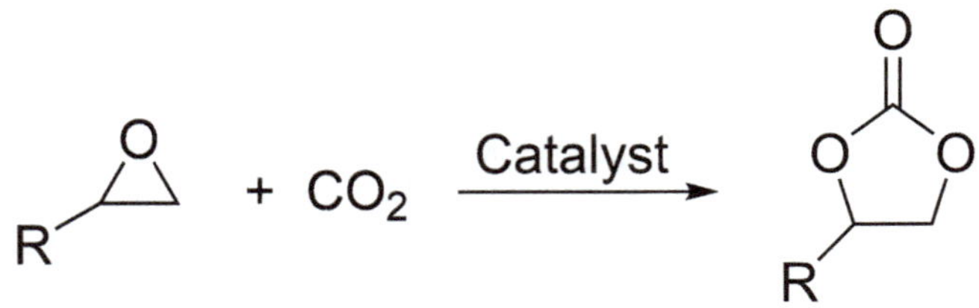

Scheme 10 General preparation of cyclic carbonates from epoxides with CO_2 via catalysis

of pyrrolidines [32]. A similar iodine-mediated approach resulted in the stereoselective formation of azetidines from α-amidomalonate and enones [33].

2.3 5-Membered Rings

Among the 5-membered rings, the synthesis of pyrrolidines attracted the most sizeable attention; with regard to other such heterocyclic systems, the synthesis of cyclic carbonates from epoxides and oxazolidinones from aziridines are the most common.

Epoxides react readily with CO_2 in the presence of a broad range of catalysts (Scheme 10). Yang et al. described the application of zinc clusters in this reaction without the use of any organic solvents and under mild conditions (1 atm, 25 °C) [34]. Other catalysts include a metal–organic framework Cr-MIL-101 [35].

Oxazolidines efficiently form when aziridines react in a catalytic reaction with CO_2 (Scheme 11). A broad variety of oxazolidines were prepared by using recyclable protic onium salts in good to excellent yields [36]. However, a recent study pointed out that the reaction, in fact, occurs without the application of any solvent or catalyst at room temperature. Although the room temperature may be an excessive claim since the mechanochemical activation generates heat as the authors used high speed ball milling, a newly

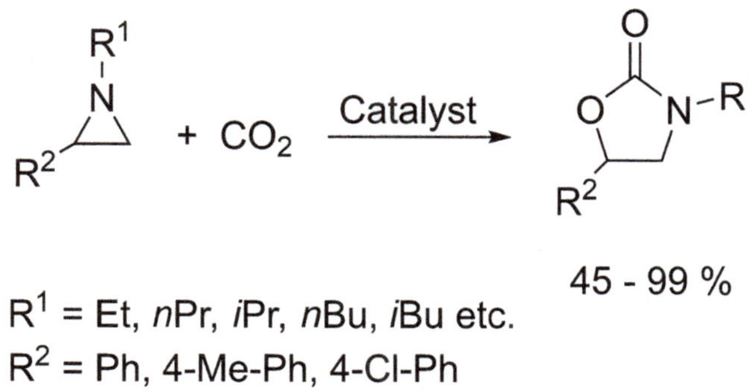

Scheme 11 General preparation of oxazolidines from aziridines with CO_2 via catalysis

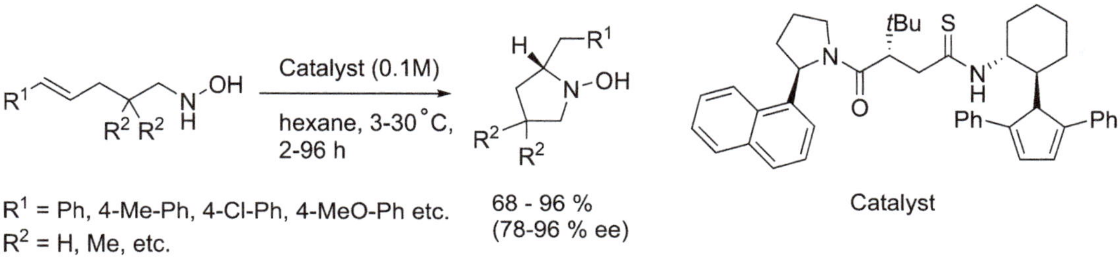

Scheme 12 Enantioselective thiourea-catalyzed intramolecular Cope-Type hydroamination

developed activation method. Nonetheless, most of the reactions occurred with quantitative yield [37].

The synthesis of pyrrolidines includes a broad variety of methods. The synthesis of chiral products dominates this area. A recent review summarizes the latest developments in the organocatalytic preparation of these compounds [38]. An effective enantioselective thiourea-catalyzed intramolecular Cope-type hydroamination reaction was described by Brown et al. (Scheme 12) [39]. The reaction provides the target compounds in good to excellent yields and high enantioselectivity (83–94%ee) using an organocatalyst. Further applications include a similar reaction catalyzed by binaphthyl diamines [40].

2.4 6-Membered Rings

Among the 6-membered one-heteroatom heterocyclic rings, piperidines enjoy the most attention. The convergent synthesis of piperidines from conjugated alkynes with imines and aldehydes is described by Chen and Micalizio [41]. Highly substituted piperidines were obtained in a two-step process. First, the imine reacted with the alkyne in a $Ti(O\textit{i}Pr)_4$-catalyzed reaction, and the intermediate was cyclized with aldehydes by the Pictet-Spengler or the aza-Prins cyclizations. The products were obtained in moderate yield (Scheme 13).

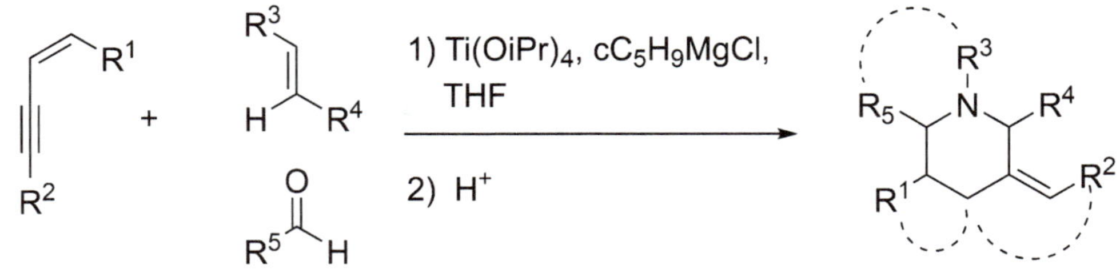

52 - 79 %

R^1 = various cycles
R^2 = $(CH_2)_2OH$
R^3 = Pr, Bn
R^4 = Ph, 3-Cl-Ph
R^5 = H, Me, various cycles

Scheme 13 Convergent synthesis of piperidines from conjugated alkynes with imines and aldehydes

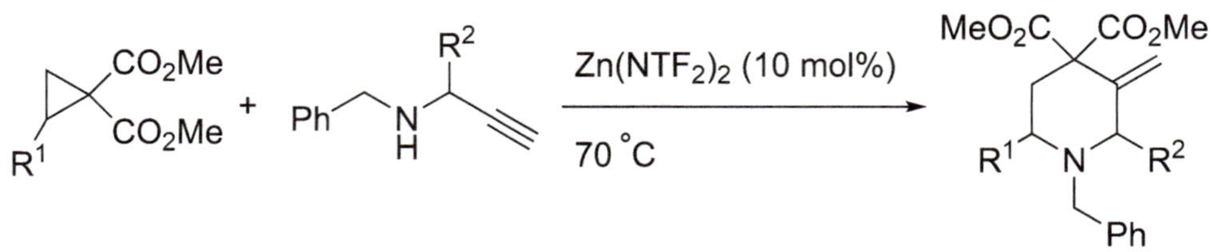

R^1 = Me, Ph, 4-MeO-Ph, 4-Br-Ph etc.
R^2 = H, Me

59 - 99 %

Scheme 14 Zn(II)-catalyzed synthesis of highly substituted piperidines from propargyl amines and cyclopropanes

The reaction of benzyl-protected propargyl amines and 1,1-cyclopropane diesters in the presence of catalytic amount of zinc bis(trifluoromethylsulfonyl)imide ($Zn(NTf_2)_2$) results in highly substituted piperidines (Scheme 14). The process occurs by a tandem cyclopropane ring opening followed by the Conia-ene cyclization and provides the products in excellent yields [42].

The Lewis acid-catalyzed cyclization of γ-amino or hydroxyl-allylic alcohol derivatives readily yields *cis*-2,6-piperidines and *cis*-2,6-tetrahydropyranes in moderate to excellent yields (Scheme 15) [43].

Enantioselective approaches including the application of chiral BINOL-based phosphoric acid derivative-catalyzed cyclization [44] are also available.

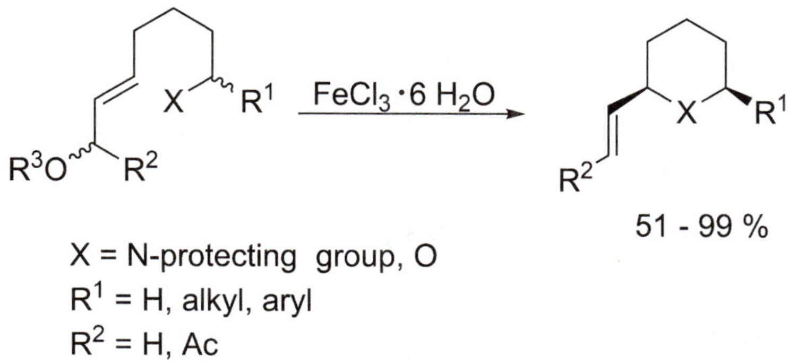

Scheme 15 $FeCl_3$-catalyzed diastereoselective cyclizaton of piperidines and tetrahydropyranes

3 Aromatic Heterocycles

3.1 5-Membered Rings

3.1.1 Pyrroles

The classical Paal-Knorr pyrrole synthesis was recently revisited by several groups. It was demonstrated that the use of an acid catalyst typically applied in this process is not necessary. Pyrroles can be formed from 1,4-diketones and amines in good to excellent yields under catalyst-free conditions in refluxing water [45] or more recently under solvent- and catalyst-free conditions (Scheme 16) [46]. The second method also utilized aq. NH_4OH as an amine equivalent at room temperature providing pyrroles underivatized on their nitrogen.

It has been shown that the synthesis of 2-aryl substituted pyrroles is possible from ketoximes using calcium carbide as a sustainable carbon source [47]. Under the reaction conditions employed, CaC_2 slowly decomposes to form acetylene which reacts with the oxime in an addition-rearrangement-cyclization-elimination sequence to form the pyrroles (Scheme 17).

Ketoxime carboxylates can be homocoupled using a simple copper(I) salt as catalyst to yield symmetrical 2,5-diaryl-pyrroles in good yields [48]. The presence of the aromatic ring appears to be of importance as reactions with an alkyl side chain do not result in the product formation (Scheme 18).

The synthesis of *N*-substituted pyrroles from 2,5-dimethoxytetrahydrofurans was recently achieved by reaction of the THF derivative with a variety of aromatic amines in a deep eutectic solvent as reaction medium (Scheme 19) [49]. The authors could show that the reaction medium could be reused up to 5 times and the reaction could easily be performed on a multigram scale. The reaction medium promotes the elimination of MeOH to form an activated enol species that can then be attacked by the amine.

In a very similar approach, solid acid catalysts under microwave irradiation [50] and ionic liquids [51] have been used for the synthesis of *N*-aryl or *N*-arylsulfonyl pyrroles, using the same starting materials.

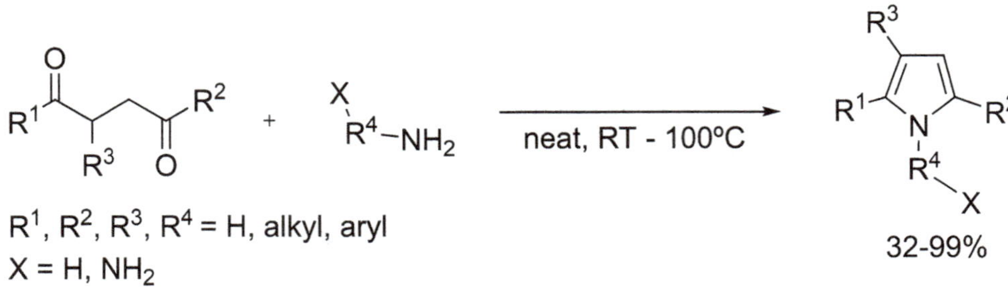

Scheme 16 A catalyst and solvent-free Paal-Knorr-type synthesis of pyrroles

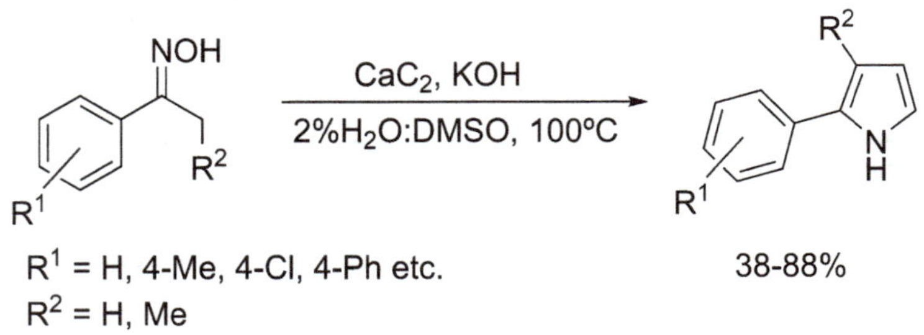

Scheme 17 Direct synthesis of aryl-substituted pyrroles from calcium carbide

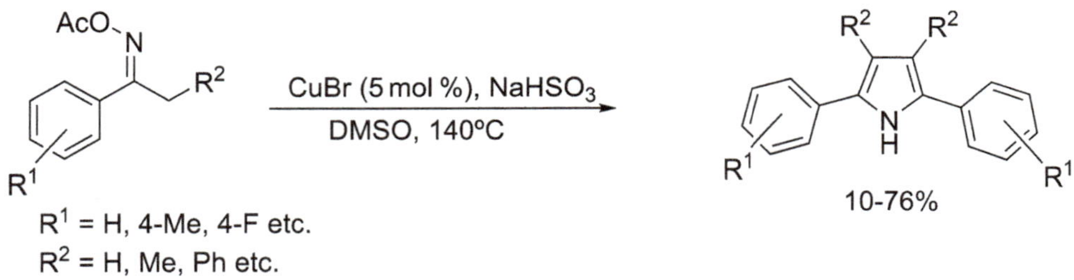

Scheme 18 Copper(I)-catalyzed homocoupling of ketoxime carboxylates for the synthesis of symmetrical pyrroles

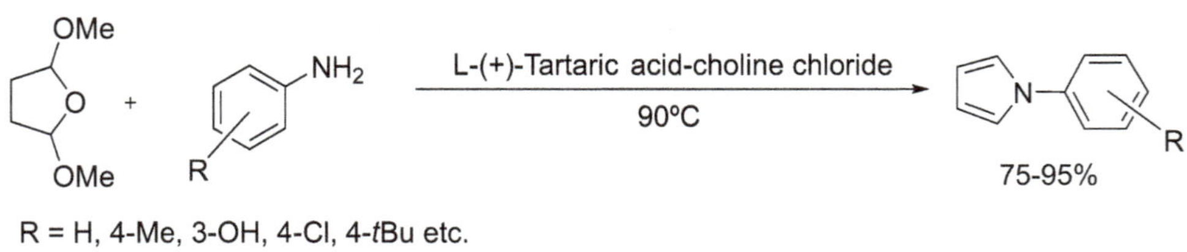

Scheme 19 Synthesis of *N*-substituted pyrrols using a THF derivative in deep eutectic solvents

A series of highly substituted pyrroles was synthesized in a multicomponent reaction catalyzed by β-cyclodextrin [52]. Aromatic amines, acetylenedicarboxylic acid esters, and glyoxal are coupled together to give the pyrroles in high yield using water as

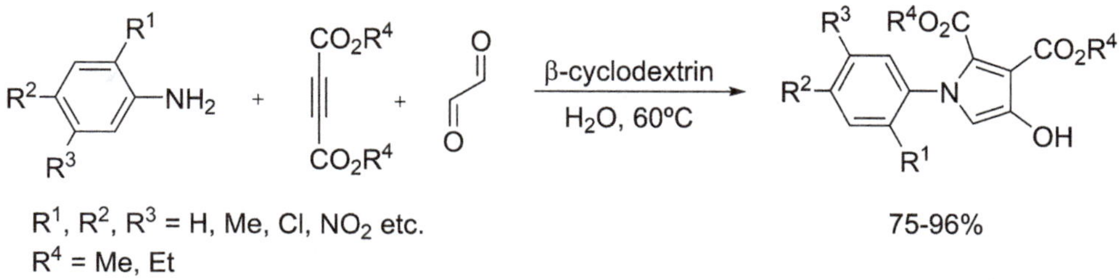

Scheme 20 Synthesis of substituted pyrrols by a multi-component reaction

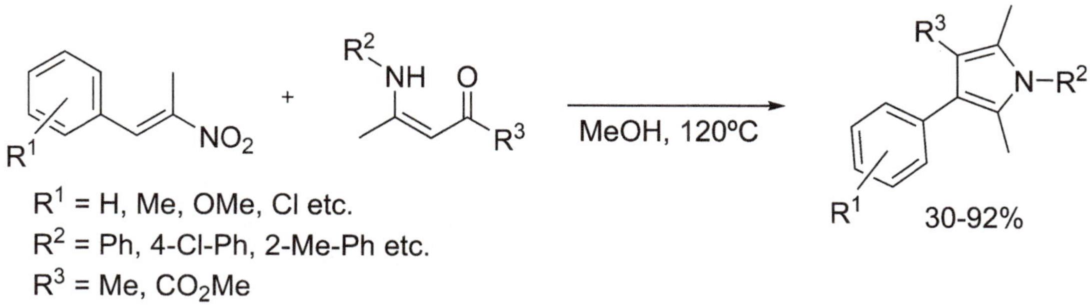

Scheme 21 Synthesis of multisubstituted pyrroles from β-amino ketones and nitroolefins

a solvent (Scheme 20). The β-cyclodextrin catalyst could be reused up to four times without loss of activity and without any additional treatment.

Coupling nitroolefins with enaminoesters also leads to the formation of multisubstituted pyrroles [53]. The reaction proceeds in MeOH at high temperatures in a closed system without the use of any additive or catalyst. The pyrroles were obtained in moderate to good yields (Scheme 21).

3.1.2 Pyrazoles

Pyrazoles are commonly obtained via ring derivatization, addition, or cyclization of acyclic precursors. The most common classical synthetic route is based on the cyclization of hydrazines with 1,3-dicarbonyl (or in general 1,3-bifunctional) compounds. However, the use of unsymmetrical 1,3-diketones gives a mixture of regioisomers. In addition, most traditional methods require two steps: a cyclization and a subsequent aromatization (oxidation).

This classical synthesis has been modified to avoid the use of harmful chemicals. The cyclization and the subsequent aromatization of chalcones with arylhydrazones provided the corresponding pyrazoles in excellent yields [54]. The simple mechanical mixture of Pd/C catalyst and K-10 montmorillonite was applied as a catalyst. While the solid acid K-10 ensured the effective cyclization, the subsequent dehydrogenation by the Pd catalyst resulted in the aromatic products (Scheme 22).

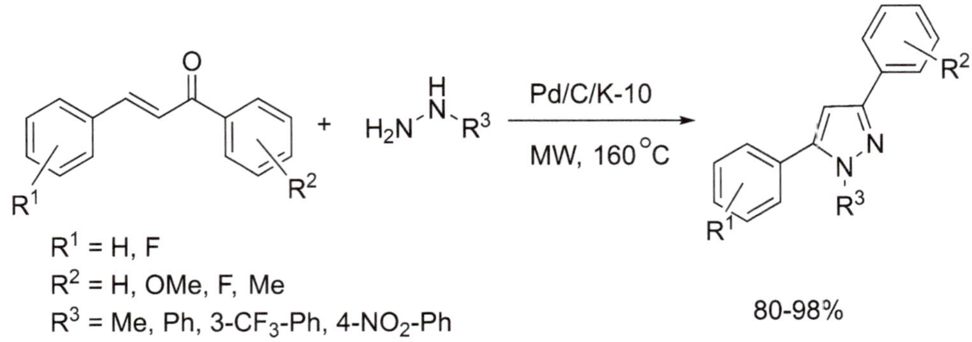

Scheme 22 Synthesis of pyrazoles from α,β-unsaturated ketones and hydrazines

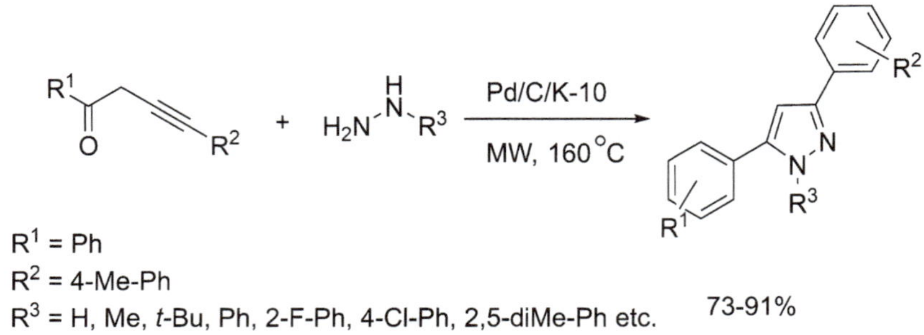

Scheme 23 Synthesis of pyrazoles from alk-3-yn-1-ones and hydrazines

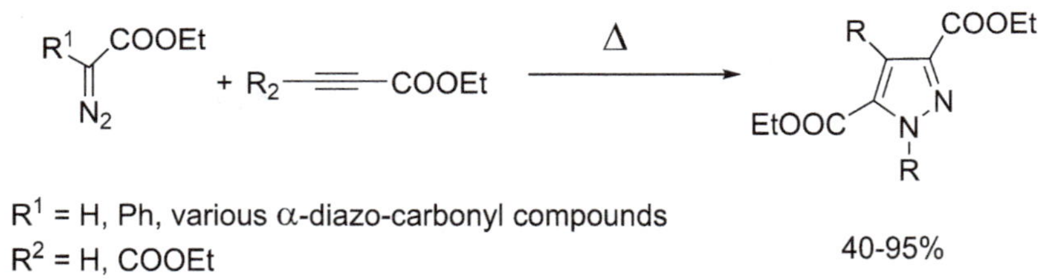

Scheme 24 Synthesis of pyrazoles from alkynes and diazo compounds

The application of alk-3-yn-1-one as a starting material allowed the elimination of the second aromatization step providing the pyrazoles in one step in good to excellent yields (Scheme 23) [55].

A catalyst-free simple thermal cycloaddition of diazo compounds to alkynes provides pyrazoles in moderate to excellent yields (Scheme 24). When the reaction is carried out with a-diazocarbonyl substrates, the reaction can be performed without solvent, and the products can be isolated in high yields without any work-up or purification [56].

Due to the practical importance of pyrazoles, new environmentally benign methods are being published frequently. Several recent papers offer a wide range of synthesis possibilities by catalytic,

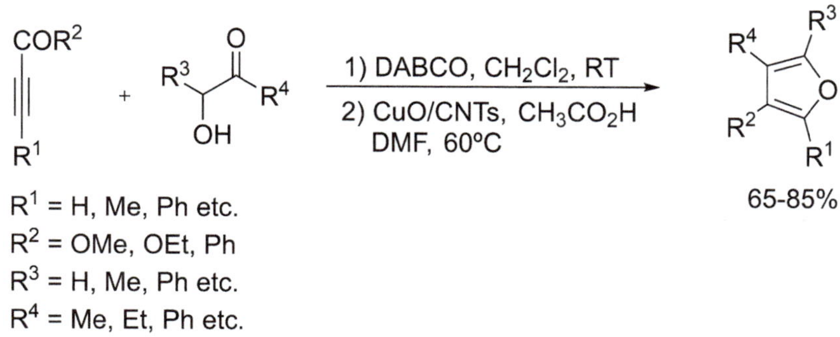

Scheme 25 Synthesis of furans from alkynes and α-hydroxyketones

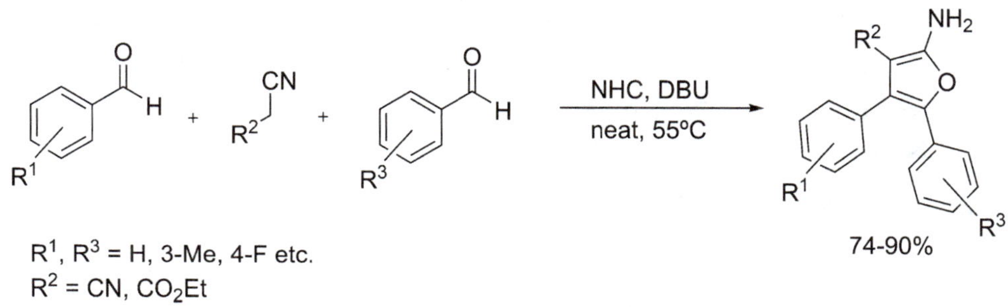

Scheme 26 NHC-catalyzed synthesis of furans

multicomponent, or solvent-free reactions [57–59]. Additional reviews also describe a broad array of contemporary pyrazole synthesis [60, 61].

3.1.3 Furans

Polysubstituted furan derivatives have been synthesized in a 2-step procedure from electron deficient alkynes and α-hydroxy ketones [62]. The first step consists of an additional reaction of the alcohol to the triple bond followed by a CuO/CNTs (CNT = carbon nanotube) catalyzed cyclization (Scheme 25). The reaction proceeds in good yields, and the heterogenous catalyst could be reused up to 5 times without notably loss of activity. A similar method was described using electron-deficient alkynes and 2-yn-1-ols as starting materials [63].

A *N*-heterocyclic carbene (NHC)-catalyzed three-component coupling between two aromatic aldehydes and a nitrile results in the formation of 2,3-diarylated, fully substituted furans in high yields [64]. The reaction proceeds without the use of a solvent and uses a thiazolium salt in combination with DBU as catalytic system (Scheme 26).

It has been shown that furans can be obtained via a poly(ethylene glycol) (PEG-400) promoted three-component coupling of isocyanides with α,β-unsaturated aldehydes and dialkyl acetylenedicarboxylates (Scheme 27) [65]. The reaction yields polysubstituted

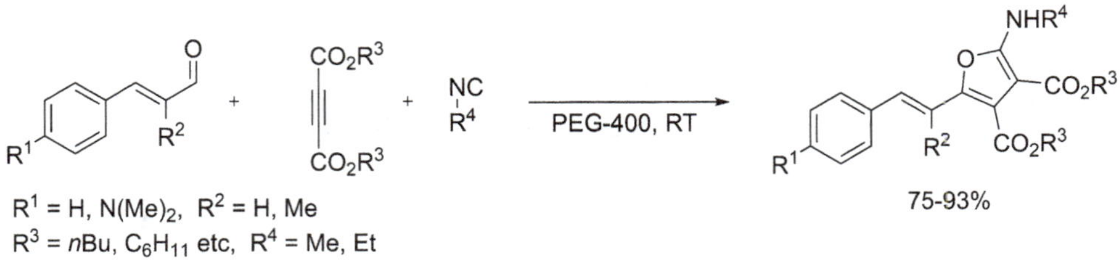

Scheme 27 PEG-400 promoted synthesis of furans

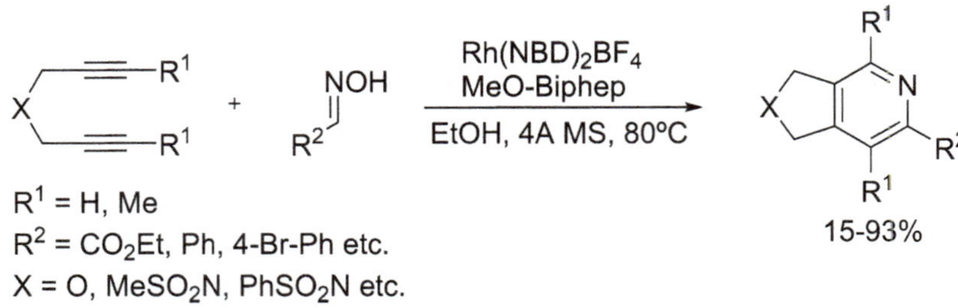

Scheme 28 Pyridine synthesis from dialkynes and oximes

furans in high yields, and the PEG-400 could be reused for further runs after separation from the product. The reaction is thought to proceed via a zwitterionic intermediate after attack of the isocyanite on the alkyne.

A similar reaction has been described using water as a solvent and benzyltriethyl ammonium chloride as a phase transfer catalysts [66, 67].

3.2 6-Membered Rings

3.2.1 Pyridines

Pyridines and their synthesis have been the subject of numerous reviews; most recently, their metal-free multicomponent synthesis has been reviewed [68].

The synthesis of highly functionalized pyridine derivatives has been recently described via a Rh-catalyzed [2 + 2 + 2] cycloaddition of diynes with oximes (Scheme 28) [69]. The reaction produces the pyridines in medium to high yields and can be run with preformed oxime or as a one-pot reaction combining the oxime formation and, after addition of the diyne and the catalyst, the cycloaddition.

The 3-component coupling of aromatic aldehydes, malonitrile, and thiophenols results in the formation of fully substituted 6-thiopyridines in high yields (Scheme 29) [70]. The reaction proceeds in PEG-400 as a reaction medium and is used as a catalytic quantity of K_2CO_3. The employed PEG-400 could be reused for up to 4 times without notably drop of product yield. The same reaction can also be performed using NH_4OMe as a catalyst in MeOH as the reaction medium [71].

The solvent- and catalyst-free synthesis of hydroxylated triarylpyridines was achieved under microwave irradiation (Scheme 30) [72]. The pyridines are obtained through the coupling of an

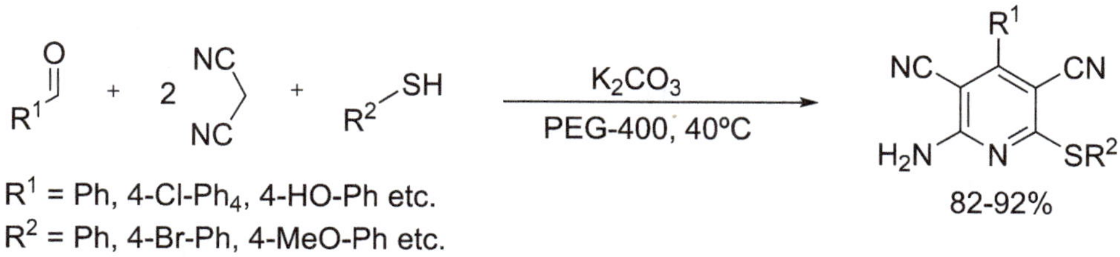

Scheme 29 Three-component coupling of aldehydes, malonitrile and thiophenols for the synthesis of pyridines

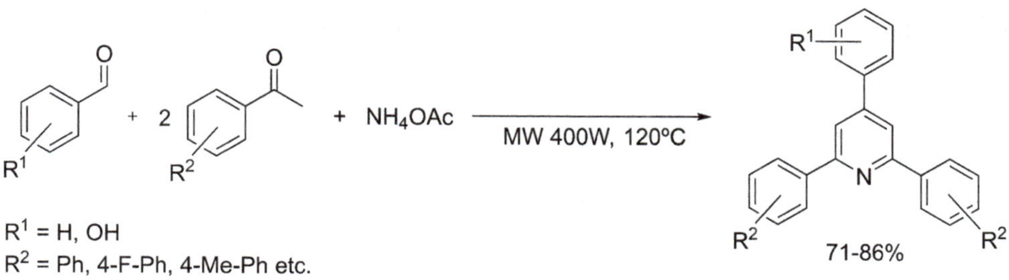

Scheme 30 Microwave-assisted synthesis of pyridines

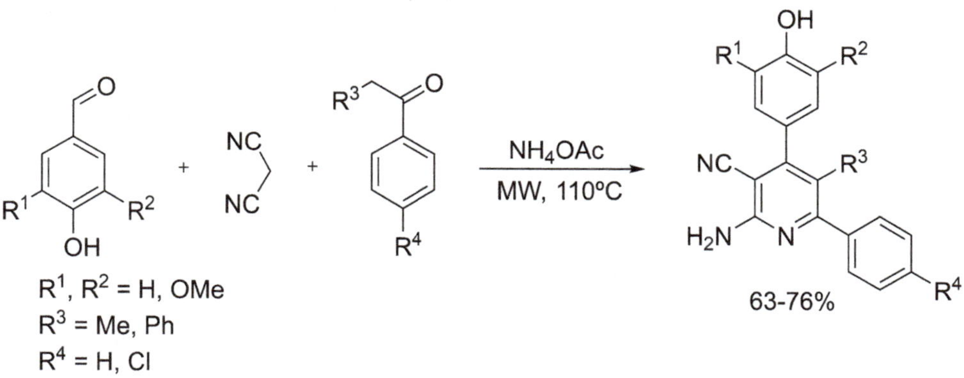

Scheme 31 Synthesis of highly substituted pyridines by microwave-assisted synthesis

aromatic aldehyde with two molecules of an aromatic ketone in the presence of NH_4OAc as a nitrogen source.

Microwave irradiation was also used in the solvent-free synthesis of fully substituted pyridines (Scheme 31) [73]. In this 4-component coupling reaction, an aromatic aldehyde, an aromatic ketone, and a malonitrile react in the presence of NH_4OAc to yield the pyridines in short reaction times in medium to good yields.

A similar approach to symmetrically substituted pyridines is described using a multicomponent domino cyclization-oxadative aromatization approach [74]. Two molecules of ethyl pyruvate react with an aldehyde and NH_4OAc as nitrogen source to form a dihydropyridine which undergoes dehydrogenative aromatization to give pyridines in moderate to high yields (Scheme 32).

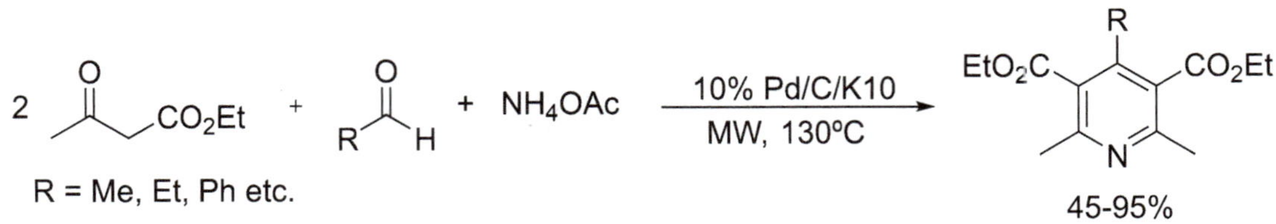

Scheme 32 Microwave-assisted multicomponent reaction for the synthesis of pyridines

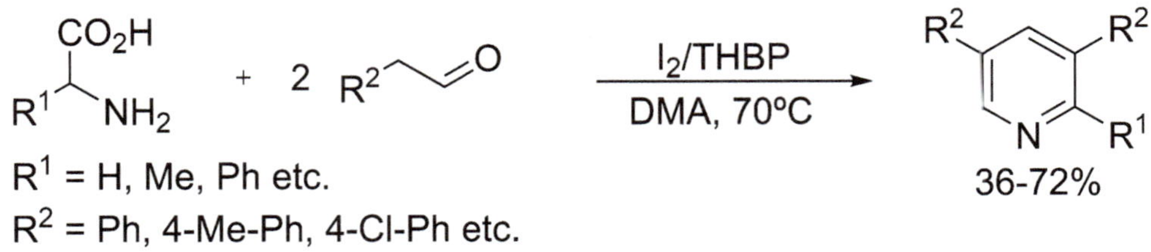

Scheme 33 Oxidative cyclization reaction for the synthesis of pyridines

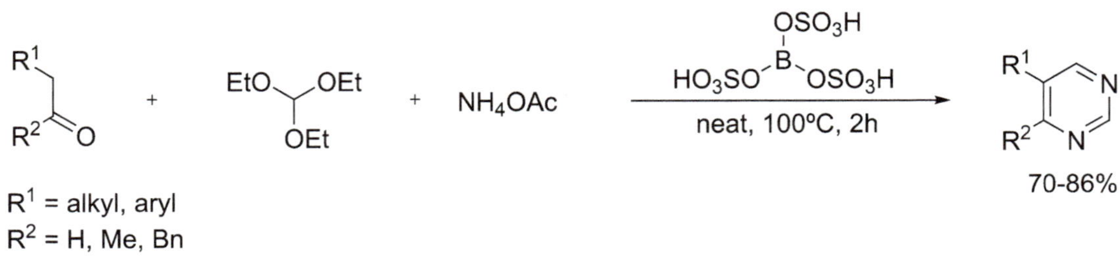

Scheme 34 Pyrimidine synthesis via multi-component reaction using a reusable acid catalyst

Pyridines containing the side chain of amino acids can be prepared through a decarboxylative cyclization method [75]. An amino acid and two equivalents of an aldehyde are coupled together under oxidative conditions using I_2/THBP as an oxidative system (Scheme 33). In contrast to previously described methods, no metal catalyst is used, and the reaction proceeds under milder conditions.

3.2.2 Pyrimidines

As the synthesis of pyrimidines was recently reviewed [76, 77], only a selection of the most recent work will be given here.

The synthesis of pyrimidines through a solvent-free multicomponent reaction has been described by Soheilizad et al. [78]. They could show that the coupling of ketones, triethylorthoformate, and ammoniumacetate is possible under solvent-free conditions (Scheme 34). The reaction proceeds with the help of a reusable acid catalyst.

In a similar approach, Sivagamisundari et al. described the synthesis of pyrimidines using a silica-bonded *S*-sulphonic acid (SBSSA) as catalyst under solvent-free conditions [79]. The reaction proceeds in short reaction times (<1 min) under microwave

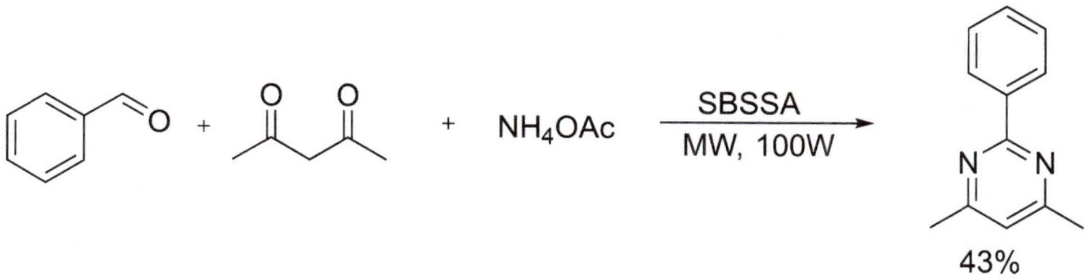

Scheme 35 Three-component synthesis of pyrimidines using a solid acid catalyst

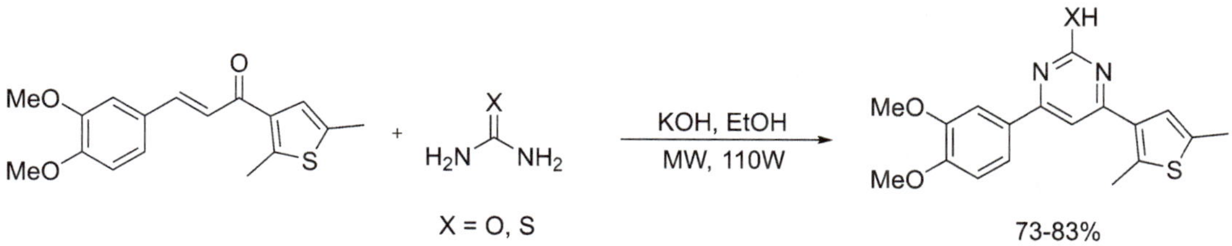

Scheme 36 Microwave-assisted condensation reaction for the synthesis of pyrimidines

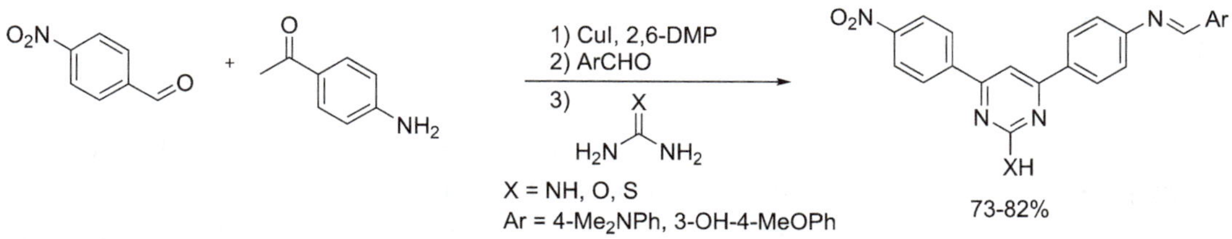

Scheme 37 Synthesis of substituted pyrimidines in a CuI and base-catalyzed multicomponent reaction

irradiation (Scheme 35). When urea is used instead of ammonium acetate as reagent, the moderate yield of only 43% can be overcome with the drawback that only the corresponding tetrahydropyrimidines are obtained.

During their investigation on the synthesis of new antibacterial compounds, Khan et al. found that chalcones are useful substrates for the synthesis of different heterocycles [80]. When urea or thiourea is reacted under basic conditions with activation by microwave irradiation, substituted pyrimidines were obtained in good yield (Scheme 36). A similar reaction forming NH_2-substituted pyrimidines was described by Ezhilarasi et al. [81], substituting urea compounds with guanidine nitrate. The reaction resulted in the formation of the desired compounds in medium to good yields.

Babu et al. reported a one-pot reaction furnishing similar substituted pyrimidines starting from benzaldehydes and acetophenones [82]. A chalcone intermediate is formed in a CuI- and base-catalyzed reaction before a urea derivate is added and the pyrimidine is formed (Scheme 37).

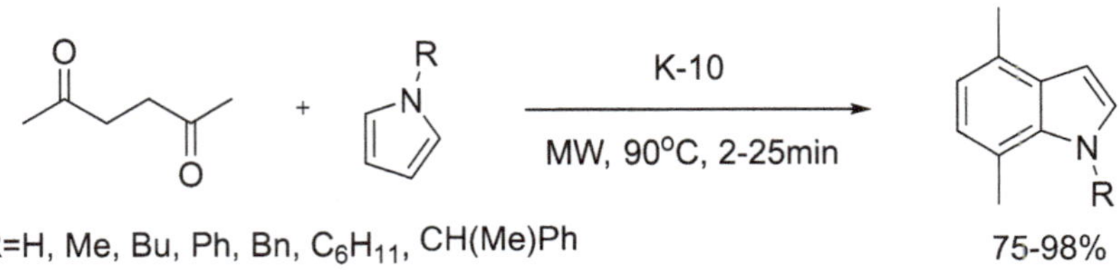

Scheme 38 Synthesis of indoles via a microwave-assisted, solid acid-catalyzed solvent-free annelation of pyrroles with 1,4-diketones

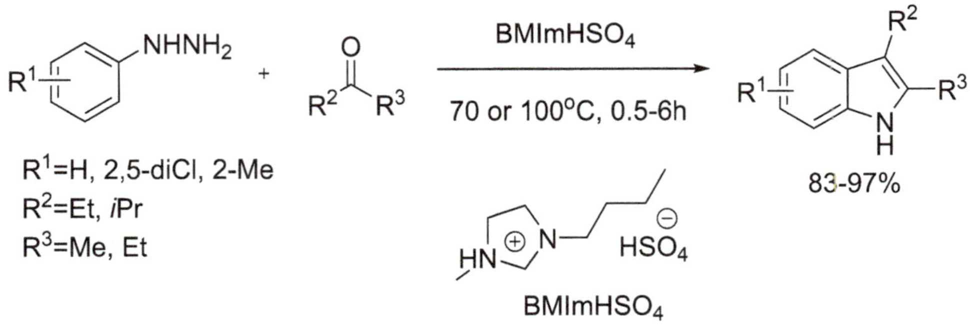

Scheme 39 Synthesis of 2,3-disubstituted indoles via the ionic liquid-catalyzed the reaction of alkyl methyl ketones and arylhydrazines

3.3 Condensed Heterocycles

The synthesis of condensed heterocycles is an important task in organic and medicinal chemistry due to the frequent occurrence of such motifs in bioactive compounds, and they often are potential lead compounds or drug candidates in the pharmaceutical industry. Therefore, the development of green and eco-friendly procedures for their synthesis has been a top priority.

3.3.1 Indoles

An effective microwave-induced, solid acid-catalyzed solvent-free annelation of pyrroles with 1,4-diketones to is substituted indoles described (Scheme 38) [83]. The new synthetic methodology is based on the use of a considerably strong solid acid, K-10 montmorillonite. The combination of solid acid catalysis, solvent-free conditions, and microwave irradiation results in excellent yields, reduces the reaction times, and represents a highly ecofriendly approach.

A novel one-pot Fischer indole synthesis approach has been developed by using Brønsted acidic ionic liquids as dual solvent-catalysts [84]. Yields of 83–97% were obtained after reaction in $BMImHSO_4$ (1-Butyl-3-methylimidazolium Hydrogen Sulfate) at 70–100 °C in 0.5–6 h, and exclusive formation of 2,3-disubstituted indoles was observed in the reaction of alkyl methyl ketones and arylhydrazines (Scheme 39). The indoles produced could be conveniently separated from the reaction mixture without any volatile

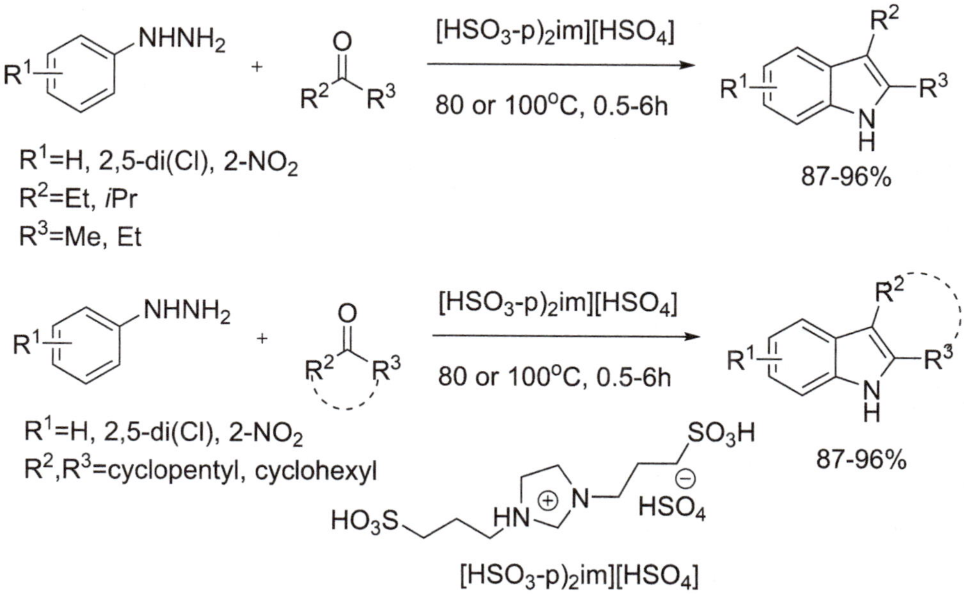

Scheme 40 An improved, ionic liquid-catalyzed Fischer indole synthesis

organic solvents, and the $BMImHSO_4$ could be readily reused without efficiency loss after simple treatment involving only 1 equiv. of HCl for reacidification followed by filtration.

The same authors improved their previous attempts [85]. Novel SO_3H-functionalized ionic liquids bearing two alkyl sulfonic acid groups in the imidazolium cations were designed and successfully applied as catalysts for the one-pot Fischer indole synthesis in aqueous medium (Scheme 40). The developed methodology offered significant improvements: (1) the whole process for the Fischer indoles was performed in water without using any organic solvents; (2) the reaction–separation–recycle process was convenient and the catalytic system of $[(HSO_3\text{-p})_2\text{im}][HSO_4]/H_2O$ (1,3-bis(3-sulfopropyl)imidazolidin-1-ium hydrogen sulfate/water) could be reused at least 12 times without an obvious decrease in catalytic activity; (3) the novel catalytic system could be successfully applied in the synthesis of indoles from a broad variety of carbonyl compounds affording good to excellent yields.

Presently, metal organic frameworks (MOFs) have attracted growing attention from both academia and industry owing to their outstanding catalytic features. In addition to the well-demonstrated catalytic activity of pure MOFs [86–90], these materials are also attractive candidates for catalyst supports because of their unique porous features. A hydrothermally stable metal-organic framework, MIL-101, was used as support for a metallic Pd nanoparticle catalyst [91]. Palladium nanoparticles (<3 nm) confined in the cages of MIL-101 could be employed as a catalyst for the domino synthesis of indole derivatives in water (Scheme 41).

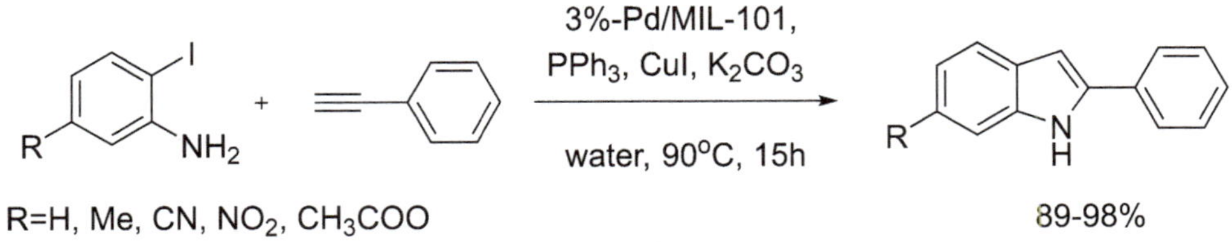

Scheme 41 An aqueous domino synthesis of indole derivatives catalyzed by MOF supported Pd nanoparticles

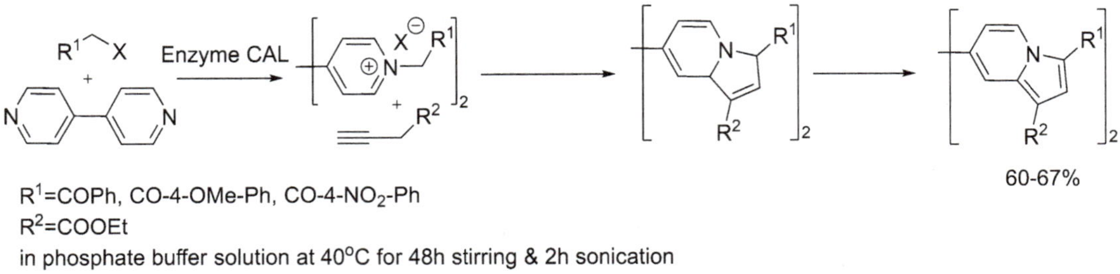

Scheme 42 A biocatalytic route to indolizines

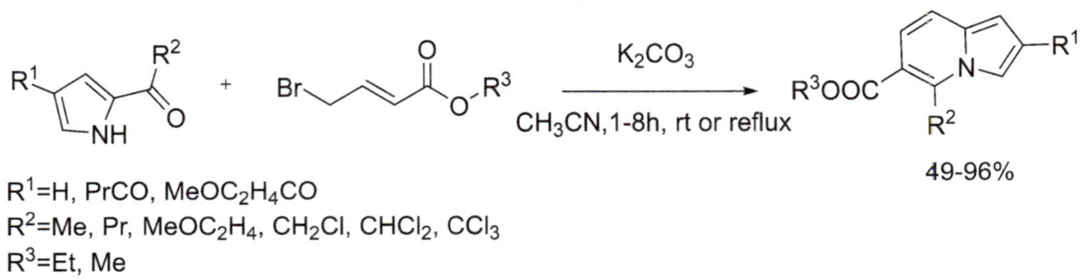

Scheme 43 A metal-free tandem reaction for the synthesis of indolizines

3.3.2 Indolizines

The first example for biocatalyzed *one-pot* synthesis for indolizines, which involves cycloadditions of ethyl propiolate with in situ generated ylides, in aqueous medium was described [92]. The best results were obtained when the *Candida antarctica* lipase (CAL) catalyzed the biotransformation of 4,4′-bipyridine, halide derivatives, and ethyl propiolate to form indolizines (Scheme 42). The catalytic reactions with lipase from *Candida antarctica* produced almost pure indolizines with good yields and in very short time under ultrasonic activation. As mentioned above, the lipase-catalyzed synthesis was performed in water, the greenest solvent in contrast to the organic solvents, usually required for the classical methodologies. The increasing importance of safety, health, and environmental issues as well as initiatives in green and sustainable chemistry and the new perspectives of molecular economy plainly justify this biocatalytic approach for indolizine synthesis using enzymes.

A series of transition metal-free tandem reactions for the synthesis of indolizines by the reaction of 4-bromobut-2-enoate with 2-acetyl pyrrole derivatives is described (Scheme 43) [93]. These tandem reactions could be used to synthesize indolizine under mild and green conditions.

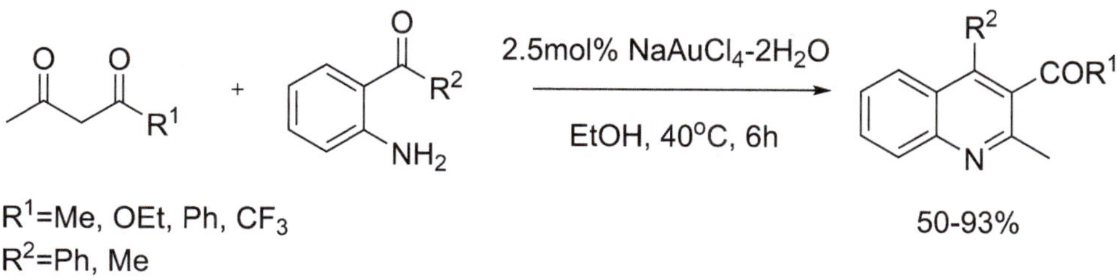

Scheme 44 A gold(III)–catalyzed sequential condensation/annulation of 2-amino substituted aromatic carbonyl compounds with β-ketoesters

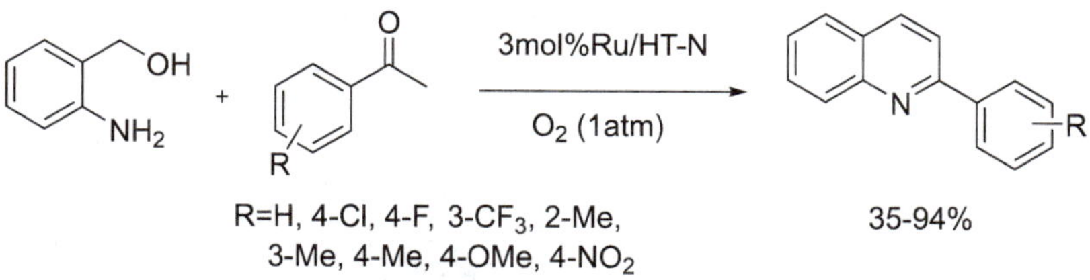

Scheme 45 A Ru-grafted hydrotalcite-catalyzed synthesis of quinolines

3.3.3 Quinolines

A green approach, developed to replace the Friedländer synthesis [94] of quinolines, requires neither harsh conditions nor the use of hazardous acids or bases. The gold(III)-catalyzed sequential condensation/annulation reaction of 2-amino substituted aromatic carbonyl compounds with ketones containing active methylene groups appears to provide an efficient new tool for the synthesis of quinolines under mild conditions (Scheme 44) [95].

Hydrotalcites (HTs) are useful solid catalysts for nano-scale design of active sites because of their unique properties such as alternating cationic and anionic layers, surface tunable basicity, and adsorption capacity [96]. A Ru-grafted hydrotalcite was found to be an excellent multifunctional catalyst for the one-pot synthesis of quinolines from 2-aminobenzyl alcohol and various carbonyl compounds (Scheme 45) [97]. These quinolines were obtained through aerobic oxidation by the Ru species, followed by aldol reaction on base sites of the hydrotalcite. Atmospheric pressure molecular oxygen was used as a green oxidant.

The preparation of substituted quinoline derivatives through a Friedländer condensation reaction utilizing the ionic liquid, 1-butyl-3-methylimidazolium tetrafluoroborate [bmim][BF_4], as reaction medium and iron chloride hexahydrate ($FeCl_3 \cdot 6H_2O$) as a catalyst is described (Scheme 46) [98]. The advantages of using this method include its environmental friendliness, simple operating process under both mild and neutral reaction conditions and good yields. The simple experimental and product isolation procedures, combined with ease of recovery and reuse of the catalyst and the reaction media, are expected to contribute to the development of greener and waste-free chemical processes for the preparation of quinolines.

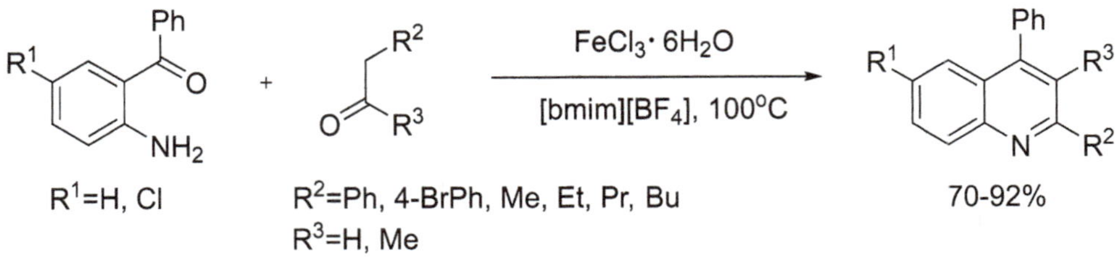

Scheme 46 An ionic liquid-catalyzed synthesis of quinolines

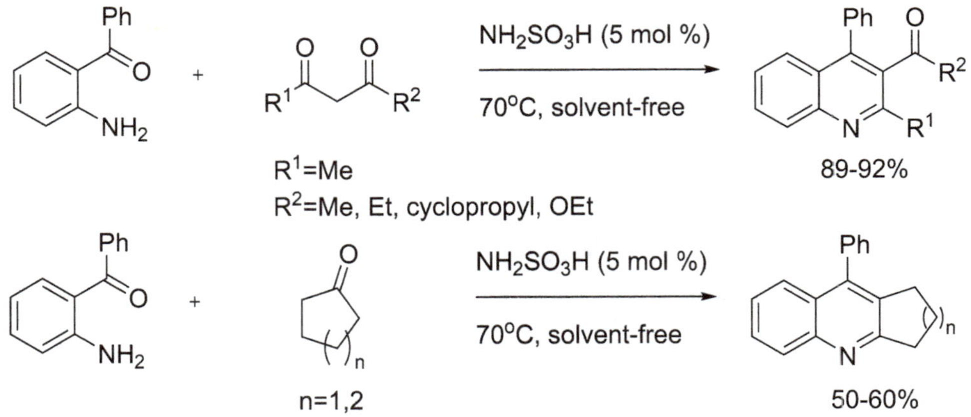

Scheme 47 Synthesis of quinolines by sulfamic acid catalysis

Heterogeneous solid acids are advantageous over conventional homogeneous acid catalysts as they can be easily recovered from the reaction mixture by simple filtration and can be re-used with or even without activation, thereby making the process economically and environmentally more viable. During the last few years, sulfamic acid (NH_2SO_3H) (SA) has emerged as a substitute for conventional acidic catalysts. Sulfamic acid is a common inorganic acid with mild acidity, is non-volatile and non-corrosive, and is insoluble in common organic solvents. A mild and efficient approach for the synthesis of polysubstituted quinolines via Friedländer annulation using a catalytic amount of SA under solvent-free conditions was reported (Scheme 47) [99].

A medium chain SO_3H-functionalized ionic imidazole-based liquid was successfully applied as a water tolerant acid catalyst for the synthesis of quinolines with a one-pot domino approach in an aqueous medium [100]. Various types of quinolines were prepared from 2-aminoaryl ketones and β-ketoesters/ketones in 85–98% yields using this catalytic system (Scheme 48). The ionic liquid was synthesized in a combinatorial fashion. The quinoline products could be conveniently separated from the reaction mixture by filtration, due to the aqueous medium. More importantly, the catalyst could be easily recycled for 5 times without loss of activity. The catalytic system possesses the advantages of both homogeneous and heterogeneous catalysts.

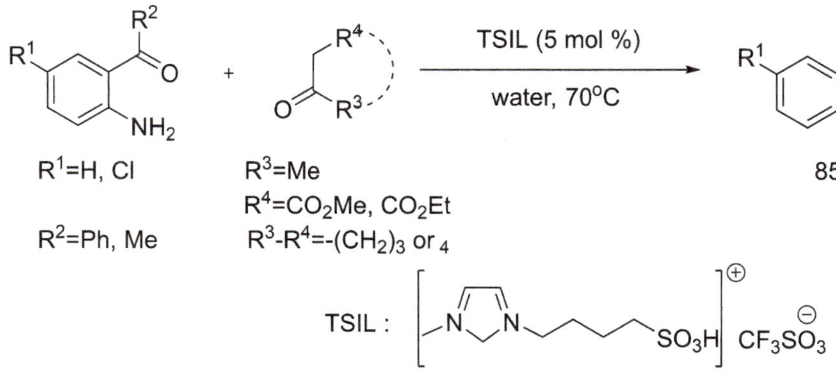

Scheme 48 Synthesis of quinolines catalyzed by an SO_3H-functionalized imidazole-based ionic liquid

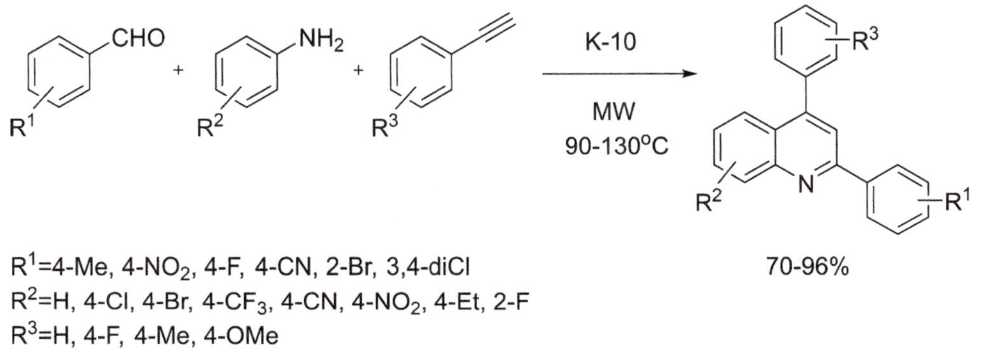

Scheme 49 A K-10 montmorillonite-catalyzed microwave-assested synthesis of quinolines

A broad variety of 2,4-disubstituted quinolines were synthesized by a three-component reaction of anilines, benzaldehydes, and terminal phenylacetylenes in the presence of K-10 montmorillonite under solvent-free conditions (Scheme 49) [101]. All attempted combinations of the three components resulted in the formation of the corresponding 2,4-diarylquinolines in good to excellent yields. The catalyst could be recycled; the catalytic activity was unaltered even after 5 cycles.

3.3.4 Carbazoles

A new, effective, solid acid-catalyzed synthesis of substituted carbazoles is described by Abid et al. [69]. The method provides the products in very high yields and excellent selectivities in short reaction times (Scheme 50). In addition to efficiency, the solvent-free environment, solid acid catalysis, very limited energy consumption, and waste-free nature make the process very attractive for the environmentally benign synthesis of the carbazoles. This work represents the first effective method that is based on a successive solid acid-catalyzed Friedel–Crafts cyclialkylation/annelation for the synthesis of the substituted carbazoles.

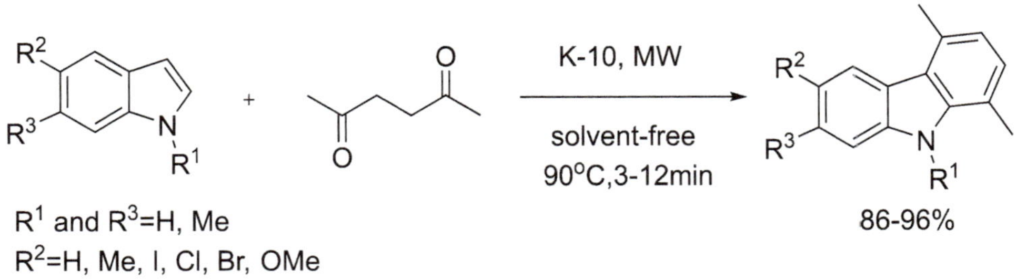

Scheme 50 Synthesis of carbazoles by a K-10-catalyzed microwave-assisted annelation of indoles with 1,4-diketones

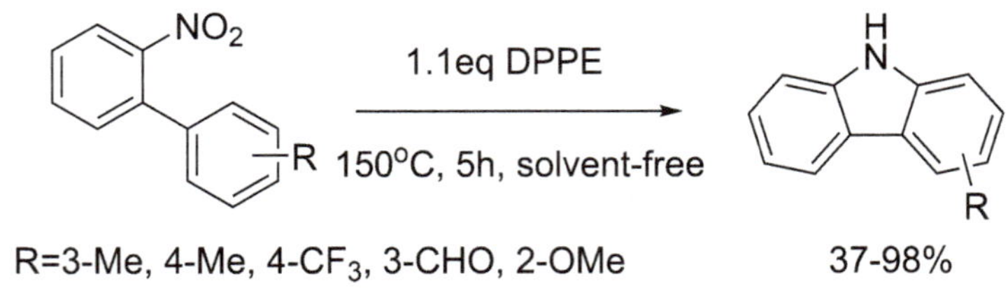

Scheme 51 Synthesis of functionalized carbazoles via reductive ring closure catalyzed by 1,2-bis(dipenylphosphino)ethane (DPPE)

A green and efficient preparation of functionalized carbazoles via reductive ring closure by 1,2-bis(dipenylphosphino)ethane (DPPE) under solvent-free conditions is described (Scheme 51) [102]. One general method for carbazole synthesis is the Cadogan cyclization [103] which involves the reductive cyclization of 2-nitrobiphenyl derivatives in the presence of trivalent organophosphorus reagents. $P(OEt)_3$ and PPh_3 are usually employed as reductants, and $P(OEt)_3$ is also used as solvent in the reactions. The reductive cyclizations with $P(OEt)_3$ suffer from several drawbacks. To overcome the drawbacks, DPPE is employed as a greener alternative. The starting 2-nitrobiphenyl derivatives are readily prepared through the Suzuki-Miyaura cross-coupling reaction from commercially available compounds. The polar by-product ethane-1,2-diylbis(diphenylphosphine oxide) is easily removed from the relatively polar reaction mixture. Various substituted carbazoles are obtained in acceptable yields. It is particularly worth mentioning that substrates with electron-withdrawing groups also give the desired products in good yield.

Metal oxide-supported palladium hydroxide ($Pd(OH)_2/MO_x$) catalysts enabled the oxidative intramolecular coupling of diarylamines to carbazoles via double aryl C–H bond functionalizations with molecular oxygen as the sole oxidant (Scheme 52) [104]. While supported PdO, Pd, and palladium acetate catalysts showed poor catalytic performance, supported $Pd(OH)_2$ exhibited remarkably high catalytic activity. Among supported $Pd(OH)_2$

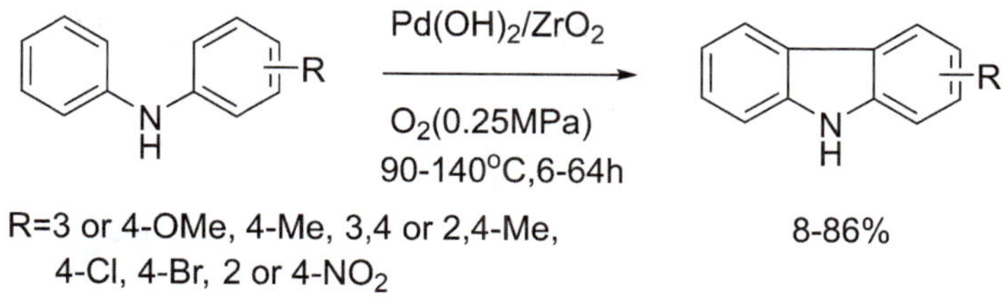

Scheme 52 A Pd-catalyzed oxidative intramolecular coupling of diarylamines to carbazoles

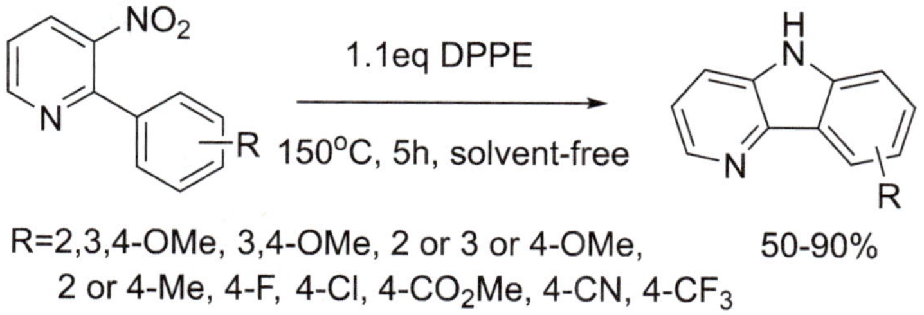

Scheme 53 Synthesis of carbolines from functionalized 3-nitro-2-phenylpyridines

catalysts, $Pd(OH)_2/ZrO_2$ was found to be the most efficient catalyst in terms of both catalytic activities and selectivities for the synthesis of carbazoles.

3.3.5 Carbolines

An efficient and practical solvent-free synthesis of novel or useful δ-carbolines has been developed [85]. A wide variety of functionalized 3-nitro-2-phenylpyridine derivatives were used to prepare the desired aza-heterocycles in good to excellent yields, and various functional groups are tolerated well (Scheme 53). The advantages of this method include its environmentally benign, simple manipulation, and high regioselectivity. This is the first time that 1,2-bis (Diphenylphosphino)-ethane (DPPE) was used as a reductant in Cadogan cyclization.

The bifunctional (metal/acid) Pd/C/K-10 catalyst was applied in a multistep synthesis of β-carbolines starting with tryptamines and carbonyl compounds such as benzaldehyhdes and aromatic glyoxals, respectively (Scheme 54) [105]. The final products are formed in a one-pot, three-step domino sequence. In the first step, the aldehyde/glyoxal is condensed with the primary amino group of the tryptamine to form an imine. In the second step, the imine underwent a K-10-catalyzed Pictet–Spengler cyclization reaction. In the third and final step, the tetrahydro-β-carboline intermediate was dehydrogenated by Pd/C to form β-carbolines in good to excellent yields.

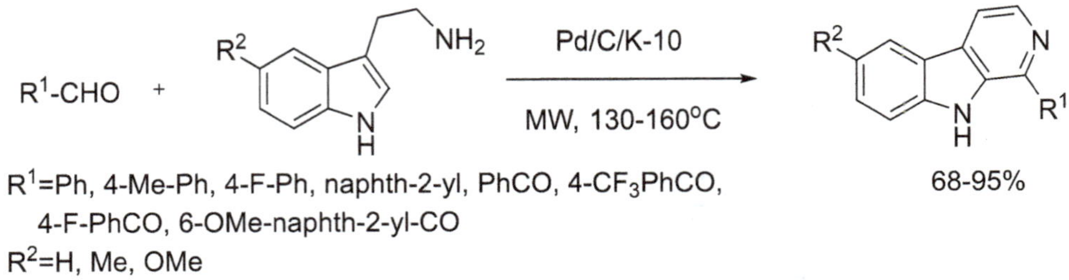

Scheme 54 A microwave-assisted synthesis of β-carbolines by a bifunctional Pd/C-K-10 catalyst

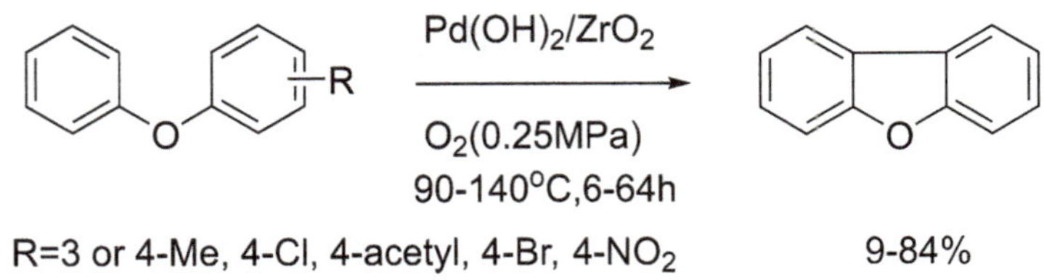

Scheme 55 A Pd-catalyzed intramolecular oxidative coupling of diaryl ethers to dibenzofurans

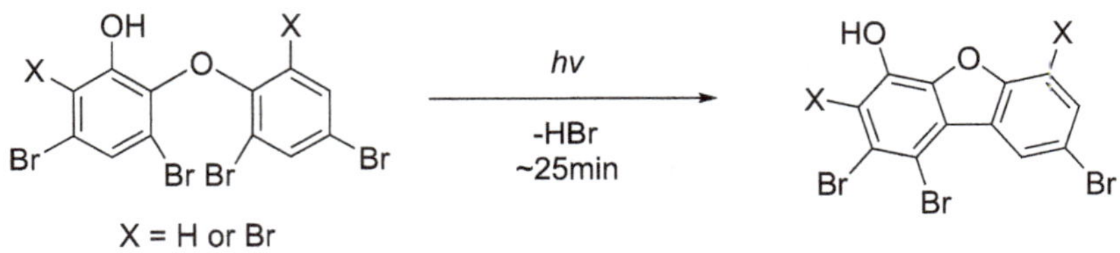

Scheme 56 Photochemical coupling of polybrominated diphenyl ethers to dibenzofurans

3.3.6 Dibenzofurans

ZrO_2 supported $Pd(OH)_2$ catalyst promoted the intramolecular oxidative coupling of C–H bonds of diarylethers to afford dibenzofurans with high selectivity using molecular O_2 as a green oxidant (Scheme 55) [91]. The yield was 71–84% except 4-bromodiphenyl ether which gave only 9% desired product. This is due to the formation of debrominated diphenyl ether. It suggested that oxidative addition of 4-bromodiphenyl ether to Pd(0), which was formed in situ, occurred to produce debrominated substrate, diphenyl ether. At the same time, the oxidative coupling of diphenyl ether was significantly inhibited. It is likely that Pd is deactivated once the Pd-Br bond is formed, although the Pd-Cl bond does not affect the catalyst activity.

The photochemical conversion of selected hydroxylated polybrominated diphenyl ethers (OH-PBDEs) to dibenzofuran and other products was also investigated (Scheme 56) [106]. The starting OH-PBDEs are both transformation products of polybrominated diphenyl ethers and naturally occurring compounds. The aim of this work was to investigate the photochemical degradation of three different OH-PBDEs. While the reported photoproducts are concerning from a toxicity standpoint (polybrominated dibenzofuran, dioxins, phenols, and dihydroxybiphenyls), this work could serve as a foundation for the development of a green synthetic pathway to dibenzofuran without any toxic catalyst and solvent.

References

1. Gilchrist TL (1997) Heterocyclic chemistry, 3rd edn. Addison, Wesley, Longman, Harlow
2. Singh GS, D'hooghe M, De Kimpe N (2007) Synthesis and reactivity of C-heteroatom-substituted aziridines. Chem Rev 107:2080–2135
3. Yudin AK (2006) Aziridines and epoxides in organic synthesis. Wiley-VCH, Weinheim
4. Chawla R, Singh AK, Yadav LDS (2013) Organocatalysis in synthesis and reactions of epoxides and aziridines. RSC Adv 13:11385–11403
5. Grigoropoulou G, Clark JH, Elings JA (2003) Recent developments on the epoxidation of alkenes using hydrogen peroxide as an oxidant. Green Chem 5:1
6. Tebandeke E, Coman C, Guillois K et al (2014) Epoxidation of olefins with molecular oxygen as the oxidant using gold catalysts supported on polyoxometalates. Green Chem 16:1586–1593
7. Tang B, Dai W, Sun X et al (2014) A procedure for the preparation of Ti-Beta zeolites for catalytic epoxidation with hydrogen peroxide. Green Chem 16:2281–2291
8. Bawaked S, Dummer NF, Bethell D et al (2011) Solvent-free selective epoxidation of cyclooctene using supported gold catalysts: an investigation of catalyst re-use. Green Chem 13:127–134
9. Doherty S, Knight JG, Ellison JR et al (2012) An efficient recyclable peroxometalate-based polymer-immobilised ionic liquid phase (PIILP) catalyst for hydrogen peroxide-mediated oxidation. Green Chem 14:925–929
10. Qiao Y, Hou Z, Li H et al (2009) Polyoxometalate-based protic alkylimidazolium salts as reaction-induced phase-separation catalysts for olefin epoxidation. Green Chem 11:1955–1960
11. Hasan K, Brown N, Kozak CM (2011) Iron-catalyzed epoxidation of olefins using hydrogen peroxide. Green Chem 13:1230–1237
12. Chen Q, Beckman EJ (2008) One-pot green synthesis of propylene oxide using in situ generated hydrogen peroxide in carbon dioxide. Green Chem 10:934–938
13. Sharpless KB (2002) Searching for new reactivity (Nobel lecture). Angew Chem Int Ed 41:2024–2032
14. De Faveri G, Ilyashenko G, Watkinson M (2011) Recent advances in catalytic asymmetric epoxidation using the environmentally benign oxidant hydrogen peroxide and its derivatives. Chem Soc Rev 40:1722–1760
15. Kawai H, Okusu S, Yuan Z et al (2013) Enantioselective synthesis of epoxides having a tetrasubstituted trifluoromethylated carbon center: methylhydrazine-induced aerobic epoxidation of α,β-disubstituted enones. Angew Chem Int Ed 52:2221–2225
16. Aouf C, Durand E, Lecomte J et al (2014) The use of lipases as biocatalysts for the epoxidation of fatty acids and phenolic compounds. Green Chem 16:1740–1754
17. Kluge M, Ullrich R, Scheibner K et al (2012) Stereoselective benzylic hydroxylation of alkylbenzenes and epoxidation of styrene derivatives catalyzed by the peroxygenase of *Agrocybe aegerita*. Green Chem 14:440–446
18. Chavan VP, Pathwardan AV, Gogate PR (2012) Intensification of epoxidation of soybean oil using sonochemical reactors. Chem Eng Process 54:22–28
19. Sato K, Aoki M, Ogawa M, Hashimoto T, Noyori R (1996) A practical method for epoxidation of terminal olefins with 30% hydrogen peroxide under halide-free conditions. J Org Chem 61:8310–8311
20. Sato K, Aoki M, Ogawa M, Hashimoto T, Panyella D, Noyori R (1997) A halide free method for olefin epoxidation with 30% hydrogen peroxide. Bull Chem Soc Jpn 70:905–915
21. Degennaro L, Trinchera P, Luisi R (2014) Recent advances in the stereoselective synthesis of aziridines. Chem Rev 114:7881–7929
22. Jin LM, Xu X, Lu H (2013) Effective synthesis of chiral N-fluoroaryl aziridines through enantioselective aziridination of alkenes with

fluoroaryl azides. Angew Chem Int Ed 52:5309–5313

23. Jenkins DM (2012) Atom-economical C2 + N1 aziridination: progress towards catalytic intermolecular reactions using alkenes and aryl azides. Synlett 2012:1267–1270
24. Cramer SA, Jenkins DM (2011) Synthesis of aziridines from alkenes and aryl azides with a reusable macrocyclic tetracarbene iron catalyst. J Am Chem Soc 133:19342–19345
25. Jung N, Bräse S (2012) New catalysts for the transition-metal-catalyzed synthesis of aziridines. Angew Chem Int Ed 51:5538–5540
26. Jat JL, Paudyal MP, Gao H, Xu QL, Yousufuddin M, Deverajan D, Ass DH, Kurti L, Falck JR (2014) Direct stereospecific synthesis of unprotected N-H and N-Me aziridines from olefins. Science 343:61–65
27. Borkin D, Carlson A, Török B (2010) K-10-catalyzed highly diastereoselective synthesis of aziridines. Synlett 2010:745–748
28. Buckley BR, Patel AP, Wijayantha KGU (2013) Observations on the modified wenker synthesis of aziridines and the development of a biphasic system. J. Org. Chem. 78:1289–1292
29. Abe M (2010) Formation of a four-membered ring: oxetanes. In: Albini A, Fagnoni M (eds) Handbook of synthetic photochemistry. Wiley-VCH, Weinheim, pp 217–239
30. Adam W, Stegmann VR (2011) Hydroxy-group directivity in the regioselective and diastereoselective [2+2] photocycloaddition (Paterno-Buechi reaction) of aromatic carbonyl compounds to chiral and achiral allylic substrates: the preparation of oxetanes with up to three stereogenic centers as synthetic building blocks. Synthesis 2001:1203–1214
31. Christlieb M, Davies JE, Eames J et al (2001) The stereoselective synthesis of oxetanes; exploration of a new, Mitsunobu-style procedure for the cyclization of 1,3-diols. J Chem Soc Perkin Trans 1(2001):2983–2996
32. Feula A, Dhillon SS, Byravan R et al (2013) Synthesis of azetidines and pyrrolidines via iodocyclisation of homoallyl amines and exploration of activity in a zebrafish embryo assay. Org Biomol Chem 11:5083–5093
33. Miao CB, Dong CP, Zhang M et al (2013) Highly stereoselective, one-pot synthesis of azetidines and 2,4-dioxo-1,3-diazabicyclo [3.2.0] compounds mediated by I_2. J. Org. Chem. 78:4329–4340
34. Yang Y, Hayashi Y, Fujii Y et al (2012) Efficient cyclic carbonate synthesis catalyzed by zinc cluster systems under mild conditions. Catal Sci Technol 2:509–513
35. Zalomaeva OV, Chibiryaev AM, Kovalenko KA et al (2013) Cyclic carbonates synthesis from epoxides and CO_2 over metal–organic framework Cr-MIL-101. J Catal 298:179–185
36. Yang ZZ, Li YN, Wei YY et al (2011) Protic onium salts-catalyzed synthesis of 5-aryl-2-oxazolidinones from aziridines and CO_2 under mild conditions. Green Chem 13:2351–2353
37. Phung C, Ulrich RM, Ibrahim M et al (2011) The solvent-free and catalyst-free conversion of an aziridine to an oxazolidinone using only carbon dioxide. Green Chem 13:3224–3229
38. Han MY, Jia JY, Wang W (2014) Recent advances in organocatalytic asymmetric synthesis of polysubstituted pyrrolidines. Tetrahedron Lett 55:784–794
39. Brown AR, Uyeda C, Brotherton CA et al (2013) Enantioselective thiourea-catalyzed intramolecular cope-type hydroamination. J Am Chem Soc 135:6747–6749
40. Chapurina Y, Ibrahim H, Guillot R (2011) Catalytic, enantioselective intramolecular hydroamination of primary amines tethered to Di- and trisubstituted alkenes. J Org Chem 76:10163–10172
41. Chen MZ, Micalizio GC (2009) Convergent synthesis of piperidines by the union of conjugated alkynes with imines: a unique regioselective bond construction for heterocycle synthesis. Org Lett 11:4982–4985
42. Lebold TP, Leduc AB, Kerr MA (2009) Zn (II)-catalyzed synthesis of piperidines from propargyl amines and cyclopropanes. Org Lett 11:3770–3772
43. Guérinot A, Serra-Muns A, Gnamm C et al (2010) $FeCl_3$-catalyzed highly diastereoselective synthesis of substituted piperidines and tetrahydropyrans. Org Lett 12:1808–1811
44. Sun Z, Winschel GA, Zimmerman PM et al (2014) Enantioselective synthesis of piperidines through the formation of chiral mixed phosphoric acid acetals: experimental and theoretical studies. Angew Chem Int Ed 53:11194–11198
45. Akbaslar D, Demirkol O, Giray S (2014) Paal-Knorr pyrrole synthesis in water. Synth Commun 44:1323–1332
46. Cho H, Madden R, Nisanci B et al (2015) The Paal-Knorr reaction revisited. A catalyst and solvent-free synthesis of underivatized and N-substituted pyrroles. Green Chem 17:1088–1099

47. Kaewchangwat N, Sukato R, Vchirawongkwin V et al (2015) Direct synthesis of aryl substituted pyrroles from calcium carbide: an underestimated chemical feedstock. Green Chem 17:460.
48. Ran L, Ren Z-H, Wang Y-Y et al (2014) Copper-catalyzed homocoupling of ketoxime carboxylates for synthesis of symmetrical pyrroles. Green Chem 16:112–115
49. Wang P, Ma F-P, Zhang Z-H (2014) *L*-(+)-Tartaric acid and choline chloride based deep eutectic solvent: an efficient and reusable medium for synthesis of N-substituted pyrroles via Clauson-Kaas reaction. J Mol Liq 198:259–262
50. Abid M, Landge SM, Török B (2006) An efficient and rapid synthesis of N-substituted pyrroles by microwave assisted solid acid catalysis. Org Prep Proced Int 38:495–500
51. Naeimi H, Dadaei M (2014) Efficient and green synthesis of N-aryl pyrroles catalyzed by ionic liquid [H-NMP][HSO_4] in water at room temperature. J Chin Chem Soc 61:1127–1132
52. Singh SB, Verma PK, Tiwari K et al (2014) Supramolecular catalysis in the synthesis of polyfunctionalised pyrroles. Supramol Chem 26:882–889
53. Guan Z-H, Li L, Ren Z-H et al (2011) A facile and efficient synthesis of multisubstituted pyrroles from enaminoesters and nitroolefins. Green Chem 13:1664–1668
54. Landge SM, Schmidt A, Outerbridge V, Török B (2007) Synthesis of pyrazoles by a one-pot tandem cyclization-dehydrogenation approach on Pd/C/K-10 catalyst. Synlett:1600–1604
55. Borkin DA, Puscau M, Carlson A, Solan A, Wheeler KA, Török B, Dembinski R (2012) Synthesis of diversely 1,3,5-trisubstituted pyrazoles via 5-exo-dig cyclization. Org Biomol Chem 10:4505–4508
56. Vuluga D, Legros J, Crousse B, Bonnet-Delpon D (2009) Synthesis of pyrazoles through catalyst-free cycloaddition of diazo compounds to alkynes. Green Chem 11:156–159
57. Li XT, Liu YH, Liu X, Zhang ZH (2015) Meglumine catalyzed one-pot, three-component combinatorial synthesis of pyrazoles bearing a coumarin unit. RSC Adv 5:25625–25633
58. Schmitt DC, Taylor AP, Flick AC, Kyne RE (2015) Synthesis of pyrazoles from 1,3-diols via hydrogen transfer catalysis. Org Lett 17:1405–1408
59. Matcha K, Antonchick AP (2014) Cascade multicomponent synthesis of indoles, pyrazoles and pyridazinones by functionalization of alkenes. Agnew Chem Int Ed 53:11960–11964
60. Dias D, Pacheco BS, Cunico W, Pizzuti L, Pereira CMP (2014) Recent advances on the green synthesis and antioxidant activities of pyrazoles. Mini-Rev Med Chem 14:1078–1092
61. Fustero S, Sanchez-Rosello M, Barrio P, Simon-Fuentes A (2011) From 2000 to mid-2010: a fruitful decade for the synthesis of pyrazoles. Chem Rev 111:6984–7034
62. Cao H, Jiang H-F, Zhou X-S et al (2012) CuO/CNTs-catalyzed heterogeneous process: a convenient strategy to prepare furan derivatives from electron-deficient alkynes and [α]-hydroxy ketones. Green Chem 14:2710–2714
63. Cao H, Jiang H, Huang H (2011) Transition-metal-catalyzed domino reactions: efficient one-pot regiospecific synthesis of highly functionalized polysubstituted furans from electron-deficient alkynes and 2-Yn-1-ols. Synthesis 2011:1019–1036
64. Yu C, Lu J, Li T et al (2011) A NHC-involved, cascade, metal-free, and three-component synthesis of 2,3-diarylated fully substituted furans under solvent-free conditions. Synlett 2011:2420–2424
65. Reddy BVS, Somashekar D, Reddy AM et al (2010) PEG 400 as a reusable solvent for 1,4-dipolar cycloadditions via a three-component reaction. Synthesis 2010:2069–2074
66. Yadav JS, Reddy BVS, Shubashree S et al (2007) Organic synthesis in water: green protocol for the synthesis of 2-amino furan derivatives. J Mol Cat A Chem 272:128–131
67. Azizian J, Mohammadizadeh MR, Mohammadi A et al (2005) A modified and green methodology for preparation of polysubstituted furans. Heteroat Chem 16:259–262
68. Allais C, Grassot J-M, Rodriguez J et al (2014) Metal-free multicomponent syntheses of pyridines. Chem Rev 114:10829–10868
69. Xu F, Wang C, Wang H et al (2015) Eco-friendly synthesis of pyridines via rhodium-catalyzed cyclization of diynes with oximes. Green Chem 17:799. https://doi.org/10.1039/C4GC01756K
70. Kidwai M, Chauhan R (2014) K_2CO_3 catalyzed green and rapid access to 2-amino-3,5-dicarbonitrile-6-thio-pyridines. J Iran Chem Soc 11:1005–1013

71. Shaikh YI, Shaikh AA, Nazeruddin GM (2012) Ammonia solution catalyzed one-pot synthesis of highly functionalized pyridine derivatives. J Chem Pharm Res 4:4953–4956
72. Yin G, Liu Q, Ma J (2012) Solvent- and catalyst-free synthesis of new hydroxylated trisubstituted pyridines under microwave irradiation. Green Chem 14:1796–1798
73. Yang X-H, Zhou Y-H, Zhang P-H et al (2013) An effective, one-pot synthesis of fully substituted pyridines under microwave irradiation in the absence of solvent. J Heterocyclic Chem 50:1346–1350
74. De Paolis O, Baffoe J, Landge SM et al (2008) Multicomponent domino cyclization-oxidative aromatization on a bifunctional Pd/C/K-10 catalyst: an environmentally benign approach toward the synthesis of pyridines. Synthesis 21:3423–3428
75. Wang Q, Wan C, Gu Y et al (2011) A metal-free decarboxylative cyclization from natural α-amino acids to construct pyridine derivatives. Green Chem 13:578–581
76. Lou S, Zhang J (2013) Pyrimidines. In: Jie JL (ed) Heterocyclic chemistry in drug discovery. John Wiley & Sons, Inc, New York, NY, pp 569–613
77. Gore RP, Rajput AP (2013) A review on recent progress in multicomponent reactions of pyrimidine synthesis. Drug Invent Today 5:148–152
78. Soheilizad M, Adib M, Sajjadifar S (2014) One-pot and three-component synthesis of substituted pyrimidines catalysed by boron sulfuric acid under solvent-free conditions. J Chem Res 38:524–527
79. Sivagamisundari G, Pushpalatha AM, Ranee SJ (2014) Silica bonded S-sulphonic acid as a green catalyst in the synthesis of functionalized pyrimidine under solvent-free microwave irradiation conditions. Int J Sci Eng Technol 3:852–855
80. Khan SA, Asiri AM, Kumar S et al (2014) Green synthesis, antibacterial activity and computational study of pyrazoline and pyrimidine derivatives from 3-(3,4-dimethoxy-phenyl)-1-(2,5-dimethylthiophen-3-yl)-propenone. Eur J Chem 5:85–90
81. Ezhilarasi MR, Prabha B, Prabakaran S (2014) Microwave assisted synthesis and spectral studies of thiophenyl pyrimidine derivatives. J Appl Chem 3:1929–1935
82. Babu KR, Paul VL, Rao VM (2014) Substituted pyridine catalysed domino synthesis of pyrazolines and pyrimidines. World J Pharm Res 3:389–398
83. Abid M, Spaeth A, Török B (2006) Solvent-free solid acid-catalyzed electrophilic annelations: a new green approach for the synthesis of substituted five-membered N-heterocycles. Adv Synth Catal 348:2191–2196
84. Xu D, Yang W, Luo S, Wang B, Wu J, Xu Z (2007) Fischer indole synthesis in Brønsted acidic ionic liquids: a green, mild, and regiospecific reaction system. Eur J Org Chem 2007:1007–1012
85. Xu D, Wu J, Luo S, Zhang J, Wu J, Du X, Xu Z (2009) Fischer indole synthesis catalyzed by novel SO_3H-functionalized ionic liquids in water. Green Chem 11:1239–1246
86. Alaerts L, Seguin E, Poelman H, Thibault-Starzyk F, Jacobs P, de Vos DE (2006) Probing the Lewis acidity and catalytic activity of the metal–organic framework $[Cu_3(BTC)_2]$ (BTC=Benzene-1,3,5-tricarboxylate). Chem Eur J 12(28):7353–7363
87. Schlichte K, Kratzke T, Kaskel S (2004) Improved synthesis, thermal stability and catalytic properties of the metal-organic framework compound $Cu_3(BTC)_2$. Microporous Mesoporous Mater 73:81–88
88. Horcajada P, Surble S, Serre C, Hong D, Seo Y, Chang JS, Greneche JM, Margiolaki I, Ferey G (2007) Synthesis and catalytic properties of MIL-100(Fe), an iron (III) carboxylate with large pores. Chem Commun:2820–2822
89. Gascon J, Hernandez-Alonso M, Ameida A, Van Klink GPM, Kapteijn F, Mul G (2009) Amino-based metal-organic frameworks as stable, highly active basic catalysts. J Catal 261:75–87
90. Savonnet M, Aguado S, Ravon U, Bazer-Bachi D, Lecocq V, Bats N, Pinel C, Farrusseng D (2009) Solvent free base catalysis and transesterification over basic functionalized metal-organic frameworks. Green Chem 11:1729–1732
91. Li H, Zhu Z, Zhang F, Xie S, Li H, Li P, Zhou X (2011) Palladium nanoparticles confined in the cages of MIL-101: an efficient catalyst for the one-pot indole synthesis in water. ACS Catal 1:1604–1612
92. Dinica R, Furdui B, Ghinea I, Bahrim G, Bonte S, Demeunynck M (2013) Novel *one-pot* green synthesis of indolizines biocatalysed by *Candida antarctica* lipases. Mar Drugs 11:431–439
93. Yang B, Huang Z, Guan H, Niu X, Li Y, Fang S, Ma C (2013) New routes for the synthesis of fused pyrrole scaffolds through

transition metal-free tandem reactions. Tetrahedron Lett 54:5994–5997

94. Cheng CC, Yan SJ (2005) The Friedländer synthesis of quinolines. Organic reactions. Wiley, New York, NY

95. Arcadi A, Chiarini M, Giuseppe SD, Marinelli F (2003) A new approach to the Friedländer synthesis of quinolines. Synlett 2003:203–206

96. Sels BF, De Vos DE, Jacobs PA (2001) Hydrotalcite-like anionic clays in catalytic organic reactions. Catal Rev 43:443–448

97. Motokura K, Mizugaki T, Ebitani K, Kaneda K (2004) Multifunctional catalysis of a ruthenium-grafted hydrotalcite: one-pot synthesis of quinolines from 2-aminobenzyl alcohol and various carbonyl compounds via aerobic oxidation and aldol reaction. Tetrahedron Lett 45:6029–6032

98. Wang J, Fan X, Zhang X, Han L (2004) Green preparation of quinoline derivatives through $FeCl_3{\cdot}6H_2O$ catalyzed Friedländer reaction in ionic liquids. Can J Chem 82:1192–1196

99. Yadav JS, Purushothama Rao P, Sreenu D, Srinivasa Rao R, Naveen Kumar V, Nagaiah K, Prasad AR (2005) Sulfamic acid: an efficient, cost-effective and recyclable solid acid catalyst for the Friedländer quinoline synthesis. Tetrahedron Lett 46:7249–7253

100. Akbari J, Heydari A, Kalhor HR, Kohan SA (2010) Sulfonic acid functionalized ionic liquid in combinatorial approach, a recyclable and water tolerant-acidic catalyst for one-pot Friedländer quinoline synthesis. J Comb Chem 12:137–140

101. Kulkarni A, Török B (2010) Microwave-assisted multicomponent domino cyclization–aromatization: an efficient approach for the synthesis of substituted quinolines. Green Chem 12:875–878

102. Peng H, Chen X, Chen Y, He Q, Xie Y, Yang C (2011) Solvent-free synthesis of δ-carbolines/carbazoles from 3-nitro-2-phenylpyridines/2-nitrobiphenyl derivatives using DPPE as a reducing agent. Tetrahedron 67:5725–5731

103. Cadogan JIG, Carmeron-Wood M, Makie RK, Searle RJG (1965) The reactivity of organophosphorus compounds. Part XIX. Reduction of nitro-compounds by triethyl phosphite: a convenient new route to carbazoles, indoles, indazoles, triazoles, and related compounds. J Chem Soc:4831–4837

104. Ishida T, Tsunoda R, Zhang Z, Hamasaki A, Honma T, Ohashi H, Yokoyama T, Tokunaga M (2014) Supported palladium hydroxide-catalyzed intramolecular double C-H bond functionalization for synthesis of carbazoles and dibenzofurans. Appl Catal B Environ 150–151:523–531

105. Kulkarni A, Abid M, Török B, Huang X (2009) A direct synthesis of b-carbolines via a three-step one-pot domino approach with a bifunctional Pd/C/K-10 catalyst. Tetrahedron Lett 50:1791–1794

106. Erickson PR, Grandbois M, Arnold WA, McNeil K (2012) Photochemical formation of brominated dioxins and other products of concern from hydroxylated polybrominated diphenyl ethers (OH-PBDEs). Environ Sci Technol 46:8174–8180

Chapter 2

Greener Methods for Amide Bond Synthesis

Nathan J. Oldenhuis, Aaron M. Whittaker, and Vy M. Dong

Abstract

The amide bond is an important linkage ubiquitous in nature and technology found in many small molecules and materials. Formation of amide bonds is most easily accomplished via condensation at high temperatures using amines and carboxylic acids, but peptide coupling agents are pervasively used to form amides under milder conditions. While these coupling agents are robust synthetic reagents, they are generally used in super stoichiometric amounts which can be disadvantageous due to their sensitizing nature and lack of atom economy. This chapter presents a brief survey of "greener" methods to form amides via the catalytic use or release of small benign molecules such as H_2 or H_2O. Both heterogenous and homogenous catalytic systems will be discussed. We will group catalytic methods by which functional moieties (e.g., alcohol, aldehyde, nitrile, azide, and acid homologs) are used as starting materials for the reaction. We hope to inform the reader of the broad scope of "green" amide bond-forming reactions which can be used in lieu of canonical peptide coupling reagents.

Key words Amides, Green, Catalysis, Hydrogen borrowing, Amidation

1 Introduction

Amide bonds stitch together life's most basic building blocks to create various peptides and proteins. Moreover, amides are common intermediates and targets in the creation of new materials and medicines. Due to the importance of these bonds, there now exists many ways to create amide linkages to access endogenous biomolecules, as well as non-endogenous pharmaceuticals, synthetic proteins, and more, by a robust set of reagents and methods [1–3]. Most commonly used are coupling reagents, which operate by activating a carboxylic acid for nucleophilic displacement by an amine nucleophile (Fig. 1). These transformations are general and high yielding, but require an excess of the coupling reagent, as well as a stoichiometric amount of base to prevent insoluble carboxylate ammonium salt formation. The excess reagent used in these reactions is wasteful and can often be toxic. A survey of pharmaceutical manufacturers in 2005 named "amide formation avoiding poor

Paul F. Richardson (ed.), *Green Chemistry in Drug Discovery: From Academia to Industry*, Methods in Pharmacology and Toxicology, https://doi.org/10.1007/978-1-0716-1579-9_2, © Springer Science+Business Media, LLC, part of Springer Nature 2022

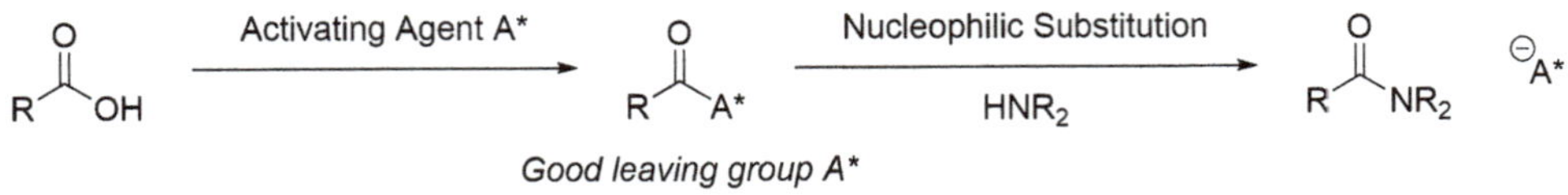

Fig. 1 Traditional approach to amide bond formation via activated carboxylic acid

atom economy reagents" the number one challenge to address in the coming years [4].

In response to this challenge, several new amide-forming reactions have emerged to improve atom economy [5–9]. Methods that form amides by catalysis via the release or use of small benign molecules, such as H_2O, O_2, N_2, H_2, etc. are targeted to decrease waste. This chapter will survey amide bond formation through the use or loss of H_2. For other promising approaches using catalysis, we point readers to recent reviews [1, 8, 10]. Relevant catalytic methods involving the transfer of other small benign molecules (water, nitrogen, etc.) will be included briefly for topical clarity, and amidation reactions involving the transfer of hydrogen from aldehydes, esters, azides, and nitriles will also be covered.

2 Dehydrogenative Amide Synthesis (DAS)

2.1 Dehydrogenative Amide Synthesis (DAS) Using Alcohols and Amines

The first example of amide synthesis from amines and alcohols was explored by Murahashi with the formation of lactams from aminoalcohols in the presence of $RuH_2(PPh_3)_4$ [11]. Using aldehydes and secondary amines, both γ and δ lactams were synthesized using this system as well as tertiary amides. In this seminal work, a stoichiometric amount of a hydrogen acceptor (benzylacetone) was required to remove the equivalent of hydrogen from the ruthenium and turn over the catalytic cycle. When the hydrogen acceptor was removed from the reaction, the authors noted that direct amination via dehydration occurred instead. Similar catalytic activity for lactam formation was observed using $Ru_3(CO)_{12}$, $RuCl_3 \cdot H_2O$, $RuCl_2(PPh_3)_3$, and $RhCl(PPh_3)_3$. Many of these metal complexes make later appearances in this chapter as precursors for more robust catalytic systems.

Milstein demonstrated the first dehydrogenative bimolecular amide formation from alcohols and amines using a novel tridentate phosphine-nitrogen-nitrogen (PNN)-type ruthenium(II) pincer complex (Fig. 2) [12]. The pyridine-based pincer complex had previously been shown to catalyze ester formation upon activation with base from alcohols via the direct elimination of hydrogen gas from the catalyst. The elimination of dihydrogen directly from the catalyst after oxidation was facilitated via a Lewis basic site formed through dearomatization of the pyridine on the ligand and the Lewis acidic ruthenium [13–15]. During their optimization studies, the authors found that amide formation was favored over ester

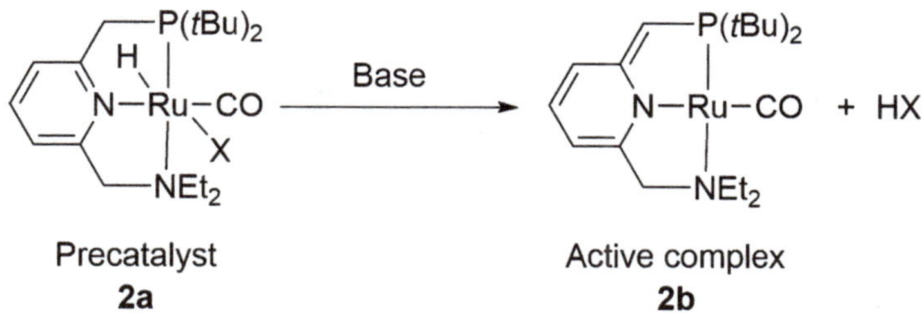

Fig. 2 Pyridine-based PNN-pincer complex for amide bond formation

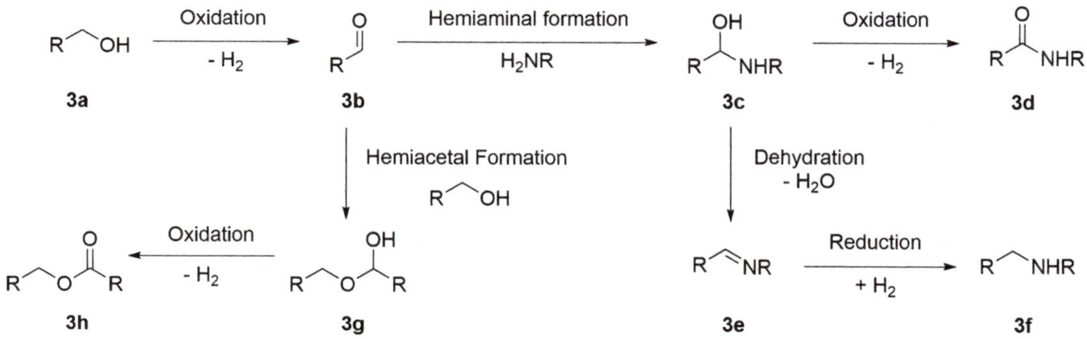

Fig. 3 Potential amidation side products

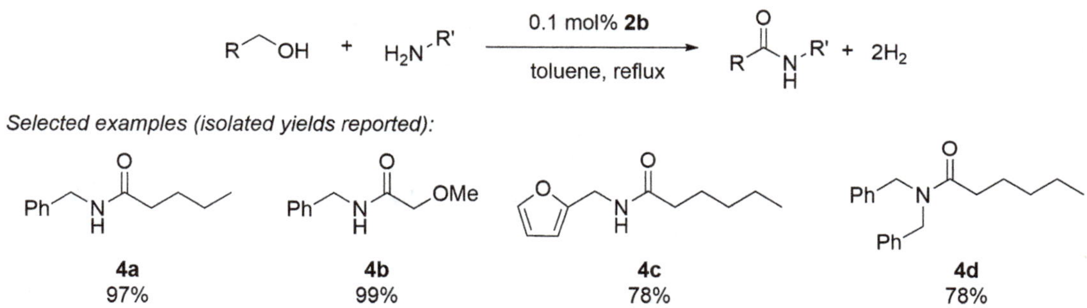

Fig. 4 Selected examples of DAS using complex **2a**

formation as well as imine and amine formation. Considering the oxidative route displayed in Fig. 3, after the alcohol (**3a**) is oxidized to the carbonyl complex (**3b**), both the amine and the alcohol could attack **3b** producing either a hemiaminal (**3c**) or a hemiacetal (**3g**). From there, a number of competitive pathways are possible. The hemiacetal can undergo oxidation to generate the ester product (**3h**). The hemiaminal (**3c**) can undergo dehydration to afford the imine (**3e**), which upon reduction by the metal-hydride would yield an amine (**3f**). Finally, the catalyst can oxidize the hemiaminal to yield the amide (**3d**). The robust pincer complex affords a variety of secondary amides, with only 0.1 mol% of the active complex **2b** formed in situ needed for the reaction to go to completion (Fig. 4). The elimination of stoichiometric reagents, production of

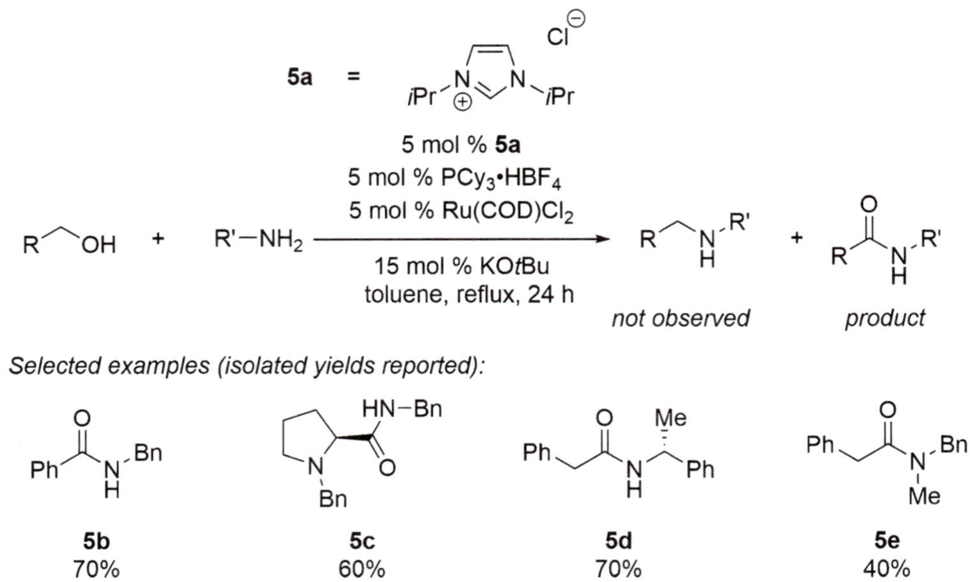

Fig. 5 Madsen's NHC complex formation for DAS

hydrogen gas as the only by-product, and eventual commercial availability demonstrated that not only was green amide bond formation possible but could become economically viable.

While looking for catalysts for amine alkylation with alcohols, Madsen found that ruthenium(II) *N*-heterocyclic carbene (NHC) phosphine complexes formed amides almost exclusively (Fig. 5) [16]. Considering the reaction pathway outlined in Fig. 3, it is easy to see that dehydrogenative amide synthesis (DAS) could compete with alkylation pathways. The use of Ru(COD)Cl_2 with imidazolium salt **5a**, tricyclopentylphosphine, and potassium *tert*-butoxide gave the highest yield of amides under reflux in toluene. A variety of primary amines and alcohols yielded secondary amides in excellent yields. Stereocenters on both the alcohol (**5c**) and amine (**5d**) were not racemized during the transformation. One example of a tertiary amide was shown using a secondary amine, but in order to obtain an appreciable amount (40% isolated yield) of amide, the mixture had to be refluxed in mesitylene (**5e**). This work showed that a ruthenium(II)-NHC complex formed in situ performed DAS but required much higher (5 mol% vs. 0.1 mol%) catalyst loadings and longer reaction times (24 h vs. 6–12 h) compared to Milstein's original report.

Milstein's report revived an interest in identifying catalytic methods for DAS. Several studies investigating the mechanism and new catalysts have since appeared. Another ruthenium(II)-based system was subsequently discovered by the Williams group [17]. [Ru(*p*-cymene)Cl_2]$_2$ was complexed with *bis* (diphenylphosphino)butane (dppb) in the presence of cesium carbonate and refluxed in *t*-BuOH to form a variety of amides from

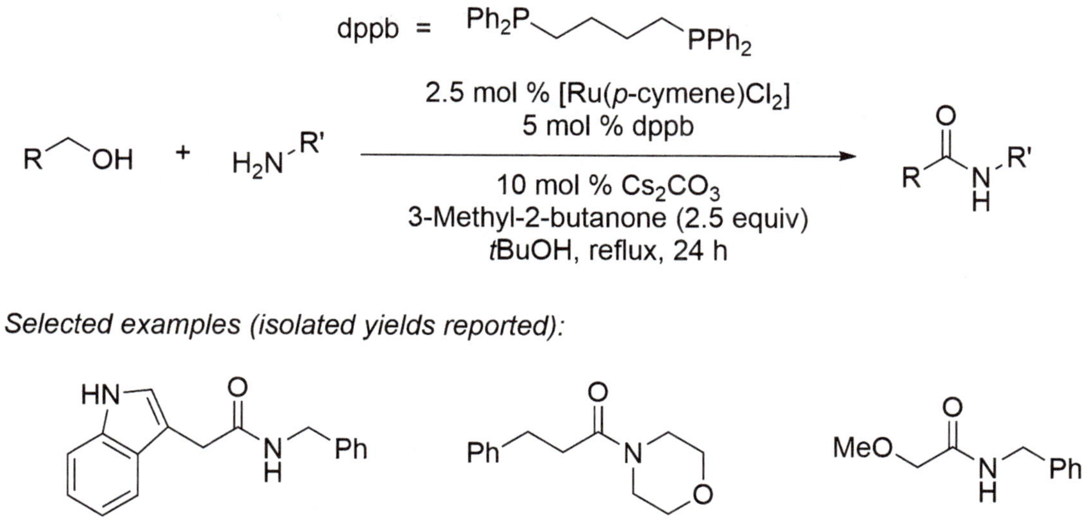

6a	**6b**	**6c**
36%	31%	73%

Fig. 6 DAS using dppb as a ligand and 2-methyl-3-butanone as a hydrogen acceptor

alcohols and amines (Fig. 6). In accordance with the work to date, the methodology was largely limited to less bulky aliphatic amines and alcohols (**6a**, **6c**). Of note, a tertiary amide was formed using morpholine, but the yield in this case was lower (**6b**). 2.5 equivalents of 3-methyl-2-butanone were used as the hydrogen acceptor, and both this ketone and the corresponding alcohol formed upon reduction can easily be removed after reaction due to their low boiling points (~110 °C). This system was considered advantageous compared to Milstein's PNN complex (**2b**) due to the convenient procedure and use of both a commercially available ligand and ruthenium source. However, lower isolated yields, higher catalyst loadings, and longer reaction times were also observed.

The Grützmacher group departed from the use of ruthenium and reported DAS from alcohols and amines using a rhodium (I) complex [18]. This work focused on reducing the high temperatures required for amide formation required by both the previously developed methods. While the [Rh(trop$_2$N)(PPh$_3$)] (Fig. 7, **7b**) looks much different than both the complexes used by Milstein and Madsen, it was designed to contain the same "cooperative ligand" effects as the Milstein catalyst (a Lewis acidic site and a Lewis basic site). According to density functional theory (DFT) calculations performed by Grützmacher, the Lewis basic nitrogen atom (HOMO) and the Lewis acidic center on the rhodium (LUMO) activate dihydrogen heterolytically. While the protocol formed carboxylic acids, esters, and amides in good yields at ambient temperatures (~25 °C), the procedure required a hydrogen acceptor (5 equivalents of cyclohexanone, or methyl methacrylate)

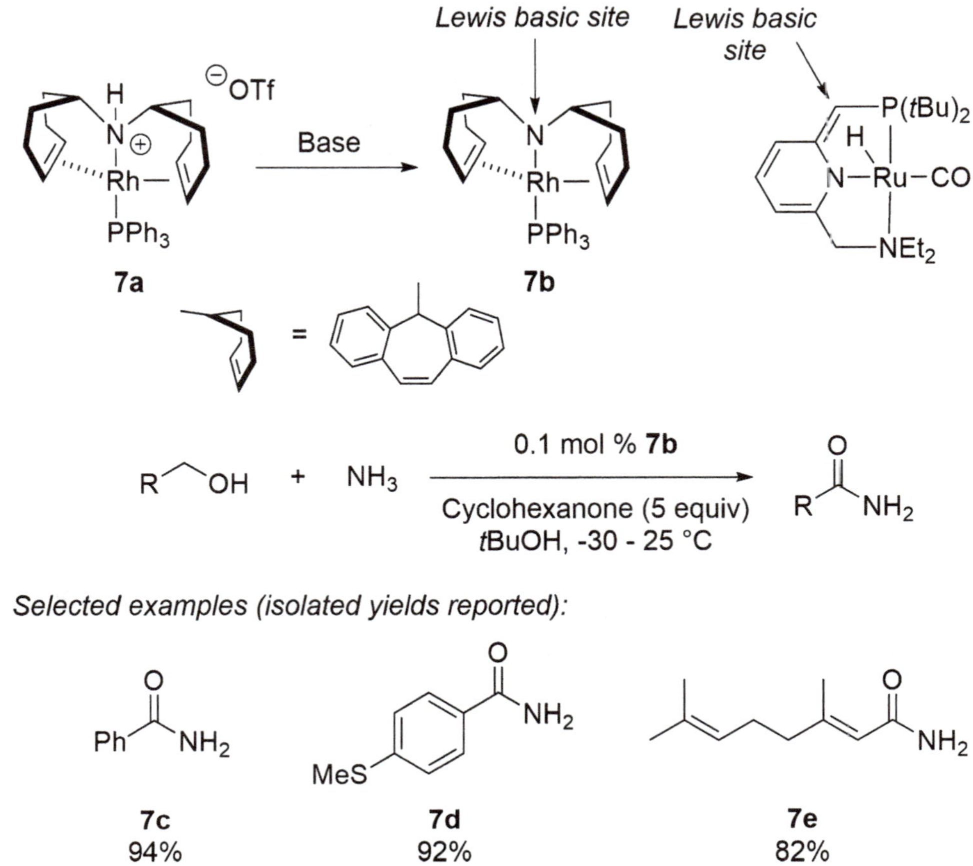

Fig. 7 Rhodium(I) trop$_2$N complex for DAS

to turn over the catalyst. Using the activated complex (**7b**) directly versus the precatalyst (**7a**) and base, one can produce primary amides by DAS from ammonia and primary alcohols (**7c**–**7e**). This result represents the first-time ammonia could be activated in a DAS reaction. Benzylamine was also utilized for the formation of secondary amides.

Until 2009, all the complexes used for catalytic amide bond formation from amines and alcohols used phosphine ligands to facilitate the reaction. The Hong group sought a phosphine-free catalytic system to decrease catalyst deactivation through the decomposition of tertiary phosphines via either heat or air [19]. Both [Ru(*p*-cymene)Cl_2]$_2$ (method **A**) and [Ru(benzene)-Cl_2]$_2$ (method **B**) in tandem with catalytic amounts of sodium hydride, an imidazolium salt (**8a**), and a ligand (pyridine or acetonitrile) produce a variety of amides under reflux conditions in toluene (Fig. 8). The reaction times were slightly longer (36 h), due to a lower catalyst TON as compared to Milstein's 2007 work. However, compared to the work by Williams ([Ru(*p*-cymene)Cl_2]$_2$ and dppb), no hydrogen acceptor was required, thus providing an

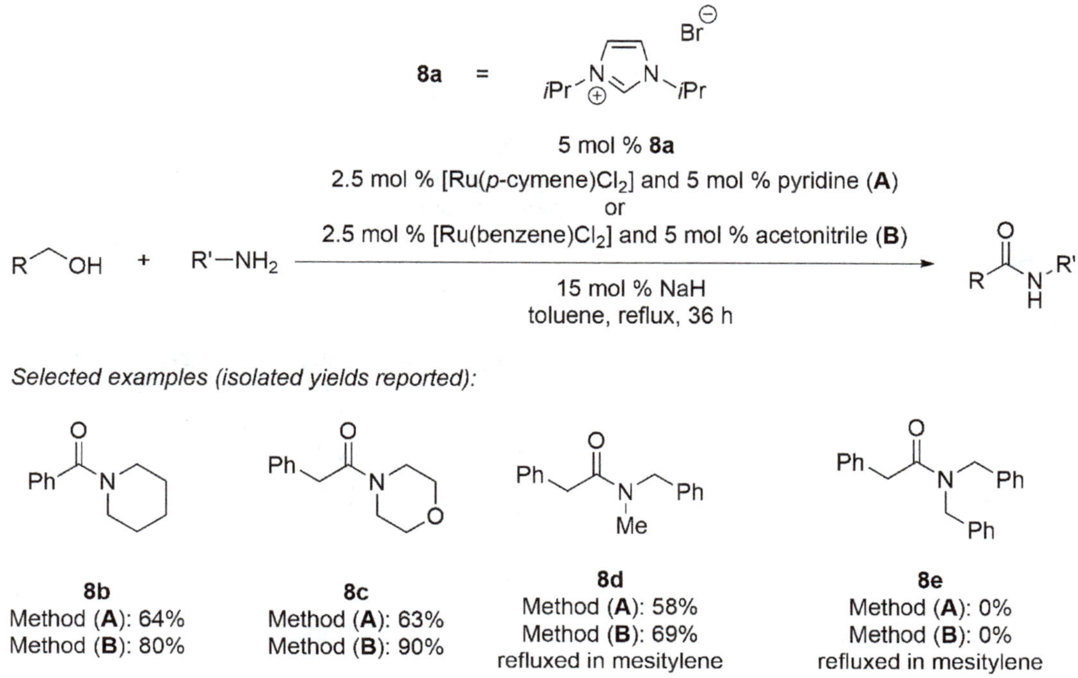

Fig. 8 Formation of tertiary amides by Ru-NHC complexes

advantage in terms of atom economy. Both catalyst systems were tolerant of aliphatic alcohols and amines to yield secondary amides and showed some effectiveness for tertiary amide formation. Piperidine, morpholine, and *N*-methylbenzylamine are coupled with various alcohols to give the corresponding tertiary amides **8b**, **8c,** and **8d** in good yields. For tertiary amide formation with *N*-methylbenzylamine, reflux in mesitylene was required. Using either method, no reactivity was observed with dibenzylamine due to the high steric encumbrance.

Several follow-up studies were published by Hong [20–22] and Madsen [23] to more thoroughly investigate the formation, mechanism and substrate scope of the ruthenium NHC based-systems for DAS. The Hong group created amidation catalysis using more economical ruthenium(II) or ruthenium (0) sources (such as $RuH_2(PPh_3)_4$, $RuCl_3$, Ru(cod)(cot), Ru Black, Ru on Al_2O_3, and $Ru_3(CO)_{12}$), an NHC precursor, base, and a ligand. A comparison of the ligands, bases, and ruthenium sources is shown in Fig. 9. Both Hong and Madsen showed that Grubb's olefin metathesis catalysts are able to mediate the formation of amide bonds from amines and alcohols as well. This work demonstrated that almost any ruthenium-based NHC catalyst showed some degree of activity for amide bond formation. An additional example was provided by Albrecht using a triazolylidene complex instead of an imidazolium-based one (Fig. 10) [24].

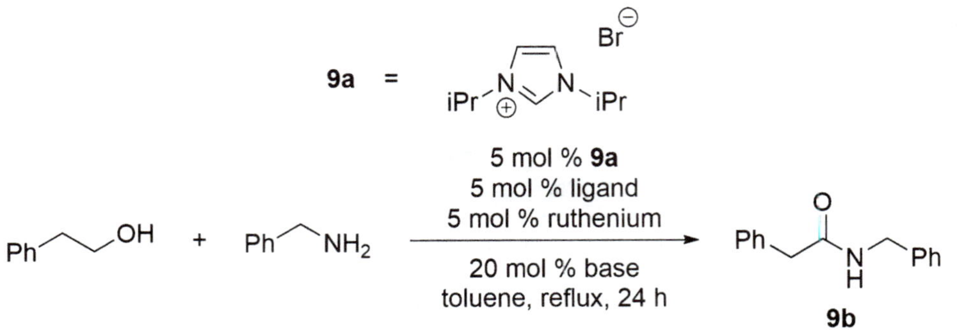

Ruthenium source	***Ligand***	***Base***	***Yield of 9b***
$RuCl_3$	Pyridine	NaH	80%
$RuH_2(PPh_3)_4$	Acetonitrile	NaH	96%
$RuH_2(PMe_3)_4$	Acetonitrile	NaH	21%
$RuH_2(CO)(PPh_3)_3$	Acetonitrile	NaH	87%
$RuH_2(CO)(PPh_3)_3$	Pyridine	NaH	74%
$RuH_2(CO)(PPh_3)_3$	PPh_3	NaH	57%
$RuHCl(CO)(PPh_3)_3$	Acetonitrile	NaH	49%
$Ru(COD)Cl_2$	$PCy_3{\bullet}HBF_4$	KO*t*Bu	92%
Ru(cod)(cot)	Acetonitrile	NaH	83%
$Ru_3(CO)_{12}$	Acetonitrile	NaH	59%
Ru Black	Acetonitrile	NaH	19%
Ru on Al_2O_3	Acetonitrile	NaH	15%

Fig. 9 Comparison of ruthenium NHC complexes for DAS (*Cy* cyclopentyl)

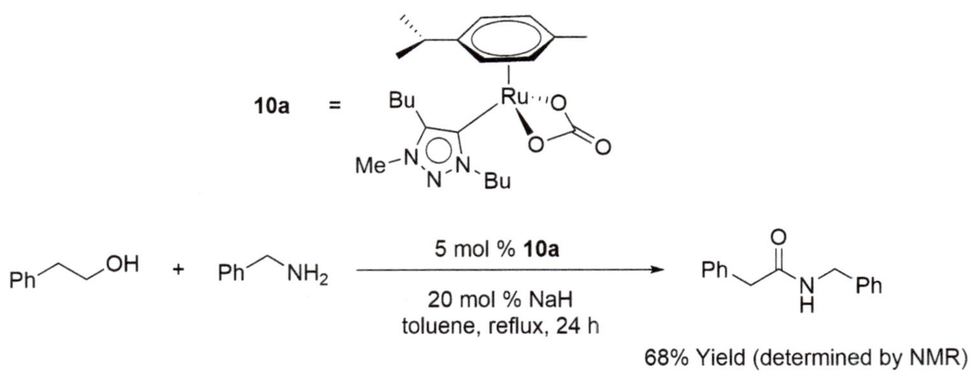

Fig. 10 Triazolylidene ruthenium(II) complex for DAS

Tertiary amide formation using primary alcohols and secondary amines has been especially challenging. Both Madsen and Hong tried to enhance the formation of tertiary amides by reducing the steric bulk of the "wingtip" groups (functional groups on the *N* position of the imidazolium) on the imidazolium salt. Unfortunately, this change did not enable the formation of tertiary amides, and in all cases to this date, *iso*-propyl substituents remain the best for amide formation. Both Madsen and Hong did not observe primary amides using ammonia (or its surrogates) with the NHC

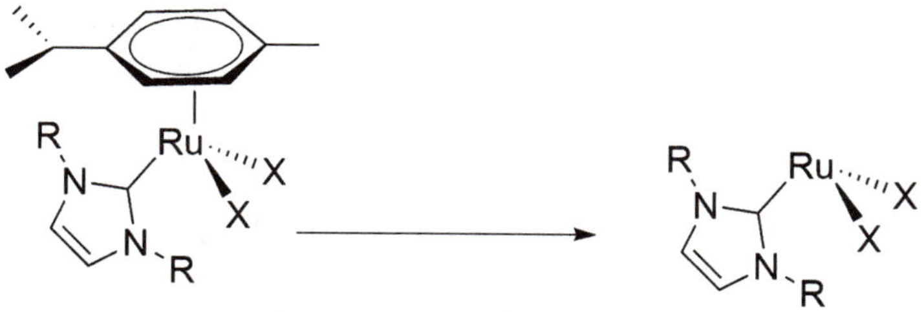

R = Me, *i*Pr, Mes
X = Cl, I

11a **11b**

Fig. 11 In situ generated ruthenium NHC catalyst species

complexes, in contrast to the Rh(I) system reported by Grützmacher. The substrate scope was still largely limited to unhindered aliphatic amines and alcohols, but both Madsen and Hong were able to show lactam formation from α,ω amino alcohols and the ability to form tertiary amides from minimally hindered secondary amines. Hong further demonstrated that using $RuH_2(PPh_3)_4$ as a precursor rather than $[Ru(p\text{-cymene})Cl_2]_2$ to generate an NHC complex greatly increased the ability of the ruthenium(II) complex to form amides directly from aldehydes and amines as well as from alcohols and amines. For more details, see Subheading 3.2.

To elucidate the structure of the active catalyst, both Madsen and Hong prepared a variety of ruthenium NHC complexes (**11a**) (Fig. 11). They postulated that **11b** is the active catalytic species as free *p*-cymene was observed while monitoring the reaction via ^1H NMR indicating disassociation. Preliminary mechanistic studies showed that the formation of a ruthenium hydride species through the action of a base was likely a key step in forming the active catalytic species. Amide bond formation was not observed when no base was added. An interesting observation by Madsen was that imine formation seemed to hinder amide formation. Amines were combined with mixtures of aldehyde (which condenses with the amine to form the imine immediately) and alcohol to measure if the amide product was still observed. Formation of the amide product decreased as the amount of aldehyde present in the system increased. This indicated that the aldehyde formed from oxidation of the alcohol must stay partially or fully coordinated to the catalyst during the reaction and remains on the metal center until a suitable nucleophile approaches.

Hong followed up with a paper addressing the formation of tertiary amides from secondary amines and alcohols using a NHC pre-catalyst derived from $[Ru(benzene)Cl_2]_2$ (Fig. 12) [25]. Much like their previous investigations, preparing the ruthenium (II) NHC "pre-catalyst" beforehand, rather than in situ, led to more efficient tertiary amide bond formation (**12b–12d**). They

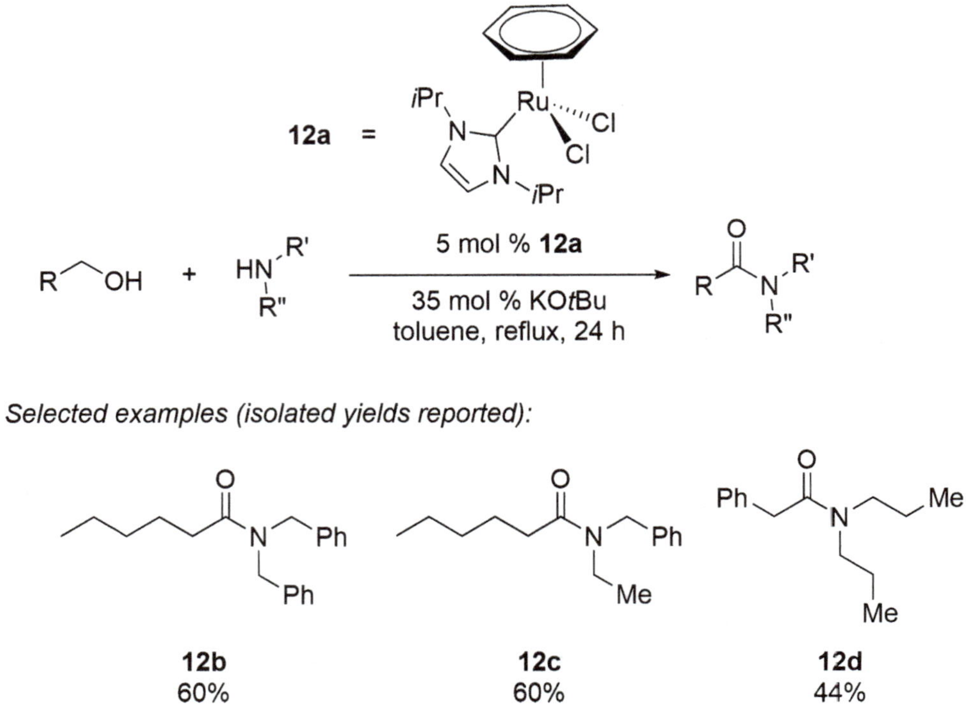

Fig. 12 Ruthenium NHC complex for tertiary amide formation

postulated that the dissociation of the arene is necessary to form the active catalytic species. An increased reaction rate was observed using the benzene complex over the *p*-cymene complex. This effect was attributed to the less sterically hindered and more electron-deficient benzene being able to disassociate faster to form the active complex. However, ester formation was observed using **12a**, and not in the case of the *p*-cymene complex. Increasing the amount of base (35 mol%) was necessary to increase tertiary amide formation, with a simultaneous decrease in ester formation. The authors did not rule out the possibility that ester formation might occur followed by transamidation with the secondary amine to yield the desired product. In summary, this work was the first to demonstrate tertiary amide synthesis in reasonable yields, as well as the ability to step away from unencumbered secondary amines to use bulkier secondary amines such as dibenzyl amine. In similar work reported by Milstein in 2013, a new dipyridine PNN (**13a**) type catalyst was shown to have activity toward tertiary amide formation (**13c–13d**) but could not react with highly sterically encumbered dibenzylamine (**13b**) (Fig. 13) [26].

To deduce what components of catalyst design were important for DAS, the Crabtree group studied a variety of diphosphine diamine catalysts (Fig. 14) based on an amidation catalyst developed by the Williams group (**14a**) in a model lactamization reaction [27]. In the cases where the diphosphine was not present, much

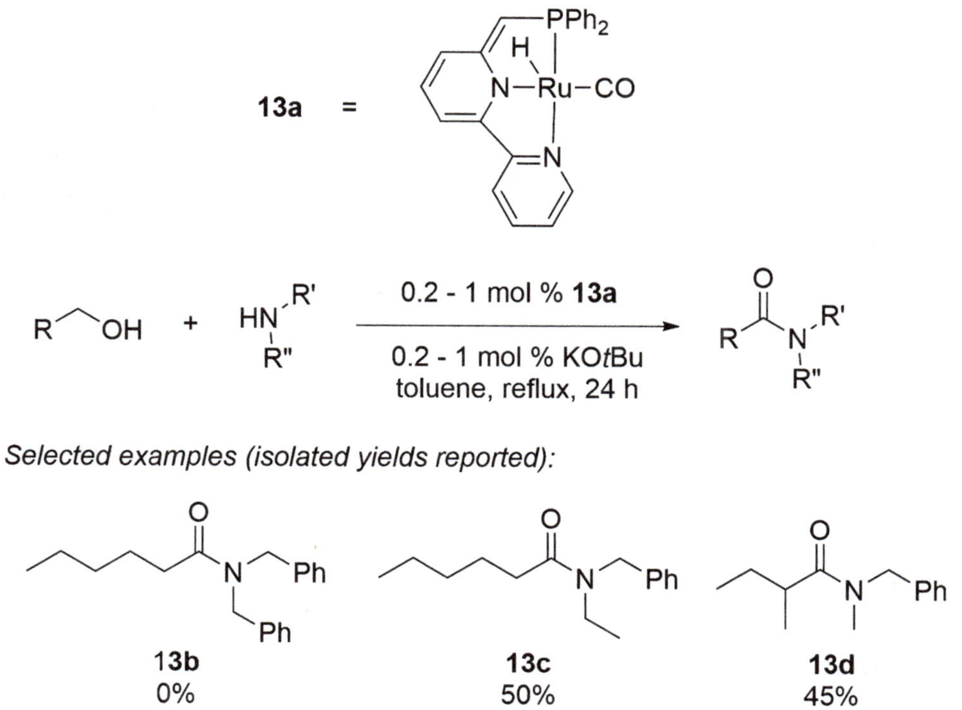

Fig. 13 Ruthenium PNN complex for tertiary amide formation

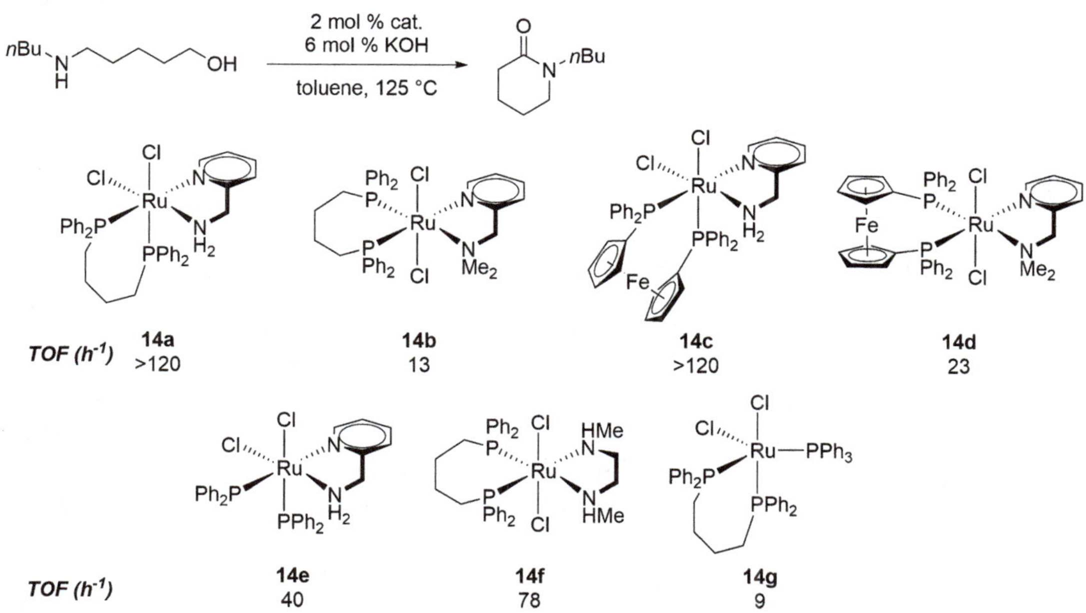

Fig. 14 Diphosphine diamine complexes prepared for study of DAS synthesis

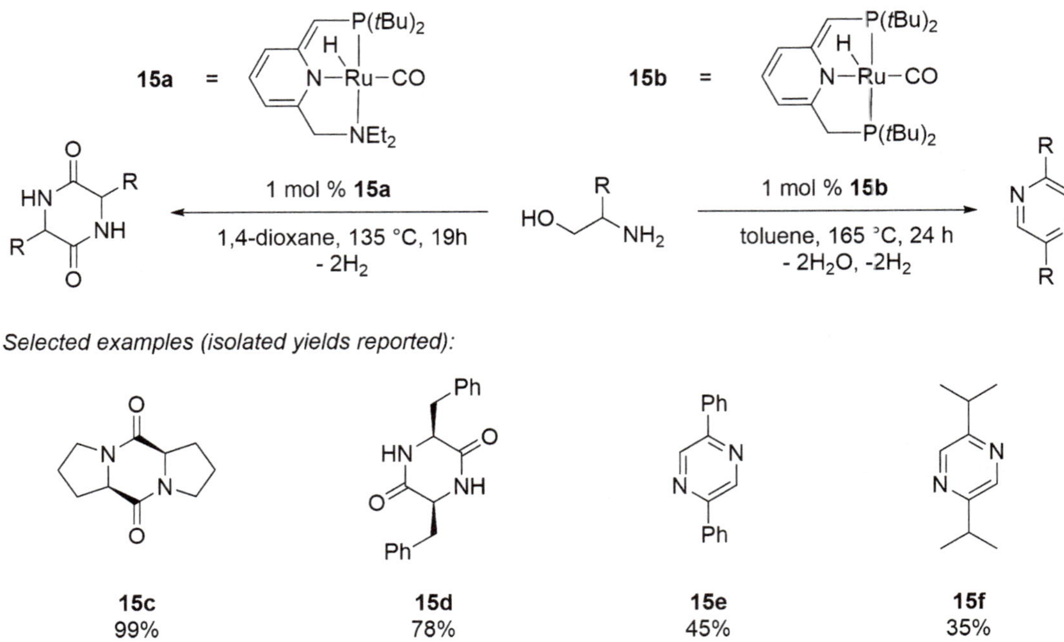

Fig. 15 Synthesis of cyclic dipeptides and pyrazines using amino alcohols and PNN or PNP Ru complexes

lower activity in lactam formation was observed (**14e**). In the diamine, if there were no N–H bonds present, TOFs were reduced (**14b**, **14d**). They reasoned through computational studies that the N–H bond helps stabilize both the aldehyde and hemiaminal intermediates through preventing them from disengaging from the catalyst and forming imines, which can interrupt and shut down the catalytic cycle (**14a**, **14c**, **14f**). While the structures of these catalysts were similar to previously mentioned systems, Crabtree's study provided insights to influence future catalyst design.

In a follow-up study, Milstein focused on making cyclic dipeptides, as well as pyrazines using their PNN- (**15a**) and PNP-based (**15b**) pincer complexes (Fig. 15) [28]. β-Amino alcohols in combination with the PNN complex were shown to favor forming the cyclic dipeptide (**15c**, **15d**). When alaninol (R = Me) was used, the polyamide was obtained instead. Pyrazine formation could be promoted with the PNP complex (**15b**) which is known to promote imine formation over amide formation. A cyclic diimine intermediate is obtained that spontaneously oxidizes to the substituted pyrazine (**15e**, **15f**). Accessing symmetrically substituted pyrazines is an important advance because these structures are present in a number of pharmacologically active complexes.

In a related study, the Glorius group used methanol as the alcohol in DAS to form primary formamides. To drive the reaction to completion, a new NHC ruthenium(II) complex was developed derived from [Ru(cod)(2-methylallyl)$_2$] and the imidazolium salt **16a** (Fig. 16). Previous catalysts were unsuccessful for this

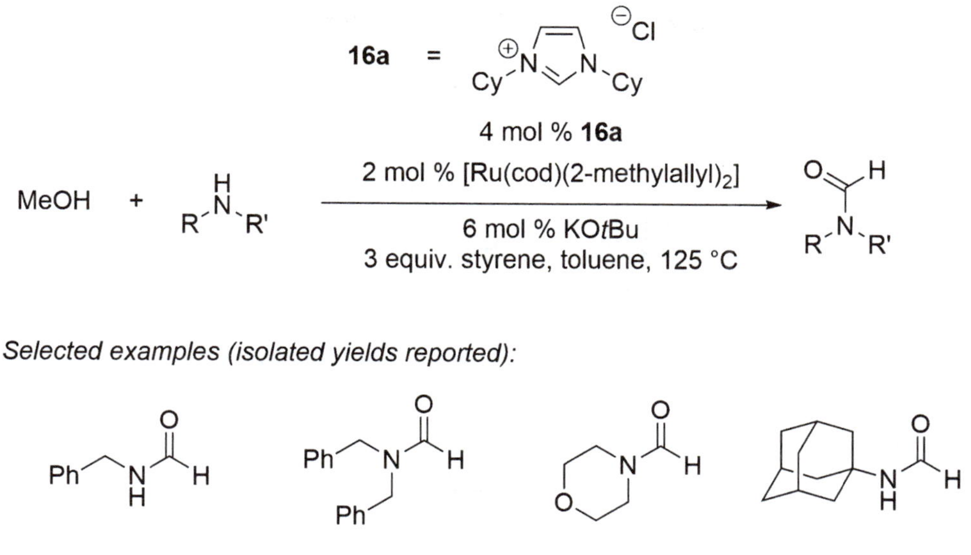

16b 83%

16c 71%

16d 74%

16e 27%

Fig. 16 Formylation of amines with methanol via DAS

transformation because the formation of formamides is thermodynamically disfavored. There was a concern that urea derivatives could be formed under the reaction conditions through oxidative amidation of the formamide product. However, formamide formation through DAS proceeded in good yields (**16b–16e**). Styrene was required as an additive in the production of the formamides in high yields because the catalyst could not turnover without the assistance of a hydrogen acceptor.

With a focus on the oxygen-source, Milstein developed an amidation methodology using water to oxidize cyclic amines to lactams [29]. With the more active acridine PNP complex **17a**, secondary cyclic amines are activated via the removal of hydrogen to form an imine and then through reversible hydration to form the hemiaminal. Finally, the catalyst oxidizes the hemiaminal to form the lactam in good yields (**17b–17e**) (Fig. 17). They also report the synthesis of lactams from diols and ammonium hydroxide as an ammonia equivalent. The oxidation of amines using water is a major advancement in green oxidation methods for DAS, but at this point requires rather harsh conditions in the form of high reaction temperatures, and only gives modest yields.

In a collaborative study, Dong and Guan reported the synthesis of secondary and tertiary amides using a commercially available PNP catalyst Ru-Macho (Fig. 18, **18a**) [30]. Originally developed for ester hydrogenolysis, the authors recognized in line with other studies [6] that these catalytic processes are often reversible and can be applied to ester formation and amide formation from amines and

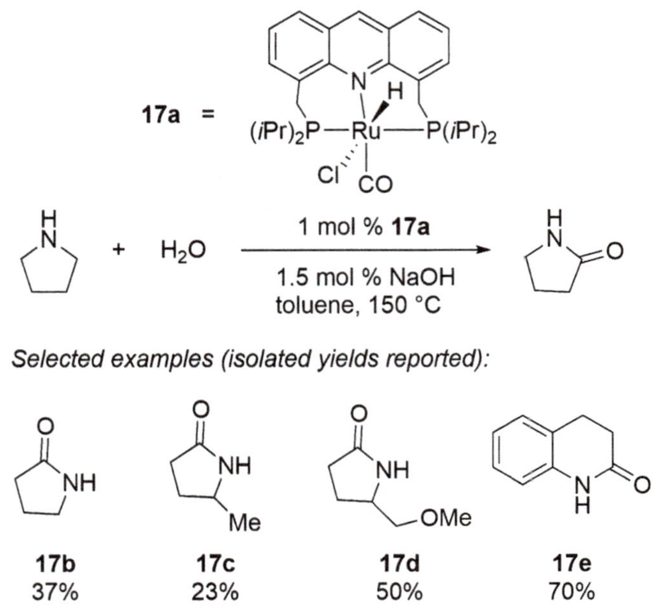

Fig. 17 DAS of formamides using water as oxidant

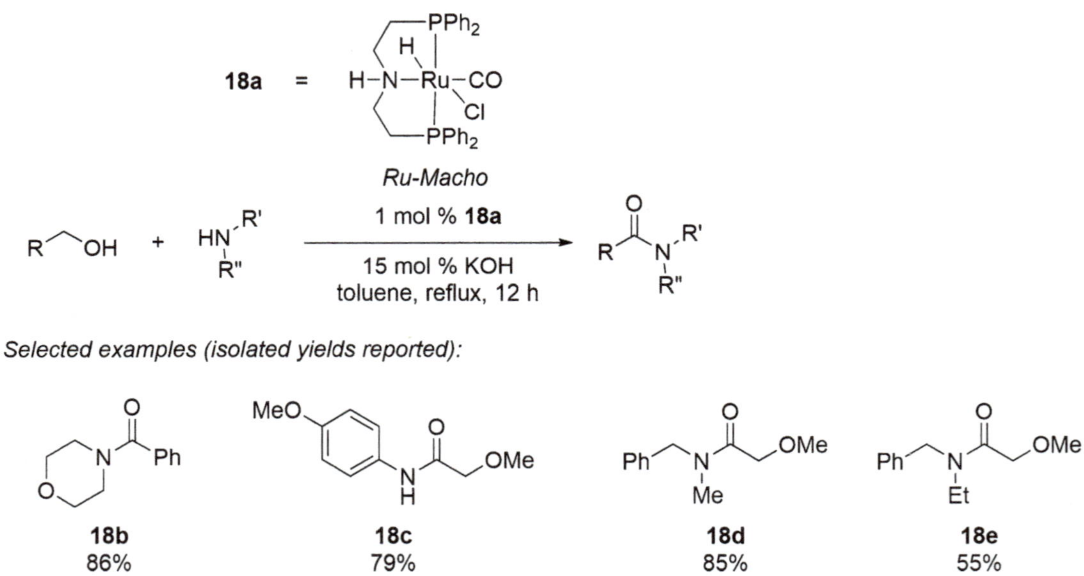

Fig. 18 DAS using a less sterically encumbered commercially available PNP catalyst

alcohols. A reduction in steric hindrance around the ligand backbone allowed for the synthesis of tertiary amides in good yields without the need for elevated reaction temperatures (**18b–18e**).

Despite the dominance of ruthenium-based catalysts for DAS, recent advances in this field have been enabled by the use of alternative transition metals. Hull and Nguyen addressed a longstanding steric limitation of DAS through the use of a rhodium

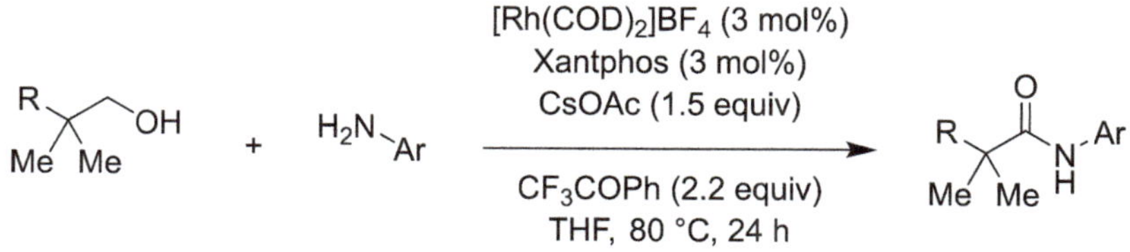

Fig. 19 Rh-catalyzed DAS for the synthesis of α,α,α-trisubstituted amides

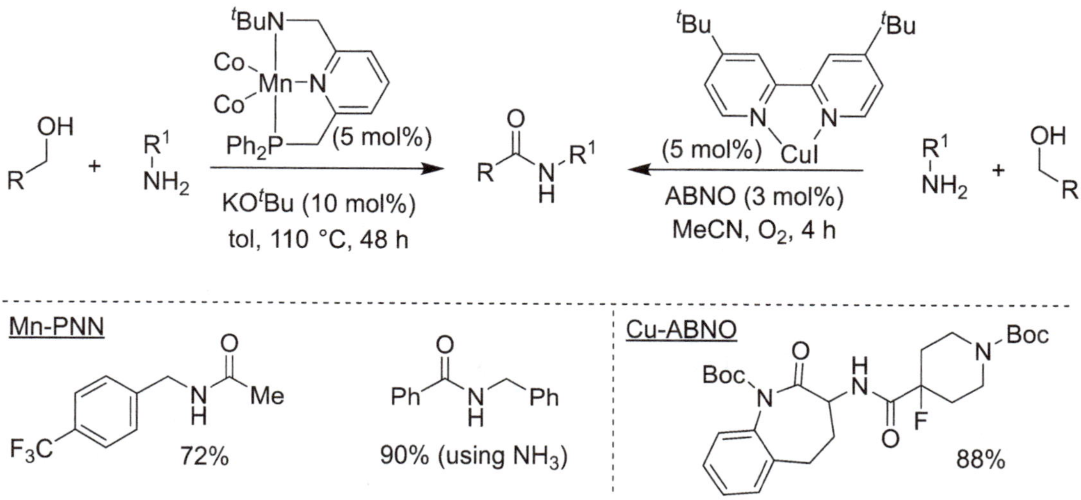

Fig. 20 Base metal-catalyzed DAS between alcohols and amines

Xantphos catalyst [31]. Although this method requires the use of trifluoroacetophenone as a hydrogen acceptor, it enables the synthesis of α,α,α-trisubstituted amides from alcohols and amines for the first time (Fig. 19).

In the interest of creating a more sustainable and cost-effective protocol, earth abundant transitions metals have also begun to gain traction. While many approaches result in large amounts of stoichiometric waste [32], two systems with no organic terminal oxidant have emerged. Milstein developed the first manganese pincer-catalyzed acceptorless dehydrogenative coupling of alcohols with benzylamines [33] or benzyl alcohols with ammonia [34] to provide the corresponding *N*-benzyl secondary amides. Separately Stahl and coworkers developed a copper-nitroxyl system with oxygen as the terminal oxidant. The mild conditions of this base metal-catalyzed transformation enabled a broad scope of drug-like molecules to be functionalized (Fig. 20) [35, 36].

2.2 Mechanistic Considerations of Homogeneous DAS

To date, several mechanistic studies have been published with a focus on DAS (dehydrogenative amide synthesis). To prevent redundancy, we direct the readers to the studies themselves for an in-depth discussion of the mechanistic considerations for each of the catalytic systems studied [37–44]. In this brief section, we will

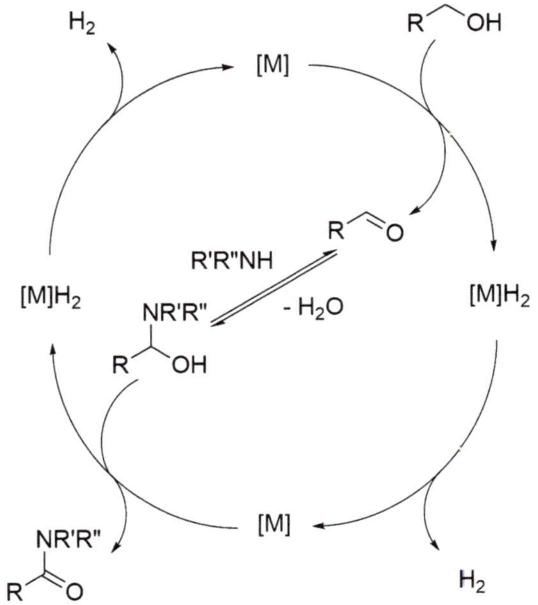

Fig. 21 Key mechanistic steps of DAS from alcohols and amines

provide what we consider to be the most important mechanistic facets of DAS from alcohols and amines across the catalysts previously surveyed. There are two mechanistic pathways that will be discussed, as well as the energetics of amide bond formation versus the competing pathways listed in Fig. 3.

In a simplified view of the reaction, DAS can be summarized in the four key steps illustrated in Fig. 21. Initially, a metal complex oxidizes an alcohol to an aldehyde via removal of hydrogen. The catalyst then eliminates the equivalent of hydrogen directly, or with the help of a hydrogen acceptor, while simultaneously the aldehyde undergoes nucleophilic attack by the amine to form the hemiaminal intermediate. In the final step, the hemiaminal is further oxidized by the catalyst to form the amide product with the catalyst regenerated via removal of hydrogen.

Upon the discovery of DAS, the operative mechanism was thought to involve β-hydride eliminations from both the alcohol and hemiaminal as the key steps to form a supposed aldehyde intermediate and product, respectively (Fig. 22). This β-hydride elimination could potentially be operative in all the above-mentioned systems but requires the amine group of the PNN ligand to be hemilabile in order to allow an open coordination site on the ruthenium center (Fig. 23). This disassociation is thought to be too high in energy to be feasible. However, pincer ligands presented by Milstein, Grutzmacher, Dong, and Guan could proceed through an additional mechanism involving a bifunctional double hydrogen transfer (BDHT) (Fig. 24). In the BDHT mechanism, no hemilabile ligand is required, which could

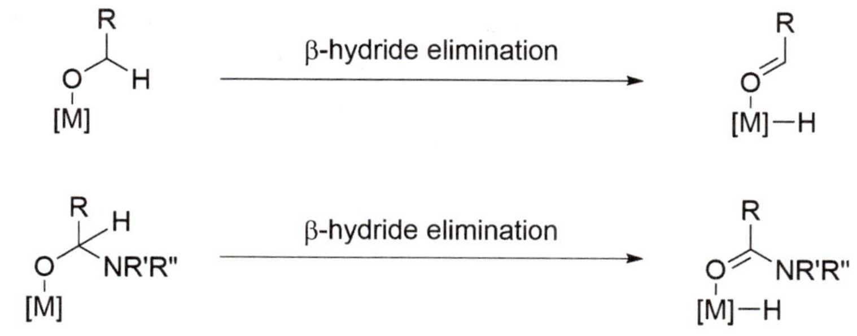

Fig. 22 Schematic of β-hydride elimination of alcohols and hemiaminals

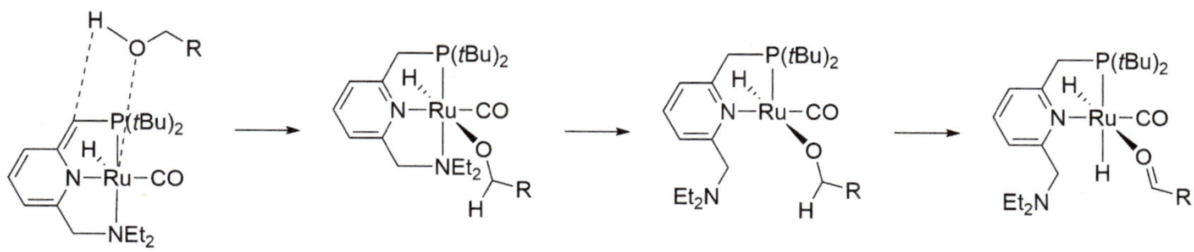

Fig. 23 Example of hemilabile ligand site required for β-hydride elimination of alcohol using PNN catalyst

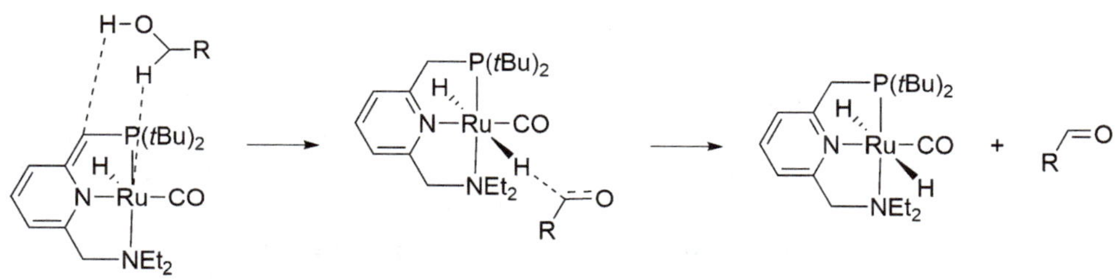

Fig. 24 BDHT mechanism operating on PNN pincer ligand. No hemilabile site is required

explain why the pincer complexes (used by the Grutzmacher group and the Dong/Guan group) facilitate DAS despite lacking this feature.

In light of the possible reaction pathways, common observations found during the initial studies help confirm certain mechanistic aspects of DAS from alcohols and amines. For example, the free aldehyde and hemiaminal tend not to be observed during the course of the reaction, despite early theories that the hemiaminal re-binds to the catalyst after being initially formed in solution. This suggests that the aldehydic intermediate must remain bound to the catalyst during the overall oxidation process from alcohol to amide. This proposal is further supported by the observation that the catalysts, which are observed to release the aldehyde after the initial oxidation, favor direct amination and imine formation as opposed to amidation. If hemiaminal formation occurs through a non-catalytic pathway, dehydration to either the imine [41, 45] or amine product out-competes sequential oxidation to form the amide. For a catalyst to be active in DAS, the oxidation of the hemiaminal species by the catalyst must out-compete dehydration to imine intermediates.

These mechanistic studies demonstrate that DAS mainly out-competes dehydrogenative ester synthesis, and computational studies [41] show that hemiaminal formation is favored over hemiacetal formation (the intermediate in dehydrogenative ester synthesis) when using primary amines. Unsurprisingly, when sterically hindered secondary amines are considered, the rate of hemiacetal formation is faster than hemiaminal formation due to the fact that the lower nucleophilicity of secondary amines makes the hemiaminal formation higher in energy than hemiacetal formation. This trend was observed in several cases, and when highly sterically encumbered amines are used in DAS, ester formation is exclusively observed.

2.3 Heterogeneous Catalysis for Amide Bond Formation

As the chemical industry initiated a push for the development of new amidation methods, the potential for heterogeneous catalysis for DAS from alcohols and amines emerged. Heterogeneous catalysts are reusable and can easily be separated from the reaction mixture for further use, a property that gives them a distinct advantage over homogeneous catalysts, which are difficult to recover, and as such can often only be used once. As a caveat, this review does not classify heterogeneous amide formation from amines and alcohols as strictly dehydrogenative. Aerobic oxidation processes are also common among the heterogeneous catalysts surveyed requiring only that the transformation be exposed to air. In these cases, water (rather than hydrogen) is the by-product. In addition to these oxygen-mediated reactions, we survey cases where the transfer of hydrogen is involved and a high atom economy observed.

Shortly after the report of bimolecular DAS by Milstein, the Christensen group reported that gold supported on TiO_2 is able to oxidize amines using molecular oxygen at high temperatures to form amides (Fig. 25) [46]. In bimolecular experiments with amines and alcohols, homocoupling of the amines was observed as a competing side reaction arising through direct oxidation of the amines. This reactivity was exploited in the conversion of 1,6-hexanediamine into caprolactam, and the authors also investigated using cyclohexyl amine and found that in this case oxime formation occurred. The study demonstrated only one intermolecular amide synthesis in relatively low yield but showed the feasibility for a dehydrogenation pathway to operate.

Mizuno reported oxidation of amines using water and an alumina-supported ruthenium hydroxide catalyst which afforded primary amides [47]. Initial investigation and optimization of the catalyst showed that many oxidative and reductive products could be obtained using different catalyst systems (Fig. 26), and changing

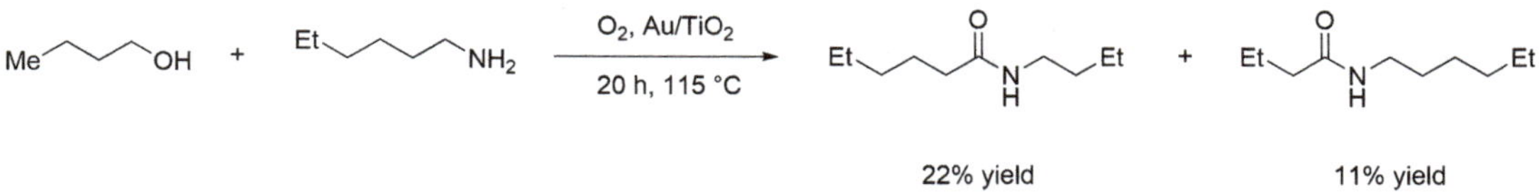

Fig. 25 Heterogeneous catalysis of amines and alcohols to form amides

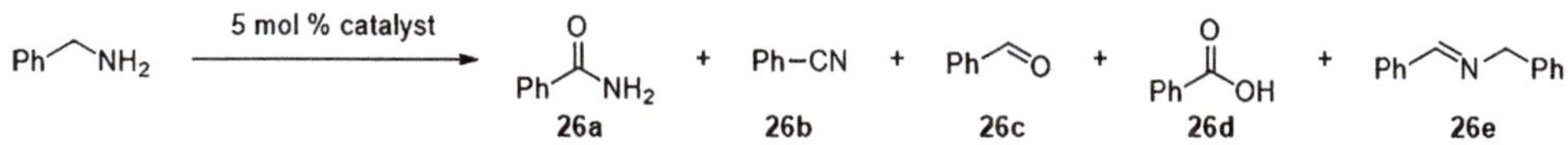

Catalyst	GC Yield %				
	Amide	Nitrile	Aldehyde	Carboxylic Acid	Imine
$Ru(OH)_x/Al_2O_3$	87	8	-	3	-
$Ir(OH)_x/Al_2O_3$	8	11	18	-	-
$Au(OH)_x/Al_2O_3$	3	3	28	3	3
$Cu(OH)_x/Al_2O_3$	1	2	25	-	9
$Rh(OH)_x/Al_2O_3$	-	1	3	-	3
$Pt(OH)_x/Al_2O_3$	-	-	21	-	1
$Pd(OH)_x/Al_2O_3$	-	-	15	-	0
$Co(OH)_x/Al_2O_3$	-	-	5	-	6
$In(OH)_x/Al_2O_3$	-	-	5	-	7
$RuCl_x/Al_2O_3$ No base	9	25	2	-	7
Ru/C	28	20	1	3	-
RuHAP	20	42	5	3	-
RuO_2 anhydrous	-	2	8	11	-
$RuCl_3 \cdot nH_2O$	38	15	6	1	-
$[Ru(acac)_3]$	25	22	7	-	-
$[RuCl_2(PPh_3)_3]$	11	21	18	2	-
$[\{RuCl_2(p\text{-cymene})\}_2]$	44	22	7	2	-
$[Ru_3(CO)_{12}]$	26	30	9	-	-
Al_2O_3 No Base	-	-	9	-	2
Al_2O_3 Base (NaOH)	-	-	2	-	2
N/A	-	-	3	-	4

Fig. 26 Mixed oxidative and reductive products obtained during catalyst screening for heterogenous DAS

the metal supported on the alumina had drastic effects on the product distribution obtained. Only ruthenium hydroxide was shown to have high catalytic activity and selectivity for amide formation. Other ruthenium sources were shown to have both less activity and little selectivity for amide bond formation. After the conditions were optimized, Mizuno and coworkers were able to demonstrate the synthesis of a variety of corresponding primary amides in excellent yields (Fig. 27). Even a coordinating pyridyl substrate (**27c**) was tolerated, which is relevant for pharmaceutical applications.

The next advance came in 2009 when Satsuma found that alumina-supported silver clusters were able to form secondary and tertiary amides from alcohols and amines [48]. A variety of different metals were complexed with alumina in differing amounts to form a variety of nanoparticles. In contrast to the report from Mizuno, the ruthenium system exhibited no activity in DAS, with silver found to be the metal displaying the highest activity. A TON of about ~130

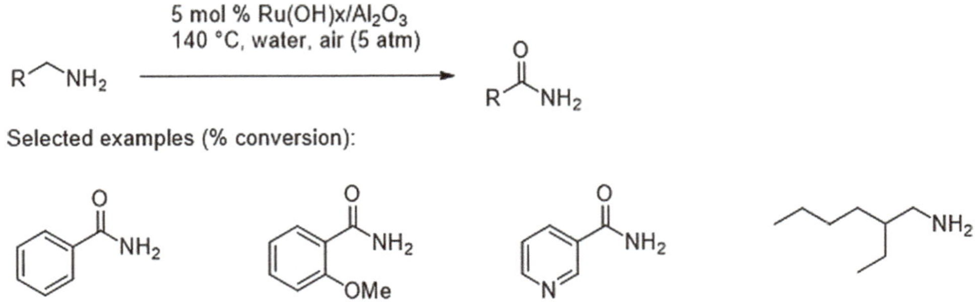

27a	27b	27c	27d
92%	90%	91%	90%

Fig. 27 Oxidation of amines with water and a solid ruthenium hydroxide catalyst supported on alumina

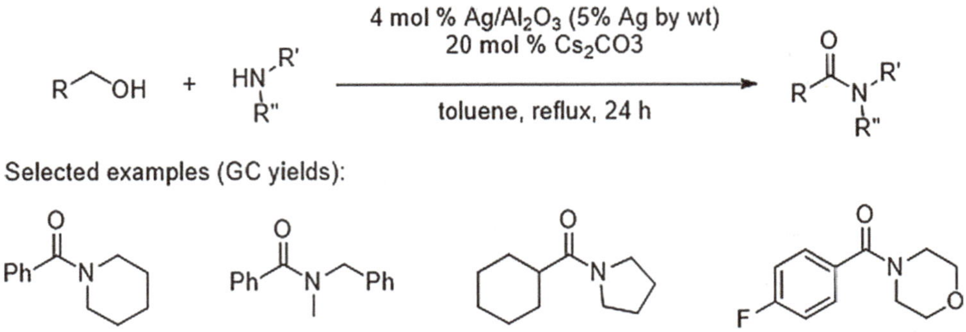

28a	28b	28c	28d
82%	78%	62%	87%

Fig. 28 Silver nanoparticle catalyzed DAS from alcohols and amines

was measured for the catalytic process using morpholine, *p*-fluorobenzyl alcohol, with Cs_2CO_3 as an additive to activate the catalyst (Fig. 28). The catalyst was recovered using centrifugation, re-generated via a series of heat treatments, and used three more times without significant loss of activity. The silver clusters were able to tolerate a variety of aliphatic amines (most notably some secondary amines, **28a–28d**) and alcohols. The reaction is believed to be facilitated by both the metal cation and the acid-base sites on the alumina support with the Lewis acidic silver atoms facilitating the abstraction of hydride from the reactants, while the acidic and basic sites on the alumina enable the pairing of the hydride with a proton and the subsequent evolution of hydrogen gas. The authors corroborate this idea by showing that silver supported on alumina is much more electronegative than the other metal species tested, with this property correlating with the observed activity of the catalyst system.

Using gold supported on nickel oxide, the Haruta group demonstrated the formylation of amines with methanol in the presence of oxygen [49]. During the optimization of the catalyst system, a range of oxidative and reductive products were obtained.

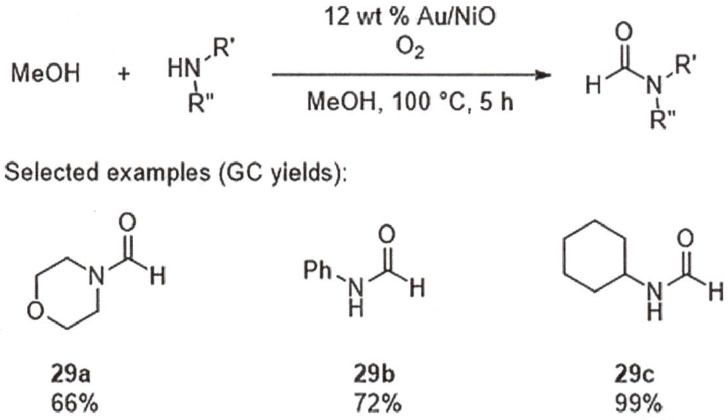

Fig. 29 Formylation of amines using methanol and oxygen

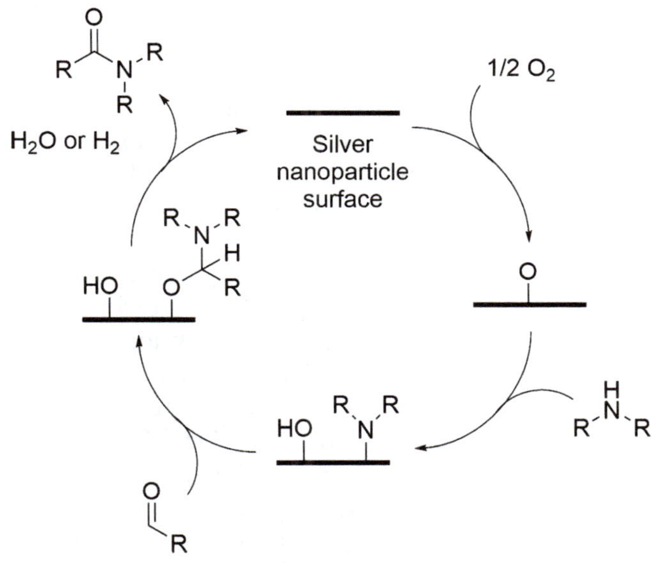

Fig. 30 Proposed mechanism of oxygen assisted amidation

Reaction of amines typically resistant to conventional formylation conditions was performed in good yields (Fig. 29, **29a–29c**). A stoichiometric amount of oxygen was shown to be required for the reaction to go to completion, indicating that it was not catalytic at the surface of the nanoparticle.

In a further development by the Friend group in late 2009, a mechanistic study into the formation of DMF from dimethylamine and formaldehyde mediated by silver surfaces yielded new information [50]. Oxygen was adsorbed onto the surface of the silver, deprotonating the amine and binding it to the catalyst surface (Fig. 30). The bound amine then attacked the aldehyde and formed a hemiaminal. Dehydration or dehydrogenation then occurred forming the amide along with an equivalent of either dihydrogen or water.

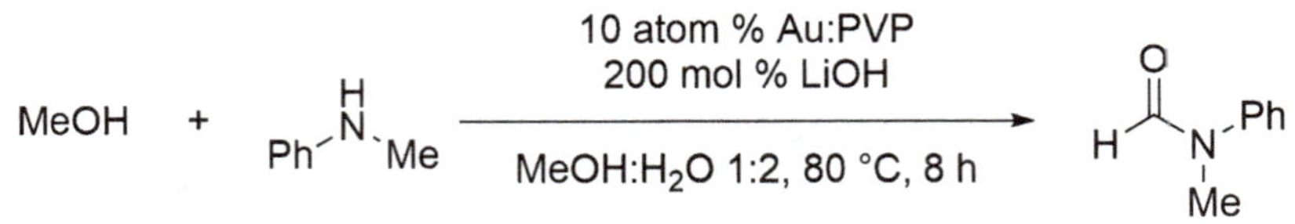

Fig. 31 Formylation of *N*-methylaniline with methanol

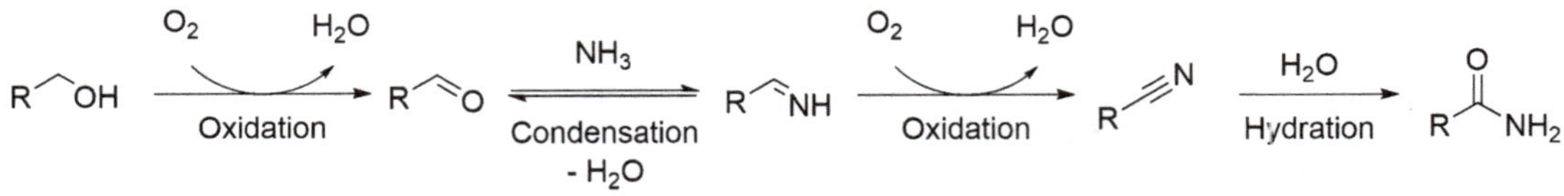

Fig. 32 Proposed reaction pathway of manganese-mediated amide synthesis

In a 2010 paper published by the Sakurai group, formylation of amines with methanol via a gold-PVP (poly(*N*-vinyl-2-pyrrolidone)) cluster was reported [51]. The reaction was thought to proceed by aerobic oxidation of the methanol to a formyl group on the catalyst surface followed by condensation with an amine to form an intermediate hemiaminal. This hypothesis was confirmed by demonstrating the oxidation of methanol to formaldehyde using the catalyst, which was then successfully utilized to formylate a variety of amines (Fig. 31).

In the following years, a variety of different heterogeneous gold catalysts were shown to form amide bonds with notable works including systems developed by Deng [52], Luque [53], Zhang [54, 55], Wang [56], and Kobayashi [57, 58]. These methods all offer an efficient way to produce amide bonds from alcohol and amines. In a number of cases, aldehydes can also be utilized to generate the desired amides (see Subheading 3).

Departing from the use of gold, silver, and ruthenium, the Mizuno group demonstrated that manganese oxide nanoparticles were able to form primary amides directly from alcohols and aqueous ammonia [59]. Manganese oxide-based octahedral molecular sieves (KMn_8O_{16}: OMS-2) are known to catalyze the dehydrogenation of alcohols to aldehydes, the dehydrogenation of aldimines to nitriles, and the hydration of nitriles independently (Fig. 32). Under aerobic conditions, all three steps were theorized to be taking place ultimately forming primary amides. This was further supported by demonstrating amidation of aldehydes and hydration of nitriles using the same catalytic conditions. A variety of primary amides were obtained in good yields (Fig. 33). Further works using the same catalyst showed amines could be oxidized aerobically to primary amides [60], as well as a further mechanistic investigation using a similar cryptomelane (a mineral) type manganese oxide-based octahedral molecular sieve [61]. A similar report of manganese oxide supported on graphene by Xiao was also highly active for

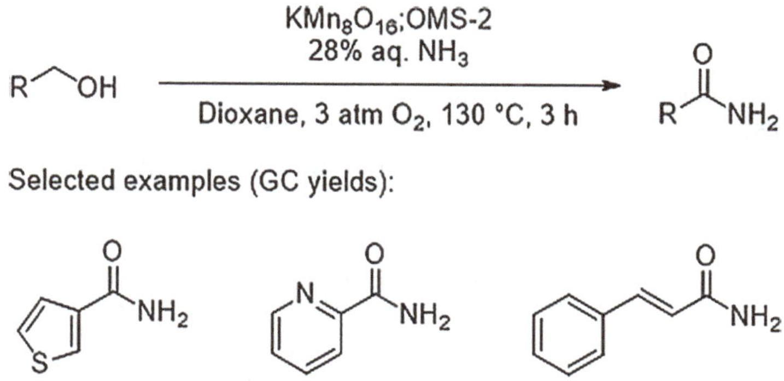

33a
87%

33b
95%

33c
99%

Fig. 33 Amide synthesis using aqueous ammonia as the nitrogen source

amide bond formation [62]. These manganese oxide-based catalysts represent an important advancement, as there are limited catalytic systems able to synthesize primary amides using ammonia.

3 Oxidative Amidation of Aldehydes

There is a propensity of naturally occurring and inexpensive commercially available aldehydes, and utilization of these in place of alcohols in the aforementioned DAS (Subheading 2) would eliminate the need for a catalyst that dehydrogenates the initial alcohol substrate. For these reasons, the direct oxidative conversion of aldehydes to amides has been heavily investigated and achieved by a myriad of mechanistic approaches, which are complementary to DAS from alcohols.

While there are certainly a large number of catalysts and oxidants that can facilitate the oxidative amidation of aldehydes, there are many additional challenges to this transformation. For example, aldehydes and amines spontaneously undergo condensation to form an imine and water without a catalyst present [63]. The following discussion will present the catalysts and procedures used to circumvent this complication as well as other limitations. For the sake of brevity, and to remain in scope of the core topic, we will focus our discussion on methods that convert aldehydes to amides through a transfer hydrogenation mechanism and reactions that proceed without the use of additional stoichiometric reagents. Where applicable, we will direct the reader to reviews that provide more detail on the topic and have partitioned the following sections according to the mechanism of oxidation.

3.1 Oxidative Amidation of Aldehydes via the Loss of H_2 Gas

All of the methods discussed in Subheading 2, which convert alcohols to amides through the loss of hydrogen gas, in theory, can convert aldehydes to amides since it is conceivable that some of these reactions proceed through an aldehyde intermediate. However, beginning with a stoichiometric amount of amine and aldehyde presents additional complications. Traditionally as mentioned, without a catalyst, aldehydes and amines spontaneously undergo condensation to form an imine and water, and in some cases, this condensation reaction may result in catalyst deactivation [64]. Throughout the course of the ruthenium-catalyzed DAS from alcohol substrates, the concentration of aldehyde remains low as it is closely associated with the catalyst after it is generated thus alleviating the potential for deleterious imine formation. Herein, we present several examples that overcame this limitation using aldehyde substrates as the primary focus of the report.

Hong and coworkers were the first to recognize that catalysts which converted alcohols to amides through DAS often were not applicable for the analogous conversion of aldehydes through DAS [22]. They postulated that a ruthenium hydride was the active catalyst and forming this from aldehyde substrates did not occur. To address this challenge, Hong used a $RuH_2(PPh_3)_4$ pre-catalyst (5 mol%) with an NHC salt (5 mol%), acetonitrile (5 mol%), and NaH (20 mol%) to generate the active ruthenium hydride catalyst in situ (Fig. 34). The corresponding $RuCl_2$ catalysts were inactive for aldehyde amidation supporting the proposal that the preformation of a ruthenium catalyst is crucial. Importantly, they discovered that the scope of the reaction was slightly different depending on the oxidation state of the initial substrate. For example, aliphatic alcohols were more successful than aliphatic aldehydes in most cases

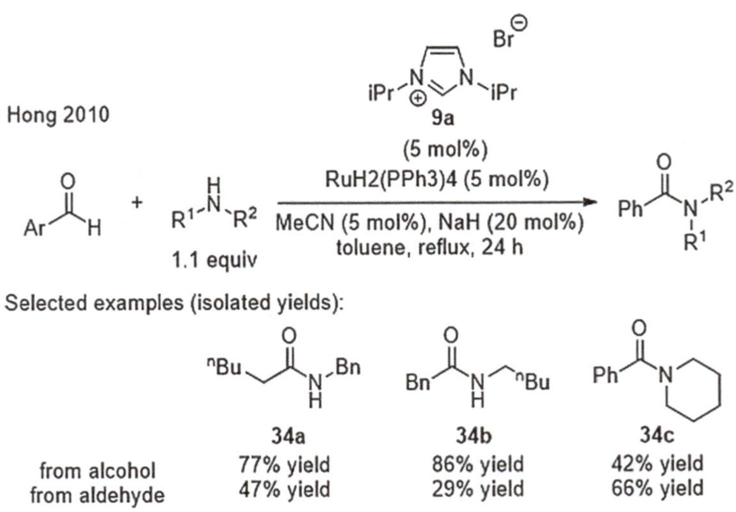

Fig. 34 Comparison of Ru-catalyzed oxidative amidation of aldehydes or alcohols

Hong 2012

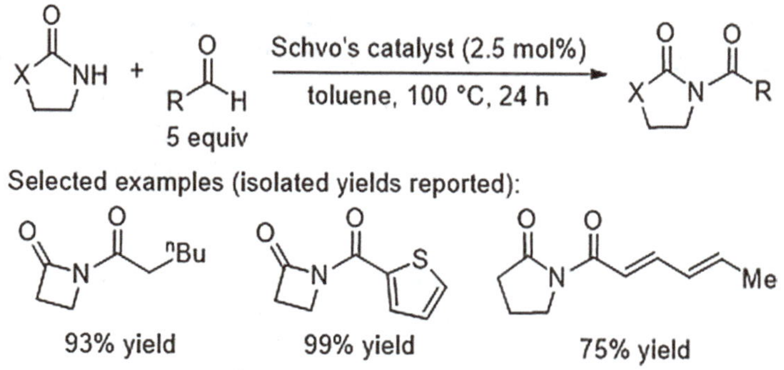

Fig. 35 Oxidative Acylation of *N*-heterocycles using Schvo's catalyst

(**34a–34c**), but secondary amines gave higher yields when starting from benzaldehyde as opposed to the corresponding alcohol (**34d**).

Another limitation in the scope of the oxidative amidation of alcohols is in the case of less nucleophilic amines. As the amine nucleophilicity approaches that of the alcohol, a competing esterification reaction can occur consuming the alcohol [14, 65]. Using Schvo's catalyst, Hong and coworkers demonstrated that less nucleophilic substrates such as lactams can be coupled with aldehydes to generate *N*-acyl lactams (Fig. 35) [66]. Using this methodology, oxazolidinones, imidazolidinones, β, γ, and δ lactams were converted to amides and ureas. The aldehyde partner could be aliphatic, heteroaromatic, or α,β-unsaturated, albeit with some competing conjugate reduction for the latter substrates. It is worth noting that the authors proposed elimination of hydrogen gas as the mechanistic pathway, though based on the established mechanistic pathways for Schvo's catalyst and the identification of alcohol side products, it is also possible that the reaction proceeds through a transfer hydrogen mechanism (see the following discussion on oxidative amidation by transfer hydrogenation) [67]. Regardless of the reaction mechanism, this example highlights the ability to overcome the limitations associated with the conversion of alcohols to amides by starting from the aldehyde oxidation state.

In summary, extending amide synthesis from alcohol substrates to aldehyde substrates by way of a dehydrogenative coupling may appear trivial, though there are additional complications that one must consider. These are to some degree compensated for as there are certain products which can be accessed in this reaction manifold that cannot be synthesized starting from the alcohol oxidation state, and because of both these factors, it is important to consider amide synthesis from alcohols and from aldehydes as distinct reactions which access unique chemical spaces.

3.2 Oxidative Amidation of Aldehydes via Transfer Hydrogenation

Transferring hydrogen from alcohols and amines to ketones has been deeply investigated [68, 69], though the analogous transfer from a hemiaminal is in comparison underdeveloped. Although reactions that fall into this category require a sacrificial hydrogen acceptor, they warrant a discussion in this chapter for three reasons: (1) transfer hydrogenation was the inspiration for direct DAS [70], (2) a broad range of catalysts can be employed including base metal catalysis, and (3) progress is being made toward the use of benign and inexpensive hydrogen acceptors, thus lowering the environmental impact of the reactions. In the following sections, we will discuss amide synthesis by transfer hydrogenation via a Cannizzaro-type mechanism and a β-hydride elimination mechanism and by direct oxidative addition into aldehyde C–H bonds.

3.2.1 Oxidative Amidation of Aldehydes via a Cannizzaro-Like Mechanism

The Cannizzaro reaction, which is the base-induced disproportionation of an aldehyde lacking an β-hydrogen, is one of the oldest known (1853) reactions in organic chemistry [71], while the related Tishchenko reaction for the conversion of aldehydes to esters has been established since 1908 [72]. Extending this classical reactivity to the synthesis of amides has been far less successful. To understand the challenges faced by amide synthesis through a Cannizzaro-type process, consideration of the plausible mechanism for this reaction is required (Fig. 36) [73]. Initial hemiaminal formation, followed by deprotonation, allows binding to a catalyst, which assists in transferring a hydride to a second equivalent of the aldehyde (**36a–36b**). Although the details of the formation of a tetrahedral intermediate and catalyst binding may vary depending on the system, the hydride transfer event occurring through a six-membered transition state remains consistent and crucial for reactivity. The limited success of amide synthesis through a

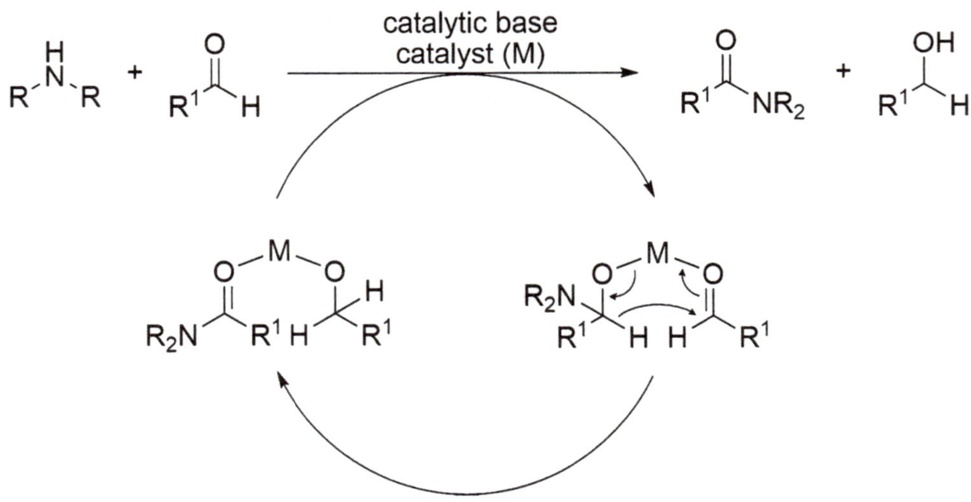

Fig. 36 Cannizzaro-like mechanism for oxidative amidation

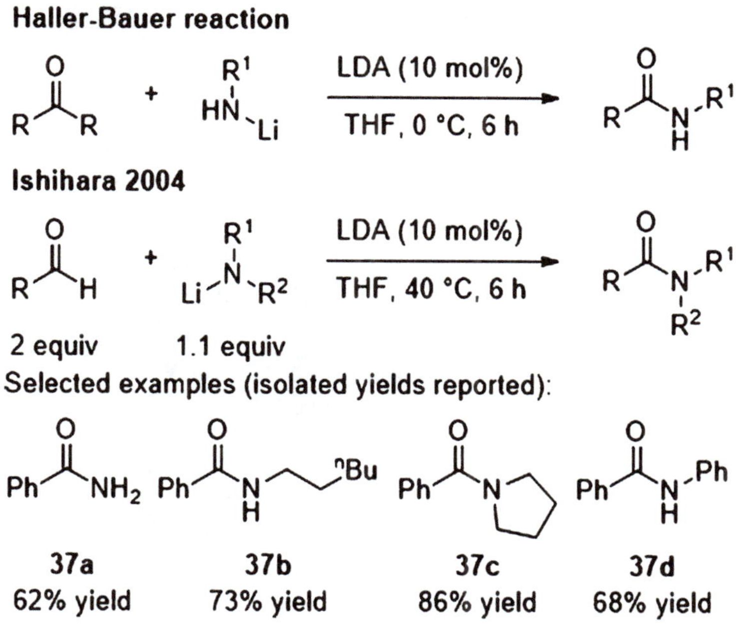

Fig. 37 LDA-catalyzed oxidative amidation using Li-amides

Cannizzaro-type mechanism stems from the need to generate a hemiaminal without generating the imine irreversibly and the difficulty of discovering a catalyst that binds to a hemiaminal preferentially over an amine.

These challenges were not addressed until the work of Ishihara in 2004 [74]. The authors of the first Cannizzaro-like reaction to produce amides stated that their inspiration was from the Haller-Bauer reaction. Their methodology used stoichiometric amounts of sodium amides or lithium amides along with a catalytic amount of lithium *N*,*N*-diisopropylamide (LDA) (Fig. 37). Although not entirely understood, the authors suggested that the role of the LDA was to chelate the lithiated hemiaminal intermediate to promote hydride transfer. Ishihara discovered that non-enolizable aldehydes give good yields of primary (**37a**), secondary (**37b**–**37c**), or tertiary (**37d**) amides at 40 °C in THF.

After the initial reports by Ishihara, alternative catalyst systems began to surface intent on eliminating the need for stoichiometric amounts of base. The first truly catalytic amide synthesis by a Cannizzaro-type mechanism was reported in 2005 by Abaee and coworkers who observed oxidative amidation of benzaldehyde with *N*,*N*-diethylamine using a catalytic amount of $MgBr_2$ in the absence of base (Fig. 38) [75]. The authors discovered this reaction while attempting to study the mechanism of a Cannizzaro reaction. Interestingly, the *N*,*N*-diethyl benzamide was considered a by-product to the desired Cannizzaro reaction, and this oxidative amidation was not developed beyond one substrate.

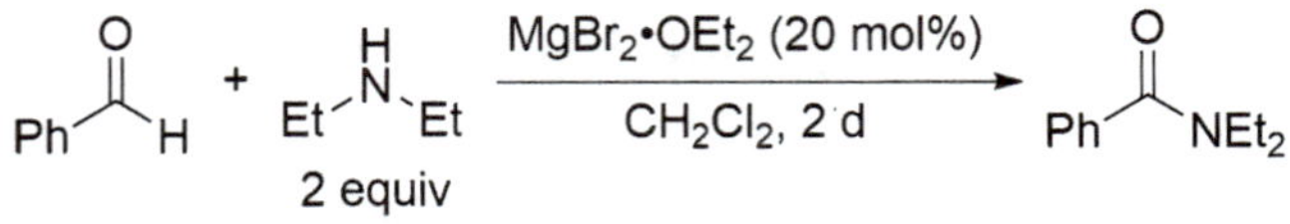

Fig. 38 $MgBr_2$-catalyzed oxidative amidation

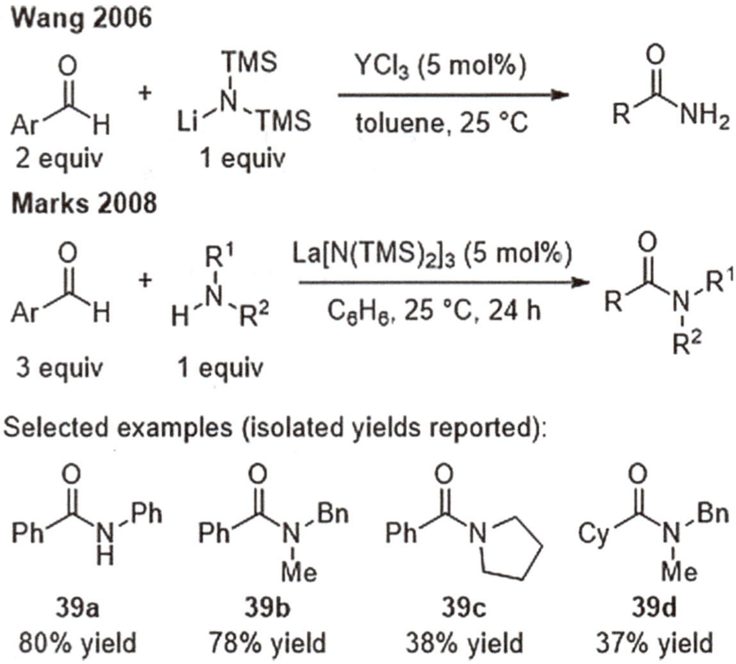

Fig. 39 Lanthanum-catalyzed oxidative amidation

Another serendipitous discovery came from the laboratory of Wang. While studying the mechanism of the Baylis-Hillman reaction, Wang and co-workers discovered that yttrium trichloride could generate amides from aromatic aldehydes and lithium amides [76]. These reactions though performed at room temperature had several drawbacks with the need to use lithium amide substrates limiting functional group compatibility as well as the reaction scope only applicable to primary amides (Fig. 39). A major breakthrough for this chemistry was published by Marks in 2008 who developed a method, which was truly catalytic in both catalyst and base through the use of a $La[N(TMS)_2]_3$ catalyst precursor [73]. The use of anilines (**39a**) and *N*-benzyl-*N*-methyl amines (**39b**) as nucleophiles was crucial, presumably because of the pK_a range making imine formation reversible under these conditions, while other amines such as pyrrolidine were found to be unreactive (**39c**). In addition, the slightly basic conditions rendered aliphatic aldehydes less effective due to competing Aldol side reactions (**39d**). Marks also demonstrated preliminary levels of reactivity with the analogous samarium and ytterbium catalysts.

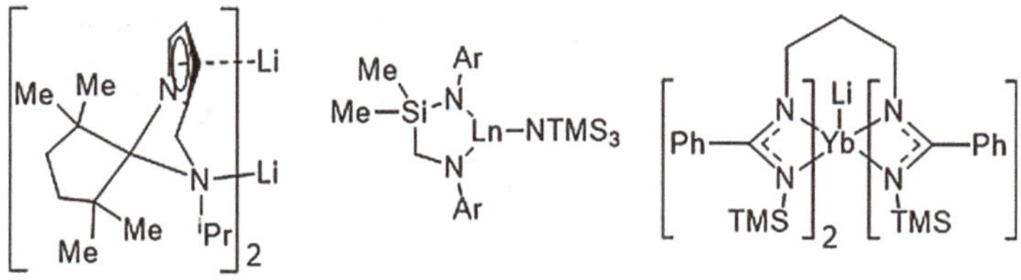

40a
Liu 2013

40b
Wang 2010

40c
Shen 2009

Fig. 40 Additional catalysts for the Cannizzaro-like oxidative amidation

Since the Marks report, oxidative amidation via a Cannizzaro-type of mechanism has been expanded to include additional aniline examples [77] and has been extended to cyclic alkyl amines. Catalysts derived from lithium (**40a**) [78], samarium (**40b**) [79], neodymium (**40b**) [80], and ytterbium (**40c**) [81] have been used with several representative examples shown in Fig. 40. The common feature to all of these recent catalyst systems is the presence of a multidentate amine ligand indicating that the identity of the ligand may be more critical than that of the metal center [82].

3.2.2 Oxidative Amidation of Aldehydes via a β-Hydride Elimination Mechanism

All of the methods listed in the previous section rely on the use of excess aldehyde, and attempts to circumvent this have resulted in a large body of research on developing methods for transferring hydrogen to various other acceptors using main group transition metal catalysis. The majority of these methods proceed through β-hydride elimination from a hemiaminal intermediate [83]. As shown in Fig. 41, a tetrahedral intermediate or a hemiaminal can coordinate to a metal center (**41a**), and β-hydride elimination from this intermediate generates the amide product (**41b**) and a metal hydride species (**41c**), which can subsequently reduce an additive to turn over the catalyst. This unique mechanistic pathway proceeding through a metal hydride allows the use of various hydride acceptors other than the excess aldehyde substrate fulfilling this role.

Since the first transition metal-catalyzed oxidative amidation of aldehydes, the mechanism shown in Fig. 41 was favored. In 1983, Yoshida and coworkers developed a triphenylphosphine stabilized palladium catalyst that could convert aldehydes to morpholine amides by using a combination of bromobenzene and potassium carbonate as the stoichiometric oxidants (Fig. 42) [84]. This method tolerated both electron deficient or neutral aromatic aldehydes (**42a**) much better than electron rich ones (**42b**), while also representing a rare example of aliphatic aldehyde amidation (**42c**). Despite the stoichiometric amount of waste produced, this report was an important advance in the synthesis of amides from aldehydes mediated by transition metal catalysis while using almost equimolar amounts of each reagent.

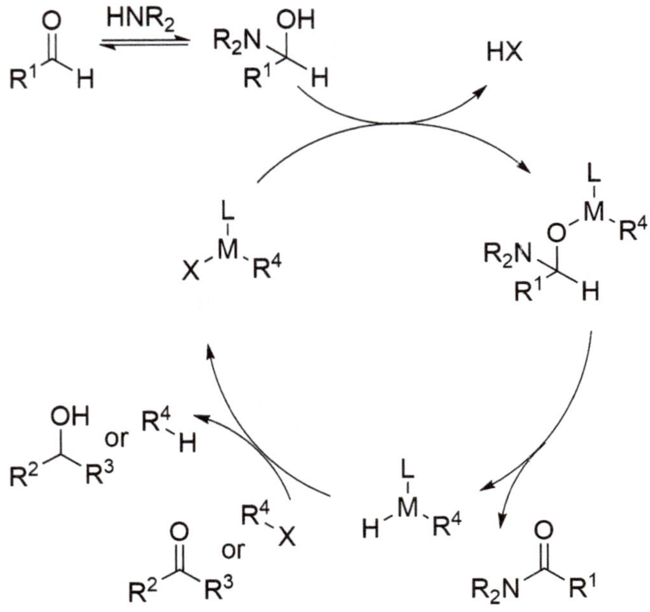

Fig. 41 Oxidative amidation via β-hydride elimination

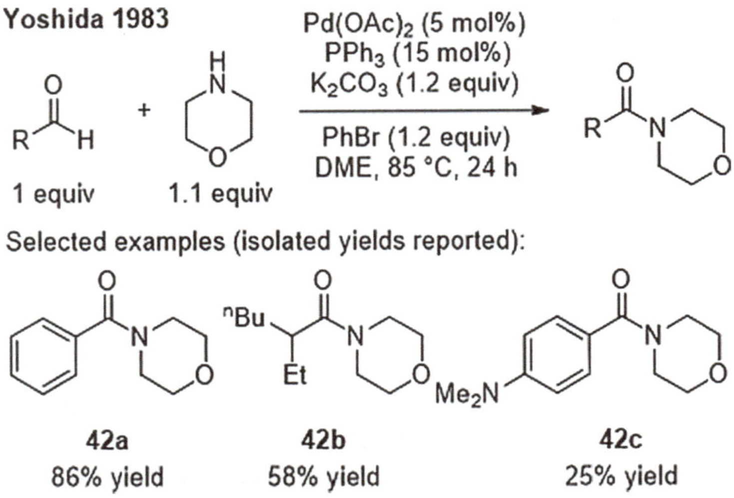

Fig. 42 Oxidative amidation utilizing Ar-Br as the oxidant

Since the initial report by Yoshida, many groups have been interested in finding alternative catalysts and acceptors for this transformation. Murahashi discovered that a ruthenium catalyst could transfer hydrogen to α,β-unsaturated compounds or to alkynes generating amides [85], while Dai used a rhodium catalyst in combination with Ag_2CO_3. The Ag_2CO_3 acts as an oxidant by promoting reductive elimination of HX from the Rh center. Beller demonstrated an oxidative amidation of aldehydes where morpholine benzamide (**43a**) could be produced by reducing an extra

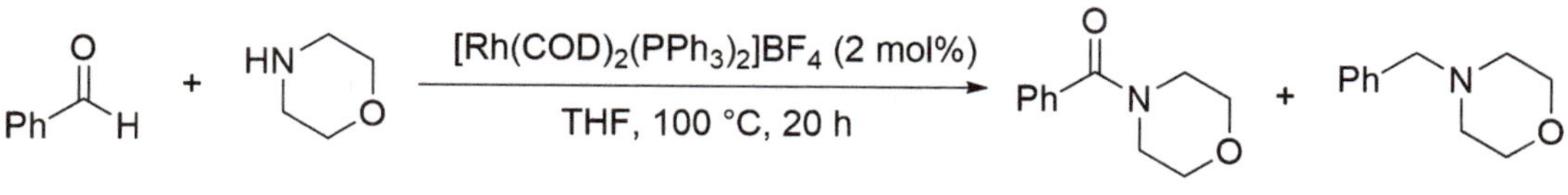

43a **43b**

Fig. 43 Rh-catalyzed oxidative amidation using the hemiaminal intermediate as an oxidant

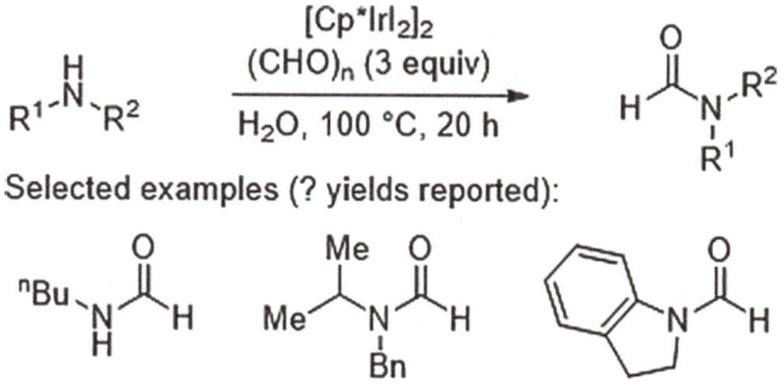

44a 90% yield **44b** 78% yield **44c** 91% yield

Fig. 44 Ir-catalyzed oxidative formamidation in water

equivalent of the hemiaminal intermediate to generate *N*-benzyl morpholine (**43b**) as the by-product (Fig. 43) [86]. Although this reaction consumes an equivalent of each reagent resulting in a maximum of 50% yield with respect to either substrate, Beller demonstrated that it was possible to improve the reaction efficiency by using *N*-methylmorpholine *N*-oxide (NMO) to re-oxidize the generated amines back to amides. The scope of the reaction was expanded to eight additional substrates, but only when NMO was used as a stoichiometric additive. This is an important example as not only was it the first Rh-catalyzed oxidative amidation reaction but also the first indication that carbonyl compounds could be potential hydrogen acceptors in amide synthesis.

Somewhat surprisingly, this chemistry remains underdeveloped with few examples of transition metal-catalyzed coupling of aldehydes and amines using carbonyl compounds as hydrogen acceptors. One notable exception was developed by Williams in 2010 [87] describing a formamide synthesis using 1 mol% of an $[Cp^*IrI_2]_2$ catalyst (Fig. 44). In this reaction, paraformaldehyde reacted as both the carbonyl source and hydrogen acceptor. The reactions, performed in water, were amenable to the synthesis of secondary and tertiary *N*-alkyl (**44a**), *N*-benzyl (**44b**), or *N*-aryl (**44c**) formamides while requiring no other additives. Unfortunately, attempts to extend this reaction to other aldehyde substrates were unsuccessful.

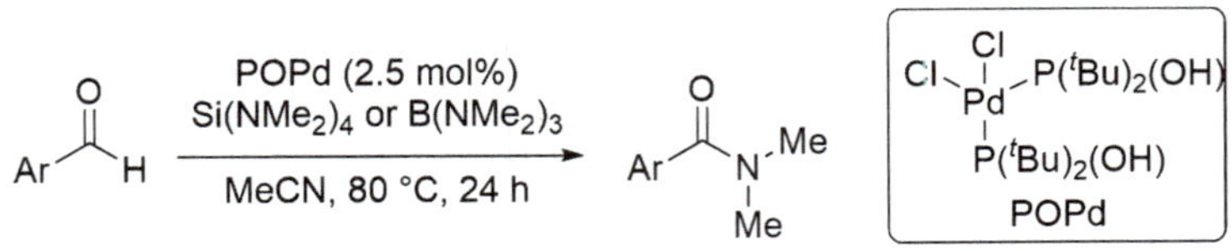

Fig. 45 Oxidative amidation using *tetra*(amino)silane or *tri*(amino)borane

A somewhat unique approach pioneered by Wolf combined the amine source with the hydrogen acceptor (Fig. 45) [88] in a method which is palladium-phosphonous acid catalyzed (**45a**) and uses either *tetrakis*(dimethylamino)silane or *tris* (dimethylamino)borane to generate *N*,*N*-dimethyl amides in good yields. Considering that one equivalent of either $Si(NMe_2)_4$ or $B(NMe_2)_3$ only generates one equivalent of amide, and these reagents need to be prepared in a discrete step, there is much work needed to enhance the practicality, the atom-efficiency, and the scope of this approach. However, the use of these amine sources is under-explored and may hold great potential if these limitations can be addressed.

3.2.3 Oxidative Amidation of Aldehydes via Direct C–H Bond Activation

The activation of an aldehyde C–H bond through oxidative addition to form an acyl metal hydride has been proposed in the hydroacylation literature for decades [89] though is rare for other aldehyde functionalization reactions, and it was not until 2015 that a nickel catalyst was proposed to insert into the aldehyde C–H bond for an oxidative amidation reaction [90]. The stoichiometric reaction between nickel oxide [91] and amido [92] complexes has been known to generate primary amides from aldehydes since 1966. However, it was not until 1997 that the first indication of a possible catalytic variant appeared in which Buchwald observed oxidative amidation as a side product of an amidative cross-coupling between *p*-chlorobenzaldehyde and pyrrolidine [93]. Although the authors did not report a yield, they proposed β-hydride elimination from a hemiaminal intermediate, presumably reducing an additional equivalent of aldehyde. This proposed mechanism was disputed by Dong in 2015, who showed that decarbonylation was a competing pathway indicating that the reaction likely proceeds through a nickel acyl hydride intermediate (Fig. 46, **46a**) [93].

Dong demonstrated a broad scope of oxidative amidation by taking advantage of this mechanism. Secondary and tertiary *N*-aryl (**46b–46d**), *N*-alkyl (**46e–46f**), or *N*-benzyl amides were synthesized using aromatic (**46b, 46e**) or aliphatic (**46c–46d, 46f**) aldehydes for the first time. The reaction was performed at 30–40 °C and employed equimolar amounts of the aldehyde and amine if α,α,α-trifluoroacetophenone (**46g**) was used as a hydrogen acceptor. Alternatively, the hydrogen acceptor could be an additional equivalent of the aldehyde or an alternative electrophilic carbonyl

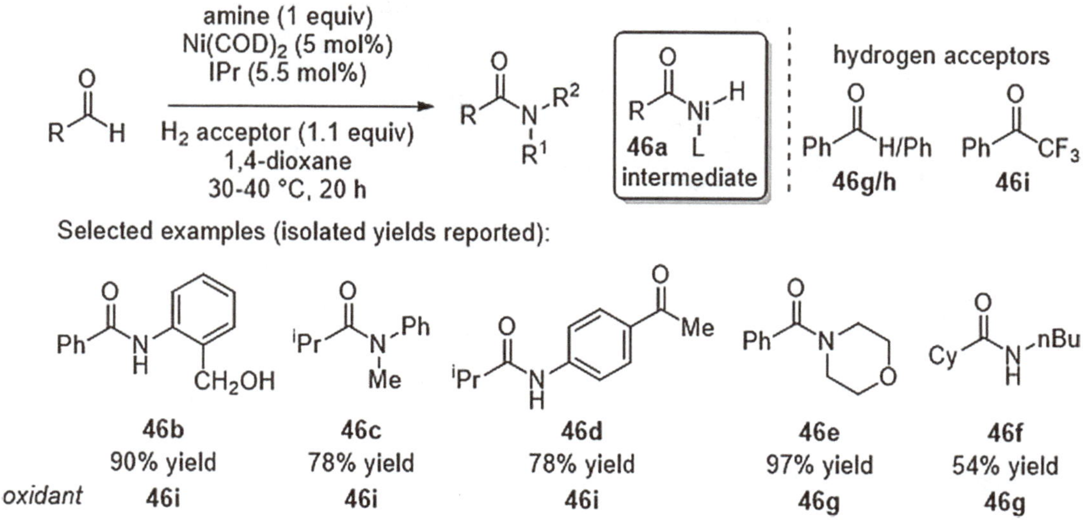

Fig. 46 Ni-catalyzed oxidative amidation via C–H activation and using ketone oxidants

compound such as benzophenone (**46g–46h**), though these acceptors were only briefly discussed for optimization. One demonstrative example employed one equivalent each of cyclohexyl carboxaldehyde, benzaldehyde, and *N*-butylamine to provide the cyclohexyl amide and benzyl alcohol indicating that benzaldehyde is an appropriate oxidant for aliphatic aldehyde amidation (**46f**). Although additional mechanistic details are needed to fully explain the selective amidation of the aliphatic aldehyde in the presence of the aromatic aldehyde, this observation is consistent with a preferential C–H activation of an aliphatic substrate and an easier reduction of benzaldehyde substrate by the generated metal hydride.

In summary, there are many promising methods for the oxidative amidation of aldehydes, which proceed through a variety of transfer hydrogenation mechanisms such as the Cannizzaro-type mechanism, β-hydride elimination, or direct oxidative addition into an aldehyde C–H bond. Recent success with carbonyl compounds as hydrogen acceptors is moving the field in a more favorable environmental direction with the future possibility to take advantage of more benign hydrogen acceptors such as acetone or carbon dioxide a significant goal in the field of green amide synthesis.

3.3 Oxidative Amidation of Aldehydes with the Solvent as an Oxidant

Many oxidative amidation reactions rely on a large excess of oxidant including the oxidant as the solvent, and although this approach may not initially appear green [94], the choice of solvent can have a major impact on how environmentally benign the reaction is. A notable example by Vishwakarma in 2014 in the synthesis of β-ketoamides showed that dimethyl sulfoxide acts as an oxidant by donating the sulfoxide oxygen generating dimethyl sulfide and water as the sole by-products (Fig. 47) [95]. Despite the stench

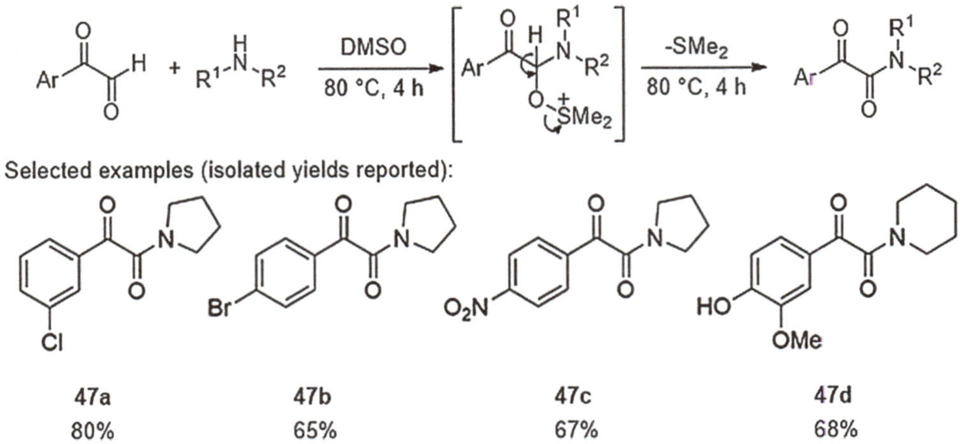

Fig. 47 Catalyst free oxidative amidation using DMSO as the oxidant

associated with this reagent, a particular appealing aspect of this chemistry is that no additional reagents or catalyst is required. This methodology is not yet applicable to inactivated aldehydes, but it demonstrates a unique and potentially more environmentally friendly approach to α-ketoamide synthesis. Additionally, the relatively mild nature of this protocol makes it compatible with a variety of functional groups including aryl chlorides (**47a**), aryl bromides (**47b**), nitro groups (**47c**), and free hydroxyls (**47d**).

3.4 Oxidative Amidation of Aldehydes Using O_2 as an Oxidant

Conceptually, the use of ambient oxygen as an oxidant for amide synthesis from aldehydes and amines is highly attractive, as in addition to requiring no stoichiometric oxidant and generating water as the only by-product, these reactions should be operationally simple since tolerance to oxygen obviates the need for Schlenk technique and should simplify scale up. However, great care is needed when combining molecular oxygen with volatile solvents, as the mixture can be highly combustible. Currently, only a few methods have achieved this goal without the super stoichiometric use of additional reagents and further due to the fact that few catalysts can tolerate exposure to oxygen at elevated temperatures.

3.4.1 Oxidative Amidation of Aldehydes Using O_2 as an Oxidant: Transition Metal Catalyzed

Of the transition metals, group 11 metals, specifically copper (II) complexes, have been known to tolerate harsh oxidative conditions [96–98] and were therefore a good choice to study in oxidative amidation of aldehydes using O_2 as the terminal oxidant. Building on the work of Jiao in 2009 who used super stoichiometric amounts of pyridine [99–101], Yamaguchi developed a method using a catalytic amount of CuI/2-pyridonate (Fig. 48) [102] with the reaction being run open to air in refluxing toluene to provide a variety of secondary benzamides. Importantly, this reaction tolerated sensitive functional groups such as nitro groups (**48a**) and heterocycles such as pyridine (**48b**). Finally, the authors

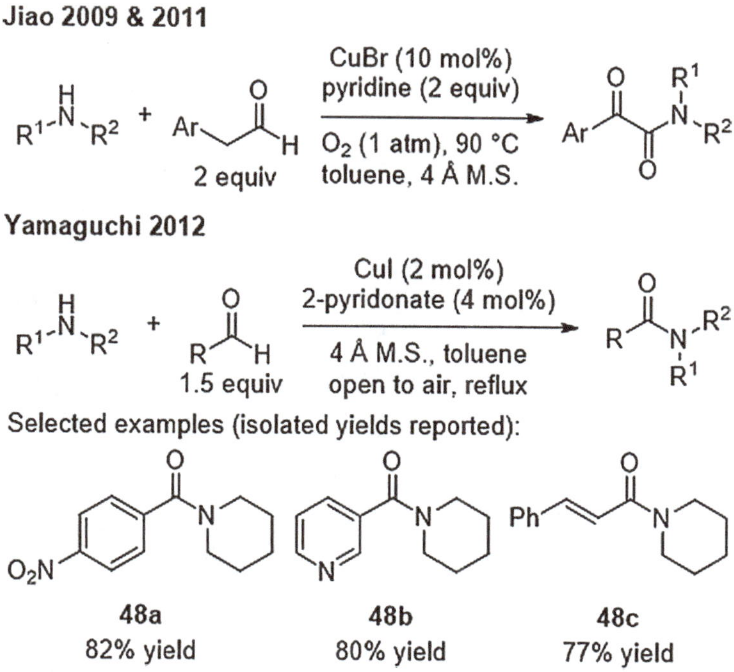

Fig. 48 Copper-catalyzed aerobic amidation of aldehydes

demonstrated a rare example of oxidative amidation of cinnamaldehydes (**48c**) for which typically competing side reactions of the α,β-unsaturated moiety would complicate the use of these substrates.

As discussed in Subheading 2.3, there is a rapidly growing field of oxygen-mediated oxidative amidation based on the other group 11 transition metals, silver, and gold. These heterogeneous catalysts have primarily focused on starting with the alcohol oxidation state, but there are a few notable exceptions. In a seminal report by Sakurai, a gold catalyst was found to formylate a variety of amines albeit with excess base, formaldehyde, and alcohol (see Subheading 2 for more details) [51]. Also in 2012, Maddix and Friend showed that silver could achieve a similar transformation, although limited in scope to the synthesis of *N*,*N*-dimethylformamide and *N*,*N*-dimethylacetamide (see Fig. 30) [103].

Detailed mechanistic studies by Kobayashi showed that extending this reactivity to include other aldehyde substrates would be challenging due to the facile formation of imines (**49a**) in the presence of nanoparticle catalysts. These reactions proved to be more successful starting from the alcohol oxidation state (see Fig. 49) [58]. Wang showed that benzaldehyde could be coupled with aniline to form *N*-phenyl benzamide (**49b**) indicating that less nucleophilic amines are the key to achieving reactivity [56]. However, the most significant contribution came from Wong who used 1 mol% of a gold catalyst and 10 mol% of K_2CO_3 as the base to

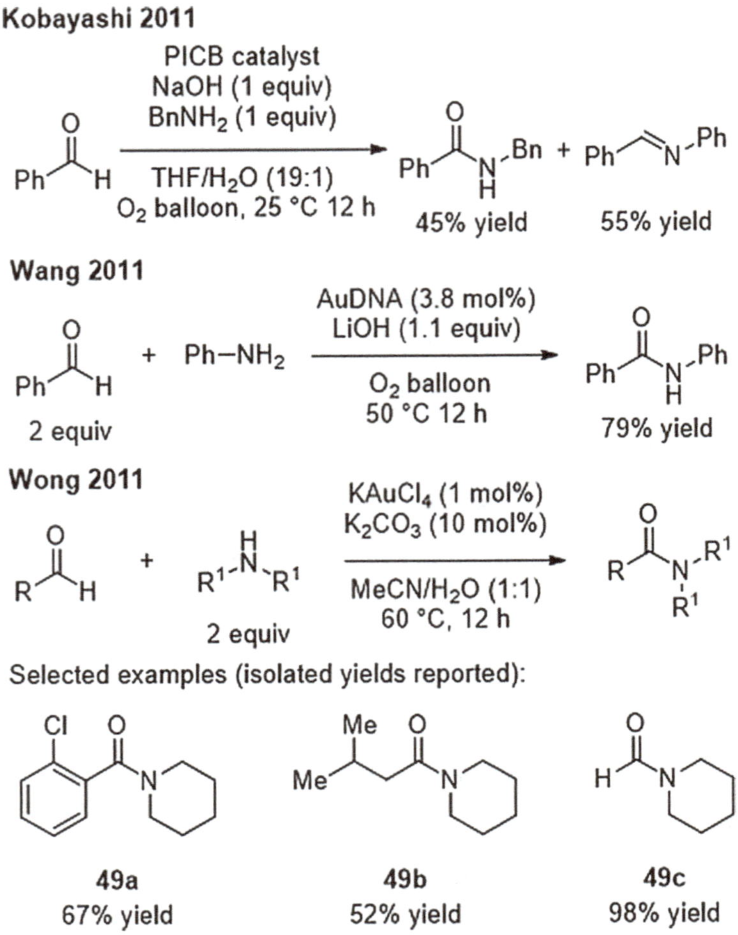

Fig. 49 Gold-catalyzed aerobic amidation of aldehydes

synthesize a variety of amides from aromatic aldehydes (**49c**), aliphatic aldehydes (**49d**), and even formaldehyde (**49e**) though many substrates gave modest yields and only cyclic amines were demonstrated [104]. This publication by Wong represents the first time that gold catalyzed amide synthesis from either alcohols or aldehydes in air takes place without a super stoichiometric amount of base. The success of this reaction as opposed to previous gold catalysts may be due to a switch to a radical based mechanism.

Besides group 11 metal catalysis, there is very little research on transition metal-catalyzed oxidative amidation using oxygen as a terminal oxidant. A manganese oxide supported on molecular sieves (KMn_8O_{16}:OMS-2) is able to convert aldehydes to nitriles which subsequently get hydrolyzed to give primary amides (Fig. 50) [59]. This reaction was also shown to be compatible with alcohol substrates (see Subheading 2.3).

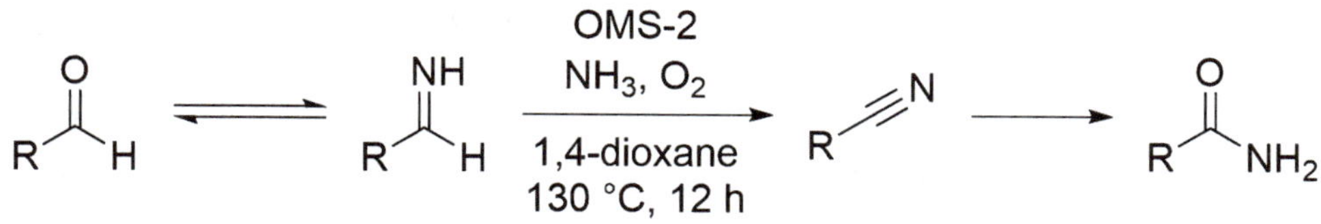

Fig. 50 OMS-catalyzed oxidative amidation via a nitrile intermediate

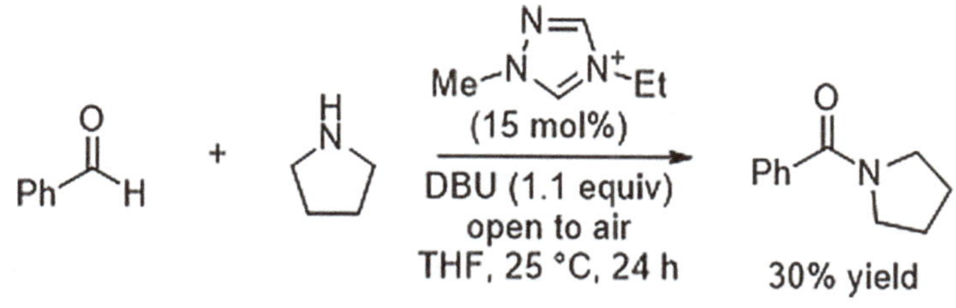

Fig. 51 NHC-catalyzed oxidative amidation of benzaldehyde

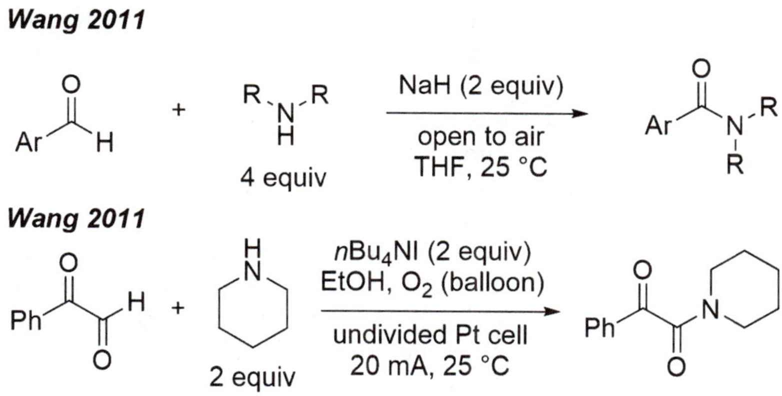

Fig. 52 Metal-free aerobic oxidative amidation

3.4.2 Oxidative Amidation of Aldehydes Using O_2 as an Oxidant: Without Transition Metals

Although typically organocatalysis is incompatible with molecular oxygen, there is one report attempting oxidative amidation of aldehydes using traditional NHC catalysis. Connon coupled benzaldehyde (**51a**) and pyrrolidine (**51b**) in 30% yield (**51c**) using a 1,2,4-triazolium catalyst (**51d**) (Fig. 51) [105]. Although this reaction requires a stoichiommetric amount of DBU, it is worth noting as it represents the first example of a challenging but important potential new avenue in amide synthesis.

Two other promising transition metal-free examples of oxidative amidation using molecular oxygen have been reported (Fig. 52). A report showing the oxidation of aldehydes in the presence of oxygen and NaH is intriguing [106], but no mechanistic details of this reaction have been given. A conceptually unique approach however is the application of electrochemistry. Wang showed amide bond formation from an α-ketoaldehyde in the presence of TBAI under oxygen [107]. The reaction only

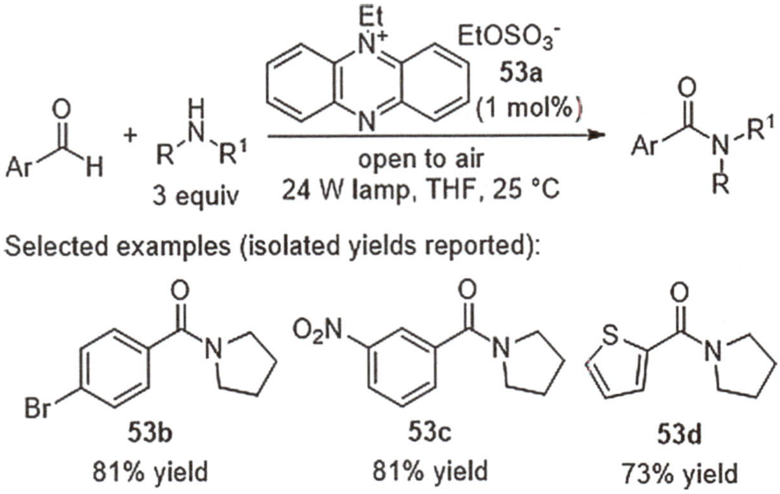

Fig. 53 Phenazinium-catalyzed aerobic oxidative amidation

proceeded when a 20 mA electric potential was applied across platinum electrodes in ethanol. In contrast to the majority of methods of oxidative amidation that require excess aldehyde as the hydrogen acceptor, this protocol uses the aldehyde as the limiting reagent.

Exchanging the electrodes for the inexpensive photocatalyst, phenazine ethosulfate (Fig. 53, **53a**), Leow synthesized a variety of tertiary benzamides at ambient temperatures [108] with the reaction displaying excellent functional group tolerance including aryl halides (**53b**), nitro groups (**53b**), and thiophenes (**53c**). Leow suggested that molecular oxygen and water together may be forming small amounts of hydrogen peroxide under the reaction conditions, and this hypothesis may provide important insight to the mechanisms of other oxidative amidation reactions using molecular oxygen. Reactions that use peroxides and other inorganic oxidants are common [109], but are outside of the scope of this chapter due to the reactive nature of these oxidants, which are typically used in excess. However, the ability to catalytically generate hydrogen peroxide directly from molecular oxygen may be important in the development of mild methods and may open new possibilities for sustainable approaches to amide synthesis within this reaction paradigm. It is worth noting that a limitation to both the Leow chemistry and the majority of aldehyde oxidative amidation reactions mediated by oxygen is the requirement for excess amine. This limitation may be difficult to overcome since amines themselves are prone to oxidation and may be getting converted to *N*-oxides in the reaction mixtures. The possibility of *N*-oxides playing a role in the oxidation step has not been investigated.

To improve catalytic efficiency, Sundén and co-workers developed a three-catalyst cooperative system in which oxidative N-acylation of oxazolidinones with aldehydes proceeds under

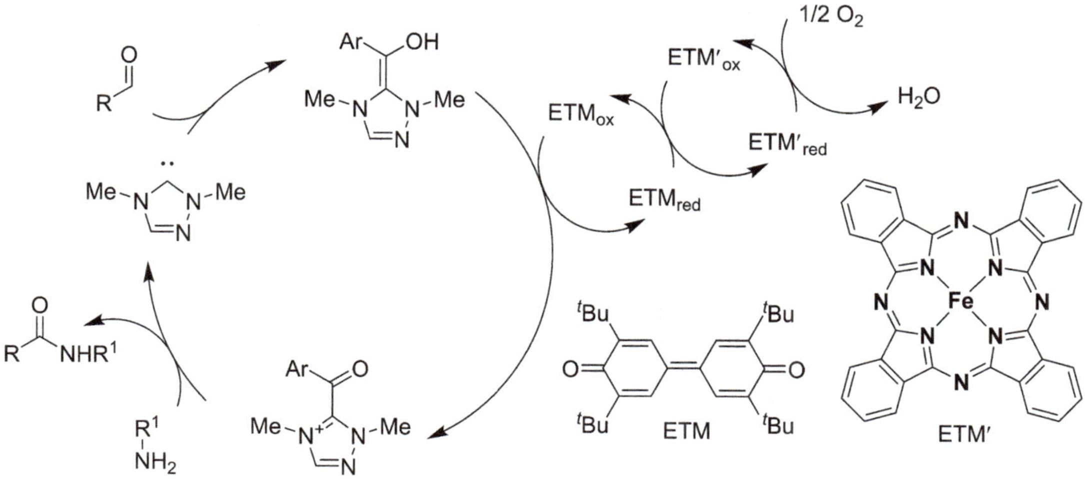

Fig. 54 NHC-catalyzed oxidative amidation of aldehydes using catalytic ETMs and O_2

mild aerobic conditions. The reaction uses an NHC-catalyzed formation of a Breslow intermediate which is then oxidized by a cooperative action of catalytic amounts of electron transfer mediators (ETM's) 3,3′,5,5′-tetra-tert-butyldiphenoquinone and iron (II)phthalocyanine. While the methodology was limited to generating acylated oxazolidinones, avoiding prefunctionalized substrates and stoichiometric oxidants is an attractive direction (Fig. 54) [110]. Replacing the ETM's with LiCl as a Lewis acid can also generate *N*-(benzo[d]thiazol-2-yl)benzamides if the imine is preformed [111].

In summary, there are a number of promising approaches toward the oxidative amidation of aldehydes using molecular oxygen as a terminal oxidant. Whereas metal catalysis is most successful with the group 11 metals that are relatively insensitive to molecular oxygen, electrochemistry and photochemistry appear to have the greatest potential for developing a metal-free oxidative amidation of aldehydes within this realm. It is possible that each of these methods rely on the catalytic generation of hydrogen peroxide that oxidizes the hemiaminal intermediate. The key challenge to overcome in the following years for these systems will be to develop methods that use equimolar ratios of both the amine and aldehyde substrates.

3.5 Isohypsic Oxidative Amidation of Aldehydes

One appealing factor of converting aldehydes or alcohols to amides using only amine nucleophiles is that all reagents are feedstock chemicals and are widely available. However, there are a few notable examples that start with aldehydes and an oxidized nitrogen source to generate amides, and alternatively, there are examples of specialized aldehyde substrates that undergo an internal redox relay in the presence of an amine to generate amides. While considering isohypsic transformations, it is important to realize that the burden of redox economy transfers to pre-functionalization steps rendering these methods less green than they may initially appear. Because of the vast array of data on this topic, including various detailed reviews [112–114], we will focus our discussion on the seminal contributions to catalytic isohypsic methods that convert aldehydes to amides while generating minimal amounts or no waste.

3.5.1 Isohypsic Oxidative Amidation of Aldehydes: Beckmann-Type Rearrangements

The most classic example of this type of isohypsic amide synthesis is the rearrangement of ketoximes to amides, or the Beckmann rearrangement [115]. Early examples of this transformation required the use of super stoichiometric amounts of a strong acid or base to convert ketones to amides by first condensing with hydroxylamine. For a more detailed discussion on these, we would like to direct the reader to the numerous reviews on the Beckmann rearrangement [116–118], as well as a recent review on the conversion of aldoximes to amides [119]. Herein, we will focus our attention to the limited number of reported methods that convert aldehydes into amides directly via in situ aldoxime generation.

While there has been much progress in developing catalytic Beckmann rearrangements during the last century, the related aldoxime rearrangement generally gives nitriles under typical Beckmann rearrangement conditions. This challenge was first addressed by Sharghi in 2002, who developed a method which utilized stoichiometric amounts of ZnO under neat conditions at high temperatures (>140 °C). The reaction was successful for the one-pot conversion of aromatic aldehydes to primary amides via an aldoxime intermediate [120]. However, it was 1 year later that Chang reported the first catalytic method using 5 mol% of Wilkinson's catalyst, which at (Fig. 55, **55a**) [121] 150 °C converted aryl (**55b**), alkenyl (**55c**), and alkyl (**55d**) aldoximes to primary amides in good yields. Chang proposed that the reaction proceeded through a nitrile intermediate (**55e**) which is in turn hydrated to form the primary amide, and this mechanistic proposal appears to hold true for many of the related reactions that have subsequently appeared. Many of these methods are highlighted in Subheading 5.

A remaining challenge for this Beckmann-type rearrangement was to synthesize amides from aldehydes in one pot. One particularly intriguing aspect to this reaction is that early attempts suggested that the ratio of *cis* to *trans* oxime was crucial for reactivity in a traditional Beckmann-type mechanism [122]. Mizuno was the first to achieve a one-pot conversion of aldehydes to primary amides by using a hydroxyl amine salt, $(NH_2OH)_2 \cdot H_2SO_4$ (Fig. 56) [123]. The reaction requires 4 mol% of a heterogeneous catalyst, $Rh(OH)_x/Al_2O_3$ at 160 °C. This one-pot reaction was greatly improved upon by Crabtree who showed an enhanced scope at lower temperatures (110 °C) and used the commercially available

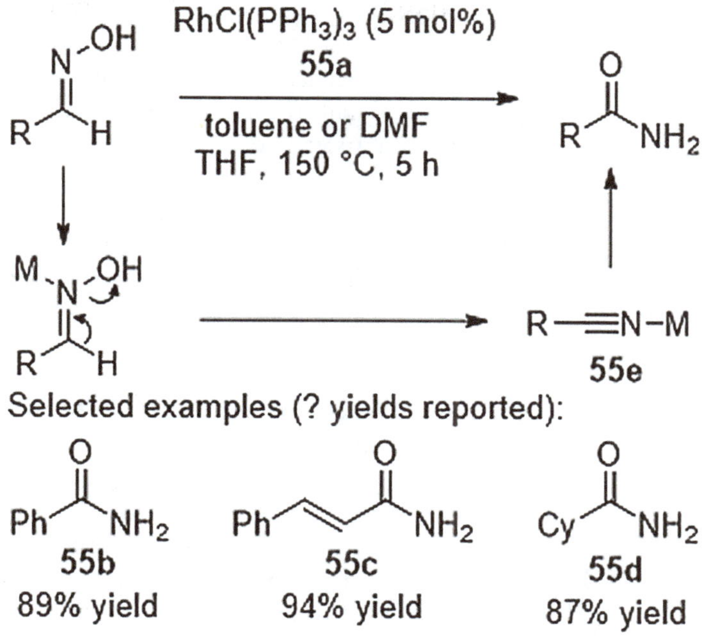

Fig. 55 The first catalytic rearrangement of aldoximes to amides

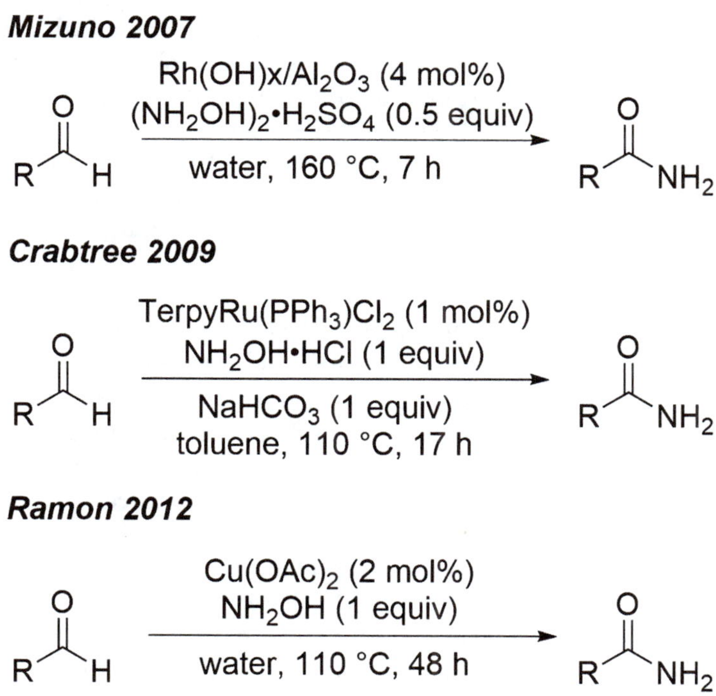

Fig. 56 Oxidative amidation via in-situ generated aldoximes

$NH_2OH \cdot HCl$ [124]. In this report, Crabtree observed the formation of a nitrile from an O-alkylated oxime indicating that the Chang mechanism discussed previously was operative. One additional detail to the mechanism was the requirement of the oxygen atom transfer to occur within the coordination sphere of the metal. Because the reaction proceeded through a nitrile intermediate, Crabtree demonstrated that the *cis-trans* isomerization of the oxime did not affect the reaction yield.

Finally, Ramón and coworkers found that they could eliminate the need for base entirely by using free hydroxylamine [125] to mediate the conversion of aldehydes to primary amides thus generating minimal amounts of waste. In addition, the Ramón report uses a base-metal catalyst ($Cu(OAc)_2$) in water at a lower temperature than many of the related transformations (110 °C). More recently, Mal and Achar removed the need for a metal catalyst entirely by running the reaction neat in a ball-mill with a chloramine precursor [126, 127]. Although outside of the scope of this chapter, it is worth noting that there are also many reactions that convert ketones to amides such as the Beckmann or the Wolff rearrangements.

3.5.2 Isohypsic Oxidative Amidation of Aldehydes: NHC-Catalyzed Methods

The desire to move away from metal catalysts entirely has driven the development of NHC catalysis. Building on the concept from Scheidt and coworkers, who added homoenolates to electrophilic nitrogen sources [128], Zhang developed a 100% atom-economical method of adding aldehydes to *N*-aryl nitroso compounds (Fig. 57) [129]. Taking advantage of the umpolung reactivity of the Breslow intermediate, nitroso compounds provide an electrophilic source of nitrogen-generating amides from aldehydes. Within 1 h at room temperature, the reaction generated hydroxamic acids, which represent a useful motif both in pharmaceutical research and metal catalysis.

Distinct from the umpolung reactivity of traditional NHC catalysis, an alternative approach to NHC-catalyzed oxidation has emerged with Corey being the first to demonstrate that the Breslow intermediate generated from an aldehyde and cyanide ion could be oxidized by manganese oxide to generate an acyl azolium intermediate that could be substituted by a nucleophile to synthesize acid derivatives [130]. Three years later, in 1971, Gilman showed that amines were compatible nucleophiles generating amides from aldehydes [131]. Performing this transformation using a catalyst and no stoichiometric oxidant however proved to be much more challenging, and the corresponding isohypsic NHC-catalyzed methods to generate esters [132] and thioesters [133] were developed long before Bode and Rovis simultaneously extended this chemistry to amide synthesis (Fig. 58).

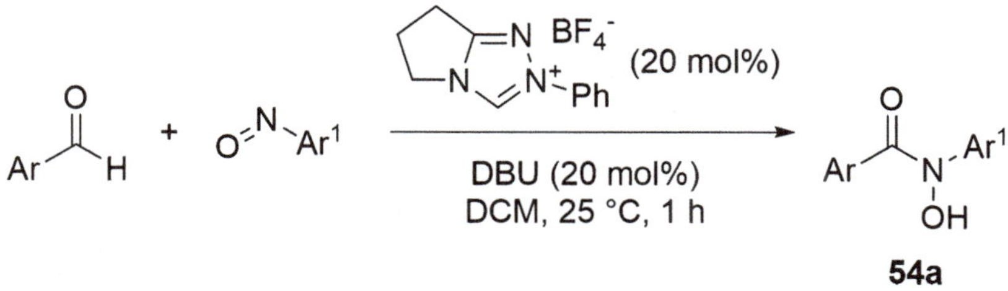

Fig. 57 NHC-catalyzed umpolung oxidative amidation

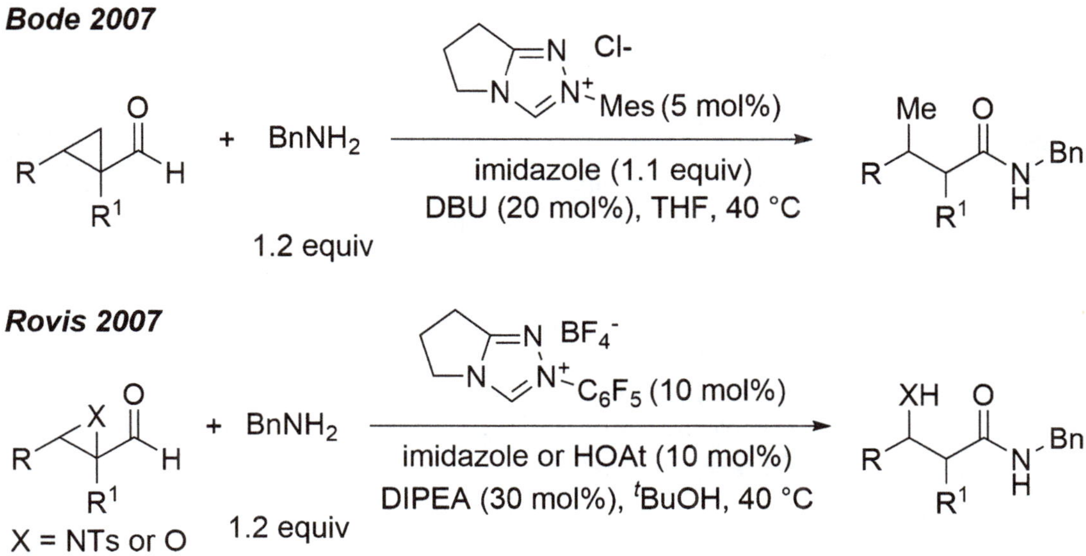

Fig. 58 NHC-catalyzed oxidative amidation via an internal redox transfer

Bode used formyl cyclopropanes with a triazole catalyst at 40 °C [134] using imidazole as a stoichiometric additive to suppress the imine formation that deactivated the catalyst. The Rovis reaction on the other hand did not require any stoichiometric additives for the majority of the substrates [135] with the improvement being achieved by using HOAt or imidazole as co-catalysts in combination with α,β-epoxides or aziridines. Interestingly, both of these reactions proved to be most successful with primary benzylic amines, while other amines gave decreased yields. The general concept of relay oxidation to achieve oxidative amidation using NHC catalysis has since been extended to polymer-supported catalysts [136], ring expansions [137], and kinetic resolutions of amines [138]. We would like to direct the reader to recent reviews that feature these methods [139].

Finally, an alternative approach to isohypsic oxidative amidation using NHC-catalysis involves generating a nucleophilic amine in situ by combining it with the redox event. Scheidt and others have achieved [3 + 2]- or [3 + 3]-cycloadditions to synthesize cyclic amides using diazines [134], or azomethine imines (Fig. 59)

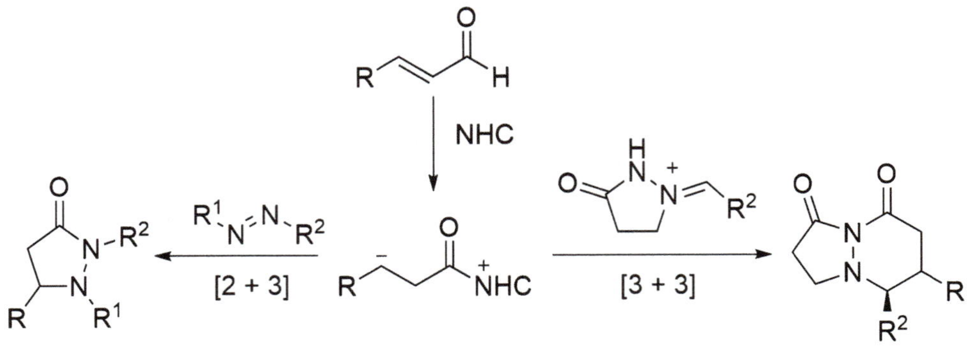

Fig. 59 Oxidative amidation using in situ generated amines

[140]. The success of these reactions is a result of the amide bond-forming event being intramolecular. The first bond-forming event (addition of the homoenolate to the diazine) generates the nucleophile (hydrazide) resulting in a low concentration of amine, thus reducing deleterious side reactions between the amine and catalyst.

In summary, isohypsic methods that rely on internal redox relays represent an atom economical synthesis of amides from aldehydes. Although the researcher has to consider the synthesis of the reagents required to achieve these transformations, there is much notable progress in both the catalytic rearrangement of oximes and NHC-catalyzed amide synthesis.

4 Direct Amidation of Carboxylic Acids and Homologs

Examples of the direct amidation of esters have been reported since the early 1990s, but generally require stoichiometric amounts of activators or metal mediators. A few catalytic methods of note are the early reports of macrolactamization [141, 142] using esters and amines, though the first methodology that focused directly on amide bond formation from esters and amines was reported by Porco in 2005 [143]. He reported that group IV metal catalysts were able to promote the formation of amides from esters and amines with after optimization, $Zr(OtBu)_4$ with a triazole (HYP, HOAt, or HOBt) additive being shown to give the best yields. This direct amidation reaction was demonstrated to give high yields with a diverse group of substrates (**60a–60c**) and show high specificity for esters (Fig. 60). Elucidation of the reaction mechanism indicated that a dimeric zirconium structure was the catalytically active species (Fig. 61).

Milstein et al. demonstrated in 2011 that their family of ruthenium(II) pincer ligands could also perform direct amidation of esters as well as DAS from alcohols and amines [144]. The PNN ligand (**2b**) was shown to have the highest activity for amide bond formation from esters and amines though unlike DAS from

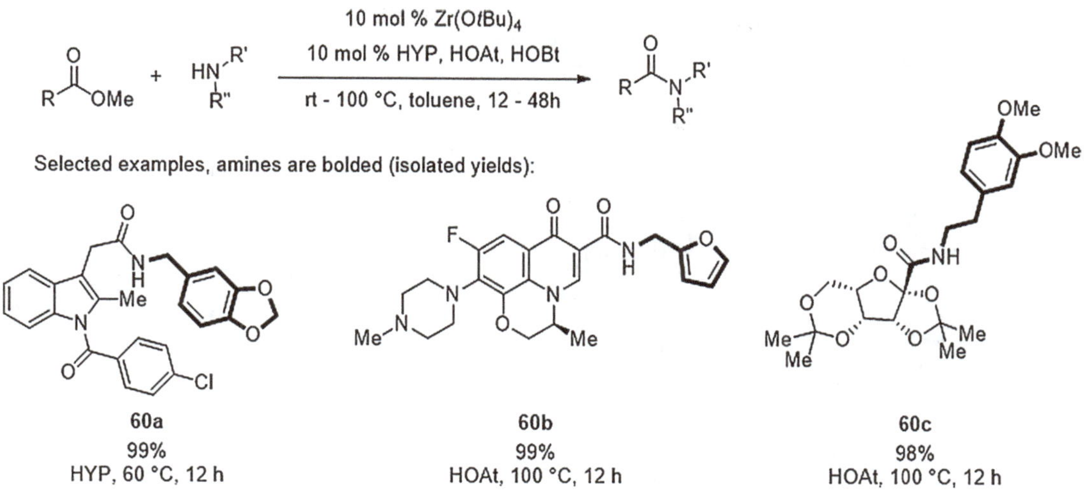

Fig. 60 Zirconium-catalyzed amidation of esters

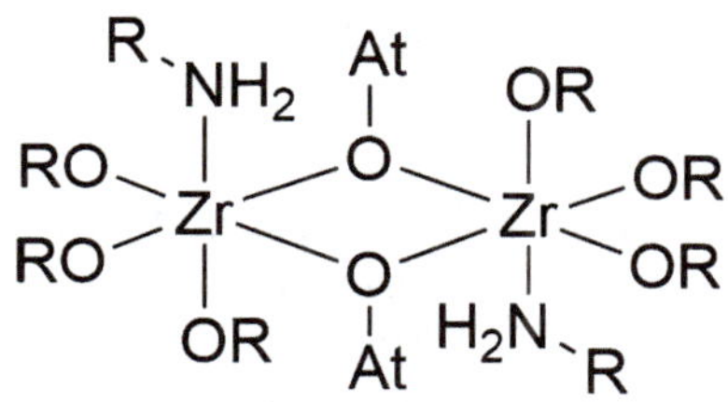

Fig. 61 Proposed transition state of zirconium-catalyzed amidation reaction

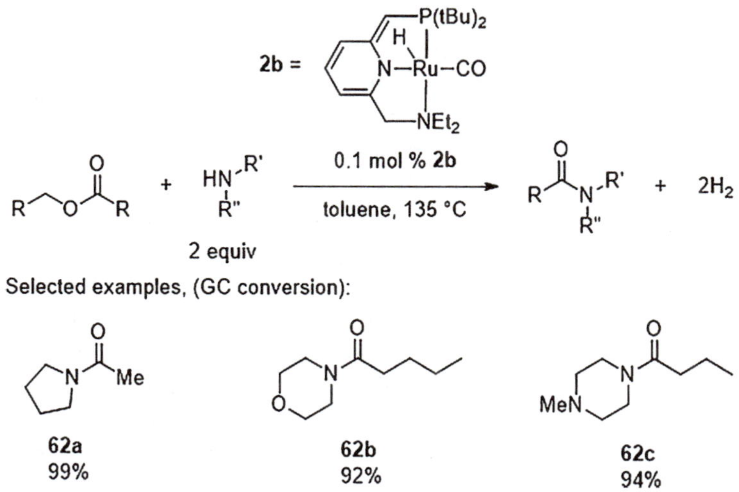

Fig. 62 Aminolysis of esters catalyzed by **2b**

alcohols and amines, no base is required for the reaction to proceed. A variety of secondary and tertiary amides (**62a–62c**) were able to be formed with a similar scope to the report using alcohols and amines (Fig. 62). The mechanism is thought to proceed

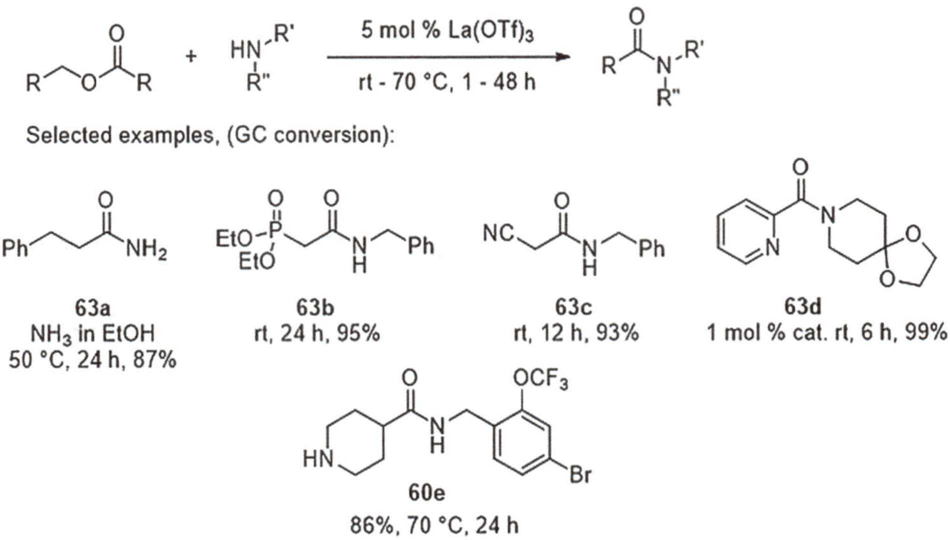

Fig. 63 Lanthanide-catalyzed amide bond formation

through an intermediate in which both the amine and ester are simultaneously coordinated to the ruthenium center at some point in the reaction pathway.

A further study in 2012 by Mashima attempted to identify benign conditions for direct ester amidation based on metals available in biological settings [145]. Catalytic NaOMe was shown to promote amide formation from esters and amines for a variety of substrates at a relatively low temperature (~50 °C). Extending this methodology to target peptide couplings lead to extensive racemization due to the basic conditions. Though by adding trifluoromethylphenol to the reaction, they were able to preclude racemization of the peptide, leading to efficient couplings.

Ohshima demonstrated the catalytic formation of several amides using a $La(OTf)_3$ catalyst at moderate temperature (70 °C) (Fig. 63) [146] with the catalyst being commercially available and showing a high tolerance to several functional groups, including phosphines (**63b**), nitriles (**63c**), sterically hindered secondary amines (**63d**), and fluorinated amines (**63e**).

Similar to the Milstein's PNN catalyst, Ru-Macho, a commercially available catalyst shown previously to promote amide formation by Dong and Guan [30], was used by Li for the formation of amides from esters and amines [147] with high activity for the formation of aliphatic secondary and tertiary amides. Since Ru-Macho was originally developed for an industrial scale reaction [148], Li tested the reaction at a larger scale (240 mmol) demonstrating no loss of activity.

Highly activated esters are a common theme in the amidation of esters, and in many cases, activation increases waste and decreases utility. Fier and Whittaker reported a rare case of atom economical amidation of activated esters in the form of *N*-carboxy anhydrides

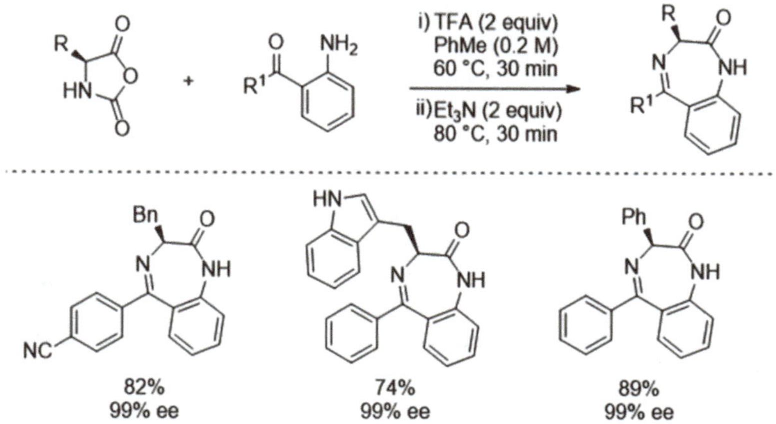

Fig. 64 *N*-carboxy anhydrides as amide coupling partners under acidic conditions

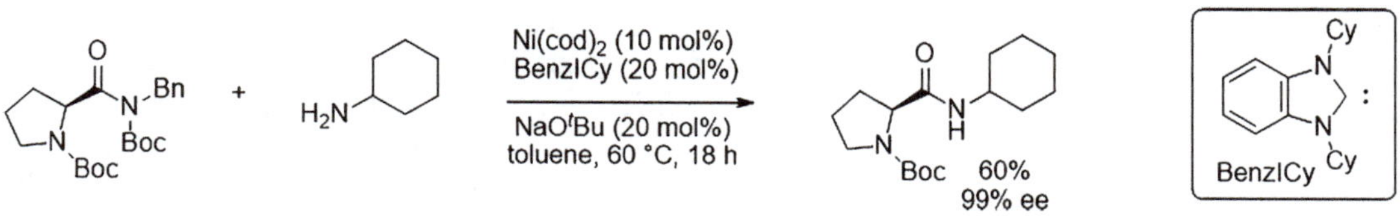

Fig. 65 Pd-catalyzed transamidation using aliphatic amines

[149]. This strategy was used for the direct formation of benzodiazepines with CO_2 and H_2O as the only byproducts. Because of the ability to run the amide formation step at low pH, no oligomerization or racemization of the was observed whatsoever (Fig. 64). Other reactive esters such as methyl benzoates or phenyl benzoates are effective electrophiles for electron-rich transition metal-NHC complexes of Pd [150] and Ni [151, 152]. Access to a Pd acyl intermediate has rendered one set of conditions applicable to both amide synthesis through activation of reactive esters and benzophenone synthesis through activation of reactive amides via a Suzuki reaction [153].

The general strategy of utilizing highly reactive amides as electrophiles has formed the basis of all existing transamidation reactions. While the methods for transition-metal-free transamidation are limited in scope to only transamidations between formamide and benzamide with anilines under high temperature or highly basic conditions [154–157], transition metal catalysis has enabled a slightly broadened scope [158]. To date there is one example of transamidation between aliphatic amides and aliphatic amines. Garg and co-workers discovered that prefunctionalization of secondary amides allows mild transamidations with minimal epimerization, a long-standing challenge to many amidation methodologies (Fig. 65) [159]. The inefficiency associated with prefunctionalization of amides limits how green this chemistry currently is, but the progress toward a more general transamidation could have major implications in the future of green peptide coupling.

A truly aspirational goal of green amide synthesis is to directly access amides from carboxylic acids without the use of stoichiometric coupling reagents. Acids and amines are some of the most widely available motifs, and the combination of the two should theoretically generate only water [8]. Ironically, waste generated from DCC, $POCl_3$, or HATU often represents a higher mass than the isolated product in addition to well-documented issues with racemization and purification. Approaches toward a direct amidation of carboxylic acids have slowly been gaining traction ever since the seminal paper on this topic in 1996 by Yamamoto [160]. Given that the first report used an aryl boronic acid catalyst, it is not surprising that the most development for direct amidation of carboxylic acids is boron based [161]. A continuing evolution from aryl boric acid catalysts [162] through diboron catalyst [163], diboronic acid anhydrides [164, 165], geminal diboron compounds [166], and finally pseudo trimeric B_3NO_2 compounds was a rational direction in light of the mechanistic understanding of the transformation (Fig. 66). The reactions all seem to proceed through a pseudo trimer containing three electronically distinct boron atoms. The most Lewis acidic boron coordinates the amine, while the least Lewis acidic boronic acid forms a hydrogen bond with the carboxylic acid promoting amidation.

Most other approaches in this arena require stoichiometric reagents such as silanes [167] or phosphites [168] to promote amidation [169] or to turn over the catalyst [170]. And while some of these catalysts offer an impressively broad scope applicable to peptide synthesis [171], the only two non-boron-based methods without a stoichiometric reagent use simple $NiCl_2$ and $ZrCl_4$. The

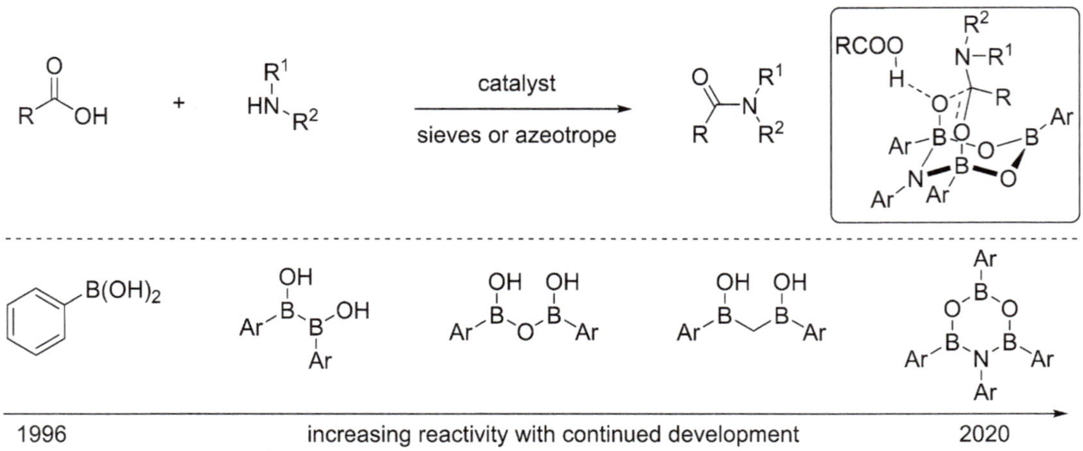

Fig. 66 Direct amidation of carboxylic acids using boronic acid catalysts

zirconium-based method relies on molecular sieves which are recyclable dehydrating aids rather than consumed reagents. In fact, many of the boron-based methods also use either sieves or a Dean-Stark trap to remove water and drive the reaction forward.

5 Formation of Amides from Nitriles

Using nitriles to form amides is known through several mechanisms, the most widely exploited being the well-established Ritter reaction [172]. More recently, nitriles have been utilized to form primary amides via hydration. Often these reactions require a strong acid or base to proceed, but there are several examples of metal catalysts being employed for the hydration of nitriles to form amides. The metal-catalyzed reactions are of particular note because they often proceed under mild conditions and exhibit maximal atom economy. Several enzymes have also been shown to be extremely active [173–179] for generating amides from nitriles and are used in many industrial processes. There are a large number of examples for the hydration of nitriles, of which we will cover the more important contributions, as there are already exhaustive reviews on this topic [119].

The first report of the catalytic hydration of nitriles to amides was reported by Taube in 1974 using $[(NH_3)_5RuCl]Cl_2$ in the presence of silver oxide, trifluoroacetic acid, air, water, and ethanol [180] (Fig. 67). They were able to form several primary amides in good yields under mild conditions. Another early report of a metal-catalyzed hydration of nitriles was made by Murahashi in 1986 using $RuH_2(PPh_3)_4$ at >100 °C [181]. They found that upon addition of ammonia to an aqueous alcoholic solution of the nitrile, secondary and tertiary amides could be obtained in good yields (**67a–67f**).

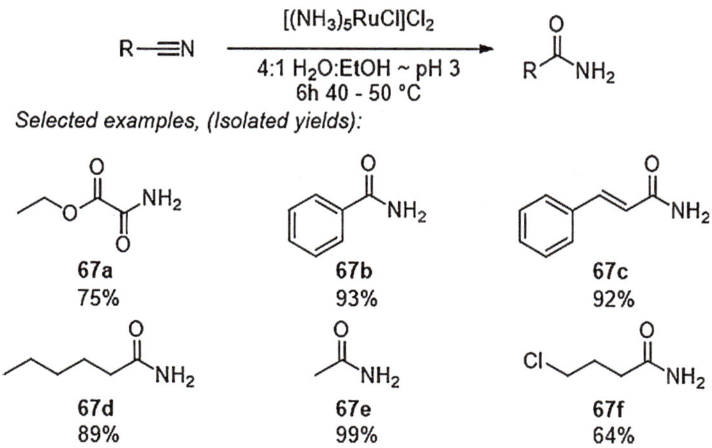

Fig. 67 Ruthenium-catalyzed primary amide formation from nitriles

In 1995, a highly active platinum catalyst [$PtH(PMe_2OH)(PMe_2O)_2H$] was reported by Parkins which gave TONs around ~5700 for this type of reaction and formed primary amides in good yields [182]. Mild conditions (rt–80 °C) in toluene allowed for thermally sensitive functionalities to be incorporated, and due to both these features, this protocol has gained popularity and has been used in the synthesis of several natural products [187].

In an attempt to limit organic solvent waste, there was a drive to switch from using water as a reagent to using water as the solvent for these processes. However, the catalyst system reported by Parkins had minimal solubility in water, and a search for catalytic systems that functioned purely in aqueous environment were sought. Systems using copper [183, 184], ruthenium [185–187], and other metals were designed using ligands specifically engineered to increase their solubility in aqueous environments.

As the hydration of nitriles is a well-established field, it was surprising that it was not until 2013 that an example of the synthesis of amides from alcohols and nitriles was reported. Using a ruthenium(II) NHC catalyst, which was similar to those developed for DAS from alcohols and amines, the Hong group demonstrated that amides (**68a–68c**) could be formed in high yield from nitriles and alcohols with complete atom economy and minimal by-product formation (Fig. 68) [188]. Using ^1H NMR to probe the reaction mechanism, the free aldimine was observed confirming that reduction of the nitrile was occurring in the reaction mixture. The reaction mechanism proposed involved initial reduction of the nitrile to the aldimine, followed by coordination and oxidation of the alcohol to the aldehyde. Interestingly, in line with previous observations, no free aldehyde was observed in the reaction. After the oxidation occurs, the aldimine was reduced to the amine, and a hemiaminal intermediate is formed. Further dehydrogenation turns over the catalyst and releases the desired amide. Using the same catalytic system though with lower loadings of both the catalyst and ligand in a subsequent paper, the Hong group also reported on the synthesis of cyclic imides (**68d–68f**, Fig. 68) [189].

6 Formation of Amides from Azides

While there is a lot of interest in finding different oxygen atom sources for amidation methods, investigation into new nitrogen atom sources is more limited. In a paper by Mizuno in 2011, while investigating the transformation of azides to nitriles using a $Ru(OH)_x/Al_2O_3$, it was noted that in the presence of water and air, azides could be transformed directly into primary amides [190]. The reaction was thought to proceed through a mechanism where the azide is transformed into a nitrile and then hydrated to

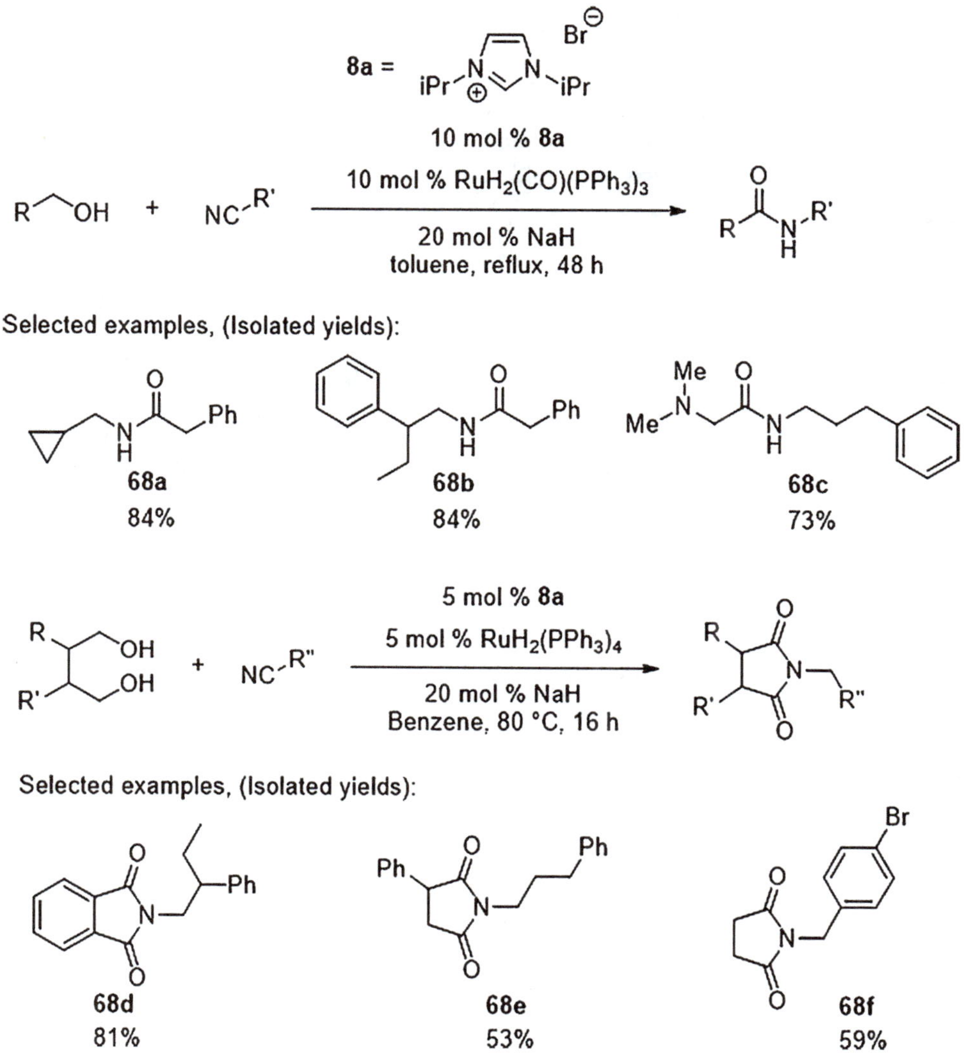

Fig. 68 Formation of amides and cyclic imides using nitriles and alcohols mediated by a ruthenium(II) NHC complex

form a primary amide. While this reaction yields amides, the use of azides does present additional challenges due to their explosive hazard.

In 2012, the Hong group reported that DAS could be performed using azides as the nitrogen source in the presence of an alcohol (Fig. 69) [191]. In a similar manner to their previous investigation into DAS using amines and alcohols, a ruthenium (II) precursor, an imidazolium salt, and base are used to form an NHC-bound ruthenium complex that catalyzes the reaction. The catalyst reduces the azide to the amine using the equivalent of hydrogen taken from the alcohol in the production of the aldehyde. The reaction is then thought to proceed in an analogous manner to

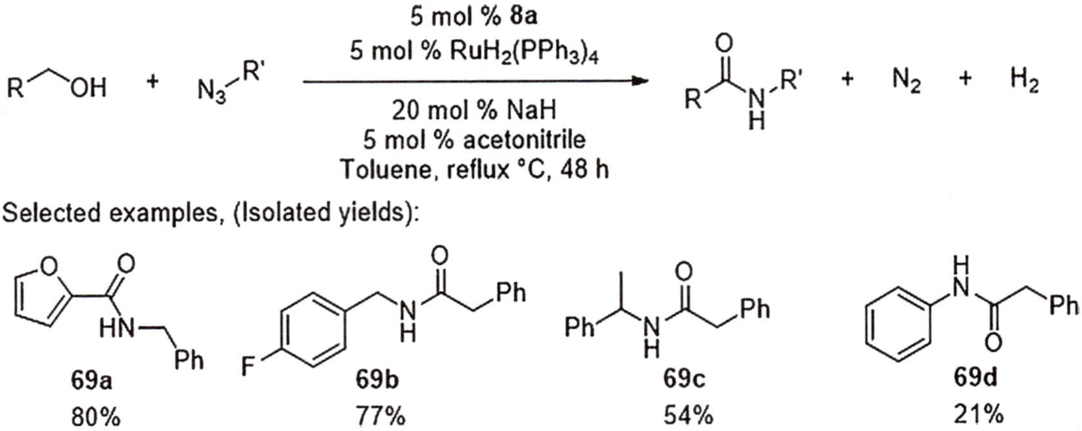

Fig. 69 Formation of amides using azides and alcohols mediated by a ruthenium(II) NHC complex

DAS between amines and alcohols and also demonstrates a similar substrate scope. Typically, both aliphatic azides and alcohols are tolerated (**69a–69d**), though in the presence of an alkyne, the ruthenium(II) NHC complex preferentially forms the triazole over the amide.

7 Conclusion

Considering the young age of this field, an impressive amount of work has been achieved over the last 10 years, yielding a wealth of methodologies for accessing amides with high atom economy. Largely dominated by both heterogeneous and homogenous transition metal-based catalysts, important groundwork has been made on both demonstrating functional tolerance and providing mechanistic insights to expand this type of reaction into a wide and useful chemical technique. Future studies should aim to find applicable systems which rely on earth abundant metals and display high activity at lower temperatures to ensure that these new methodologies can compete with canonical peptide coupling reagents. Further improving the reaction conditions and substrate tolerance could also help access important industrial applications of amide bond formation such as peptide synthesis and polyamide formation.

Although outside of the scope of this chapter, biocatalysis is an emerging technology with high potential for greener syntheses of amide bonds. Peptiligases [192], ligases [193], transaminases [194] amide synthetases, sortases, proteases, and transpeptidase [195] have all shown the ability to selectively form amides in the presence of traditionally incompatible functional groups while also operating in water. For more information on this topic, see biocatalysis discussed in Chap. 12.

References

1. Valeur E, Bradley M (2009) Amide bond formation: beyond the myth of coupling reagents. Chem Soc Rev 38:606–631. https://doi.org/10.1039/B701677h
2. Bode JW (2006) Emerging methods in amide- and peptide-bond formation. Curr Opin Drug Discov Dev 9:765–775. https://doi.org/10.1002/chin.200721229
3. Montalbetti CAGN, Falque V (2005) Amide bond formation and peptide coupling. Tetrahedron 61:10827–10852. https://doi.org/10.1016/J.Tet.2005.08.031
4. Constable DJC, Dunn PJ, Hayler JD, Humphrey GR, Leazer JL, Linderman RJ, Lorenz K, Manley J, Pearlman BA, Wells A, Zaks A, Zhang TY (2007) Key green chemistry research areas - a perspective from pharmaceutical manufacturers. Green Chem 9:411–420. https://doi.org/10.1039/B703488c
5. Allen CL, Williams JMJ (2011) Metal-catalyzed approaches to amide bond formation. Chem Soc Rev 40:3405–3415. https://doi.org/10.1039/c0cs00196a
6. Gunanathan C, Milstein D (2013) Applications of acceptorless dehydrogenation and related transformations in chemical synthesis. Science 341:249. https://doi.org/10.1126/science.1229712
7. Chen C, Hong SH (2011) Oxidative amide synthesis directly from alcohols with amines. Org Biomol Chem 9:20–26. https://doi.org/10.1039/c0ob00342e
8. Pattabiraman VR, Bode JW (2011) Rethinking amide bond synthesis. Nature 480:471–479. https://doi.org/10.1038/nature10702
9. Trost BM (1991) The atom economy - a search for synthetic efficiency. Science 254:1471–1477. https://doi.org/10.1126/science.1962206
10. Lundberg H, Tinnis F, Selander N, Adolfsson H (2014) Catalytic amide formation from non-activated carboxylic acids and amines. Chem Soc Rev 43:2714–2742. https://doi.org/10.1039/c3cs60345h
11. Naota T, Murahashi SI (1991) Ruthenium-catalyzed transformations of amino-alcohols to lactams. Synlett:693–694. https://doi.org/10.1021/jo00228a032
12. Gunanathan C, Ben-David Y, Milstein D (2007) Direct synthesis of amides from alcohols and amines with liberation of H-2. Science 317:790–792. https://doi.org/10.1126/Science.1145295
13. Zhang J, Gandelman M, Shimon LJW, Rozenberg H, Milstein D (2004) Electron-rich, bulky ruthenium PNP-type complexes. Acceptorless catalytic alcohol dehydrogenation. Organometallics 23:4026–4033. https://doi.org/10.1021/Om049716j
14. Zhang J, Leitus G, Ben-David Y, Milstein D (2005) Facile conversion of alcohols into esters and dihydrogen catalyzed by new ruthenium complexes. J Am Chem Soc 127:10840–10841. https://doi.org/10.1021/Ja052862b
15. Zhang J, Leitus G, Ben-David Y, Milstein D (2006) Efficient homogeneous catalytic hydrogenation of esters to alcohols. Angew Chem Int Ed 45:1113–1115. https://doi.org/10.1002/Anie.200503771
16. Nordstrøm LU, Vogt H, Madsen R (2008) Amide synthesis from alcohols and amines by the extrusion of dihydrogen. J Am Chem Soc 130:17672–17673. https://doi.org/10.1021/ja808129p
17. Watson JA, Maxwell AC, Williams JMJ (2009) Ruthenium-catalyzed oxidation of alcohols into amides. Org Lett 11:2667–2670. https://doi.org/10.1021/Ol900723v
18. Zweifel T, Naubron J-V, Grützmacher H (2009) Catalyzed dehydrogenative coupling of primary alcohols with water, methanol, or amines. Angew Chem Int Ed 48:559–563. https://doi.org/10.1002/anie.200804757
19. Ghosh SC, Muthaiah S, Zhang Y, Xu X, Hong SH (2009) Direct amide synthesis from alcohols and amines by phosphine-free ruthenium catalyst systems. Adv Synth Catal 351:2643–2649. https://doi.org/10.1002/adsc.200900482
20. Zhang Y, Chen C, Ghosh SC, Li Y, Hong SH (2010) Well-defined N-heterocyclic carbene based ruthenium catalysts for direct amide synthesis from alcohols and amines. Organometallics 29:1374–1378. https://doi.org/10.1021/om901020h
21. Ghosh SC, Hong SH (2010) Simple RuCl3-catalyzed amide synthesis from alcohols and amines. Eur J Org Chem 2010:4266–4270. https://doi.org/10.1002/ejoc.201000362
22. Muthaiah S, Ghosh SC, Jee J-E, Chen C, Zhang J, Hong SH (2010) Direct amide synthesis from either alcohols or aldehydes with amines: activity of Ru(II) hydride and Ru (0) complexes. J. Org. Chem. 75:3002–3006. https://doi.org/10.1021/jo100254g

23. Dam JH, Osztrovszky G, Nordstrom LU, Madsen R (2010) Amide synthesis from alcohols and amines catalyzed by ruthenium N-heterocyclic carbene complexes. Chem Eur J 16:6820–6827. https://doi.org/10.1002/chem.201000569
24. Prades A, Peris E, Albrecht M (2011) Oxidations and oxidative couplings catalyzed by triazolylidene ruthenium complexes. Organometallics 30:1162–1167. https://doi.org/10.1021/om101145y
25. Chen C, Zhang Y, Hong SH (2011) N-heterocyclic carbene based ruthenium-catalyzed direct amide synthesis from alcohols and secondary amines: involvement of esters. J. Org. Chem. 76:10005–10010. https://doi.org/10.1021/jo201756z
26. Srimani D, Balaraman E, Hu P, Ben-David Y, Milstein D (2013) Formation of tertiary amides and dihydrogen by dehydrogenative coupling of primary alcohols with secondary amines catalyzed by ruthenium bipyridine-based pincer complexes. Adv Synth Catal 355:2525–2530. https://doi.org/10.1002/adsc.201300620
27. Schley ND, Dobereiner GE, Crabtree RH (2011) Oxidative synthesis of amides and pyrroles via dehydrogenative alcohol oxidation by ruthenium diphosphine diamine complexes. Organometallics 30:4174–4179. https://doi.org/10.1021/Om2004755
28. Gnanaprakasam B, Balaraman E, Ben-David Y, Milstein D (2011) Synthesis of peptides and pyrazines from ss-amino alcohols through extrusion of H2 catalyzed by ruthenium pincer complexes: ligand-controlled selectivity. Angew Chem Int Ed 50:12240–12244. https://doi.org/10.1002/Anie.201105876
29. Khusnutdinova JR, Ben-David Y, Milstein D (2014) Oxidant-free conversion of cyclic amines to lactams and H_2 using water as the oxygen atom source. J Am Chem Soc 136:2998–3001. https://doi.org/10.1021/ja500026m
30. Oldenhuis NJ, Dong VM, Guan Z (2014) From racemic alcohols to enantiopure amines: Ru-catalyzed diastereoselective amination. J Am Chem Soc 136:12548–12551. https://doi.org/10.1021/ja5058482
31. Nguyen TT, Hull KL (2016) Rhodium-catalyzed oxidative amidation of sterically hindered aldehydes and alcohols. ACS Catal 6:8214–8218. https://doi.org/10.1021/acscatal.6b02541
32. Ovian JM, Kelly CB, Pistritto VA, Leadbeater NE (2017) Accessing N-acyl azoles via oxoammonium salt-mediated oxidative amidation. Org Lett 19:1286–1289. https://doi.org/10.1021/acs.orglett.7b00060
33. Kumar NA, Espinosa-Jalapa GL, Diskin-Posner Y, Avram L, Milstein D (2017) Direct synthesis of amides by dehydrogenative coupling of amines with either alcohols or esters: manganese pincer complex as catalyst. Angew Chem Int Ed 56:14992–14996. https://doi.org/10.1002/anie.201709180
34. Daw P, Kumar A, Espinosa-Jalapa NA, Ben-David Y, Milstein D (2019) Direct synthesis of amides by acceptorless dehydrogenative coupling of benzyl alcohols and ammonia catalyzed by a manganese pincer complex: unexpected crucial role of base. J Am Chem Soc 141:12202–12206. https://doi.org/10.1021/jacs.9b05261
35. Zultanski SL, Zhao J, Stahl SS (2016) Practical synthesis of amides via copper/ABNO-catalyzed aerobic oxidative coupling of alcohols and amines. J Am Chem Soc 138:6416–6419. https://doi.org/10.1021/jacs.6b03931
36. Piszel PE, Vasilopoulos A, Stahl SS (2019) Oxidative amide coupling from functionally diverse alcohols and amines using aerobic copper/nitroxyl catalysis. Angew Chem Int Ed 58:12211–12215. https://doi.org/10.1002/anie.201906130
37. Gruetzmacher H (2008) Cooperating ligands in catalysis. Angew Chem Int Ed 47:1814–1818. https://doi.org/10.1002/anie.200704654
38. Nova A, Balcells D, Schley ND, Dobereiner GE, Crabtree RH, Eisenstein O (2010) An experimental-theoretical study of the factors that affect the switch between ruthenium-catalyzed dehydrogenative amide formation versus amine alkylation. Organometallics 29:6548–6558. https://doi.org/10.1021/om101015u
39. Gunanathan C, Milstein D (2011) Metal-ligand cooperation by aromatization-dearomatization: a new paradigm in bond activation and "green" catalysis. Acc Chem Res 44:588–602. https://doi.org/10.1021/ar2000265
40. Li H, Wang X, Huang F, Lu G, Jiang J, Wang Z-X (2011) Computational study on the catalytic role of pincer Ruthenium(II)-PNN complex in directly synthesizing amide from alcohol and amine: the origin of selectivity of amide over ester and imine. Organometallics 30:5233–5247. https://doi.org/10.1021/om200620n
41. Zeng G, Li S (2011) Insights into dehydrogenative coupling of alcohols and amines

catalyzed by a (PNN)-Ru(II) hydride complex: unusual metal-ligand cooperation. Inorg Chem 50:10572–10580. https://doi.org/10.1021/ic200205e

42. Cho D, Ko KC, Lee JY (2013) Catalytic mechanism for the Ruthenium-complex-catalyzed synthesis of amides from alcohols and amines: a DFT study. Organometallics 32:4571–4576. https://doi.org/10.1021/om4005324
43. Mielby J, Riisager A, Fristrup P, Kegnaes S (2013) Mechanistic investigation of the one-pot formation of amides by oxidative coupling of alcohols with amines in methanol. Catal Today 203:211–216. https://doi.org/10.1016/j.cattod.2012.04.026
44. Montag M, Zhang J, Milstein D (2012) Aldehyde binding through reversible C-C coupling with the pincer ligand upon alcohol dehydrogenation by a PNP-ruthenium catalyst. J Am Chem Soc 134:10325–10328. https://doi.org/10.1021/Ja303121v
45. Gunanathan C, Gnanaprakasam B, Iron MA, Shimon LJW, Milstein D (2010) "Long-range" metal-ligand cooperation in H2 activation and ammonia-promoted hydride transfer with a ruthenium-acridine pincer complex. J Am Chem Soc 132:14763–14765. https://doi.org/10.1021/ja107770y
46. Klitgaard SK, Egeblad K, Mentzel UV, Popov AG, Jensen T, Taarning E, Nielsen IS, Christensen CH (2008) Oxidations of amines with molecular oxygen using bifunctional gold-titania catalysts. Green Chem 10:419–423. https://doi.org/10.1039/b714232c
47. Kim JW, Yamaguchi K, Mizuno N (2008) Heterogeneously catalyzed efficient oxygenation of primary amines to amides by a supported ruthenium hydroxide catalyst. Angew Chem Int Ed 47:9249–9251. https://doi.org/10.1002/anie.200802464
48. Shimizu K, Ohshima K, Satsuma A (2009) Direct dehydrogenative amide synthesis from alcohols and amines catalyzed by gamma-alumina supported silver cluster. Chem Eur J 15:9977–9980. https://doi.org/10.1002/Chem.200901896
49. Ishida T, Haruta M (2009) N-formylation of amines via the aerobic oxidation of methanol over supported gold nanoparticles. ChemSusChem 2:538–541. https://doi.org/10.1002/cssc.200800260
50. Zhou L, Freyschlag CG, Xu B, Friend CM, Madix RJ (2010) Direct selective oxygen-assisted acylation of amines driven by metallic silver surfaces: dimethylamine with formaldehyde. Chem Commun 46:704–706. https://doi.org/10.1039/B921066K
51. Preedasuriyachai P, Kitahara H, Chavasiri W, Sakurai H (2010) N-Formylation of amines catalyzed by nanogold under aerobic oxidation conditions with MeOH or formalin. Chem Lett 39:1174–1176. https://doi.org/10.1246/cl.2010.1174
52. Zhu J, Zhang Y, Shi F, Deng Y (2012) Dehydrogenative amide synthesis from alcohol and amine catalyzed by hydrotalcite-supported gold nanoparticles. Tetrahedron Lett 53:3178–3180. https://doi.org/10.1016/j.tetlet.2012.04.048
53. Pineda A, Gomez L, Balu AM, Sebastian V, Ojeda M, Arruebo M, Romero AA, Santamaria J, Luque R (2013) Laser-driven heterogeneous catalysis: efficient amide formation catalyzed by Au/SiO2 systems. Green Chem 15:2043–2049. https://doi.org/10.1039/c3gc40166a
54. Zhang L, Wang W, Wang A, Cui Y, Yang X, Huang Y, Liu X, Liu W, Son J-Y, Oji H, Zhang T (2013) Aerobic oxidative coupling of alcohols and amines over Au-Pd/resin in water: Au/Pd molar ratios switch the reaction pathways to amides or imines. Green Chem 15:2680–2684. https://doi.org/10.1039/c3gc41117f
55. Wang W, Cong Y, Zhang L, Huang Y, Wang X, Zhang T (2014) Aerobic oxidative transformation of primary alcohols and amines to amides promoted by a hydroxyapatite-supported gold catalyst in water. Tetrahedron Lett 55:124–127. https://doi.org/10.1016/j.tetlet.2013.10.133
56. Wang Y, Zhu D-p, Tang L, Wang S-J, Wang Z-Y (2011) Highly efficient amide synthesis from alcohols and amines by virtue of a water-soluble gold/DNA catalyst. Angew Chem Int Ed 50:8917–8921. https://doi.org/10.1002/anie.201102374
57. Soule J-F, Miyamura H, Kobayashi S (2011) Powerful amide synthesis from alcohols and amines under aerobic conditions catalyzed by gold or gold/iron, -nickel or -cobalt nanoparticles. J Am Chem Soc 133:18550–18553. https://doi.org/10.1021/ja2080086
58. Miyamura H, Kobayashi S (2014) Tandem oxidative processes catalyzed by polymer-incarcerated multimetallic nanoclusters with molecular oxygen. Acc Chem Res 47:1054–1066. https://doi.org/10.1021/ar400224f
59. Yamaguchi K, Kobayashi H, Oishi T, Mizuno N (2012) Heterogeneously catalyzed synthesis of primary amides directly from primary alcohols and aqueous ammonia. Angew

Chem Int Ed 51:544–547. https://doi.org/10.1002/anie.201107110

60. Wang Y, Kobayashi H, Yamaguchi K, Mizuno N (2012) Manganese oxide-catalyzed transformation of primary amines to primary amides through the sequence of oxidative dehydrogenation and successive hydration. Chem Commun 48:2642–2644. https://doi.org/10.1039/c2cc17499e
61. Yamaguchi K, Kobayashi H, Wang Y, Oishi T, Ogasawara Y, Mizuno N (2013) Green oxidative synthesis of primary amides from primary alcohols or aldehydes catalyzed by a cryptomelane-type manganese oxide-based octahedral molecular sieve, OMS-2. Catal Sci Technol 3:318–327. https://doi.org/10.1039/c2cy20178j
62. Nie R, Shi J, Xia S, Shen L, Chen P, Hou Z, Xiao F-S (2012) MnO2/graphene oxide: a highly active catalyst for amide synthesis from alcohols and ammonia in aqueous media. J Mater Chem 22:18115–18118. https://doi.org/10.1039/c2jm34652d
63. Dobereiner GE, Crabtree RH (2009) Dehydrogenation as a substrate-activating strategy in homogeneous transition-metal catalysis. Chem Rev 110:681–703. https://doi.org/10.1021/cr900202j
64. Ito T, Horino H, Koshiro Y, Yamamoto A (1982) Selective dimerization of aldehydes to esters catalyzed by hydridoruthenium complexes. Bull Chem Soc Jpn 55:504–512. https://doi.org/10.1246/bcsj.55.504
65. Nielsen M, Junge H, Kammer A, Beller M (2012) Towards a green process for bulk-scale synthesis of ethyl acetate: efficient acceptorless dehydrogenation of ethanol. Angew Chem Int Ed 51:5711–5713. https://doi.org/10.1002/anie.201200625
66. Zhang J, Hong SH (2012) Direct N-acylation of lactams, oxazolidinones, and imidazolidinones with aldehydes by Shvo's catalyst. Org Lett 14:4646–4649. https://doi.org/10.1021/ol302087z
67. Comas-Vives A, Ujaque G, Lledós A (2007) Hydrogen transfer to ketones catalyzed by Shvo's Ruthenium hydride complex: a mechanistic insight. Organometallics 26:4135–4144. https://doi.org/10.1021/om7004832
68. Noyori R (2002) Asymmetric catalysis: science and opportunities (nobel lecture). Angew Chem Int Ed 41:2008–2022. https://doi.org/10.1002/1521-3773(20020617)41:12<2008::AID-ANIE2008>3.0.CO;2-4
69. Knowles WS (2002) Asymmetric hydrogenations (nobel lecture). Angew Chem Int Ed 41:1998–2007. https://doi.org/10.1002/1521-3773(20020617)41:12<1998::AID-ANIE1998>3.0.CO;2-8
70. Blum Y, Shvo Y (1985) Catalytically reactive (η4-tetracyclone)(CO)2(H)2Ru and related complexes in dehydrogenation of alcohols to esters. J Organomet Chem 282:C7–C10. https://doi.org/10.1016/0022-328X(85)87154-0
71. Cannizzaro S (1853) Ueber den der Benzoësäure entsprechenden Alkohol. Justus Liebigs Annalen der Chemie 88:129–130. https://doi.org/10.1002/jlac.18530880114
72. Seki T, Nakajo T, Onaka M (2006) The Tishchenko reaction: a classic and practical tool for ester synthesis. Chem Lett 35:824–829. https://doi.org/10.1246/cl.2006.824
73. Seo S, Marks TJ (2008) Mild amidation of aldehydes with amines mediated by lanthanide catalysts. Org Lett 10:317–319. https://doi.org/10.1021/ol702788j
74. Ishihara K, Yano T (2004) Synthesis of carboxamides by LDA-catalyzed Haller–Bauer and Cannizzaro reactions. Org Lett 6:1983–1986. https://doi.org/10.1021/ol0494459
75. Abaee MS, Sharifi R, Mojtahedi MM (2005) Room-temperature Cannizzaro reaction under mild conditions facilitated by magnesium bromide ethyl etherate and triethylamine†. Org Lett 7:5893–5895. https://doi.org/10.1021/ol052506y
76. Zhang L, Wang S, Zhou S, Yang G, Sheng E (2006) Cannizzaro-type disproportionation of aromatic aldehydes to amides and alcohols by using either a stoichiometric amount or a catalytic amount of lanthanide compounds. J Org Chem 71:3149–3153. https://doi.org/10.1021/jo060063l
77. Qian C, Zhang X, Li J, Xu F, Zhang Y, Shen Q (2009) Trisguanidinate lanthanide complexes: syntheses, structures, and catalytic activity for mild amidation of aldehydes with amines. Organometallics 28:3856–3862. https://doi.org/10.1021/om900120v
78. Li J, Xu F, Zhang Y, Shen Q (2009) Heterobimetallic lanthanide/sodium phenoxides: efficient catalysts for amidation of aldehydes with amines. J Org Chem 74:2575–2577. https://doi.org/10.1021/jo802617d
79. Zhang L, Su S, Wu H, Wang S (2009) Synthesis of amides through the Cannizzaro-type reaction catalyzed by lanthanide chlorides. Tetrahedron 65:10022–10024. https://doi.org/10.1016/j.tet.2009.09.101
80. Wu Y, Wang S, Zhang L, Yang G, Zhu X, Zhou Z, Zhu H, Wu S (2010)

Cyclopentadienyl-free rare-earth metal amides [{(CH2SiMe2){(2,6-iPr2C6H3)N}2} Ln{N(SiMe3)2} (THF)] as highly efficient versatile catalysts for C–C and C–N bond formation. Eur J Org Chem 2010:326–332. https://doi.org/10.1002/ejoc.200901015

81. Guo Z, Liu Q, Wei X, Zhang Y, Tong H, Chao J, Guo J, Liu D (2013) 2-Aminopyrrolyl dilithium compounds: synthesis, structural diversity, and catalytic activity for amidation of aldehydes with amines. Organometallics 32:4677–4683. https://doi.org/10.1021/om4006609
82. Cao Y, Sun W, Luo G, Yu Y, Zhou Y, Zhao Y, Yang J, Luo Y (2020) Mechanistic insights into La-catalyzed amidation of aldehyde with amines. Org Lett 22:705–708. https://doi.org/10.1021/acs.orglett.9b04445
83. Fujita K-i, Takahashi Y, Owaki M, Yamamoto K, Yamaguchi R (2004) Synthesis of five-, six-, and seven-membered ring lactams by Cp*Rh complex-catalyzed oxidative N-heterocyclization of amino alcohols. Org Lett 6:2785–2788. https://doi.org/10.1021/ol0489954
84. Tamaru Y, Yamada Y, Yoshida Z-i (1983) Direct oxidative transformation of aldehydes to amides by palladium catalysis. Synthesis 1983:474–476. https://doi.org/10.1055/s-1983-30388
85. Dai W, Liu Y, Tong T, Li X, Luo F (2014) Rh (III)-catalyzed oxidative amidation of aldehydes: an efficient route to N-pyridinamides and imides. Chin J Catal 35:1012–1016. https://doi.org/10.1016/S1872-2067(14) 6014
86. Tillack A, Rudloff I, Beller M (2001) Catalytic amination of aldehydes to amides. Eur J Org Chem:523–528. https://doi.org/10.1002/1099-0690(200102)
87. Saidi O, Bamford MJ, Blacker AJ, Lynch J, Marsden SP, Plucinski P, Watson RJ, Williams JMJ (2010) Iridium-catalyzed formylation of amines with paraformaldehyde. Tetrahedron Lett 51:5804–5806. https://doi.org/10.1016/j.tetlet.2010.08.106
88. Ekoue-Kovi K, Wolf C (2008) One-pot oxidative esterification and amidation of aldehydes. Chemistry 14:6302–6315. https://doi.org/10.1002/chem.200800353
89. Willis MC (2009) Transition metal catalyzed alkene and alkyne hydroacylation. Chem Rev 110:725–748. https://doi.org/10.1021/cr900096x
90. Whittaker M, Dong VM (2015) Nickel-catalyzed dehydrogenative cross-coupling: direct transformation of aldehydes into esters and amides. Angew Chem Int Ed 54:1312–1315. https://doi.org/10.1002/anie.201410322
91. Nakagawa K, Onoue H, Minami K (1966) Oxidation with nickel peroxide. A new synthesis of amides from aldehydes or alcohols. Chem Commun:17–18. https://doi.org/10.1039/C19660000017
92. Fukuoka S, Ryang M, Tsutsumi S (1971) Reactions of lithium dimethylcarbamoylnickel carbonylate. J Org Chem 36:2721–2723. https://doi.org/10.1021/jo00817a036
93. Wolfe JP, Buchwald SL (1997) Nickel-catalyzed amination of aryl chlorides. J Am Chem Soc 119:6054–6058. https://doi.org/10.1021/ja964391m
94. Pandey G, Koley S, Talukdar R, Sahani PK (2018) Cross-dehydrogenating coupling of aldehydes with amines/ROTBS ethers by visible-light photoredox catalysis: synthesis of amides, esters, and ureas. Org Lett 20:5861–5865. https://doi.org/10.1021/acs.orglett.8b02537
95. Mupparapu N, Khan S, Battula S, Kushwaha M, Gupta AP, Ahmed QN, Vishwakarma RA (2014) Metal-free oxidative amidation of 2-oxoaldehydes: a facile access to α-ketoamides. Org Lett 16:1152–1155. https://doi.org/10.1021/ol5000204
96. Allen SE, Walvoord RR, Padilla-Salinas R, Kozlowski MC (2013) Aerobic copper-catalyzed organic reactions. Chem Rev 113:6234–6458. https://doi.org/10.1021/cr300527g
97. Jamalifard S, Mokhtar J, Mirjafary Z (2019) One-pot synthesis of amides via the oxidative amidation of aldehydes and amines catalyzed by a copper-MOF. RSC Adv 9:22749–22752. https://doi.org/10.1039/c9ra04216d
98. Patel OPS, Anand D, Maurya RK, Yadav PP (2015) Copper-catalyzed highly efficient oxidative amidation of aldehydes with 2-aminopyridines in aqueous micellar system. Green Chem 17:3728–3732. https://doi.org/10.1039/c5gc00628g
99. Álvarez-Pérez A, Esteruelas MA, Izquierdo S, Varela JA, Saá C (2019) Ruthenium-catalyzed oxidative amidation of alkynes to amides. Org Lett 21:5346–5350. https://doi.org/10.1021/acs.orglett.9b01993. For a related method proceeding through a common intermediate, but starting from an alkyne see this references
100. Zhang C, Xu Z, Zhang L, Jiao N (2011) Copper-catalyzed aerobic oxidative coupling of aryl acetaldehydes with anilines leading to α-ketoamides. Angew Chem Int Ed

50:11088–11092. https://doi.org/10.1002/anie.201105285

101. Zhang C, Jiao N (2009) Dioxygen activation under ambient conditions: Cu-catalyzed oxidative amidation–diketonization of terminal alkynes leading to α-ketoamides. J Am Chem Soc 132:28–29. https://doi.org/10.1021/ja908911n
102. Zhu M, Fujita K-i, Yamaguchi R (2012) Aerobic oxidative amidation of aromatic and cinnamic aldehydes with secondary amines by CuI/2-pyridonate catalytic system. J Org Chem 77:9102–9109. https://doi.org/10.1021/jo301553v
103. Siler CGF, Xu B, Madix RJ, Friend CM (2012) Role of surface-bound intermediates in the oxygen-assisted synthesis of amides by metallic silver and gold. J Am Chem Soc 134:12604–12610. https://doi.org/10.1021/ja303178z
104. Li G-L, Kung KK-Y, Wong M-K (2012) Gold-catalyzed amide synthesis from aldehydes and amines in aqueous medium. Chem Commun 48:4112–4114. https://doi.org/10.1039/C2CC17689K
105. Delany EG, Fagan C-L, Gundala S, Zeitler K, Connon SJ (2013) Aerobic oxidation of NHC-catalysed aldehyde esterifications with alcohols: benzoin, not the Breslow intermediate, undergoes oxidation. Chem Commun 49:6513–6515. https://doi.org/10.1039/C3CC42597E
106. Wang X, Wang DZ (2011) Aerobic oxidation of secondary benzylic alcohols and direct oxidative amidation of aryl aldehydes promoted by sodium hydride. Tetrahedron 67:3406–3411. https://doi.org/10.1016/j.tet.2011.03.052
107. Zhang Z, Su J, Zha Z, Wang Z (2013) A novel approach for the one-pot preparation of [small alpha]-ketoamides by anodic oxidation. Chem Commun 49:8982–8984. https://doi.org/10.1039/C3CC43685C
108. Leow D (2014) Phenazinium salt-catalyzed aerobic oxidative amidation of aromatic aldehydes. Org Lett 16:5812–5815. https://doi.org/10.1021/ol5029354
109. Yoo W-J, Li C-J (2006) Highly efficient oxidative amidation of aldehydes with amine hydrochloride salts. J Am Chem Soc 128:13064–13065. https://doi.org/10.1021/ja064315b
110. Ta L, Axelsson A, Sundén H (2018) *N*-acylation of oxazolidinones via aerobic oxidative NHC catalysis. J. Org. Chem 83:12261–12268. https://doi.org/10.1021/acs.joc.8b01723
111. Wang G, Fu Z, Huang W (2017) Access to amide from aldimine via aerobic oxidative carbene catalysis and LiCl as cooperative lewis acid. Org Lett 19:3362–3365. https://doi.org/10.1021/acs.orglett.7b01195
112. De Sarkar S, Biswas A, Samanta RC, Studer A (2013) Catalysis with N-heterocyclic carbenes under oxidative conditions. Chemistry 19:4664–4678. https://doi.org/10.1002/chem.201203707
113. Maki BE, Chan A, Phillips EM, Scheidt KA (2009) N-heterocyclic carbene-catalyzed oxidations. Tetrahedron 65:3102–3109. https://doi.org/10.1016/j.tet.2008.10.033
114. Crochet P, Cadierno V (2015) Catalytic synthesis of amides via aldoximes rearrangement. Chem Commun 51:2495–2505. https://doi.org/10.1039/C4CC08684H
115. Beckmann E (1886) Zur Kenntniss der Isonitrosoverbindungen. Ber Dtsch Chem Ges 19:988–993. https://doi.org/10.1002/cber.188601901222
116. Lachman A (1925) The Beckman rearrangement. II. J Am Chem Soc 47:260–265. https://doi.org/10.1021/ja01678a038
117. Blatt H (1933) The Beckmann rearrangement. Chem Rev 12:215–260. https://doi.org/10.1021/cr60042a002
118. Jefferies LR, Weber SR, Cook SP (2015) Iron-catalyzed C–N bond formation via the Beckmann rearrangement. Synlett 26:331. https://doi.org/10.1055/s-0034-1379540
119. Garcia-Alvarez R, Crochet P, Cadierno V (2013) Metal-catalyzed amide bond forming reactions in an environmentally friendly aqueous medium: nitrile hydrations and beyond. Green Chem 15:46–66. https://doi.org/10.1039/C2GC36534K
120. Sanz R, Aguado R, Pedrosa MR, Arnáiz FJ (2002) Simple and selective oxidation of thiols to disulfides with dimethylsulfoxide catalyzed by dichlorodioxomolybdenum(VI). Synthesis 2002:0856–0858. https://doi.org/10.1055/s-2002-28520
121. Park S, Choi Y-a, Han H, Ha Yang S, Chang S (2003) Rh-catalyzed one-pot and practical transformation of aldoximes to amides. Chem Commun:1936–1937. https://doi.org/10.1039/B305268K
122. Allen L, Lawrence R, Emmett L, Williams JMJ (2011) Mechanistic studies into metal-catalyzed aldoxime to amide rearrangements. Adv Synth Catal 353:3262–3268. https://doi.org/10.1002/adsc.201100650
123. Fujiwara H, Ogasawara Y, Yamaguchi K, Mizuno N (2007) A one-pot synthesis of

primary amides from aldoximes or aldehydes in water in the presence of a supported rhodium catalyst. Angew Chem Int Ed 46:5202–5205. https://doi.org/10.1002/anie.200701273

124. Gnanamgari R, Crabtree H (2009) Terpyridine ruthenium-catalyzed one-pot conversion of aldehydes into amides. Organometallics 28:922–924. https://doi.org/10.1021/om8010678

125. Martínez-Asencio A, Yus M, Ramón DJ (2012) Copper(II) acetate-catalyzed one-pot conversion of aldehydes into primary amides through a Beckmann-type rearrangement. Tetrahedron 68:3948–3951. https://doi.org/10.1016/j.tet.2012.03.085

126. Achar TK, Mal P (2015) Radical-induced metal and solvent-free cross-coupling using TBAI–TBHP: oxidative amidation of aldehydes and alcohols with *N*-chloramines via C–H activation. J. Org. Chem. 80:666–672. https://doi.org/10.1021/jo502464n

127. Mondal A, Subaramanian M, Nandakumar A, Balaraman E (2018) Manganese-catalyzed direct conversion of ester to amide with liberation of H2. Org Lett 20:3381–3384. https://doi.org/10.1021/acs.orglett.8b01305

128. Chan K, Scheidt A (2008) Direct amination of homoenolates catalyzed by N-heterocyclic carbenes. J Am Chem Soc 130:2740–2741. https://doi.org/10.1021/ja711130p

129. Wong FT, Patra PK, Seayad J, Zhang Y, Ying JY (2008) N-heterocyclic carbene (NHC)-catalyzed direct amidation of aldehydes with nitroso compounds. Org Lett 10:2333–2336. https://doi.org/10.1021/ol8004276

130. Corey J, Gilman NW, Ganem BE (1968) New methods for the oxidation of aldehydes to carboxylic acids and esters. J Am Chem Soc 90:5616–5617. https://doi.org/10.1021/ja01022a059

131. Gilman NW (1971) The preparation of carboxylic amides from aldehydes by oxidation. J Chem Soc D Chem Commun:733–734. https://doi.org/10.1039/C29710000733

132. Raj VP, Sudalai A (2005) A facile direct conversion of aldehydes to esters and amides using acetone cyanohydrin. Tetrahedron Lett 46:8303–8306. https://doi.org/10.1016/j.tetlet.2005.09.173

133. Alcaide B, Almendros P, Redondo MC (2004) Novel N1–C4 β-lactam bond breakage. Synthesis of enantiopure α-alkoxy-γ-keto acid derivatives†. Org Lett 6:1765–1767. https://doi.org/10.1021/ol049549j

134. Bode W, Sohn SS (2007) N-heterocyclic carbene-catalyzed redox amidations of α-functionalized aldehydes with amines. J Am Chem Soc 129:13798–13799. https://doi.org/10.1021/ja0768136

135. Vora HU, Rovis T (2007) Nucleophilic carbene and HOAt relay catalysis in an amide bond coupling: an orthogonal peptide bond forming reaction. J Am Chem Soc 129:13796–13797. https://doi.org/10.1021/ja0764052

136. Gondo A, Bode JW (2013) Catalytic redox amidations of aldehydes with a polymer-supported peptide–N-heterocyclic carbene multifunctional catalyst. Synlett 24:1205–1210. https://doi.org/10.1055/s-0033-1338956

137. Domingo R, Aurell MJ, Arnó M (2009) Understanding the mechanism of the N-heterocyclic carbene-catalyzed ring-expansion of 4-formyl-β-lactams to succinimide derivatives. Tetrahedron 65:3432–3440. https://doi.org/10.1016/j.tet.2009.02.030

138. Binanzer M, Hsieh S-Y, Bode JW (2011) Catalytic kinetic resolution of cyclic secondary amines. J Am Chem Soc 133:19698–19701. https://doi.org/10.1021/ja209472h

139. Vora HU, Wheeler P, Rovis T (2012) Exploiting acyl and enol azolium intermediates via N-hetero- cyclic carbene-catalyzed reactions of α-reducible aldehydes. Adv Synth Catal 354:1617–1639. https://doi.org/10.1002/adsc.201200031

140. Chan K, Scheidt A (2007) Highly stereoselective formal [3 + 3] cycloaddition of enals and azomethine imines catalyzed by N-heterocyclic carbenes. J Am Chem Soc 129:5334–5335. https://doi.org/10.1021/ja0709167

141. Ishihara K, Kuroki Y, Hanaki N, Ohara S, Yamamoto H (1996) Antimony-templated macrolactamization of tetraamino esters. Facile synthesis of macrocyclic spermine alkaloids, (+/-)-buchnerine, (+/-)-verbacine, (+/-)-verbaskine, and (+/-)-verbascenine. J Am Chem Soc 118:1569–1570. https://doi.org/10.1021/Ja953541a

142. Kuroki Y, Ishihara K, Hanaki N, Ohara S, Yamamoto H (1998) Metal-templated macrolactamization of triamino and tetramino esters. Facile synthesis of macrocyclic spermidine and spermine alkaloids, (S)-(+)-dihydroperiphylline, (+/-)-buchnerine, (+/-)-verbacine, (+/-)-verbaskine, and (+/-)-verbascenine. Bull Chem Soc Jpn

71:1221–1230. https://doi.org/10.1246/Bcsj.71.1221

143. Han J, Lee P, Lobkovsky E, Porco JA (2005) Catalytic ester-amide exchange using group (IV) metal alkoxide-activator complexes. J Am Chem Soc 127:10039–10044. https://doi.org/10.1021/Ja0527976

144. Gnanaprakasam B, Milstein D (2011) Synthesis of amides from esters and amines with liberation of H2 under neutral conditions. J Am Chem Soc 133:1682–1685. https://doi.org/10.1021/ja109944n

145. Ohshima T, Hayashi Y, Agura K, Fujii Y, Yoshiyama A, Mashima K (2012) Sodium methoxide: a simple but highly efficient catalyst for the direct amidation of esters. Chem Commun 48:5434–5436. https://doi.org/10.1039/C2cc32153j

146. Morimoto H, Fujiwara R, Shimizu Y, Morisaki K, Ohshima T (2014) Lanthanum (III) triflate catalyzed direct amidation of esters. Org Lett 16:2018–2021. https://doi.org/10.1021/ol500593v

147. Han Q, Xiong X, Li S (2015) An efficient, green and scale-up synthesis of amides from esters and amines catalyzed by Ru-MACHO catalyst under mild conditions. Catal Commun 58:85–88. https://doi.org/10.1016/j.catcom.2014.08.036

148. Kuriyama W, Matsumoto T, Ogata O, Ino Y, Aoki K, Tanaka S, Ishida K, Kobayashi T, Sayo N, Saito T (2011) Catalytic hydrogenation of esters. development of an efficient catalyst and processes for synthesising (R)-1,2-propanediol and 2-(l-menthoxy)ethanol. Org Process Res Dev 16:166–171. https://doi.org/10.1021/op200234j

149. Fier PS, Whittaker AM (2017) An atom-economical method to prepare enantiopure benzodiazepines with N-carboxyanhydrides. Org Lett 19:1454–1457. https://doi.org/10.1021/acs.orglett.7b00417

150. Halima TB, Vandavasi JK, Shkoor M, Newman SG (2017) A cross-coupling approach to amide bond formation from esters. ACS Catal 7:2176–2180. https://doi.org/10.1021/acscatal.7b00245

151. Halima TB, Masson-Makdissi J, Newman SG (2018) Nickel-catalyzed amide bond formation from methyl esters. Angew Chem Int Ed 57:12925–12929. https://doi.org/10.1002/anie.201808560

152. Zheng Y-L, Newman SG (2019) Methyl esters as cross-coupling electrophiles: direct synthesis of amide bonds. ACS Catal 9:4426–4433. https://doi.org/10.1021/acscatal.9b00884

153. Shi S, Nolan SP, Szostak M (2018) Well-defined palladium(II)–NHC precatalysts for cross-coupling reactions of amides and esters by selective N–C/O–C cleavage. Acc Chem Res 51:2589–2599. https://doi.org/10.1021/acs.accounts.8b00410

154. Ghosh T, Jana S, Dash J (2019) KOtBu-promoted transition-metal-free transamidation of primary and tertiary amides with amines. Org Lett 21:6690–6694. https://doi.org/10.1021/acs.orglett.9b02306

155. Yin J, Zhang J, Cai C, Deng G-J, Gong H (2019) Catalyst-free transamidation of aromatic amines with formamide derivatives and tertiary amides with aliphatic amines. Org Lett 21:387–392. https://doi.org/10.1021/acs.orglett.8b03542

156. Li G, Ji C-L, Hong X, Szostak M (2019) Highly chemoselective, transition-metal-free transamidation of unactivated amides and direct amidation of alkyl esters by N–C/O–C cleavage. J Am Chem Soc 141:11161–11172. https://doi.org/10.1021/jacs.9b04136

157. Li G, Szostak M (2018) Highly selective transition-metal-free transamidation of amides and amidation of esters at room temperature. Nat Commun 9:4165–4173. https://doi.org/10.1038/s41467-018-06623

158. Baker L, Yamano MM, Zhou Y, Anthony SM, Garg NK (2018) A two-step approach to achieve secondary amide transamidation enabled by nickel catalysis. Nat Commun 7:11554–11558. https://doi.org/10.1038/ncomms11554

159. Dander JE, Baker EL, Garg NK (2017) Nickel-catalyzed transamidation of aliphatic amide derivatives. Chem Sci 8:6433–6439. https://doi.org/10.1039/c7sc01980g

160. Ishihara K, Ohara S, Yamamoto H (1996) 3,4,5-Trifluorobenzeneboronic acid as an extremely active amidation catalyst. J. Org. Chem 61:4196–4197. https://doi.org/10.1021/jo9606564

161. de Figueiredo RM, Suppo J-S, Campagne J-M (2016) Nonclassical routes for amide bond formation. Chem Rev 116:12029–12122. https://doi.org/10.1021/acs.chemrev.6b00237

162. Sabatini T, Boulton LT, Sneddon HF (2019) A green chemistry perspective on catalytic amide bond formation. Nat Catal 2:10–17. https://doi.org/10.1038/s41929-018-0211-5

163. Sawant N, Bagal DB, Ogawa S, Selvam K, Saito S (2018) Diboron-catalyzed

dehydrative amidation of aromatic carboxylic acids with amines. Org Lett 20:4397–4400. https://doi.org/10.1021/acs.orglett.8b01480

164. Shimada N, Hirata M, Koshizuka M, Ohse N, Kaito R, Makino K (2019) Diboronic acid anhydrides as effective catalysts for the hydroxy-directed dehydrative amidation of carboxylic acids. Org Lett 21:4303–4308. https://doi.org/10.1021/acs.orglett.9b01484

165. Du Y, Barber T, Lim SE, Rzepa HS, Baxendale IR, Whiting A (2019) A solid-supported arylboronic acid catalyst for direct amidation. Chem Commun 55:2916–2919. https://doi.org/10.1039/c8cc09913h

166. Michigami K, Sakaguchi T, Takemoto Y (2020) Catalytic dehydrative peptide synthesis with gem-diboronic acids. ACS Catal 10:683–688. https://doi.org/10.1021/acscatal.9b03894

167. Muramatsu W, Hattori T, Yamamoto H (2019) Substrate-directed lewis-acid catalysis for peptide synthesis. J Am Chem Soc 141:12288–12295. https://doi.org/10.1021/jacs.9b03850

168. Akondi SM, Gangireddy P, Pickel TC, Liebeskind LS (2018) Aerobic, diselenide-catalyzed redox dehydration: amides and peptides. Org Lett 20:538–541. https://doi.org/10.1021/acs.orglett.7b03620

169. Liebeskind LS, Gangireddy P, Lindale MG (2016) Benzoisothiazolone organo/copper-cocatalyzed redox dehydrative construction of amides and peptides from carboxylic acids using (EtO)3P as the reductant and O2 in air as the terminal oxidant. J Am Chem Soc 138:6715–6718. https://doi.org/10.1021/jacs.6b03168

170. Braddock DC, Lickiss PD, Rowley BC, Pugh D, Purnomo T, Santhakumar G, Fussell SJ (2018) Tetramethyl orthosilicate (TMOS) as a reagent for direct amidation of carboxylic acids. Org Lett 20:950–953. https://doi.org/10.1021/acs.orglett.7b03841

171. Muramatsu W, Yamamoto H (2019) Tantalum-catalyzed amidation of amino acid homologues. J Am Chem Soc 141:18926–18931. https://doi.org/10.1021/jacs.9b08415

172. Ritter JJ, Minieri PP (1948) A new reaction of nitriles. 1. Amides from alkenes and mononitriles. J Am Chem Soc 70:4045–4048. https://doi.org/10.1021/Ja01192a022

173. Yamada H, Kobayashi M (1996) Nitrile hydratase and its application to industrial production of acrylamide. Biosci Biotechnol Biochem 60:1391–1400. https://doi.org/10.1271/bbb.60.1391

174. Kobayashi M, Shimizu S (2000) Nitrile hydrolases. Curr Opin Chem Biol 4:95–102. https://doi.org/10.1016/S1367-5931(99)00058-7

175. Endo I, Nojiri M, Tsujimura M, Nakasako M, Nagashima S, Yohda M, Odaka M (2001) Fe-type nitrile hydratase. J Inorg Biochem 83:247–253. https://doi.org/10.1016/S0162-0134(00)00171-9

176. Kovacs A (2004) Synthetic analogues of cysteinate-ligated non-heme iron and non-corrinoid cobalt enzymes. Chem Rev 104:825–848. https://doi.org/10.1021/cr020619e

177. Prasad S, Bhalla TC (2010) Nitrile hydratases (NHases): at the interface of academia and industry. Biotechnol Adv 28:725–741. https://doi.org/10.1016/j.biotechadv.2010.05.020

178. Somorjai A, Li Y (2010) Selective nanocatalysis of organic transformation by metals: concepts, model systems, and instruments. Top Catal 53:832–847. https://doi.org/10.1007/s11244-010-9511-y

179. Sanchez S, Demain AL (2011) Enzymes and bioconversions of industrial, pharmaceutical, and biotechnological significance. Org Process Res Dev 15:224–230. https://doi.org/10.1021/op100302x

180. Diamond SE, Grant B, Tom GM, Taube H (1974) The mild selective conversion of nitriles to amides. Tetrahedron Lett 15:4025–4028. https://doi.org/10.1016/S0040-4039(01)92074-X

181. Murahashi S, Naota T, Saito E (1986) Ruthenium-catalyzed amidation of nitriles with amines. A novel, facile route to amides and polyamides. J Am Chem Soc 108:7846–7847. https://doi.org/10.1021/ja00284a066

182. Ghaffar T, Parkins AW (1995) A new homogeneous platinum-containing catalyst for the hydrolysis of nitriles. Tetrahedron Lett 36:8657–8660. https://doi.org/10.1016/0040-4039(95)01785-G

183. Arévalo J, García J (2010) Bond activation with low-valent nickel in homogeneous systems. Eur J Inorg Chem 2010:4063–4074. https://doi.org/10.1002/ejic.201000621

184. Crestani MG, Arévalo A, García JJ (2006) Catalytic hydration of benzonitrile and acetonitrile using nickel(0). Adv Synth Catal 348:732–742. https://doi.org/10.1002/adsc.200505382

185. Šmejkal T, Breit B (2007) Self-assembled bidentate ligands for ruthenium-catalyzed hydration of nitriles. Organometallics 26:2461–2464. https://doi.org/10.1021/om0611047
186. Cadierno V, Francos J, Gimeno J (2008) Selective ruthenium-catalyzed hydration of nitriles to amides in pure aqueous medium under neutral conditions. Chemistry 14:6601–6605. https://doi.org/10.1002/chem.200800847
187. Cadierno V, Díez J, Francos J, Gimeno J (2010) Bis(allyl)ruthenium(IV) complexes containing water-soluble phosphane ligands: synthesis, structure, and application as catalysts in the selective hydration of organonitriles into amides. Chemistry 16:9808–9817. https://doi.org/10.1002/chem.201001152
188. Kang B, Fu Z, Hong SH (2013) Ruthenium-catalyzed redox-neutral and single-step amide synthesis from alcohol and nitrile with complete atom economy. J Am Chem Soc 135:11704–11707. https://doi.org/10.1021/ja404695t
189. Kim S, Hong H (2014) Synthesis of cyclic imides from nitriles and diols using hydrogen transfer as a substrate-activating strategy. Org Lett 16:4404–4407. https://doi.org/10.1021/ol501835t
190. He JL, Yamaguchi K, Mizuno N (2011) Aerobic oxidative transformation of primary azides to nitriles by ruthenium hydroxide catalyst. J. Org. Chem. 76:4606–4610. https://doi.org/10.1021/Jo2004956
191. Fu Z, Lee J, Kang B, Hong SH (2012) Dehydrogenative amide synthesis: azide as a nitrogen source. Org Lett 14:6028–6031. https://doi.org/10.1021/ol302915g
192. Schmidt S, Toplak A, Rozeboom HJ, Wijma HJ, Quaedflieg PJLM, van Maarseveen JH, Janssen DB, Nuijens T (2018) Design of a substrate-tailored peptiligase variant for the efficient synthesis of thymosin-α1. Org Biomol Chem 16:609–613. https://doi.org/10.1039/c7ob02812a
193. Petcheya R, Grogana G (2019) Enzyme-catalysed synthesis of secondary and tertiary amides. Adv Synth Catal 361:3895–3914. https://doi.org/10.1002/adsc.201900694
194. Adams JP, Brown MJB, Diaz-Rodriguez A, Lloyd RC, Roiban G-D (2019) Biocatalysis: a pharma perspective. Adv Synth Catal 361:2421–2432. https://doi.org/10.1002/adsc.201900424
195. Goswami A, Van Lanen SG (2015) Enzymatic strategies and biocatalysts for amide bond formation: tricks of the trade outside of the ribosome. Mol BioSyst 11:338–353. https://doi.org/10.1039/c4mb00627e

Chapter 3

Mitsunobu Reactions in Medicinal Chemistry and Development of Practical Modifications

Tsuyoshi Taniguchi

Abstract

The Mitsunobu reaction is widely used in synthetic chemistry and has contributed to the discovery and scalable synthesis of drugs because the reaction is a reliable method for the stereoselective functionalization of organic molecules. However, the Mitsunobu reaction has some serious problems from the viewpoint of green chemistry. The reaction requires hazardous reagents and produces large amounts of undesired by-products that sometimes complicate purification of the desired products. These drawbacks often hamper the practical application of the Mitsunobu reaction in synthesis. Therefore, the development of novel concepts to overcome these drawbacks is needed for wider utility of the Mitsunobu reaction to be realized. As the first topic in this chapter, recent applications of the Mitsunobu reaction in medicinal and process chemistry will be described, while recent modifications of the Mitsunobu reaction will be addressed in the second section with a focus on "greener" methodologies.

Key words Azo compounds, Green chemistry, Medicinal chemistry, Mitsunobu reactions, Phosphines

1 Introduction

Condensation reactions are fundamental methods to connect two components in synthetic chemistry. Acid-catalyzed condensation reactions are typical methods to obtain esters from carboxylic acids and alcohols (Fischer esterification) [1]. Amides can be similarly synthesized from carboxylic acids and amines, and this classical reaction has been used in industrial synthesis of simple compounds because the reaction is conceptually atom-economical and produces little waste. On the other hand, the method is not sometimes suitable for efficient synthesis of complex molecules because the reaction is reversible and often requires harsh conditions. In short, a lot of energy is needed for the process on a large scale, and a decrease in product yield is often caused under such harsh conditions. There are some examples of highly active acid catalysts that can overcome this problem [2–4], though generally this problem is

Paul F. Richardson (ed.), *Green Chemistry in Drug Discovery: From Academia to Industry*, Methods in Pharmacology and Toxicology, https://doi.org/10.1007/978-1-0716-1579-9_3,

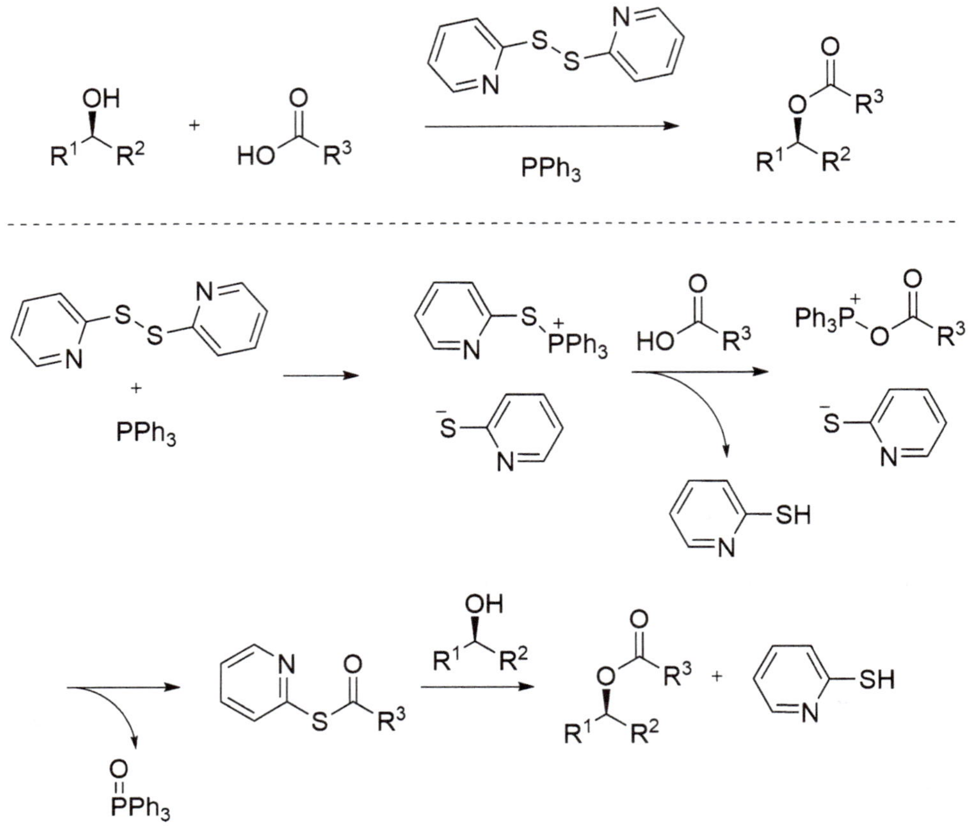

Fig. 1 Typical oxidation-reduction condensation

avoided by using condensation reagents to activate carboxylic acids [5]. Currently, many convenient reagents for this purpose are commercially available, though these lead to significant by-products, which need to be removed after the reaction.

In the 1960s, Mukaiyama and co-workers developed "oxidation-reduction condensation" as a new concept of condensation reactions [6], in which water (H_2O) is formally removed as two hydrogen atoms and an oxygen atom as a result of redox processes. For instance, condensation between alcohols and carboxylic acids proceeds to give the corresponding esters in the presence of 2,2-′-dipyridyldisulfide as an oxidant and triphenylphosphine as a reductant (Fig. 1) [6]. Carboxylic acids react with triphenylphosphine activated by the reaction with 2,2′-dipyridyldisulfide to form the corresponding acyloxyphosphonium ions. After the intermediates have reacted with 2-pyridylthiolate, the thioester intermediates further react with alcohols to give the desired esters. 2-Thiopyridone and triphenylphosphine oxide are formed as by-products in the reaction, which become unavoidable waste. The reaction occurs irreversibly under neutral conditions unlike the aforementioned acid-catalyzed condensation reactions.

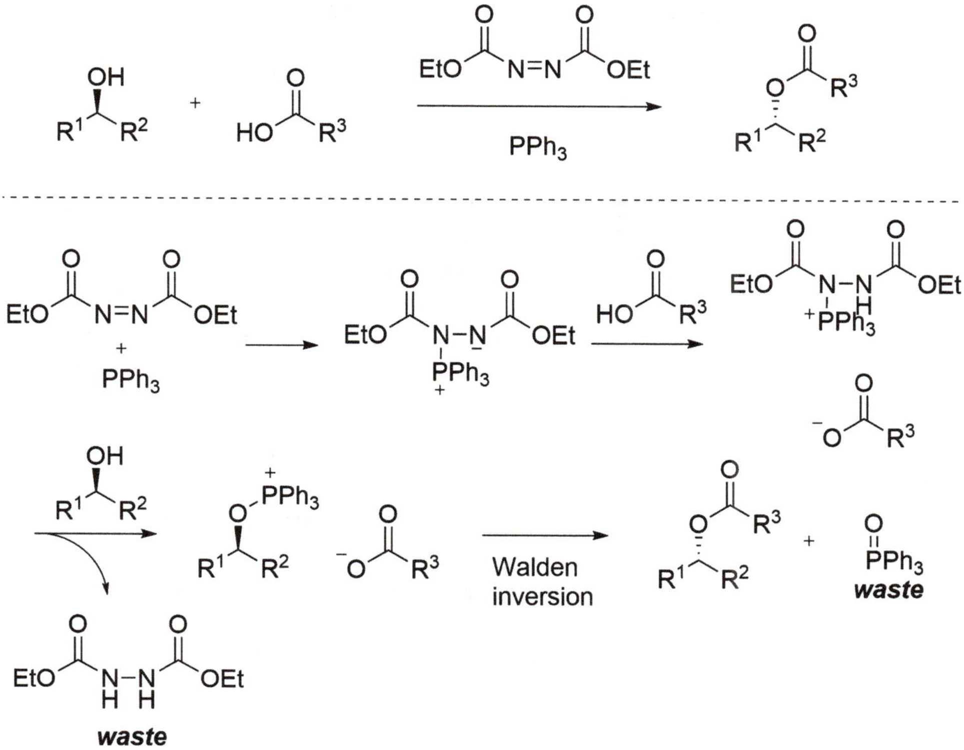

Fig. 2 Typical Mitsunobu reactions

In 1967, Mitsunobu and co-workers reported a new example of an oxidation-reduction condensation [7, 8]. Typically, diethyl azodicarboxylate (DEAD) is employed as an oxidant in the reaction, and the reaction mechanism fundamentally differs from the mechanisms of the previously reported oxidation-reduction condensation reactions (Fig. 2) [9, 10]. Triphenylphosphine reacts with the azo group of DEAD to form an N–P bond and gives rise to the reactive phosphonium intermediates, which subsequently react with alcohols to give the corresponding alkoxyphosphonium ions in contrast to the Mukaiyama protocol. The intermediates undergo an S_N2 reaction with carboxylic acids to afford the corresponding esters. Thus, since Walden inversion of alcohols occurs, the reaction is frequently utilized as a method for inversion of the stereochemistry of chiral secondary alcohols. In addition, alcohols can be converted to various functional groups by the reaction with nucleophiles such as phenols, azide, phthalimide, sulfonamides, sulfides, or active methylenes instead of carboxylic acids (Fig. 3). Generally, the reactivity in the Mitsunobu reaction depends on the acidity of nucleophiles ($pK_a < 13$ is required) because the basicity of a betaine intermediate generated from DEAD is modest. This limitation can be overcome by using modified reagents and procedures [11–15]. Owing to its reliability and predictable properties, the Mitsunobu reaction has become one of

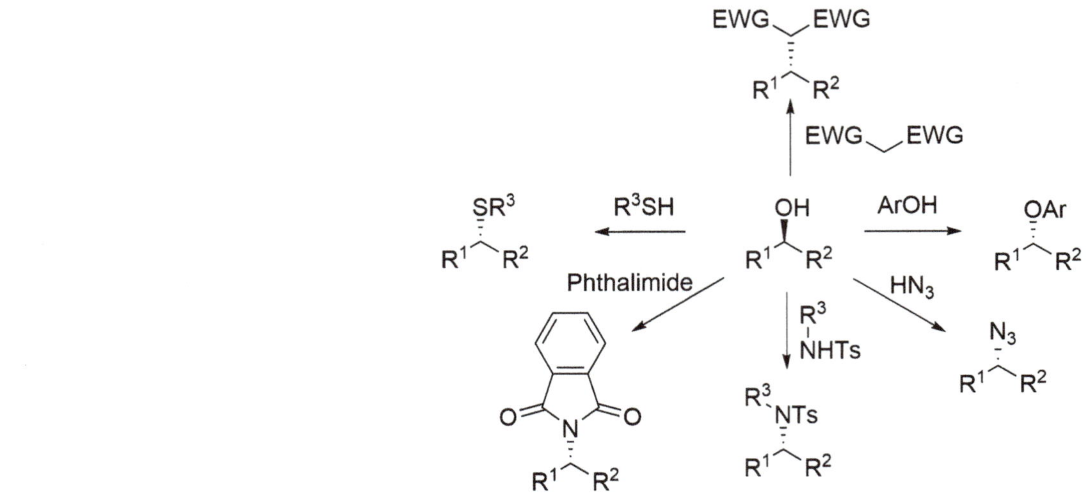

Fig. 3 Possible partners in Mitsunobu reactions

the most well-known reactions in synthetic chemistry. The experimental procedure of the Mitsunobu reaction is usually convenient. Typically, DEAD is added to a solution of substrates and triphenylphosphine in an appropriate solvent. Toluene or THF is preferably used as a solvent, though the reaction usually works well in various solvents.

Despite wide utility and large potential of the Mitsunobu reaction, its application to practical synthesis on a large scale is often limited with one of the major drawbacks being the production of a large amount of waste based on the reagents used [16]. In this respect, azodicarboxylates and organophosphines are converted to the corresponding hydrazinedicarboxylates and organophosphine oxides by redox reactions, and removal of both by-products from the crude products is often troublesome, and in some cases, they also contaminate the desired products. The mass of waste is 176.17 g/mol for diethyl 1,2-hydrazinedicarboxylate and 278.29 g/mol for triphenylphosphine oxide. Indeed, waste of more than 450 g/mol is totally produced through the condensation process. This is inefficient from a viewpoint of atom economy. Another serious drawback of the Mitsunobu reaction is that DEAD is toxic and sensitive to light and shock, which presents a challenge to performing the reaction on a large scale. Typically nowadays and in particular for scale-up, DEAD is often replaced by the more convenient alternative, diisopropyl azodicarboxylate (DIAD), or other related reagents. However, these derivatives tend to be thermally instable like DEAD. The differential scanning calorimetry (DSC) analysis indicates that most of the azodicarboxylates release a large amount of exothermic energy (ca. 1000 J/g) with decomposition at typically around 200 °C, even if the compound is a crystalline solid having a higher molecular weight than DEAD

[17]. The Koenen test on DEAD and DIAD, in which a neat form are directly heated in a sealed tube, results in explosion of the tube [17]. Therefore, only a diluted solution of DEAD (typically 40% in toluene) is currently commercially available from most of the suppliers.

Thus, the Mitsunobu reaction is a well-established method, but many practical aspects of the reaction are still under active development. Indeed, there are many reports of the application of the Mitsunobu reaction in medicinal chemistry on a laboratory scale, while on the other hand, extensive efforts have often been required to solve the problems of waste removal and product purification in these synthetic studies. Therefore, the development of practical modifications for reducing the waste generated in the Mitsunobu reaction is an important subject from the viewpoints of both process and green chemistry. This chapter mainly covers representative applications of the Mitsunobu reaction to medicinal and green chemistry. More examples have been displayed in recent reviews on the Mitsunobu reaction [18, 19].

2 Applications of the Mitsunobu Reaction in Medicinal Chemistry

The Mitsunobu reaction is widely utilized in the discovery of drug candidates because it provides a reliable method for introducing various functional groups to target compounds in a stereospecific manner. In addition, application of the Mitsunobu reaction in a parallel manner not only can enable medicinal chemists to efficiently add libraries of compounds to their collection for SAR purposes but also enables diversification and potential further elaboration of the compounds through the employment of different nucleophiles [20]. Although Mitsunobu chemistry has been demonstrated in such a parallel manner, novel solutions have to be implemented to facilitate the purification [21]. In this section, we provide examples of the utilization of the Mitsunobu reaction across a range of drug discovery programs.

The Mitsunobu reaction was utilized for efficiently obtaining several optically active derivatives to study the structure–activity relationships of inhibitors of adrenal steroid 11β-hydroxylations (Fig. 4) [22]. Since the starting chiral alcohols **1** are easily obtained by an enzymatic optical resolution method [23, 24], the chiral alkylated derivatives **3** could be systematically synthesized by employing the Mitsunobu alkylation of imidazole **2** with a series of chiral alcohols **1**, which predictably proceeds with complete inversion. Alkylation predominantly occurs at more nucleophilic N-1 of imidazoles. This is a classical example demonstrating the broad utility of the Mitsunobu reaction in medicinal chemistry. The azo reagent used here is di-*tert*-butyl azodicarboxylate (DTBAD), which is more convenient than DEAD. It is solid, and waste from the reagent can be easily removed (*vide infra*).

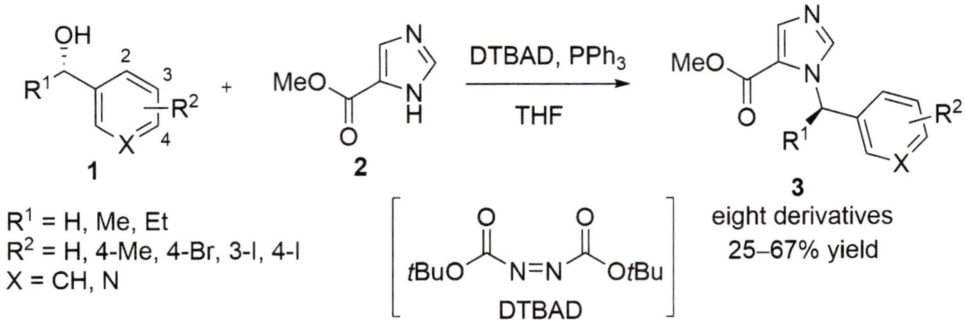

Fig. 4 Systematic synthesis of inhibitors of adrenal steroid 11β-hydroxylations

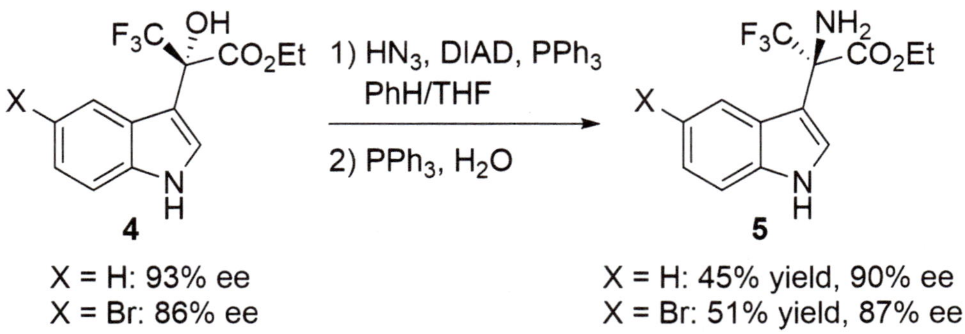

Fig. 5 Synthesis of chiral α-amino acid derivatives by the Mitsunobu reaction

The indole compounds **4** are potent HIV-1 non-nucleoside reverse transcriptase inhibitors, and the optically active compounds are obtained by an enantioselective Friedel-Crafts reaction between indoles and ethyl 3,3,3-trifluoropyruvate with an organocatalyst [25]. The Mitsunobu reaction was utilized to efficiently access the corresponding α-amino acid derivatives **5** for a compound library [26]. Treatment of the enantiopure tertiary alcohols **4** with hydrazoic acid, diisopropyl azodicarboxylate (DIAD), and triphenylphosphine produced the corresponding azides, and subsequent treatment of the reaction mixture with additional triphenylphosphine and water provided the amines **5** by reduction of azides in moderate yield with complete inversion of stereochemistry (Fig. 5). Although the use of highly toxic hydrogen azide is hazardous, this is a rare example of a successful Mitsunobu reaction of tertiary alcohols as sterically hindered alcohols are usually less reactive.

The Mitsunobu reaction served as a method for ensuring stereocontrol in the synthesis of carbocyclic nucleoside analogues that show potent anti-HIV activity, such as *carba*-BVdU [27, 28] (Fig. 6). A typical hydroboration-oxidation sequence of cyclopentene derivative **6** afforded deoxyribose mimics **7** and **8**, with the *R*-configured alcohol **7** always being the major product under a range of conditions tried. A sufficient amount of the *S*-configured alcohol

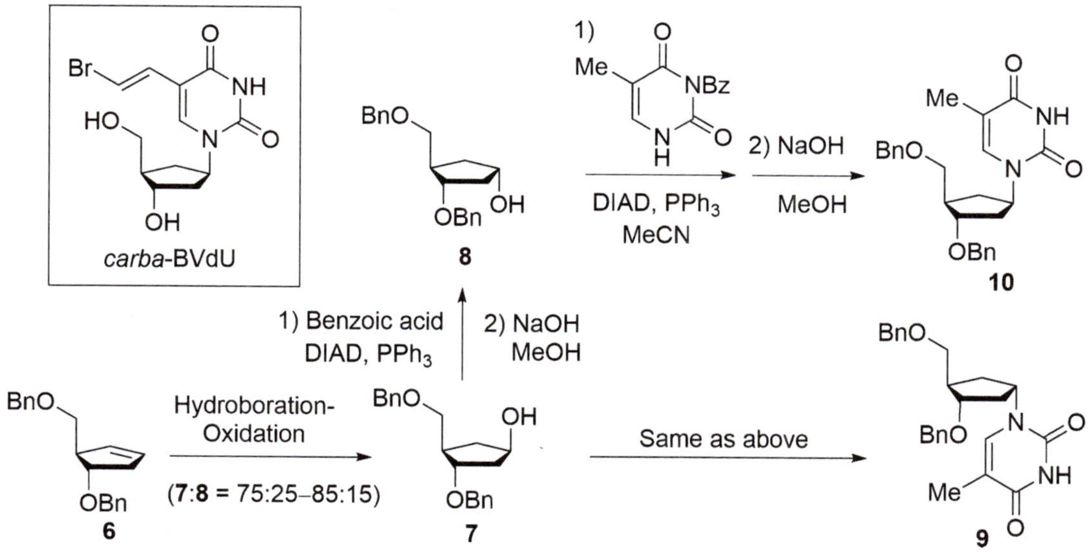

Fig. 6 Divergent synthesis of carbocyclic nucleosides with the Mitsunobu reaction

8 could be obtained by inversion of the stereochemistry of 7 employing the Mitsunobu reaction. Subsequently, a thymine unit was introduced into 7 and **8** by a second Mitsunobu reaction to obtain of the protected versions of the nucleosides, **9** and **10**. Thus, the Mitsunobu reaction has become a powerful tool to obtain multiple enantiomers of a drug candidate from a common key intermediate even when conventional transformations show a strong intrinsic bias toward a single enantiomer.

2-Nitrobenzenesulfonamides (nosylamides) function as good nucleophiles under Mitsunobu conditions because they show higher acidity than that of the corresponding *p*-toluenesulfonylamides (tosylamides) and can also be readily removed under mild conditions using thiolates [29, 30]. Therefore, the Mitsunobu reaction with nosylamides is a useful method for synthesis of amine compounds, and this is illustrated in a flexible synthesis of heterocyclic scaffolds in the search for potent peptidomimetic drug candidates [31]. When nosylamide **13** synthesized by the condensation between the nosyl-protected tryptophan **11** and tetrahydroisoquinoline derivative **12** was exposed to Mitsunobu conditions, intramolecular C–N bond formation took place to afford piperazinone compound **14**, which can be easily be elaborated to the non-peptide somatostatin analogue **15** (Fig. 7). Compound **14** was directly used for the next transformation without purification. This example demonstrates that the Mitsunobu reaction can be used as a method to provide rapid access to heterocyclic cores of drug candidates.

The chemistry of nosylamides has also proven its merits in the synthesis of secondary amines, and an improved Mitsunobu method was applied to the synthesis of tenascin lipopeptide **20**

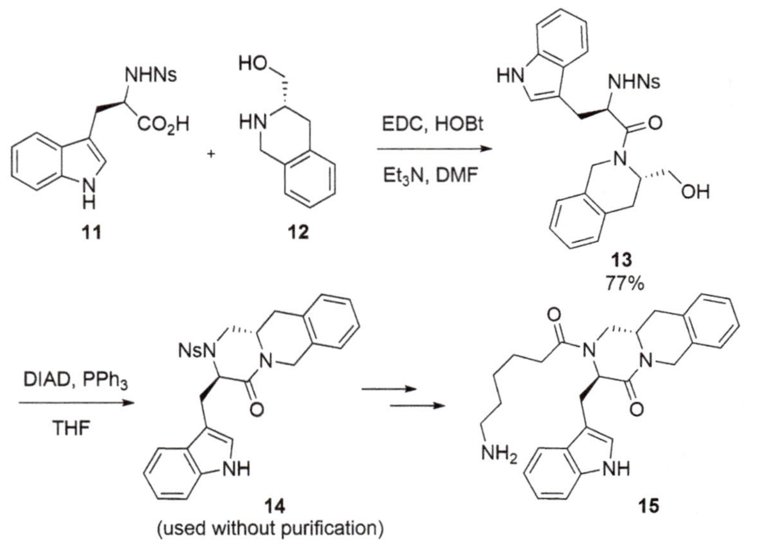

Fig. 7 Synthesis of heterocyclic scaffolds for peptidomimetic drug design with the Fukuyama-Mitsunobu reaction

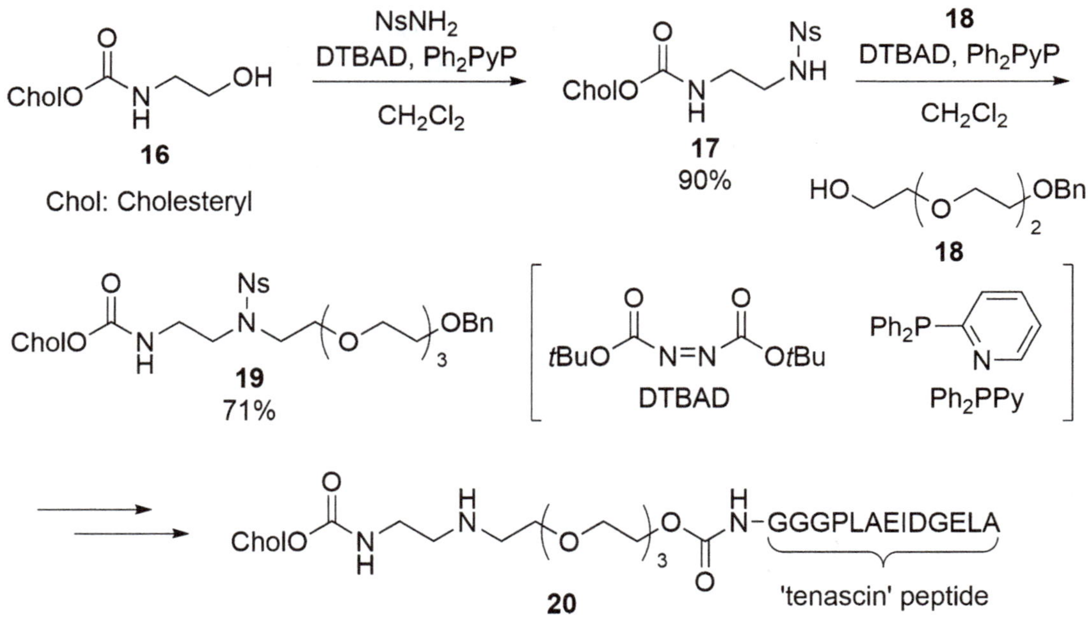

Fig. 8 Synthesis of a secondary amine having a tenascin lipopetide by the Fukuyama-Mitsunobu procedure

(Fig. 8) [32]. The Mitsunobu reaction between alcohol **16** and nosylamide under standard conditions (DEAD/PPh_3) complicated the purification and only provided compound **17** in moderate yield (43%). When the standard reagents were replaced with a combination of di-*tert*-butyl azodicarboxylate (DTBAD) and diphenyl-2-pyridylphosphine under slow addition of DTBAD or the use of excess amounts of reagents (2 equiv), the product **17** was isolated

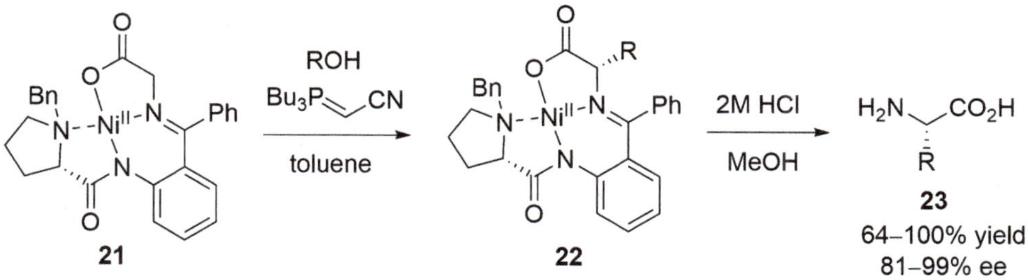

Fig. 9 Synthesis of chiral amino acids with the Tsunoda reagent

in a significantly improved yield (90%). The by-products could be easily removed by acidic quenching followed by extraction as di-*tert*-butyl 1,2-hydrazinedicarboxylate generated from DTBAD decomposes under acidic conditions to *iso*-butene, carbon dioxide, and hydrazine, which along with diphenyl-2-pyridylphosphine oxide can be removed by acidic extraction [33]. The second Mitsunobu reaction of **17** with alcohol **18** under the same conditions gave the corresponding protected amine **19**, which was converted to tenascin lipopeptide **20** by a series of subsequent chemical transformations.

The Tsunoda reagent [(cyanomethylene)tributylphosphorane] can promote the Mitsunobu reaction even if nucleophiles with poor acidity are employed because the reaction with this reagent can be performed at high temperature (80–120 °C), which generates the highly basic α-cyanomethyl anion [13, 14]. Even carbon nucleophiles such as active methylene compounds can partner with alcohols in the reaction utilizing this reagent [15, 16]. For example, the Mitsunobu reaction with this reagent was applied to the synthesis of various amino acids from a glycine derivative and a series of alcohols [34]. Specifically, the optically active amino acids **23** could be synthesized by the reaction of the nickel–(*S*)-2-[*N*-(*N*′-benzylprolyl)amino]benzophenone complex **21** with alcohols via the alkylated complexes **22** (Fig. 9), which provides a reliable method for obtaining non-proteinogenic amino acids.

The Mitsunobu reaction is highly practical for medicinal chemistry on a laboratory scale, though chemists often hesitate to use the Mitsunobu reaction in the industrial field because the method is environmentally and atom-economically unfavorable chiefly due to production of a large amount of waste, as described above [35, 36]. Nevertheless, the Mitsunobu reaction is often chosen for the preparation of drug candidates on a large scale as it is often superior to other methods [16]. The typical Mitsunobu reaction conditions are usually used, with the procedure for purification being individually modified to facilitate isolation of the desired compound.

The Mitsunobu reaction was employed to connect sulfamide **24** with alcohol **25** at a late stage in the synthesis of an active

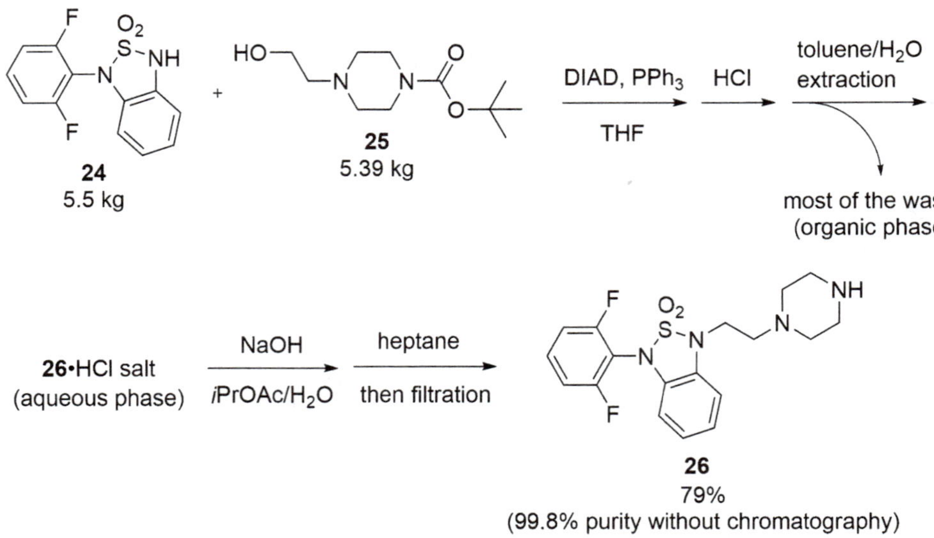

Fig. 10 Pilot-plant scale synthesis of an active pharmaceutical ingredient by the Mitsunobu reaction

pharmaceutical ingredient (API) **26** on pilot-plant scale (Fig. 10) [37]. Apparently, alkylation of **24** with alkyl halides corresponding to **25** is preferred because the reaction is more atom-economical, though such halides bear structures of toxic nitrogen mustard reagents, and therefore contamination of the final product with this becomes a serious problem. Therefore, a Mitsunobu procedure with a modified work-up eliminating chromatography was adopted. After the Mitsunobu reaction between **24** and **25**, the *tert*-butoxycarbonyl group was removed by treatment of the reaction mixture with hydrochloric acid. When the mixture was extracted with toluene/water, the desired product was included in the aqueous phase because the hydrochloride salt was formed. Therefore, most of the waste such as the hydrazinedicarboxylate and triphenylphosphine oxide remaining in the organic phase could be separated from the product in this step. Finally, highly pure product **26** was obtained by basification with sodium hydroxide followed by extraction into *i*PrOAc and addition of heptane to precipitate the desired compound.

A Mitsunobu azidation was employed for the synthesis of drug candidate **30**, which is one of the hepatitis C virus (HCV) NS5B-polymerase allosteric inhibitors, on a pilot-plant scale (Fig. 11) [38]. After the Mitsunobu reaction of **27** with diphenylphosphoryl azide (DPPA) to give the crude azide compound **28**, another equivalent of triphenylphosphine and water was added to the reaction mixture to provide the corresponding amine via a Staudinger-reduction. Treatment of the crude product with hydrochloric acid provided the corresponding hydrochloride salt **29**, which could be isolated as a solid by simple filtration.

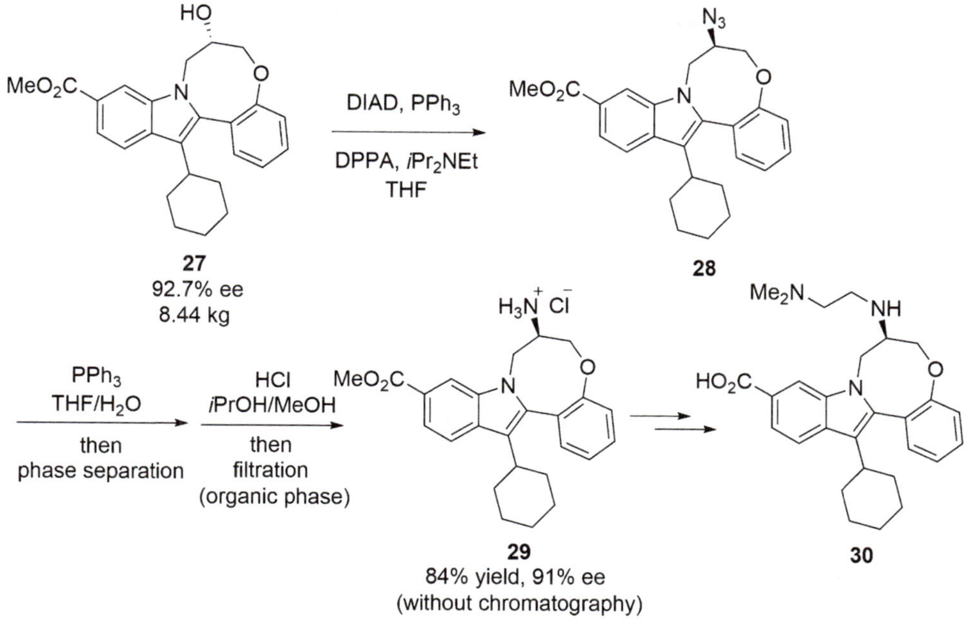

Fig. 11 Multikilogram synthesis of an HCV polymerase inhibitor featuring a Mitsunobu azidation

A thorough safety evaluation of the reaction is critical when the step includes the reaction with azide reagents in the synthesis on a plant scale [39, 40] in particular to stringently monitor for the potential formation of toxic and explosive hydrazoic acid (HN_3). Through development of the overall process shown in Fig. 11, it was confirmed that formation of hydrazoic acid was suppressed below 10 ppm, though concentration of hydrazoic acid temporally increased during addition of DPPA to the mixture.

The Mitsunobu reaction was adopted as the first step in the synthesis of crizotinib (PF-02341066) (**34**), which is an anti-cancer drug for the treatment of NSCLC functioning as an ALK (anaplastic lymphoma kinase) and is currently being evaluated as a ROS1 (c-ros oncogene 1) inhibitor, on a pilot-plant scale because mild reaction conditions for connecting **31** with **32** were required due to the potential thermal instability of **32** (Fig. 12) [41]. Again, a modified procedure was developed in order to simplify the purification of **33**. After the typical Mitsunobu reaction between **31** and **32**, most of the waste mainly including triphenylphosphine oxide and diisopropyl 1,2-hydrazinedicarboxylate was removed by addition of water and subsequent filtration of the complex formed. Solubility profiling of the desired product leading to replacement of the solvent with ethanol represented a key step as this allowed direct precipitation product **33**, which could be isolated in high purity by filtration, and used directly in the subsequent Ni-catalyzed reduction of a nitro group.

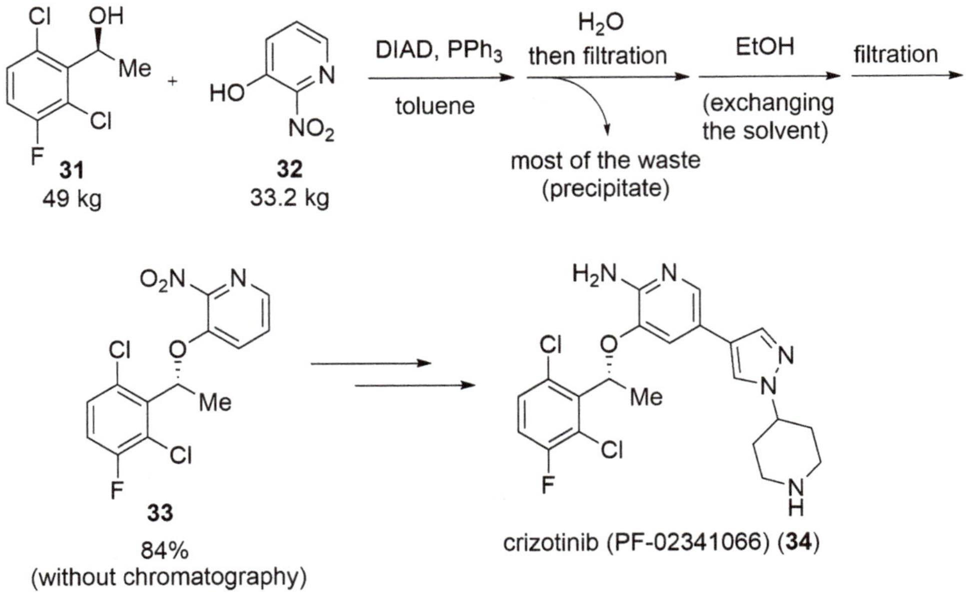

Fig. 12 Application of the Mitsunobu reaction to the pilot-plant scale synthesis of crizotinib

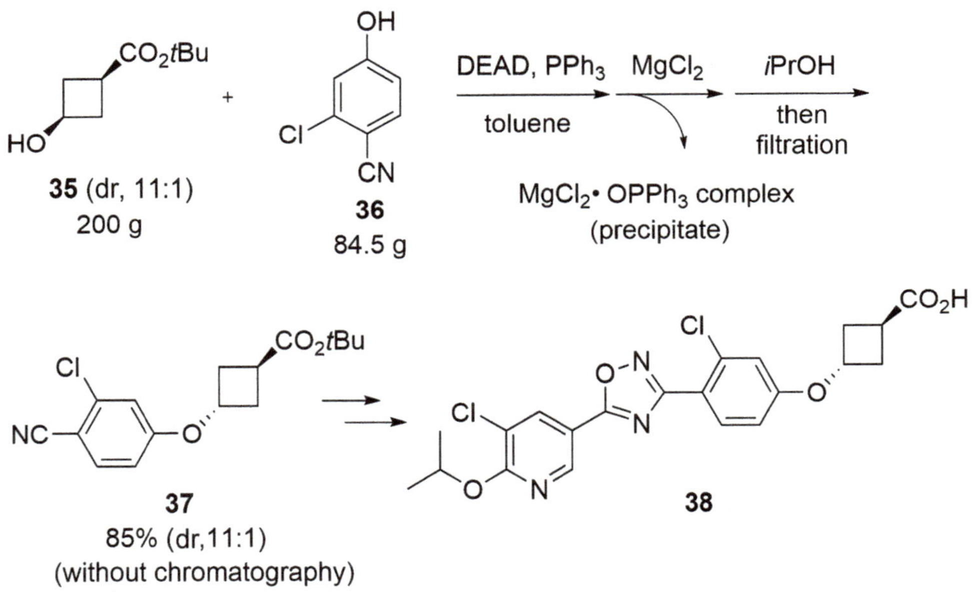

Fig. 13 Mitsunobu reaction with modified purification in the scalable synthesis of an $S1P_1$ receptor agonist

It has been established that triphenylphosphine oxide forms stable complexes with some metal halides, and this unique phenomenon was applied to remove triphenylphosphine oxide when the Mitsunobu reaction was used for the scalable synthesis of the $S1P_1$ receptor agonist **38** (Fig. 13) [42]. When magnesium chloride was added to the reaction mixture after completion of the Mitsunobu reaction between alcohol **35** and phenol **36**, an insoluble complex of magnesium chloride with triphenylphosphine oxide was formed in the toluene solution. Most of triphenylphosphine (>95%) oxide could be removed by filtration of this suspension. Subsequently, product **37** could be isolated in high purity by filtration as **37** precipitated from 2-propanol enabling the hydrazinedicarboxylate by-product formed to be removed in this step.

There are numerous similar examples of the Mitsunobu reactions on a large scale [43–46] with a focus on for modifying work-up procedures to facilitate purification and avoid chromatography in the majority of cases.

3 Modifications of the Mitsunobu Reaction

As described in the previous section, the production of a large amount of waste seriously hampers application of the Mitsunobu reaction to the practical synthesis of drugs and their intermediates. Many interesting concepts for improving the Mitsunobu reaction have recently been reported with many of these modifications aimed at the facile removal of waste after the reaction.

Separation of waste by extraction operative after the reaction represents one of the simplest modifications with the standard example being the Mitsunobu reaction promoted by a combination of diphenyl-2-pyridylphosphine and di-*tert*-butyl azodicarboxylate (DTBAD), which can be extracted/decomposed by an aqueous acidic work-up (Fig. 8) [29].

The crown ether tag attached to the triphenylphosphine derivative **39** facilitates removal of the corresponding phosphine oxide with an ArgoPore™ ammonium trifluoroacetate column (Fig. 14) [47]. When the crude product was purified by the column with dichloromethane after the Mitsunobu reaction between 7-hydroxycoumarin and benzyl alcohol, no phosphine oxide was detected from the elution leading to the benzylated product, though further

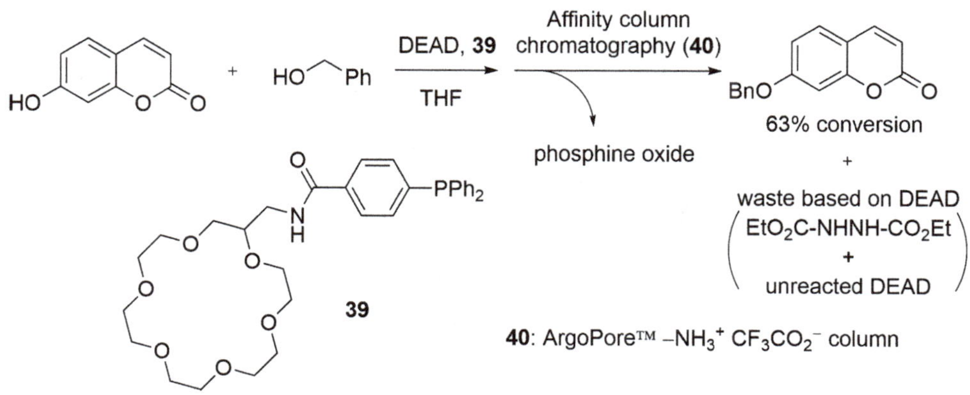

Fig. 14 Mitsunobu reaction with a phosphine reagent bearing a crown ether tag

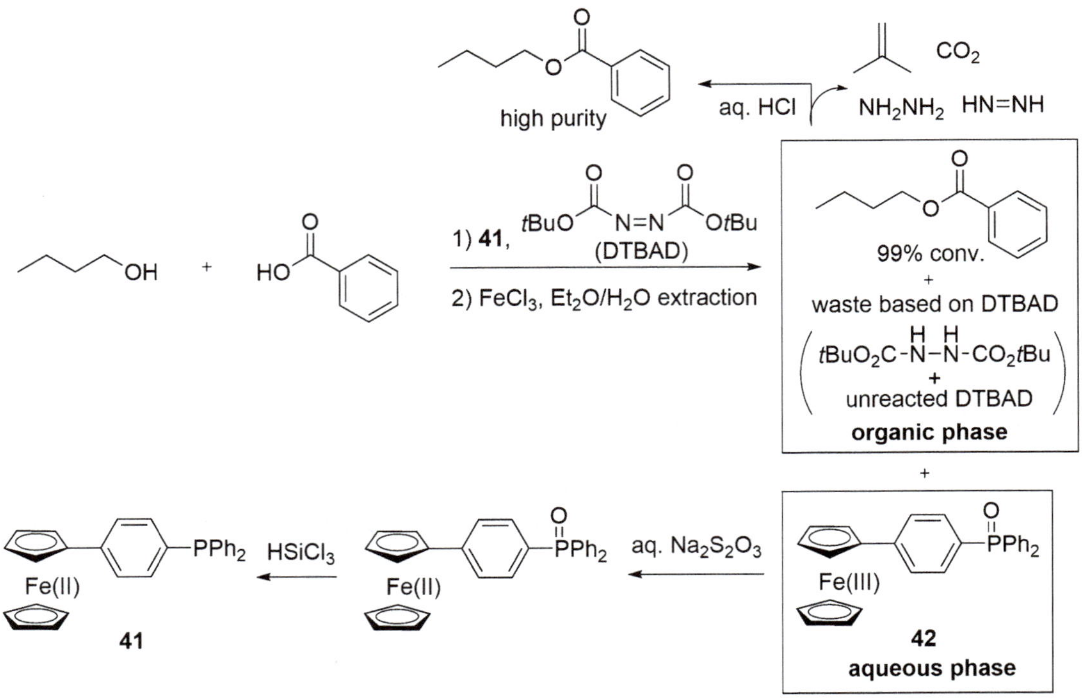

Fig. 15 Mitsunobu reaction with a phosphine reagent bearing a redox-switching tag

purification to remove waste based on DEAD is required for in order to obtain highly pure desired product. This methodology may not be practical on a large scale due to the use of toxic dichloromethane in chromatography, but would be useful on a laboratory scale when typical phosphine oxides become problematic waste in purification of the desired products.

Plenio and co-workers developed phosphine **41** having a redox-switching tag (Fig. 15) [48] and utilized a combination of this reagent and di-*tert*-butyl azodicarboxylate (DTBAD) to promote the Mitsunobu reaction. A treatment of the crude reaction mixture with iron(III) chloride followed by ether/water extraction provided an organic phase, which includes the product and waste based on the azo reagent, and an aqueous phase including the phosphine oxide **42** possessing an oxidized ferrocene core as high-valent ferrocene has a higher solubility in water. The separated phosphine oxide **42** can be reduced in a stepwise manner with sodium thiosulfate and trichlorosilane to recover the starting phosphine **41**, while the pure product is obtained after removal of waste based on the azo reagent by treatment with hydrochloric acid followed by ether extraction.

Curran and co-workers have applied their fluorous chemistry demonstrating that [49–51] the Mitsunobu reaction with azo reagents **43–46** and phosphine **47** having fluorous tags smoothly proceeds in THF, and highly pure products could be obtained by

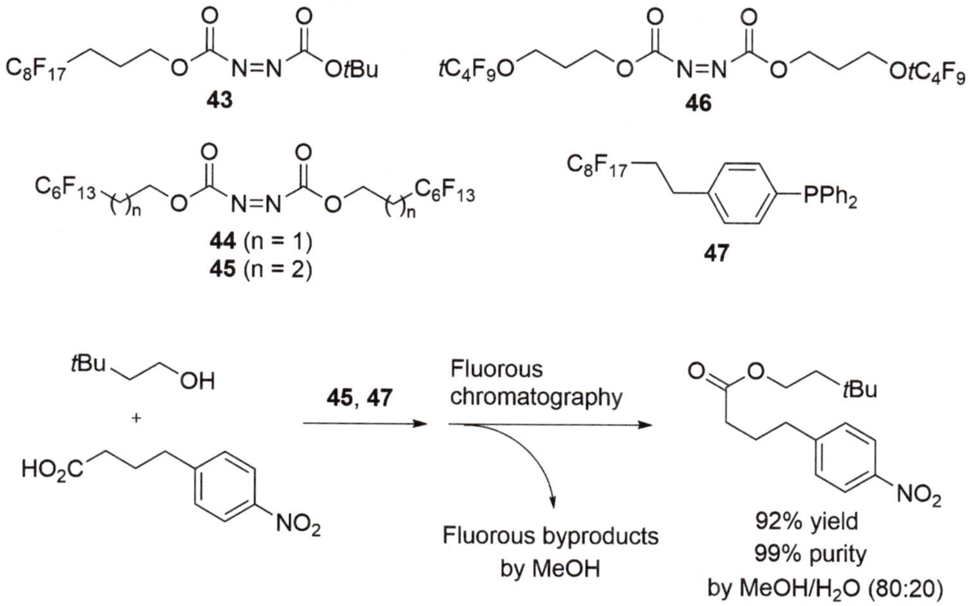

Fig. 16 Fluorous Mitsunobu reagents

solid-phase extraction with fluorous silica gel (Fluoro*Flash*®) (Fig. 16). For instance, in the reaction between 4-fluorobenzyl alcohol and 4-cyanophenol with azo compound **45** and phosphine **47**, methanol/water (80:20) eluted only the corresponding ether product in high purity, whereas the fluorous side products could be recovered by switching the solvent to absolute methanol. Although a recovered fluorous mixture needs to be further purified by usual silica gel chromatography, fluorous hydrazine compounds are reoxidized to the azo reagents with combination of bromine and pyridine. In a similar context, a fluorous phosohine oxide is recyclable by reconversion to the fluorous phosphine with alane–*N*,*N*-dimethylamine complex. Azo reagent **45** having a propylene spacer tends to provide better results than azo reagent **44** having an ethylene spacer when steric hindered alcohols or weakly acidic nucleophiles are used in the reaction. Azo reagents **43** and **46** are more atom-economical from a standpoint of fluorine atoms. However, more fluorous silica gel is needed to get sufficient separation in the reactions with these reagents, compared with the cases of **45** and **46**.

Di-2-methoxyethyl azodicarboxylate (**48**) (DMEAD®) is a new entry among modified azo reagents (Fig. 17) [52] and is convenient for handling due to being a crystalline solid. **48** is easily prepared by acylation between commercially available hydrazine hydrate and 2-methoxyethyl chloroformate followed by oxidation with *N*-bromosuccinimide, and it is obtained as a pure crystalline solid by recrystallization. The by-product from the reduction after reaction of DMEAD®, di-2-methoxyethyl hydrazinedicarboxylate

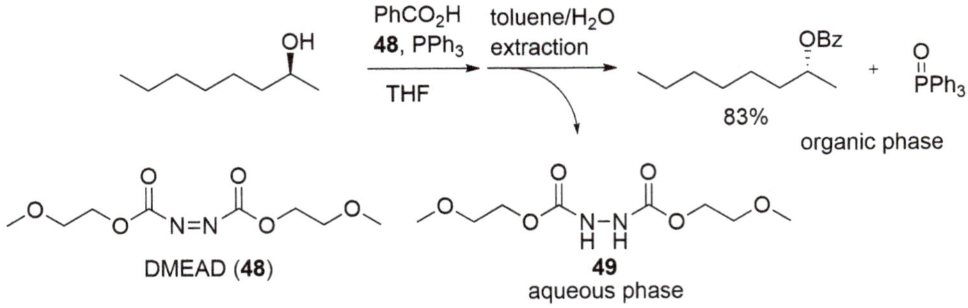

Fig. 17 Mitsunobu reaction with DMEAD

(**49**) can be easily removed by simple extraction with water. Since most of triphenylphosphine oxide can often be removed by trituration with hexanes followed by filtration, obtaining a highly pure product is often possible after short silica gel column chromatography. In a similar manner, utilizing azodicarbonyl dimorpholide (ADDM) allows for the reduced by-product to also be removed after reaction by aqueous extraction, and in this case, it was shown that the reagent can be easily recycled by oxidation with silver (I) oxide and re-used [53].

Solid-phase synthesis facilitates removal of waste in organic synthesis because the desired products can be isolated simply by filtration. As with typical solid-phase synthesis, there are a very large number of examples of the Mitsunobu reaction using polymer-supported substrates (alcohols, carboxylic acids, or other nucleophiles) [54]. Such a methodology is utilized for construction of medicinal libraries [55–57]. In this respect, Schultz and co-workers reported a unique solid-phase method in the construction of combinatorial libraries of 2,6,9-substituted purines (Fig. 18) [58]. The solid-phase Mitsunobu reaction with substrates possessing the polymer tag proved to be problematic due to incomplete alkylation with secondary alcohols and the use of excessive amounts of reagents. This is a typical problem in solid-phase reactions, which is often attributed to an issue of solubility. As such, the reactions were carried out in solution phase, and after the reaction of purine **50** had been performed, the crude products were captured by resin-bound amines **52** prior to purification. Subsequent filtration of the resultant suspension gave products **53** in high purity with this strategy enabling both the reliable solution-phase Mitsunobu reaction to occur and the easy removal of waste from the desired products. Numerous purine derivatives **54** could be synthesized through additional solid-phase reactions, with final isolation occurring through cleavage of the resin. Construction of a 1000 compound library is possible employing this methodology in a 96-well format.

On the other hand, there are examples of solution-phase Mitsunobu reactions employing polymer-supported reagents, which

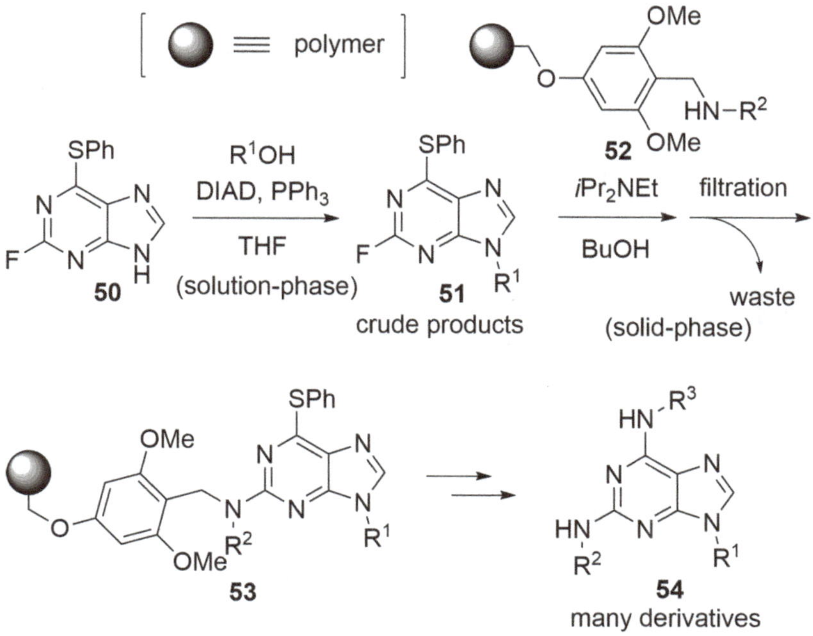

Fig. 18 Synthesis of purine derivatives through a combination of the solution-phase Mitsunobu reaction and solid-phase reactions

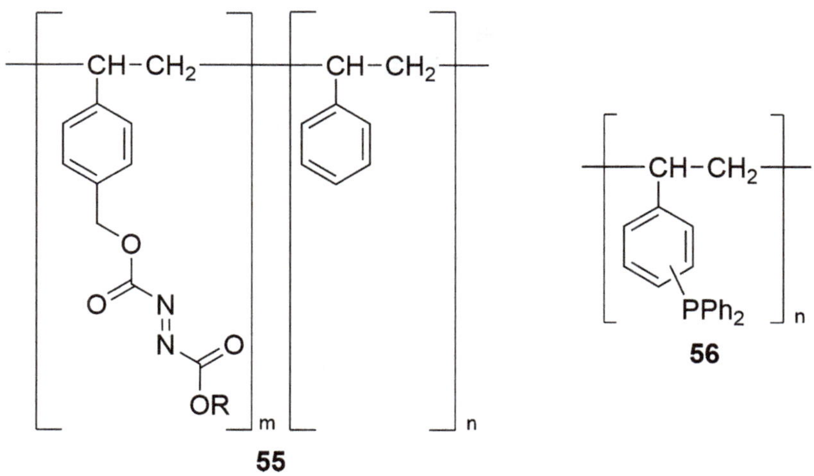

Fig. 19 Typical examples of polymer-supported Mitsunobu reagents

enables both the reagents and waste based upon them to be removed by filtration of the reaction mixture. This method is more convenient than the solid-phase method described previously as it is applicable to a diverse range of substrates. Both polystyrene-bounded azo **55** and phosphine **56** have been reported (Fig. 19) [59–62], though typically either the azo reagent or the phosphine reagent is used as a polymer-supported reagent in the Mitsunobu reaction because if both are employed as polymers both reaction

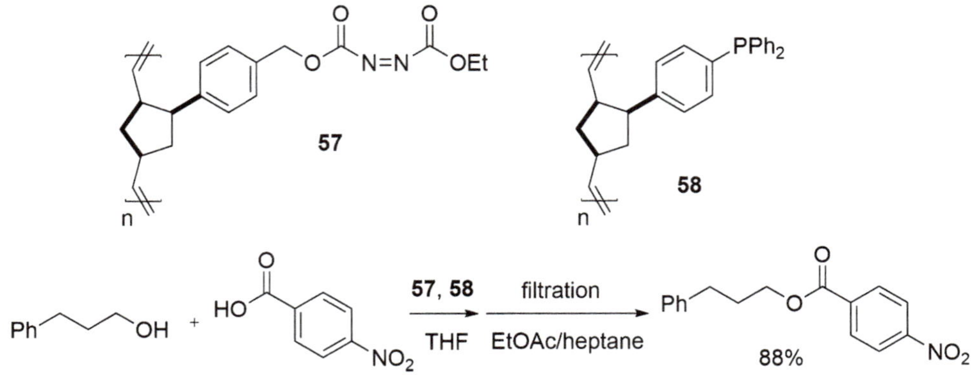

Fig. 20 Mitsunobu reaction using "soluble" polymer-supported Mitsunobu reagents

rate and the amount of solids in the reaction medium often become a problem.

Recently, Hanson and co-workers reported a solution-phase Mitsunobu reaction with azo and phosphine reagents **57** and **58** featuring ring-opening metathesis (ROM) oligomerization of a norbornenyl tag, which are soluble in THF (Fig. 20) [63]. After the Mitsunobu reaction has been conducted with both polymer reagents, the polymers can be easily removed by two filtration processes through a short pad of silica gel eluting with EtOAc/heptane, enabling the product is readily obtained in high purity. Monomer reagents bearing a norbornenyl tag also work well in the Mitsunobu reaction. As an interesting example, after completion of the reaction, oligomerization with norbornenyl-tagged magnetic nanoparticles enables removal of by-products by a magnet [64]. There are similar examples using other polymers or dendrimers such as polystyrene, polyglycerol, and polyethylene glycol [65, 66].

Although these methods facilitate the purification of Mitsunobu reactions, application of these on a large scale may not be practical owing to either prohibitive cost, or large-scale availability of the necessary reagents.

An innovative but fundamental improvement of the Mitsunobu reaction is to avoid the stoichiometric use of azo and/or phosphine reagents. Recycling reagents in the reaction mixture is a feasible concept, and thus this system should consist of catalytic amounts of Mitsunobu reagents, which are recycled by stoichiometric amounts of convenient sacrificial reagents, of which the generated by-products are easy to remove after reaction.

The concept has already been realized in the case of hazardous azo reagents. In 2006, Toy and co-workers developed the first example of catalytic Mitsunobu reaction using iodobenzene diacetate as a terminal oxidant (Fig. 21) [67] with diethyl hydrazinecarboxylate generated in the Mitsunobu reaction being reoxidized to the starting DEAD in the presence of iodobenzene diacetate, and

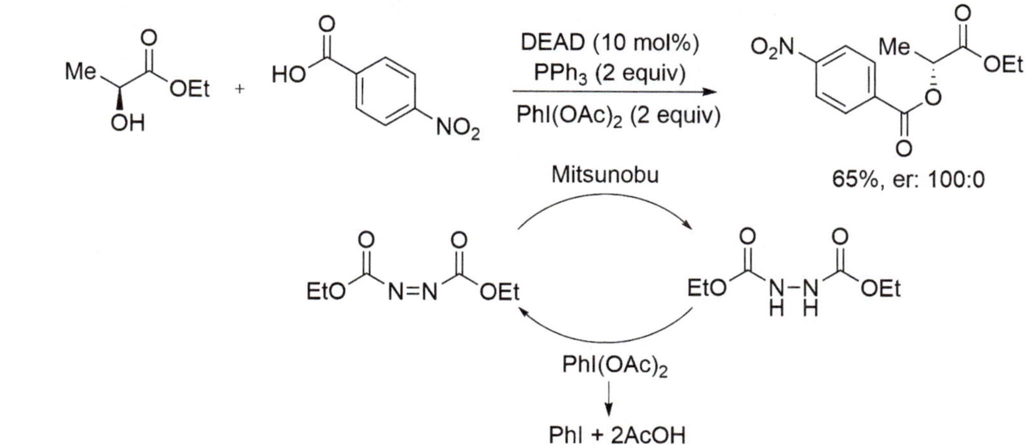

Fig. 21 Catalytic Mitsunobu reaction with iodosobenzene diacetate as a sacrificial oxidant

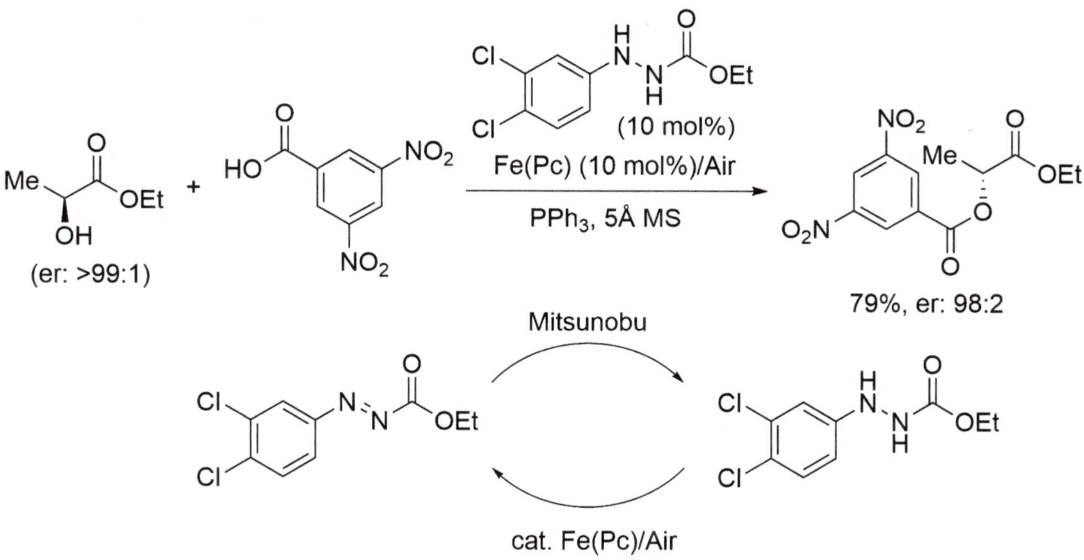

Fig. 22 Catalytic Mitsunobu reaction with azo-reagent recycling through aerobic oxidation

loading 10 mol% of DEAD is thus sufficient to obtain the product in good yield. The sacrificial reagent produces iodobenzene and acetic acid as the bulk of the waste, and a loading of 10 mol% of DEAD is sufficient to obtain the desired products in good yield. Although only examples with carboxylic acids have been shown in scope of reactions, the concept of the reaction is very impressive.

A second example of a catalytic Mitsunobu reaction was reported in 2013 (Fig. 22) [68] with the outstanding feature being the use of atmospheric oxygen as a terminal oxidant for recycling the azo reagent with the aid of an iron catalyst. Since the standard by-product, diethyl 1,2-hydrazinedicarboxylate, was intact under the aerobic iron-catalyzed oxidation conditions, a new azo reagent that is recyclable under these conditions was developed. 2-Phenylhydrazinecarboxylate derivatives are easily oxidized to the corresponding 2-phenylazocarboxylate derivatives by aerobic oxidation with a catalytic amount of iron phthalocyanine [Fe(Pc)] because an aromatic ring can stabilize intermediates such as a cation or a radical. Phenylazocarboxylate derivatives induced the Mitsunobu reaction like DEAD, and 2-phenylhydrazinecarboxylate derivatives should work as a catalyst in Mitsunobu reaction in the presence of iron phthalocyanine and air. Ethyl 2-(3,4-dichlorophenyl)hydrazinecarboxylate has been identified as an efficient catalyst in this catalytic process. Iron phthalocyanine is commercially available as an inexpensive and non-toxic reagent. In addition, recently, a 2-(3,4-dichlorophenyl)hydrazinecarboxylate has been also commercially available from a supplier. The experimental procedure is very simple because the mixture of all materials and reagents was just stirred under air (typically a balloon technique was used). This catalytic system is conceptually close to the ideal catalytic Mitsunobu reaction as the stoichiometric waste generated from the terminal oxidant is water. Addition of molecular sieves is essential to sequester water formed from oxygen in the system. Although reactivity is still moderate compared with the typical Mitsunobu reaction, this catalytic method is applicable to reactions with several nucleophiles such as phenol, phthalimide, and a nosylamide.

Catalytic Mitsunobu reactions for phosphine reagents have been recently reported [69]. Cyclic phosphine oxides efficiently work as a catalyst combining with phenylsilane as a sacrificial reducing reagent in the Mitsunobu reaction with a stoichiometric amount of DIAD (Fig. 23). This method also has relatively general scope of substrates. The catalytic concept is based on catalytic Wittig reactions reported in 2009 [70].

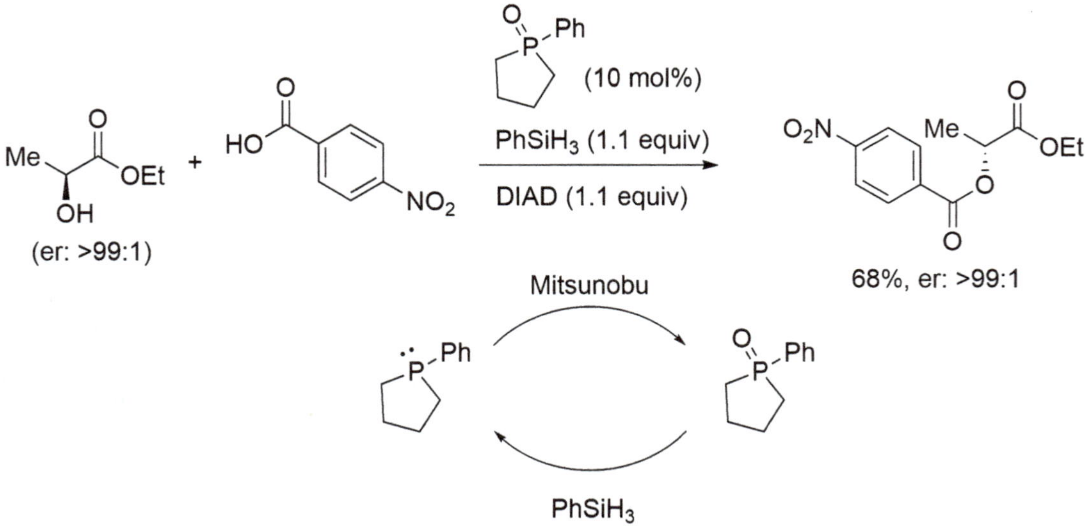

Fig. 23 Catalytic Mitsunobu reaction with a phosphine reagent recycled through reduction with phenylsilane

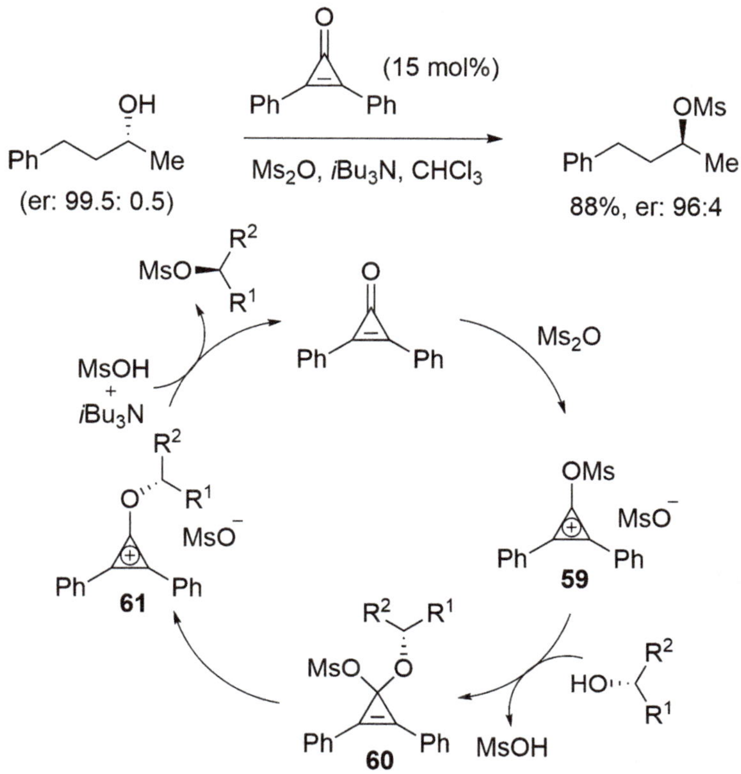

Fig. 24 Cyclopropenone-catalyzed S_N2 reaction with the methanesulfonate anion

The azo and phosphine reagent-free Mitsunobu reaction is a further goal in sustainable chemistry to avoid the significant drawbacks of these reagents, and there have been a few promising examples reported in this area. Diphenylcyclopropenone can work as a catalyst to promote elimination of the hydroxyl groups of alcohols (Fig. 24) [71] with the activated form of this catalyst being the stable cyclopropenium cation **59** generated by the reaction with methanesulfonic anhydride (Ms_2O). When an alcohol is added to cation **59**, cyclopropene intermediate **60** is rapidly formed and then converted to cation **61** by elimination of the methanesulfonate anion (MsO^-). The methanesulfonate anion then attacks the activated alcohol from the back side to provide a mesylate with inversion of stereochemistry. The reaction is applicable to various secondary alcohols bearing functional groups due to the mild conditions, but the applicable nucleophile is only the methanesulfonate anion at the present conditions. If mesylates are used for precursors of S_N2 reactions with other nucleophiles, overall stereochemistry is retained from alcohols. Thus, this catalytic method cannot be full replacement for the Mitsunobu reaction at the moment, though the concept is very attractive. This catalytic method has also been demonstrated for the direct chlorination of alcohols, but it is rather a replacement for Appel reactions [72].

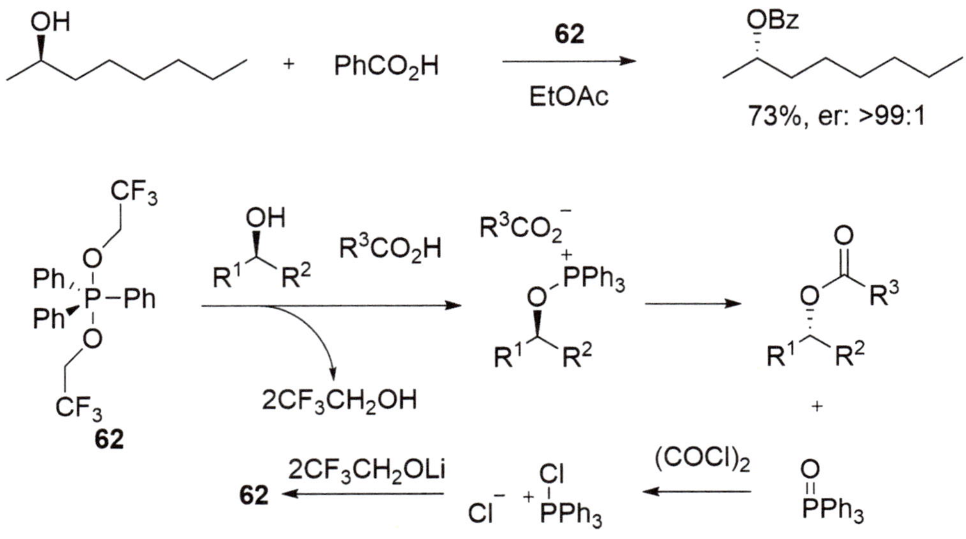

Fig. 25 Azo reagent-free Mitsunobu reaction

An azo reagent-free Mitsunobu reaction has also been achieved by the use of an activated phosphorus(V) reagent **62** (Fig. 25) [73]. Treatment of a mixture of an alcohol and a carboxylic acid with reagent **62** directly provides the corresponding Mitsunobu product. The phosphorus(V) reagent **62** is prepared from triphenylphosphine oxide by treatment with oxalyl chloride followed by addition of lithium 1,1,1-trifluoroethoxide. The overall reaction mechanism does not involve a redox process in contrast to the conventional Mitsunobu reaction, and reagent **62** is also sensitive to moisture though can be stocked as an ethyl acetate solution.

4 Conclusion

Comprehensive application of the Mitsunobu reaction to medicinal chemistry has been illustrated. The typical Mitsunobu reaction has matured into a reliable and efficient method on a laboratory scale, whereas serious problems often appear on a plant scale because of the use of hazardous reagents and generation of much waste. Many impressive approaches to solve such the problems have been reported from the academia, but further efforts will be needed to realize their practical application. It is expected that safe, economic, and ecological Mitsunobu methods will be developed in the future [74, 75].

Acknowledgment

Mr. Daisuke Hirose and Mr. Takuma Hashimoto are acknowledged for their partial assistance in the preparation of the manuscript.

References

1. Fischer E, Speier A (1895) Darstellung der Ester. Chem Ber 28:3252–3258
2. Wakasugi K, Misaki T, Yamada K, Tanabe Y (2000) Diphenylammonium triflate (DPAT): efficient catalyst for esterification of carboxylic acids and for transesterification of carboxylic esters with nearly equimolar amounts of alcohols. Tetrahedron Lett 41:5249–5252
3. Ishihara K, Ohara S, Yamamoto H (2000) Direct condensation of carboxylic acids with alcohols catalyzed by hafnium(IV) salts. Science 290:1140–1142
4. Ishihara K, Nakagawa S, Sakakura A (2005) Bulky diarylammonium arenesulfonates as selective esterification catalysts. J Am Chem Soc 127:4168–4169
5. Valeur E, Bradley M (2009) Amide bond formation: beyond the myth of coupling reagents. Chem Soc Rev 38:606–631
6. Mukaiyama T (1976) Oxidation-reduction condensation. Angew Chem Int Ed Eng 15:94–103
7. Mitsunobu O, Yamada M, Mukaiyama T (1967) Preparation of esters of phosphoric acid by the reaction of trivalent phosphorus compounds with diethyl azodicarboxylate in the presence of alcohols. Bull Chem Soc Jpn 40:935–939
8. Mitsunobu O, Yamada M (1967) Preparation of esters of carboxylic and phosphoric acid *via* quaternary phosphonium salts. Bull Chem Soc Jpn 40:2380–2382
9. Mitsunobu O (1981) The use of diethyl azodicarboxylate and triphenylphosphine in synthesis and transformation of natural products. Synthesis:1–28
10. Hughes DL (1992) The Mitsunobu reaction. Org React 42:335–656
11. Tsunoda T, Yamamiya Y, Itô S (1993) 1,1-′-(Azodicarbonyl)dipiperidine-tributylphosphine, a new reagent system for Mitsunobu reaction. Tetrahedron Lett 34:1639–1642
12. Tsunoda T, Otsuka J, Yamamiya Y, Itô S (1994) N,N,N′,N-′-tetramethylazodicarboxamide (TMAD), a new versatile reagent for mitsunobu reaction. its application to synthesis of secondary amines. Chem Lett:539–542
13. Tsunoda T, Ozaki F, Itô S (1994) Novel reactivity of stabilized methylenetributylphosphorane: a new Mitsunobu reagent. Tetrahedron Lett 35:5081–5082
14. Tsunoda T, Nagaku M, Nagino C, Kawamura Y, Ozaki F, Hioki H, Itô S (1995) Carbon-carbon bond formation with new Mitsunobu reagents. Tetrahedron Lett 36:2531–2534
15. Tsunoda T, Nagino C, Ogure M, Itô S (1996) Mitsunobu-type alkylation with active methine compounds. Tetrahedron Lett 37:2459–2462
16. Constable DJC, Dunn PJ, Hayler JD, Humphrey GR, Leazer JL Jr, Linderman RJ, Lorenz K, Manley J, Pearlman BA, Wells A, Zaks A, Zhang TY (2007) Key green chemistry research areas–a perspective from pharmaceutical manufacturers. Green Chem 9:411–420
17. Berger A, Wehrstedt KD (2010) Azodicarboxylates: explosive properties and DSC measurements. J Loss Prev Process Ind 23:734–739
18. Kumara Swamy KC, Kumar Bhuvan NN, Balaraman E, Pavan Kumar KVP (2009) Mitsunobu and related reactions: advances and applications. Chem Rev 109:2551–2651
19. Fletcher S (2015) The Mitsunobu reaction in the 21st century. Org Chem Front 2:739–752
20. Pelletier J, Kincaid S (2000) Mitsunobu reaction modifications allowing product isolation without chromatography: application to a small parallel library. Tetrahedron Lett 41:797–800
21. Proctor AJ, Beautement K, Clough JM, Knight DW, Li Y (2006) Chromatography-free product separation in the Mitsunobu reaction. Tetrahedron Lett 47:5151–5154
22. Zolle IM, Berger ML, Hammerschmidt F, Hahner S, Schirbel A, Peric-Simov B (2008) New selective inhibitors of steroid 11-hydroxylation in the adrenal cortex. Synthesis and structure–activity relationship of potent etomidate analogues. J Med Chem 51:2244–2253
23. Laumen K, Schneider MP (1988) A highly selective ester hydrolase from *Pseudomonas sp.* for the enzymatic preparation of enantiomerically pure secondary alcohols; chiral auxiliaries in organic synthesis. J Chem Soc Chem Commun 598–600
24. Seemayer R, Schneider MP (1992) Preparation of optically pure pyridyl-1-ethanols. Tetrahedron Asymmetry 3:827–830
25. Han X, Liu B, Zhou H-B, Dong C (2012) Enhanced efficiency of recyclable C_3-symmetric cinchonine-squaramides in the asymmetric Friedel–Crafts reaction of indoles with alkyl trifluoropyruvate. Tetrahedron Asymmetry 23:1332–1337
26. Han X, Wu H, Wang W, Dong C, Tien P, Wu S, Zhou H-B (2014) Synthesis and SARs of

indole-based α-amino acids as potent HIV-1 non-nucleoside reverse transcriptase inhibitors. Org Biomol Chem 12:8308–8317
27. Ludek OR, Meier C (2003) New convergent synthesis of carbocyclic nucleoside analogues. Synthesis:2101–2109
28. Ludek OR, Krämer T, Balzarini J, Meier C (2006) Divergent synthesis and biological evaluation of carbocyclic α-, *iso*- and 3'-*epi*-nucleosides and their lipophilic nucleotide prodrugs. Synthesis:1313–1324
29. Fukuyama T, Jow C-K, Cheung M (1995) 2- and 4-Nitrobenzenesulfonamides: exceptionally versatile means for preparation of secondary amines and protection of amines. Tetrahedron Lett 36:6373–6374
30. Kan T, Fukuyama T (2004) Ns strategies: a highly versatile synthetic method for amines. Chem Commun 353–359
31. Zapf CW, Del Valle JR, Goodman M (2005) Utilizing the intramolecular Fukuyama–Mitsunobu reaction for a flexible synthesis of novel heterocyclic scaffolds for peptidomimetic drug design. Bioorg Med Chem Lett 15:4033–4036
32. Guisado C, Waterhouse JE, Price WS, Jorgensen MR, Miller AD (2005) The facile preparation of primary and secondary amines *via* an improved Fukuyama–Mitsunobu procedure. Application to the synthesis of a lung-targeted gene delivery agent. Org Biomol Chem 3:1049–1057
33. Kiankarimi M, Lowe R, McCarthy JR, Whitten JP (1999) Diphenyl 2-pyridylphosphine and di-*tert*-butylazodicarboxylate: convenient reagents for the Mitsunobu reaction. Tetrahedron Lett 40:4497–4500
34. Noisier AF, Harris CS, Brimble MA (2013) Novel preparation of chiral α-amino acids using the Mitsunobu–Tsunoda reaction. Chem Commun 49:7744–7746
35. Girardin M, Dolman SJ, Lauzon S, Ouellet SG, Hughes G, Fernandez P, Zhou G, O'Shea PD (2011) Development of a practical synthesis of stearoyl-CoA desaturase (SCD1) inhibitor MK-8245. Org Process Res Dev 15:1073–1080
36. Schmidt G, Reber S, Bolli MH, Abele S (2012) Practical and scalable synthesis of $S1P_1$ receptor agonist ACT-209905. Org Process Res Dev 16:595–604
37. Connolly TJ, Auguscinski W, Fung P, Galante R, Liu W, McGovern L, Sebastian A, Shen X, Shi X, Wilk B, Varssalona R, Zhong H (2010) Development of a pilot-plant-scale synthesis of an alkylated dihydrobenzothiadiazole *S,S*-dioxide: incorporation of a late-stage mitsunobu reaction. Org Process Res Dev 14:868–877
38. Scott JP, Alam M, Bremeyer N, Goodyear A, Lam T, Wilson RD, Zhou G (2011) Mitsunobu inversion of a secondary alcohol with diphenylphosphoryl azide. Application to the enantioselective multikilogram synthesis of a HCV polymerase inhibitor. Org Process Res Dev 15:1116–1123
39. Wiss J, Fleury C, Onken U (2006) Safety improvement of chemical processes involving azides by online monitoring of the hydrazoic acid concentration. Org Process Res Dev 10:349–353
40. Wiss J, Fleury C, Heuberger C, Onken U (2007) Explosion and decomposition characteristics of hydrazoic acid in the gas phase. Org Process Res Dev 11:1096–1103
41. Koning PD, McAndrew D, Moore R, Moses IB, Boyles DC, Kissick K, Stanchina CL, Cuthbertson T, Kamatani A, Rahman L, Rodriguez R, Urbina A, Sandoval (née Accacia) A, Rose PR (2011) Fit-for-purpose development of the enabling route to crizotinib (PF-02341066). Org Process Res Dev 15:1018–1026
42. Lukin K, Kishore V, Gordon T (2013) Development of a scalable synthesis of oxadiazole based $S1P_1$ receptor agonists. Org Process Res Dev 17:666–671
43. Wallace M, McGuire MA, Yu MS, Goldfinger L, Liu L, Dai W, Shilcrat S (2004) Multi-kiloscale enantioselective synthesis of a vitronectin receptor antagonist. Org Process Res Dev 8:738–743
44. Anelli PL, Brocchetta M, Lattuada L, Manfredi G, Morosini P, Murru M, Palano D, Sipioni M, Visigalli M (2009) Scale-up of trisodium [(3β,5β,12α)-3-[[4(*S*)-4-[*Bis*[2-[*bis*[(carboxy-*kO*)methyl]amino-*kN*]ethyl]amino-*kN*]-4-(carboxy-*kO*)-1-oxobutyl]amino]-12-hydroxycholan-24-oato(6-)]gadolinate(3-)], a Gd(III) complex under development as a contrast agent for MRI coronary angiography. Org Process Res Dev 13:739–746
45. Bellingham R, Buswell AM, Choudary BM, Gordon AH, Moore SO, Peterson M, Sasse M, Shamji A, Urquhart MWJ (2010) Discovery and development of an efficient, scalable, and robust route to the novel CENP-E inhibitor GSK923295A. Org Process Res Dev 14:1254–1263
46. Beliaev A, Wahnon J, Russo D (2012) Process research for multikilogram production of etamicastat: a novel dopamine β-hydroxylase inhibitor. Org Process Res Dev 16:704–709

47. Jackson T, Routledge A (2003) Synthesis and application of crown ether tagged triarylphosphines. Tetrahedron Lett 44:1305–1307
48. Fleckenstein CA, Plenio H (2006) Redox-switchable phase tags-facile Mitsunobu reactions using ferrocenyl-tagged triphenylphosphine. Adv Synth Catal 348:1058–1062
49. Dandapani S, Curran DP (2002) Fluorous Mitsunobu reagents and reactions. Tetrahedron 58:3855–3864
50. Dandapani S, Curran DP (2004) Second generation fluorous DEAD reagents have expanded scope in the Mitsunobu reaction and retain convenient separation features. J Org Chem 69:8751–8757
51. Chu Q, Henry C, Curran DP (2008) Second-generation tags for fluorous chemistry exemplified with a new fluorous Mitsunobu reagent. Org Lett 10:2453–2456
52. Sugimura T, Hagiya K (2007) Di-2-methoxyethyl azodicarboxylate (DMEAD): an inexpensive and separation-friendly alternative reagent for the Mitsunobu reaction. Chem Lett 36:566–567
53. Lanning ME, Fletcher S (2013) Azodicarbonyl dimorpholide (ADDM): an effective, versatile, and water-soluble Mitsunobu reagent. Tetrahedron Lett 54:4624–4628
54. Aronov AM, Gelb MH (1998) Phthalimide resin reagent for efficient Mitsunobu aminodehydroxylation. Tetrahedron Lett 39:4947–4950
55. Ryoo S-J, Kim J, Kim J-S, Lee Y-S (2002) Efficient methods of converting hydroxyl groups into amino groups in poly(ethylene glycol)-grafted polystyrene resin. J Comb Chem 4:187–190
56. Guo G, Arvanitis EA, Pottorf RS, Player MR (2003) Solid-phase synthesis of a tyrphostin ether library. J Comb Chem 5:408–413
57. Arya P, Wei C-Q, Barnes ML, Daroszewska M (2004) A solid phase library synthesis of hydroxyindoline-derived tricyclic derivatives by Mitsunobu approach. J Comb Chem 6:65–72
58. Ding S, Gray NS, Ding Q, Schultz PG (2001) Concise and traceless linker strategy toward combinatorial libraries of 2,6,9-substituted purines. J Org Chem 66:8273–8276
59. Arnold LD, Assil HI, Vederas JC (1989) Polymer-supported alkyl azodicarboxylates for Mitsunobu reactions. J Am Chem Soc 111:3973–3976
60. Amos RA, Emblidge RW, Havens N (1983) Esterification using a polymer-supported phosphine reagent. J Org Chem 48:3598–3600
61. Tunoori AR, Dutta D, Georg GI (1998) Polymer-bound triphenylphosphine as traceless reagent for Mitsunobu reactions in combinatorial chemistry: synthesis of aryl ethers from phenols and alcohols. Tetrahedron Lett 39:8751–8754
62. Lizarzaburu ME, Shuttleworth SJ (2003) Convenient preparation of aryl ether derivatives using a sequence of functionalized polymers. Tetrahedron Lett 44:4873–4876
63. Harned AM, He HS, Toy PH, Flynn DL, Hanson PR (2005) Multipolymer solution-phase reactions: application to the Mitsunobu reaction. J Am Chem Soc 127:52–53
64. Maity PK, Rolfe A, Samarakoon TB, Faisal S, Kurtz RD, Long TR, Schätz A, Flynn DL, Grass RN, Stark WJ, Reiser O, Hanson PR (2011) Monomer-on-monomer (MoM) Mitsunobu reaction: facile purification utilizing surface-initiated sequestration. Org Lett 13:8–10
65. Roller S, Zhou H, Haag R (2005) High-loading polyglycerol supported reagents for Mitsunobu- and acylation-reactions and other useful polyglycerol derivatives. Mol Divers 9:305–316
66. Figlus M, Wellaway N, Cooper AWJ, Sollis SL, Hartley RC (2011) Synthesis of arrays using low molecular weight MPEG-assisted Mitsunobu reaction. ACS Comb Sci 13:280–285
67. But TYS, Toy PH (2006) Organocatalytic Mitsunobu reactions. J Am Chem Soc 128:9636–9637
68. Hirose D, Taniguchi T, Ishibashi H (2013) Recyclable Mitsunobu reagents: catalytic mitsunobu reactions with an iron catalyst and atmospheric oxygen. Angew Chem Int Ed 52:4613–4617
69. Buonomo JA, Aldrich CC (2015) Mitsunobu reactions catalytic in phosphine and a fully catalytic system. Angew Chem Int Ed 54:13041–13044
70. O'Brien CJ, Tellez JL, Nixon ZS, Kang LJ, Carter AL, Kunkel SR, Przeworski KC, Chass GA (2009) Recycling the waste: the development of a catalytic Wittig reaction. Angew Chem Int Ed 48:6836–6839
71. Nacsa ED, Lambert TH (2013) Cyclopropenone catalyzed substitution of alcohols with mesylate ion. Org Lett 15:38–41
72. Vanos CM, Lambert TH (2011) Development of a catalytic platform for nucleophilic substitution: cyclopropenone-catalyzed chlorodehydration of alcohols. Angew Chem Int Ed 50:12222–12226
73. Tang X, Chapman C, Whiting M, Denton R (2014) Development of a redox-free

Mitsunobu reaction exploiting phosphine oxides as precursors to dioxyphosphoranes. Chem Commun 50:7340–7343

74. Beddoe RH, Sneddon HF, Denton RM (2018) The catalytic Mitsunobu reaction: a critical analysis of the current state-of-the-art. Org Biomol Chem 16:7774–7781

75. Beddoe RH, Andrews KG, Magné V, Cuthbertson JD, Saska J, Shannon-Little AL, Shanahan SE, Sneddon HF, Denton RM (2019) Redox-neutral organocatalytic Mitsunobu reactions. Science 365:910–914

Chapter 4

Direct Nucleophilic Substitution of Alcohols by Brønsted or Lewis Acids Activation: An Update

Pier Giorgio Cozzi, Andrea Gualandi, Luca Mengozzi, Elisabetta Manoni, and Claire Margaret Wilson

Abstract

Typically substitution of alcohols involves a two-step activation/displacement pathway thus leading to the generation of additional waste. The current chapter considers an alternative reaction manifold with the displacement taking place directly through activation of the alcohol by either a Lewis or Brønsted acid. With particular focus on the literature since 2011, an initial overview of carbenium ion reactivity is provided followed by a survey of displacement reactions grouped by the nature of the nucleophile. Finally, advances in both diastereoselective and enantioselective variants of the reaction are discussed.

Key words Alcohols, Direct nucleophilic substitution, Lewis acids, Brønsted acids, Mayr scale, Carbenium ions, Piancatelli rearrangement, Meyer–Schuster rearrangement, Cascade reactions, Enantioselectivity, Diastereoselectivity

1 Introduction

"In 2005, the ACS Green Chemistry Institute® in conjunction with pharmaceutical corporations formed the ACS GCI Pharmaceutical Roundtable to encourage innovation while catalysing the integration of green chemistry and green engineering in the pharmaceutical industry. The activities of the Roundtable reflect its members' shared belief that the pursuit of green chemistry and engineering is imperative for business and environmental sustainability" [1]. These statements cleanly summarize the on-going shift to the development of sustainable and green chemical processes with problems related to pollutants and the costly use of toxic metals or reagents changing the way in which new chemical processes are realized. In addition, changes in industrial policy are also influencing the development and invention of new chemical methodologies. The pharmaceutical roundtable has developed a list of imperative research topics, with the vision that improved methodologies to realize these will lead to a significant reduction in

Paul F. Richardson (ed.), *Green Chemistry in Drug Discovery: From Academia to Industry*, Methods in Pharmacology and Toxicology, https://doi.org/10.1007/978-1-0716-1579-9_4,

impact in producing APIs and fine chemicals. Within that list, the nucleophilic substitution of alcohols is considered one of the most prominent topics. Alcohols are used in universities and industry as starting materials for important synthetic transformations. However, they need to be activated by the introduction of a good leaving group in order to accomplish many of these transformations. This activation is normally conducted in a discrete single step reaction, often followed by a nucleophilic displacement, thus producing additional waste.

In recent years, the use of Brønsted or Lewis acids has been proposed as a possible solution to this shortcoming to enable the direct nucleophilic substitution of alcohols [2]. In this emerging protocol, relatively stable carbenium ions are formed, and the by-product generated by the nucleophilic displacements is water (Scheme 1). Although generally considered green processes, these transformations are typically carried out in the presence of an excess of the nucleophile which in itself is "wasteful." Given this, these reactions need to be critically accessed, taking into account the conditions, solvent, and promoters used. Although the scope is sometimes limited and conditions are not always "green," these new methodologies are valuable processes and should be considered in synthetic planning discussions. In 2011, we presented a general review [2] of the direct nucleophilic substitution of alcohols, in which a series of Brønsted and Lewis acids were examined as catalysts or promoters in nucleophilic S_N1-type reactions. In the last 4 years, various new processes, methods, and reactions have been developed in this area, and a more up-to-date review is presented herein to cover these developments [3–7]. In this chapter, we briefly highlight and summarize the key principles and understanding necessary to use, develop, and design new S_N1-type reactions of alcohols, highlighting particularly the progress made in the last several years in the direct nucleophilic substitution of alcohols.

2 Alcohols as Electrophiles: Generation of Carbenium Ions by Brønsted or Lewis Acids

A common mechanism underlies the direct substitution of alcohols in the presence of Brønsted or Lewis acids. Fundamental studies of S_N1-type reactions have shown that carbocations are key intermediates in many organic reactions [8] including Friedel–Crafts (FC) reactions and polymerizations. The generation of carbocations as suitable precursors by laser flash photolysis provides information regarding their reactivity with many different nucleophiles, and if the lifetime of the photogenerated species is sufficient to observe reactivity towards the nucleophiles of interest, then kinetic data can also be generated [9]. And if the lifetime of the photogenerated carbocations is sufficient to observe reactivity toward the

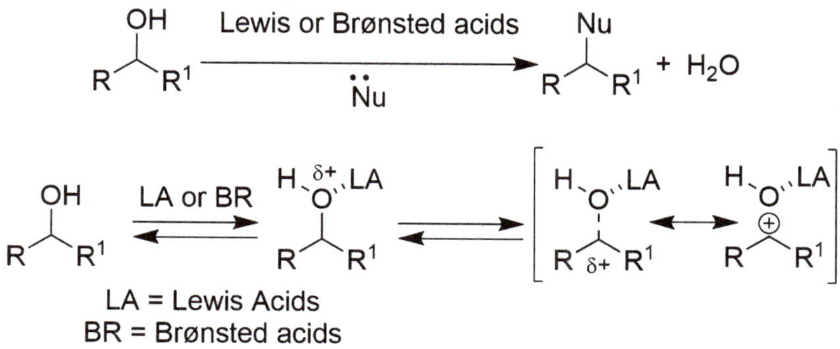

Scheme 1 Mechanism involved in the direct nucleophilic substitution of alcohols

nucleophiles of interest, then kinetic data can be generated. The combination of conventional UV–Vis spectroscopy, stop-flow methods, and laser flash photolysis allows for the investigation of a reaction series across a wide range of rate constants from 10^{-4} M^{-1} s^{-1} up to the diffusion limit (10^9 M^{-1} s^{-1}). Using carbenium ions of varying stability, Mayr was able to build up a comprehensive reactivity scale, which is designed to consider carbenium ions with reactivity differences over 30 orders of magnitude in reactions with the same nucleophile [10]. The Mayr scale of reactivity gives a general idea about the stability and the possible use of different carbenium ions in chemical processes. Mayr was able to express this relationship in the logarithmic equation $\log k = s\,(E + N)$, with k as the rate constant, E representing electrophilicity, N representing nucleophilicity, and s representing specific nucleophilicity. From this equation, it is possible to measure the reaction rate of substrates covering more than 10 units of electrophilicity (E) or 14 units of nucleophilicity (N), and it is also possible to establish a "rule of thumb" for a successful reaction between a nucleophile and electrophile, allowing a rapid assessment of a nucleophile-electrophile combination to be made to see if a positive reaction is likely to occur [11]. Specifically, a successful reaction of nucleophile and electrophile will have an $E + N > -5$. The equation is easily understood and rationalized and can further be extended to predict the potential success of reactions that takes place in the direct nucleophilic substitution of alcohols. It is important to note that in this process the alcohols generate carbenium ions (Scheme 1) with the stability of this intermediate species representing an important factor. The different methodologies discussed in this chapter generate *stabilized* carbenium ions with the majority of the alcohols discussed in the literature being allylic, benzylic, benzhydrylic, or propargylic. The stability of a carbenium ion derived from an alcohol can be easily determined from a simplified Mayr scale, as depicted in Fig. 1.

The use of secondary and tertiary alcohols bearing only alkyl groups is quite uncommon in these transformations as the electrophilicity of the generated carbenium ion will be too high. Carbenium ions of high electrophilicity are not only difficult to generate

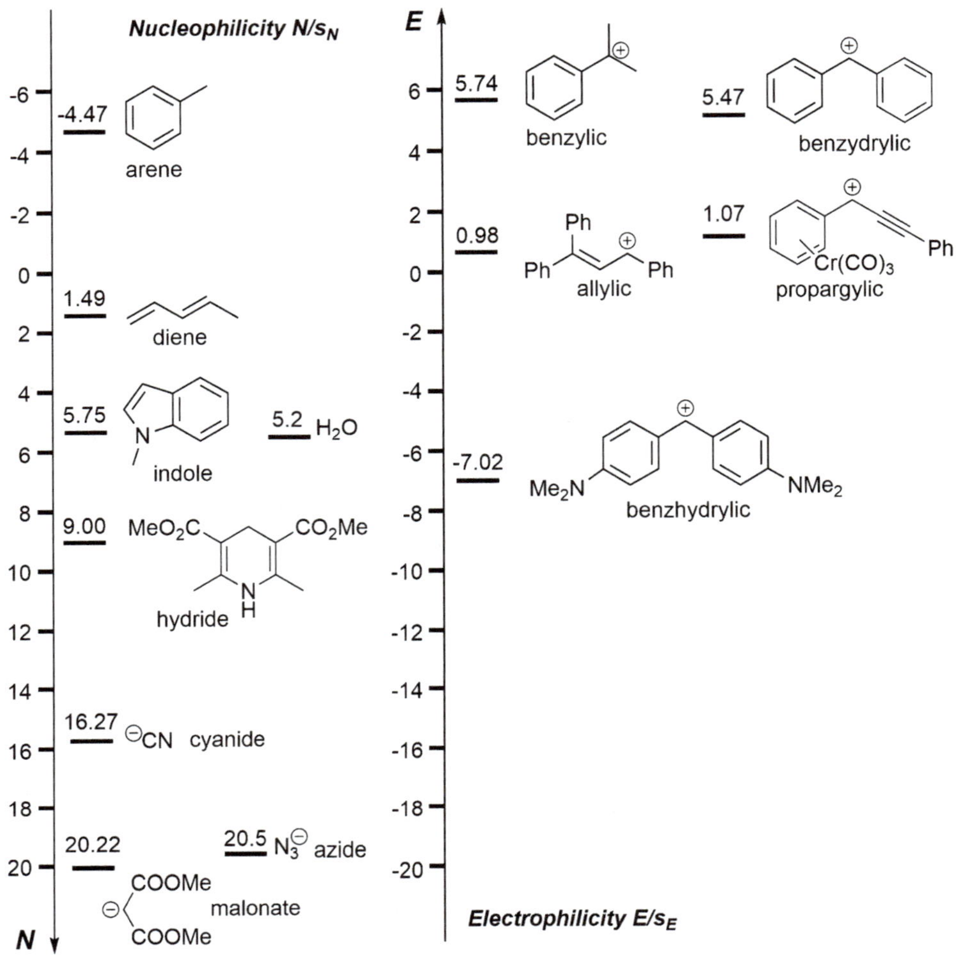

Fig. 1 Mayr's scale with common nucleophiles and selected carbenium ion generated by the action of Lewis or Brønsted acids on alcohols

but also are prone to side reactions such as rearrangements or eliminations. Additionally, the Brønsted and Lewis acids utilized to generate the carbenium ion along with the nucleophilicity (according to the Mayr equation), and concentration of the reaction partners also influences the likelihood of the reaction having a positive outcome. Normally, these considerations are not the key points of focus in reports on the direct nucleophilic substitution of alcohols, as authors are more interested in highlighting the advantages such as the mild reaction conditions, catalytic use of Brønsted or Lewis acid, and reaction scope. As discussed, carbenium ions are common intermediates generated in these reactions, and given this, we will try to discuss the recent developments in the field by unifying the types of alcohols and considering the various acidic conditions used for the generation of the reactive species. Initially, allylic, benzylic, benzhydrylic, and propargylic alcohols will be discussed together as starting materials with comments and discussion of the different acidic compounds used for the generation of the common intermediate. This forms a core area of research with often the same model reaction partners being described by various research groups, with different Brønsted or Lewis acids employed to promote the desired transformation.

2.1 Brønsted and Lewis Acids for the Generation of Carbenium Ions

The direct use of alcohols in S_N1-type reactions results in the formation of a stoichiometric amount of water as a by-product, therefore requiring the Brønsted or Lewis acid catalyst employed in this transformation to be both water-tolerant and not deactivated by the hydroxyl group. Kobayashi has nicely summarized the role and features of water-tolerant Lewis acids in organic transformations and was able to link two physical parameters to the successful use of such acids under these conditions [12]. Firstly, the stability constant (hydrolysis) of the Lewis acid needs to be sufficient to allow the activation of the hydroxyl group via coordination of the Lewis acid to the alcohol, and particularly oxophilic Lewis acids can facilitate this coordination event. Secondly, after formation of the carbenium ion and the successive reaction with the nucleophile, the Lewis acid needs to be stable in the presence of the water by-product to drive the catalytic cycle. In addition, the alcohol must be able to successfully out-compete water in terms of coordination to the Lewis acidic center, as coordination of the latter will diminish the efficiency of the Lewis acid. The overall degree of alcohol coordination is related to the exchange rate of water coordinated to the Lewis acidic center. To some extent, these considerations can be applied also to Brønsted acids with the acidity being particularly relevant, as this shifts the equilibria of the reaction to the protonated alcohol form. With Brønsted acids, it is also important to consider the basicity of the species generated by this protonation. As the Brønsted acids employed in direct nucleophilic substitution are normally highly acidic, the basicity of the anionic species is low. When the basicity of this species is increased, elimination and other side reactions could be favored. One further consideration arises when cationic Lewis acid catalysts, such as metal chlorides, triflates, or perchlorates, are utilized in the reactions, as these tend not to be stable in the presence of water being hydrolyzed to produce strong Brønsted acids often resulting in side reactions. In fact, this hydrolysis has raised questions as to whether a Lewis acid is the true catalyst in many S_N1-type reactions.

3 Classes of Nucleophiles for S_N1-Type Reaction of Alcohols: C, N, O, and S Nucleophiles

Generally, the direct nucleophilic substitution of alcohols allows a great diversity in the scope of nucleophiles that can be used, with acidic ketoesters, diketones, or malonates commonly employed as carbon-nucleophiles. With acidic substrates and particularly Lewis acids, it is possible to generate the metal enolate, which has been widely successfully employed in direct nucleophilic substitutions. In addition, the employment of potentially bidentate nucleophiles such as CN is sometimes described in these transformations, though less acidic organometallic compounds are not commonly used due to the competitive deprotonation of alcohols. Alkynes are employed when the Lewis acid is able to activate the alkyne toward the formation of the corresponding metal acetylide. Another class of carbon nucleophiles extensively used are aromatic or heteroaromatic compounds. The Mayr scale can be utilized to indicate the nucleophilicity of aromatic compounds, and both pyrrole and indole find widespread employment in a large number of direct nucleophilic substitutions given that they are 5–7 orders of magnitude more nucleophilic than aromatic compounds substituted with electron-donating groups EDG (EWG = electron withdrawing group) [10, 11]. Typically, azoles, pyridine, and aromatic compounds substituted with EWG groups are not suitable nucleophiles for direct substitution due to their reduced nucleophilicity. Nitrogen-nucleophiles have been extensively examined in direct nucleophilic substitution, as amines and their derivatives are important intermediates in synthesis. The direct nucleophilic substitution of alcohols to form amines represents a cornerstone transformation and has been described using highly nucleophilic azides or the less nucleophilic (though not basic) electron-deficient amine derivatives (carbamates, amides). Direct substitution with oxygen nucleophiles can be considered in order to avoid a deprotonation step, and low molecular weight alcohols (including acidic phenols) are sometimes utilized for this purpose. Acidic thiol (RSH) derivatives are also suitable for direct nucleophilic substitution chemistry, though care must be exercised to avoid oxidation of both the thiols and the highly thiophilic Lewis acids which lead to a significant reduction in yields. Samec has investigated the effect of the electrophiles and nucleophiles with different catalysts in the catalytic S_N1-type substitution of alcohols with varying degrees of activation using a series of C-, N-, S-, and O-centered nucleophiles [13]. As carbenium ions play a key role, the use of different catalysts does not produce a significant difference in reactivity, though the electrophilicity of the carbenium ion is a key factor. However, the catalyst employed is important as it exhibits different reactivity when varying the nucleophile. For example, Bi^{III}-Lewis acidic catalysts are highly efficient with all nucleophiles, whereas Fe^{III}- and Au^{III}-based catalysts show higher conversions for S-, C-, and N-centered nucleophiles, while Rh^{VII}-, Pd^{II}-, and Pt^{II}-derived systems are the catalysts of choice for O-centered nucleophiles though lead to by-products (typically ethers formed by the alcohols reacting with itself) when C, N, and S nucleophiles are employed. With some Lewis acids (e.g., those based on indium), certain ethers (e.g., allylic ethers) have been shown to form reversibly, though can revert to give the desired carbenium ion. However, if the ether is more stable, the formation of this by-product (which is favored with oxophilic Lewis acids) will reduce both the yield and the selectivity of the reaction. Although theoretical investigations have been conducted on the catalytic steps involved in the direct Lewis or Brønsted acid-promoted nucleophilic addition reaction [14], a clear and deep understanding of the reactivity trends involved mechanistically has not yet emerged.

4 Development of Direct Nucleophilic Substitution with C-Nucleophiles

4.1 Addition of Carbonyl Derivatives: Use of Lewis Acids

Many Lewis acids have been reported as promoters for the addition of malonates, ketoesters, and diketones to alcohols, the general scheme for which is indicated in Scheme 1. Considering oxophilic Lewis acids, the mode of activation of these catalysts involves the coordination of the hydroxyl group to the Lewis acid leading to the formation of a hydroxo–anionic complex. $Fe_2(SO_4)_3 \cdot H_2O$ on silica has been described as a novel and efficient catalyst for the direct nucleophilic substitution of alcohols under solvent-free conditions [15]. The reaction with benzhydrylic, benzylic, and allylic alcohols has been shown to give high yields of the desired compounds using either symmetrical or unsymmetrical diketones. By selecting *bis*(4-methoxyphenyl)methanol ($E = 0$), it was possible to confirm the intermediacy of the carbenium ion. Despite the electrophilicity of this carbenium ion, its stability is increased by the presence of SiO_2, and as such the cation's presence could be observed (as indicated by the red color) under an atmosphere of air for ca. 1 h. Many S_N1-type reactions of acidic diketones and ketoesters were described in our previous review [2], though it is interesting to report the possibility of using less acidic substrates for the direct α-alkylation, as described by Gu, who reported the direct nucleophilic substitution of aryl methyl ketones **2** with benzhydrylic alcohols **1** and xanthinol derivatives (Scheme 2) [16].

With chlorobenzene as the reaction solvent and $Fe(OTf)_3$ as the catalyst at high temperature (130 °C), it proved possible to use ketones **2** as substrates due to their equilibrium with their corresponding enol form, with yields for products **3** reported in the range of 34–70%. It is important to note that highly electrophilic alcohols cannot be used, as illustrated in the case of benzhydrols (with $E > 0$) where oxidation of the alcohol took place through an internal redox reaction, with the concomitant loss of a hydride.

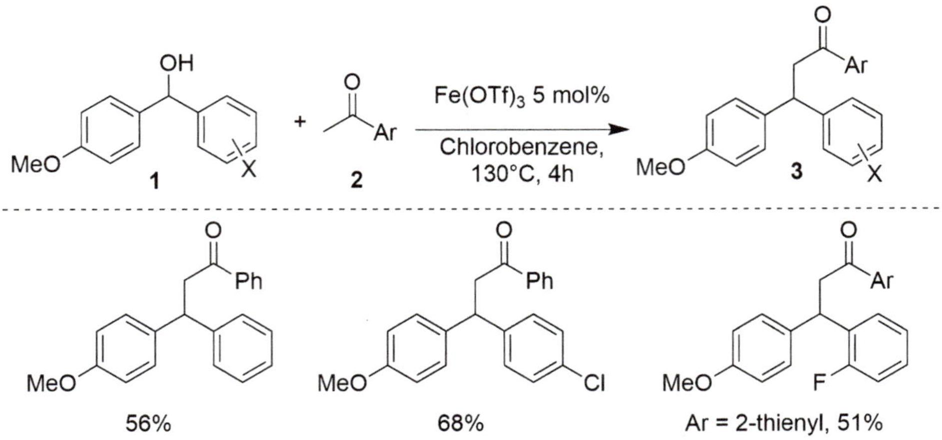

Scheme 2 Direct nucleophilic substitution of aryl methyl ketones with Benzhydrylic alcohols

4.1.1 Addition of Carbonyl Derivatives: Use of Brønsted Acids

The direct α-alkylation of unactivated ketones is also possible when Brønsted acids are employed as catalysts using benzylic alcohols **4** as the electrophiles (Scheme 3) [17]. The reaction is performed in triflic acid with trimethyl orthoformate as an additive, which plays an important role as it forms the corresponding methyl vinyl ether of the ketone, a nucleophilic species that intercepts the carbenium ion generated under the reaction conditions. Diverse aryl ketones **3** could be alkylated with diarylmethanols, cinnamyl alcohols, and phenyl propargyl alcohols of varying electrophilicities (Fig. 1).

4.2 Friedel-Crafts Reactions of Alcohols: Use of Lewis Acids

Numerous publications on the direct FC (FC = Friedel-Crafts) reaction have appeared since 2011, as this transformation gives rapid access to various important intermediates. Different Lewis and Brønsted acids have been described for the promotion of the FC reaction, with a diverse array of alcohol starting materials employed. Many recent works have described the use of boron-based Lewis acids for the generation of the carbenium ions. In addition, $Bi(OTf)_3$ has been used in the catalytic intramolecular FC reaction of tertiary alcohols to prepare disubstituted tetrahydronapthalenes, chromans, thiochromans, tetrahydroquinolines, and tetrahydroisoquinolines. The reactions were conducted in $CHCl_3$ at reflux and in general gave good yields [18]. More recently, an FC reaction involving benzylic alcohols with nucleophilic anilines has been reported [19] with Re_2O_7 catalyst (5 mol%) at 80 °C in CH_3CN providing good yields of the desired products. As herein non-protected anilines are employed as the nucleophile, questions can be raised about the nature of the effective kinetic adduct. The reaction first involves the formation of a mixture of the N-alkylated (the kinetically controlled product) as well as the C-alkylated products (the thermodynamically controlled product). Re_2O_7 is able to further activate the *N*-alkyl product thus

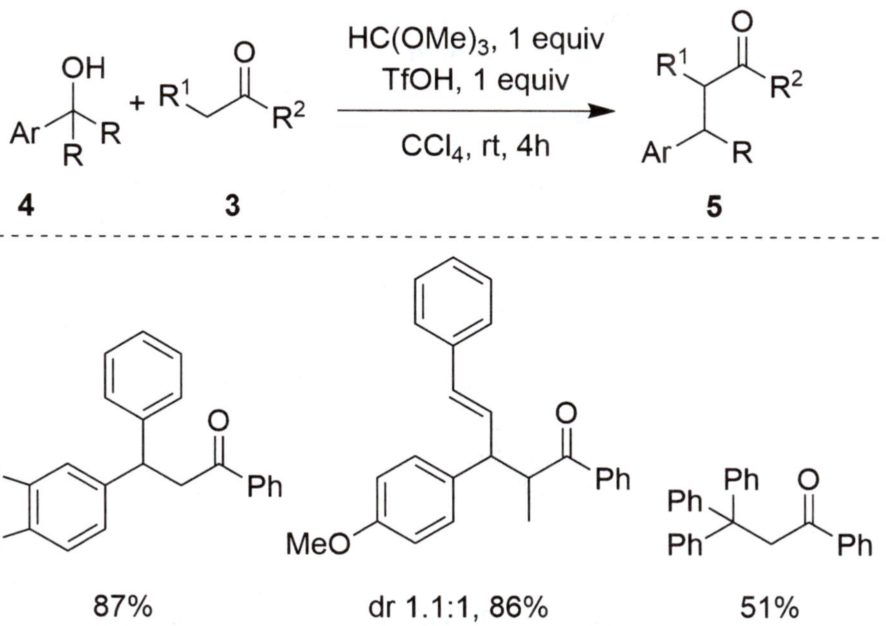

Scheme 3 α-Alkylation of unactivated ketones

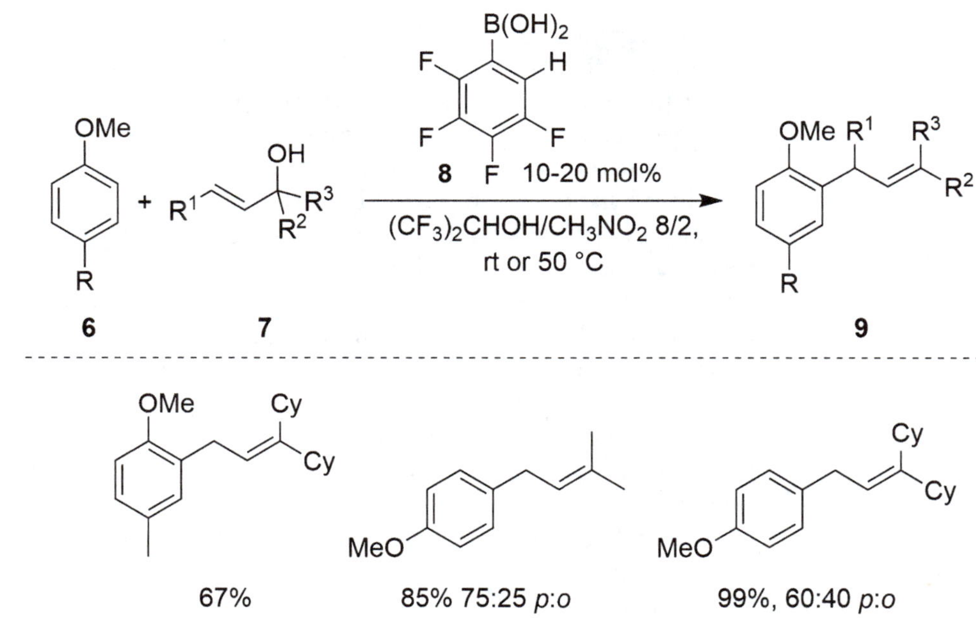

Scheme 4 Tetrafluorophenylboronic acid as an air- and moisture-tolerant catalyst for the reaction of alcohols

regenerating the carbocation. Bisai and co-workers conducted FC reactions of 3-hydroxy-2-oxindoles with electron-rich aromatics (phenols) in order to access the core structure of azonazine indole alkaloids [20]. A variety of Lewis acids were shown to successfully promote the ionization of the starting oxindole lea3ding to the FC product, with $In(OTf)_3$, $Cu(OTf)_2$, $Bi(OTf)_3$, and $Sn(OTf)_2$ all providing yields of greater than 85%. Several recent reports published on the direct nucleophilic substitution of alcohols in FC reactions have focused on the activation of allylic and propargylic alcohols. For allylic alcohols, Hall has reported a boronic acid catalyst, which was initially applied to a sequential Nazarov cyclization of divinyl alcohols/Diels–Alder cycloaddition for the preparation of densely functionalized bicyclic compounds in a highly diastereoselective fashion [21]. More recently, Hall has described the use of 2,3,4,5-tetrafluorophenylboronic acid **8** as an air- and moisture-tolerant catalyst for the reaction of both allylic **7** and benzylic alcohols with a variety of relatively non-nucleophilic arene compounds **6** (Scheme 4) [22].

It is hypothesized that the presence of hydrogen in the *ortho*-position of **8** could influence the stability of the transition state for hydroxyl-activation through hydrogen bonding and that the fluorine groups enhance the Lewis acidity. Propargylic alcohols have also been successfully employed in FC reactions with nucleophilic aromatic and heteroaromatic (indole and pyrrole) compounds using either $Ce(OTf)_3$ [23] or $SnCl_2$ [24] as the catalysts with nitromethane as the solvent.

4.2.1 Friedel-Crafts Reactions of Alcohols: Use of Brønsted Acids

An FC reaction promoted by TsOH was used by Heo and co-workers for the synthesis of laetevirenol A **10** (Scheme 5) [25]. Interestingly, this reaction involved the formation of the tertiary carbenium ion, which rearranged to the less stable benzylic carbenium ion, which in turn lead to the desired product. An α-amidoalkylation reaction, catalyzed by very low levels of HCl produced in situ from 1,1,2,2-tetrachloroethane at high temperature, was recently described by Dalla and co-workers [26]. This work draws into question the role played by Lewis acids given that chlorinated solvents are used at high temperature. The reaction is described in both an intramolecular and intermolecular fashion with a broad scope of nucleophilic partners. In particular, the skeletal framework obtained in the intramolecular FC reaction provides rapid access to useful intermediates for the synthesis of alkaloids.

Other highly electrophilic benzylic alcohols can be activated toward nucleophiles using the strong Brønsted acid, *o*-benzenedisulfonimide [27]. In this case, the reaction can be performed under solvent-free conditions using 10 mol% of the Brønsted acid, and importantly, it was possible to perform the alkylation with challenging highly electrophilic substrates ($E = 5$) and achieve yields equal to or higher than those described in the literature using Lewis acids. Finally, acidic medium behaving as a Brønsted acid has been for S_N1-type reactions. For example, allylic alcohols can be activated toward the direct nucleophilic substitution reaction using 1,1,1,3,3,3-hexafluoroisopropanol (HFIP) and 2,2,2-

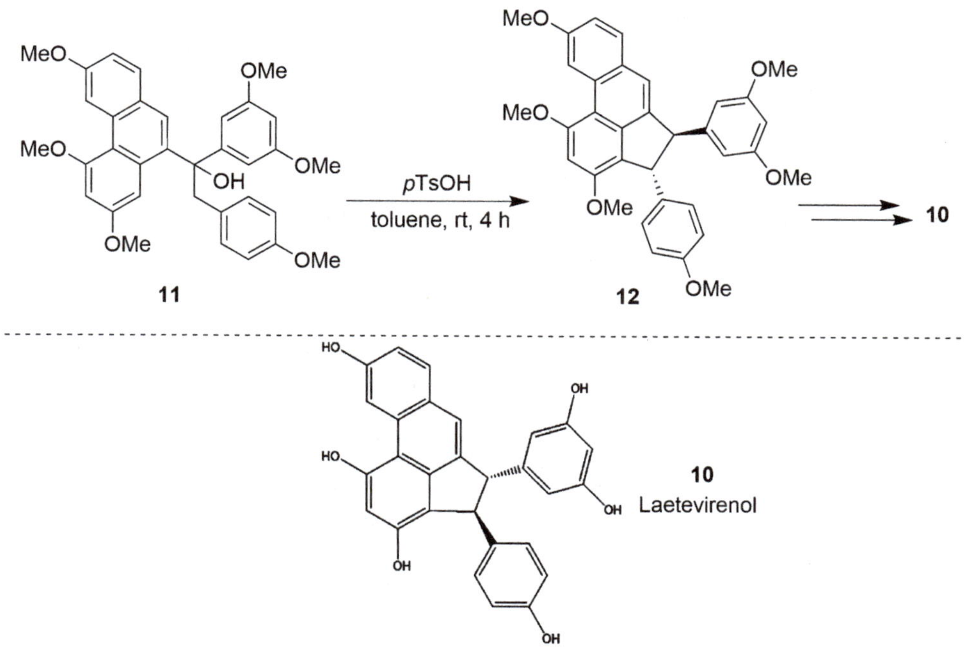

Scheme 5 Synthesis of a natural product by a Friedel-Crafts reaction of a tertiary alcohol

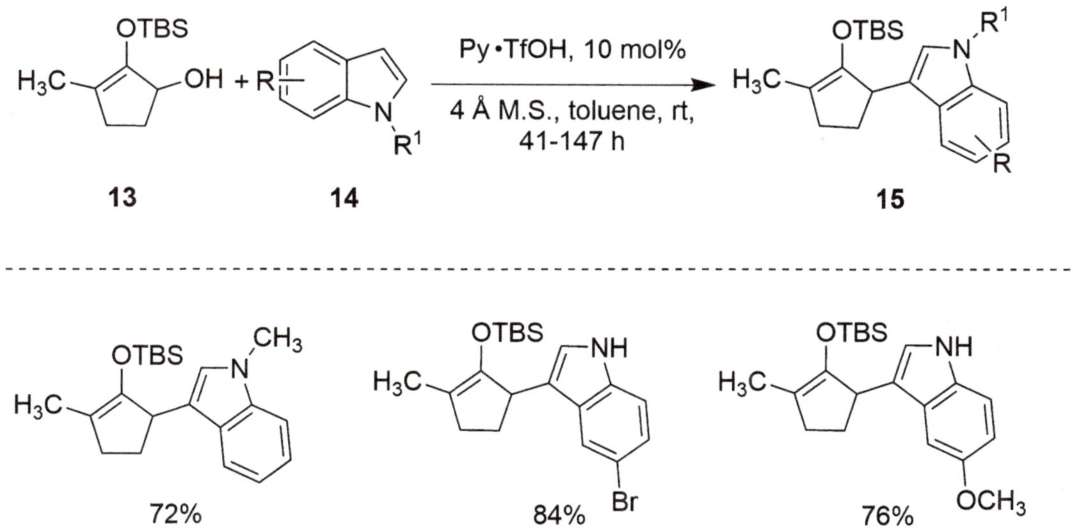

Scheme 6 A Facile formation of an oxallyl cation

trifluoroethanol (TFE) as reaction solvents [28]. This procedure is general and works under mild conditions with the key properties of HFIP and TFE for generating carbenium ions being disclosed by Mayr [29]. Essentially, the low nucleophilicity of the solvent ($N = 0$–1) combined with the acidity of the proton enables the facile generation of the carbenium ion possible. As well as the generality shown for the FC reaction of nucleophilic arenes and allylic alcohols, these reaction conditions are also suitable for the S_N1 reaction of different nucleophiles, and it is important to keep in mind that 1,3-dicarbonyl compounds (acetylacetone and Meldrum's acid) are also suitable substrates for this transformation. Cyclic silyl enol ethers **13** bearing a hydroxyl group in the α-′-position can be treated with Brønsted acids like pyridinium triflate to afford a stable oxallyl cation, which can undergo FC reactions with nucleophilic indole derivatives **14** (Scheme 6) [30].

4.3 Nitriles, Cyanide, and Other Nucleophiles

A simple iron catalyst has been described for a Ritter-type reaction allowing the synthesis of primary, secondary, and tertiary amides [31]. The catalytic system also permits the addition of other nucleophiles, such as phenylacetylene, to produce substituted phenyl ketones. The yields are good with benzylic alcohol substrates, and for reactions with yields below 40% under the standard conditions, it is possible to improve them using cationic Fe(III) complexes generated in situ in particular where the activation of secondary allylic alcohols is required (normally difficult substrates for S_N1-type reactions). Zhou has reported the preparation of functionalized 2-oxindoles by the Ritter reaction using $HClO_4$ (10–20 mol%) in the presence of alkyl, vinyl, or aryl nitriles to give amides [32]. The reactions of nitrones with allenyl cations generated by triflic acid or by $Sc(OTf)_3$ were reported by Liu and co-workers in the preparation of dihydro-β-carbolines [33].

5 Development of Direct Nucleophilic Substitution with N-Nucleophiles

In our previous review, we reported the use of many Lewis and Brønsted acidic conditions for the S_N1-type directed nucleophilic substitution of alcohols, using 4-nitroaniline, tosylamide, methyl and benzyl carbamate, simple amides, and *tert*-butoxycarbonyl (Boc) amines as substrates [2]. An update on the recent progress both in terms of the conditions used and the scope of nucleophile-electrophile combination will be illustrated here.

5.1 The Use of Lewis Acids with N-Nucleophiles

Although various expensive Lewis acids have been reported for the direct amination of alcohols, it has been demonstrated that $Al(OTf)_3$ is a reliable, cheap alternative catalyst for this transformation [34]. The scope of the reaction is broad with allylic-, benzylic-alcohols, and benzhydrols all being shown to be suitable substrates to participate in the reaction with various amides. The reaction, typically performed in CH_3NO_2 with 1 mol% of the catalyst at room temperature is extremely fast (max 1 h) and provides the desired products in good to excellent yields. A particularly notable feature of this transformation is the possibility to use highly electrophilic carbenium ions derived from benzhydrols ($E = 1$–4). The more stable ferrocenyl alcohols ($E = -3.5$) have been employed in

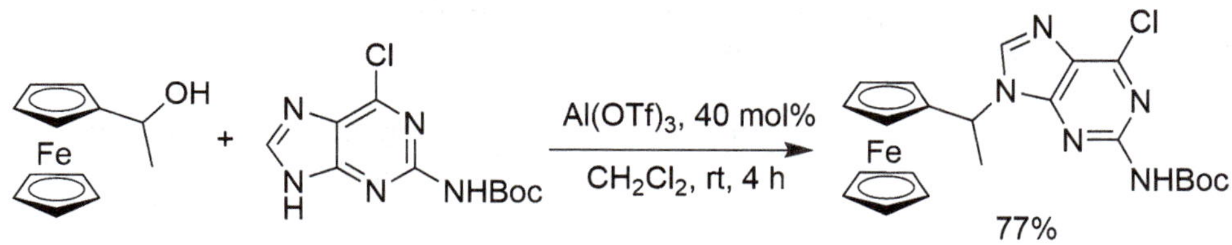

16 **17** **18**

Scheme 7 Direct nucleophilic reaction of a guanine derivative with ferrocenyl alcohol

the synthesis of guanine and guanosine derivatives by a direct nucleophilic substitution of either racemic or enantioenriched ferrocenylethanol derivatives in the presence of the same catalyst, (Al $(OTf)_3$) [35]. The lower Lewis acidity of $Al(OTf)_3$ allowed the direct introduction of ferrocenylethanol **16** at the 9-position of the Boc-protected guanine derivative **17**. Through the judicious choice of Lewis acid derivatives (indium vs. aluminum), it was also possible to use Boc protected guanine derivatives (Scheme 7).

Another suitable Lewis acid for the direct introduction of N-based nucleophiles to alcohols is iron(III). Najera has investigated the role played by iron in these reactions [36, 37] with an investigation using allylic alcohols in water at 80–90 °C. It is remarkable that the direct substitution reaction takes place in water (or "on water"), as the carbenium ion formed preferentially reacts with the nucleophilic species present forming the desired amine derivative, which is stable under the reaction conditions. The hydrate of iron has been demonstrated to be the active catalyst in these reactions. Re_2O_7 is also capable of activating alcohols in direct nucleophilic reactions [38], and various amides have been successfully used with both benzylic and propargylic alcohols. A typical protocol uses CH_2Cl_2 as the reaction solvent at room temperature, with catalyst loadings of less than 2% used. The Zhan group has developed a method to prepare acrylonitriles via a domino reaction involving initially propargylic substitution followed by rearrangement using $FeCl_3$ (10 mol%) as the Lewis acid catalyst [39]. The transformation of benzylic alcohols into the corresponding azides [40] (using 1 mol% of $Bi(OTf)_3$ in CH_2Cl_2—caution when using azides, and especially in solvents such as CH_2Cl_2) or into the corresponding sulfonamides (by the use of BF_3) has also been demonstrated [41].

5.2 The Use of Brønsted Acids with N-Nucleophiles

Many examples of the use of Brønsted acids in direct substitution with N-nucleophiles have been reported in our previous review [2]. However, a recent report has described the use of functionalized sulfonic acid ionic liquids as efficient catalysts for the direct amination of alcohols [42].

6 Development of Direct Nucleophilic Substitution with O-Nucleophiles

The direct nucleophilic substitution of alcohols can be used to access a variety of different cyclic and acyclic ethers, and either Lewis or Brønsted acids can be used to promote these transformations.

6.1 The Use of Lewis Acids with O-Nucleophiles

More recent results reported in this area rely on the use of aluminum-, boron-, or gold-based Lewis acids as catalysts, with a typical example being the use of $Al(OTf)_3$ as a new efficient catalyst for the direct nucleophilic substitution of ferrocenyl alcohols reported by Mamane as a useful method for the synthesis of ferrocene-containing PEG (polyethylene glycol) derivatives [43]. 2,2-Disubstituted-2*H*-chromenes and 2,2-disubstituted-2*H*-1-benzopyrans are present in many natural products and pharmaceutically relevant molecules demonstrating a broad range of activity, including antitumor, anti-HIV, antibacterial, and antifungal. Propargylic alcohols can be used for the formation of chromenes in the presence of BF_3 [44] with the reaction involving an initial direct nucleophilic substitution of a phenol with the propargyl carbenium ion, which is formed by reaction with 15 mol% of BF_3 in CH_2Cl_2 at room temperature, followed by ring closure through a Claisen rearrangement to the desired chromene. Although primary and secondary propargylic alcohols are relatively unreactive substrates for this transformation, tertiary propargylic alcohols react with numerous differently substituted phenols to provide the chromenes in reaction times ranging from 5 to 7 h. Moran has reported the use of $B(C_6F_5)_3$ in the formation of both cyclic ethers and pyrrolidines **20** (Scheme 8) [45].

In this reaction, in addition to using allylic and propargylic alcohols as suitable substrates, it was also demonstrated that aliphatic primary alcohols could be activated though at higher temperatures (100 °C in CH_3NO_2). Normally cyclization of primary aliphatic alcohols occurs only when stoichiometric amounts of dehydrating agents or superacids are used at significantly elevated temperatures (>160 °C). Iron Lewis acids have also been shown to be compatible for direct O-substitution reactions with the use of $FeCl_3$ (25 mol%) in dichloroethane at 80 °C reported for the direct nucleophilic displacement by alcohols in the formation of benzoin derivatives [46], while 2*H*-chromenes can be obtained in a similar manner by iron(III) complex-catalyzed cyclization of 2-(1-hydroxyallyl) phenols [47]. An interesting mechanistic analysis was disclosed by Widenhoefer [48] using gold carbene complexes in the direct nucleophilic addition of alcohols to allylic alcohols to form allylic ethers. These regio- and stereo-specific transformations were shown to take place exclusively via γ-addition *syn* to the departing hydroxyl group of the nucleophile on the double bond, followed by Au-complexation of the allylic alcohol. The successive elimination

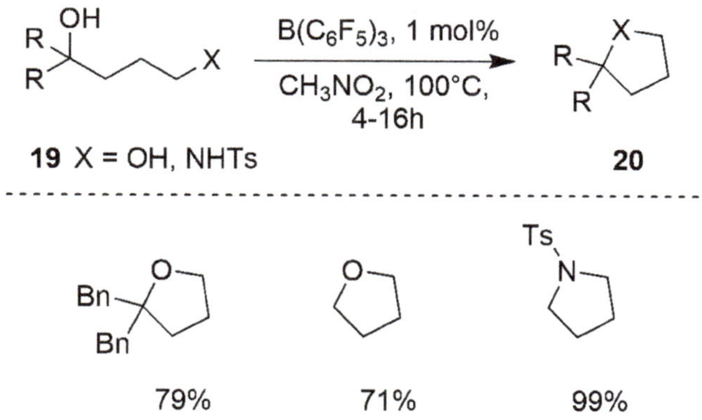

Scheme 8 Synthesis of cyclic ethers and pyrrolidines by the use of a boron-based Lewis acid

of the hydroxy group and de-complexation of Au then provides the *E*-olefin. It is important to consider this mechanistic study when new transformations based on either Au(I) or Au(III) Lewis acids are studied.

6.2 The Use of Brønsted Acids with O-Nucleophiles

In the S_N1-type direct reaction of nucleophiles with alcohols, nucleophilic solvents like water are often avoided though interestingly it has recently been shown that if the nucleophilicity of the nucleophiles present in the reaction mixture is greater than the nucleophilicity of the solvent, the desired reaction will occur much faster than the competing reaction of the solvent (though naturally the relative concentration of nucleophile and solvent must also be considered). Therefore, it is possible to carry out many S_N1-type transformations in water. For example, hot water displays acidic properties and has been reported to efficiently promote the intramolecular direct nucleophilic substitution reaction of unsaturated alcohols to form cyclic ethers [49], as well as the 1,*n*-rearrangement ($n = 3, 5, 7, 9$) of allylic alcohols taking advantage of the direct nucleophilic substitution of solvent water [50].

7 Non-stabilized Carbenium Ions from Alcohols: The Use of Calcium-Based Lewis Acids

Niggeman and co-workers have published several interesting developments for the direct nucleophilic activation of alcohols using calcium Lewis acids [51]. Not only is calcium inexpensive, non-toxic, and abundant (as the fifth most frequent element in the Earth's crust), but Lewis acidic calcium complexes also display remarkable properties and have the ability to activate numerous alcohols that cannot be activated by other common and widely employed Lewis acids. In particular, calcium complexes can be used with secondary and tertiary alcohols leading to carbenium ions that are so electrophilic; they are not incorporated on the Mayr scale. Typically, alcohols are activated by calcium complexes surrounded by weakly coordinating, non-basic anions such as TfO^-, F^-, or 1:1 combination of triflimidate $(NTf_2)^-$ and hexafluorophosphate $(PF_6)^-$ as these salts are particularly stable to both air and moisture. The Lewis acid catalyst formed by the combination of $Ca(NTf_2)_2$ and Bu_4NPF_6 has proven to be highly active for S_N1-type reactions of alcohols, operating under mild conditions at room temperature. It has been suggested [51] that calcium is particularly active in these transformations due to its ability to coordinate oxygen though another interesting property is its ability to enhance the activity of acids. When the carbenium ion is formed from the alcohol in these transformations, the alkoxide ion coordinating to the calcium atom is basic and able to act as strong base while calcium hydroxide also acts as a base and restores the calcium complex through the liberation of water. The combination of oxophilicity, Lewis acid activity, and basicity of the coordinate hydroxide makes the calcium Lewis acid particularly suitable for use in S_N1-type reactions, and although primary alcohols are unsuitable substrates, the S_N1-type reactions with calcium ions have broad scope and unique reactivity activating many substrates that are not activated under more conventional conditions. Again considering the Mayr scale to compare reactivity of the corresponding carbenium ions generated, it is easy to discern that alcohols located above $E = 3$ or 4 scale can be used in S_N1-type reactions (Fig. 1). Several key findings and applications of calcium-catalyzed S_N1-type reactions were summarized in our previous review article in 2011 [2]. A further example is presented in the direct functionalization of alkynes with carbon-centered electrophiles. After the nucleophilic attack of the alkyne, a vinyl cation intermediate is generated, which can be intercepted by a water molecule, and after keto-enol tautomerization, the corresponding ketone is generated. Alcohols are also suitable substrates for this transformation, with water being generated in the activation step. Taking advantage of this, Niggemann recently disclosed a carbohydroxylation of alkynes with allyl and propargyl alcohols using a calcium Lewis acid at 5 mol% loading [52]. To stabilize the intermediate carbenium ion and to promote the transformation, a weakly Lewis basic electron-pair donor additive, cyclopentanone, was also used. These conditions were shown to be highly efficient for the stabilization of either allyl- or vinyl-cationic intermediates in combination with a $Ca(NTf_2)_2$ and Bu_4NPF_6-based catalyst system with a cascade reaction based on this concept being developed, involving the formation of a vinyl cation that is intercepted by a series of arene nucleophiles (Scheme 9) [53].

Recently, an interesting transformation was described for the formal intermolecular [2 + 2 + 2] cycloaddition reaction of enynes

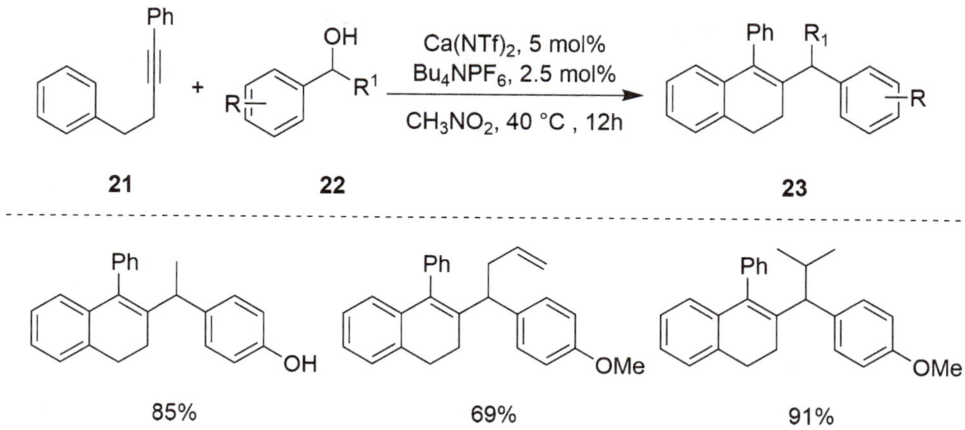

Scheme 9 Combination of $Ca(NTf_2)_2$ and Bu_4NPF_6 as active catalysts for the activation of highly electrophilic alcohols

involving the formation of a propargylic cation (which can also be considered an allenyl-cation), a calcium-catalyzed dehydration, followed by nucleophilic attack of an olefin. The reaction proceeded with excellent regio- and chemoselectivity, producing bicyclic building blocks containing up to three stereocenters in a highly diastereoselective manner [54].

8 Complex Rearrangements and Cascade Reactions: Aza- and Oxa-Piancatelli Cascade Reactions

Alcohols are involved in the well-established Pinacol and Prins-Pinacol rearrangements, and these transformations have been used in numerous complex syntheses, with the alcohol starting materials often being treated with Brønsted or Lewis acids. Unprotected glycals can also be considered as useful starting materials for allylic S_N1-type reactions as they readily undergo catalyzed substitution reactions at C-1, accompanied by allylic rearrangement, with this reaction commonly known as the Ferrier rearrangement or the Ferrier Type-I reaction [55].

The classical Meyer–Schuster rearrangement involves heating an alcohol with a strong acid though these somewhat harsh conditions are often incompatible with the presence of a number of functional groups [56]. However, this transformations has recently been described under milder conditions using either Lewis or Brønsted acids enabling a wide range of primary, secondary, and tertiary propargylic alcohols to undergo the Meyer–Schuster rearrangement to give enones at room temperature typically using Au (I)-based catalysts in the presence of small quantities of either MeOH or 4-methoxyphenylboronic acid (Scheme 10) [57].

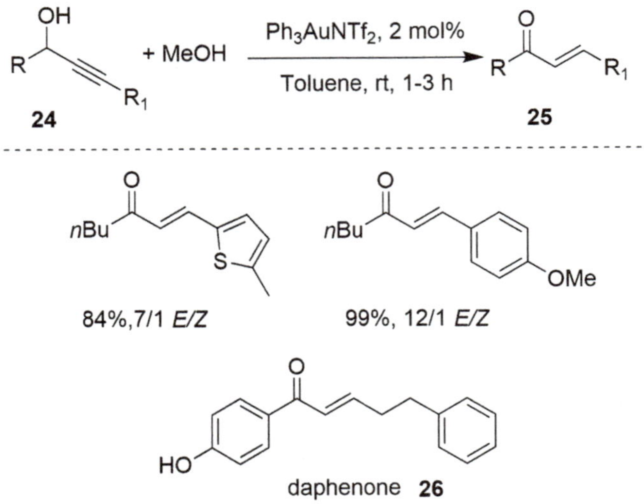

Scheme 10 Meyer–Schuster rearrangement to give enones at room temperature

This transformation enabled the synthesis of the enone-containing natural product daphenone **26**, and in addition, the rearrangement of primary propargylic alcohols was combined in a one-pot procedure with various nucleophiles to give access to β-aryl, β-alkoxy, and β-sulfido ketones. A similar transformation was described by Yanada using $Bi(OTf)_3$ [58] and Amberlyst-15 immobilized in a [Bmim][PF_6] ionic liquid [59]. In another example of a reaction mediated by an initial Meyer–Schuster rearrangement [60], various 1,1-diaryl-2-propyn-1-ols are coupled with indan-1,3-dione. The reaction proceeds in water under microwave irradiation in the presence of catalytic amounts of the water-tolerant $InCl_3$. Allenes are also versatile synthons in organic chemistry that can be obtained in a facile manner reacting a Au (III)-cationic complex with propargylic alcohols in the presence of nucleophilic arene compounds. Similar to the Mayer-Schuster rearrangement, the propargylic cation generated under the reaction conditions is in equilibrium with the allenylic cation, which is intercepted by the nucleophilic aromatic compound in an FC-type reaction. When furylcarbinols are treated with acids, a rearrangement can take place, giving access to 4,5-substituted cyclopentenones, and this transformation was recently reviewed [61] along with other rearrangements that involve propargylic alcohols, in which carbenium ions are formed in the presence of Lewis acids capable of activating alkynes (Au, Ag, Yb) [62]. Common to all these transformations is the fact that after the activation step of the carbenium ion generates the allenyl cation, which is then attacked by either an internal or external nucleophile, often in a FC-type reaction leading cyclic or bicyclic substrates. The Piancatelli rearrangement has been exploited to allow access to a series of novel aza-spirocycles and spirocyclic ethers [63]. For example, 1,2-Dihydropyridine-5,6-dicarboxylates were efficiently obtained by a cascade reaction in which primary amines and 2-butynedioate are reacted with carbenium ions obtained from propargylic alcohols [64] with allenic carbocations being key intermediates in this reaction, being intercepted by the nucleophilic enamines which are formed in situ. An interesting series of reactions was described by Yeh and co-workers in which homoallylic cations reacted with an adjacent tethered-alkyne (Scheme 11) [65].

As seen in Scheme 11, alcohol **27** is reacted with $FeCl_3$, with the cationic promoted ring-opening reaction producing a homoallylic cation that undergoes ring-expansion and trapping with chloride leading to the desired product **28**. In a similar manner, the Nazarov rearrangement reaction can be performed as a type of cyclization reaction, with a variant (although not formally a Nazarov cyclization) recently described in which allylic alcohols undergo $FeCl_3$-catalyzed cyclization to form functionalized indenes [66].

9 Stereoselective Reactions

Stereoselective reactions using alcohols as intermediates have several advantages, as in particular functionalization or activation steps are avoided. In the direct nucleophilic substitution of alcohols, carbenium ion intermediates are generated. Due to their planar structure, nucleophiles can react at either of the two pro-stereogenic faces, and in order to react at one face preferentially, three strategies have been developed (Fig. 2).

Firstly, if the alcohol contains a stereogenic center, the formation of a preferential diastereoisomer is possible, whereas other strategies consider the formation of an intimate chiral pair [67]. The third strategy was developed in recent years with the renewed interest in organocatalytic activation modes, and this has been shown to be particularly successful with the development of a large family of chiral Brønsted acids [68]. This approach also considers the formation of a chiral nucleophile using a promoter in sub-stoichiometric amounts. For example, chiral enamines (can be generated in situ employing chiral primary or secondary amines) subsequently react with the carbenium ions generated from alcohols in a stereoselective fashion. In our previous review, this approach was introduced, highlighting the methodology, scope, and possibilities offered by such a strategy, and as such herein, the focus will be on summarizing more recent results [2].

9.1 Diastereoselective Reactions with Chiral Alcohols

After carrying out a detailed investigation on the reaction of chiral benzylic cations [69], Bach reported on a series of tyrosine derivatives focusing on the stereo-controlled addition of nucleophiles to the carbenium ion, generated using HBF_4 (3 equiv.). The protecting groups on the tyrosine were varied, while the

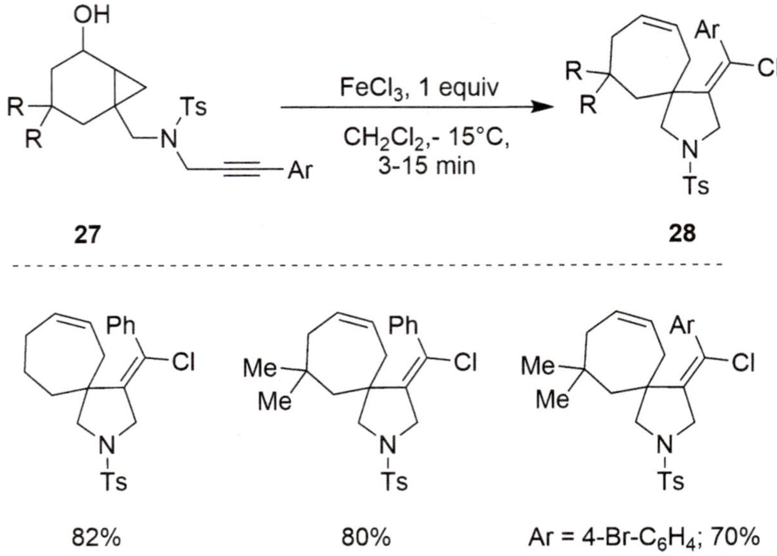

Scheme 11 Reaction of homoallylic cations with an adjacent tethered alkyne

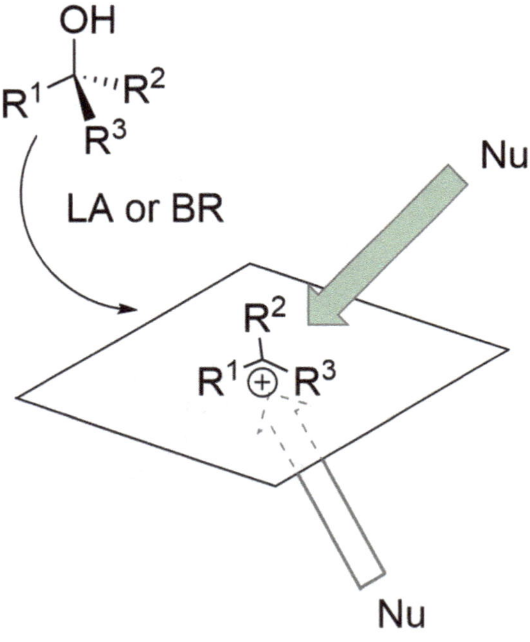

Fig. 2 Generation of a new stereocenter in the reaction involving the activation of alcohols by Lewis or Brønsted acids

diastereoselectivities obtained in the FC reaction examined. The reactions were shown to be fully stereo-convergent and proceeded under kinetic control. In order to explain the observed stereoselectivity, the proposed model suggested a conformationally locked benzylic cation intermediate, in which the conformation is controlled by allylic strain. By introducing specific protecting groups, it

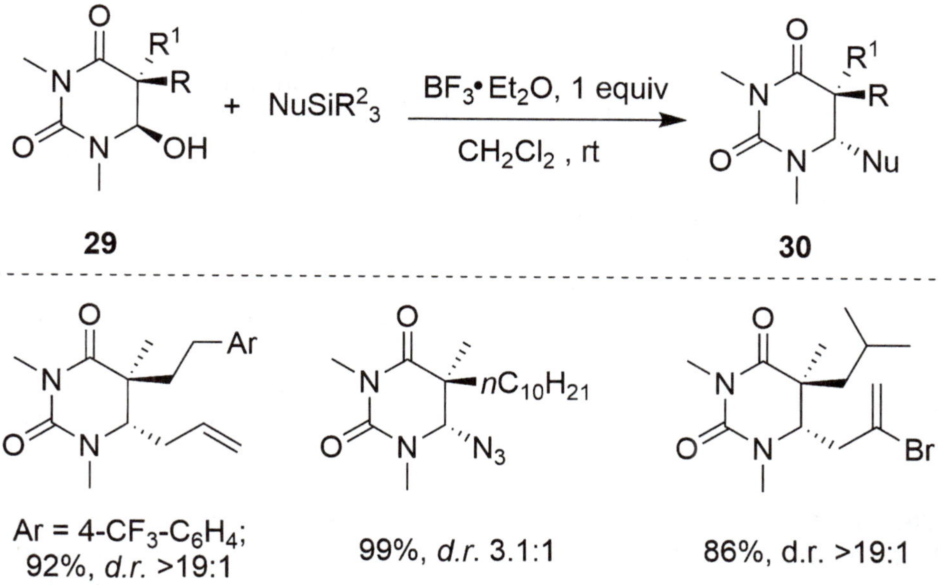

Scheme 12 Acyliminium ions obtained from barbituric acids by activation of alcohols

was possible to achieve high levels of stereocontrol in the FC reaction, employing various nucleophilic aromatic substrates. A similar pattern was exploited by Hanessian in order to access the 2-amino-1,1-diarylalkanes chiral unit, through introduction of a phenylamine moiety [70]. In this report, the starting materials are electron-rich benzylic alcohols carrying adjacent α-nitro or α-azido groups with the reaction promoted by either $AuCl_3$ or $Bi(OTf)_3$. In both cases, a *syn*-diastereoisomer was kinetically favored based on the minimization of allylic strain in the intermediate resonance-stabilized benzylic carbenium ions on which the nucleophilic attack takes place.

N-Acyliminium ions are important intermediates in synthesis and are readily accessible by a number of methods [71]. For example, acyliminium ions can be obtained from stable α-amino alcohols derived from barbituric acids (uracils) and have been utilized in the direct nucleophilic displacement of alcohols to produce useful uracil derivatives **30** (Scheme 12) [72, 73].

This methodology relies on an initial chemoselective mono-reduction of barbituric acids using a SmI_2–H_2O complex, after which the stable amino alcohols obtained can be reacted with several nucleophiles in the presence of a Lewis acid. The reaction features a very broad substrate scope and proceeds with excellent stereoselectivity (dr up to >95:5), which is primarily a function of the stereogenic center present in the starting compounds. Among all the Lewis acids tested, BF_3 gave high yields, excellent stereoselectivity, and a broad reaction scope with regard to the nucleophile employed.

A diastereoselective alkylation of chiral nitro derivatives was reported by Cozzi and co-workers [74]. Given that the simple alkylation of nitro derivatives is known to be difficult as normally electrophilic reagents have the intrinsic preference for attack on the oxygen atom of the nitronate anion, therefore α-alkylation of such species is quite rare. In this study, trifluoroethanol, a non-nucleophilic and acidic alcohol [29], acts as both the reaction medium and activator for the formation of carbenium ions from various benzylic, benzhydrylic, and propargylic alcohols with enantio-enriched substrates obtained from nitro-Michael or Henry reactions utilized as the starting materials. Various novel chiral pyrrolidines, which have been shown to be useful secondary amines in the promotion of organocatalytic reactions, can be readily obtained using this methodology.

9.2 Stereoselective Reactions with Chiral Counterions

In 2012, the Gong group described the highly enantioselective substitution reaction of 3-substituted 3-hydroxy-oxindoles **31** with ene-carbamates using 10 mol% of chiral phosphoric acid **32**, allowing access to enantioenriched oxindoles **33** in 72–92% yield with 90–96% ee (Scheme 13) [75].

In the same year, the use of a series of simpler cyclic and acyclic ketones as nucleophiles was reported by Guo and Peng [76] in the direct α-alkylation of 3-indolyl 3-hydroxy-3-indolyloxindoles. The ability of 3-hydroxy-indole to readily form carbenium ions is largely due to the high stability of the resulting species, and although their electrophilicity has not been fully investigated, these diarylcarbenium tetrafluoroborates containing *N*-heteroarenes are bench stable and can be isolated [77]. In a similar manner, 3-indolyl 3-hydroxyoxindole fragments can be used for the formal asymmetric [3 + 3] cycloaddition with an azomethine ylide generated in situ, with the nucleophilic species being formed through an interaction of a phosphoric acid with the activated alcohol [78].

Allylic alcohols can be alkylated through the activation of leaving groups such as halides, carbonates, or acetates in Tsuji-Trost type-reactions [79], though the stability of the carbenium ions derived from allylic substrates can be also exploited in the presence of Brønsted acids or Lewis acids, as we have briefly illustrated in previous sections. Chiral nucleophiles have also been utilized in the presence of Lewis acids or transition metals in order to facilitate the formation of π-allylic cationic species from allylic alcohols [80], though due to their relative high electrophilicity, in order to allow the reaction to proceed, highly nucleophilic species need to be employed. Chiral phosphoric acids of moderated acidity [81], which are capable of catalyzing a variety of such reactions [80], are also able to protonate the allylic alcohol to generate a chiral ion pair, and this was exploited in a highly enantioselective organocatalytic intermolecular allylic alkylation of allylic alcohols with 1,3-dicarbonyls described by Gong (Scheme 14) [82].

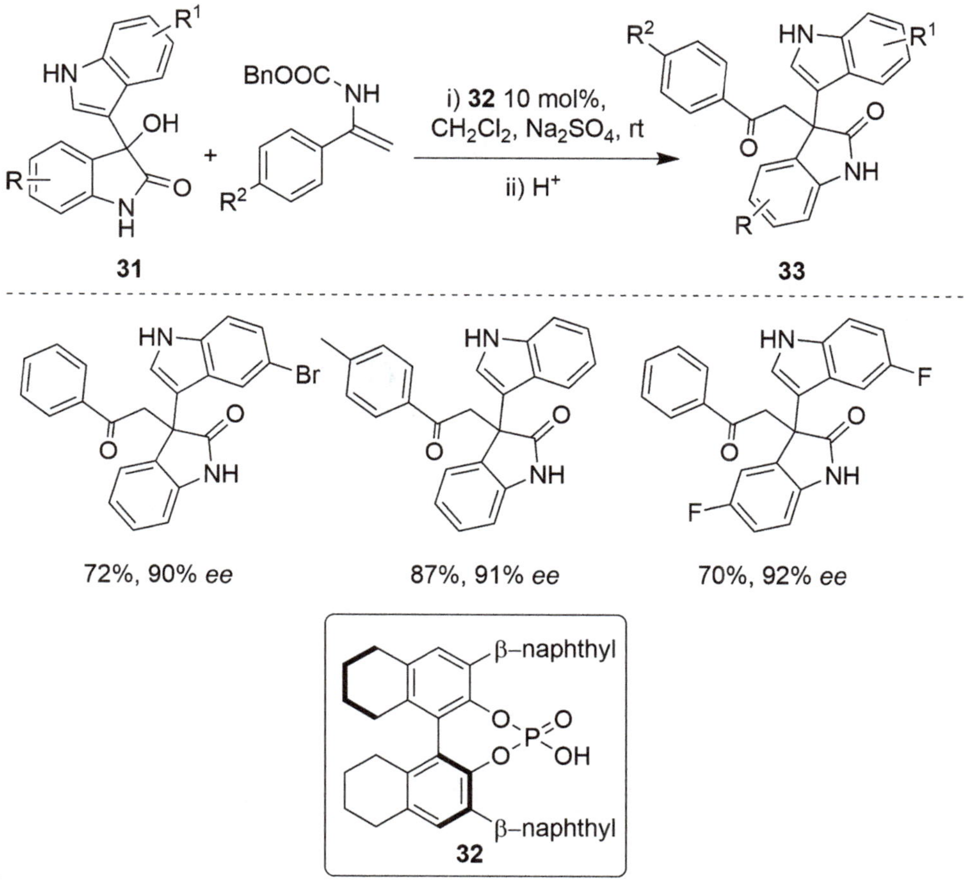

Scheme 13 Synthesis of enantioenriched oxindoles

In this study, cyclic and acyclic diketones **35** gave very high enantiomeric excesses when the reaction was performed in DCE at −20 °C, while acetoacetate also gave good results. This reaction was exploited in the total synthesis of hydroxymetasequirin-A tetramethyl ether diacetate (9 steps, 37% overall yield), though has only been described for the less electrophilic allylic cations. The use of less electrophilic alcohols in the enantioselective alkylation of carbonyls was described by Najera and Baeza [83], who utilized a dual hydrogen bond activation in an S_N1-type reaction for methylene compounds, using benzydrylic alcohols as the alkylating agents and employing *bis*(2-aminobenzoimidazole) **40** in combination with an achiral Brønsted acid (TFA or TfOH) as the catalysts (Scheme 15).

In this case, enantiomeric excesses of up to 90% were obtained using various alcohols, and it is interesting to note that the methodology was also applicable to allylic alcohols, although the results obtained were not as good.

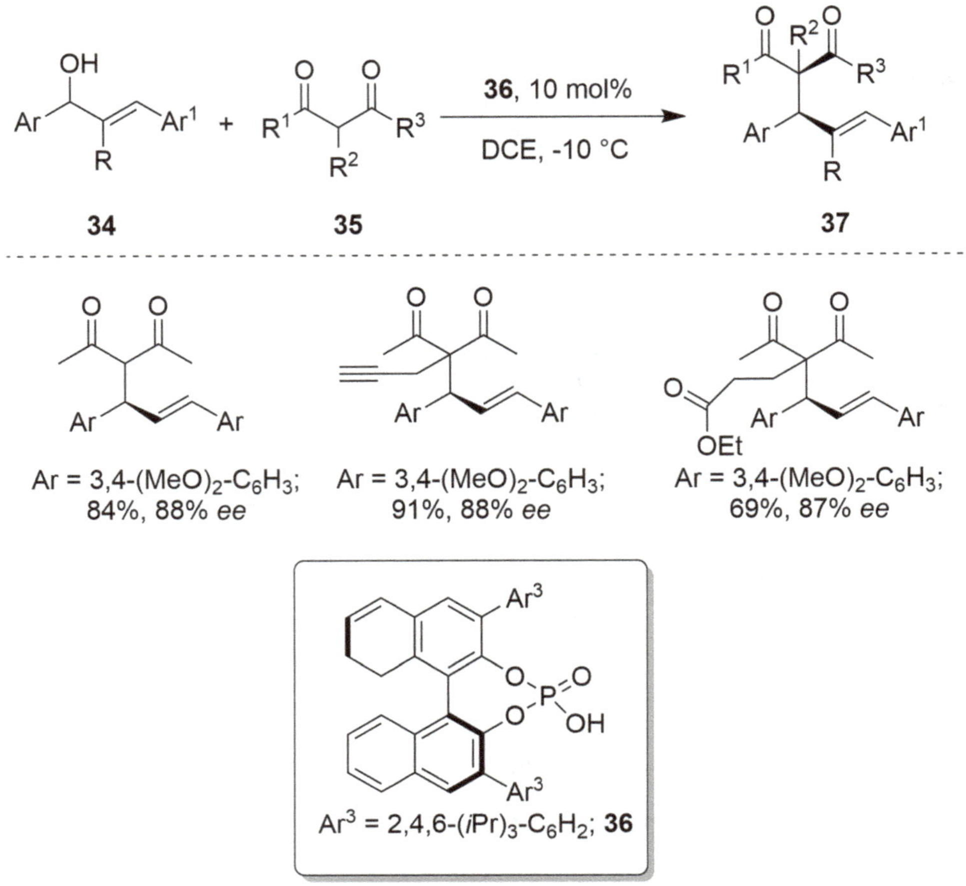

Scheme 14 Intermolecular allylic alkylation of allylic alcohols with 1,3-dicarbonyl compounds

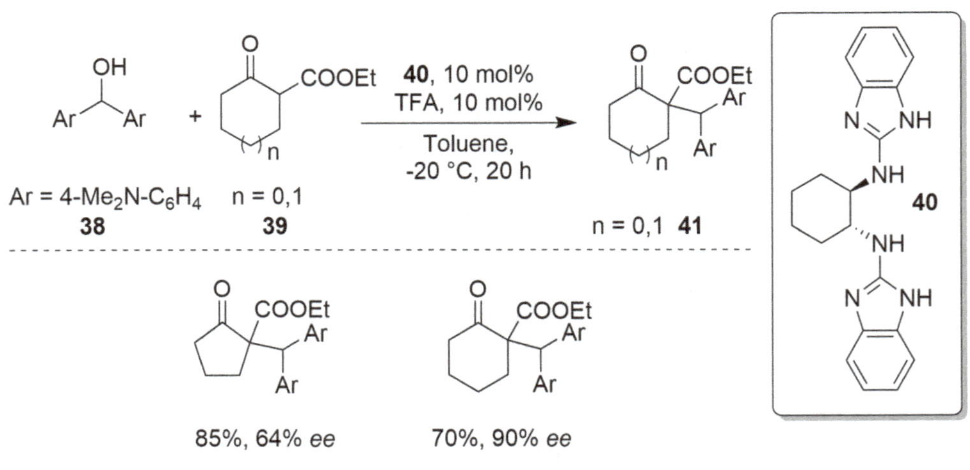

Scheme 15 A dual hydrogen bond activation in an S_N1-type reaction

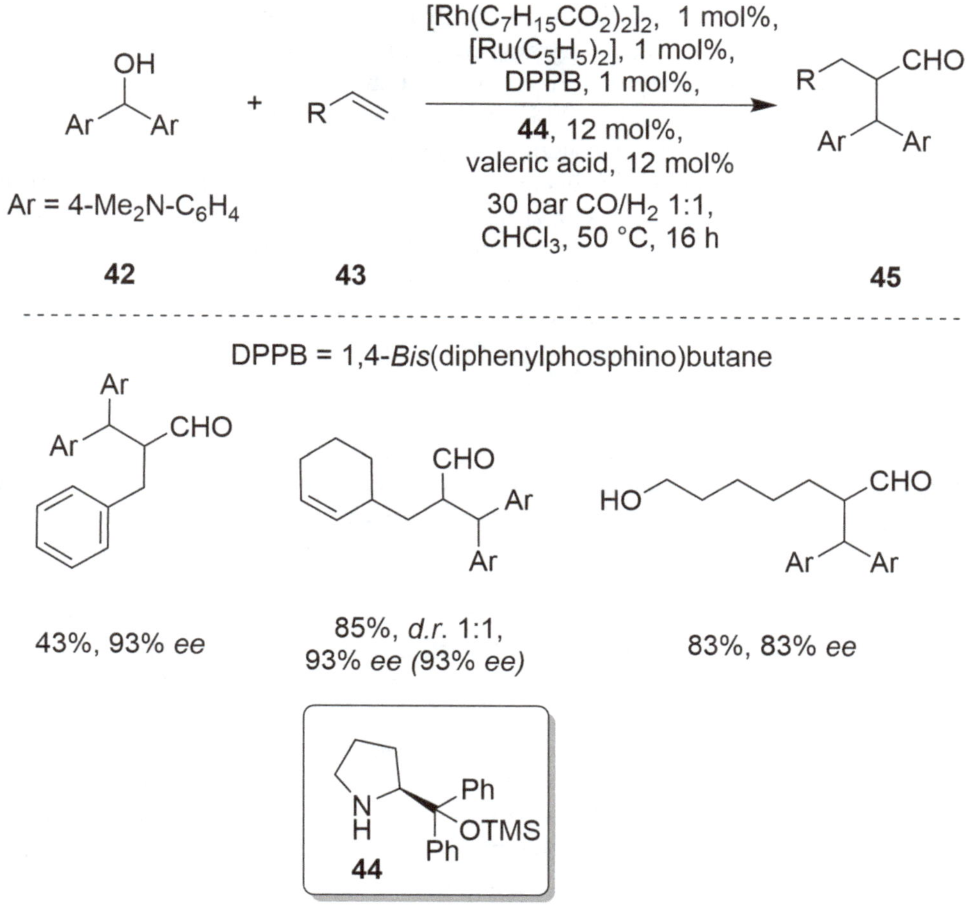

Scheme 16 Tandem hydroformylation/asymmetric organocatalytic alkylation reaction

9.3 Stereoselective Reactions with Chiral Nucleophiles

Following on from Cozzi's asymmetric organocatalytic α-alkylation of aldehydes [84], Cristmann and co-workers reported an orthogonal tandem hydroformylation/asymmetric organocatalytic α-alkylation of both simple and functionalized olefins **43** with benzhydrylic alcohols, though this chemistry was limited to the use of highly stable carbenium ions ($E = -7$) (Scheme 16) [85].

Similarly, the same stable benzhydrylic alcohols were used by Kokotos in the alkylation of cyclic ketones, promoted by 10 mol% of a thioxotetrahydropyrimidinone prolinol-derived catalyst [86]. The use of primary amines derived from quinine was reported by Gong and co-workers, in combination with an amine and 30% of the chiral phosphoric acid **6**, to mediate the highly enantioselective alkylation of 3-indolyl-3-hydroxyoxindoles with *O*-protected-acetaldehyde to give a quaternary oxindole product with both high dr values and excellent ee [87]. The 3-indolyl-3-hydroxyoxindole framework was also a viable substrate using asymmetric enamine catalysis in aqueous media as demonstrated by Wang and Ji [88]. Guo and co-workers have described an efficient α-alkylation reaction of α-amino aldehydes with 3-indolylmethanols [89], with use of a primary amine thiourea-based catalyst allowing the formation of the corresponding enamine of α-amino aldehydes in high yields (up to 99%), with good diastereoselectivities (up to 88:12), and with excellent enantioselectivities (up to 96% ee). The products obtained in this reaction can be readily converted into tryptophan derivatives. Cozzi has described an efficient prolinol-derived catalyst containing a ferrocene framework for the stereoselective alkylation of aldehydes [90]. In contrast to the Hayashi-Jørgensen catalyst which contains a sensitive OTMS group, the pyrrolidine ferrocene can be used in the presence of $InBr_3$ or a range of other Lewis acids.

9.4 Stereoselective Reactions with Chiral Metal Complexes

A detailed mechanistic analysis was performed by Bandini and Miscione using experiments and DFT calculations on an enantioselective intramolecular FC reaction of indole with allylic alcohols [91]. In this publication, both the role played of gold catalysis and the effective generation of a carbenium allylic intermediate are discussed. Although apparently these reactions are often viewed as Lewis acid mediated reactions, with the gold complexes activating the alcohols, this mechanistic study shows that the reaction is a stepwise S_N2'-type in which the auration of the C–C double bond of the allylic alcohols takes place due to FC reaction of indole. After rearomatization of the indole moiety, a subsequent β-elimination of [Au(I)]-H_2O occurs, thus demonstrating that this reaction does not involve direct nucleophilic displacement and formation of a carbenium ion. Nevertheless, it is important to illustrate these results relative to other gold-mediated nucleophilic displacements of allylic alcohols, where the mechanism could probably be similar to the S_N2' mechanism determined for the intramolecular FC reaction. It is also worth mentioning that the counterion on the gold complex (a triflate in this case) plays a central role throughout the mechanism, with hydrogen bonds formed in the preferred transition state between both the indolic proton and the allylic hydroxyl group with the counterion leading to the favored orientation of reagents. The counterion in this example is also involved in the shuttling of the indolic protons to the hydroxyl group, leading to the formation of a water molecule, which is subsequently eliminated.

9.5 Stereoselective S_N1-Type Reactions with Enantioenriched Alcohols

Gong has reported an Au(I)-catalyzed multiple cascade reaction between 2-alkynyl arylazides and propargylic alcohols, generating pyrroloindolone derivatives with a great degree of structural diversity [92], with the gold complex participating in a tandem triple catalysis cascade when enantioenriched propargylic alcohols **47** are used (Scheme 17).

This leads to enantioenriched pyrroloindolone derivatives **48** with two stereogenic centers, including a quaternary one, with excellent stereocontrol, with the transformation only being

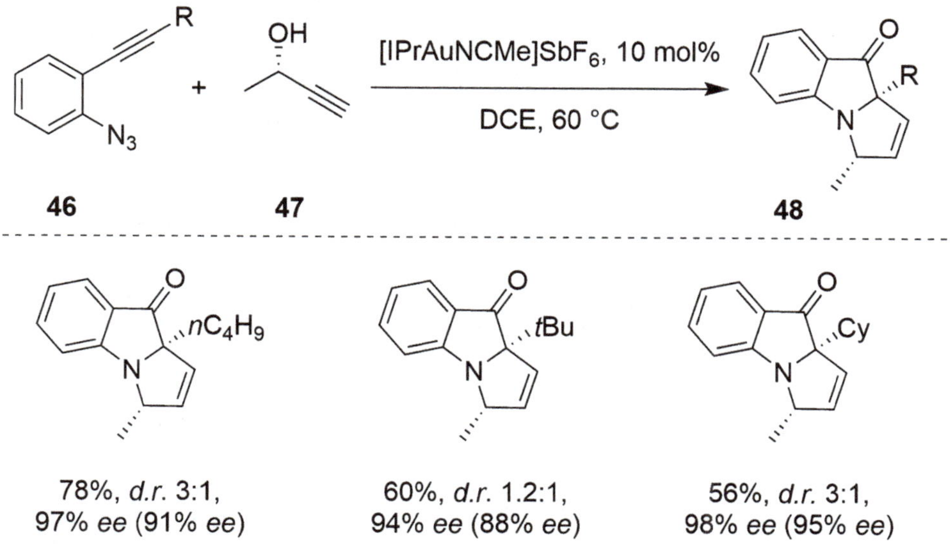

Scheme 17 Triple tandem catalysis with enantioenriched propargylic alcohols

possible because the reaction does not occur via a carbenium ion but instead through a carbenoid gold complex, which reacts with the alcohol formed after a Saucy-Marbet rearrangement involving the propargylic indole, to give the observed pyrroloindolone product.

10 Miscellaneous

Morita–Baylis–Hillman (MBH) adducts [93] are allylic alcohols that can be used as S_N1 nucleophilies, and to demonstrate this, Sasai has reported an S_N1'-type reaction using Deoxo-Fluor [94] for the preparation of alkenes bearing four different C-substituents, with high regio- and stereo-control. The formation of a C(sp^3)–F bond is realized in this transformation, with the reaction showing a fairly broad scope.

The direct reaction of hydrides with carbenium ions is a useful strategy for the transformation of alcohols to the corresponding alkanes. With the nucleophilicity of hydride reagents being recently established by Mayr [95], different electrophilic carbenium ions generated using Bronsted or Lewis acids can be used in hydrogenolysis. In this regard, Samec has reported a Pd-catalyzed direct nucleophilic reduction of primary, secondary, and tertiary benzylic alcohols with formic acid [96], while the reductive deoxygenation of allylic or benzylic alcohols with excess Ti(III) reagent has been reported by Arteaga and Jaraíz [97]. This latter reaction proceeds via an allyl(benzyl)–radical and an allyl(benzyl)–Ti species, which is then selectively protonated. The capability of titanium to perform de-oxygenation is well established and observed in the classical

McMurry reaction [98]. In similar transformations, tertiary and secondary benzylic alcohols are transformed into the corresponding alkanes in good to quantitative yields by use of tin (IV) ion-exchanged montmorillonite (Sn-Mont) as a solid acid catalyst and Et_3SiH as the hydride source.

11 Conclusion and Future Perspectives

In this update to our previous comprehensive review [2], we have presented new reactions and methodologies in the S_N1-type reaction of alcohols, reported in the last few years. Progress has been made, particularly in the application of chiral Brønsted acids in stereoselective reactions in which carbenium ions are formed. We envision that future developments in the field will consider the use of solid embedded Lewis acids and organic-inorganic solid combinations for selective S_N1-type reactions, with the possible recovery and reuse of the catalysts. Although secondary and primary alcohols are still not suitable substrates for such transformations, the invention of strong acids with stabilizing effects would allow reactions to be extended to these substrates, further enhancing the scope of S_N1-type transformations. New photocatalytic reactions, in which mild conditions are used to generate radical species, can be coupled with alcohols and carbenium ions, allowing for the development of interesting novel cascade and multicomponent transformations in which activation of the alcohols will be more facile. It is important to consider that alcohols will be valuable materials for use as biodiesel and for the preparation of key compounds within the remit renewable and green chemistry, and S_N1-type direct nucleophilic substitution can potentially play a pivotal role to achieve this goal.

References

1. http://www.acs.org/content/acs/en/greenchemistry/industry-business/pharmaceutical.html
2. Emer E, Sinisi R, Cozzi PG et al (2011) Direct nucleophilic S_N1-type reactions of alcohols. Eur J Org Chem:647–666
3. Zhu Y, Sun L, Lu P et al (2014) Recent advances on the Lewis acid-catalyzed cascade rearrangements of propargylic alcohols and their derivatives. ACS Catal 4:1911–1925
4. Chen L, Yin X-P, Wang C-H et al (2014) Catalytic functionalization of tertiary alcohols to fully substituted carbon centres. Org Biomol Chem 12:6033–6048
5. Chaskar A, Murugan K (2014) Direct allylation of alcohols using allyltrimethylsilane: a move towards an economical and ecological protocol for C-C bond formation. Catal Sci Tech 4:1852–1868
6. Baeza A, Najera C (2014) Recent advances in the direct nucleophilic substitution of allylic alcohols through S_N1-type reactions. Synthesis 46:25–34
7. Bandini M, Cera G, Chiarucci M (2012) Catalytic enantioselective alkylations with allylic alcohols. Synthesis 44:504–512
8. Naredla R, Klumpp DA (2013) Contemporary carbocation chemistry: applications in organic synthesis. Chem Rev 113:6905–6948
9. Nigst TA, Ammer J, Mayr H (2012) Photogeneration of benzhydryl cations by near-UV laser flash photolysis of pyridinium salts. J Phys Chem 116:8494–8499

10. Mayr H, Kempf B, Ofial AR (2003) π-Nucleophilicity in carbon-carbon bond-forming reactions. Acc Chem Res 36:66–77
11. Mayr H, Ofial AR (2005) Kinetics of electrophile-nucleophile combinations: a general approach to polar organic reactivity. Pure Appl Chem 77:1807–1821
12. Kobayashi S (2013) The new world of organic reaction in water. Pure Appl Chem 85:1089–1101
13. Biswas S, Samec JSM (2013) The efficiency of the metal catalysts in the nucleophilic substitution of alcohols is dependent on the nucleophile and not on the electrophile. Chem Asian J 8:974–981
14. Olah GA, Surya Prakash GK, Bach T et al (2014) Chiral propargylic cations as intermediates in S_N1-type reactions: substitution pattern, nuclear magnetic resonance studies, and origin of the diastereoselectivity. J Am Chem Soc 136:2851–2857
15. Li L, Zhu A, Zhang Y et al (2014) $Fe_2(SO_4)_3$.x. H_2O on silica: an efficient and low-cost catalyst for the direct nucleophilic substitution of alcohols in solvent-free conditions. RSC Adv 4:4286–4291
16. Pan X, Li M, Gu Y (2014) Fe $(OTf)_3$-catalyzed α-benzylation of aryl methyl ketones with electrophilic secondary and aryl alcohols. Chem Asian J 9:268–274
17. Koppolu SR, Naveen N, Balamurugan R (2014) Triflic acid promoted direct α-alkylation of unactivated ketones using benzylic alcohols via in situ formed acetals. J Org Chem 79:6069–6078
18. Nammalwar B, Bunce RA (2013) Friedel--Crafts cyclization of tertiary alcohols using bismuth(III) triflate. Tetrahedron Lett 54:4330–4332
19. Nallagonda R, Rehan M, Ghorai P (2014) Chemoselective C-benzylation of unprotected anilines with benzyl alcohols using Re_2O_7 catalyst. J Org Chem 79:2934–2943
20. Ghosh S, Kinthada LK, Bhunia S, Bisai A (2012) Lewis acid-catalyzed Friedel–Crafts alkylations of 3-hydroxy-2-oxindole: an efficient approach to the core structure of azonazine. Chem Commun 48:10132–10134
21. Zheng H, Lejkowski M, Hall DG (2013) Mild boronic acid catalyzed Nazarov cyclization of divinyl alcohols in tandem with Diels–Alder cycloaddition. Tetrahedron Lett 54:91–94
22. Ricardo CL, Mo X, Hall DG (2015) A surprising substituent effect provides a superior boronic acid catalyst for mild and metal-free direct Friedel–Crafts alkylations and prenylations of neutral arenes. Chem Eur J 21:4218–4223
23. Silveira CC, Mendes SR, Martins GM (2012) Propargylation of aromatic compounds using $Ce(OTf)_3$ as catalyst. Tetrahedron Lett 53:1567–1570
24. Masuyama Y, Hayashi H, Suzuki N (2013, 2013) $SnCl_2$-catalyzed propargylic substitution of propargylic alcohols with carbon and nitrogen nucleophiles. Eur J Org Chem:2914–2921
25. Choi YL, Kim BT, Heo J-N (2012) Total synthesis of Laetevirenol A. J. Org. Chem. 77:8762–8767
26. Hamon M, Dickinson N, Dalla V et al (2014) Intra- and intermolecular alkylation of N, O-acetals and π-activated alcohols catalyzed by in situ generated acid. J Org Chem 79:1900–1912
27. Barbero M, Cadamuro S, Dughera S et al (2014) Solvent-free Brønsted acid catalysed alkylation of arenes and heteroarenes with benzylic alcohols. Tetrahedron 70:1818–1826
28. Trillo P, Baeza A, Nájera C (2012) Fluorinated alcohols as promoters for the metal-free direct substitution reaction of allylic alcohols with nitrogenated, silylated, and carbon nucleophiles. J Org Chem 77:7344–7354
29. Westermaier M, Mayr H (2008) Regio- and stereoselective ring-opening reactions of epoxides with indoles and pyrroles in 2,2,2-trifluoroethanol. Chem Eur J 14:1638–1647
30. Ayala CE, Dange NS, Kartika R et al (2015) Brønsted acid catalysed α′-functionalization of silylenol ethers with indole. Angew Chem Int Ed 54:4641–4645
31. Jefferies LR, Cook SP (2014) Alcohols as electrophiles: iron-catalyzed Ritter reaction and alcohol addition to alkynes. Tetrahedron 70:4204–4207
32. Zhou F, Ding M, Zhou J (2012) A catalytic metal-free Ritter reaction to 3-substituted 3-aminooxindoles. Org Biomol Chem 10:3178–3181
33. Wang L, Xie X, Liu Y (2012) Facile synthesis of fully substituted dihydro-β-carbolines via Brønsted acid promoted cascade reactions of α-indolyl propargylic alcohols with nitrones. Org Lett 14:5848–5851
34. Ohshima T, Ipposhi J, Mashima K (2012) Aluminum triflate as a powerful catalyst for direct amination of alcohols, including electron-withdrawing group-substituted benzhydrols. Adv Synth Catal 354:2447–2452
35. Masiero S, Cozzi PG, Paolucci F (2014) New approaches toward ferrocene–guanine

conjugates: synthesis and electrochemical behavior. Organometallics 33:4986–4993
36. Trillo P, Baeza A, Nájera C (2013) Direct nucleophilic substitution of free allylic alcohols in water catalyzed by $FeCl_3 \cdot 6H_2O$: which is the real catalyst. ChemCatChem 5:1538–1542
37. Trillo P, Baeza A, Nájera C (2012) $FeCl_3 \cdot 6H_2O$ and TfOH as catalysts for allylic amination reaction: a comparative study. Eur J Org Chem:2929–2934
38. Das BG, Nallagonda R, Ghorai P (2012) Direct substitution of hydroxy group of π-activated alcohols with electron-deficient amines using Re_2O_7 catalyst. J Org Chem 77:5577–5583
39. Hao L, Wu F, Zhan Z-P et al (2012) Synthesis of acrylonitriles through an $FeCl_3$-catalyzed domino propargylic substitution/aza-Meyer–Schuster rearrangement sequence. Chem Eur J 18:6453–6456
40. Tummatorn J, Thongsornkleeb C, Ruchirawat S et al (2015) Convenient and direct azidation of sec-benzyl alcohols by trimethylsilyl azide with bismuth(III) triflate catalyst. Synthesis 47:323–329
41. Pan J, Li J-Q, Xiong Y et al (2015) Metal-free direct N-benzylation of sulfonamides with benzyl alcohols by employing boron trifluoride--diethyl ether complex. Synthesis 47:1101–1108
42. Han F, Yang L, Xia C et al (2012) Sulfonic acid-functionalized ionic liquids as metal-free, efficient and reusable catalysts for direct amination of alcohols. Adv Synth Catal 354:1052–1060
43. Allali N, Mamane V (2012) $Al(OTf)_3$ as a new efficient catalyst for the direct nucleophilic substitution of ferrocenyl alcohol substrates. Convenient preparation of ferrocenyl-PEG Compounds. Tetrahedron Lett 53:2604–2607
44. Madabhushi S, Jillella R, Godala KR et al (2012) An efficient and simple method for synthesis of 2,2-disubstituted-2H-chromenes by condensation of a phenol with a 1,1-disubstituted propargyl alcohol using $BF_3.Et_2O$ as the catalyst. Tetrahedron Lett 53:5275–5279
45. Hellal M, Falk FC, Moran J et al (2014) Breaking the dichotomy of reactivity vs. chemoselectivity in catalytic S_N1 reactions of alcohols. Org Biomol Chem 12:5990–5994
46. Mirzaei A, Biswas S, Samec JSM (2012) Iron (III)-catalyzed nucleophilic substitution of the hydroxy group in benzoin by alcohols. Synthesis 44:1213–1218
47. Calmus L, Corbu A, Cossy J (2015) 2H-Chromenes generated by an iron(III) complex-catalyzed allylic cyclization. Adv Synth Catal 57:1381–1386
48. Mukherjee P, Widenhoefer RA (2013) The regio- and stereospecific intermolecular dehydrative alkoxylation of allylic alcohols catalyzed by a gold(I) *N*-heterocyclic carbene complex. Chem Eur J 19:3437–3444
49. Zhang F-Z, Tian Y, Qu J et al (2015) Intramolecular etherification and polyene cyclization of π-activated alcohols promoted by hot water. J Org Chem 80:1107–1115
50. Li P-F, Wang H-L, Qu J (2014) 1, n-rearrangement of allylic alcohols promoted by hot water: application to the synthesis of navenone B, a polyene natural product. J Org Chem 79:3955–3962
51. Begouin J-M, Niggemann M (2013) Calcium-based Lewis acid catalysts. Chem Eur J 19:8030–8041
52. Stopka T, Niggemann M (2015) Cyclopentanone as a cation-stabilizing electron-pair donor in the calcium-catalyzed intermolecular carbohydroxylation of alkynes. Org Lett 17:1437–1440
53. Fu L, Niggemann M (2015) Calcium-catalyzed carboarylation of alkynes. Chem Eur J 21:6367–6370
54. Meyer VJ, Ascheberg C, Niggemann M (2015) Calcium-catalyzed formal [2+2+2] cycloaddition. Chem Eur J 21:6371–6374
55. Gómez AM, Lobo F, López JC et al (2013) Recent developments in the Ferrier rearrangement. Eur J Org Chem:7221–7262
56. Wen J-J, Zhu Y, Zhan Z-P (2012) The synthesis of aromatic heterocycles from propargylic compounds. Asian J Org Chem 1:108–129
57. Pennell MN, Turner PG, Sheppard TD (2012) Gold-and silver-catalyzed reactions of propargylic alcohols in the presence of protic additives. Chem Eur J 18:4748–4758
58. Okamoto N, Sueda T, Yanada R (2014) Bi $(OTf)_3$-catalyzed tandem Meyer–Schuster rearrangement and 1,4-addition to the resulting vinyl ketone. J Org Chem 79:9854–9859
59. Wagh KV, Bhanage BM (2015) Synthesis of substituted aryl ketones by addition of alcohols to alkynes using amberlyst-15/ionic liquid as a recyclable catalytic system. Synthesis 26:759–764
60. Francos J, Borge J, Cadierno V et al (2015) Easy entry to donor/acceptor butadiene dyes through a MW-assisted $InCl_3$-catalyzed coupling of propargylic alcohols with indan-1,3-dione in water. Catal Commun 63:10–14

61. Palmer LI, de Alaniz JR (2014) Acid catalyzed rearrangement of furylcarbinols: the aza- and oxa-piancatelli cascade reaction. Synlett 25:8–11
62. Ayers BJ, Chan PWH (2015) Harnessing the versatile reactivity of propargyl alcohols and their derivatives for sustainable complex molecule synthesis. Synlett 26. https://doi.org/10.1055/s-0034-1380402
63. Palmer LI, Read de Alaniz J (2013) Rapid and stereoselective synthesis of spirocyclic ethers via the intramolecular Piancatelli rearrangement. Org Lett 15:476–479
64. Yin G, Lu P, Wang Y et al (2013) Lewis acid-promoted cascade reaction of primary amine, 2-butynedioate, and propargylic alcohol: a convenient approach to 1,2-dihydropyridines and 1H-pyrrolo[3,4-b]pyridine-5,7(2H,6H)-diones. Tetrahedron 69:8353–8359
65. Yeh M-C P, Liang C-J, Fan C-W et al (2012) Synthesis of 2-azaspiro[4.6]undec-7-enes from N-tosyl-N-(3-arylpropargyl)-tethered 3-methylcyclohex-2-en-1-ols. J Org Chem 77:9707–9717
66. Dethe DH, Murhade G (2013) $FeCl_3$ catalyzed prins-type cyclization for the synthesis of highly substituted indenes: application to the total synthesis of (±)-jungianol and epi-jungianol. Org Lett 15:429–431
67. Phipps RJ, Hamilton GL, Toste FD (2012) The progression of chiral anions from concepts to applications in asymmetric catalysis. Nat Chem 4:603–614
68. Parmar D, Sugiono E, Rueping M et al (2014) Complete field guide to asymmetric BINOL-phosphate derived Brønsted acid and metal catalysis: history and classification by mode of activation; Brønsted acidity, hydrogen bonding, ion pairing, and metal phosphates. Chem Rev 114:9047–9153
69. Cozzi PG, Benfatti F (2010) Stereoselective reactions with stabilized carbocations. Angew Chem Int Ed 49:256–259
70. Chénard E, Hanessian S (2014) Kinetic diastereomer differentiation in Au(III)- and Bi(III)-catalyzed benzylic arylation: concise and stereocontrolled synthesis of 2-Amino-1,1 diarylalkanes. Org Lett 16:2668–2671
71. Gualandi A, Mengozzi L, Cozzi PG et al (2015) Stereosective catalytic addition of nucleophiles to isoquinolinium and 3,4-dihydroisoquinolinium ions: a simple approach to the synthesis of isoquinoline alkaloids. Catal Lett 145:398–419
72. Szostak M, Sautier B, Procter DJ (2014) Stereoselective capture of *N*-acyliminium ions generated from α-hydroxy-*N*-acylcarbamides: direct synthesis of uracils from barbituric acids enabled by SmI_2 reduction. Org Lett 16:452–455
73. Szostak M, Sautier B, Procter DJ (2014) Structural analysis and reactivity of unusual tetrahedral intermediates enabled by SmI_2-mediated reduction of barbituric acids: vinylogous N-acyliminium additions to α-hydroxy-N-acyl-carbamides. Chem Commun 50:2518–2521
74. Petruzziello D, Gualandi A, Cozzi PG et al (2012) Direct and stereoselective alkylation of nitro derivatives with activated alcohols in trifluoroethanol. Eur J Org Chem:6697–6701
75. Guo C, Song J, Gong L-Z et al (2012) Core-structure-oriented asymmetric organocatalytic substitution of 3-hydroxyoxindoles: application in the enantioselective total synthesis of (+)-folicanthine. Angew Chem Int Ed 51:1046–1048
76. Song L, Guo Q-X, Peng Y-G et al (2012) The direct asymmetric α-alkylation of ketones by Brønsted acid catalysis. Angew Chem Int Ed 51:1899–1902
77. Barbero M, Buscaino R, Cozzi PG et al (2015) Synthesis of bench-stable diarylmethylium tetrafluoroborates. J Org Chem 80:4791–4796
78. Shi F, Zhu R-Y, Tu S-J et al (2014) Catalytic asymmetric formal [3+3] cycloaddition of an azomethine ylide with 3-indolylmethanol: enantioselective construction of a six-membered piperidine framework. Chem Eur J 20:2597–2604
79. Trost BM, Crawley ML (2003) Asymmetric transition-metal-catalyzed allylic alkylations: applications in total synthesis. Chem Rev 103:2921–2934
80. Gualandi A, Mengozzi L, Cozzi PG et al (2014) Synergy, compatibility, and innovation: merging Lewis acids with stereoselective enamine catalysis. Chem Asian J 9:984–995
81. Christ P, Berkessel A, O'Donaghue AC et al (2011) pKa Values of chiral Brønsted acid catalysts: phosphoric acids/amides, sulfonyl/sulfuryl imides, and perfluorinated TADDOLs (TEFDDOLs). Chem Eur J 17:8524–8528
82. Wang P-S, Zhou X-L, Gong L-Z (2014) An organocatalytic asymmetric allylic alkylation allows enantioselective total synthesis of hydroxymetasequirin-A and metasequirin-B tetramethyl ether diacetates. Org Lett 16:976–979
83. Trillo P, Baeza A, Nájera C (2014) (2-Aminobenzoimidazole)-organocatalyzed asymmetric alkylation of activated methylene compounds with benzylic and allylic alcohols. Synthesis 46:3399–3414

84. Cozzi PG, Benfatti F, Zoli L (2009) Organocatalytic asymmetric alkylation of aldehydes by S_N1-type reaction of alcohols. Angew Chem Int Ed 48:1313–1316
85. Stiller J, Behr A, Christmann M et al (2012) Enantioselective tandem reactions at elevated temperatures: one-pot hydroformylation/S_N1 alkylation. Chem Eur J 18:9496–9499
86. Trifonidou M, Kokotos CG (2012) Enantioselective organocatalytic α-alkylation of ketones by S_N1-type reaction of alcohols. Eur J Org Chem:1563–1568
87. Song J, Guo C, Gong L-Z (2013) Enantioselective organocatalytic construction of hexahydropyrroloindole by means of α-alkylation of aldehydes leading to the total synthesis of (+)-gliocladin C. Chem Eur J 19:3319–3323
88. Zhang Y, Wang S-Y, Ji J (2013) Asymmetric α-alkylation of aldehydes with 3-hydroxy-3-indolylox-indoles in aqueous media. Org Biomol Chem 11:1933–1937
89. Guo Z-L, Xue J-H, Guo Q-X et al (2014) The direct asymmetric alkylation of α-amino aldehydes with 3-indolylmethanols by enamine catalysis. Org Lett 16:6472–6475
90. Petruzziello D, Stenta M, Cozzi PG et al (2014) A rational approach towards a new ferrocenyl pyrrolidine for stereoselective enamine catalysis. Chem Eur J 19:7696–7000
91. Bandini M, Bottoni A, Miscione GP (2012) Mechanistic insights into enantioselective gold-catalyzed allylation of indoles with alcohols: the counterion effect. J Am Chem Soc 134:20690–20700
92. Li N, Wang T-Y, Gong L-Z et al (2015) Gold-catalyzed multiple cascade reaction of 2-alkynylphenylazides with propargyl alcohols. Chem Eur J 21:3585–3588
93. Basavaiah D, Reddy BS, Badsara SS (2010) Recent contributions from the Baylis–Hillman reaction to organic chemistry. Chem Rev 110:5447–5674
94. Takizawa S, Arteaga Arteaga F, Sasai H et al (2014) Facile regio- and stereoselective metal-free synthesis of all-carbon tetrasubstituted alkenes bearing a C(sp_3)–F unit via dehydroxyfluorination of Morita–Baylis–Hillman (MBH) adducts. Org Lett 16:4162–4165
95. Horn M, Mayr H, Ofial AR et al (2013) Towards a comprehensive hydride donor ability scale. Chem Eur J 19:249–263
96. Sawadjoon S, Lundstedt A, Samec JS (2013) Pd-catalyzed transfer hydrogenolysis of primary, secondary, and tertiary benzylic alcohols by formic acid: a mechanistic study. ACS Catal 3:635–642
97. Prieto C, Arteaga JF, Jaraíz M et al (2015) Homocoupling versus reduction of radicals: an experimental and theoretical study of Ti (III)- mediated deoxygenation of activated alcohols. Org Biomol Chem 13:3462–3469
98. McMurry JE (1989) Carbonyl-coupling reactions using low-valent titanium. Chem Rev 89:1513–1524

Chapter 5

Friedel-Crafts Reactions

Grigoriy Sereda

Abstract

The history, mechanism, diversity, industrial applications, and perspectives of the Friedel-Crafts reactions have been discussed in terms of "green" chemical processes. The rational design of reaction conditions and correct choice of the catalyst provide a wide variety of options to conduct one of the most fundamental groups of organic reactions in a responsible and environmentally friendly manner. Utilization of now available methods of nanotechnology and non-traditional sources of energy paves the way for the further implementation of the principles of "green" chemistry for industrial Friedel-Crafts reactions.

Key words Friedel-Crafts reactions, Alkylation, Acylation, Catalysis, Drug discovery, Green chemistry, Selectivity, Mechanisms, Toxicity, Photochemistry

1 Types and Mechanisms of Friedel-Crafts Alkylation and Acylation

In 1873, Grucareviz and Mertz [1] reported the preparation of ketones from acyl chlorides and aromatic hydrocarbons in the presence of Zn metal or ZnO. Four years later, in 1877, Charles Friedel and James Crafts employed a Lewis acid (aluminum chloride, $AlCl_3$) for introduction of an amyl group to the molecule of benzene, using 1-chloropentane as the source of the alkyl group (alkylation agent) [2]. The paper of Friedel and Crafts explicitly recognized the role of a metal halide in the process of acylation and alkylation of aromatic hydrocarbons. This type of reaction as such was named after Friedel and Crafts. Later on, many other alkylation (alkyl halides, alkenes, alcohols) and acylation (acid halides, anhydrides, acids, esters, amides) agents in combination with various catalysts and substrates have been reported to carry out the preparation of functionalized aromatic compounds, which remain extremely important in the chemical industry today [3, 4]. Friedel-Crafts reactions can be accompanied by cyclizations, rearrangements of substrates, reagents, and products (particularly in alkylation). It is important to note that alkylation can also be a reversible process [3].

Paul F. Richardson (ed.), *Green Chemistry in Drug Discovery: From Academia to Industry*, Methods in Pharmacology and Toxicology, https://doi.org/10.1007/978-1-0716-1579-9_5, © Springer Science+Business Media, LLC, part of Springer Nature 2022

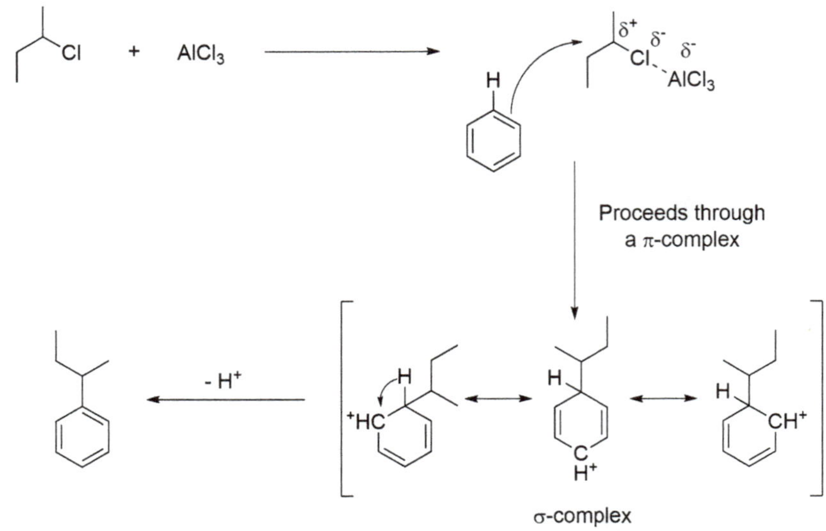

Scheme 1 Mechanism of the Friedel-Crafts alkylation

The **Friedel-Crafts alkylation** is the introduction of an alkyl group to an aromatic compound by the mechanism of electrophilic aromatic substitution (Scheme 1). The electrophilic agent is generated by the coordination of a Lewis acid (in this example $AlCl_3$) with the halogen of the alkyl halide (herein an alkyl chloride), which polarized the C–Cl bond, increases the partial positive charge on carbon, and, most importantly, facilitates further polarization of this bond during the following electrophilic attack. The step of the electrophilic attack leads to the σ-complex, which produces the product after deprotonation. Removal of HCl regenerates the $AlCl_3$ catalyst. Each of the variety of alkylation methods discussed in this chapter utilizes some form of positive charge induction on the carbon, up to generation of the carbocation by protonation of a tertiary alcohol or alkene.

The **Friedel-Crafts acylation** is the introduction of an acyl group to an aromatic compound by the mechanism of electrophilic aromatic substitution. The initial hypothesis of the reaction mechanism suggested a complex between the aromatic substrate and aluminum chloride as the reactive intermediate, which next reacts with the acylation reagent [4]. Now, it is generally accepted that Friedel-Crafts reactions proceed by a classical aromatic electrophilic substitution mechanism in which the catalyst activates the alkylation or herein the acylation agent thereby inducing and/or increasing its electrophilicity [3, 4] (Scheme 2).

However, interaction of the Lewis acid catalyst with the aromatic substrate should not be ignored, especially when the substrate already contains an acyl group (Scheme 3).

From the viewpoint of atom economy, a promising acylating agent is a combination of an alkyl halide and CO, which is

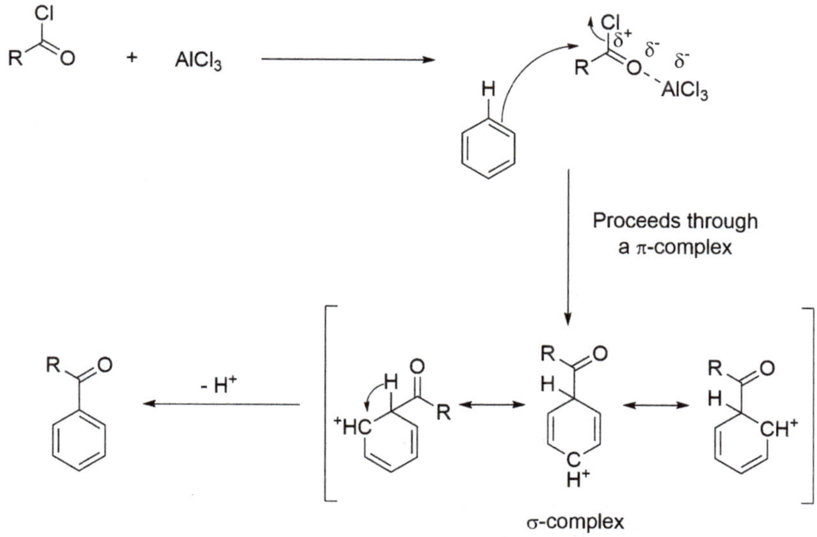

Scheme 2 Mechanism of the Friedel-Crafts acylation

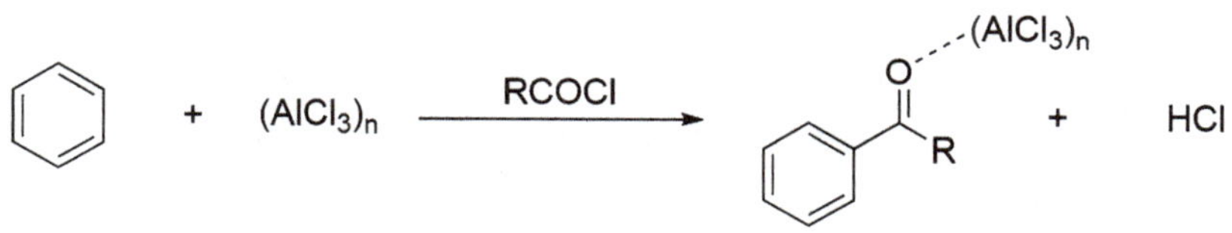

Scheme 3 Interaction of $AlCl_3$ with the product of acylation

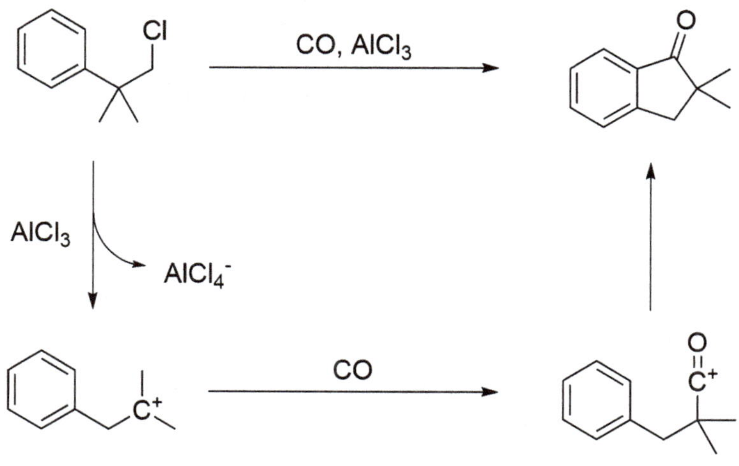

Scheme 4 Aromatic cycloacylation by an alkyl halide and CO

presumably first converted to an acyl halide. This cyclization reaction is often accompanied by carbocation rearrangements [3] (Scheme 4).

After initial product formation, a tight complex forms between the acylated product and the Lewis acid catalyst (aluminum

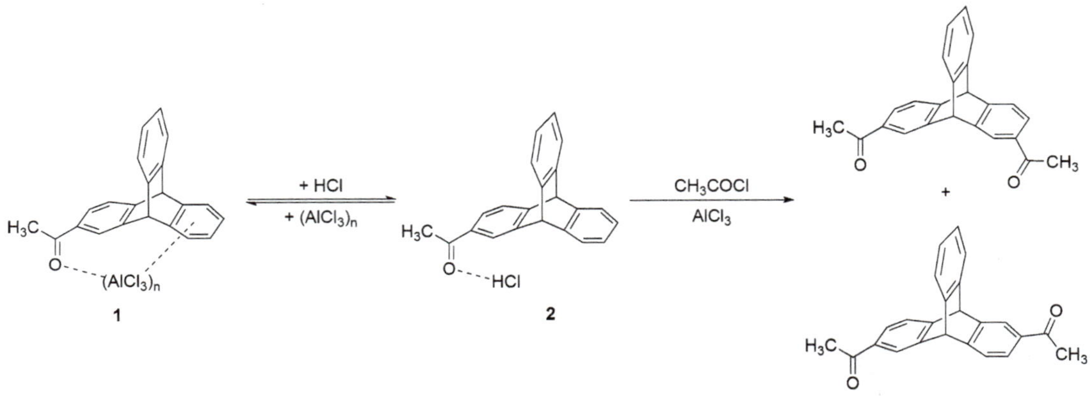

Scheme 5 Activation of complex 1 by HCl

chloride is most commonly used), which deactivates the Lewis acid and thus often requires its presence in stoichiometric amounts. This distinguishes the role of aluminum chloride in acylation reactions, wherein the Lewis acid is required in stoichiometric amounts from purely alkylation reactions where theoretically only a catalytic quantity is required. Second, the product/aluminum chloride complex becomes inert toward further acylation of cage-like aromatic structures such as triptycene even in the presence of excess of an acylating agent [5]. As reported by Sereda and Skvarchenko in 1994, introduction of the next acetyl group into the triptycene system is possible only when the complex of the intermediate diacetyltriptycene with aluminum chloride **1** is broken up by HCl formed in the reaction mixture (Scheme 5) [5]. The released ketone **2** then becomes susceptible to further acetylation.

The presence in the molecule of triptycene of two acetyl groups enhances the deactivating effect of complexation to such an extent that in order to introduce the third acetyl group into the molecule, the ketone-$AlCl_3$ complex must be broken up by excess of HCl introduced from an external source [5]. Alternatively, multiple acylations of triptycene can be achieved by using 1,1-dichlorodimethyl ether as an alkylating agent [6] (Scheme 6).

Although mechanistically this is an alkylation reaction, the overall transformation is in practice an acylation as hydrolysis of the initially formed alkylation product readily produces the carbonyl group. The major drawbacks in many classical Friedel-Crafts acylations are the necessity to use aluminum chloride (or a similar Lewis Acid) in stoichiometric excess leading to the generation of large quantities of waste as well as complicating purification. Within addition, the evolution of noxious HCl in the majority of cases makes the search for cleaner and more sustainable routes especially important for the Friedel-Crafts acylation [7].

Friedel-Crafts reactions are usually understood in terms of the electrophilic attack on an aromatic substrate. However, elevated

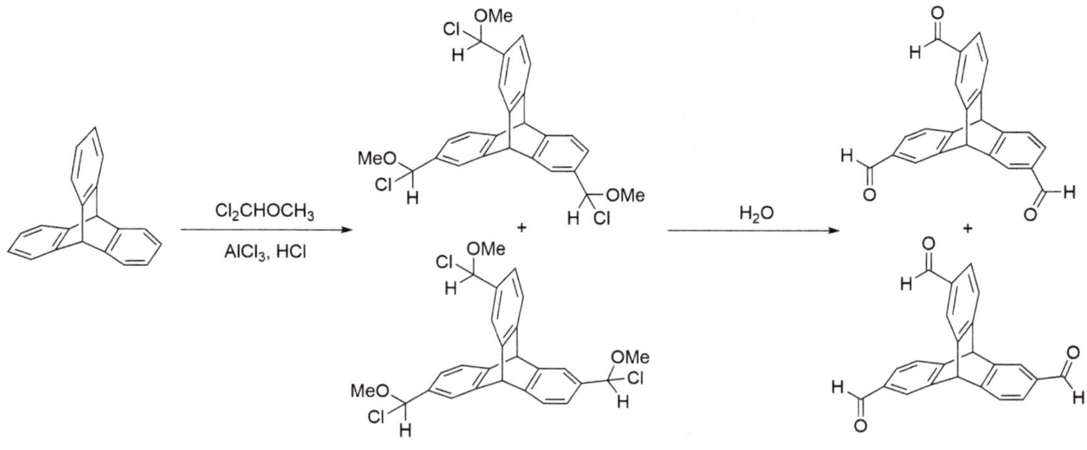

Scheme 6 Overall acylation triptycene by an alkylating agent

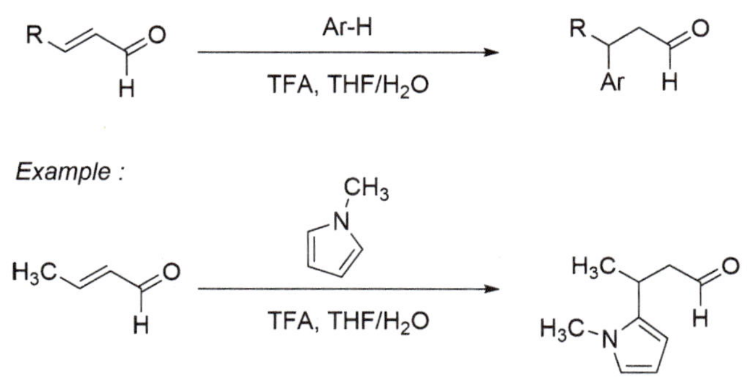

Scheme 7 Michael-type Friedel-Crafts reaction

nucleophilicity of some heterocyclic substrates combined with the inherent electrophilicity of the β-position in α,β-unsaturated carbonyl compounds makes the line between electrophilic aromatic substitution and nucleophilic Michael addition in these cases somewhat blurred. Thus, introduction of the γ-carbonyl group to phenols, pyrroles, and indoles are examples of the "Michael-type Friedel-Crafts" reaction [8] (Scheme 7).

While leading to a significantly different synthetic outcome, the hydroxyalkylation of aromatic compounds with carbonyl compounds and similar processes are often also referred to as Friedel-Crafts reactions. The most practical of these for the pharmaceutical industry are acyliminium Friedel-Crafts reaction [9] and those using epoxides as alkylating agents [10]. It is also worthwhile to note that Friedel-Crafts reactions are often used as steps in a more complex sequence of reactions, such as tandem cyclopropane/ring opening/Friedel-Crafts [11] (Scheme 8) or organocatalytic cascade reactions [12] (Scheme 9). In the tandem cyclopropane ring opening/Friedel-Crafts alkylation, the cyclopropane ring hyperconjugated with an ester group opens and subsequently acts as an

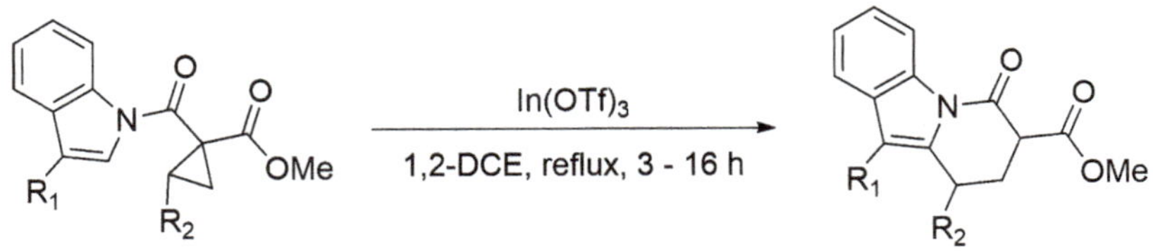

Scheme 8 Tandem Friedel-Crafts reaction—cyclopropane ring opening

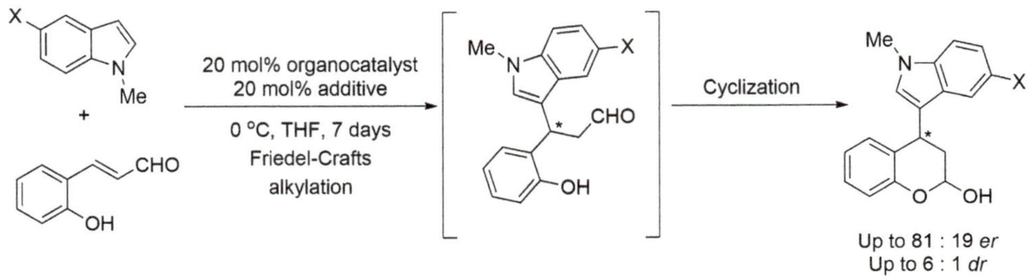

Scheme 9 Organocatalytic cascade reaction in the presence of a chiral organocatalyst and an acid

electrophile adding to the N-substituted indole fragment of the molecule (Scheme 8).

2 Significance of Friedel-Crafts Reactions for Drug Discovery

The wide diversity of available reagents, substrates, catalysts, and pathways of Friedel-Crafts reactions open many opportunities for the development of atom efficient and environmentally friendly processes to access structurally complex molecules for the chemical industry in general and drug discovery in particular.

Friedel-Crafts reactions are often the method of choice to elaborate the carbon skeleton of a variety of aromatic compounds. Therefore, these reactions have routinely been used for the industrial production of many aromatic pharmaceuticals such as the analgesics aspirin (the acetylation or carboxylation of phenol) and acetaminophen (again the acetylation of phenol) (Scheme 10), the anticoagulants 4-hydroxycoumarin and warfarin (again the acetylation of a phenol) (Scheme 11), and a series of antioxidant flavonoids (acylation of phenol) [4]. The acylation of a pyrrole moiety by a range of acid chlorides is a key step in the synthesis of the antiparasitic indanomycin, the antibiotic calcimycin, and the analgesic zomax (Scheme 12) [13].

Due to the high reactivity of pyrrole toward electrophilic substitution, this transformation can be catalyzed by reusable ZnO (at least 3 times without a significant drop in the yield) under solvent-free conditions [13]. A similar ZnO-catalyzed Benzoylation of indoles performed in an ionic liquid stereoselectively proceeds to the 3-position [14] and thus opens access to a variety of compounds

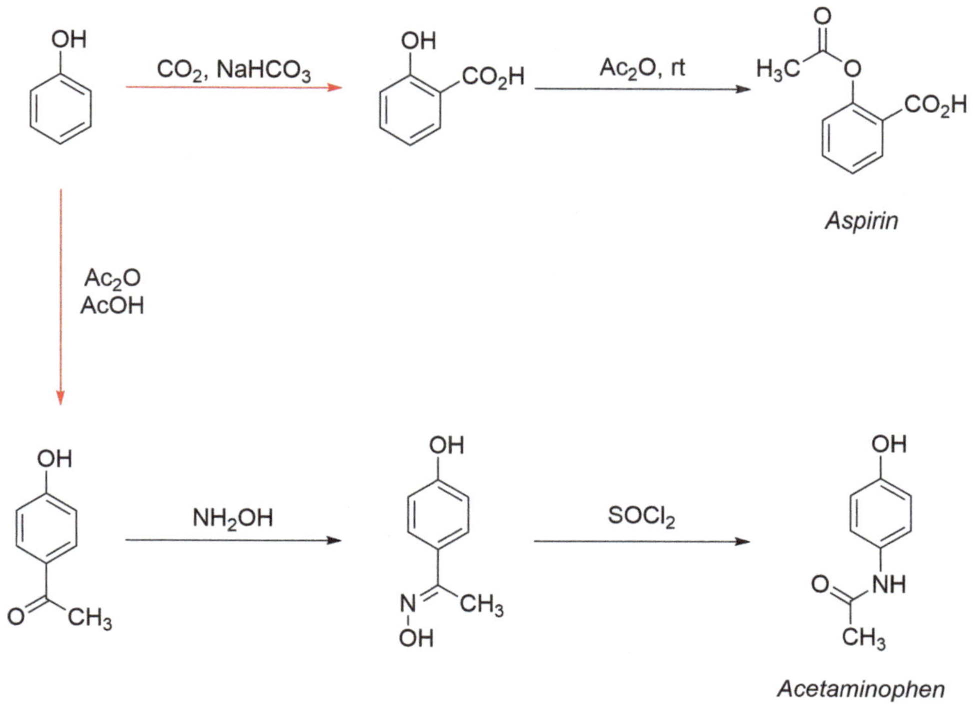

Scheme 10 Synthesis of selected pharmaceuticals using a Friedel-Crafts reaction step—1

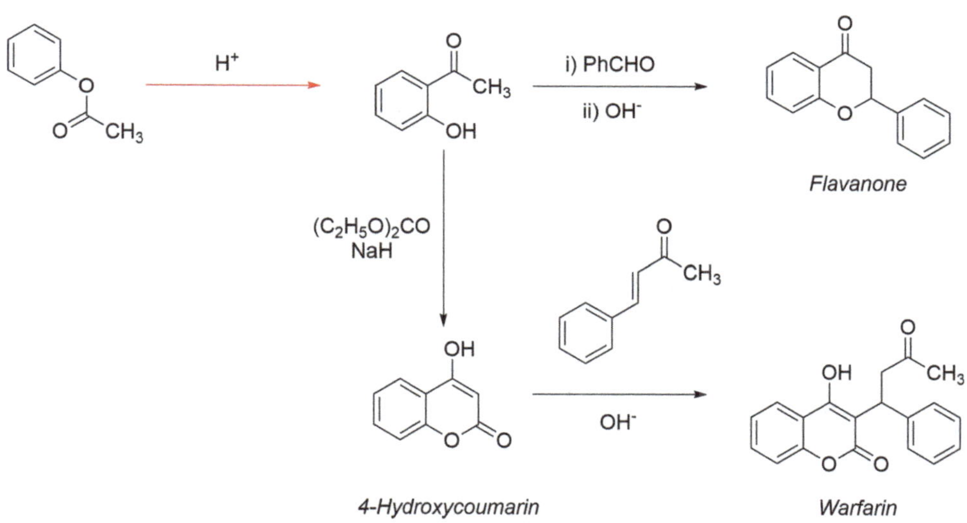

Scheme 11 Synthesis of selected pharmaceuticals using a Friedel-Crafts reaction step—2

whose anti-cancer activity is comparable with that of colchicine and similar alkaloids [15] (Scheme 13).

Hydroxymethylation of aromatic compounds with aldehydes or ketones leads to the benzyl alcohols, which in turn can act as an alkylating agent for another equivalent of the aromatic substrate (Scheme 14). Such a "double Friedel-Crafts reaction" was

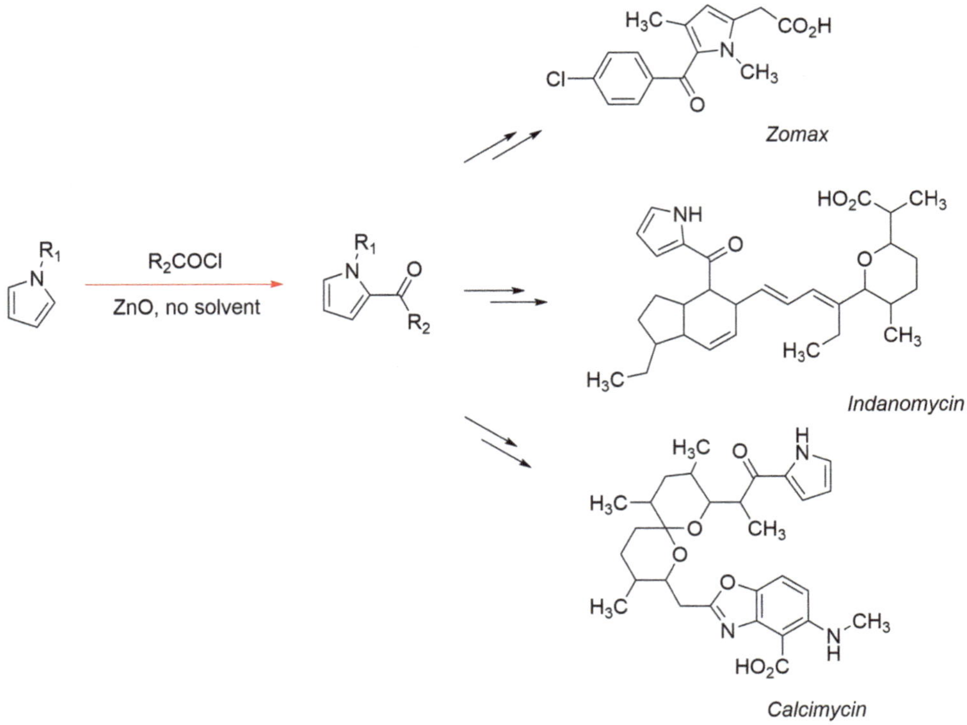

Scheme 12 Synthesis of selected pharmaceuticals using a Friedel-Crafts reaction step—3

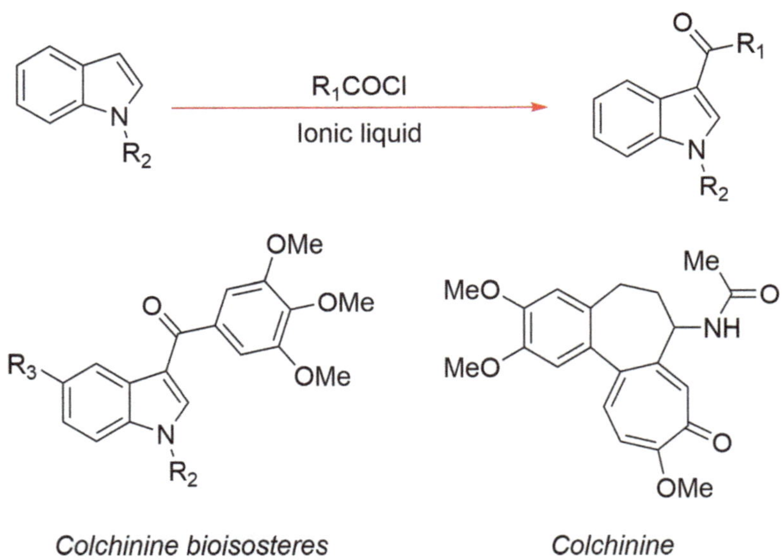

Scheme 13 Synthesis of indole derivatives bioisosteric to colchicine

successfully employed to alkylate two equivalents of 2-substituted furan with an aldehyde in the presence of 10 mol% of molecular iodine [16]. The resulting *bis*(heteroaryl)alkanes have been found to significantly inhibit nitric oxide (NO) production in a

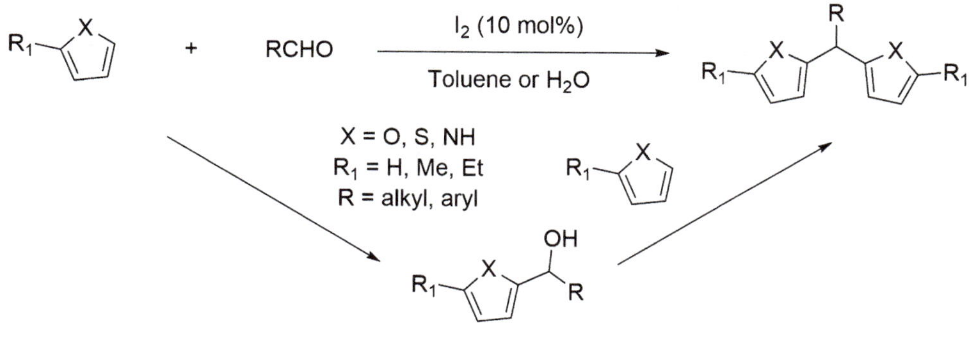

Scheme 14 Tandem hydroxyalkylation—alkylation

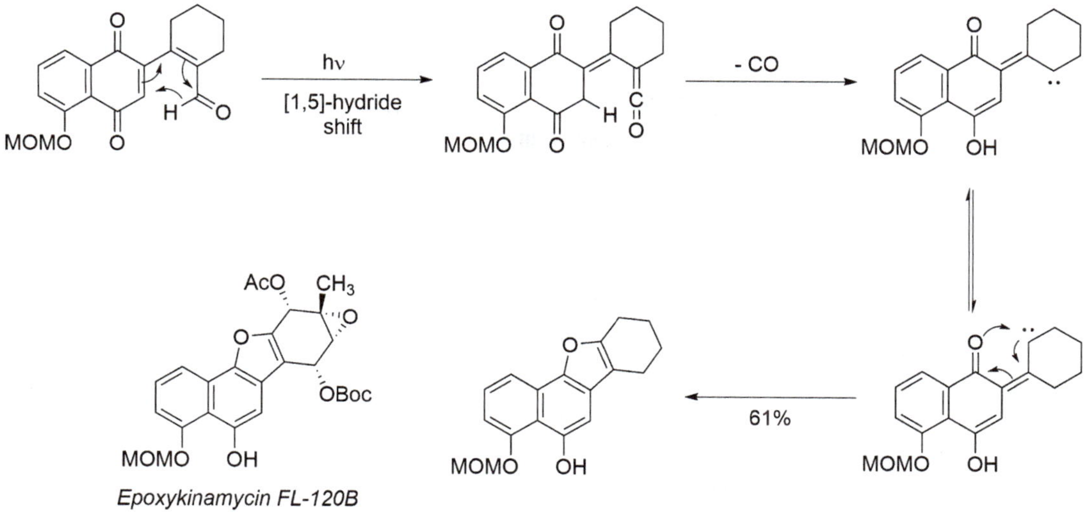

Scheme 15 Mechanism of the decarbonylative "Photo-Friedel-Crafts reaction"

macrophage model and are therefore promising leads for new non-steroidal anti-inflammatory agents [16].

Interestingly, photoacylation of quinones with aldehydes, which is mechanistically very different from electrophilic aromatic substitution, is still referred to as a "photo-Friedel-Crafts reaction" [17]. The reaction involves hydrogen transfer from the aldehyde group to the excited state of quinone followed by the free radical acylation step. An intramolecular version of this reaction, which involved decarbonylation followed by carbene addition, was employed to develop a new approach to the total synthesis of the natural anticancer compound and antibiotic, epoxykinamycin FL-120B (Scheme 15) [17]. Somewhat surprisingly herein, the intermediate ketene does not act as an acylating agent but rather expels carbon monoxide (CO) and produces a carbene.

In 2011, Benites, Rios, Diaz, and Valderrama reported the heteroacylation of 1,4-quinones with aldehydes derived from furan, thiophene, pyrrole, and pyridine induced by solar light

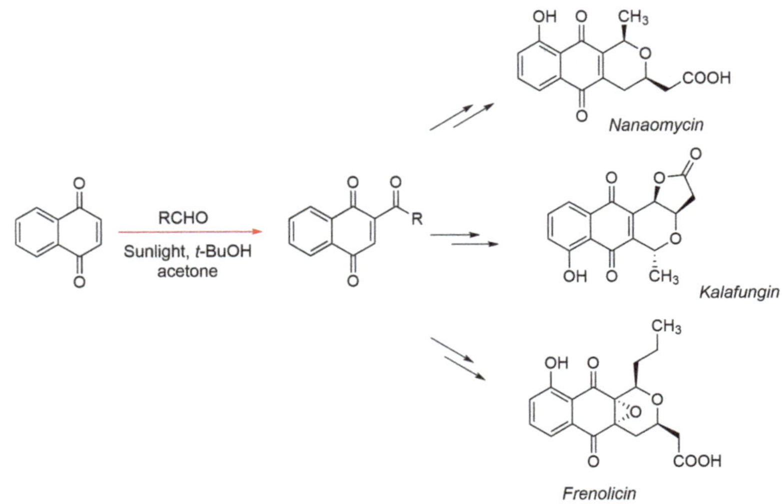

Scheme 16 The use of the photoacylation of quinones in the synthesis of pyranonaphthoquinone antibiotics

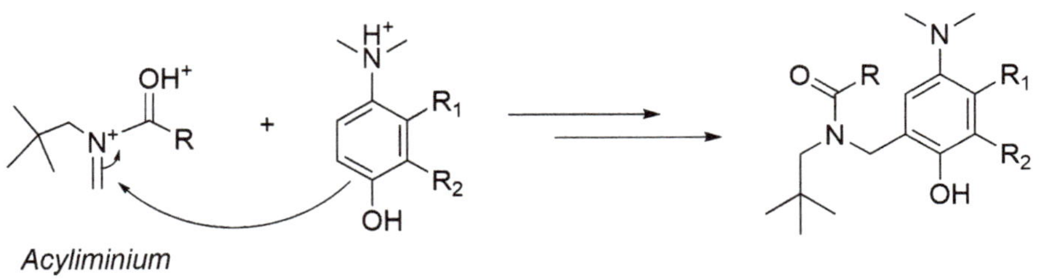

Scheme 17 Aminoalkylation by acyliminium reagents

[18]. This type of acylation has proven to be useful for constructing polycyclic structures with antiproliferative activity [19] and later for a wider variety of biologically active compounds such as pyranonaphthoquinone antibiotics (Scheme 16) [20, 21].

A new type of aminoalkylating Friedel-Crafts reagents (acyliminiums) were introduced in 2008 by Chung et al. (Scheme 17) [9]. Typically generated in situ, the acyliminium reagent reacts with an activated aromatic system, starting a synthetic sequence to produce the advanced antibiotic PTK0796, which belongs to the class of aminomethylcyclines [9] (Scheme 18). This study from Merck evaluates a series of acyliminium precursors in terms of their reactivity, the resulting product profile from the reaction as well as the subsequent ease of removal of the protecting group from the final desired compound.

In 2013, a more classical Friedel-Crafts reaction—intramolecular alkylation using an allylic bromide—was employed to form the substituted 3-benzazepine motif essential for the synthesis of a series of natural products including a number of dopamine agonists (Scheme 19) [22]. In this study, a range of Lewis acids were

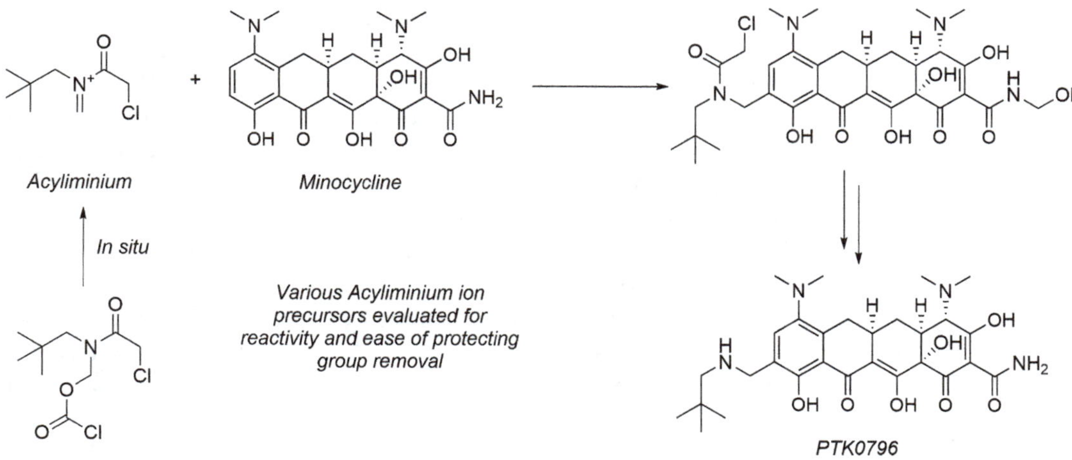

Scheme 18 Merck's acyliminium-based approach to the antibiotic PTK0796

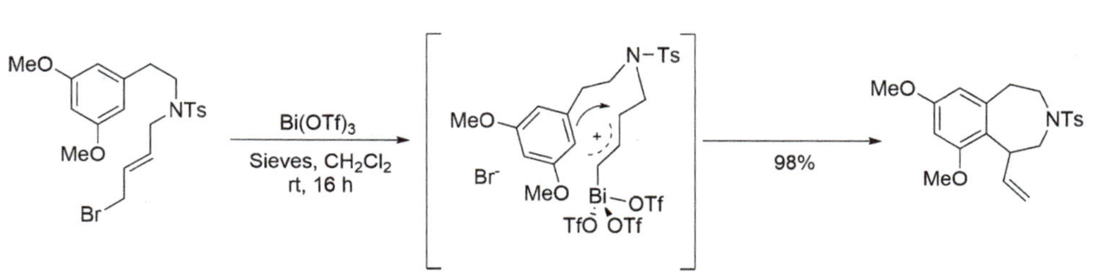

Scheme 19 Bi(OTf)$_3$-mediated synthesis of 3-benazazepines

evaluated for their ability to mediate this transformation with Bi(OTf)$_3$ shown to be optimal.

In another example, which we have previously seen (Scheme 9), the C=C bond of a substituted *ortho*-hydroxycinnamaldehyde alkylates an N-substituted indole, with aldehyde group then forming a cyclic hemiacetal with the phenolic group. This alkylation-cyclization cascade reaction gives access to a wide variety of substituted indoles known for their high-affinity binding to a number of receptors and, therefore, important in drug discovery [12].

3 Minimizing Quantity of the Catalyst or Co-reactant

In 2000, Anastas and coworkers proposed the 12 Principles of Green Chemistry listed in Appendix 1 [23]. The major effort to evolve the Friedel-Crafts acylation to a "green" industrial process is focused on the replacement of the Lewis acid co-reactant usually used in a stoichiometric excess, with a true catalyst. As previously mentioned in contrast to the catalytic Friedel-Crafts alkylation, the classical implementation of its acylation counterpart requires a stoichiometric amount of $AlCl_3$, due to the formation of an inactive complex of the Lewis acid with the ketone product. Therefore,

$AlCl_3$ often referred to as a catalyst actually plays the role of a co-reactant. The first example of electrophilic acylation was reported in 1876 by Crucareviz and Merz, who isolated benzophenone from the unsuccessful coupling of benzoyl chloride using Zn with benzene as the solvent [1]. One year later, Friedel and Crafts attributed the catalytic effect to $ZnCl_2$ formed in situ from Zn [2]. The weaker binding affinity to ketones of $ZnCl_2$ compared to $AlCl_3$ has now opened the gateway to the wide range of catalysts available for aromatic acylation. Zn, In, and Fe used in amounts ranging from 20% to 100% with respect of the acyl chloride have all been reported to catalyze acylation of activated and deactivated benzenes under microwave irradiation [4]. Catalytic amounts of $HClO_4$, $AlH(SO_4)_2$, or molecular I_2 promote the reaction of acid chlorides or anhydrides with activated aromatic substrates [4]. Molecular I_2 is also able to promote the alkylation of furan derivatives using aldehydes [16]. While switching from stoichiometric $AlCl_3$ to catalytic amounts of an alternative substance definitely contributes to both the reduction of waste and hazards involved with the chemical synthesis (Green Chemistry Principles 1, 2, 3, and 9), most of the listed reactions also proceed at room temperature, which is critical for energy efficient processes (Principle 6). The benefits of catalytic acylation have been further refined by using numerous inorganic and organic heterogeneous compositions as catalysts, thus further simplifying isolation of the desired products and enabling the processes to be run in a continuous flow manner [24].

As an example of this, *Zeolites* are natural microporous aluminoosilicates comprised of SiO_4^- and AlO_4^- tetrahedra, which are capable of accommodating a wide variety of cations. Due to the lower valence of Al compared to Si, AlO_4^- tetrahedra impart the negative charge to the inorganic covalent framework. This negative charge can be balanced by protons thus providing Brønsted acidic catalytic sites. Their density can be tuned by the Al/Si ratio in the material, while their distribution is determined by the zeolite's morphology [25]. A range of available controlled processes for the synthesis of mesoporous materials such as MCM-41 (Mobile Composition of Matter-41) and their doping with other metals have led to the development of a wide variety of tunable catalysts for Friedel-Crafts reactions. Thus, the BEA (Beta-Polymorph A) zeolite activates both acid chlorides and anhydrides toward the acylation of toluene [4]. However, better atom economy (Green Chemistry Principle 2) and fewer synthetic steps (Green Chemistry Principle 8) are achieved when a carboxylic acid acts as the acylating agent as the whole organic component of the carboxylic acid ends up in the reaction product, while only half of an anhydride is incorporated. In addition, a derivatization step is removed from the overall synthetic sequence. Excellent results have been reported by Chiche and coworkers, when a Ce-doped zeolite (CeY) was used to catalyze the acylation of toluene by acetic acid [26]. Alternatively, methanesulfonic acid was employed for broader reaction scope in the acylation of aromatic substrates with carboxylic acids under solvent-free conditions [27].

Clays: are layered silicates with a negatively charged framework consisting of various arrays of tetrahedral and octahedral layers. While clay-catalyzed acylation of most aromatic compounds by acyl chlorides requires elevated temperatures (up to 300 °C), acylation of activated aromatics (benzo-crown ethers, pyrroles, furans, and thiophenes) with anhydrides and acyl chlorides proceeds under 80 °C without unwanted acid-catalyzed degradation of the substrate [4]. Especially efficient are Sn-doped clays due to the ability of Sn^{2+} to abstract Cl^- from an acyl chloride and generate a reactive acylium cation [4].

Graphite: In 1997, Kodomari and coworkers have reported that commercial graphite catalyzes the acylation of moderately activated aromatic compounds with both acyl chlorides and acyl bromides upon refluxing in either 1,2-dichloroethane or benzene [28]. In 2004, the use of the graphite catalyst was expanded to the alkylation of toluene and *p*-xylene by benzyl and alkyl halides [29]. Interestingly, graphite-catalyzed alkylation proceeded very smoothly for a primary alkylating agent and a phenyl ester substrate, which would typically rearrange in the presence of traditional strong Lewis acids [30]. Scheme 20 shows the graphite-catalyzed alkylation of a primary bromoalkyl-substituted benzene to the *para*-position by an ester-functionalized benzyl chloride.

In order to promote "green" synthetic methods in the chemical education, alkylation and acylation on graphite have been introduced both to the laboratory chemistry curriculum [31] and to a workshop for undergraduate students [32].

Naflon: is a perfluorinated polymer, which is considered as a "superacid" (pK_a ~ −5 to −9) due to the presence of sulfonic acid groups. Catalysis by Naflon provides practical results for the benzoylation of toluene with benzoyl chloride. However, acylation by alkanoyl chlorides is problematic due to the formation of ketone degradation products [4]. Much better results have been achieved by using Teflon-silica composites where either nanoparticles of

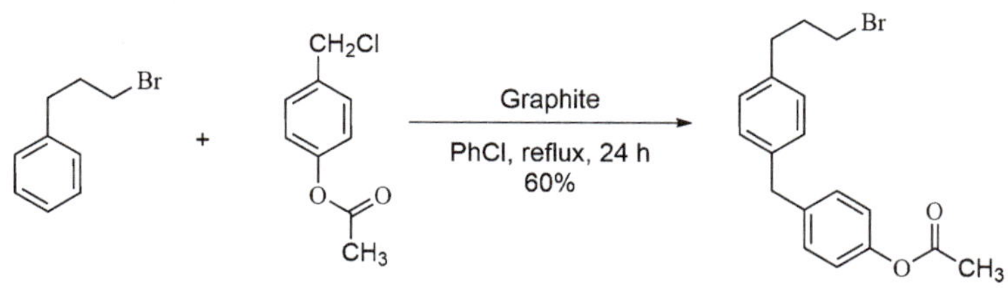

Scheme 20 Chemoselective alkylation on commercial graphite

Naflon were embedded into a silica matrix, or perfluoroalkylsulfonic acid chains were anchored to the surface of silica. These composite catalysts require milder conditions to effect the reaction, produce much less unwanted degradation products, and are able to activate acyl chlorides, anhydrides, and carboxylic acids [4].

Light-driven Friedel-Crafts reactions: The ultimate result of minimizing catalyst amount is obviously to use no catalyst at all, and this can be achieved if the necessary energy is supplied to the reaction mixture by light. For example, the Photo-Fries rearrangement, in which a phenyl ester acts as an intramolecular source of the acylating agent, can be carried out at room temperature upon irradiation at 254 nm [4]. The overall outcome of the reaction is aromatic acylation, which is characteristic for a Friedel-Crafts type of reaction. An even further shift toward an eco-friendly acylation is demonstrated by using solar light as the source of energy, which is keeping up with the sixth Principle of Green Chemistry (Design for energy efficiency). Exposure to sunlight initiated the alkylation of quinones by aromatic [20] and heteroaromatic [18] aldehydes, leading to a series of anti-proliferative diaryl ketones [19]. The attempted intramolecular version of this reaction was unexpectedly accompanied by decarbonylation, which led to a previously discussed new method for the synthesis of the natural anticancer antibiotic Epoxykinamycin FL-120B in the presence of solar light (Scheme 15) [17].

Besides promoting photochemical reactions, solar light can be also used simply for heating. Thus, the classical acylation of aromatic compounds with acyl chlorides in the presence of $AlCl_3$ has been carried out at elevated temperatures created by solar light concentrated at the reactor by parabolic mirrors, which generated temperatures over 300 °C [33].

4 Reducing Hazardous Effects of the Process

The widespread use of acyl chlorides for Friedel-Crafts acylation and alkyl chlorides for the corresponding alkylation, brings up the issue of dealing with the noxious HCl by-product and disposal of the corrosive $AlCl_3$. Options for alternative acylating agents such as carboxylic acids and their anhydrides have been recently expanded to phenylamides with an electron-withdrawing substituent on the aromatic ring. In the presence of triflic acid, these amides have been shown to acylate benzene almost quantitatively within 3 h [34]. In a similar manner, N-nitroisocyanate acylates two molecules of benzene, producing benzophenone in 90% yield. The alkylation equivalent of this reaction can be significantly improved when alcohols or ethers can be used for the key alkylation. Thus, $Sc(OTf)_3$ catalyzes alkylation of aromatic compounds by secondary benzylic alcohols,

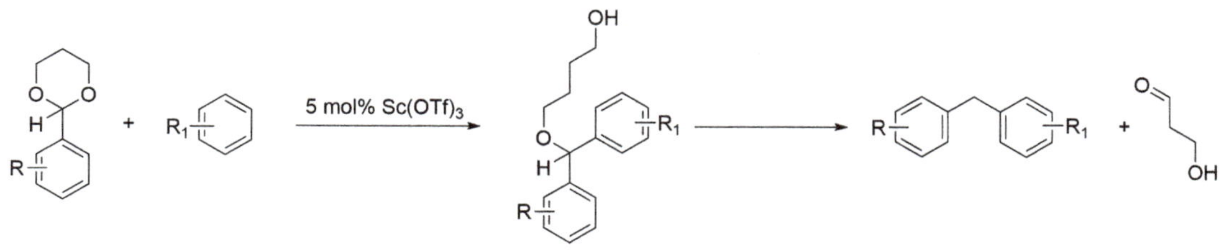

Scheme 21 Redox-alkylation using acetals

and $FeCl_3$ enables alkylation by secondary benzylic ethers [35]. A very interesting alkylation occurs upon activation of 1,3-propaneacetals promoted by $Sc(OTf)_3$ (Scheme 21), with Rueping and co-workers [35] suggesting the involvement of a hydride shift in this process.

Using these alternative acylating and alkylated reagents enables the reaction to be run under less hazardous conditions through utilizing safer reagents and also prevents the generation of acidic waste (Principles 3 and 5 of Green Chemistry: less hazardous chemical synthesis and using safer solvents and auxiliaries).

For a given reaction, there are three major ways to reduce the inherent toxicity of the overall catalytic process: (1) use of non-toxic catalysts, (2) immobilization of catalysts to a solid support thus decreasing their volatility and tendency to form aerosols, and (3) utilization of a non-toxic or non-hazardous solvent, or no solvent at all. Thus, the alpha-acylation of pyrrole was catalyzed by non-toxic, reusable ZnO as was mentioned above [13]. In addition, this reaction was performed without any solvent, which is consistent with Principle 5 of Green Chemistry (using safer solvents or auxiliaries, or no solvent at all). However, one should consider that solvents act as a heat sink, which limits their safe use to reactions with the appropriate heat effects. Another solvent-free reaction catalyzed by low-toxic graphene oxide was reported for the beta-alkylation of indole by epoxides [10]. Non-toxic commercial graphite catalyzes both Friedel-Crafts acylation [28] and alkylation [28–32] reactions. It was found that catalytic activity of commercial graphite toward alkylation is accounted for by the presence of Fe in the material [30]. A material prepared by growing magnetite nanoparticles on carboxy-functionalized graphite has demonstrated excellent catalytic activity for the alkylation of aromatic compounds by various organic halides [36]. This "hybrid catalyst" is an example of a catalyst immobilized on a solid support. These types of materials are more dispersed, reactive, and hinder release of the catalyst to the environment, which is critical both to enable recycling thus easing isolation of products and for less hazardous chemical synthesis (Principle 3 of Green Chemistry). Graphite-supported tosic acid is able to catalyze acylation of substituted benzenes by carboxylic acids at 90 °C [4]. Another strong acid, triflic acid, supported by poly(4-vinylpyridine), was employed to catalyze Friedel-Crafts

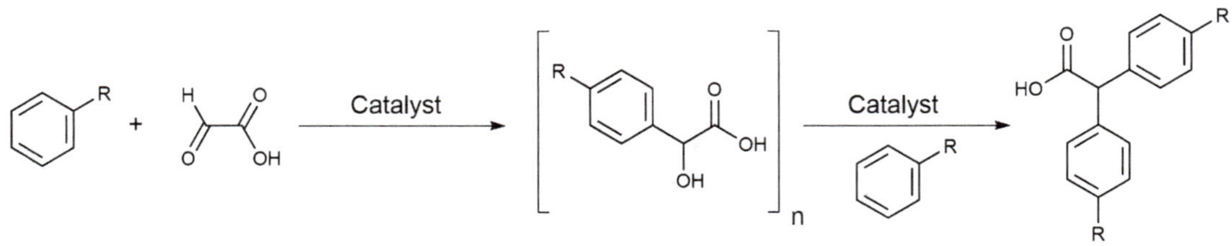

Scheme 22 Hydroxylation followed by alkylation catalyzed by solid-supported triflic acid

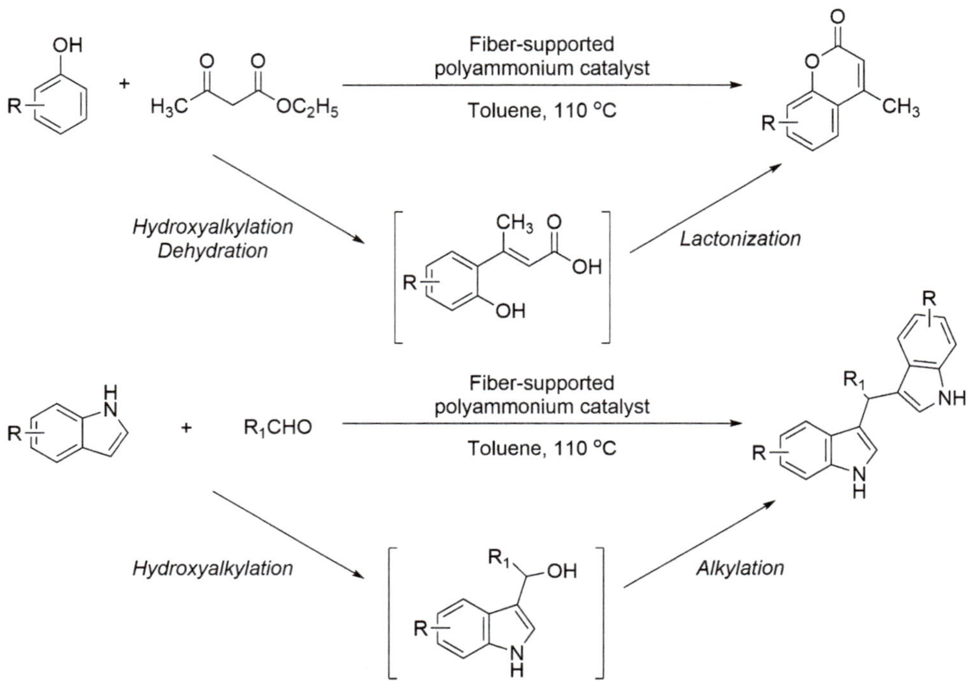

Scheme 23 One-pot solid-support catalyzed synthesis of coumarins from phenols and hydroxymethylation of indoles

acylation and hydroxymethylation followed by alkylation (Scheme 22) [37].

Acylation of anisole by a carboxylic acid (benzoic acid) can also be catalyzed by a weaker (tungstophosphoric) acid supported by a mesoporous MCM-41 material [38]. However, the reaction was complicated by demethylation that led to the minor product phenylbenzoate. In 2014, Zhang and coworkers have covalently attached polyammonium chains to polyacrylonitrile fibers [39]. This environmentally benign material catalyzed the conversion of phenols to coumarins, which features a hydroxymethylation step using acetoacetic ester and the hydroxymethylation of indoles by aldehydes (Scheme 23) [39].

A variety of zeolite-supported Lewis acids have been explored for Friedel-Crafts reactions [4] with the most utilized zeolites being BEA, Y, ZSM-5, MOR, and MCM-41. An interesting feature of

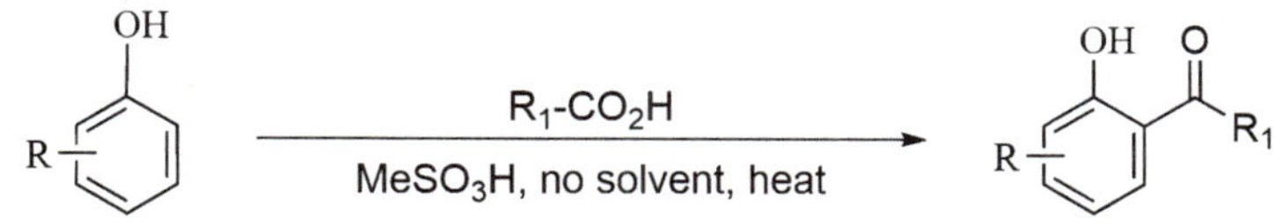

Scheme 24 Solvent-free *ortho*-acylation of phenols

zeolites is both their ability to catalyze acylation by acids rather than their anhydrides or halides and their utility for application for continuous flow processes in fixed bed reactors [4]. As previously demonstrated herein, 12 metal salts deposited on 4 Å molecular sieves were tested to catalyze an intramolecular aromatic allylation featuring an in situ isomerization of the key reactive center (see Scheme 19) [22] with an almost quantitative yield being obtained using 20 mol% of $Bi(OTf)_3$ [22].

In 2010, a group of French chemists demonstrated the use of supercritical CO_2 as a suitable solvent for the acylation of benzofurans using acyl chlorides in the presence of various catalysts [7]. This technique has eliminated concerns about toxicity or flammability typically associated with organic solvents. Two years later, the ionic liquid [BMM][PF_6] (1-butyl-2,3-dimethylimidazolium hexafluorophosphate) was utilized as a solvent for the acylation of indoles with acyl chlorides [14]. Due to their very low vapor pressures, ionic liquids do not emanate any toxic fumes to the environment, which is in keeping with the third Principle of Green Chemistry. As an additional advantage, they can often be easily recycled and re-used without any adverse effects on the efficiency of the process. The Michael-type alkylation of indoles allows for the use of a water-compatible organocatalyst *D*-camphor-10-sulfonic acid, described by Chimni [40]. In addition to other aspects of a clean technology, conducting the reaction in water reduces the probability of a chemical accident involving release of toxic gases or liquids to the environment (Principle 12 of Green Chemistry). In addition, the methanesulfonic acid-catalyzed acylation of phenols by carboxylic acids themselves can be carried out under solvent-free conditions [41] (Scheme 24), with interestingly acylation occurring primarily at the *ortho*-position.

5 Application of Friedel-Crafts Reactions to the Synthesis of "Green" Products

Though the role of the Friedel-Crafts reaction in "green" technology has so far been the focus of this chapter, with a variety of examples of environmentally friendly reaction conditions, it should not be overlooked that a reaction that can be used as a platform for the synthesis of biodegradable, non-toxic, or otherwise "green" products also contributes to "green" technology even if in its direct application in these processes, it is not "green" itself. An example of this is the $AlCl_3$-catalyzed acylation of the renewable feedstock cinnamic acid by maleic anhydride, which efficiently leads to a pivotal intermediate for the production of a "green" epoxy resin (Scheme 25) [42].

6 Increasing Selectivity of Alkylation and Acylation

The benefits of selectivity for a "green" organic reaction often overweigh all the previous factors that we have discussed, simply because increasing the yield of the target compound directly improves the atom economy (Principle 2 of Green Chemistry) and greatly reduces both the toxic exhausts and energy costs in the process of isolation and purification, especially with regard to column chromatography. Thus, in the presence of a chiral catalyst and an acidic co-catalyst, Michael-type alkylation of indoles by *ortho*-hydroxycinnamic aldehydes followed by cyclization produces, in a variety of solvents, chiral coumarins with modest to high enantiomeric excess (Scheme 26) [12]. Depending on the acid (trifluoroacetic acid, dichloroacetic acid, trichloroacetic acid), solvent (dichloromethane, acetonitrile, toluene, THF, ethyl acetate, 1,4-dioxane), and reaction time, the yields ranged from 35% to 91%

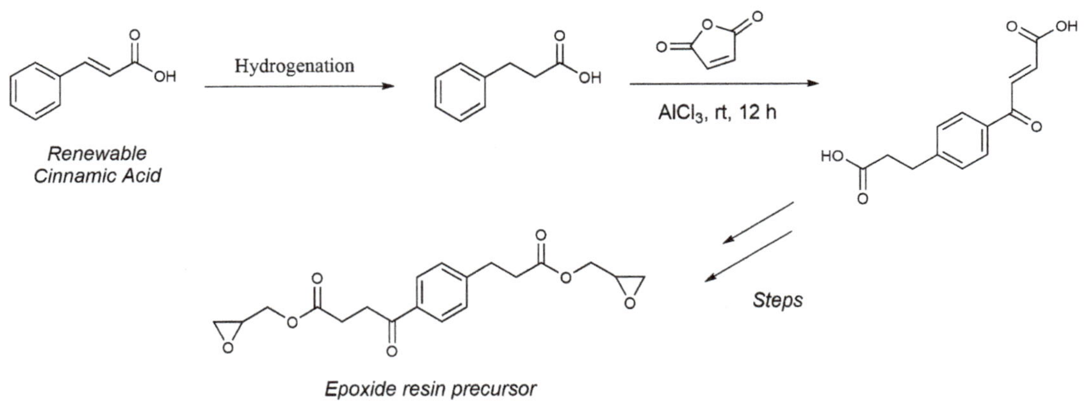

Scheme 25 Synthesis of a "green" epoxy-resin precursor via Friedel-Crafts acylation

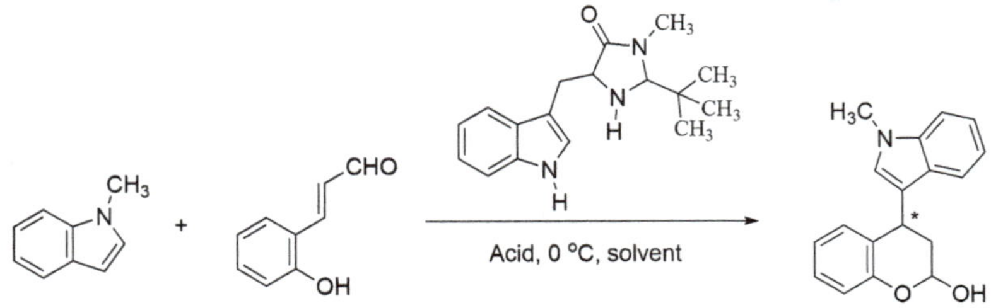

Scheme 26 Chiral Michael-type alkylation of indoles

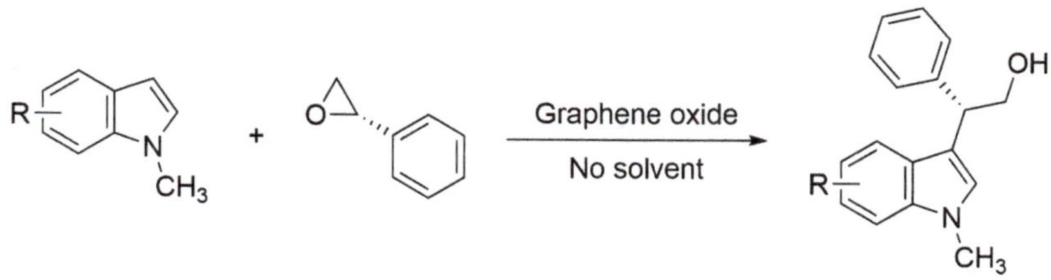

Scheme 27 Unexpected selectivity of alkylation on graphene oxide

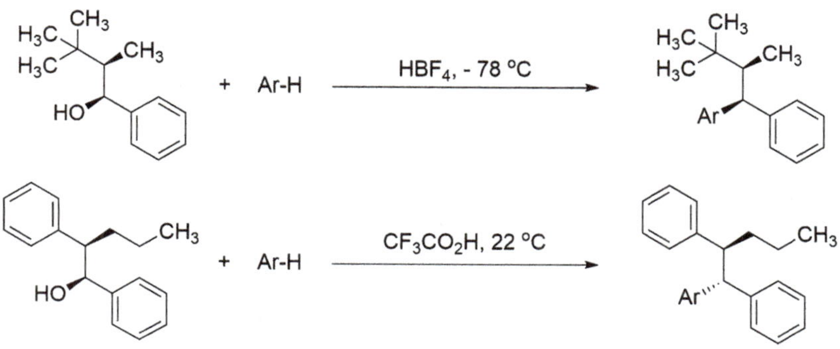

Scheme 28 Diastereoselective alkylation of chiral benzylic alcohols

and er from 51:49 to 73:27. Similar results have been achieved for substituted indoles and *ortho*-hydroxycinnamates [12].

This Michael-type alkylation is, perhaps, the most refined example of enantioselective Friedel-Crafts reactions and is discussed in detail in a 2012 review [8]. A chiral alkylating agent can also act as the source of chirality for the products of electrophilic substitution reactions. For instance, alkylation of a variety of indoles by chiral phenyloxiranes in the presence of graphene oxide produces regioselectively and stereoselectively primary alcohols with high enantiomeric excess (87–99) (Scheme 27) [10].

This high level of regioselectivity is not usually observed under acidic conditions as significant racemization occurs due to the intermediary of a benzyl carbocation. However, this graphene oxide-catalyzed reaction under solvent-free conditions results in retention to a large degree of the enantioselectivity [10], perhaps due to the increased nucleophilicity of the poorly solvated indole. Alkylation of a variety of heterocyclic compounds (thiophenes, pyrroles, and benzofurans) by chiral benzyl alcohols in the presence of a Lewis acid also proceeds with high diastereoselectivity (Scheme 28) [35]. The diastereoselectivity is achieved as the bulky chiral neighbor shields one side of the intermediate benzyl carbocation thus making it more sterically hindered than the other.

Two interesting cases of selective Friedel-Crafts acylation of a bicyclic system by acetyl chloride have been reported by Sereda and co-workers in 2012 [43]. First, the acetyl group is preferentially

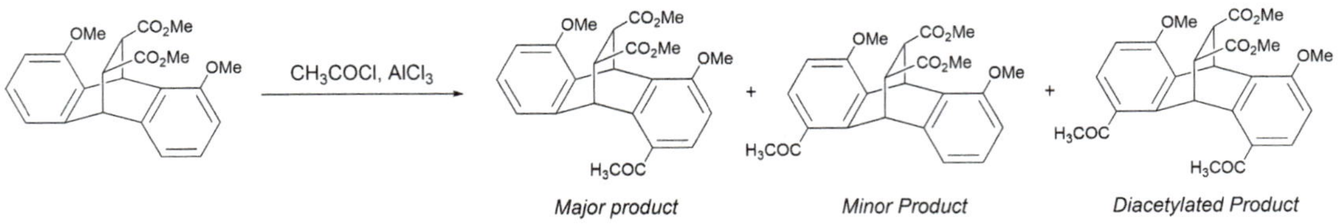

Scheme 29 Stereoselective acetylation directed by pendant ester groups

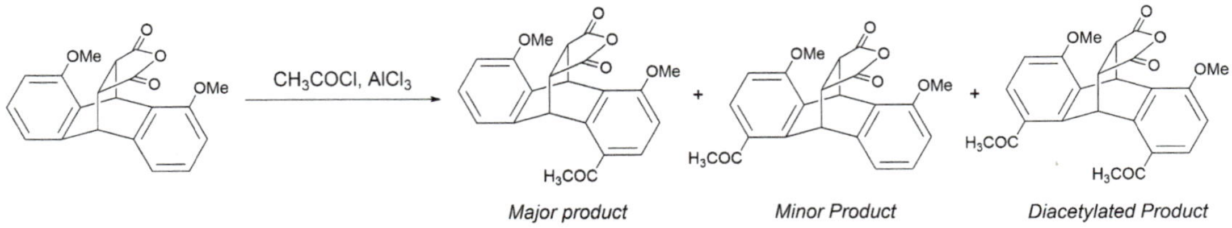

Scheme 30 Stereoselective acetylation directed by an unreactive pendant anhydride moiety

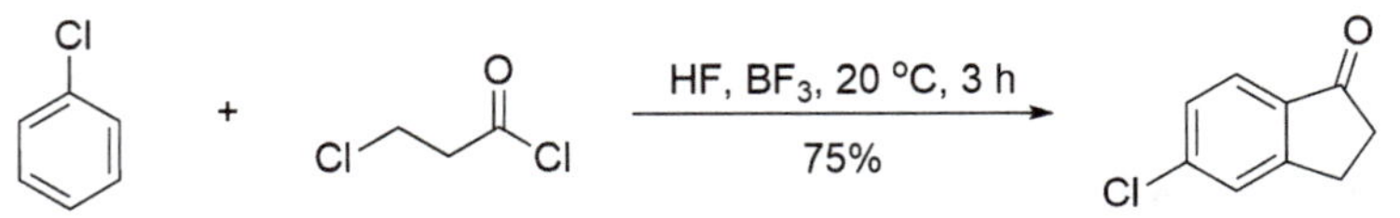

Scheme 31 Chemoselective sequential acylation/alkylation

introduced to one of the two benzene rings due to the directing effect of the pendant ester groups coordinating with $AlCl_3$ (Scheme 29).

In addition, the pendant anhydride group exhibits a similar directing effect. In this case, the sterically hindered anhydride also remains inactive as a potential acylating agent. Therefore, this stereo- and chemo-selective reaction affords practical yields of the target aromatic ketones (Scheme 30) [43].

While the hybrid catalyst prepared by growing magnetite nanoparticles on carboxylated graphite has shown superior activity for many types of alkylating agents, the less active commercial graphite allows for chemoselective alkylation of aromatic compounds in the presence of primary alkyl and allyl bromides and *para*-nitrobenzyl chloride, which remain unreactive under the same conditions [36].

The sequential acylation-alkylation of chlorobenzene by the bifunctional reagent 3-chloropropanoyl chloride further demonstrates chemoselectivity (Scheme 31) [4]. The bifunctional acylating-alkylating agent 3-chloropropanoyl chloride is more reactive for Friedel-Crafts acylation, which proceeds first to the *para*-position of chlorobenzene. Next, the cycloalkylation process converts the intermediate to the cyclized final product. The observed reaction product can be explained by the higher reactivity of the acylating part of the reagent directed to the *para*-position by the halogen substituent. In the next step, alkylation to the *meta*-position with respect to Cl despite being contrary to the classical expected selectivity of an electrophilic aromatic substitution is driven by the stability of the five-membered bicyclic intermediate as opposed to its prohibitively strained six-membered counterpart.

7 Conclusion

Friedel-Crafts aromatic electrophilic substitution was one of the earliest discovered organic reactions to be performed in the presence of a Lewis acid and is now routinely employed for the synthesis of many pharmaceutically active compounds and approved drugs. Originally applied to alkylation and acylation, the term "Friedel-Crafts reaction" is now used for a wide variety of electrophilic reactions that form a bond between a carbon and an aromatic ring, including hydroxyalkylation, aminoalkylation, and Michael-type reactions. A broad spectrum of possible mechanisms, electrophilic agents, and aromatic substrates open up a range of opportunities to perform these processes in a "green" fashion. First, classical aromatic acylation by acyl chlorides involves $AlCl_3$ as a reactant rather than a catalyst due to its ability to form complexes with the final ketone. The development of new non-toxic catalytic systems with lower affinity toward the acylation product allows one to perform the process in a truly catalytic manner, thus minimizing waste and increasing the atom economy. In some cases, the catalyst is active enough to replace acyl chlorides or anhydrides with the carboxylic acid itself, which produces water as the only by-product. The last decade has been marked by a clear trend toward exploring solvents not producing toxic or flammable fumes (ionic liquids) and non-toxic solvents (water) and even performing reactions under solvent-free conditions. The reports of using solar energy for either photochemical Friedel-Crafts reactions or for mere heating of the reaction mixture are also on the rise. The progress of modern organic chemistry has revealed a frequent occurrence of Friedel-Crafts reactions as steps performed in one-pot synthetic cascade sequences, which have been designed to allow them to perform with high selectivity. This aspect significantly cuts the number of necessary procedures for performing the process and isolating the product and, therefore, reduces exponentially adverse environmental effects associated with single-step synthetic chemistry. Last but not least is the utilization of aromatic electrophilic reactions to produce biodegradable materials from renewable feedstocks. While the current state of the art presented here of performing Friedel-Craft reactions in an environmentally friendly fashion is impressive, further improvements are still possible due to the continuous discovery of less toxic and more efficient catalytic materials, safer solvents, and energy-efficient sources.

Appendix 1: Twelve Principles of Green Chemistry [23]

1. Prevention (it is better to prevent waste than to clean it up)
2. Atom economy (as many atoms as possible should be included in the final product rather than in undesired by-products)
3. Less hazardous chemical synthesis (minimizing generation of compounds toxic to human health and the environment)
4. Designing safer chemicals (preference should be given to chemicals with similar functions but lower toxicity)
5. Safer solvents and auxiliaries (use of auxiliary materials should be avoided or minimized)
6. Design for energy efficiency (if possible, synthesis should be conducted at ambient temperature and pressure)
7. Use of renewable feedstocks
8. Reduce derivatives (minimize the number of synthetic steps)
9. Catalysis (selective catalysts are superior to stoichiometric reagents)
10. Design for degradation (chemical products should break down into safe degradation products at the end of their function)
11. Real-time analysis for pollution prevention (formation of hazardous substances should be detected as soon as possible)
12. Inherently safer chemistry for accident prevention (the possibility for chemical accidents should be minimized)

References

1. Grucarevic S, Merz V (1873) Zur synthese der ketone. Chem Ber 6:60
2. Friedel C, Crafts J (1877) A new general synthetic method of producing hydrocarbons. J Chem Soc 32:725–791
3. Roberts R, Khalaf A (1984) Friedel crafts alkylation chemistry. A century of discovery. Marcel Dekker Inc, New York, NY
4. Sartori G, Maggi R (2010) Advances in Friedel-Crafts acylation reactions. Catalytic and green processes. CRC Press, Boca Baton, FL
5. Sereda G, Skvarchenko V (1994) The regulating action of hydrogen chloride on the acetylation and formylation of triptycene and acetyltriptycenes. Russ J Org Chem 30 (3):442–444
6. Sereda G, Lapteva V, Skvarchenko V (1994) Triformyltriptycene. Russ J Org Chem 30 (7):1137–1138
7. Aribert N, Camy S, Lucchese Y, Condoret J-S, Cognet P (2010) Cleaner routes for Friedel-Crafts acylation. Int J Chem Eng 8:A53
8. Chuhan P, Chimni S (2012) Recent advances in asymmetric organocatalytic conjugate addition of arenes and hetero-arenes. RSC Adv 2:6117–6134
9. Chung J, Hartner F, Cvedovich R (2008) Synthesis development of an aminomethylcycline antibiotic *via* an electronically tuned acyliminium Friedel-Crafts reaction. T Lett 49:6095–6100
10. Acocella M, Mauro M, Guerra G (2014) Regio- and enantioselective Friedel-Crafts reactions of indoles to epoxides catalyzed by graphene oxide: a green approach. ChemSusChem 7:3279–3283
11. Patil D, Cavitt M, Grzybowski P, France S (2011) An efficient synthesis of hydropyrido [1,2-*a*]indole-6(7H)-ones *via* an In(III)-catalyzed tandem cyclopropane ring-opening/

Friedel-Crafts alkylation sequence. Chem Commun 47:10278–10280

12. Gwon S, Kim S, Kim S-G (2011) Assymetric organocatalytic Friedel-Crafts alkylation-cyclization cascade reaction of indoles with *o*-hydroxyaromatic α,β-unsaturated aldehydes. Bull Kor Chem Soc 32(12):4163–4164
13. Zhang S, Feng C, Cai J, Chen J, Hu H, Ji M (2013) Solvent free synthesis of 2-acetylpyrroles and its derivatives catalyzed by reuseable zinc oxide. J Chem Res:480–482
14. Zhang L-R, Yi F-P, Zou J-Z, Zhang X, Wang Z (2012) Regioselective Friedel-Crafts acylation of indoles catalyzed by zinc oxide in an ionic liquid. J Chem Res 36:600–602
15. La Regina G, Sarkar T, Bai R, Edler M, Saletti R, Coluccia A, Piscitelli F, Minelli L, Gatti V, Mazzoccoli C, Palermo V, Mazzoni C, Falcone C, Scovassi A, Giansanti V, Campiglia P, Porta A, Maresca B, Hamel E, Brancale A, Novellino E, Silvestri R (2009) New arylthioindoles and related bioisosteres at the sulfur bridging group. 4. Synthesis, tubulin polymerization, cell growth inhibition, and molecular modeling studies. J Med Chem 52:7512–7537
16. Jaratjaroonphong J, Tuengpanya S, Saeeng R, Udompong S, Srisook K (2014) Green synthesis and anti-inflammatory studies of a series of 1,1-bis(heteroaryl)alkane derivatives. Eur J Med Chem 83:561–568
17. Scully S, Porco J (2012) Studies toward the synthesis of the Epoxykinamycin FL-120B: discovery of a decarboxylative photocyclization. Org Lett 14(10):2646–2649
18. Benites J, Rios D, Diaz P, Valderrama J (2011) The solar-chemical photo-Friedel-Crafts heteroacylation of 1,4-quinones. Tetrahedron Lett 52:609–611
19. Arenas P, Peda A, Rios D, Benites J, Muccioli G, Calderon P, Valderrama JA (2013) Eco-friendly synthesis and antiproliferative evaluation of some oxygen substituted diaryl ketones. Molecules 18:9818–9832
20. Mitchell L, Lewis W, Moody C (2013) Solar photochemistry: optimisation of the photo Friedel-Crafts acylation of naphthoquinones. Green Chem 15:2830–2842
21. Uno H (1986) Allylation of 2-alkanoyl 1,4-quinones with allylsilanes and allylstannanes. Efficient synthesis of pyranonaphthoquinone antibiotics. J Org Chem 51:350–358
22. Kargbo R, Sajjadi-Hashemi Z, Roy S, Jin X, Herr R (2013) Synthesis of 3-benzazepines and azepino[4,5-*b*]heterocyclic ring systems via intramolecular Friedel-Crafts cyclization. Tetrahedron Lett 54:2018–2021
23. Anastas P, Bartlett L, Kirchhoff M, Williamson T (2000) The role of catalysis in the design, development, and implementation of green chemistry. Catal Today 55:11–22
24. Aribert N, Camy S, Lucchese Y, Condoret J, Cognet P (2010) Cleaner routes for Friedel-Crafts acylation. Int J Chem React Eng 8:A53
25. Coma A (1995) Inorganic solid acids and their use in acid-catalyzed hydrocarbon reactions. Chem Rev 95:559–614
26. Chiche B, Finiels A, Gauthier C, Geneste P, Graille J, Pioch D (1986) Friedel-Crafts acylation of toluene and *p*-xylene with carboxylic acids catalyzed by zeolites. J Org Chem 51:2128–2130
27. Wilkinson M (2011) "Greener" friedel-crafts acylations: a metal- and halogen-free methodology. Org Lett 13:2232–2235
28. Kodomari M, Suzuki Y, Yoshida K (1997) Graphite as an effective catalyst for Friedel-Crafts acylation. Chem Commun:1567–1568
29. Sereda G (2004) Alkylation on graphite in the absence of Lewis acids. Tetrahedron Lett 45:7265–7627
30. Sereda G, Rajpara V, Slaba R (2007) The synthetic potential of graphite-catalyzed alkylation. Tetrahedron 63:8351–8357
31. Sereda G, Rajpara V (2007) A green alternative to aluminum chloride alkylation of xylene. J Chem Ed 84(4):692–693
32. Boote B, Cartmill L, Meier E, Parker N, Pinkelman R, Pudenz M, Weber A, Sereda G (2007) Exploration of graphite-catalyzed alkylation and acylation at an REU workshop. J Undergrad Chem Res 1:7–9
33. Agee B, Mullins G, Swartling D (2013) Friedel-Crafts acylation using solar irradiation. ACS Sustain Chem Eng 1:1580–1583
34. Raja E, DeSchepper D, Lill S, Klumpp D (2012) Friedel-Crafts acylation with amides. J Org Chem 77:5788–5793
35. Rueping M, Nachtsheim B (2010) A review of new development in the Friedel-Crafts alkylation – from green chemistry to asymmetric catalysis. Beilstein J Org Chem 6(6). https://doi.org/10.3762/bjoc.6.6
36. Rajpara V, Banerjee S, Sereda G (2010) Iron oxide nanoparticles grown on carboxy-functionalized graphite: an efficient reusable catalyst for alkylation of arenes. Synthesis 16:2835–2840
37. Prakash G, Paknia F, Kulkarni A, Narayanan A, Wang F, Rasul G, Mathew T, Olah GA (2015) Taming of superacids: PVP-triflic acid as an effective solid triflic acid equivalent for Friedel-Crafts hydroxyalkylation and acylation. J Fluor Chem 171:102–11234

38. Zhang H, Han J, Tian F, Chen Q, Wang C, Jin H, Bai G (2014) Green synthesis of 4-methoxybenzophenone from anisole and benzoic acid catalyzed by tungstophosphoric acid supported on MCM-41. Res Chem Intermed 41:6731. https://doi.org/10.1007/s11164-014-1772-9
39. Shi X, Xing X, Lin H, Zhang W (2014) Synthesis of polyacrylonitrile fiber-supported poly (ammoniummethanesulfonate)s as active and recyclable heterogeneous Bronsted acid catalysts. Adv Synth Catal 356:2349–2354
40. Chauhan P, Singh S, Chimni S (2013) *D*-Camphor-10-sulfonic acid: a water compatible organocatalyst for Friedel-Crafts reaction of indoles with electron deficient olefins. Indian J Chem 52B:245–251
41. Naeimi H, Raesi A, Moradian M (2011) Iranian J Catal 1(2):65–70
42. Xin J, Zhang P, Huang K, Zhang J (2014) Study of green epoxy resins derived from renewable cinnamic acid and dipentene: synthesis, curing and properties. RSC Adv 4:8525–8532
43. Sereda G, Moorthy R, Basa P, Sykes A, Purohit K, Rajpara V, Becht D (2012) Unexpected diastereoselective acetylation of 1,8-dimethoxyanthracene adducts with maleic anhydride and dimethyl maleate. Eur J Org Chem 4:810–816

Chapter 6

Ionic Liquids: Design and Applications

Arturo Obregón-Zúñiga and Eusebio Juaristi

Abstract

In this chapter, which is divided in two parts, several aspects concerning ionic liquids are discussed. Part one contains a brief history of the development of ionic liquids, their main physical properties, the most convenient ways to manipulate them, the most salient synthetic strategies for the preparation of ionic liquids, and several aspects related with their toxicity and biodegradability. Part two covers current applications of ionic liquids, mainly in catalysis (asymmetric and supported catalysis), including their use as solvents on different "green" applications, their use as electrolytes, and their pharmaceutical applications in drug production, as well as the potential of ionic liquids as drug candidates.

Key words Ionic liquid, Imidazolium ionic liquid, Low-melting salt, Green solvent, Protic ionic liquid, Recoverable catalyst, Asymmetric organocatalysis, Chiral ionic liquid

1 Introduction and Design of Ionic Liquids

Ionic liquids (ILs) are defined as ionic compounds that present a melting point (mp) below 100 °C. This physical property differentiates ILs and ionic salts, such as sodium chloride, with mp = 801 °C. Owing to this property, ILs are also known by the name of "molten salts," and they are a special type of compounds usually constituted by an organic cation (the most common are imidazolium, pyridinium, ammonium and phosphonium cation) and an anion (most commonly a halogen such as chloride, bromide, or iodide, or larger inorganic anions such as $AlCl_4^-$, BF_4^-, PF_6^-, or even organic anions like carboxylic acids or sulfonic acids). Figure 1 shows the most common ions found in ILs.

The first IL was synthesized 100 years ago, in 1914, by Paul Walden [1]. This IL, ethylammonium nitrate (mp = 12–14 °C), is therefore a room temperature ionic liquid (RTIL), which means an ionic liquid with mp below 25 °C. However, and despite this discovery, it was not until 1982 that the first applications of ILs were reported by Wilkes et al. [2]. Wilkes's group synthesized 1-ethyl-3-methylimidazolium tetrachloroaluminate ([EMIM]

Paul F. Richardson (ed.), *Green Chemistry in Drug Discovery: From Academia to Industry*, Methods in Pharmacology and Toxicology, https://doi.org/10.1007/978-1-0716-1579-9_6,

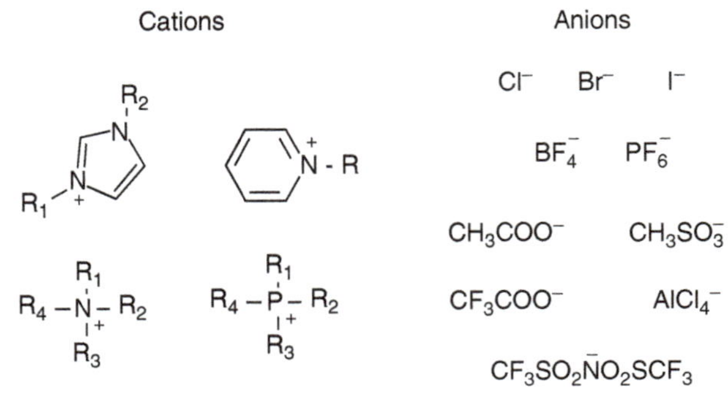

Fig. 1 Common ions found in ionic liquids

$AlCl_4$) by the reaction between 1-ethyl-3-methylimidazolium chloride and $AlCl_3$. This IL is a liquid at room temperature and was initially used as a battery electrolyte, but later was tested in the Friedel-Crafts reaction proving to be both an excellent catalyst and solvent [3]. Although several tetrachloroaluminate ILs were synthesized, they had the drawback of being air and moisture sensitive, which limited their use. Several years later, in 1992, Zaworotko and Wilkes took the task to develop water stable imidazolium ionic liquids, with the strategy to switch the tetrachloroaluminate ion for tetrafluoroborate, hexafluorophosphate, nitrate, sulfate, and acetate ions [4]. Zaworotko and Wilkes accomplished their goal and opened a new application field for ILs as non-volatile organic solvents. This discovery further established imidazolium ILs as the most widely used ILs. More recently, in 2001, Visser et al. described rational modifications of ILs leading to more specific applications [5]. These researchers synthesized several imidazolium ILs with urea and thiourea groups as side chains in order to extract and remove Hg^{2+} and Cd^{2+} from aqueous solutions. This publication marked the beginning of task-specific ionic liquids (TSILs) and also established another name to ionic ILs as "designer solvents," owing to their capacity of being modified, both in the cation and anion, to achieve a desired application. Some authors also call TSILs "iLiquids," drawing an analogy between ILs and their applications with an iPhone and its apps [6].

The history of the development of ILs can be characterized as suggested by Rogers et al. [7] in three generations: the first generation is readily associated with tetrachloroaluminate ILs, whose development was mostly based on their physical properties. The second generation includes Zaworotko's air and water stable ILs, whose development was based on both their physical and chemical properties. Finally, the third generation comprises TSILs. This generation also includes ILs with biological activity, which can be used as active pharmacological ingredients (APIs) (see Subheading 2.4.2). In Fig. 2, a representative molecule from each generation is shown.

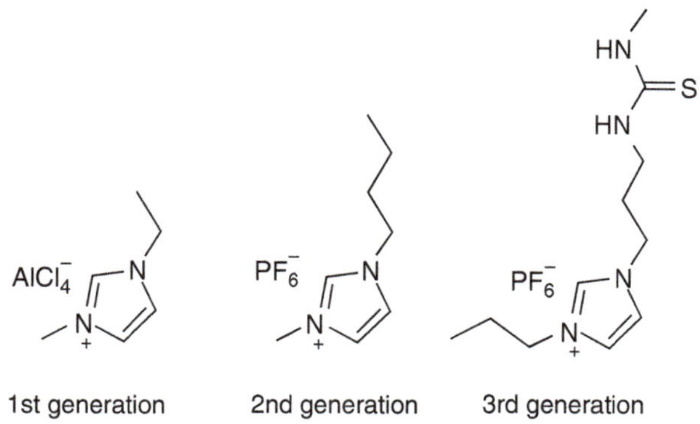

Fig. 2 Development of ILs

As it can be appreciated from this introduction, the possibilities to create ILs are limitless, and indeed with many new ILs waiting to be discovered, one can only wonder at the novel applications that these might enable.

1.1 Physical Properties of Ionic Liquids

As previously stated, the structure of ILs can be tuned by varying the nature of their cation and/or anion. These modifications of course bring changes in their physical properties making each IL unique. In particular, some of their physical properties make them attractive as "green solvents," owing to their high thermal stability, large liquid range, negligible vapor pressure, and the possibility that they can be reused. In this regard, their solubility in polar solvents facilitates the separation of products from the crude reaction media.

1.1.1 Melting Point

It was already mentioned in the introduction that mp is an important physical property that is useful to distinguish an IL from an ionic salt. In particular, any ionic compound (salt) needs to present a mp below 100 °C to be classified as an IL. Furthermore, when one refers to ionic "liquids," it is expected to have a liquid compound to work with. In this case, when we think of ILs we really have in mind RTILs. With this in mind, lowering the mp of the ionic substance is desirable in most cases.

The influence of the cation on the melting behavior is most relevant. It has been found that in order to get a low-melting salt, a large cation with low symmetry, weak intermolecular interactions, and a good charge distribution is required [8]. For example, the sterically large tetraethylammonium cation ($[N_{2222}]^+$) gives rise to ionic compounds with much lower mp when compared with the smaller and spherical sodium cation (Table 1, entries 1 and 2). By the same token, if one compares the ammonium cations $[N_{5555}]^+$ and $[N_{8444}]^+$, both containing 20 carbons, the latter one will afford ILs with lower mp due to its lower symmetry (Table 1, entries 3 and 4). Regarding the distribution of charge, heterocyclic cations

Table 1
Melting points of selected IL bromide salts

Entry	Cation	Melting point (°C)	Ref.
1	Na	755	[9]
2	N_{2222}	284	[10]
3	N_{5555}	101	[10]
4	N_{84444}	67	[10]
5	EMIM	81	[10]
6	C_8MIM	2	[11]
7	C_{10}MIM	10	[11]

usually present a better charge distribution than aliphatic cations due to electron delocalization and/or aromaticity. For instance, imidazolium cation $[EMIM]^+$ produces ILs with lower mp relative to bigger aliphatic ammonium ions like $[N_{5555}]^+$ (Table 1, entries 3 and 5). Finally, considering weak intermolecular interactions, when one increases the *N*-alkyl chain length of an imidazolium cation, a decrease in mp is observed, owing to increased cation dissymmetry. Nevertheless, this is only true up to a certain limit because of the increased participation of van der Waals intermolecular interactions for longer chain lengths, which may lead to an increase in the mp. This effect can be observed when comparing imidazolium cations $[C_8MIM]^+$ and $[C_{10}MIM]^+$ in Table 1, entries 6 and 7.

On the other hand, in order to obtain RTILs, the use of large anions with low coordination ability (large polarizability) is recommended. For example, perfluorinated anions, like tetrafluoroborate, are an attractive option to prepare RTILs due to their size and low coordination properties. Table 2 provides the mp values of EMIM salts with different anions.

1.1.2 Liquid Range and Thermal Stability

The liquid range of a substance is defined by two limits: the lower limit corresponds to its mp and the upper limit corresponds to its boiling point. In the case of ILs, the upper limit usually corresponds to the decomposition temperature, as their ionic nature renders them non-volatile compounds. In the previous section, the structural features required to obtain a low mp IL were discussed. This section will focus on some characteristics of the ions present in the IL, which lead to higher decomposition temperatures thus providing ILs with wider liquid ranges.

From a mechanistic perspective, the most accepted decomposition pathways of ILs are the E2 and S_N2 reactions [14, 15]. Both these degradation mechanisms depend on the particular structure

Table 2
Melting points of selected EMIM ILs

Entry	Anion	Melting point (°C)	Ref.
1	Cl	87	[2]
2	NO_3	38	[4]
3	BF_4	6	[12]
4	CF_3SO_3	−9	[13]
5	CF_3CO_2	−14	[13]

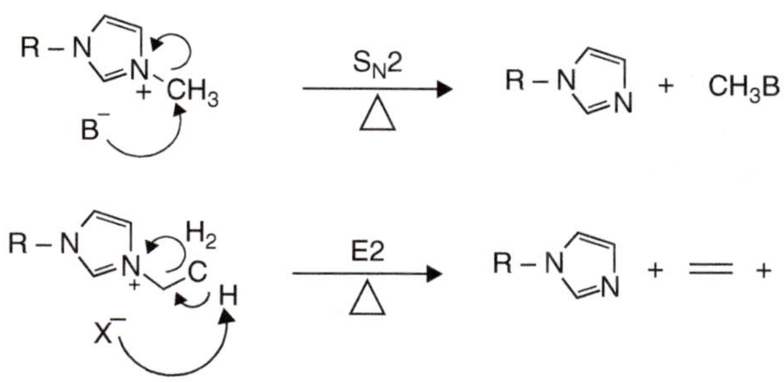

Scheme 1 Plausible thermal decomposition mechanisms

of the anion; thus the thermal stabilities of ILs are mainly dictated by the nature of the anion. If the anion is basic, the S_N2 mechanism will predominate. On the other hand, when the anion is non-coordinative, the E2 mechanism becomes the main decomposition route [16]. In Scheme 1, an example of each decomposition mechanism is displayed.

From an alternative viewpoint, the thermal stability of an IL is inversely proportional to its anion's coordinating ability, because an anion with low coordinating property is neither basic nor nucleophilic, which are both required characteristics for the two decomposition mechanisms. Therefore, the thermal stability of analogous ILs would increase in the order $Cl^- < [BF_4]^- \sim [PF_6]^- < [NTf_2]^-$. For example, $[EMIM]BF_4$ has a mp = 6 °C and a decomposition temperature of 300 °C [12], while $[EMIM]NTf_2$ has a mp of −3 °C and a decomposition point at 400 °C, thus showing a liquid range 100 °C wider than the former [13]. For comparison purposes, water has a liquid range of 100 °C (0–100 °C), whereas the ILs mentioned previously have a liquid range of 300 °C and 400 °C, respectively. This high thermal stability is one of the characteristics that make ILs excellent candidates as green solvents.

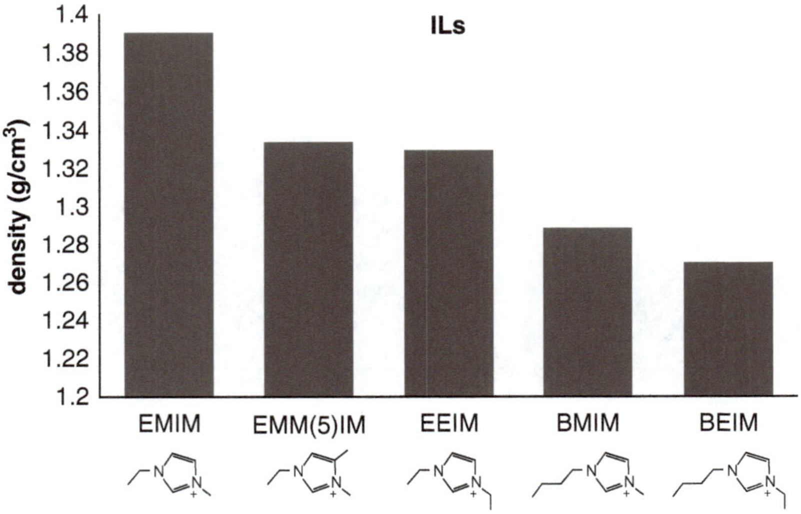

Fig. 3 Densities of different triflate imidazolium ILs

1.1.3 Density

Density, as well as melting point, is a physical property which is affected by both the cation and the anion.

For the case of the anions, the density of ILs displays a trend that is directly proportional to the anion's mass: the heavier the anion, the denser the ionic liquid. As an example, see the density trends according to anion mass in the following series: $[CH_3SO_3]^- \approx [BF_4]^- < [CF_3CO_2]^- < [CF_3SO_3]^- < [C_3F_7CO_2]^- < [(CF_3SO_2)_2N]^-$ [17].

By contrast, cations show the opposite trend relative to anions with the density of an IL decreasing as the cation size increases. For example, as depicted in Fig. 3, the density value exhibits essentially a linear tendency when different triflate imidazolium ILs are compared. As can be seen, density value decreases when the *N*-alkyl chain length increases [17].

1.1.4 Viscosity

In contrast to density, viscosity is a property drastically affected by several external factors like temperature and the potential presence of contaminants. Therefore, it is very important to have access to ILs with high purity when determining their viscosity. Because of this difficulty, less data is available to establish reliable trends for this property.

Despite these analytical problems, it has been observed that viscosity is related to the strength of anion–cation interactions, such as hydrogen bonding and van der Waals forces [13]. In the case of the anion, size is not important and does not influence viscosity as can be seen in the following anionic series: $[(CF_3SO_2)_2N]^- \leq [BF_4]^- \leq [CF_3CO_2]^- \leq [CF_3SO_3]^- < [(C_2H_5SO_2)_2N]^- < [C_3F_7CO_2]^- < [CH_3CO_2]^- \leq [CH_3SO_3]^- < [C_4F_9SO_3]^-$ [17]. For instance, although the anion $[(CF_3SO_2)_2N]^-$ presents a larger size, and also has a more extensive network of van der Waals

$$V_m = V_{ion}(A^+) + V_{ion}(X^-) \qquad \text{equation 1}$$

$$\eta = ae^{bV_m} \qquad \text{equation 2}$$

Where: η = viscosity (cP), a = pre-exponential factor (cP)

b = empirical constant (nm^{-3}), V_m = molecular volume (nm^3)

$$\sigma = ce^{-dV_m} \qquad \text{equation 3}$$

Where: σ = conductivity ($mS\ cm^{-1}$), c = pre-exponential factor ($mS\ cm^{-1}$)

d = empirical constant (nm^{-3})

Fig. 4 Equations that correlate V_m with viscosity and conductivity

interactions relative to $[CH_3CO_2]^-$, the former affords ILs with lower viscosities as a consequence of its inability to form effective hydrogen bonds with the cation because of substantial negative charge distribution through resonance effects.

In contrast, for the cation, size is an important factor to be considered, especially because of the potential van de Waals interactions. For example due to this, imidazolium cations present higher viscosities as the length of their *N*-alkyl chain is increased [17].

The exact physical properties of any new IL must be determined experimentally, though alternatively approximate prediction of some physical properties such as viscosity may be feasible. In particular, Slattery et al. developed equations that correlate molecular volume (V_m), which is the sum of the ionic volumes of the cation and the anion (Fig. 4, Eq. 1) and viscosity (Fig. 4, Eq. 2) [18]. This equation shows that the viscosity of the ILs increases exponentially with increasing V_m. However, investigating a specific series of ILs indicates that the change in viscosity is largely independent of the molecular structure of the cation, with the correlations being strongly anion dependent. Application of these equations in general showed an excellent correlation between the calculated and experimental viscosity values. Finally, a similar relationship between V_m and conductivity, which is another important physical property for electrochemical applications (see Subheading 2.2), was also observed (Fig. 4, Eq. 3).

1.2 Synthesis of Ionic Liquids

In this section, several synthetic methodologies for the preparation of ILs will be presented. There are two main strategies: one option is to create the cationic core by a quaternization reaction (Subheading 1.2.1) and the second option consists of replacement of the anionic segment in the IL via an exchange reaction (Subheading 1.2.2). In addition, it is important to mention in this section the multicomponent Radziszewski reaction, which is a unique method to obtain homosubstituted 1,3-dialkylimidazolium ILs (Subheading 1.2.3).

1.2.1 Quaternization Reactions

In these reactions, the construction of the IL takes place from a molecule containing the heteroatom—more commonly nitrogen or phosphorus—that will become the cation. This is combined with a second suitable reagent, which is usually an acid (protonating agent) or an alkyl halide (alkylating agent), that becomes the anion.

Quaternization reactions are the most common synthetic methods for the production of ILs and can be divided into two further categories: protonation (Subheading 1.2.1.1) and alkylation reactions (Subheading 1.2.1.2).

Protonation Reactions

These reactions have the advantage that they produce ILs with different anions by simply mixing an aliphatic or aromatic acid with an aliphatic or aromatic amine. The reactions showed in Scheme 2 are illustrative of the ILs that can be obtained by this method. In this context, the first reported IL, ethylammonium nitrate, was prepared by a protonation reaction.

ILs prepared by this method are known as "protic ionic liquids" (PILs). Their thermal stability is usually low, and their liquid range tends to be fairly narrow. Some PILs are distillable and thus recoverable via distillation (see Subheading 2.1.1 for applications) [19–21].

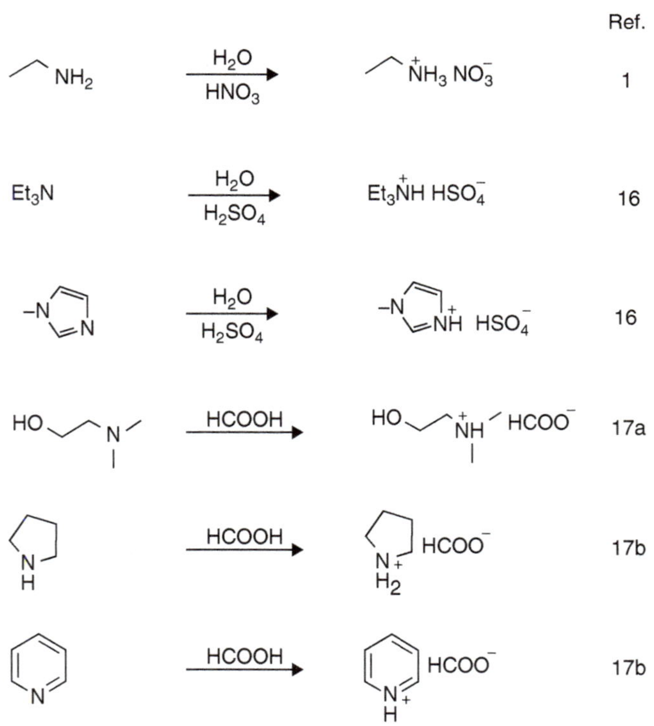

Scheme 2 Examples of the preparation of ILs via protonation reactions

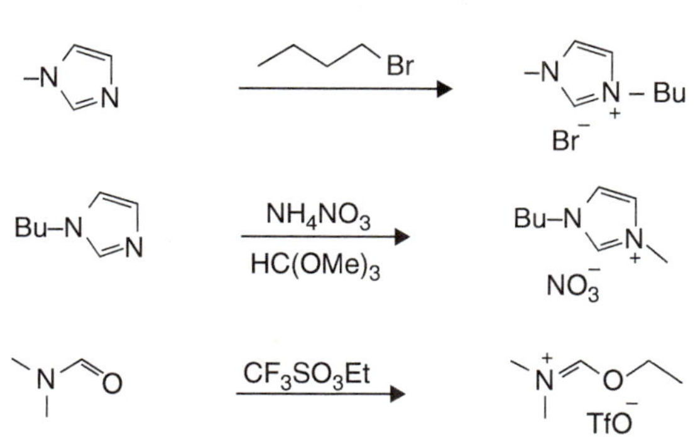

Scheme 3 Examples of alkylation reactions in the preparation of ILs

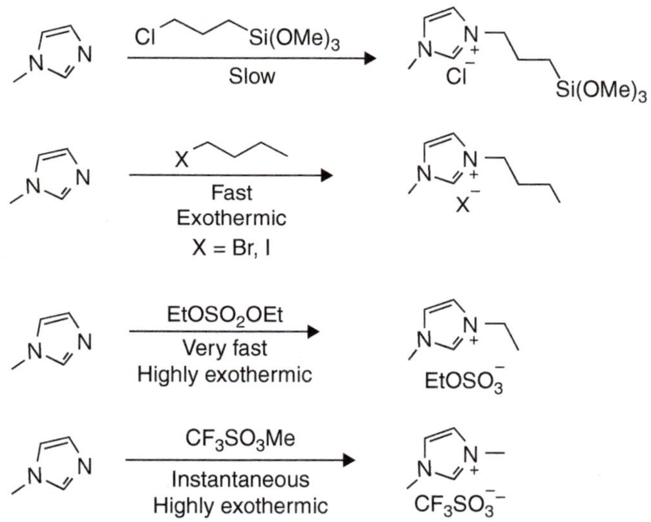

Scheme 4 Relative reaction rates and exothermicity in alkylation reactions

Alkylation Reactions

Alkylation reactions are the most common quaternization reactions for the preparation of ILs, and these can often be extended to access a range of diverse ILs through anion exchange.

Alkylation reactions are performed via reaction of a nucleophilic molecule (mainly amines, phosphines, and sulfides) with an electrophile such as an alkyl halide or a sulfonate ester. As shown in Scheme 3, a wide range of electrophiles can be used. The outcome of this reaction depends on the electrophile being utilized; in particular, substrates with better leaving groups tend to react faster and in a more exothermic manner [22]. This trend in reactivity is in line with anticipation based on the Hammond's postulate, i.e., lower activation energies associated to more exothermic reactions (see Scheme 4).

Alkylation reactions are also an excellent option to access TSILs by using heterocyclic compounds as the electrophiles, enabling different functional groups to be introduced as side chains in a

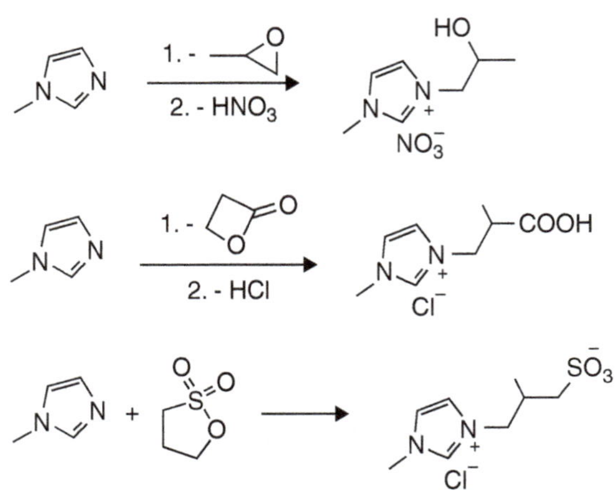

Scheme 5 Examples of TSILs synthesized by an alkylation reaction

single reaction. Scheme 5 presents examples where different heterocycles (oxiranes, lactones, and sultones) are used in order to get TSILs.

1.2.2 Anion Exchange Reactions

Unlike quaternization reactions, which usually require amines as starting materials, anion exchange utilizes an organic salt (often a halide) as the starting material to afford the desired ILs. This reaction is commonly employed to synthesize RTILs and/or hydrophobic ILs, for instance by exchanging a halide for a perfluorinated anion.

In a similar manner to quaternization reactions, there are two types of anion exchange reactions: reactions with Lewis acids (Subheading 1.2.2.1) and anion metathesis reactions (Subheading 1.2.2.2).

Reactions with Lewis Acids

This reaction has been used to synthesize the first generation of ILs (see Introduction). The methodology consists of the reaction of a halide IL with a Lewis acid under an inert atmosphere due to moisture sensitivity of both the Lewis acids and the products (Scheme 6).

Anion Metathesis Reactions

Anion metathesis has been used to synthesize second generation of ILs (see Introduction). Illustrative examples are provided in Scheme 7.

1.2.3 Multicomponent Radziszewski Reaction

The so-called Radziszewski reaction is a synthetic method initially developed to obtain substituted imidazoles, which involves a multicomponent reaction between ammonia and amines, a dicarbonyl compound, and an aldehyde (Scheme 8). Like most multicomponent reactions, the Radziszewski reaction presents high atom economy with water being formed as the only by-product.

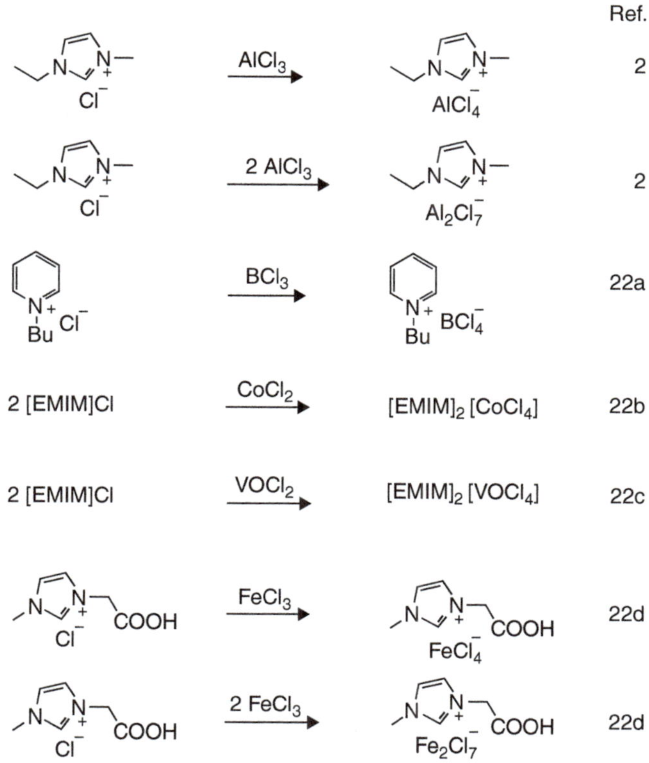

Scheme 6 Representative reactions of halide IL with Lewis acids

Very recently, the Radziszewski reaction has been modified to obtain homosubstituted 1,3-dialkylimidazolium salts in high yields and good purity. In this case, ammonia is substituted by two equivalents of an amine, which become the *N*-alkyl chains, with the addition of an acid necessary both to catalyze the reaction and to provide the counter ion in the IL.

Among recent examples of this methodology, Zimmermann et al. examined various parameters that influence the modified Radziszewski reaction in continuous-flow synthesis using a microreactor setup. Several water soluble imidazolium ILs were prepared in this fashion, with good to excellent yields and high purity (>95%) being obtained (Scheme 9) [23].

Employing this multicomponent reaction, Esposito and co-workers developed a green process to obtain ILs from renewable starting materials [24, 25]. Initially, they synthesized zwitterionic imidazolium derivatives of amino acids and dicarbonyl compounds that can both be produced through bioprocesses. Subsequently, the zwitterionic imidazolium compounds were decarboxylated in the presence of acetic acid in a continuous flow process in order to get imidazolium ILs in moderate to excellent yield (Scheme 10).

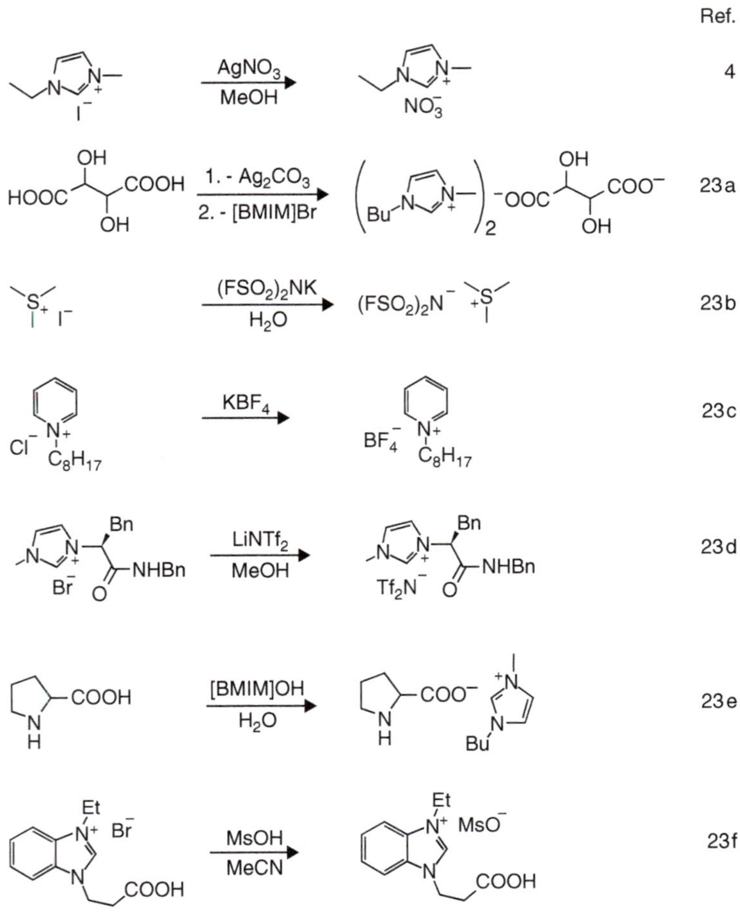

Scheme 7 Selected anion exchange reactions

1.3 Environmental Aspects of Ionic Liquids

Ionic liquids are frequently classified as green solvents and considered non-toxic. Indeed, ILs present negligible vapor pressure and thus are not volatile. Nevertheless following their use, highly stable ILs may be present in water or soil. Therefore, with the increasing use of ILs on industrial processes, their presence in the environment will accumulate and persist, and as a consequence, studies about their impact on the environment must be carried out.

1.3.1 Design of Ionic Liquids with Low Toxicity

Recently, Sosnowska et al. analyzed the toxicity of 375 ILs, containing six different cations and 64 different anions, in various living organisms and found that the nature of the cation (size and degree of branching) has a higher effect on the toxicity of an IL than the anion's structure [26]. Nevertheless, the use of anions such as saccharinate or acesulfamate has also been found to reduce the toxicity of some ILs [27].

In the case of imidazolium cations, a study conducted by Romero and co-workers concluded that the toxicity of this ion increases with chain length [28]. There is a limit however, with the upper limit lying between 14 and 16 carbon atoms [29, 30].

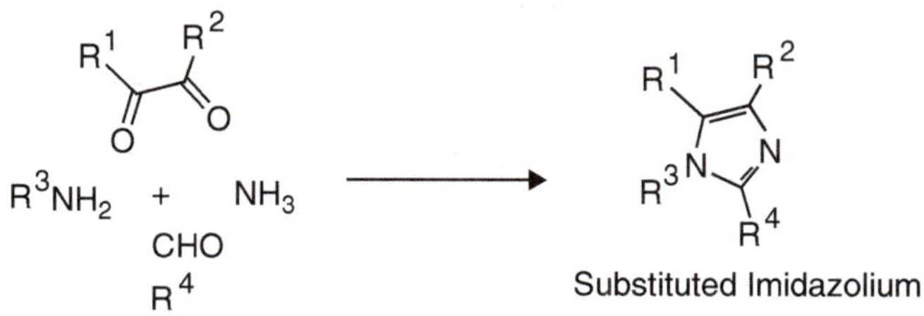

Scheme 8 Multicomponent Radziszewski reaction

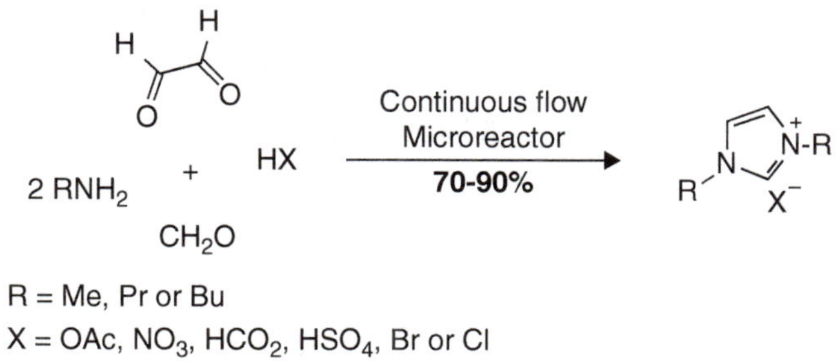

Scheme 9 Modified Radziszewski reaction in the preparation of ILs

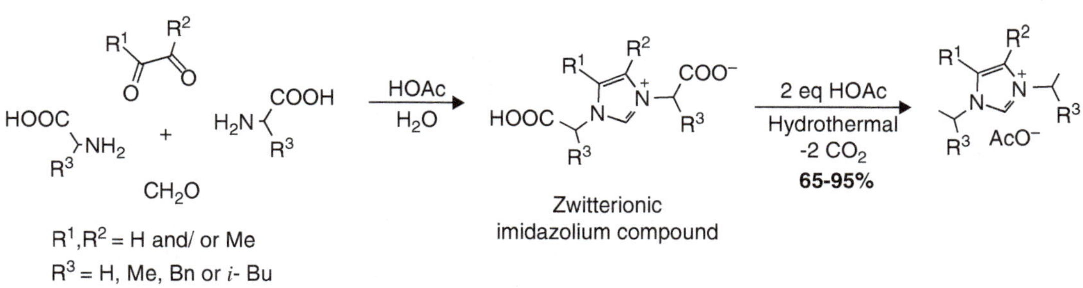

Scheme 10 Green process to obtain imidazolium ILs from renewable feedstocks

A similar pattern is observed in pyridinium cations, with their toxicity increasing with chain length [31, 32]. Furthermore, it has been observed that the introduction of methyl groups in the aromatic ring, mainly at positions 3 and 5, increases the toxicity of ILs containing the pyridinium ion [33].

In the case of alicyclic cations (morpholinium, piperidinium, and pyrrolidinium), it has been observed that toxicity increases with ring size (e.g., piperidinium > pyrrolidinium) [34]. Also, for identical ring sizes, it was found that aromatic rings are more toxic than their alicyclic congeners (e.g., pyridinium > piperidinium) [35].

With regard to aliphatic ammonium cations, their toxicity, as in the case of the cyclic cations, depends on the alkyl chain length [36]. In this context though, the toxicity can be reduced with the introduction of a hydroxyl group on the alkyl chain [37, 38]. A

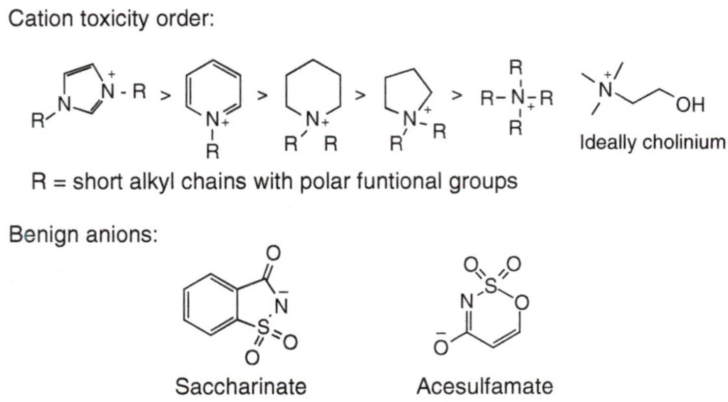

Fig. 5 Structural requirements to design ILs with low toxicity

special ammonium cation with low toxicity is the cholinium cation that combines a series of short alkyl chains with the presence of a terminal hydroxyl group in one of them [39–41].

Finally, Stolte et al. studied the toxicity of aromatic cations (imidazolium and pyridinium), alicyclic cations (morpholinium, piperidinium and pyrrolidinium), and aliphatic ammonium cation discussed previously [30]. It was concluded that lipophilicity is a key factor in connection with toxicity, which explains why long alkyl chains and aromaticity confer higher toxicity to the cations. Figure 5 summarizes these main characteristics impacting the toxicity of ILs.

1.3.2 Biodegradability of Ionic Liquids

Biodegradability is defined as the chemical decomposition of a substance by a microorganism in the environment and is a property usually related to toxicity, because if a compound is very toxic towards microorganisms, which are responsible for the compound degradation, they will not be able to perform their function, and the compound will persist in the environment.

Although there is very little information about the biodegradability in comparison with the toxicity of ILs, some structural characteristics have been examined in order to design environmentally friendly ILs. In principle, biodegradability depends both on the structure of the cation and the anion. It has been observed, for instance, that the incorporation of an ester group in the alkyl chain of the cation increases its potential for biodegradability, probably through initial ester hydrolysis [42]. Other functionalities examined such as ethers do not influence the biodegradability rate of the cation [43]. Regarding alkyl chain length, ILs with long chains are more susceptible to biodegradation [42].

On the other hand, pyridinium ILs show a higher capacity for biodegradation compared with the imidazolium ILs [44]. However, the aliphatic cholinium cation, which also presents low toxicity, is more biodegradable than both these cyclic cations [45, 46].

In the case of the anions, it is important to note that the fluorinated and the nitrile-containing anions ($[(CF_3SO_2)_2N]^-$, $[(C_2F_5)_3PF_3]^-$, $[N(CN)_2]^-$, $[C(CN)_3]^-$, $[B(CN)_4]^-$) are the most persistent species. In contrast, anions derived from carboxylic acids are highly biodegradable [47, 48]. Also, benign anions as saccharinate and acesulfamate have proven to be highly biodegradable [49].

As demonstrated in this section, there are key structural characteristics that can help in the design of more sustainable ILs which are both safer and easier to biodegrade in line with 2 of the 12 principles of Green Chemistry.

2 Current Applications of Ionic Liquids

2.1 Ionic Liquids as Solvents

Of the many applications developed for ILs, perhaps the best known is their use as sustainable solvents, the following sections highlight two current uses of ILs as a solvent within a green chemistry context. These are the extraction of natural polymers with particular emphasis on lignin and keratin (Subheading 2.1.1) and ILs used as solvents for some specific organic transformations (Subheading 2.1.2).

2.1.1 Ionic Liquids in the Extraction of Natural Polymers

The dissolution and processing of natural polymers prior to their degradation to afford useful organic molecules has become an important research field in the area of green chemistry. Indeed, from abundant and renewable natural sources such as wood and bagasse, it is possible to obtain valuable fuels and synthetic precursors. In particular, with advances and discoveries in this area known as biorefinery, it is envisioned that oil dependence as an energy source will diminish.

Since the year 2002, when Rogers et al. reported the facile dissolution of cellulose in imidazolium ILs [50], much research has been carried out in the field of cellulose processing and its degradation to simple sugars using ILs as the solvent [51–53]. Following these advances, the next development consisted of investigating lignocellulosic biomass degradation in ILs [54]. From this, cellulose—a glucose polymer—is the major component, with hemicellulose (a polysaccharide formed by hexoses and pentoses) and lignin (an aromatic, water-insoluble polymer) also being isolated. From these three main components, lignin has been the least exploited from a biorenewable feedstock perspective because it does not contain carbohydrates. Nevertheless, lignin can be used as a source of aromatic substances, which are the most desirable for biorefinery needs [55]. Therefore, efficient lignin separation from the other two components is a desirable objective. In this context, Henderson and co-workers used a low-cost PIL, pyrrolidinium acetate ([Pyrr]Ac), to dissolve lignin selectively out of lignocellulosic biomass [20]. Furthermore, the PIL could be separated from the lignin extract by vacuum distillation and thus fully recovered and re-used. In addition, Itoh et al. have developed a complete separation process for the three components of lignocellulosic mass [56]. These researchers used the IL, *N*-methyl-*N*-(2-methoxyethyl)-pyrrolidin-1-ium 2,6-diaminohexanoate ([P_{1ME}][Lys]) and found that lignin is selectively soluble at 60 °C, while cellulose becomes soluble at 80 °C. With these variations in solubility at different temperatures, lignin, cellulose, and hemicellulose could be separated from wood powder and the IL simply recovered by washing with water and re-used. The whole process is highly sustainable with the IL representing an example of an optimally designed compound presenting low toxicity owing to the anion, which is derived from an amino acid and an alicyclic cation (see Subheading 1.3.1).

Another natural polymer of potential commercial interest is keratin, which is a protein constituent in bird feathers, human hair, finger nails, and wool. This biopolymer is a by-product from poultry production and is generated in large quantities. From a commercial perspective, keratin can be used as a source of polyamide polymers; nevertheless despite this potential, the work done in the field of keratin isolation is somewhat scarce.

MacFarlane and co-workers tested several ILs for keratin extraction from turkey feathers. In one recent publication, they used two imidazolium ILs, [BMIM]Cl and [AMIM]Cl, as well as choline thioglycolate. These three ILs were able to dissolve 45% by weight keratin when heating to 130 °C [57], and then the pure keratin could be recovered through precipitation from the IL by adding water, leading to a recovery of up to 51% of the initial mass. In recent work, MacFarlane and co-workers used the PIL, *N,N*-dimethylethanolammonium formate ([DMEA][HCOO]), which dissolved 150 mg of keratin per gram of IL at 100 °C [19]. This development allows for the recovery of the IL (up to 99%), following keratin precipitation from methanol, by simple distillation to 122 °C.

These findings indicate that the field of biopolymer extraction via ILs is open to many potentially very valuable applications. For example, chitin may become a source of energy in the not too distant future, and as such investigations into its isolation through selective extraction into ILs are highly valuable.

2.1.2 Ionic Liquids as Reaction Media

The use of ILs as solvent for organic transformations is a rapidly growing field, and as such this section discusses only a few recent examples with a focus on their sustainable nature.

Galy et al. developed a green methodology for peptide synthesis employing an ionic liquid, 1,1′-carbonylbis(3-ethylimidazolium) triflate (CBEIT), as a substitute for the toxic

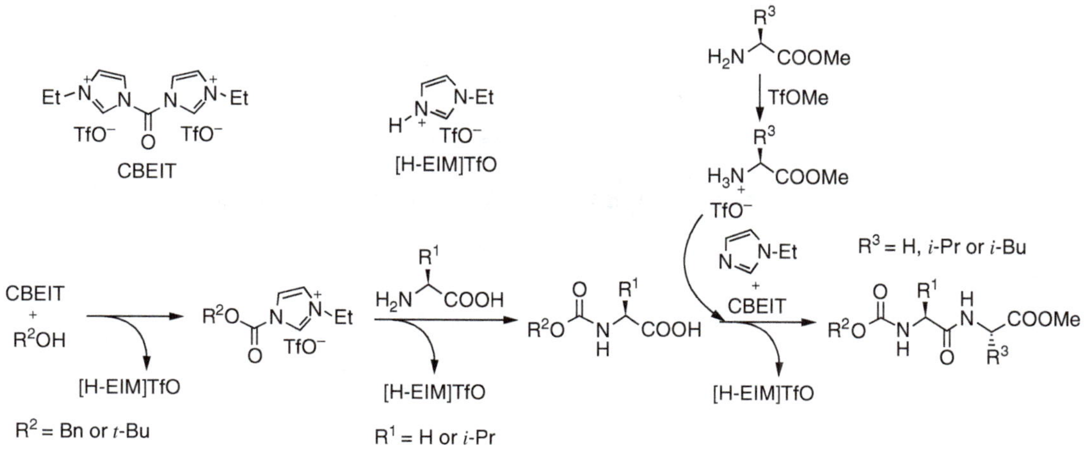

Scheme 11 Waste-free peptide synthesis

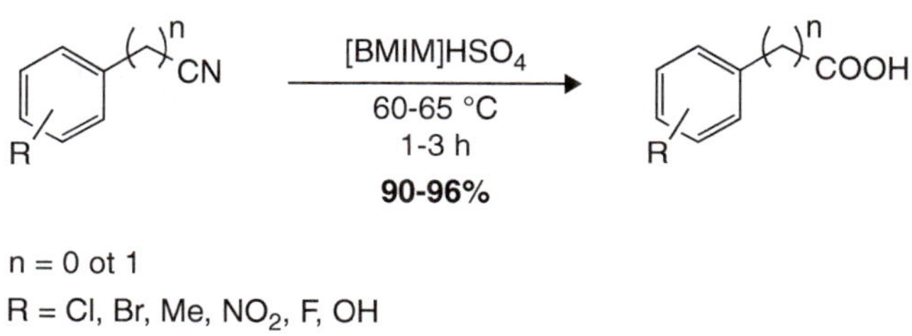

Scheme 12 Hydrolysis of nitriles using [BMM]HSO$_4$

1,1′-carbonyldiimidazole [58]. Once the coupling of the amino acids is completed, the only by-products of the reaction are CO_2 and ethylimidazolium triflate ([H-EIM]TfO), a PIL that works as the reaction solvent and can also be separated and reused. Furthermore, a waste-free protocol was developed (Scheme 11) through the in situ synthesis of the desired amino esters. Finally, the coupling was performed by reaction of both C- and N-protected amino acids in the presence of CBEIT and ethylimidazole to provide the expected couplings in yields ranging from 55% to 80%.

Recently, Kumar and co-workers have synthesized carboxylic acids from nitriles via hydrolysis using the ionic liquid [BMIM] HSO_4 as both the solvent and catalyst (Scheme 12) [59]. The method produces nitriles with excellent purity under very mild conditions, and the IL can be reused up to 5 times without any significant loss in its activity.

ILs can also be used as oxidizing agents when a suitable anion is used [60]. For example, [PMIM]IO_4 was used to oxidize a variety of sulfides to sulfones with good yields and in the absence of organic solvent (Scheme 13).

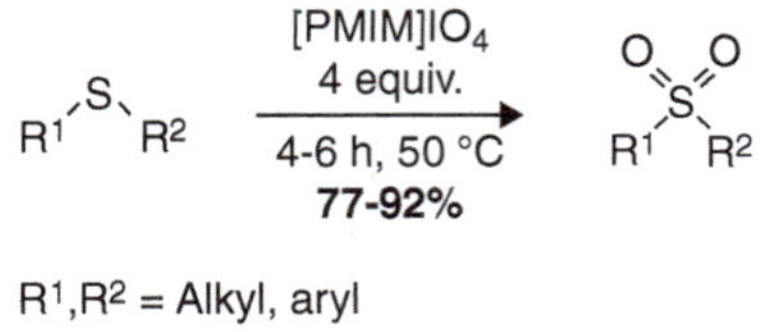

Scheme 13 Oxidation of sulfides to sulfones by an oxidant IL

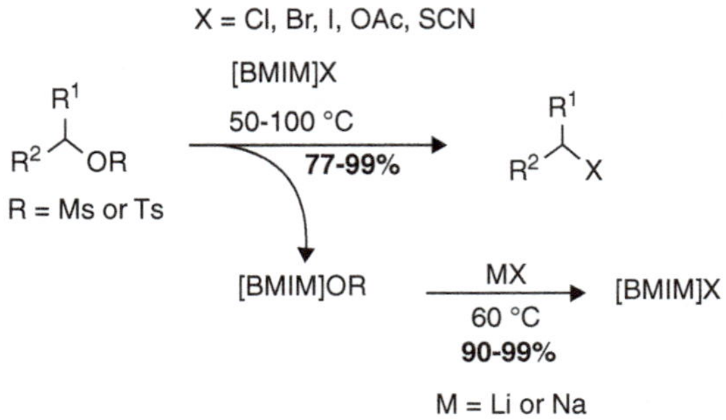

Scheme 14 Green protocol for S_N2 reaction on alkyl sulfonates

In the area of substitution reactions, a green protocol for the S_N2 displacement of sulfonate esters derived from primary and secondary alcohols was developed by Liu et al. [61]. Alkyl mesylates and tosylates were treated with [BMIM]X (X = Cl, Br, I, OAc, SCN) ILs to afford the substitution products (RX) and the corresponding [BMIM] sulfonates. Moreover, the reaction is performed without solvent and [BMIM]X can be regenerated from the [BMIM] sulfonates by anion exchange with simple alkaline salts (Scheme 14).

2.2 Ionic Liquids as Electrolytes

The use of ILs as electrolytes was among the first applications of these liquid salts. In fact, the first generation of ILs (see the Introduction) was developed specifically for this purpose. Although those ILs presented several drawbacks particularly in terms of their moisture sensitivity, the development of second-generation ILs enabled their efficient use as electrolytes. In the past 20 years, several applications of ILs in electrochemical devices have been developed including batteries, capacitors, fuel cells, biosensors, and photovoltaic cells [62–64].

Although many ILs fulfill the required properties for efficient application as electrolytes in electrochemical devices (reasonable conductivity and a large electrochemical window in terms of applied voltage), they cannot be used in liquid state due to the present architecture of commonly used electrochemical devices. Therefore, solid electrolytes are preferred, and this led to the emergence of polymeric conducting materials (PCM), which are ILs coupled to polymers acting as supports [65–67].

2.2.1 Ion Jelly®

Although IL-PCMs have replaced traditional electrolytes in a number of applications, these solid electrolytes usually involve complex operating conditions that result in a significant increase in their cost thus limiting their bulk scale application. In 2008, Vidinha et al. had the idea to create gelatins with ILs instead of water simply through the dissolution of the gelatin in the IL followed by cooling, which leads to formation of the solid. Surprisingly, this new jelly exhibited the characteristic conductivity of ILs and also the mechanical flexibility of a polymer, so Ion jelly® as an ideal PCM was born [68]. Gelatin is a widely available, inexpensive, and biodegradable protein obtained through the thermal denaturation of collagen. This accessibility combined with the green characteristics of ILs turns Ion jelly® into an environmentally friendly PCM. Furthermore, it is possible to cut or mold the Ion jelly® into any desired shape, which makes processing fairly facile. The specific Ion jelly® properties depend on the IL used; for instance, the optimal conducting Ion jelly® discovered at this time is the one derived from [BMIM][$N(CN)_2$].

In contrast, because a small amount of water is necessary for the formation of Ion jelly®, ILs with hydrophobic fluorinated anions (PF_6^-, BF_4^- and Tf_2N^-) do not form Ion jelly®. Moreover, Ion jelly® derived from ILs with long alkyl chains showed low conductivities.

Recently, Vidinha's group expanded on this concept with the idea to impregnate silk-fabrics with Ion jelly® derived from [BMIM][$N(CN)_2$] [69]. In this way, they created Silk-Ion jelly®, which is a thin, light-weight PCM showing excellent mechanical stability and high room temperature conductivity (2.9×10^{-3} S cm^{-1}), which is very similar to the pure IL. Therefore, there is no doubt that Silk-Ion jelly® will find potential applications in electrochemical devices due to both its facile and eco-friendly method of manufacture technique (use of only biomaterials and green solvents), as well as its high conductivity, mechanical stability, safety (non-explosive), and thermal stability (non-volatile). In addition, using the impregnation process, Vidinha's group applied Ion jelly® to a cellulose support to create Ion jelly membranes [70]. These membranes were tested in separation processes using supercritical CO_2 and demonstrated permeabilities similar to commercial reverse osmosis membranes. In addition, the permeability to pure gases (H_2, N_2, O_2, CO_2, and CH_4) was very similar to other IL membranes. As can be appreciated, the use of Ion jelly® is not limited to electrochemical devices, so its potential applications will no doubt continue to expand in the upcoming years.

2.3 Ionic Liquids as Catalysts

Just as in the case of their use as solvents, the application of ILs as catalysts is a field under constant expansion. Since ILs are organic salts, many of them are considered as organocatalysts (small organic

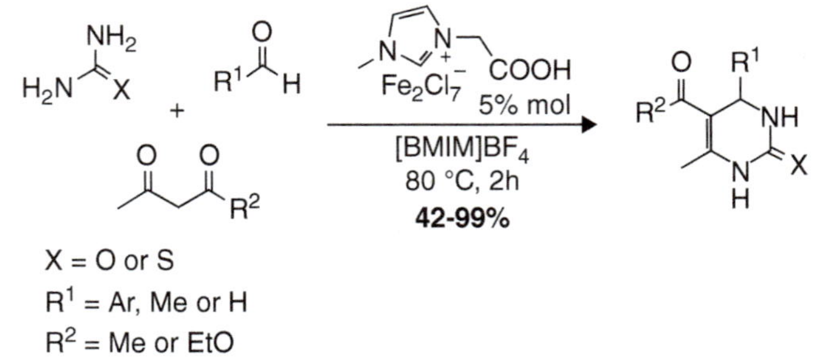

Scheme 15 Biginelli reaction catalyzed by an iron-based IL

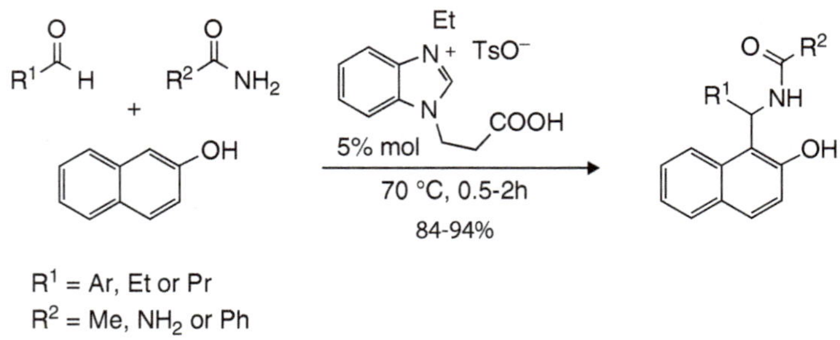

Scheme 16 Synthesis of 1-amidoalkyl Naphthols by benzilimidazolium IL

molecules which catalyze organic reactions in the absence of metals). Furthermore, their recyclability establishes ILs as excellent candidates to fulfill the principles of green chemistry. In this section, some examples of current literature will be presented to illustrate the multiple uses and benefits that can be obtained with ILs as catalysts.

Very recently, Neto et al. used an iron-based IL which acts both as catalyst and solvent for the multicomponent Biginelli reaction [71]. The reaction product could be easily removed by filtration, and the iron-based IL was re-used up 8 times without any loss in efficiency (Scheme 15).

Recently, Muskawar and co-workers developed benzimidazolium ILs as catalysts for the synthesis of 1-amidoalkyl naphthol derivatives in high yields [72]. This reaction was performed under neat conditions with low IL loading, and the IL could be reused up to 7 times without significant loss of activity (Scheme 16).

Song et al. designed and prepared the first aza-crown ether complex cation ILs, which were tested as catalysts in several organic reactions with excellent results [73]. In these reactions, the ILs exhibited cooperative catalytic activity between the anion and cation. As shown in Scheme 17, depending on the reaction being catalyzed, certain specific anion/cation combinations proved to be optimal.

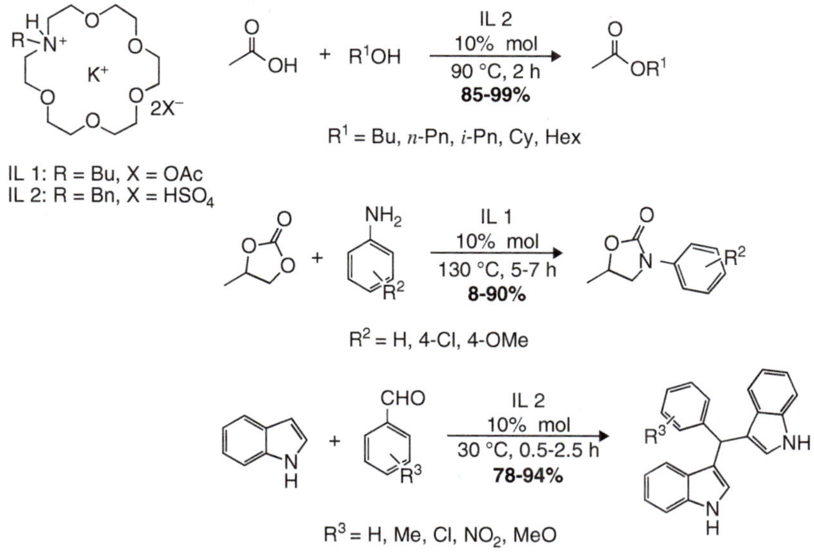

Scheme 17 Organic reactions in aza-crown ether cation complex ILs

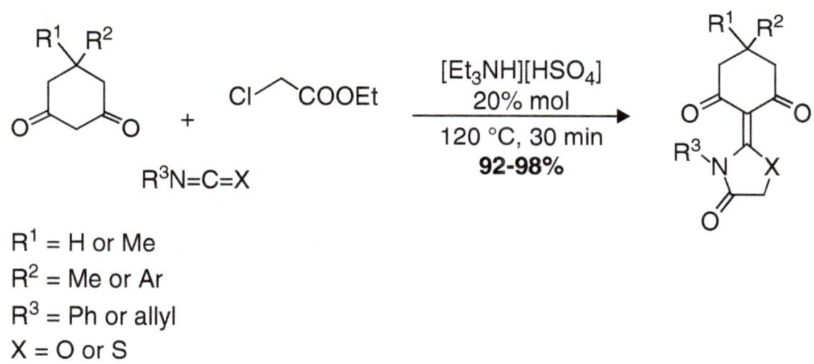

Scheme 18 Green one-pot thiazolidine/oxazolidine synthesis

Most recently, Malla and co-workers developed a green, one-pot three-component reaction for the synthesis of thiazolidine/oxazolidine derivatives in high yield and purity [74]. The IL used was the PIL, $[Et_3NH][HSO_4]$, which was used neat and reused up to 5 times without any notable deterioration in activity (Scheme 18).

2.3.1 Ionic Liquids as Catalysts in Asymmetric Reactions

Enantioselectivity is an important aspect to be considered when designing a molecule presenting centers of chirality, especially when the objective molecule is a drug. Since the thalidomide incident [75–77], determination of the absolute configuration and biological effect of each stereoisomer turned a mandatory job.

In this context, 15 years ago and essentially simultaneously, asymmetric organocatalysis with chiral ionic liquids (CILs) was reported for the first time [78, 79]. Since then, CILs have become

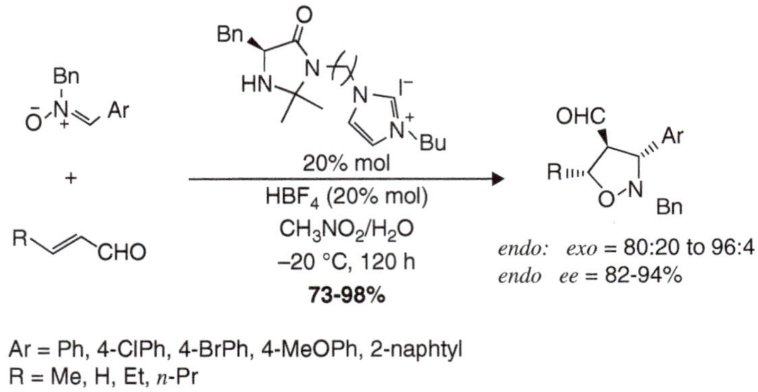

Scheme 19 Enantioselective 1,3-dipolar cycloadition

important green catalysts in asymmetric synthesis and furthermore have made it possible for many organic enantioselective transformations to be carried out in water.

In 2012, Shen et al. synthesized and applied a recyclable IL-supported chiral imidazolidinone in the enantioselective 1,3-dipolar cycloaddition between nitrones and α,β-unsaturated aldehydes to set three stereocenters in the oxazolidine product [80]. The reaction proceeded in good yield and high enantioselectivities preferentially forming the *endo* diastereomer (Scheme 19). Finally, the CIL could be recovered and recycled up to 5 times without any observable decrease in catalytic activity.

The aldol reaction has probably been the most studied reaction in asymmetric organocatalysis, following the pioneering report by List, Lerner, and Barbas III [79]. It is well known that the aldol reaction is one of the main synthetic strategies for building C–C bonds. Furthermore, this methodology can give rise to two new stereocenters. In 2012, Zlotin et al. developed a C_2-*bis*-prolinamide supported on an IL as an organocatalyst for an asymmetric aldol reaction in water [81]. This IL showed excellent results both in yields and enantioselectivities and was recycled 15 times without significant loss in selectivity (Scheme 20).

The Michael addition reaction has also been widely explored in asymmetric organocatalysis. In 2013, Headley and co-workers synthesized and tested a quaternary ammonium CIL derived from (*S*)-proline, to be used in the asymmetric Michael reaction under aqueous conditions [82]. These researchers also employed an IL-supported benzoic acid as an additive and obtained excellent yields and stereoselectivities across a range of substrates (Scheme 21). In addition, the catalyst was used at loadings of 5 mol%, and the reaction media could be recycled up to 10 times without any significant loss of enantioselectivity (Scheme 21).

More recently, Zlotin et al. have developed a primary amine supported on an *N*-(carboxyalkyl)imidazolium IL for the asymmetric Michael reaction between 4-hydroxy-2*H*-chromen-2-one and

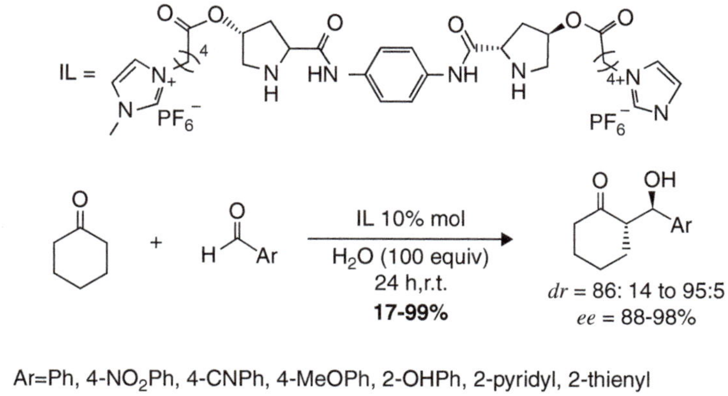

Scheme 20 Efficient asymmetric aldol reaction in water catalyzed by a CIL

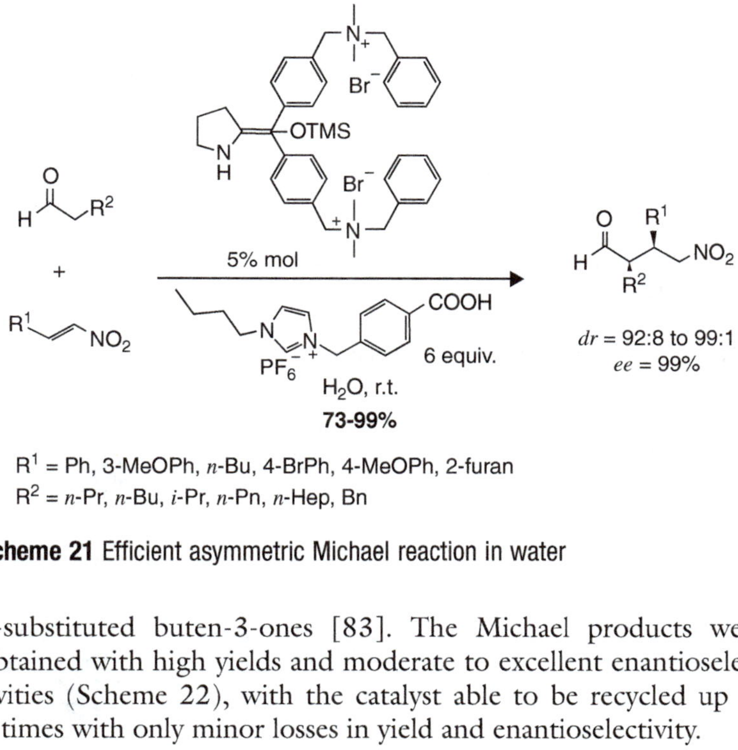

Scheme 21 Efficient asymmetric Michael reaction in water

1-substituted buten-3-ones [83]. The Michael products were obtained with high yields and moderate to excellent enantioselectivities (Scheme 22), with the catalyst able to be recycled up to 5 times with only minor losses in yield and enantioselectivity.

2.3.2 Ionic Liquids as Supported Catalysts

Although ILs are very efficient in catalysis, it is normal to lose some amount of catalyst with each reuse cycle, which translates to a decrease in the reaction performance. With this in mind, supported ionic liquids (SILs) were developed in order to overcome this problem.

An interesting SIL was developed by Yin et al. supporting an (*S*)-proline imidazolium IL on magnetic nanoparticles (SiO_2@-Fe_2O_4) in order to form an easy to recover magnetic asymmetric catalyst [84]. This SIL was used in the asymmetric aldol reaction in

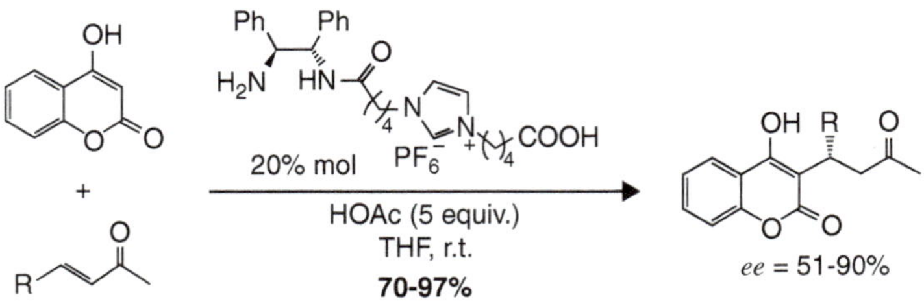

R = Ph, 4-MeOPH, 3-pyridyl, 2-thi ebyl, Cy, ferrocenyl

Scheme 22 Enantioselective synthesis of 3-substituted chromenones by Michael reaction

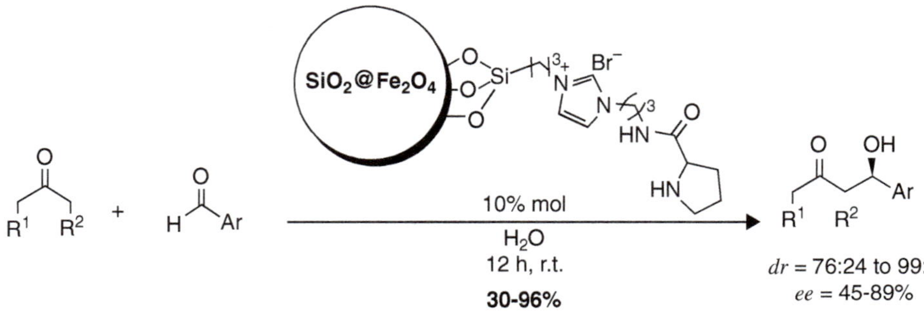

Ar = Ph, 4-NO_2Ph, 2-NO_2Ph, 2-ClPh, 4-BrPh, 2-naphtyl
$R^1 = R^2$ = H, -$(CH_2)_3$-, -$(CH_2)_2$-

Scheme 23 Magnetically recoverable chiral SIL used in the asymmetric aldol reaction

water giving promising results with the catalyst being easily recovered using a permanent magnet and reused up to 5 times without significant loss in activity (Scheme 23).

Recently, Shinde and co-workers synthesized a polystyrene-supported solid catalyst with one active molecule of a *tert*-butyl alcohol imidazolium on the polymer. This catalyst was examined in the aliphatic nucleophilic substitution of primary and secondary substrates using alkali metal salts as the nucleophiles [85]. The SIL acts as a phase transfer catalyst, enhancing the reactivity of the alkali metal salts and making product isolation very simple by filtration. Furthermore, the catalyst can be reused up to 5 cycles without any decrease in reaction yield (Scheme 24).

Finally, Sharma et al. developed a novel silica-supported sulphonic acid catalyst which was coated with [BMIM]PF_6 [86]. This SIL was used in the one-pot synthesis of 1,4-dihydropyridines, affording the desired products with excellent yields in short reaction times (Scheme 25). Moreover, the reaction was performed without solvent, and the catalyst was recycled up to 7 times with only a negligible decrease in yield.

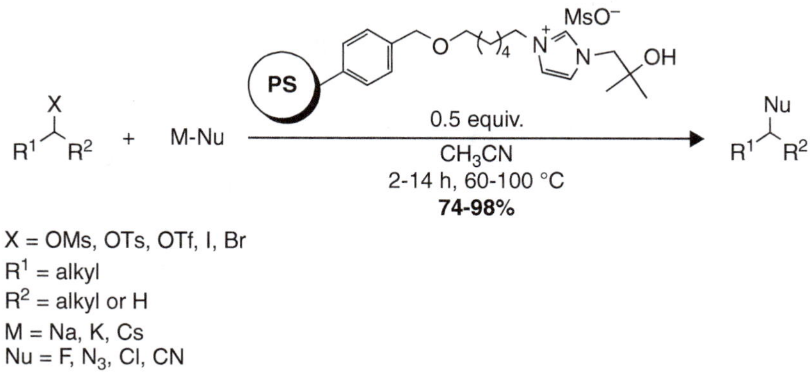

Scheme 24 SIL for efficient S_N2 reaction

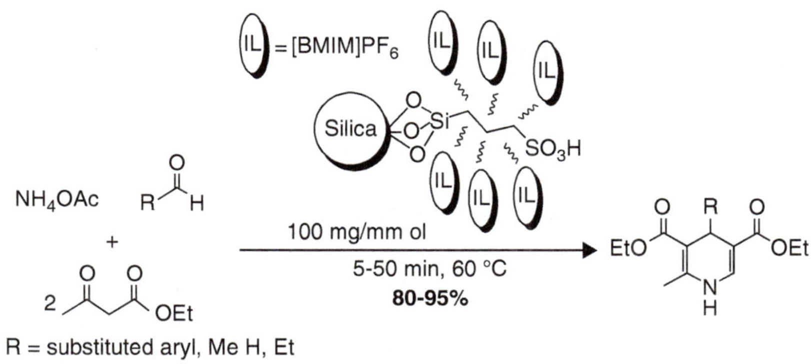

Scheme 25 One-pot synthesis of 1,4-dihydropyridines using a IL coated solid supported catalyst

As it can be appreciated, ILs have evolved up to a point where they are closely associated to the fields of organocatalysis and asymmetric synthesis and by definition green chemistry and continue to find applications in numerous different organic reactions and transformations.

2.4 Pharmaceutical Applications of Ionic Liquids

The use of ILs in pharmaceutical applications is at an early stage that originates with the use of these molecules as green solvents for synthetic processes. Later, it was found that ILs can also be employed as active pharmaceutical ingredients (APIs). In this final section, current applications of ILs in the pharmaceutical industry will be presented.

2.4.1 Ionic Liquids in Drug Production

As it was already mentioned, in the pharmaceutical field the use of ILs as reaction solvent during drug production is quite attractive, owing to their green properties. Furthermore (see Subheading 2.3), reactions carried out in IL media are usually faster and can give products with higher purities and which are more easily isolated. An example is the synthesis of pravadoline, a

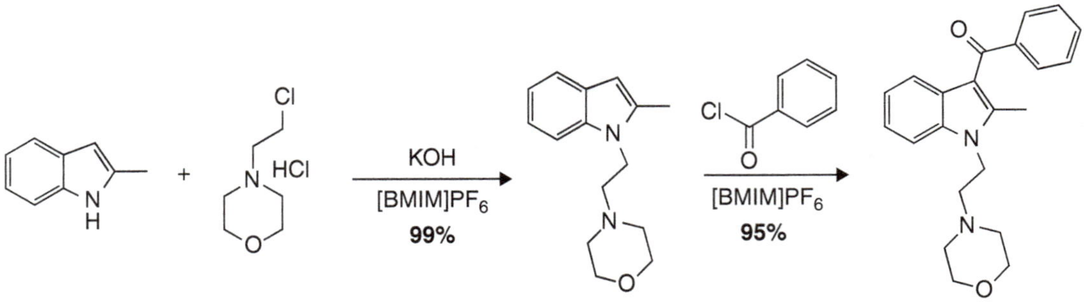

Scheme 26 Green synthesis of Pravadoline in [BMIM]PF$_6$

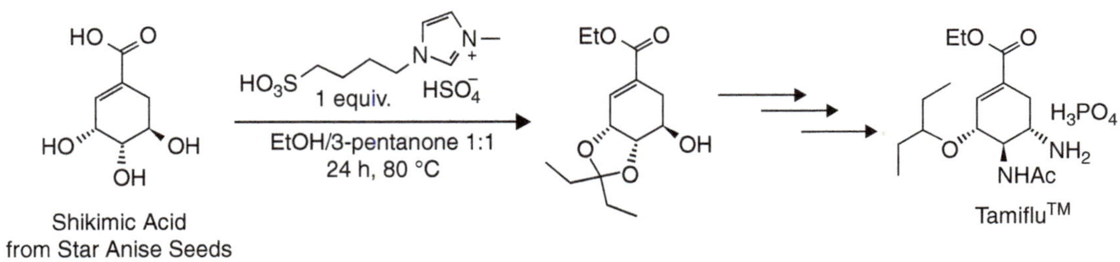

Scheme 27 Reactive dissolution of shikimic acid in sulfonic acid IL

non-steroidal anti-inflammatory drug in [BMIM]PF$_6$ [87]. The product was obtained in an overall yield of 94% over two steps with the IL easily recovered and reused (Scheme 26).

In this context, Kumar and Malhotra compared the outcome of the reactions involved in the synthesis of nucleoside anti-viral drugs, comparing ILs (imidazolium ILs) with other polar solvents (pyridine, acetonitrile, TFA, and DMA) [88]. ILs turned out to be better solvents (in particular [BMIM]TFA) in terms of both their solubility characteristics and reaction rate.

In the year 2011, Bica et al. used a reactive dissolution with a sulfonic IL to isolate from star anise seeds, and to react, a shikimic acid derivative in the synthesis of the anti-influenza drug Tamiflu™ [89] (Scheme 27).

ILs can be used in the preparation of pharmaceutical formulations or as drug carriers. In this context, Goto et al. prepared a microemulsion system consisting of oil and an imidazolium IL, which was effective to both dissolve drugs that are poorly soluble in water, and be used as a drug carrier [90]. In another report, Mahkam and co-workers developed a pH-responsive controlled drug release system for insulin by modifying silica nanoparticles with an imidazolium IL (Fig. 6) [91]. This system works in the following manner: as a consequence of the positive charges surrounding the silica nanoparticles, anionic molecules can be efficiently adsorbed inside of the nanochannels with minimal release under acidic pH conditions. At neutral or slightly basic

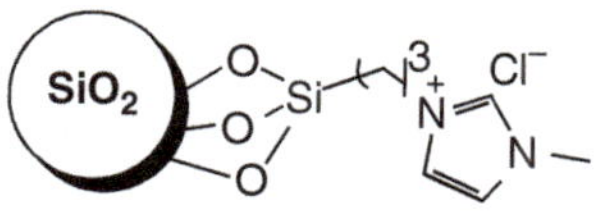

Fig. 6 IL functionalized silica nanoparticles for drug delivery

physiological pH, because of the deprotonation of surface silanol groups, the strong electrostatic repulsion opens the nanochannels and increases the release rate of the adsorbed drug molecules.

2.4.2 Ionic Liquids as Potential Drug Candidates

With the emergence of third-generation ILs, their biological properties were evaluated in addition to their chemical and physical properties (see the Introduction). API-ILs have higher solubility in water relative both to the neutral API and the ionic form of API. Because of this increased solubility, it is expected that API-ILs will also exhibit higher bioavailability. Finally, an important advantage when using an API-IL instead of a neutral API drug is the elimination of the possibility of polymorphism, which is an undesirable property that can cause a solid to crystallize in different crystalline forms and hence modify the pharmacological profile of the drug [92].

In 2007, Rogers et al. synthesized three API-ILs: lidocaine docusate, ranitidine docusate, and didecyldimethylammonium ibuprofen, which exhibited improved properties relative to the original drugs [7]. For example, lidocaine docusate (LD) is a hydrophobic RTIL which, when compared to lidocaine hydrochloride, exhibits better solubility, increased thermal stability, and a significant enhancement in efficacy as a topical analgesic due to increased adsorption.

Although not a drug, the IL 1-methyl-3-[3,4-bis(dodecyloxy)-benzyl]-4*H*-imidazolium chloride is biologically relevant as it can act as a vector for siRNA transfection, thus enabling the introduction of RNA into a living cell [93]. This IL showed high transfection efficacy with more than 80% inhibition of the targeted gene. With these results, these type of IL can become the new generation of transfection reagents for RNA interference.

Petrovski et al. synthesized ampicillin ILs containing various cations and evaluated their physical and thermal properties [94]. Cholinium ampicillin was the most interesting example of the ILs prepared because it presented a low melting point, very high water solubility, as well as good biocompatibility and low toxicity.

In 2013, Feder-Kubis and Tomczuk developed ILs (ammonium, imidazolium, alkoxymethylimidazolium, and pyridinium) and chlorides incorporating a menthol group and tested their antimicrobial activity against different microorganisms [95]. It was concluded that the ILs with the higher potency and broad spectrum activity were the ammonium ILs, followed by the imidazoliums and the alkoxymethylimidazoliums, with the least active being the pyridinium ILs (Fig. 7).

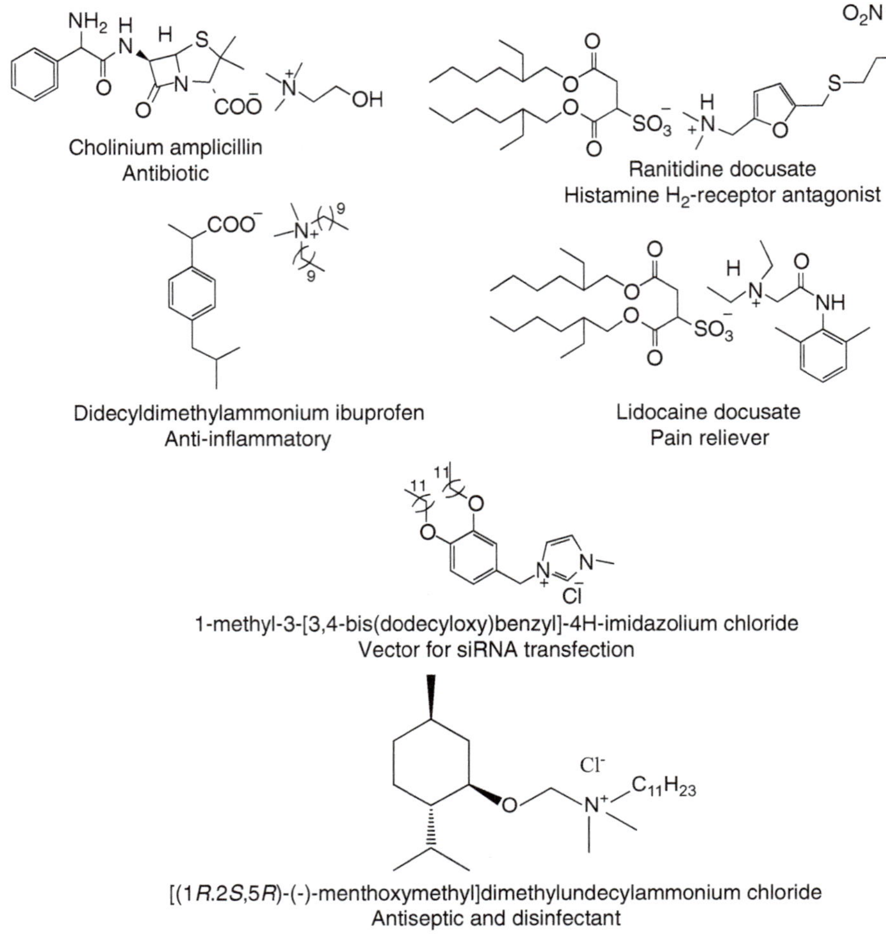

Fig. 7 Biologically active ILs

In conclusion, ILs are versatile, sustainable compounds that can be applied in many fields. The present text emphasizes some current applications, with a focus on green chemistry.

References

1. Walden P (1914) Molecular weights and electrical conductivity of several fused salts. Bull Acad Imp Sci St Pétersbourg 8:405–422
2. Wilkes JS, Levisky JA, Wilson RA, Hussey CL (1982) Dialkylimidazolium chloroaluminate melts: a new class of room-temperature ionic liquids for electrochemistry, spectroscopy, and synthesis. Inorg Chem 21:1263–1264
3. Boon J, Levisky JA, Pflug JL, Wilkes JS (1986) Friedel-crafts reactions in ambient-temperature molten salts. J Org Chem 51:480–483
4. Wilkes JS, Zaworotko MJ (1992) Air and water stable 1-ethyl-3-methylimidazolium based ionic liquids. J Chem Soc Chem Commun:965–967
5. Visser AE, Swatloski RP, Reichert WM, Mayton R, Sheff S, Wierzbicki A, Davis JH, Rogers RD (2001) Task-specific ionic liquids for the extraction of metal ions from aqueous solutions. Chem Commun:135–136
6. Fei Z, Dyson PJ (2013) The making of iLiquids – the chemist's equivalent of the iPhone. Chem Commun 49:2594–2596
7. Hough WL, Smiglak M, Rodríguez H, Swatloski RP, Spear SK, Daly DT, Pernak J, Grisel JE, Carliss RD, Soutullo MD, Davis JH,

Rogers RD (2007) The third evolution of ionic liquids: active pharmaceutical ingredients. New J Chem 31:1429–1436

8. Wasserscheid P, Keim W (2000) Ionic liquids-new "solutions" for transition metal catalysis. Angew Chem Int Ed 39:3772–3789
9. O'Neil MJ (2006) The Merck Index - an encyclopedia of chemicals, drugs, and biologicals. Merck and Co., Whitehouse Station, NJ, p 1480
10. Holbrey JD, Rogers RD (2002) Melting points and phase diagrams. In: Wasserscheid P, Welton T (eds) Ionic liquids in synthesis. Wiley-VCH Verlag GmbH & Co KGaA, Weinheim
11. Vitz J, Erdmenger T, Haensch C, Schubert US (2009) Extended dissolution studies of cellulose in imidazolium based ionic liquids. Green Chem 11:417–424
12. Holbrey JD, Seddon KR (1999) The phase behaviour of 1-alkyl-3-methylimidazolium tetrafluoroborates; ionic liquids and ionic liquid crystals. J Chem Soc Dalton Trans:2133–2140
13. Bonhôte P, Dias A-P, Papageorgiou N, Kalyanasundaram K, Grätzel M (1996) Hydrophobic, highly conductive ambient-temperature molten salts. Inorg Chem 35:1168–1178
14. Chan BKM, Chang NH, Grimmett MR (1977) The synthesis and thermolysis of imidazole quaternary salts. Aust J Chem 30:2005–2013
15. Clough MT, Geyer K, Hunt PA, Mertes J, Welton T (2013) Thermal decomposition of carboxylate ionic liquids: trends and mechanisms. Phys Chem Chem Phys 15:20480–20495
16. Maton C, De Vos N, Stevens CV (2013) Ionic liquid thermal stabilities: decomposition mechanisms and analysis tools. Chem Soc Rev 42:5963–5977
17. Mantz RA, Trulove PC (2002) Viscosity and density of ionic liquids. In: Wasserscheid P, Welton T (eds) Ionic liquids in synthesis. Wiley-VCH Verlag GmbH & Co KGaA, Weinheim
18. Slattery JM, Daguenet C, Dyson PJ, Schubert TJS, Krossing I (2007) How to predict the physical properties of ionic liquids: a volume-based approach. Angew Chem Int Ed 46:5384–5388
19. Idris A, Vijayaraghavan R, Patti AF, MacFarlane DR (2014) Distillable protic ionic liquids for keratin dissolution and recovery. ACS Sustain Chem Eng 2:1888–1894
20. Achinivu EC, Howard RM, Li G, Graczb H, Henderson WA (2014) Lignin extraction from biomass with protic ionic liquids. Green Chem 16:1114–1119
21. King AWT, Asikkala J, Mutikainen I, Järvi P, Kilpeläinen I (2011) Distillable acid-base conjugate ionic liquids for cellulose dissolution and processing. Angew Chem Int Ed 50:6301–6305
22. Löwe H, Axinte RD, Breuch D, Hofmann C, Petersen JH, Pommersheim R, Wang A (2010) Flow chemistry: imidazole-based ionic liquid syntheses in micro-scale. Chem Eng J 163:429–437
23. Zimmermann J, Ondruschka B, Stark A (2010) Efficient synthesis of 1,3-dialkylimidazolium-based ionic liquids: the modified continuous Radziszewski reaction in a microreactor setup. Org Process Res Dev 14:1102–1109
24. Esposito D, Kirchhecker S, Antonietti M (2013) A sustainable route towards imidazolium building blocks based on biomass molecules. Chem Eur J 19:15097–15100
25. Kirchhecker S, Antonietti M, Esposito D (2014) Hydrothermal decarboxylation of amino acid derived imidazolium zwitterions: a sustainable approach towards ionic liquids. Green Chem 16:3705–3709
26. Sosnowska A, Barycki M, Zaborowska M, Rybinska A, Puzyn T (2014) Towards designing environmentally safe ionic liquids: the influence of the cation structure. Green Chem 16:4749–4757
27. Hough-Troutman WL, Smiglak M, Griffin S, Reichert WM, Mirska I, Jodynis-Liebert J, Adamska T, Nawrot J, Stasiewicz M, Rogers RD, Pernak J (2009) Ionic liquids with dual biological function: sweet and anti-microbial, hydrophobic quaternary ammonium-based salts. New J Chem 33:26–33
28. Romero A, Santos A, Tojo J, Rodríguez A (2008) Toxicity and biodegradability of imidazolium ionic liquids. J Hazard Mater 151:268–273
29. Wells AS, Coombe VT (2006) On the freshwater ecotoxicity and biodegradation properties of some common ionic liquids. Org Process Res Dev 10:794–798
30. Stolte S, Matzke M, Arning J, Böschen A, Pitner W-R, Welz-Biermann U, Jastorff B, Ranke J (2007) Effects of different head groups and functionalised side chains on the aquatic toxicity of ionic liquids. Green Chem 9:1170–1179
31. Costello DM, Brown LM, Lamberti GA (2009) Acute toxic effects of ionic liquids on zebra mussel (Dreissena polymorpha) survival and feeding. Green Chem 11:548–553
32. Pham TPT, Cho C-W, Min J, Yun Y-S (2008) Alkyl-chain length effects of imidazolium and pyridinium ionic liquids on photosynthetic

response of Pseudokirchneriella subcapitata. J Biosci Bioeng 105:425–428

33. Docherty KM, Kulpa CF (2005) Toxicity and antimicrobial activity of imidazolium and pyridinium ionic liquids. Green Chem 7:185–189
34. Kumar RA, Papaïconomou N, Lee J-M, Salminen J, Clark DS, Prausnitz JM (2009) In vitro cytotoxicities of ionic liquids: effect of cation rings, functional groups, and anions. Environ Toxicol 24:388–395
35. Stolte S, Arning J, Bottin-Weber U, Müller A, Pitner W-R, Welz-Biermann U, Jastorff B, Ranke J (2007) Effects of different head groups and functionalised side chains on the cytotoxicity of ionic liquids. Green Chem 9:760–767
36. Ying G-G (2006) Fate, behavior and effects of surfactants and their degradation products in the environment. Environ Int 32:417–431
37. Pernak J, Chwala P (2003) Synthesis and antimicrobial activities of choline-like quaternary ammonium chlorides. Eur J Med Chem 38:1035–1042
38. Couling DJ, Bernot RJ, Docherty KM, Dixon JK, Maginn EJ (2006) Assessing the factors responsible for ionic liquid toxicity to aquatic organisms via quantitative structure-property relationship modeling. Green Chem 8:82–90
39. Pernak J, Syguda A, Mirska I, Pernak A, Nawrot J, Pradzynska A, Griffin ST, Rogers RD (2007) Choline-derivative-based ionic liquids. Chem Eur J 13:6817–6827
40. Petkovic M, Ferguson JL, Bohn A, Trindade J, Martins I, Carvalho MB, Leitão MC, Rodrigues C, Garcia H, Ferreira R, Seddon KR, Rebelo LPN, Silva Pereira C (2009) Exploring fungal activity in the presence of ionic liquids. Green Chem 11:889–894
41. Dipeolu O, Green E, Stephens G (2009) Effects of water-miscible ionic liquids on cell growth and nitro reduction using Clostridium sporogenes. Green Chem 11:397–401
42. Coleman D, Gathergood N (2010) Biodegradation studies of ionic liquids. Chem Soc Rev 39:600–637
43. Morrissey S, Pegot B, Coleman D, Garcia MT, Ferguson D, Quilty B, Gathergood N (2009) Biodegradable, non-bactericidal oxygen-functionalised imidazolium esters: a step towards 'greener' ionic liquids. Green Chem 11:475–483
44. Grabinska-Sota E, Kalka J (2003) An assessment of the toxicity of pyridinium chlorides and their biodegradation intermediates. Environ Int 28:687–690
45. Boethling RS, Sommer E, DiFiore D (2007) Designing small molecules for biodegradability. Chem Rev 107:2207–2227
46. Petkovic M, Ferguson JL, Gunaratne HQN, Ferreira R, Leitão MC, Seddon KR, Rebelo LPN, Silva Pereira C (2010) Novel biocompatible cholinium-based ionic liquids-toxicity and biodegradability. Green Chem 12:643–649
47. Neumann J, Cho C-W, Steudte S, Köser J, Uerdingen M, Thöming J, Stolte S (2012) Biodegradability of fluoroorganic and cyano-based ionic liquid anions under aerobic and anaerobic conditions. Green Chem 14:410–418
48. Deng Y, Beadham I, Ghavre M, Costa Gomes MF, Gathergood N, Husson P, Légeret B, Quilty B, Sancelme M, Besse-Hoggan P (2015) When can ionic liquids be considered readily biodegradable? biodegradation pathways of pyridinium, pyrrolidinium and ammonium-based ionic liquids. Green Chem 17:1479–1491
49. Stasiewicz M, Mulkiewicz E, Tomczak-Wandzel R, Kumirska J, Siedlecka EM, Golebiowski M, Gajdus J, Czerwicka M, Stepnowski P (2008) Assessing toxicity and biodegradation of novel, environmentally benign ionic liquids (1-alkoxymethyl-3-hydroxypyridinium chloride, saccharinate and acesulfamates) on cellular and molecular level. Ecotoxicol Environ Saf 1:157–165
50. Swatloski RP, Spear SK, Holbrey JD, Rogers RD (2002) Dissolution of cellose with ionic liquids. J Am Chem Soc 124:4974–4975
51. Remsing RC, Swatloski RP, Rogers RD, Moyona G (2006) Mechanism of cellulose dissolution in the ionic liquid 1-*n*-butyl-3-methylimidazolium chloride: a ^{13}C and $^{35/37}Cl$ NMR relaxation study on model systems. Chem Commun:1271–1273
52. Shill K, Padmanabhan S, Xin Q, Prausnitz JM, Clark DS, Blanch HW (2011) Ionic liquid pretreatment of cellulosic biomass: enzymatic hydrolysis and ionic liquid recycle. Biotechnol Bioeng 108:511–520
53. Hauru LKJ, Hummel M, King AWT, Kilpeläinen I, Sixta H (2012) Role of solvent parameters in the regeneration of cellulose from ionic liquid solutions. Biomacromolecules 13:2896–2905
54. Brandt A, Gräsvik J, Halletta JP, Welton T (2013) Deconstruction of lignocellulosic biomass with ionic liquids. Green Chem 15:550–583
55. Zakzeski J, Bruijnincx PCA, Jongerius AL, Weckhuysen BM (2010) The catalytic valorization of lignin for the production of renewable chemicals. Chem Rev 110:3552–3599
56. Hamada Y, Yoshida K, Asai R-I, Hayase S, Nokami T, Izumib S, Itoh T (2013) A possible

means of realizing a sacrifice-free three component separation of lignocellulose from wood biomass using an amino acid ionic liquid. Green Chem 15:1863–1868

57. Idris A, Vijayaraghavan R, Rana UA, Fredericks D, Patti AF, MacFarlane DR (2013) Dissolution of feather keratin in ionic liquids. Green Chem 15:525–534
58. Galy N, Mazières MR, Plaquevent J-C (2013) Toward waste-free peptide synthesis using ionic reagents and ionic liquids as solvents. Tetrahedron Lett 54:2703–2705
59. Kumar S, Dixit SK, Awasthi SK (2014) An efficient one pot method for synthesis of carboxylic acids from nitriles using recyclable ionic liquid [bmim]HSO_4. Tetrahedron Lett 55:3802–3804
60. Ahammeda S, Kundu D, Siddiqui MN, Ranu BC (2015) Metal and solvent free selective oxidation of sulfides to sulfone using bifunctional ionic liquid [pmim]IO_4. Tetrahedron Lett 56:335–337
61. Liu Y, Xu Y, Jung SH, Chae J (2012) A facile and green protocol for nucleophilic substitution reactions of sulfonate esters by recyclable ionic liquids [bmim][X]. Synlett 23:2692–2698
62. Byrne N, Howlett PC, MacFarlane DR, Forsyth M (2005) The zwitterion effect in ionic liquids: towards practical rechargeable lithium-metal batteries. Adv Mater 17:2497–2501
63. Fernicola A, Scrosati B, Ohno H (2006) Potentialities of ionic liquids as new electrolyte media in advanced electrochemical devices. Ionics 12:95–102
64. Galinski M, Lewandowski A, Stępniak I (2006) Ionic liquids as electrolytes. Electrochim Acta 51:5567–5580
65. Matsumi N, Sugai K, Miyake M, Ohno H (2006) Polymerized ionic liquids via hydroboration polymerization as single ion conductive polymer electrolytes. Macromolecules 39:6924–6927
66. Tiyapiboonchaiya C, Pringle JM, MacFarlane DR, Forsyth M, Sun J (2003) Polyelectrolyte-in-ionic-liquid electrolytes. Macromol Chem Phys 204:2147–2154
67. Stephan AM (2006) Review on gel polymer electrolytes for lithium batteries. Eur Polym J 42:21–42
68. Vidinha P, Lourenço NMT, Pinheiro C, Brás AR, Carvalho T, Santos-Silva T, Mukhopadhyay A, Romão MJ, Parola J, Dionisio M, Cabral JMS, Afonso CAM, Barreiros S (2008) Ion jelly: a tailor-made conducting material for smart electrochemical devices. Chem Commun:5842–5844
69. Rana S, Carvalho T, Fangueiro R, Vidinha P (2013) Silk-Ion Jelly: a novel ion conducting polymeric material with high conductivity and excellent mechanical stability. Polym Adv Technol 24:191–196
70. Couto RM, Carvalho T, Neves LA, Ruivo RM, Vidinha P, Paiva A, Coelhoso IM, Barreiros S, Simões PC (2013) Development of Ion-Jelly® membranes. Sep Purif Technol 106:22–31
71. Ramos LM, Guido BC, Nobrega CC, Corrêa JR, Silva RG, De Oliveira HCB, Gomes AF, Gozzo FC, Neto BAD (2013) The Biginelli reaction with an imidazolium-tagged recyclable iron catalyst: kinetics, mechanism, and antitumoral activity. Chem Eur J 19:4156–4168
72. Muskawar PN, Kumar SS, Bhagat PR (2013) Carboxyl-functionalized ionic liquids based on benzimidazolium cation: study of Hammett values and catalytic activity towards one-pot synthesis of 1-amidoalkyl naphthols. J Mol Catal A Chem 380:112–117
73. Song Y, Cheng C, Jing H (2014) Aza-crown ether complex cation ionic liquids: preparation and applications in organic reactions. Chem Eur J 20:12894–12900
74. Malla AM, Parveen M, Ahmad F, Azaz S, Alam M (2015) [Et_3NH][HSO_4]-catalyzed eco-friendly and expeditious synthesis of thiazolidine and oxazolidine derivatives. RSC Adv 5:19552–19569
75. Mcbride WG (1961) Thalidomide and congenital abnormalities. Lancet 278:1358
76. Lenz W, Pfeiffer RA, Kosenow W, Hayman DJ (1962) Thalidomide and congenital abnormalities. Lancet 279:45–46
77. Speirs AL, Aberd MD (1962) Thalidomide and congenital abnormalities. Lancet 279:303–305
78. Earle MJ, McCormac PB, Seddon KR (1999) Diels–Alder reactions in ionic liquids. A safe recyclable alternative to lithium perchlorate-diethyl ether mixtures. Green Chem 1:23–25
79. List B, Lerner RA, Barbas CF III (2000) Proline-catalyzed direct asymmetric aldol reactions. J Am Chem Soc 122:2395–2396
80. Shen Z-L, Goh KKK, Wong CHA, Loo W-Y, Yang Y-S, Lu J, Loh T-P (2012) Synthesis and application of a recyclable ionic liquid-supported imidazolidinone catalyst in enantioselective 1,3-dipolar cycloaddition. Chem Commun 48:5856–5858
81. Kochetkov SV, Kucherenko AS, Kryshtal GV, Zhdankina GM, Zlotin SG (2012) Simple ionic liquid supported C_2-symmetric bisprolinamides as recoverable organocatalysts for the asymmetric aldol reaction in the presence of water. Eur J Org Chem:7129–7134

82. Ghosh SK, Qiao Y, Ni B, Headley AD (2013) Asymmetric Michael reactions catalyzed by a highly efficient and recyclable quaternary ammonium ionic liquid-supported organocatalyst in aqueous media. Org Biomol Chem 11:1801–1804
83. Kucherenko AS, Lisnyak VG, Chizhov AO, Zlotin SG (2014) Primary amine attached to an *N*-(carboxyalkyl)imidazolium cation: a recyclable organocatalyst for the asymmetric Michael reaction. Eur J Org Chem:3808–3813
84. Kong Y, Tan R, Zhao L, Yin D (2013) L-Proline supported on ionic liquid-modified magnetic nanoparticles as a highly efficient and reusable organocatalyst for direct asymmetric aldol reaction in water. Green Chem 15:2422–2433
85. Shinde SS, Patil SN (2014) One molecule of ionic liquid and *tert*-alcohol on a polystyrene-support as catalysts for efficient nucleophilic substitution including fluorination. Org Biomol Chem 12:9264–9271
86. Sharma P, Gupta M (2015) Silica functionalized sulphonic acid coated with ionic liquid: an efficient and recyclable heterogeneous catalyst for the one-pot synthesis of 1,4-dihydropyridines under solvent-free conditions. Green Chem 17:1100–1106
87. Earle MJ, McCormac PB, Seddon KR (2000) The first high yield green route to a pharmaceutical in a room temperature ionic liquid. Green Chem 2:261–262
88. Kumar V, Malhotra SV (2008) Synthesis of nucleoside-based antiviral drugs in ionic liquids. Bioorg Med Chem Lett 18:5640–5642
89. Ressmann AK, Gaertner P, Bica K (2011) From plant to drug: ionic liquids for the reactive dissolution of biomass. Green Chem 13:1442–1447
90. Moniruzzaman M, Kamiya N, Goto M (2010) Ionic liquid based microemulsion with pharmaceutically accepted components: formulation and potential applications. J Colloid Interface Sci 352:136–142
91. Mahkam M, Hosseinzadeh F, Galehassadi M (2012) Preparation of ionic liquid functionalized silica nanoparticles for oral drug delivery. J Biomater Nanobiotechnol 3:391–395
92. Marrucho IM, Branco LC, Rebelo LPN (2014) Ionic liquids in pharmaceutical applications. Annu Rev Chem Biomol Eng 5:527–546
93. Dobbs W, Heinrich B, Bourgogne C, Donnio B, Terazzi E, Bonnet M-E, Stock F, Erbacher P, Bolcato-Bellemin A-L, Douce L (2009) Mesomorphic imidazolium salts: new vectors for efficient siRNA transfection. J Am Chem Soc 131:13338–13346
94. Ferraz R, Branco LC, Marrucho IM, Araújo JMM, Rebelo LPN, da Ponte MN, Prudêncio C, Noronha JP, Petrovski Z (2012) Development of novel ionic liquids based on ampicillin. Med Chem Commun 3:494–497
95. Feder-Kubis J, Tomczuk K (2013) The effect of the cationic structures of chiral ionic liquids on their antimicrobial activities. Tetrahedron 69:4190–4198

Part II

Synthetic Strategy

Chapter 7

Designing Efficient Cascade Reactions in Drug Discovery

Chenguang Yu, He Huang, Chunquan Sheng, and Wei Wang

Abstract

This chapter focuses on the exploration and application of cascade reactions in the total synthesis of both biologically active natural products and their analogues and synthetic substances for drug discovery. The high synthetic efficiency of such processes enables the facile and rapid construction of bioactive molecules and furthermore streamlines the study of structure-activity relationships to accelerate the identification of potential drug-like compounds for further biological evaluation. In addition, the power of cascade strategies is also demonstrated in the pilot plant-scale synthesis of highly valued, structurally complex drugs and drug candidates.

Key words Cascade reaction, Organocatalysis, Transition metal catalysis, Drug discovery and development, Pilot plant-scale synthesis, Structure-activity relationships

1 Introduction

The development of efficient synthetic technologies for the cost-effective preparation of bioactive natural products, therapeutic agents, and drug candidates continues to be of considerable significance. The use of cascade reactions involving multiple-bond-forming events occurring in a single vessel can lead to a dramatic increase in the overall synthetic efficiency for a process. In addition, cascade reactions by their nature embody high efficiency particularly in terms of atom, time, energy, and labor economy, which meets the requirement of a green and sustainable synthesis. Due to their inherent unrivaled merits, a number of powerful cascade reactions have been developed for numerous syntheses including both natural products and synthetic substances with complex architectures [1–3], as well as the increased implementation of this powerful strategy in the field of drug discovery, development, and production [4]. Therefore, this chapter chiefly focuses on the exploration of cascade reactions for the construction of natural products and their analogues and bioactive synthetic compounds for drug discovery with examples of synthesis of therapeutics on pilot plant

Paul F. Richardson (ed.), *Green Chemistry in Drug Discovery: From Academia to Industry*, Methods in Pharmacology and Toxicology, https://doi.org/10.1007/978-1-0716-1579-9_7, © Springer Science+Business Media, LLC, part of Springer Nature 2022

scale also included. It is important to note that, in order to be concise, this isn't a comprehensive review, and as such only selected examples are highlighted to illustrate the recent advances and developments of cascade reactions in drug discovery and development. Interesting readers may refer to a number of excellent reviews and books for more detail and an overview on the general developments of cascade reactions [3, 5–14].

2 Cascade Reactions for the Synthesis of Natural Products/Analogues

Natural products have proven and continue to be a validated source for drug development with a number of currently clinically used therapeutics originally arising from their analogues and derivatives [15–20]. However, in many cases, the low natural abundance of the natural product limits its biological evaluation, and furthermore many derivatives cannot be accessed from nature. Thus, the production of sufficient quantities of a compound for both biological studies and drug discovery relies on synthesis. These natural products often have highly complex structures featuring multiple functionalities and/or stereogenic centers, therefore presenting significant synthetic challenges leading to their successful total synthesis requiring numerous steps and significant resources. To improve upon synthetic efficiency, chemists have designed and implemented elegant cascade reactions into these syntheses [1, 21], with several successful studies related to drug discovery and development discussed herein.

The cortistatins were originally isolated by Kobayashi from the marine sponge *Corticium simplex* in 2006 and showed selective antiproliferative properties against human umbilical vein endothelial cells (HUVECs) [22], with cortistatins A and J being the most potent members of the series (IC_{50} = 1.8 nM and 8.0 nM against HUVECs, respectively), displaying a selectivity index of greater than 3000- and 300–1100-fold over normal human dermal fibroblasts (NHDF) and other tumor cell lines (KB3–1, K562, and Neuro2A). The cortistatin family shares a complex pentacyclic core structure, the construction of which presents the most critical and challenging aspect of a total synthesis. In 2008, Nicolaou and co-workers successfully accomplished the synthesis of cortistatins A (**4**) and J (**5**) [23], with their approach featuring a powerful base-promoted oxa-Michael addition/Aldol/dehydration cascade reaction to facially construct the two bridged rings of core **3** (Scheme 1). In this process, potassium carbonate triggers the initial intramolecular oxa-Michael addition of the hydroxyl group of **2** to the α,β-unsaturated cyclohexanone to form a transient enolate, which undergoes an Aldol condensation with the aldehyde followed by a dehydration to afford the desired core, **3**. Cortistatins A **4** and J **5** were obtained from **3** after a number of synthetic

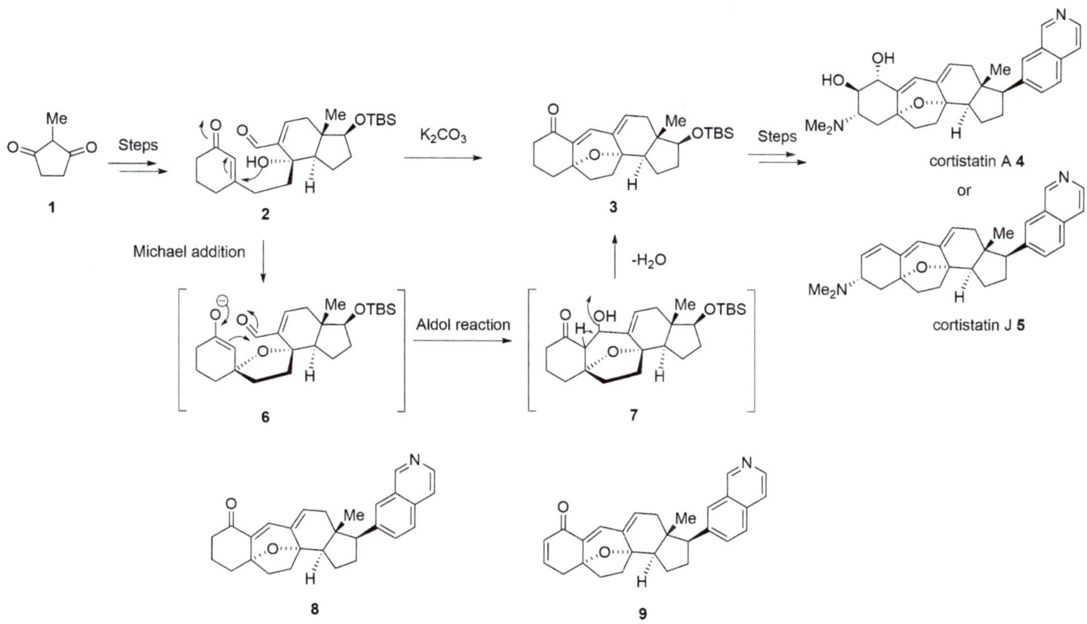

Scheme 1 K_2CO_3-promoted oxa-Michael addition/Aldol/dehydration cascade reaction as the key step in the total synthesis of cortistatins A and J

transformations. Importantly, the versatile cascade reaction enabled the synthesis of numerous analogues among which **8** and **9**, both lacking the dimethylamino and hydroxyl groups in the A ring, showed similar antiproliferative activities to these natural products [21], thus providing important information for the design of structurally simplified cortistatin analogues for drug discovery.

Calcitriol, hormone 1α,25-dihydroxyvitamin D_3, is the active form of vitamin D_3 and is involved in many biological activities, such as cell proliferation, immune response, and bond mineralization [24–27], which make it a valuable target for drug discovery. Such compounds can be potentially used for the treatment of hyperproliferative diseases and immune disorders, though clinical studies on the parent compound show that it exhibits side effects including significant hypercalcemia and increased bone resorption. Accordingly, a great deal of effort has been devoted to finding analogues with improved specificity such as the work of Mouriño and co-workers, whom designed and synthesized 1α,25-dihydroxyvitamin D_3 **15** and the analogues **13a–c** (Scheme 2) [28]. Their synthetic route features a novel palladium-catalyzed carbocyclization/Negishi cross-coupling cascade reaction as the key step for the construction of the triene core system. Herein, the chiral intermediate **11**, prepared from (*R*)-carvone **10**, underwent a $Pd(PPh_3)_4$-promoted intramolecular cyclization between the vinyl triflate and the alkyne to give the palladated alkene complex **14**, which subsequently reacted with the zinc complex **12b** via a Negishi

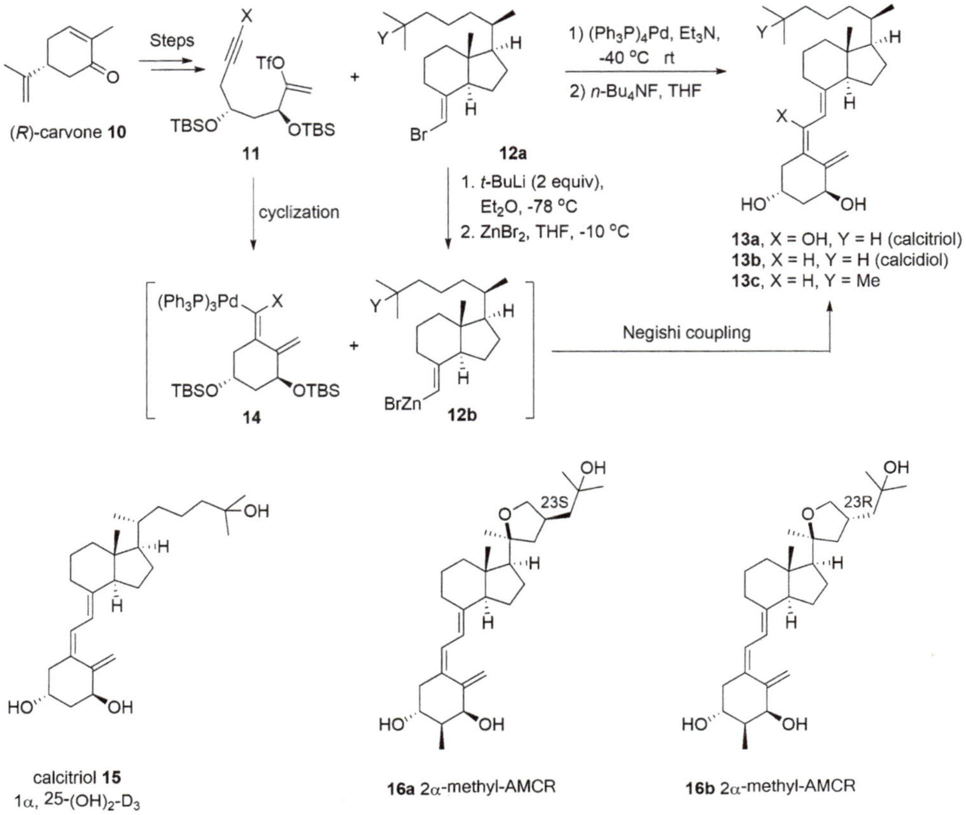

Scheme 2 Total synthesis of 1α,25-dihydroxyvitamin D_3 analogues through a palladium-promoted cyclization/Negishi cross-coupling cascade reaction

cross-coupling reaction to produce the desired products **13a–c**. Using the same approach, Rochel and co-workers prepared the analogues **16a** and **16b** [29], and it was shown that while **16a** showed super-agonistic and cellular antiproliferation/pro-differentiation properties but with similar calcemic side effects to calcitriol **15**, **16b** exhibited a similar bioactivity profile but without the side effects.

An alternative report to similar analogues avoided using the unstable in situ-derived zinc complex **12b** by employing the stable vinyl boronic esters **18a–c** to assemble the core structure (Scheme 3) [30]. This led to the preparation of three new analogues (**20a–c**) featuring two different side chains at C-12 and C-17 prepared in 15%, 6%, and 3% overall yields, respectively, through a related Pd (0)-catalyzed cyclization/Suzuki cross-coupling cascade process. Compounds **20a–c** were assayed against human colon and breast cancer cell lines and in mice and were demonstrated to have similar pro-differentiating, antiproliferative, and transcriptional actions to calcitriol in the in vitro studies. Critically, they exhibited markedly lower calcemic effects than calcitriol in vivo, thus making **20a–c** promising lead molecules for further development.

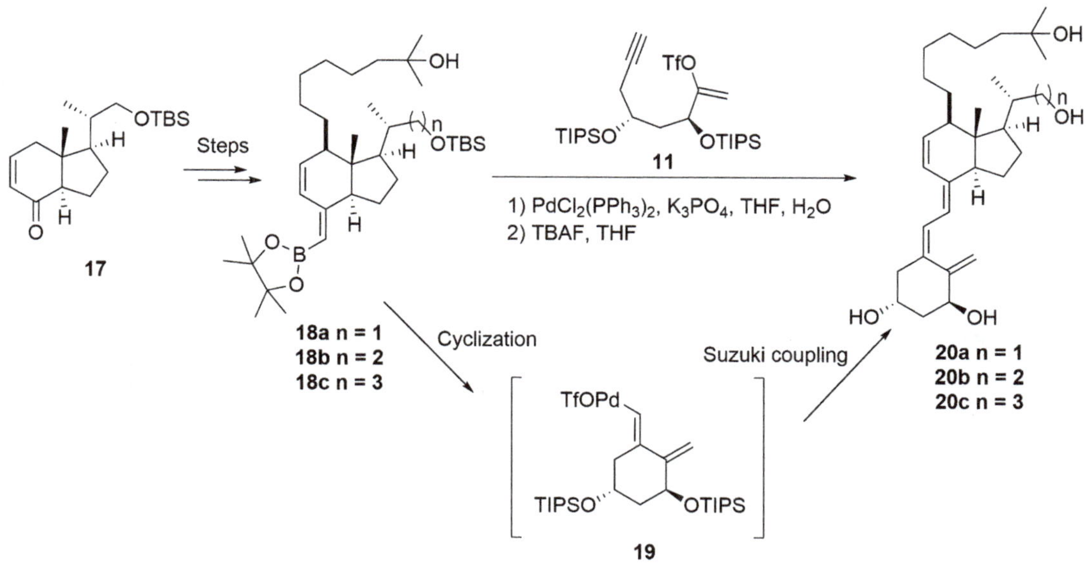

Scheme 3 Synthesis of 1α,25-dihydroxyvitamin D_3 analogues through a palladium-promoted cyclization-Suzuki cross-coupling cascade reaction

Actinophyllic **24** acid, a biologically active indole alkaloid, which possesses an indole core featuring six rings with five contiguous chiral centers, was isolated in 2005 by Carroll and co-workers from the leaves of *Alstonia actinophylla* [31]. Despite the complexity of the structure presenting a synthetic challenge, Martin and co-workers devised an elegant ten-step method for the total synthesis of (±)-actinophyllic acid from readily available chemicals (Scheme 4) [32]. The synthesis features a facial assembly of the key hexacyclic intermediate **23** through a powerful Lewis acid-promoted cascade reaction of an *N*-stabilized carbocation with a π-nucleophile. In this one pot protocol, the carbocation **25** generated by the treatment of indolyl acetate **21** with TMSOTf is immediately trapped by the cyclic enamidodiene **22** to produce the *N*-acyliminium ion **26**, which undergoes intramolecular Mannich-like reaction to give **27** with high diastereoselectivity due to the inherent geometric constraints of the system. Desilylation with TBAF in the same pot provides access to the key intermediate **27** in an impressive 92% yield. The synthetically derived actinophyllic acid **24** and various analogues of **27** were evaluated for their ability to inhibit the human breast cancer Hs578t cell line, and from these studies it was determined that while **28** was the most potent (IC_{50} = 8 nM), actinophyllic acid **24** itself was totally inactive. Further studies also demonstrated that **28** exhibited anticancer activity against human lymphoma (U937), human lung cancer (A549), and human glioblastoma cell lines (U87) with IC_{50} values ranging between 5 and 8μM. This study provides an example

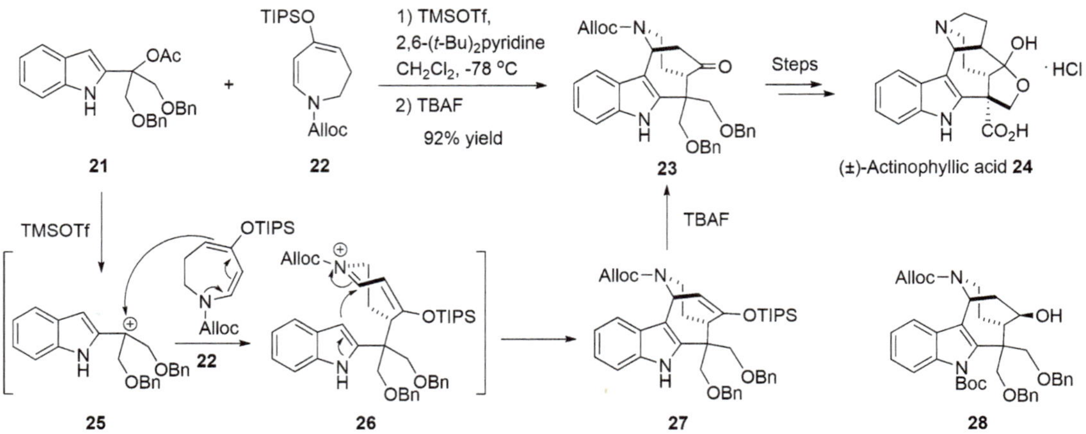

Scheme 4 Total synthesis of (±)-actinophyllic acid involving a Lewis acid-promoted cascade reaction of an N-stabilized carbocation with a π-nucleophile

wherein a simplified natural product analogue has been identified through synthesis with more potent biological activities than the parent from which it is derived.

Aculeatins, isolated from rhizome of *Amomun aculeatm*, have been used as a folk medicine to treat both fever and malaria in Papua New Guinea [33, 34]. This class of compounds feature three successive spirocyclic rings, and in 2002, Wong and co-workers developed an efficient PIFA-mediated dithiane deprotection/phenolic oxidative dearomatization cascade process to construct this spirocyclic core to give both (±)-aculeatine A (**31**) and (±)-aculeatine B (**32**) in 44% and 15% yield, respectively (Scheme 5) [35]. Based on this synthetic strategy, additional analogues (±)-**33a–e**, (±)-**34**, (±)-**35**, and (±)-**36** were synthesized, and through the design of a specific precursor, **37**, the access to more complicated compounds containing five spirocyclic rings (±)-**38a–c** and (±)-**40a–c** was also enabled [36]. Compounds (±)-**38b** and (±)-**40b** showed potent antiparasitic activities against two widespread Apicomplexa with IC_{50}'s of 81 nM and 92 nM, respectively, against the *P. falciparum* 3D7 strain, while displaying low toxicities toward the mammalian cells (TI > 100 for *P. falciparum*), thus making them promising molecules for the treatment of parasitic infections.

Neovibsanins A and B, isolated from the leaves of *Viburnum awabuki* by Fukuyama in 1996, exhibit potent neurotrophic activity and very low cytotoxicity [37, 38]. This family of natural products features fused tricyclic rings including six chiral centers and three chiral quaternary centers, and therefore how to assemble this core with the correct configuration is the crux of the synthetic challenge. Williams et al. successfully developed an expedient approach for the preparation of (±)-4,5-*bis*-epi-neovibsanins A (**45**) and B (**46**) in only 12 steps (Scheme 6) [39], developing a one-pot acid-promoted 5-step cascade reaction to assemble the

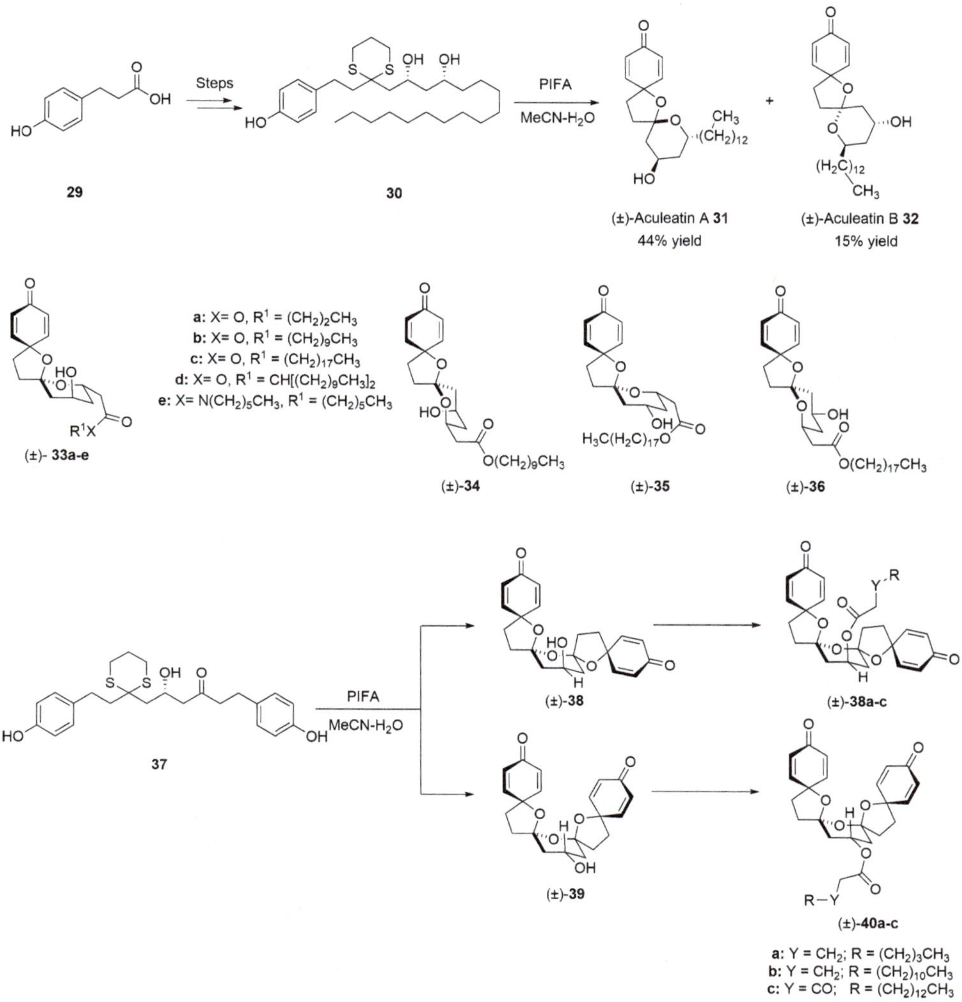

Scheme 5 Total synthesis of aculeatins A, B, and their analogues by dithiane deprotection/phenolic oxidative dearomatization cascade reaction

tricyclic core to afford the key intermediates **43** and **44** in 73% yield and with a 5:1 ratio of diasteromers. The desilylation of **42** occurred under acidic conditions to initiate the cascade process, which involved an intramolecular oxo-Michael addition to produce **48**, which spontaneously underwent lactone ring-opening to yield **49**, thus generating a new hydroxyl functionality, which was able to attack the ketone to generate hemiacetal **50** that was then converted to ketals **43** and **44** under the strongly acidic reaction conditions. Compounds **43** and **44** were converted to (±)-4,5-*bis*-epi-neovibsanin A (**45**) and B (**46**) in two steps with both compounds being demonstrated to strongly potentiate neurite outgrowth in NGF-stimulated PC12 cells.

Isolated from the heartwood of *Chamaecyparis obtuse* var. *fromosana*, (+)-chamaecypanone C (**57**) shows potent cytotoxicity against several human cancer cells [40]. Taking inspiration from

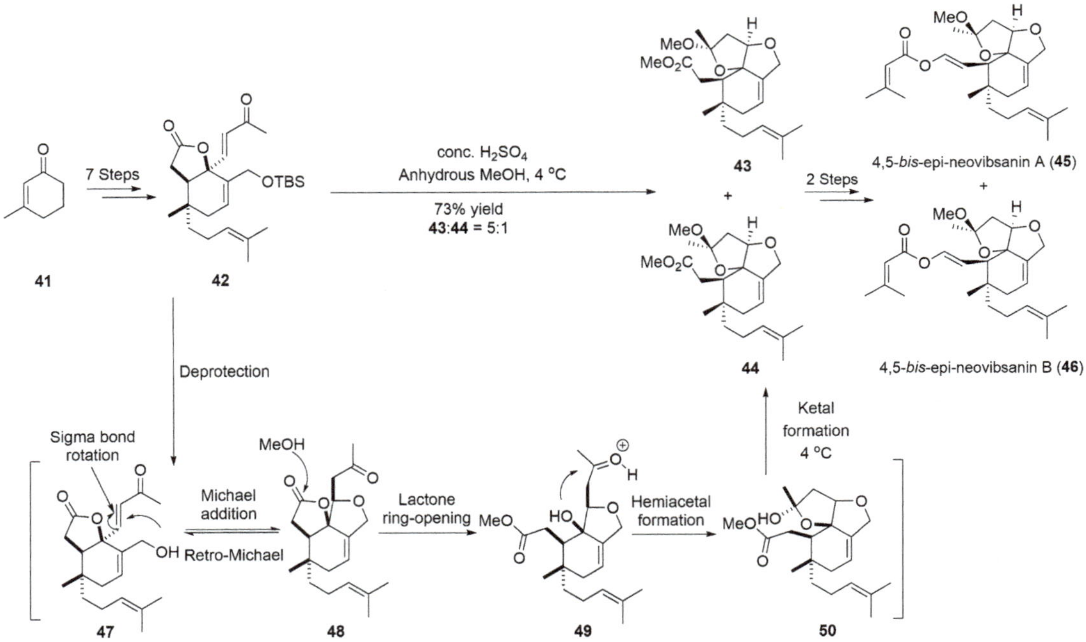

Scheme 6 Total synthesis of 4,5-*bis*-epi-neovibsanin A and B involving an acid-catalyzed five-step cascade reaction

the biosynthetic pathway [41], a dehydrogenation/retro-Diels-Alder/Diels-Alder cascade sequence was designed to form the tricyclic core (Scheme 7a) [42]. Initially the Diels-Alder reaction between the diaryl cyclopentadiene generated in situ from **52** and chiral cyclohexa-2-4-diene **55** (also generated in situ) was attempted but proved to be unsuccessful. Switching to using cyclopentadienone as the dienophile led to initial problems obtaining pure material although the compound did form. Fortunately, in situ formation of cyclopentadienone from cyclopentenone **53** using DDQ as oxidizing reagent enabled the Diels-Alder reaction with the chiral diene **55** formed in situ to proceed to give the desired product **56** in 61% yield, which could be demethylated under standard conditions to afford (−)-chamaecypanone C (**57**). To access the naturally occurring enantiomer, (+)-chamaecypanone C, the same route was utilized though starting with the enantiomer of **55**. Biological studies demonstrated the importance of the chiral center with (+)-chamaecypanone C being shown to inhibit tumor cell growth by an average of 71%, while its enantiomer had no observed inhibitory effects. Guided by this chemistry, though replacing DDQ with Rh/C as the dehydrogenation reagent, the same cascade reaction was employed for the synthesis of a series of (+)-chamaecypanone C analogues (Scheme 7b) [43], with one compound (+)-**61** found to have higher activity as both a microtubule inhibitor and cytotoxic agent than the parent itself (IC_{50} 1.5μM vs. 2.2μM against tubulin assembly). Structure-

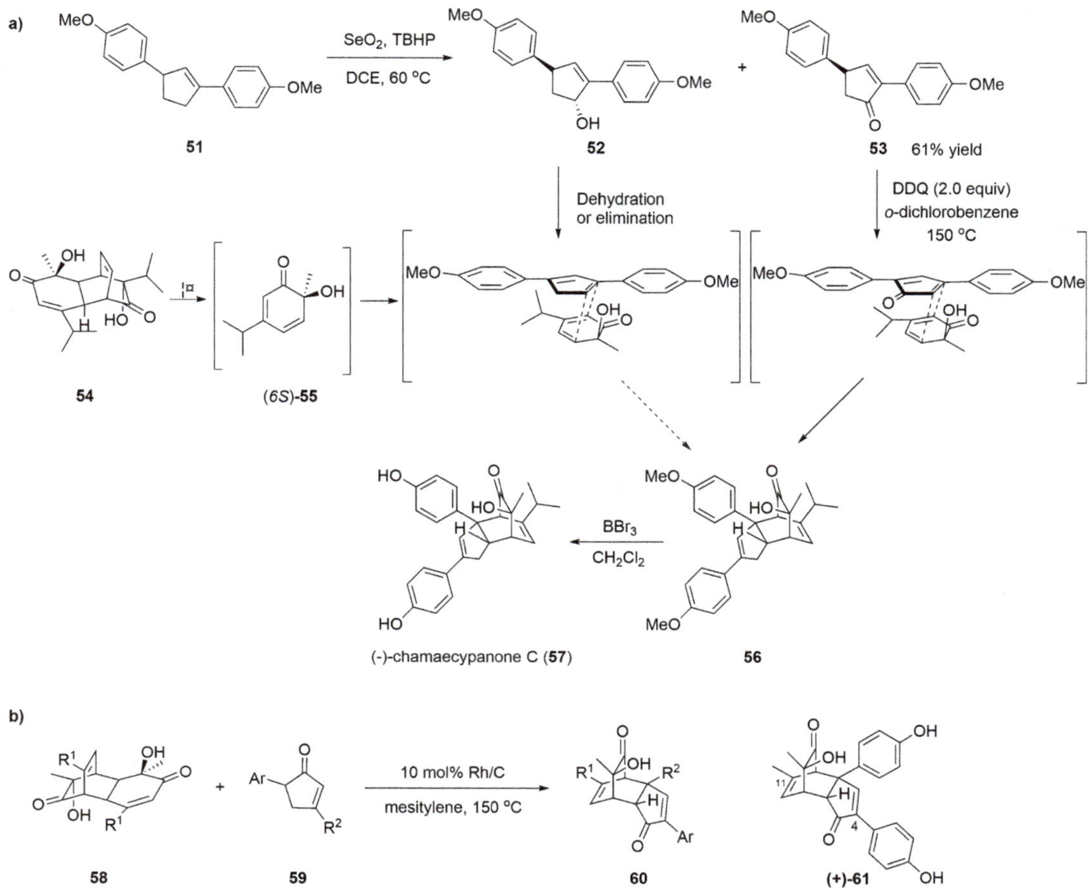

Scheme 7 (a, b) Total synthesis of (−)- and (+)-chamaecypanone C and their analogues using a dehydrogenation/retro-Diels-Alder/Diels-Alder cascade reaction

activity relationship studies demonstrated that the 4-*p*-hydroxyphenyl group is crucial for interactions with tubulin, and a smaller C-11-substitution improved the bioactivity (methyl group is more potent than *iso*propyl group, while *t*-Bu group leads to a loss of activity), thus providing guidance for the further design of more potent analogues.

(+)-Ricciocapin A, a furanquiterpene lactone, was isolated from an axenic culture of the European liverwort *Ricciocarpos natans* [44] and shows high molluscicidal activity against the water snail *Biomphalaria glabrata*, a vector of schistosomiasis [45, 46]. In 2009, List and co-workers devised an efficient one-pot, three-step, organocatalytic reductive Michael/Tishchenko cascade, affording not only both enantiomers of ricciocarpin A **68** and **69** but also the analogues **70–73** (Scheme 8) [47]. In this one-pot operation, a previously reported asymmetric reductive Michael cascade reaction developed by the same group was used to synthesize the precursor *cis*-**66** in the presence of the chiral secondary amine

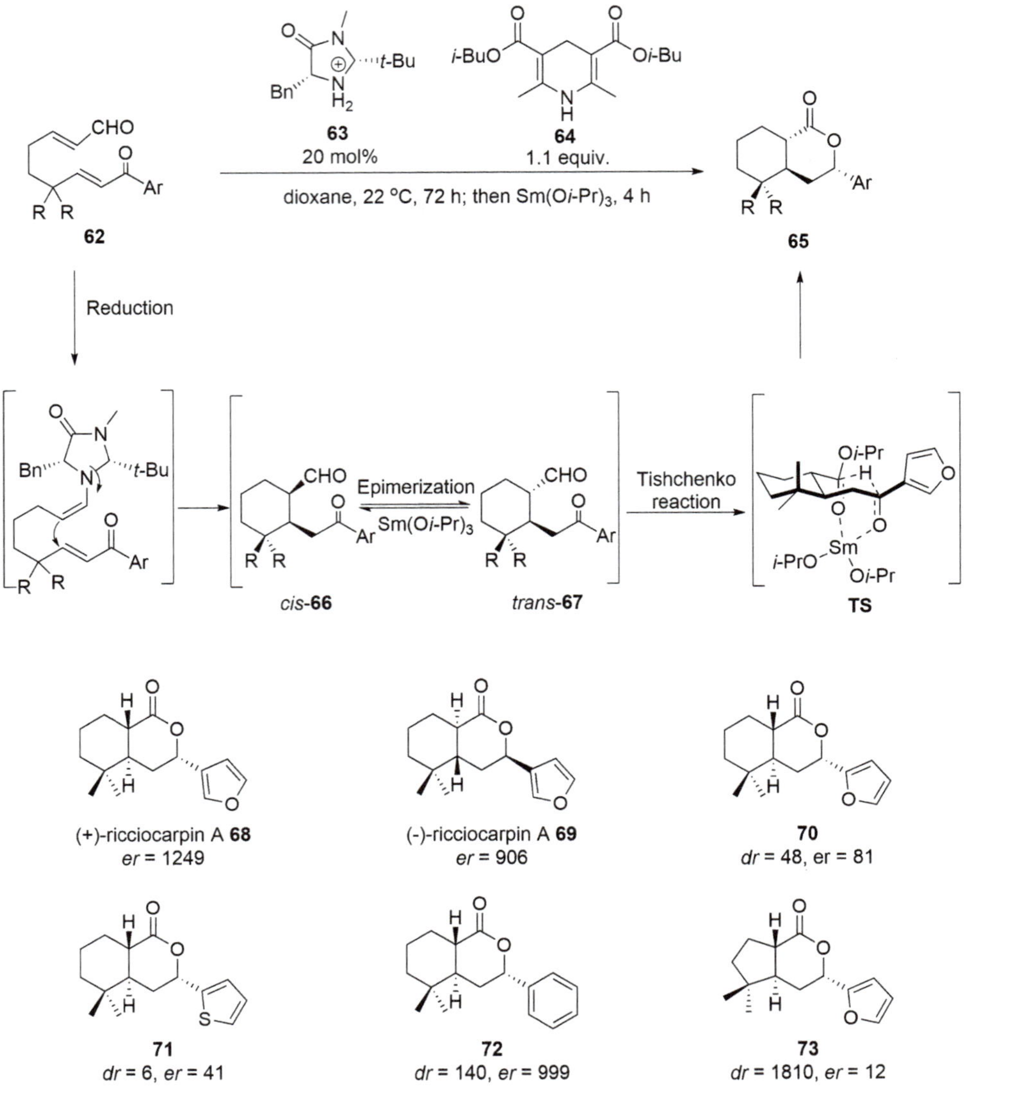

Scheme 8 Total synthesis of (+)-ricciocarpin A and analogues through a one-pot, three-step, organocatalytic reductive Michael/Tishchenko reaction cascade

catalyst **63** [48]. After 72 h, Sm(O*i*-Pr)$_3$ was added to the reaction mixture to promote an epimerization to provide *trans*-**67**, of which the configuration was critical in order for the subsequent Tishchenko reaction to yield the desired lactone **65**. Due to the good substrate tolerance of this protocol, the synthesis of a number of analogues could be easily achieved enabling the evaluation of several structure-activity relationship trends of the ricciocarpin family, which lead to the finding that **71** is more potent than the parent in terms of molluscicidal activity against *B. glabrata*.

3 Cascade Reactions for the Synthesis of Bioactive Small Molecules

In a complimentary manner to natural products and their derivatives, synthetic small molecules are rich sources for the discovery and development of therapeutic agents. Both scaffold diversity and complexity have been widely regarded as the main characteristics for the creation of collections of high-quality compound libraries [49–53]. Compounds with a range of molecular shapes and high percentages of sp_3-character, critical for broad biological activity, are primarily dependent on the diversity and nature of the central scaffold. However, the assembly of skeletally different complex drug-like small-molecule architectures and the rapid identification of useful lead compounds represent the most challenging aspects of diversity-oriented (DOS) and combinatorial library synthesis [49–53]. Cascade reactions with the capacity to enable the facile construction of complex natural product-like molecular frameworks may offer a solution and thus facilitate the drug discovery process.

Heterocyclic furo-[2,3-*c*]-quinolones **76** are a medicinally important class of compounds displaying a broad spectrum of bioactivities (Scheme 9) [54], and a palladium-catalyzed Sonogashira coupling/intramolecular 5-*endo-dig* cyclization cascade reaction has been developed to enable their synthesis [54]. Subsequent transformations were utilized to convert **76** into furo-[2,3-*c*]-quinolone-4-amines **77**, while a similar strategy was employed for the synthesis of furo-[3,2-*c*]-quinolone-4-amines **80–82** using different substrates. On profiling these series of compounds, several of them were recognized to possess a pure TLR8-agonistic activity profile with **79**, with an *n*-butyl substituent, displaying maximal potency ($EC_{50} = 1.6\mu M$).

An analogous catalytic system has been applied for the construction of 1*H*-pyrrolo-[3,2-*c*]-pyridines by changing the nucleophile from the OH group to a tosyl-protected amine (Scheme 10) [55]. Utilizing structure-based design within this series, **86** was found to be a highly potent inhibitor of MPS1 that is aberrantly overexpressed in many human cancer cells (P-MPS1 $IC_{50} = 0.04\mu M$ and HCT116 $GI_{50} = 0.16\mu M$) while also exhibiting a good oral pharmacokinetic profile in mouse and rat as well as the ability to inhibit MPS1 in vivo after oral administration.

Pierre and co-workers have developed a palladium-mediated Suzuki cross-coupling/intramolecular amide formation cascade, with the resultant amides **89** being further transformed to target molecules such as **90** (Scheme 11) [56]. The inhibitory ability of this series of compounds was evaluated against CK2, which is a prime cancer drug target due to its roles of being either deregulated or overexpressed in numerous cancer-promoting pro-survival and anti-apoptotic pathways. From these studies, compound **91** was found to be both a potent and selective inhibitor and was evaluated

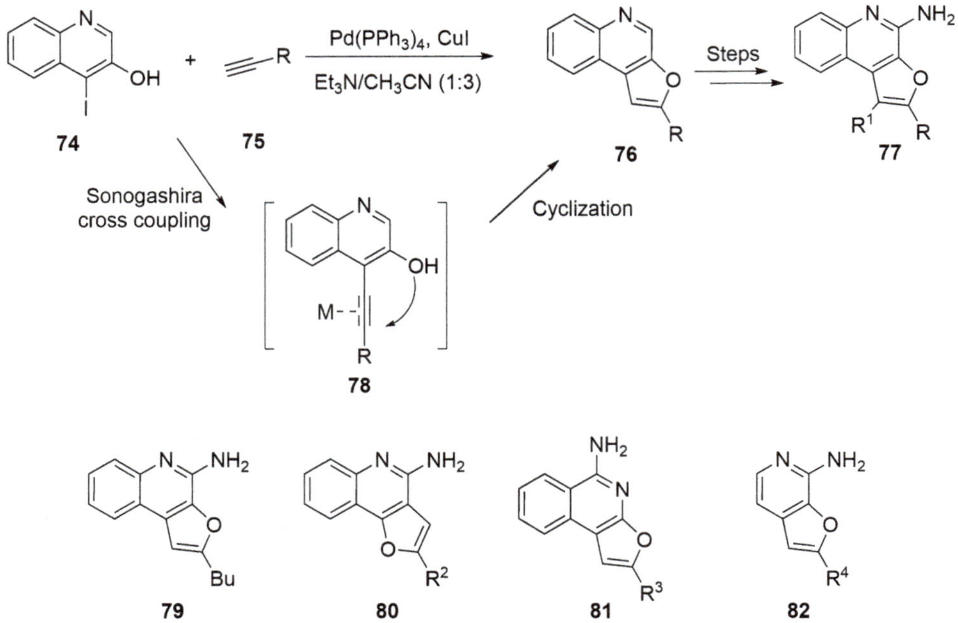

Scheme 9 Synthesis of furo[2,3-*c*]quinolone-4-amines and furo[3,2-*c*]quinolone-4-amines by a palladium-catalyzed Sonogashira cross-coupling/cyclization cascade process

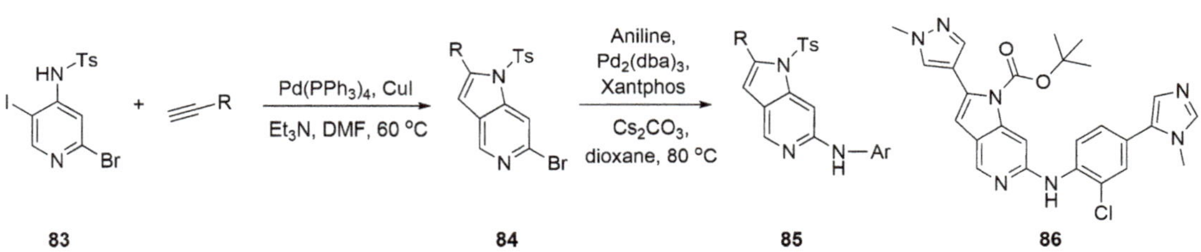

Scheme 10 Synthesis of 1*H*-pyrrolo-[3,2-*c*]-pyridines involving a palladium-catalyzed Sonogashira cross-coupling/cyclization cascade reaction

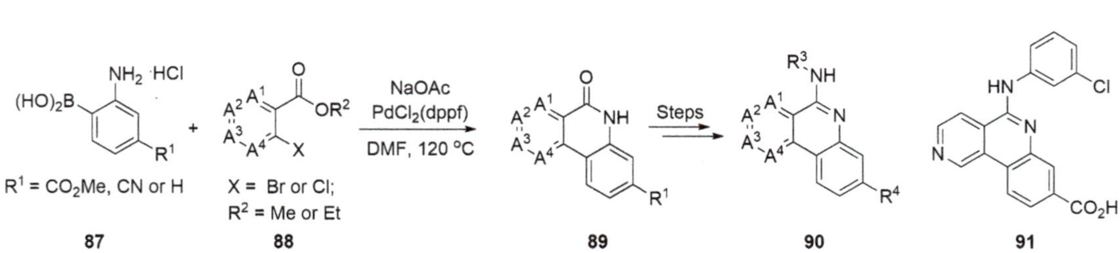

Scheme 11 Palladium-catalyzed Suzuki cross-coupling/intramolecular amide formation cascade reaction

in the clinic. A molecular binding study indicated that the strong interaction of **91** with the ATP-binding pocket of the kinase came from a combination of a series of hydrophobic interactions, an ionic bridge with Lys68, and hydrogen bonding with the hinge region, with these findings providing insightful guidance for the design and optimization of new small-molecule CK2 inhibitors.

Gold complexes, which are well established as π-Lewis acids, have been successfully exploited in a number of cascade reactions [57], and Wang and co-workers have employed $Ph_3PAuNTf$ as the catalyst in an intramolecular cyclization/nucleophilic addition cascade between the alkyne and pendent amide functionalities to afford tricyclic indolines as exemplified by **94**, **95**, and **96** (Scheme 12) [58]. The simple ring opening of the tricyclic indolines delivered, respectively, **97**, **98**, and **99**, with this powerful method allowing for the rapid access to 120 polycyclic indolines with 26 distinct skeletons and a variety of functional groups. These compounds were screened for their ability as resistance-modifying agents in combination with antibiotics to overcome the challenging issue of drug resistance with compound **100** found to selectively potentiate the activity of β-lactam antibiotics in multidrug-resistant methicillin-resistant *Staphylococcus aureus* (MRSA) but not in methicillin-sensitive *S. aureus*. It was also shown to resensitize several MRSA strains to the β-lactam antibiotic amoxicillin/clavulanic acid as well as the first-generation cephalosporin cefazolin. Further structure-activity relationship studies based on **100** led to the discovery of a more potent but less mammalian toxic analogue **101** [59].

Tropanes are known to possess high binding affinity for both the dopamine and serotonin transporter sites and are hence viewed as promising lead compounds for the development of therapeutics for the treatment of cocaine addiction [60–62]. Although much effort has been devoted to the synthesis of numerous tropane derivatives, a significant limitation exists in that many of the traditional synthetic methods rely on either the use of (−)-cocaine or tropinone as essential starting materials for structural modification [63], thus hampering attempts to create novel scaffolds beyond these materials. In 1991, Davies and co-workers devised a rhodium-catalyzed cyclopropanation/Cope rearrangement cascade as the key step for the facial access to various previously inaccessible tropane derivatives as exemplified **106** (Scheme 13) [64]. Rhodium-initiated cyclopropanation of pyrroles **102** with the vinylcarbenoids **103** produced the highly strained **105**, which undergoes a spontaneous Cope rearrangement to provide the bridged bicyclics **104**. Subsequently, the authors developed an asymmetric cascade version utilizing a chiral rhodium catalyst though only obtained moderate enantioselectivities [65]. Using this strategy, a number of novel tropane analogues were efficiently synthesized and subjected to biological evaluation, leading to the finding that a ketone functionality at the C-2 position instead of an ester could dramatically enhance the metabolic stability. Notably **107** from this series has been widely used as a probe to study the neurobiology of cocaine reinforcement and as a potential pharmacotherapeutic agent for the treatment of cocaine addiction [66–68]. In addition, compound

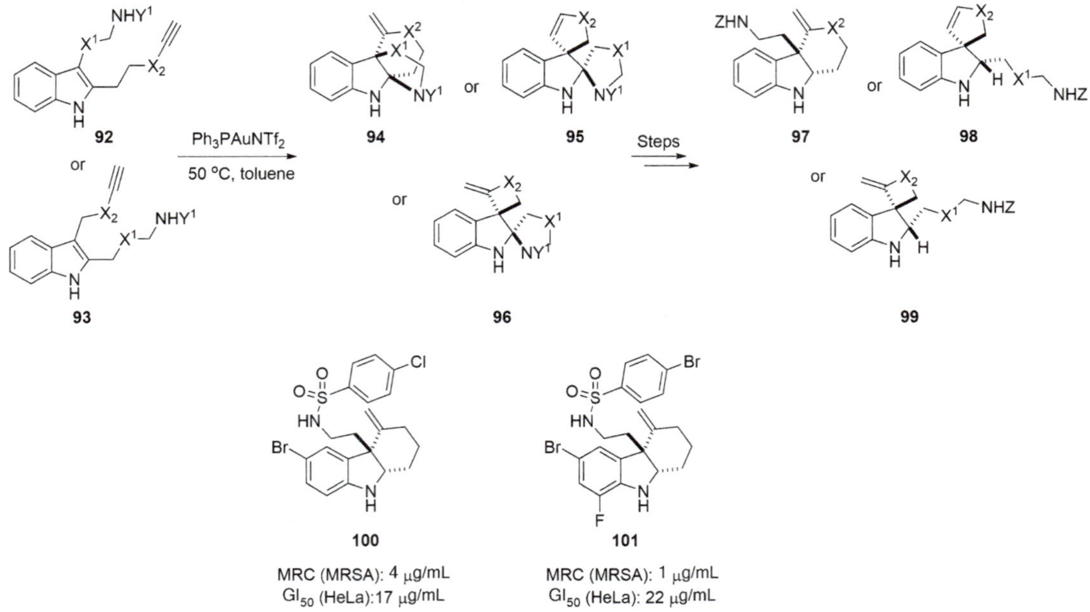

Scheme 12 Synthesis of tricyclic indolines through a gold-catalyzed cyclization/nucleophilic addition cascade reaction

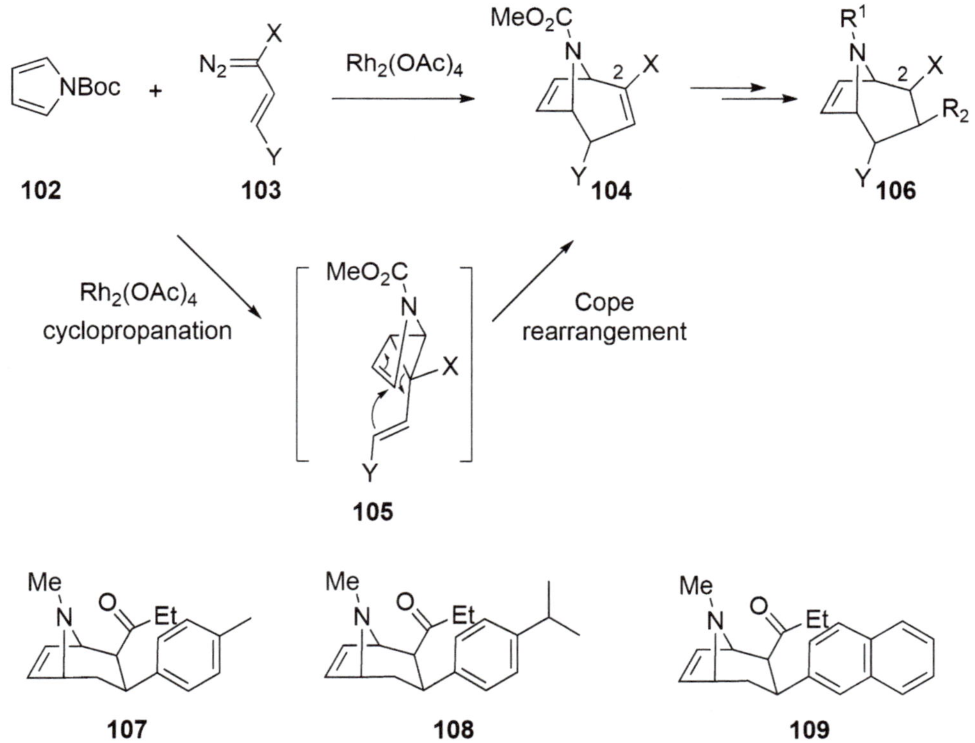

Scheme 13 Synthesis of tropane analogues through a rhodium-catalyzed cyclopropanation/Cope rearrangement cascade reaction

108 is the first potent tropane, which shows high selectivity for the SERT [69–71], while **109** exhibits the best potency among tropanes so far ($IC_{50} = 0.12$ nM at DAT) and has an extremely long duration of action in vivo [68, 72].

With an explosive development in the number of applications over the past decade, organocatalysis has become an important tool complementary to both organometallic and enzymatic catalysis in synthesis. Furthermore, the mild reaction conditions and excellent stereocontrol afforded by an organocatalyst has enabled a number of unprecedented cascade reactions to be developed [7–13, 73, 74]. In addition, the combinations of two different organocatalysts [75–77] or an organocatalyst and a metal catalyst [78–83] represents a powerful approach for the promotion of new cascade reactions to produce chiral complex molecules. In order to evaluate a novel series of molecules for biological properties, Wang and co-workers developed a novel aminocatalytic asymmetric Michael/Michael/Aldol cascade reaction between cinnamaldehydes (**110**) and thiazolidinediones (**111**) or rhodanines (**112**), producing medicinally relevant spirothiazolidinediones **113** or spirorhodanines **114**, respectively (Scheme 14a) [84]. In contrast, under similar reaction conditions, a Michael/cyclization cascade reaction between isorhodanines (**115**) and cinnamaldehydes (**110**) proceeded to yield fused thiopyranoids **116** (Scheme 14b). The further synthetic elaboration of this series of scaffolds was carried out to rapidly create a number of diverse complex structures such as **119**, **123**, and **124**, which were evaluated in both antifungal and antitumor bioassays. Herein, compounds **118**, **123**, and **124** exhibited potent antifungal activities against *Candida albicans* and *Cryptococcus neoformans*, while **117**, **118**, and **119** showed strong inhibitory activity against human breast cancer cells.

Diverse chiral pyranocoumarins **129** and 2-amino-4*H*-chromenes **130** were synthesized using the chiral thiourea-tertiary amine bifunctional catalyst **128** through an enantioselective Michael/cyclization/tautomerization cascade reaction of **125** with either 4-hydroxycoumarins **126** or cyclohexane-1,3-dione **127** (Scheme 15) [85]. Preliminary biological screening indicated that these novel heterocyclic compounds exhibited some promising antibacterial activity.

An unprecedented 1-pot 12-step cascade reaction involving nine different reactions mediated by two organocatalysts was reported for the synthesis of complex indoloquinolizines **146** (Scheme 16) [86]. Initially, triphenylphosphine catalyzes a two-step cascade reaction between an electron deficient alkyne **133** and a chromone **134** to produce a tricyclic benzopyrone **136**, which then undergoes a cascade reaction mediated by the second organocatalyst camphorsulfonic acid with the tryptamine **137**, involving a conjugate *N*-addition, conjugate *O*-addition, imine formation/isomerization, chromone ring opening, nucleophilic aromatic addition, aza-Claisen rearrangement, Pictet-

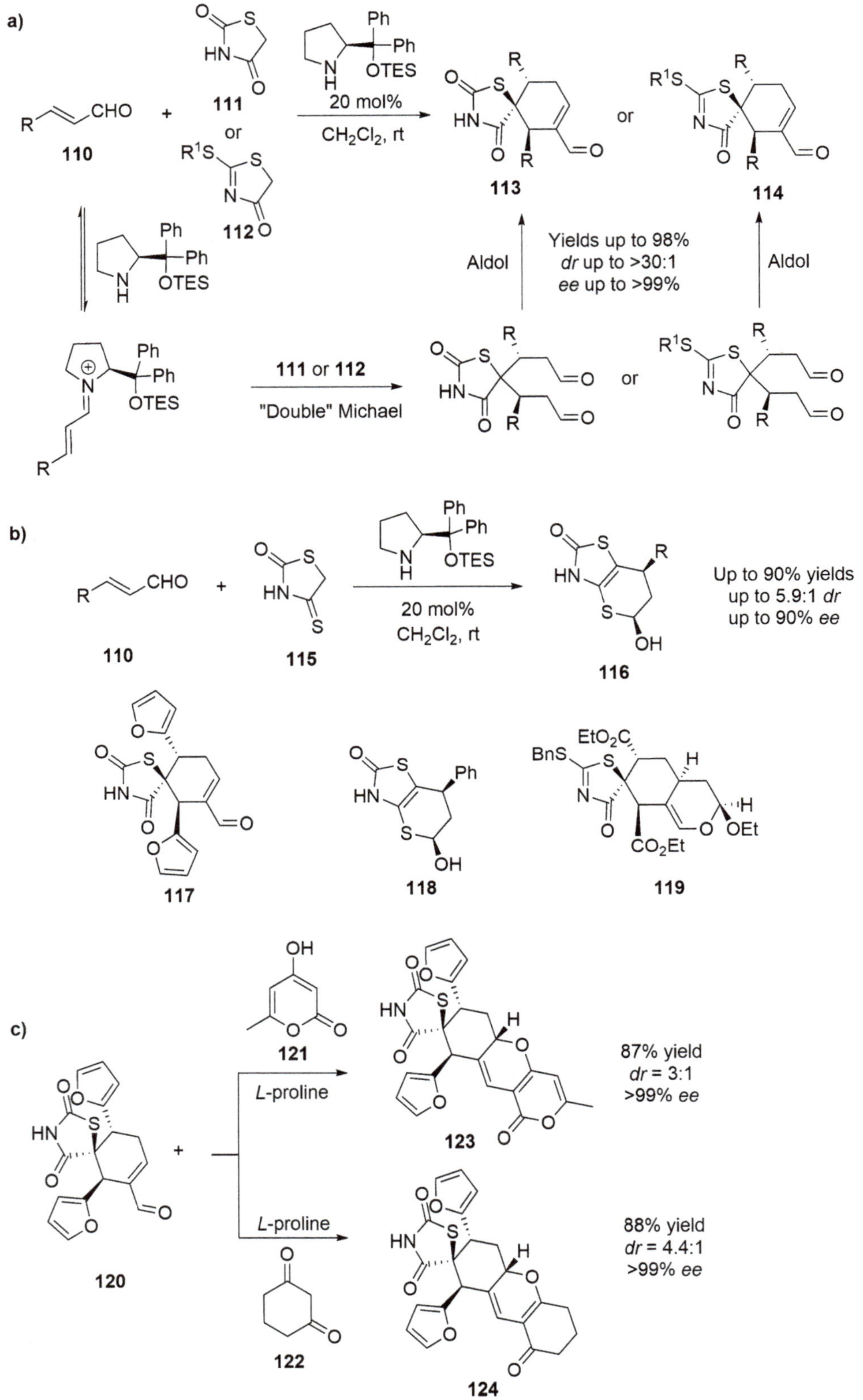

Scheme 14 (**a–c**) Synthesis of chiral spirothiazolidinediones, spirorhodanines, and fused thiopyranoids through chiral secondary amine-catalyzed asymmetric cascade reactions

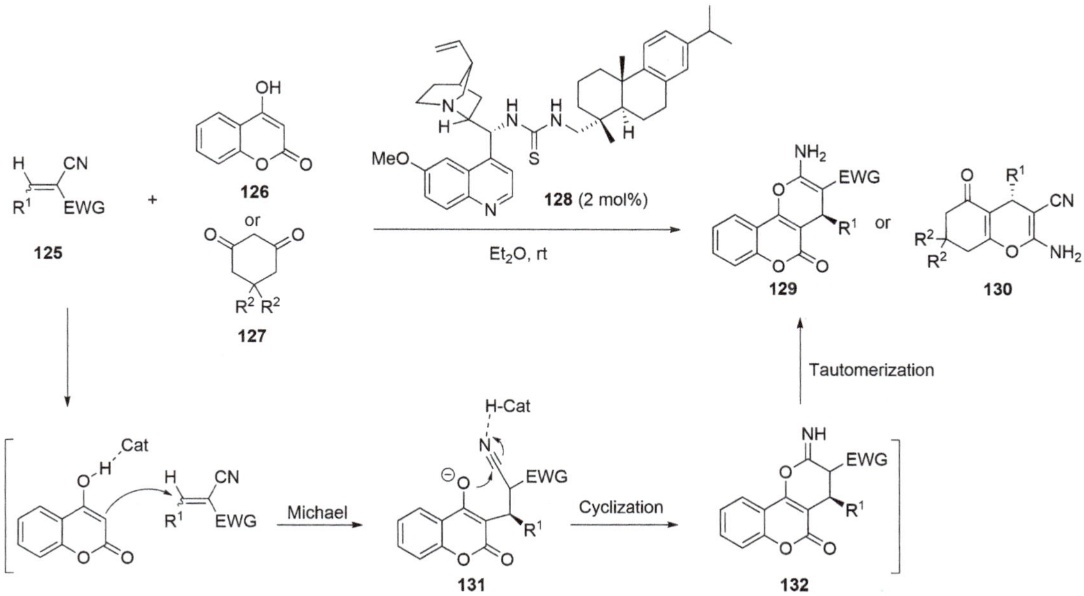

Scheme 15 Synthesis of chiral pyranocoumarins and 2-amino-4*H*-chromenes through a thiourea-tertiary amine bifunctional catalyst-promoted asymmetric Michael/cyclization/tautomerization cascade reactions

Spengler cyclization, and a retro-Michael addition. Notably, the construction of the complex indoloquinolizines **146** only takes 10–30 min and proceeds in 20–88% yield. The prepared compounds were probed for the possible modulation of cell division with *R*-**147** and *R*-**148** found to be the most potent by simultaneously targeting both NMP and Crm1, which are associated with cancer development.

4 Cascade Reactions for the Pilot Plant Synthesis of Drugs

The high efficiency of cascade reactions makes them attractive for the large-scale production of drugs as they offer a means to potentially significantly reduce cost, the amount of chemical waste, and the hazards associated with the process. Surprisingly though, relatively few cascade reactions have been exploited on scale possibly due to the difficulties in tracking the numerous intermediates in such a process, or the potential for various by-products to be formed from the truncated steps in the cascade.

Galanthamine, an *Amaryllidaceae* alkaloid, has entered the European market and been approved by the FDA for the treatment of Alzheimer's disease [87, 88], though, as a result of only limited availability from the natural source, the development of a practical and efficient synthetic strategy is critical to meet the large demand for clinical use. Jordis and co-workers developed an efficient nine-step route for the large-scale synthesis of galanthamine in 12.4%

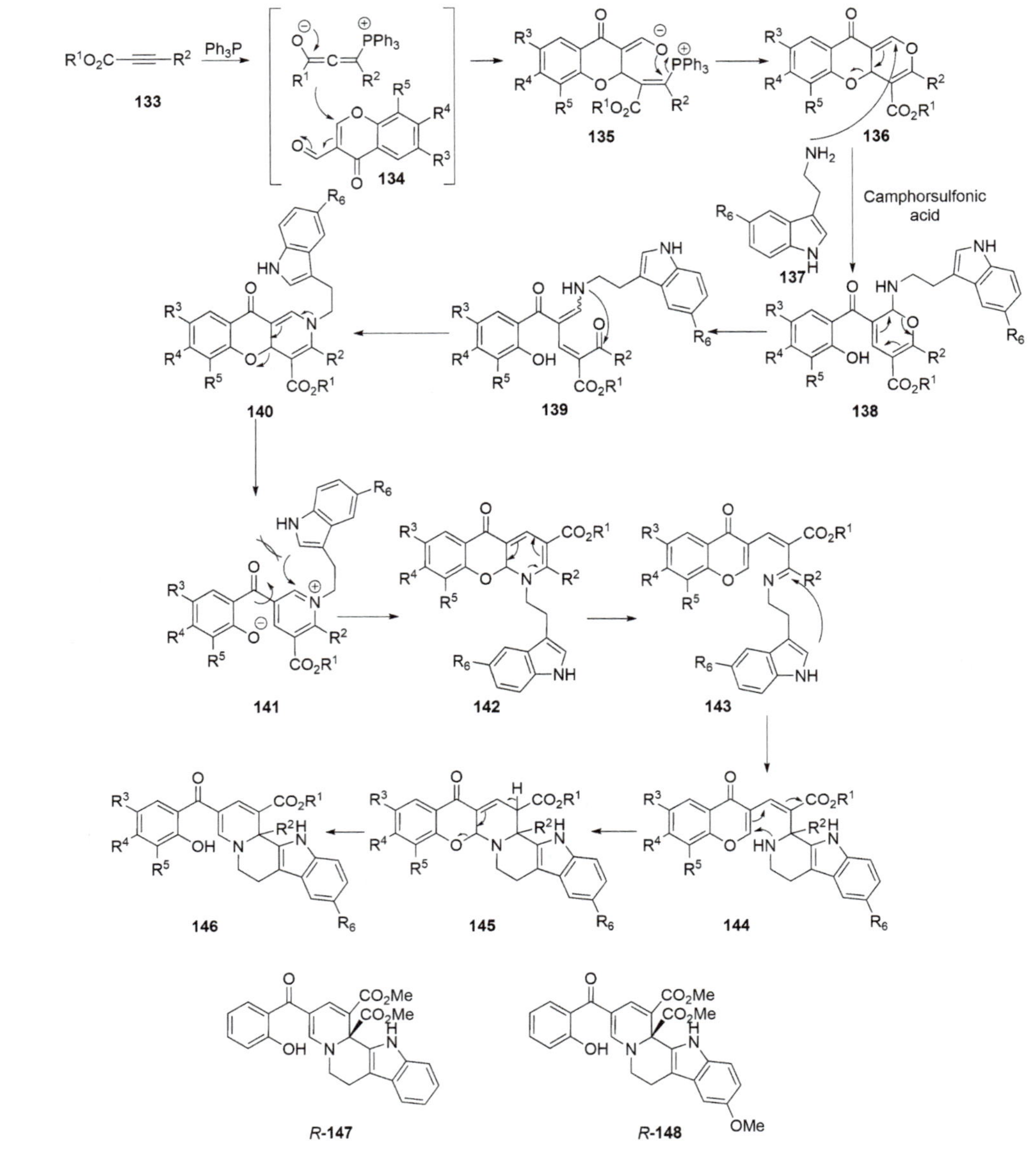

Scheme 16 Synthesis of indoloquinolizines through a dual organocatalytic 12-step cascade reaction

overall yield (Scheme 17) [89], with the key step in the process involving an oxidative phenol coupling/oxo-Michael addition cascade reaction mediated by $K_3[Fe(CN)_3]$ and K_2CO_3. The protocol enabled the construction of the core **151** with the formation of four fused rings in one pot and has been scaled up to 12 kg with reproducible yields (40–42%) and is the basis for industrial galanthamine production.

Ascomycin (**153**), isolated from *Streptomyces hygroscopicus* fermentation cultures, possesses strong immunosuppressant properties [90–92], and its derivative **154** has been proposed for both the topical and oral treatments of allergic contact dermatitis, psoriasis,

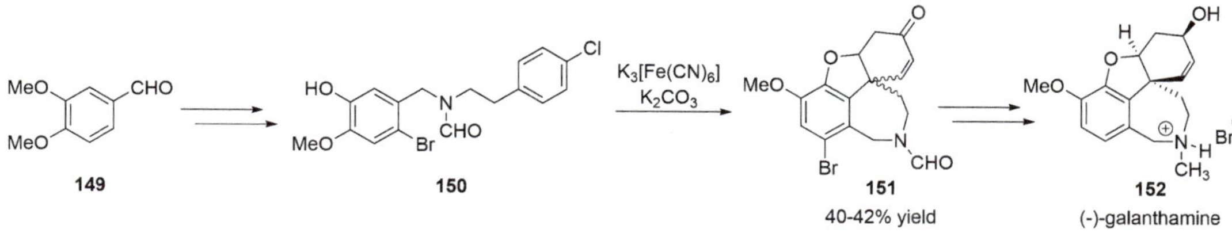

Scheme 17 Pilot plant synthesis of (−)-galanthamine through an oxidative phenol coupling/oxo-Michael addition cascade reaction

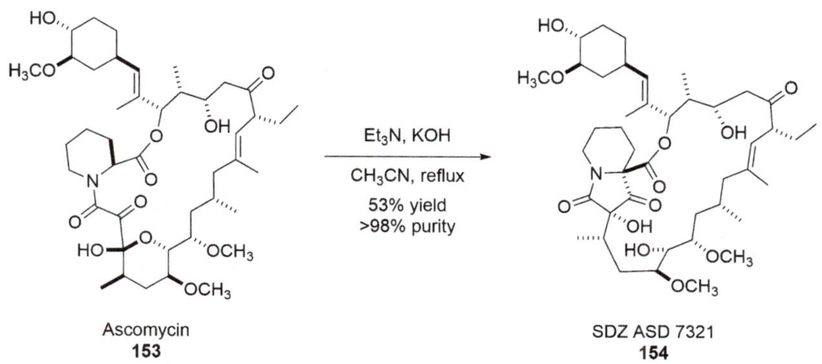

Scheme 18 Synthesis of ascomycin analogue SDZ ASD 7321 through rearrangement-epimerization cascade reaction

and atopic dermatitis [93]. Koch and co-workers demonstrated that semisynthetic **154** can be obtained in 53% overall yield and with a final purity of >98% through a base-promoted rearrangement/epimerization cascade reaction of ascomycin **153** (Scheme 18) [94], and furthermore this protocol can be scaled up to multikilogram to provide material for biological studies.

5 Conclusion

As described in this chapter, cascade reactions serve as highly powerful tools for the rapid assembly of complex natural products, their analogues, biologically interesting synthetic substances, and even drugs and drug candidates on pilot plant scale. Furthermore, they enable the production of bioactive compounds with novel natural product-like complex structures, which would otherwise be difficult to access by traditional stepwise processes. Therefore, employing these synthetic strategies can significantly facilitate drug discovery, development, and production with the benefits of potentially reducing waste, cost, energy, time, and labor. It should be stressed that although a number of highly efficient cascade reactions have been developed, their application in the drug discovery setting is still in its infancy. However, it is clear that the value and significance of cascade reactions as applied to the synthesis of pharmaceuticals will rise exponentially as new processes are developed.

Acknowledgments

The financial support from the NSF (CHE-1057569) is gratefully acknowledged.

References

1. Nicolaou KC, Edmonds DJ, Bulger PG (2006) Cascade reactions in total synthesis. Angew Chem Int Ed 45:7134–7186
2. Nicolaou KC, Chen JS (2009) The art of total synthesis through cascade reactions. Chem Soc Rev 38:2993–3009
3. Xu P-F, Wang W (2013) Catalytic cascade reactions. John Wiley & Sons, New York, NY
4. Andrushko V, Andrushko N (2013) Stereoselective synthesis of drugs and natural products. Wiley, Hoboken, NJ
5. Tietze LF, Brasche G, Gericke KM (2006) Domino reactions in organic synthesis. Wiley-VCH, Weinheim
6. Pellissier H (2006) Asymmetric domino reactions. Part A: Reactions based on the use of chiral auxiliaries. Tetrahedron 62:1619–1665
7. Yu X, Wang W (2008) Organocatalysis: asymmetric cascade reactions catalysed by chiral secondary amines. Org Biomol Chem 6:2037–2046
8. Pellissier H (2006) Asymmetric domino reactions. Part B: reactions based on the use of chiral catalysts and biocatalysts. Tetrahedron 62:2143–2173
9. Enders D, Grondal C, Huttl MRM (2007) Asymmetric organocatalytic domino reactions. Angew Chem Int Ed 46:1570–1581
10. Vlaar T, Ruijter E, Orru RV (2011) A recent advances in palladium-catalyzed cascade cyclizations. Adv Synth Catal 353:809–841
11. Ramachary DB, Jain S (2011) Sequential one-pot combination of multi-component and multi-catalysis cascade reactions: an emerging technology in organic synthesis. Org Biomol Chem 9:1277–1300
12. Lu L-Q, Chen J-R, Xiao W-J (2012) Development of cascade reactions for the concise construction of diverse heterocyclic architectures. Acc Chem Res 45:1278–1293
13. Pellissier H (2012) Recent developments in asymmetric organocatalytic domino reactions. Adv Synth Catal 354:237–294
14. Tietze LF (2014) Domino reactions: concepts for efficient organic synthesis. Wiley-VCH, Weinheim
15. Newman DJ, Cragg GM, Snader KM (2003) Natural products as sources of new drugs over the period 1981–2002. J Nat Prod 66:1022–1037
16. Kingston DG, Newman DJ (2005) The search for novel drug leads for predominately antitumor therapies by utilizing mother nature's pharmacophoric libraries. Curr Opin Drug Discov Dev 8:207–227
17. Newman DJ, Cragg GM (2007) Natural products as sources of new drugs over the last 25 years. J Nat Prod 70:461–477
18. Saklani A, Kutty SK (2008) Plant-derived compounds in clinical trials. Drug Discov Today 13:161–171
19. Ojima I (2008) Modern natural products chemistry and drug discovery. J Med Chem 51:2587–2588
20. Butler MS (2008) Natural products to drugs: natural product-derived compounds in clinical trials. Nat Prod Rep 25:475–516
21. Nicolaou KC, Peng X-S, Sun Y-P, Polet D, Zou B, Lim CS, Chen DYK (2009) Total synthesis and biological evaluation of cortistatins A and J and analogues thereof. J Am Chem Soc 131:10587–10597
22. Aoki S, Watanabe Y, Sanagawa M, Setiawan A, Kotoku N, Kobayashi M (2006) Cortistatins A, B, C, and D, anti-angiogenic steroidal alkaloids, from the marine sponge corticium simplex. J Am Chem Soc 128:3148–3149
23. Nicolaou KC, Sun Y-P, Peng X-S, Polet D, Chen DYK (2008) Total synthesis of (+)-cortistatin A. Angew Chem Int Ed 47:7310–7313
24. Norman AW, Bouillon R, Thomasset M (2000) Vitamin D endocrine system: structural, biological, genetic and clinical aspects. University of California, Riverside, CA

25. Nagpal S, Na S, Rathnachalam R (2005) Non-calcemic actions of vitamin D receptor ligands. Endocr Rev 26:662–687
26. Campbell MJ, Adorini L (2006) The vitamin D receptor as a therapeutic target. Expert Opin Ther Targets 10:735–748
27. Takahashi T, Morikawa K (2006) Vitamin D receptor agonists: opportunities and challenges in drug discovery. Curr Top Med Chem 6:1303–1316
28. Gomez-Reino C, Vitale C, Maestro M, Mourino A (2005) Pd-catalyzed carbocyclization-Negishi cross-coupling cascade: a novel approach to 1alpha,25-dihydroxyvitamin D3 and analogues. Org Lett 7:5885–5887
29. Antony P, Sigüeiro R, Huet T, Sato Y, Ramalanjaona N, Rodrigues LC, Mouriño A, Moras D, Rochel N (2010) Structure–function relationships and crystal structures of the vitamin D receptor bound 2α-methyl-(*20S, 23S*)- and 2α-methyl-(*20S, 23R*)-epoxymethano-1α,25-dihydroxyvitamin D3. J Med Chem 53:1159–1171
30. Carballa DM, Seoane S, Zacconi F, Pérez X, Rumbo A, Alvarez-Díaz S, Larriba MJ, Pérez-Fernández R, Muñoz A, Maestro M, Mouriño A, Torneiro M (2012) Synthesis and biological evaluation of 1α,25-dihydroxyvitamin D_3 analogues with a long side chain at C12 and short C17 side chains. J Med Chem 55:8642–8656
31. Carroll AR, Hyde E, Smith J, Quinn RJ, Guymer G, Forster PI (2005) Actinophyllic acid, a potent indole alkaloid inhibitor of the coupled enzyme assay carboxypeptidase U/hippuricase from the leaves of Alstonia actinophylla (Apocynaceae). J Org Chem 70:1096–1099
32. Granger BA, Jewett IT, Butler JD, Hua B, Knezevic CE, Parkinson EI, Hergenrother PJ, Martin SF (2013) Synthesis of (±)-actinophyllic acid and analogs: applications of cascade reactions and diverted total synthesis. J Am Chem Soc 135:12984–12986
33. Heilmann J, Mayr S, Brun R, Rali T, Sticher O (2000) Antiprotozoal activity and cytotoxicity of novel 1,7-dioxadispiro[5.1.5.2]pentadeca-9,12-dien-11-one derivatives from Amomum aculeatum. Helv Chim Acta 83:2939–2945
34. Řezanka T, Guschina IA (2001) Macrolactone glycosides of three lichen acids from Acarospora gobiensis, a lichen of central Asia. Phytochemistry 58:1281–1287
35. Wong Y-S (2002) Synthesis of (±)-aculeatins A and B. Chem Commun:686–687
36. Peuchmaur M, Saïdani N, Botté C, Maréchal E, Vial H, Wong Y-S (2008) Enhanced antimalarial activity of novel synthetic aculeatin derivatives. J Med Chem 51:4870–4873
37. Fukuyama Y, Minami H, Takeuchi K, Kodama M, Kawazu K (1996) Neovibsanines A and B, unprecedented diterpenes from Viburnum awabuki. Tetrahedron Lett 37:6767–6770
38. Fukuyama Y, Esumi T (2007) Chemistry and biological activity of vibsane-type diterpenes. J Syn Org Chem Jpn 65:585–597
39. Chen AP, Muller CC, Cooper HM, Williams CM (2009) Total synthesis of (±)-4,5-*bis*-epi-neovibsanin A and B: a neurite outgrowth comparison study. Org Lett 11:3758–3761
40. Chien S-C, Chang J-Y, Kuo C-C, Hsieh C-C, Yang N-S, Kuo Y-H (2007) Cytotoxic and novel skeleton compounds from the heartwood of Chamaecyparis obtusa var. formosana. Tetrahedron Lett 48:1567–1569
41. Oikawa H, Tokiwano T (2004) Enzymatic catalysis of the Diels-Alder reaction in the biosynthesis of natural products. Nat Prod Rep 21:321–352
42. Dong S, Hamel E, Bai R, Covell DG, Beutler JA, Porco JA Jr (2009) Enantioselective synthesis of (+)-chamaecypanone C: a novel microtubule inhibitor. Angew Chem Int Ed 48:1494–1497
43. Dong S, Qin T, Hamel E, Beutler JA, Porco JA Jr (2012) Synthesis of chamaecypanone C analogues from *in situ*-generated cyclopentadienones and their biological evaluation. J Am Chem Soc 134:19782–19787
44. Wurzel G, Becker H (1990) Sesquiterpenoids from the liverwort Ricciocarpos natans. Phytochemistry 29:2565–2568
45. Wurzel G, Becker H, Eicher T, Tiefensee K (1990) Molluscicidal properties of constituents from the liverwort Ricciocarpos natans and of synthetic lunularic acid derivatives. Planta Med 56:444–445
46. Zinsmeister HD, Becker H, Eicher T (1991) Bryophytes, a source of biologically active, naturally occurring material? Angew Chem Int Ed 30:130–147
47. Michrowska A, List B (2009) Concise synthesis of ricciocarpin A and discovery of a more potent analogue. Nat Chem 1:225–228
48. Yang JW, Hechavarria Fonseca MT, List B (2005) Catalytic asymmetric reductive Michael cyclization. J Am Chem Soc 127:15036–15037
49. Burke MD, Berger EM, Schreiber SL (2003) Generating diverse skeletons of small molecules combinatorially. Science 302:613–618

50. Tan DS (2005) Diversity-oriented synthesis: exploring the intersections between chemistry and biology. Nat Chem Biol 1:74–84
51. O'Connor CJ, Beckmann HSG, Spring DR (2012) Diversity-oriented synthesis: producing chemical tools for dissecting biology. Chem Soc Rev 41:4444–4456
52. Galloway WRJD, Isidro-Llobet A, Spring DR (2010) Diversity-oriented synthesis as a tool for the discovery of novel biologically active small molecules. Nat Commun 1:Gal1/1–Gal1/13
53. Trabocchi A, Schreiber SL (2013) Diversity-oriented synthesis: basics and applications in organic synthesis, drug discovery, and chemical biology. Wiley, Hoboken, NJ
54. Kokatla HP, Sil D, Malladi SS, Balakrishna R, Hermanson AR, Fox LM, Wang X, Dixit A, David SA (2013) Exquisite selectivity for human toll-like receptor 8 in substituted furo [2,3-*c*]quinolines. J Med Chem 56:6871–6885
55. Naud S, Westwood IM, Faisal A, Sheldrake P, Bavetsias V, Atrash B, Cheung KM, Liu M, Hayes A, Schmitt J, Wood A, Choi V, Boxall K, Mak G, Gurden M, Valenti M, de Haven BA, Henley A, Baker R, McAndrew C, Matijssen B, Burke R, Hoelder S, Eccles SA, Raynaud FI, Linardopoulos S, van Montfort RL, Blagg J (2013) Structure-based design of orally bioavailable 1*H*-pyrrolo[3,2-*c*]pyridine inhibitors of mitotic kinase monopolar spindle 1 (MPS1). J Med Chem 56:10045–10065
56. Pierre F, Chua PC, O'Brien SE, Siddiqui-Jain A, Bourbon P, Haddach M, Michaux J, Nagasawa J, Schwaebe MK, Stefan E, Vialettes A, Whitten JP, Chen TK, Darjania L, Stansfield R, Anderes K, Bliesath J, Drygin D, Ho C, Omori M, Proffitt C, Streiner N, Trent K, Rice WG, Ryckman DM (2011) Discovery and SAR of 5-(3-chlorophenylamino) benzo[*c*][2,6]naphthyridine-8-carboxylic acid (CX-4945), the first clinical stage inhibitor of protein kinase CK2 for the treatment of cancer. J Med Chem 54:635–654
57. Ohno H (2013) Gold-catalyzed cascade reactions of alkynes for construction of polycyclic compounds. Israel J Chem 53:869–882
58. Podoll JD, Liu Y, Chang L, Walls S, Wang W, Wang X (2013) Bio-inspired synthesis yields a tricyclic indoline that selectively resensitizes methicillin-resistant Staphylococcus aureus (MRSA) to beta-lactam antibiotics. Proc Natl Acad Sci U S A 110:15573–15578
59. Chang L, Podoll JD, Wang W, Walls S, O'Rourke CP, Wang X (2014) Structure–activity relationship studies of the tricyclic indoline resistance-modifying agent. J Med Chem 57:3803–3817
60. Carroll FI, Howell LL, Kuhar MJ (1999) Pharmacotherapies for treatment of cocaine abuse: preclinical aspects. J Med Chem 42:2721–2736
61. Smith MP, Hoepping A, Johnson KM, Trzcinska M, Kozikowski AP (1999) Dopaminergic agents for the treatment of cocaine abuse. Drug Discov Today 4:322–332
62. Singh S (2000) Chemistry, design, and structure–activity relationship of cocaine antagonists. Chem Rev 100:925–1024
63. Clarke R, Daum S, Gambino A, Aceto M, Pearl J, Levitt M, Cuminskey W, Bogado E (1973) Compounds affecting the central nervous system. 4. 3.beta.-phenyltropane-2-carboxylic esters and analogs. J Med Chem 16:1260–1267
64. Davies HML, Saikali E, Young WB (1991) Synthesis of (±)-ferruginine and (±)-anhydroecgonine methyl-ester by a tandem cyclopropanation/Cope rearrangement. J Org Chem 56:5696–5700
65. Davies HML, Matasi JJ, Hodges LM, Huby NJS, Thornley C, Kong N, Houser JH (1997) Enantioselective synthesis of functionalized tropanes by rhodium(II) carboxylate-catalyzed decomposition of vinyldiazomethanes in the presence of pyrroles. J Org Chem 62:1095–1105
66. Hemby SE, Co C, Reboussin D, Davies HM, Dworkin SI, Smith JE (1995) Comparison of a novel tropane analog of cocaine, 2 beta-propanoyl-3 beta-(4-tolyl) tropane with cocaine HCl in rats: nucleus accumbens extracellular dopamine concentration and motor activity. J Pharmacol Exp Ther 273:656–666
67. Porrino LJ, Davies HM, Childers SR (1995) Behavioral and local cerebral metabolic effects of the novel tropane analog, 2 beta-propanoyl-3 beta-(4-tolyl)-tropane. J Pharmacol Exp Ther 272:901–910
68. Bennett BA, Hollingsworth CK, Martin RS, Childers SR, Ehrenkaufer RE, Porrino LJ, Davies HM (1998) Prolonged dopamine and serotonin transporter inhibition after exposure to tropanes. Neuropharmacology 37:123–130
69. Daunais JB, Hart SL, Hedgecock-Rowe A, Matasi JJ, Thornley C, Davies HM, Porrino LJ (1997) Alterations in behavior and opioid gene expression induced by the novel tropane analog WF-31. Mol Brain Res 50:293–304
70. Hemby SE, Lucki I, Gatto G, Singh A, Thornley C, Matasi J, Kong N, Smith JE, Davies HM, Dworkin SI (1997) Potential antidepressant effects of novel tropane compounds, selective for serotonin or dopamine transporters. J Pharmacol Exp Ther 282:727–733

71. Porrino LJ, Miller M, Hedgecock AA, Thornley C, Matasi JJ, Davies HM (1997) Local cerebral metabolic effects of the novel cocaine analog, WF-31: comparisons to fluoxetine. Synapse 27:26–35
72. Daunais JB, Hart SL, Smith HR, Letchworth SR, Davies HM, Sexton T, Bennett BA, Childers SR, Porrino LJ (1998) Long-acting blockade of biogenic amine transporters in rat brain by administration of the potent novel tropane 2beta-propanoyl-3beta-(2-naphthyl)-tropane. J Pharmacol Exp Ther 285:1246–1254
73. Grondal C, Jeanty M, Enders D (2010) Organocatalytic cascade reactions as a new tool in total synthesis. Nat Chem 2:167–178
74. Volla CMR, Atodiresei I, Rueping M (2013) Catalytic C–C bond-forming multi-component cascade or domino reactions: pushing the boundaries of complexity in asymmetric organocatalysis. Chem Rev 114:2390–2431
75. Lan Y-B, Zhao H, Liu Z-M, Liu G-G, Tao J-C, Wang X-W (2011) Chiral counteranion synergistic organocatalysis under high temperature: efficient construction of optically pure spiro [cyclohexanone-oxindole] backbone. Org Lett 13:4866–4869
76. Wang Y, Han R-G, Zhao Y-L, Yang S, Xu P-F, Dixon DJ (2009) Asymmetric organocatalytic relay cascades: catalyst-controlled stereoisomer selection in the synthesis of functionalized cyclohexanes. Angew Chem Int Ed 48:9834–9838
77. Wang Y, Yu D-F, Liu Y-Z, Wei H, Luo Y-C, Dixon DJ, Xu P-F (2010) Multiple-organocatalyst-promoted cascade reaction: a fast and efficient entry into fully substituted piperidines. Chem Eur J 16:3922–3925
78. Hashmi ASK, Hubbert C (2010) Gold and organocatalysis combined. Angew Chem Int Ed 49:1010–1012
79. Zhong C, Shi X (2010) When organocatalysis meets transition-metal catalysis. Eur J Org Chem:2999–3025
80. Zhou J (2010) Recent advances in multicatalyst promoted asymmetric tandem reactions. Chem Asian J 5:422–434
81. Allen AE, MacMillan DWC (2012) Synergistic catalysis: a powerful synthetic strategy for new reaction development. Chem Sci 3:633–658
82. Cohen DT, Scheidt KA (2012) Cooperative Lewis acid/N-heterocyclic carbene catalysis. Chem Sci 3:53–57
83. Du Z, Shao Z (2013) Combining transition metal catalysis and organocatalysis - an update. Chem Soc Rev 42:1337–1378
84. Zhang Y, Wang S, Wu S, Zhu S, Dong G, Miao Z, Yao J, Zhang W, Sheng C, Wang W (2013) Facile construction of structurally diverse thiazolidinedione-derived compounds via divergent stereoselective cascade organocatalysis and their biological exploratory studies. ACS Comb Sci 15:298–308
85. Zhang G, Zhang Y, Yan J, Chen R, Wang S, Ma Y, Wang R (2012) One-pot enantioselective synthesis of functionalized pyranocoumarins and 2-amino-4H-chromenes: discovery of a type of potent antibacterial agent. J Org Chem 77:878–888
86. Dückert H, Pries V, Khedkar V, Menninger S, Bruss H, Bird AW, Maliga Z, Brockmeyer A, Janning P, Hyman A, Grimme S, Schürmann M, Preut H, Hübel K, Ziegler S, Kumar K, Waldmann H (2012) Natural product–inspired cascade synthesis yields modulators of centrosome integrity. Nat Chem Biol 8:179–184
87. Bores GM, Kosley RW (1996) Galanthamine derivatives for the treatment of Alzheimer's disease. Drugs Future 21:621–625
88. Raskind MA, Peskind ER, Wessel T, Yuan W (2000) Galantamine in AD: a 6-month randomized, placebo-controlled trial with a 6-month extension. The Galantamine USA-1 Study Group. Neurology 54:2261–2268
89. Küenburg B, Czollner L, Fröhlich J, Jordis U (1999) Development of a pilot scale process for the anti-Alzheimer drug (-)-galanthamine using large-scale phenolic oxidative coupling and crystallisation-induced chiral conversion. Org Process Res Dev 3:425–431
90. Tanaka H, Kuroda A, Marusawa H, Hatanaka H, Kino T, Goto T, Hashimoto M, Taga T (1987) Structure of FK506, a novel immunosuppressant isolated from Streptomyces. J Am Chem Soc 109:5031–5033
91. Hatanaka H, Kino T, Miyata S, Inamura N, Kuroda A, Goto T, Tanaka H, Okuhara M (1988) FR-900520 and FR-900523, novel immunosuppressants isolated from a Streptomyces. II. Fermentation, isolation and physico-chemical and biological characteristics. J Antibio 41:1592–1601
92. Morisaki M, Arai T (1992) Identity of immunosuppressant FR-900520 with ascomycin. J Antibio 45:126–128
93. Baumann K (1993) Preparation of tricyclic macrolides as drugs. European Patent 0569337, 10 Nov 1993
94. Koch G, Jeck R, Hartmann O, Küsters E (2001) Selective synthesis of a new ascomycin rearrangement product (SDZ ASD732) on a pilot plant scale. Org Process Res Dev 5:211–215

Chapter 8

Multicomponent Synthesis: Cohesive Integration of Green Chemistry Principles

Razvan Cioc, Eelco Ruijter, and Romano V. A. Orru

Abstract

The application of multicomponent reactions (MCRs) in the generation of compound libraries has long been recognized as a key strategy for the development of lead matter in drug discovery. Given the numerous advantages that these processes possess not only from the ability to generate large numbers of diverse compounds but also from a sustainability perspective, it is somewhat surprising that MCRs have not enjoyed such a widespread application in downstream development and the later scale-up of pharmaceutical candidates. The current chapter provides an overview of the most common MCRs with a critical focus on their perceived benefits as a sustainable technology. In addition, an overview of the combination of MCRs with other "green" technologies such as both biocatalysis and flow chemistry is also provided.

Key words Multicomponent reactions, Biocatalysis, Asymmetric synthesis, Flow chemistry, Drug discovery, Green chemistry, Selectivity, Mechanisms, Strecker, Hantzsch, Biginelli, Mannich, Passerini, Ugi, Petasis, Isocyanide

1 Introduction

Multicomponent reactions (MCRs) [1] feature the combination of three or more reagents in a single operation to selectively form a product that retains the majority of the atoms of the starting materials. Rather than high-order collisions, MCRs are mechanistically (and sometimes also operationally) sequences of sub-reactions occurring successively. For this reason we consider that the simultaneous addition of all inputs is not essential, and therefore the definition of multicomponent processes should be regarded in a broader sense, including sequential one-pot transformations and tandem/domino/cascade reactions.

Multicomponent chemistry dates back to the early days of organic synthesis, with the discovery of reactions such as the Strecker 3CR [2], the Hantzsch 4CR [3], and the Biginelli 3CR [4] only shortly following Wöhler's synthesis of urea [5]. However, with the exception of some pivotal moments (e.g., the discovery of

Paul F. Richardson (ed.), *Green Chemistry in Drug Discovery: From Academia to Industry*, Methods in Pharmacology and Toxicology, https://doi.org/10.1007/978-1-0716-1579-9_8, © Springer Science+Business Media, LLC, part of Springer Nature 2022

the Ugi 4CR in the 1960s), it is only since the mid-1990s, in the age of combinatorial chemistry and diversity-oriented synthesis, that multicomponent reactions have seen a truly major expansion in development and application. The list of multicomponent name reactions contains numerous recent entries, and it can only be anticipated to grow further in the future (Table 1).

The single most important area of application of multicomponent reactions is drug discovery and design. With the use of MCRs, large compound libraries can be constructed from simple building blocks in short sequences or even in a single step. Furthermore, multicomponent synthesis can be readily automated resulting in important resource savings. Given the convergent and selective nature of the chemical transformations involved, the purity of the end products is typically high. All these arguments promote multicomponent reactions as efficient tools for the highly productive synthesis of collections of compounds to be utilized for high-throughput screening against various biological targets, structure-activity relationship studies, and lead optimization.

In view of their diversity and variability, multicomponent reactions have the unmatched ability to cover vast domains of chemical space. For instance, the medicinally relevant space of nitrogen heterocycles can be systematically explored according to the ring size (azetidinones via the Staudinger reaction, pyrroles and pyridines via the Hantzsch reaction) and the number of N-atoms and unsaturation (imidazolines according to the Orru reaction, imidazoles by the van Leusen 3CR, dihydropyrimidinones via the Biginelli 3CR), as shown in Scheme 1. Furthermore, numerous MCRs come in different variants. The classical Ugi 4CR involves aldehydes, amines, isocyanides, and carboxylic acids. However, the latter can be substituted by other acidic components, such as water, carbonic acid, thiocarboxylic acids, phenols, HN_3, H_2S, HNCO, HNCS, etc., leading to an entire range of primary scaffolds (α-acylaminocarboxamides, carbamates, α-aminoamides, tetrazoles, hydantoins, etc.). Moreover, this diversity can be expanded much further according to several general strategies, such as the use of bifunctional inputs, secondary reactions, union of MCRs, etc. Finally, the exploitation of the different variation points in the input materials dramatically condenses the synthetic effort to assemble compound libraries of the desired dimension (Fig. 1).

Importantly, the large chemical space that multicomponent products span has considerable relevance for medicinal applications. For instance, cyclic systems occupy a privileged position in multicomponent chemistry; from non-flat scaffolds to planar aromatics, including various polycyclic systems (fused, spiro, bridged), heterocycles, macrocycles, etc., all are well represented in the MCR space. Another distinguishing feature is the dense functionalization of MCR-derived molecules: typically an important number of functional groups (originating from the primary scaffold and from the

Table 1
Representative examples of multicomponent chemistry

Reaction	Year	Scheme	AE	By-product
Strecker	1850	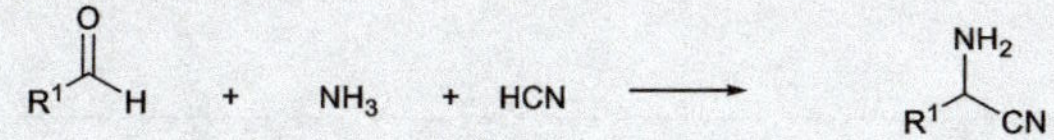	0.80	H_2O
Hantzsch	1881	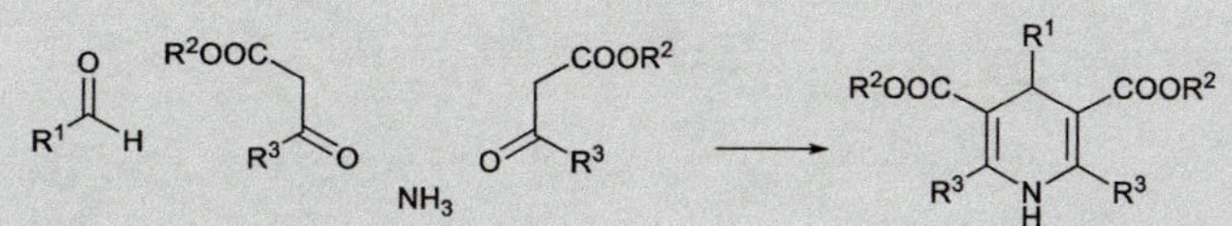	0.82	$3H_2O$
Biginelli	1891	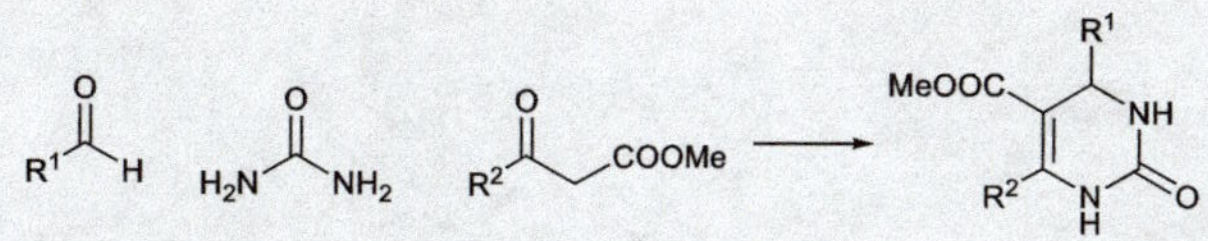	0.84	$2H_2O$
Mannich	1912	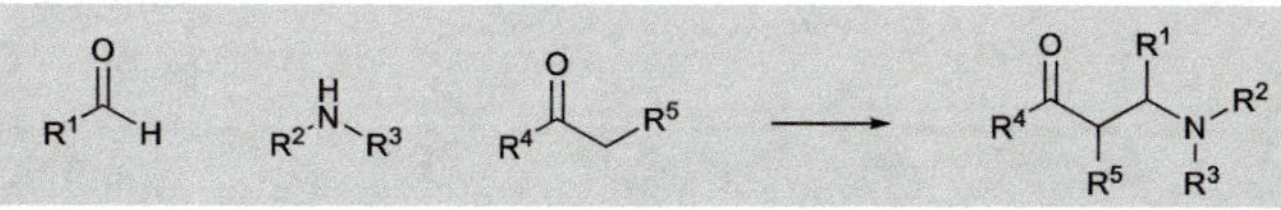	0.89	H_2O
Passerini	1921	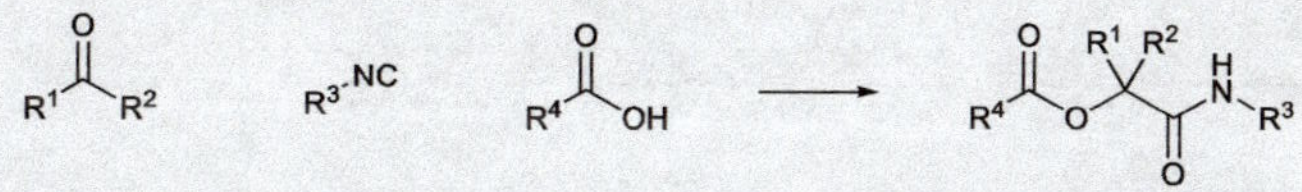	1.00	None
Ugi	1959	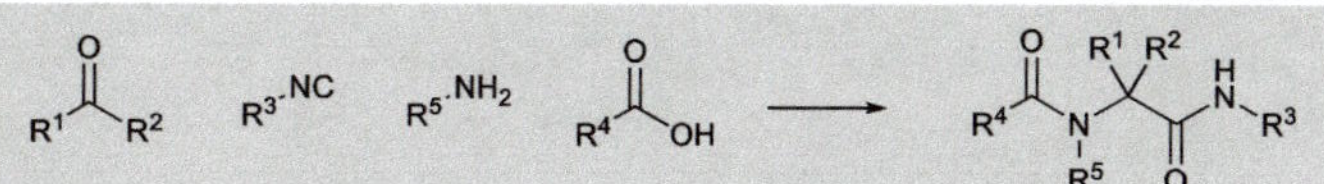	0.91	H_2O
Petasis	1993	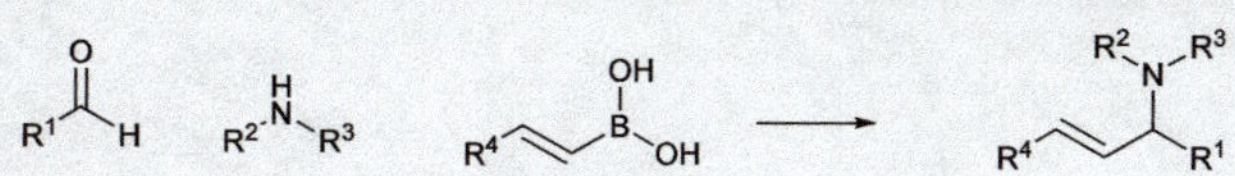	0.65	$B(OH)_3$

(continued)

Table 1 (continued)

Reaction	Year	Scheme	AE	By-product
Groebke-Blackburn-Bienaymé	1998	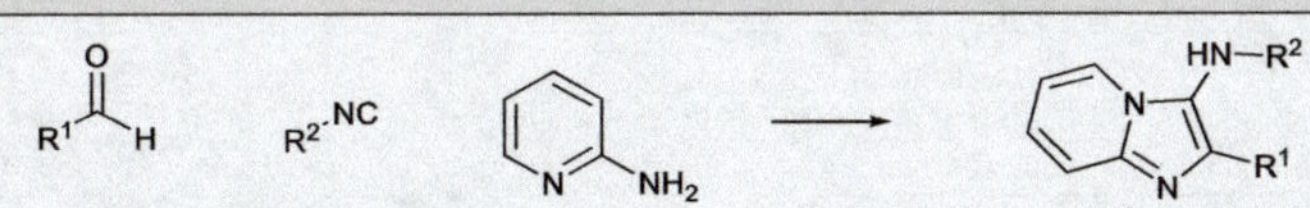	0.90	H_2O
(GBB)Passerini-Dömling	2000	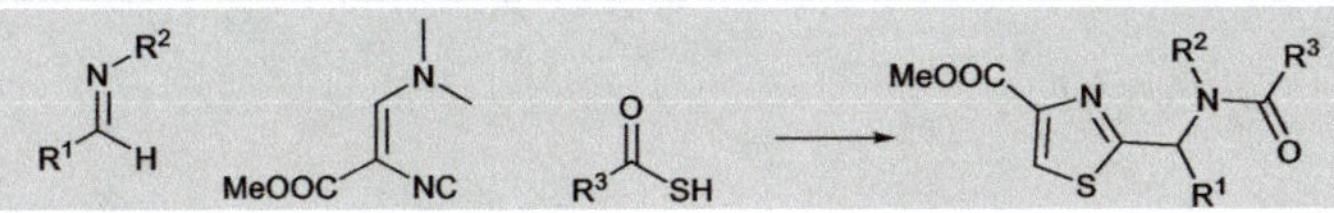	0.84	Me_2NH
A^3 coupling	2001	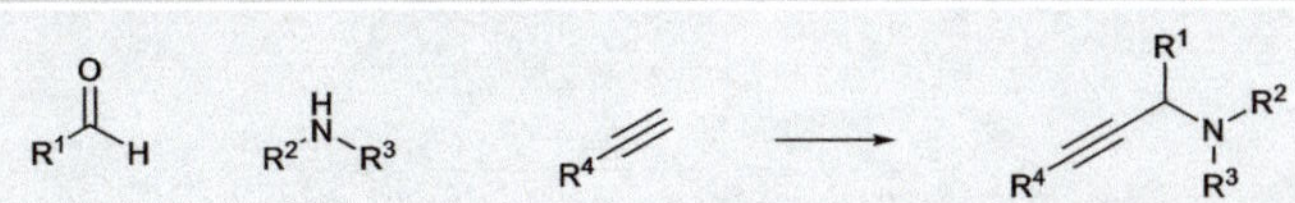	0.86	H_2O
Orru	2003	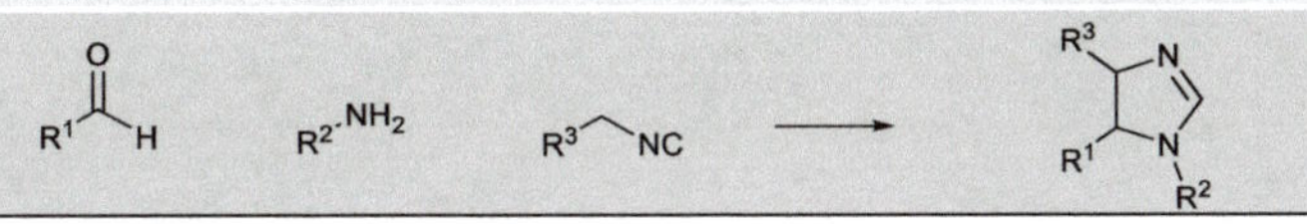	0.86	H_2O

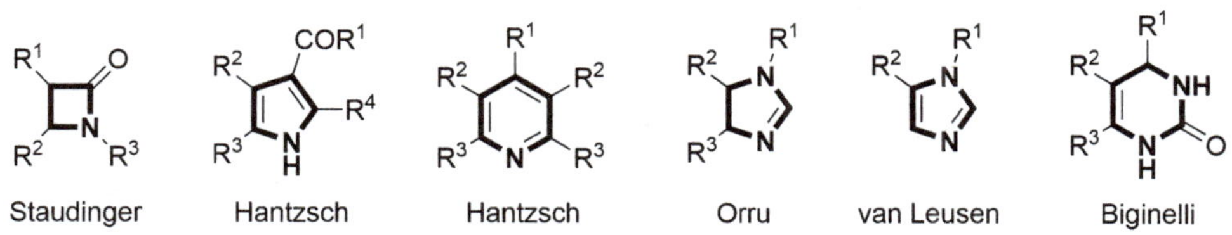

Scheme 1 Exploring the space of N-containing heterocycles with MCRs

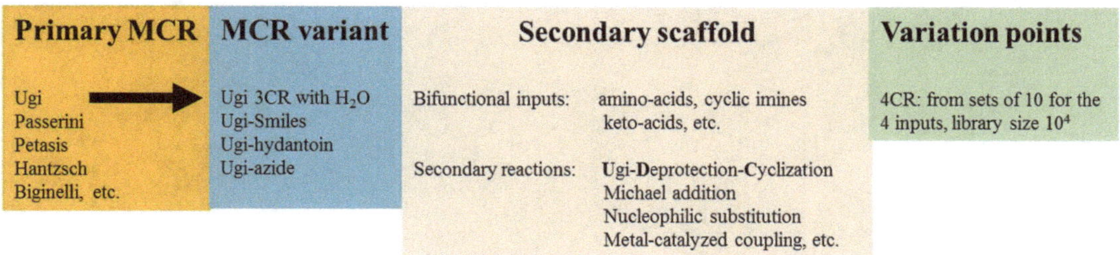

Fig. 1 Diversity generation with multicomponent reactions

generally broad functional group tolerance of the multicomponent condensation) decorate the hydrocarbon framework. This facilitates the embedding of the desired physicochemical properties in the products to suit the requirements for most medicinal chemistry applications. Equally noteworthy is that multicomponent chemistry can rapidly supply completely novel scaffolds to bioactivity screening projects [6], or extensively functionalize bioactive molecules, including natural product-based leads [7, 8]. Furthermore, once validated in natural product synthesis, the convergent multicomponent approach features unrivaled productivity in generating libraries of analogues with improved bioactivity [9–14]. A comprehensive review highlighting the power of MCRs in drug discovery was published by Dömling in 2012 [6], and various aspects of this topic have also been discussed in other reports [15–19].

Moving from the drug discovery phase to the production stage, the pharmaceutical industry can profit considerably from the implementation of multicomponent approaches in API synthesis. In the sustainability of chemical processes, MCRs play a particularly important part as they generally score high in most areas of green chemistry [20]. The successful application of multicomponent reactions in the manufacturing of pharmaceuticals is highlighted in Scheme 2, with examples ranging from simple small molecules to relatively complex structures.

Importantly, together with their intrinsic green character, multicomponent reactions are compatible with other environmentally friendly technologies (flow chemistry, biocatalysis, green solvents, etc.). The cohesive integration of these concepts in process chemistry represents a unique opportunity to adapt the industrial

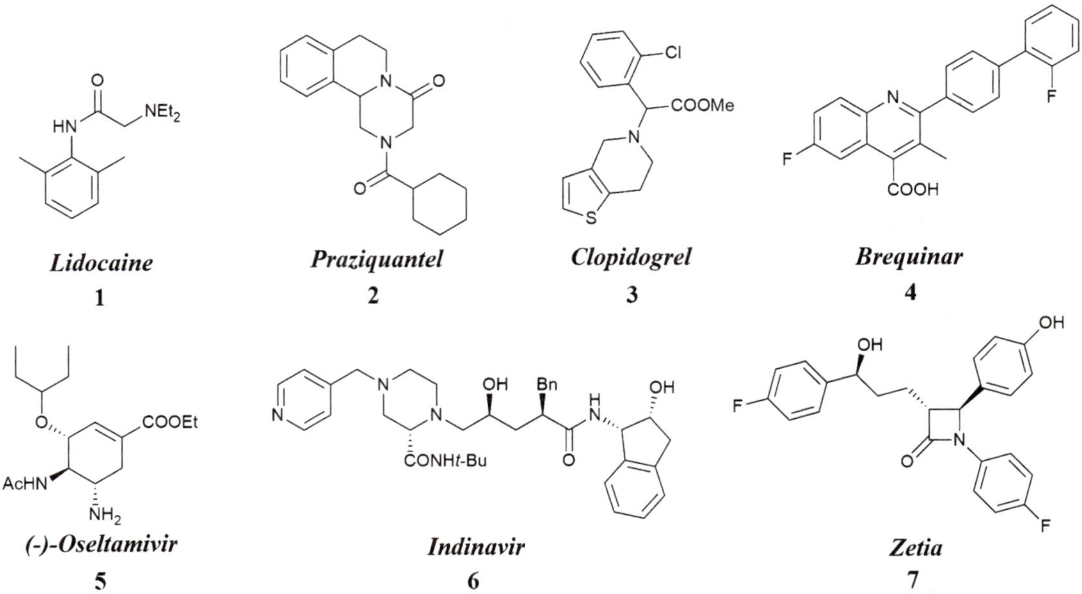

Scheme 2 Multicomponent reactions for the synthesis of APIs

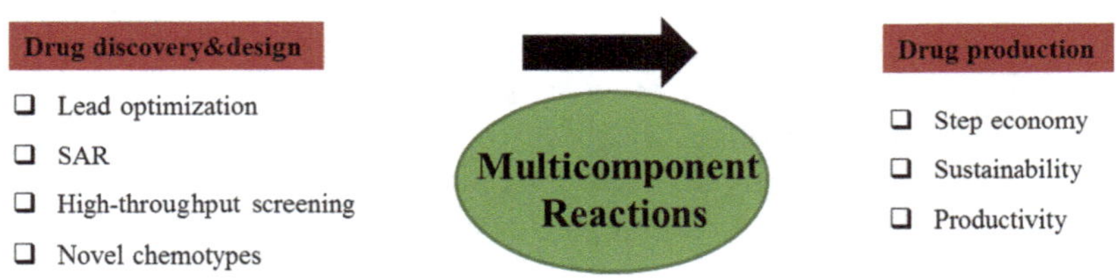

Fig. 2 Multicomponent reactions in the pharmaceutical industry

production of pharmaceuticals to the economic, societal, and regulatory demands of the twenty-first century and take a major step toward a truly sustainable synthesis.

This chapter reviews the possibilities of applying multicomponent reactions in drug design and production with a strong emphasis given to the sustainable dimension of this chemistry (see Fig. 2).

2 Atom Economy and Waste Prevention

Atom economy is perhaps the greatest virtue of multicomponent reactions from the sustainability point of view: the majority of the landmark MCRs exhibit high atom economy (at least 80%, Table 1). This important ecological benefit is the consequence of complex reaction mechanisms consisting primarily of addition and/or condensation steps, which incorporate in the product most of the atoms of the starting materials, except for typically a small molecule like water. Indeed, imine formation followed by the addition of a nucleophile (cyanide, enols, isocyanide, boronic acid, alkyne, etc.) represents the most common reaction manifold encountered in multicomponent chemistry (Strecker, Mannich, Biginelli, Ugi, Petasis, A^3 reaction, respectively). Nevertheless, the diversity of scaffolds that are accessible via MCR condensations is remarkable, as shown in Table 1.

Despite this general mechanistic resemblance, the optimal reaction conditions for these processes are quite particular. This usually enables a good control over the desired reaction pathway, and consequently the selectivity of MCRs is high. For example, the addition of boronic acids to imines is efficient in ethanol, whereas the addition of alkynes requires a nonpolar solvent (e.g., toluene) and transition-metal catalysis (most typically copper salts). Similarly, when performing the Ugi 4CR, the Passerini product, corresponding to a condensation of only three of the four inputs (aldehyde, *amine*, isocyanide, and carboxylic acid), is not formed. The Passerini reaction is very sluggish in the polar protic solvents typically used in Ugi-type chemistry and as such cannot compete with isocyanide addition to the rapidly forming imine under these conditions. The judicious choice of reaction parameters can therefore lead to high degrees of chemoselectivity in MCRs. An interesting manipulation of the reaction outcome by tuning conditions is shown in Scheme 3, where the selective formation of imidazolines or oxazoles from the same starting inputs can be promoted by the use of suitable additives [21]. Silver acetate is an efficient catalyst for the three component imidazoline synthesis, whereas triethylammonium hydrochloride is used in equimolar amounts as a Brønsted acid imine activator. The stoichiometry of the inputs is also important for the efficiency and selectivity of the reaction: the **8**:**9**:**10** ratio is 1:2:1.5 for the imidazoline synthesis and 1:1:1 for the oxazole synthesis.

The good selectivity of multicomponent processes can in many cases be attributed to thermodynamic factors. Often these reactions are performed under equilibrium conditions, and a certain product is substantially favored over other outputs in view of its lower energetic content, cyclic and/or aromatic nature etc. In this

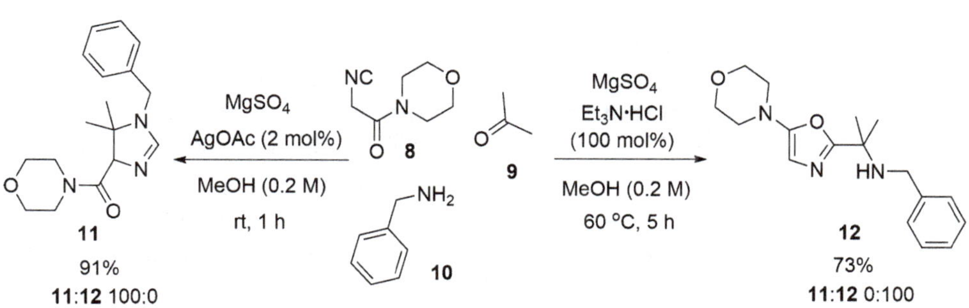

Scheme 3 Tuning the course of MCR with additives

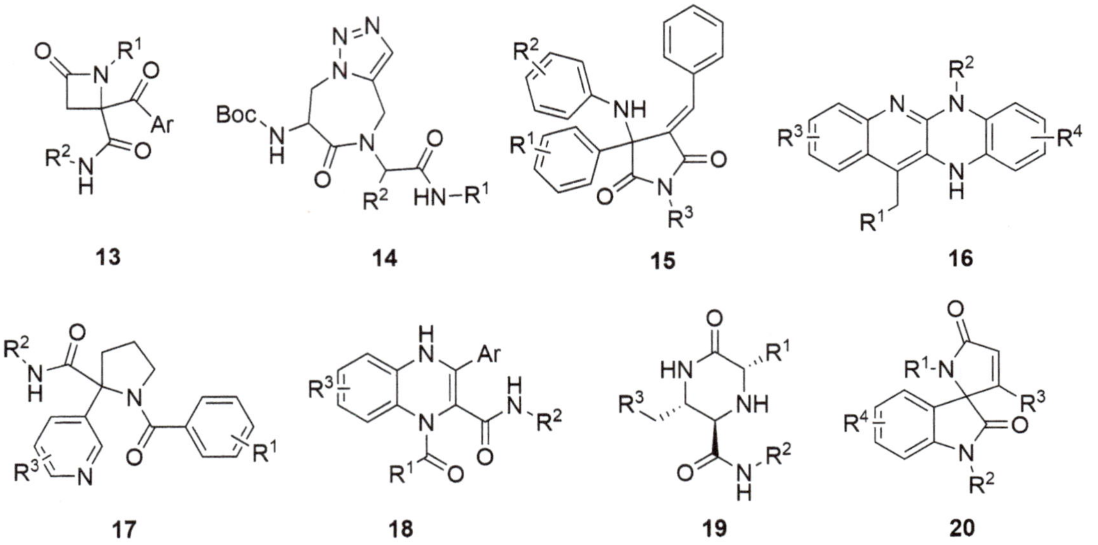

Scheme 4 Secondary isocyanide-based MCR scaffolds

respect, the century-old Hantzsch reaction is noteworthy as numerous reaction pathways can be imagined between the three inputs required, but the stability of the highly conjugated dihydropyridine product dictates the outcome. Similarly, the formation of two new σ-bonds involving the formally divalent carbon atom of isocyanides represents a powerful driving force that has been extensively exploited in the synthesis [22–24].

When it comes to drug discovery applications, these isocyanide-based MCRs are important tools since they allow the rapid generation of large libraries of compounds to be tested for bioactivity. Isocyanide chemistry is remarkably versatile, and the chemical space that it can cover is considerable. Most importantly, this chemistry is compatible with various atom-economical follow-up reactions that can be effected in the same pot to further increase the chemical diversity. Great focus has been given to this strategy in recent years, and Scheme 4 comprises a selected set of scaffolds that result from post-multicomponent condensation functionalization [25–32].

Therefore, isocyanide-based MCRs have unquestionable value in green synthesis. However, the isocyanide generation in an eco-friendly way remains an important issue to be addressed. The typical strategies to access isocyanides (dehydration of formamides [33], Hoffmann procedure [34]), albeit generally efficient, do not go hand-in-hand with the principles of green chemistry. These reactions have poor atom economy, use hazardous reagents and problematic solvents, generate waste, etc. Finding a green solution to isocyanide preparation requires intensive creative efforts, as the isocyanide functionality is commonly generated via substitution or elimination rather than addition, which is by definition contrary to a green approach. An unrefined alternative would be the so-called isocyanide-free multicomponent reactions, in which the isocyanide is prepared in situ and used directly in the synthetic application [35, 36]. Another recent conceptual advance in this direction involves the synthesis of isocyanides from readily available tertiary alcohols and TMSCN under Lewis acid catalysis [37]. These approaches have important limitations (e.g., "isocyanide-free" MCRs employ $KAg(CN)_2$, generating silver waste) but serve as a testimony to the engagement of the scientific community to deliver a sustainable solution to isocyanide synthesis.

Having discussed the atom-economical nature of multicomponent reactions, it is just as important to note that the multicomponent approach brings benefits also in the form of step economy. MCRs produce high levels of complexity in a single step, often without the need of protection-deprotection stages, which can lead to a substantial shortening of a reaction sequence. Particularly in the area of peptidomimetics synthesis, MCRs can be highly productive [38]. For instance, the novel Ugi-type 5C-4CR reaction discovered by Dömling [39] generates a dipeptide fragment in a single step from α-amino acids, secondary amines, aldehydes or ketones, and isocyanides (Scheme 5). Conventionally, this structural motif is assembled from protected building blocks using stoichiometric amounts of activating agents and bases, in a multistep highly wasteful process. Since this protocol is general, its advantages (in terms of both productivity and sustainability) in combinatorial synthesis and applications in lead-optimization are evident. Operationally, this reaction is extremely simple. The equimolar mixture of the four components is stirred at room temperature in a methanol-water solution for 1–3 days.

Furthermore, step economy of MCRs can have an even larger impact in generic drug production. By the strategic use of a biocatalytic desymmetrization and two multicomponent reactions, the synthesis of the anti-hepatitis C drug Telaprevir could be achieved in only 11 steps compared to 24 as reported previously [40] in 45% overall yield and advanced purity [41]. The highlight is a diastereoselective Ugi-Joullié 3CR (dr 6.4:1) between imine **23**, isocyanide **24**, and carboxylic acid **28** (used in 1.4:1.4:1 ratio) performed

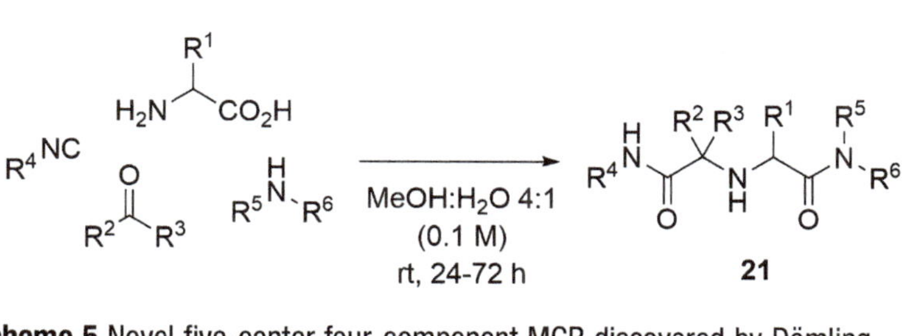

Scheme 5 Novel five-center four-component MCR discovered by Dömling

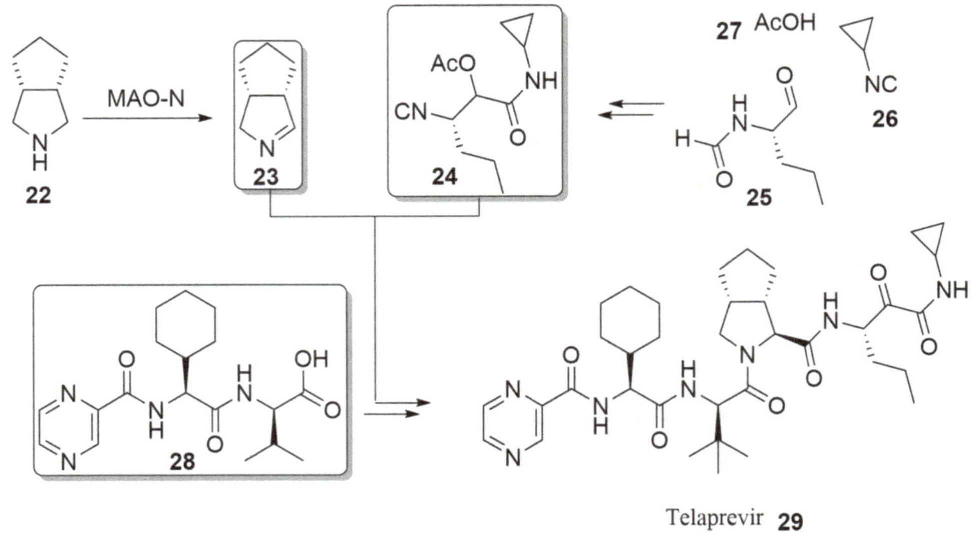

Scheme 6 Multicomponent synthesis of Telaprevir

in dichloromethane (0.34 M) at room temperature over 3 days (from a sustainability perspective, dichloromethane is viewed as an undesirable solvent, though there is no discussion herein of whether the multicomponent reaction would work in an alternative solvent. However Ichikawa et al. have demonstrated that not only can this 3CR be performed in a variety of solvents but that the choice of solvent plays a pivotal role in the mechanistic pathway**). Upon basic hydrolysis of the acetate and oxidation of the resulting alcohol group with Dess-Martin periodinane, Telaprevir **29** was obtained in a 61% yield as single diastereoisomer after chromatographic purification (prior to chromatography, Telaprevir is obtained as a 83:13:4 mixture of diastereomers, with one minor diastereomer derived from the incomplete stereoinduction of the Ugi-type 3CR and the other from the minor enantiomer of imine) (Scheme 6).

To conclude, multicomponent reactions are ideal tools to effect the synthesis of diverse complex molecules with limited generation of waste, both in combinatorial chemistry applications and on process scale.

3 Green Solvents

Solvents are perhaps the most active area of research in green chemistry, since they make up not only by far the majority of the waste [42, 43] but are also a significant component of the hazards and energy intensity of a process. In multicomponent chemistry, as shown in the previous section, solvents may markedly influence the course, rate, and selectivity of a reaction. The multicomponent product is the result of a complex sequence of mechanistic events with potentially contradicting solvent requirements, which often makes the solvent choice far from trivial. For instance, for imine-based MCRs, often preformation of the imine in a different solvent (typically with the use of a drying agent, such as Na_2SO_4, $MgSO_4$, molecular sieves) has a beneficial effect on the productivity of the reaction. For thermodynamically controlled MCRs where the precipitation of the product out of the solution phase is the driving force, solubility issues are also critical. For these reasons solvent screening is most often included in MCR development. Even the robust Ugi reaction, which in the vast majority of cases proceeds well in methanol, is found in certain applications to perform better in alternative solvents [44] with trifluoroethanol having being reported to offer improvements [45, 46]. Despite a good general understanding of solvent requirements for most MCRs, in some cases the reaction is very substrate-specific. The Petasis reaction, for instance, is applied in a broad spectrum of solvents such as CH_2Cl_2 [47], EtOH [48], dioxane [49], DMF [50], etc., depending on the specific structures of the substrates. From the sustainability perspective, in general a good compatibility of multicomponent chemistry with environmentally benign solvents is either observed or can be reached. While the use of less desirable solvents like dichloromethane is acceptable on lab-scale, in industry a green compromise solution can be determined in most cases, since MCRs are quite versatile with respect to solvent requirements.

Moreover, it is typical for MCRs to be run in concentrated solutions (1 M), keeping the consumption of solvents as reaction media at a minimum. This statement remains valid for the downstream processing and purification: since the multicomponent products have a high molecular weight and are typically crystalline, they can be recovered by precipitation and purified via recrystallization, thus avoiding extractive and/or chromatographic workups.

The application of MCRs in eco-friendly solvents has received a great deal of interest in the last decade. A comprehensive review on this topic was published in 2012 [51]. The surveyed list of novel MCRs or improved variants running in water, ionic liquids, polyethylene glycol polymers (PEGs), $scCO_2$, bio-derived solvents, and neat systems is impressive. Furthermore, multicomponent chemistry is well represented in another 2012 review summarizing the progress of organic synthesis in water [52]. The general compatibility of MCRs with ionic liquids [53] and solvent-free conditions [54] has also been discussed in detail in a number of recent reviews. This interest devoted to multicomponent reactions in green chemistry surveys shows the privileged position that this chemistry occupies in the repertoire of sustainable synthetic tools.

Table 2
Multicomponent reactions in green solvents

Reaction	Water	Ionic liquids	PEGs	Bio-derived	Neat
Strecker	Ref. 55	Ref. 56	Ref. 57		Ref. 58
Hantzsch	Ref. 59	Ref. 60	Ref. 61		Ref. 62
Biginelli	Ref. 63	Ref. 64	Ref. 65	Ref. 66	Ref. 67
Mannich	Ref. 68	Ref. 69			Ref. 70
Passerini	Ref. 71	Ref. 72	Ref. 72		Ref. 73
Ugi	Ref. 71	Ref. 74	Ref. 75		Ref. 76
Petasis	Ref. 77	Ref. 78			Ref. 79
GBB		Ref. 80			Ref. 81
A^3 coupling	Ref. 82	Ref. 83	Ref. 84		Ref. 85

The majority of most established MCRs have been successfully applied in the various categories of recognized sustainable solvents, and only a few additions are still required to complete the matrix of multicomponent chemistries in water, ionic liquids, PEGs, bio-based solvents, or solvent-free systems (Table 2). However, despite this growing interest in combining MCRs with eco-friendly solvents, it should be underlined that in typical synthetic applications, the use of conventional solvents still remains the more popular option.

4 Energy Efficiency

Multicomponent reactions are characteristically energy-efficient with the majority of processes being performed at ambient temperatures and pressures, often with the aid of catalysis. Extreme reaction temperatures are required only in specific cases, such as either cryogenic temperatures for organocatalytic enantioselective MCRs, or temperatures above 100 °C for metal-initiated cascades. Microwave irradiation is often found to be beneficial for multicomponent chemistry [86, 87], and in recent years attention has been devoted to developing protocols that employ alternative sources of energy, which are preferable to thermal heating from a sustainability point of view [88, 89], such as ultrasound irradiation [90–96] and mechanical energy [97, 98]. Moreover, the demonstrated compatibility of multicomponent chemistry with continuous operation in microreactors opens new avenues for the optimization of reactions from a process intensification standpoint [99].

5 Asymmetric Synthesis

Asymmetric synthesis represents a major focus area in multicomponent chemistry [100–103]. From a sustainability perspective, asymmetric synthesis is clearly preferable over the wasteful and often difficult separation of diastereoisomers or enantiomers, while in drug discovery, facile access to the various pure stereoisomers is highly desirable to expedite the bioactivity profiling of drug candidates. Together with these important practical benefits, the quest for asymmetry in multicomponent chemistry also finds inspiration in the intrinsic challenge of controlling selectivity in complex reaction sequences with often a limited initial mechanistic understanding of the overall process and/or the pivotal steps involved.

Thus, highly efficient diastereoselective variants of MCRs are still relatively rare in asymmetric synthesis, though a number of examples have been reported. An impressive case is the work of van der Eycken et al., who showed that the use of 2-′-formylbiphenyl-2-carboxylic acid **30** in an Ugi 3CR affords the dibenzo[*c*,*e*]azepinone products as single diastereoisomers (Scheme 7) [104]. Higher dilution than usual for Ugi chemistry was required in this system for the selective formation of the seven-membered ring over competing intermolecular condensations. Remarkably, this methodology proved very general, and a fairly large library of derivatives, of which several showed antiproliferative activity against tumor cell lines, were prepared with good overall yields.

A comparably high control in diastereoselectivity was demonstrated also for the Petasis 3CR reaction involving sulfinamides [105]. Intensive optimization efforts showed that $InBr_3$ was the best catalyst for this reaction that upon cleavage of the *tert*-butyl sulfinamide yields synthetically valuable β,γ-unsaturated amino-acids as the products (Scheme 8).

Amine-based chiral auxiliaries have also found successful application in the diastereoselective Povarov-type condensation of O-pivaloylated β-D-galactosylimines with cyclic enol ethers and alcohols (Scheme 9) [106]. Excellent yields ranging from 73% to

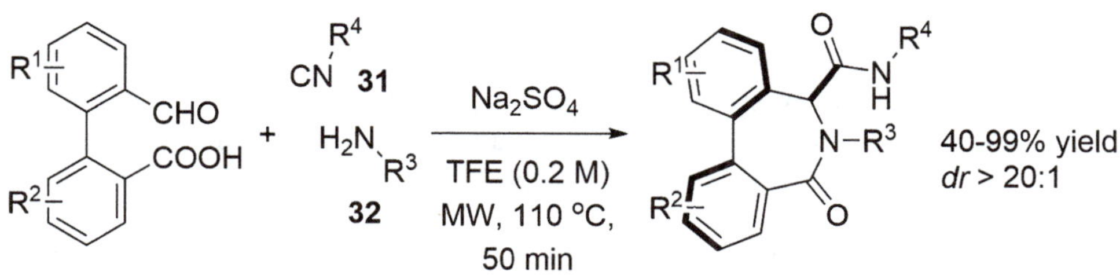

30

33

Scheme 7 Diastereoselective Ugi 3CR reaction

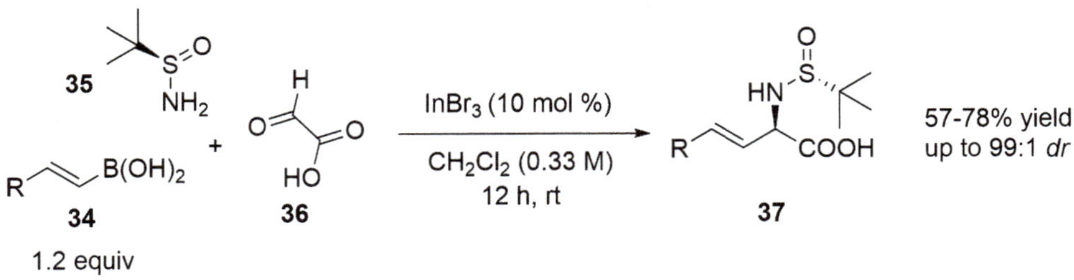

Scheme 8 Diastereoselective Petasis reaction

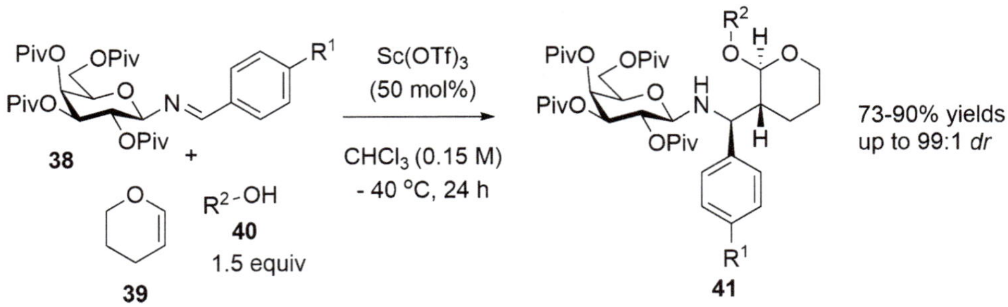

Scheme 9 Diastereoselective Povarov-type reaction

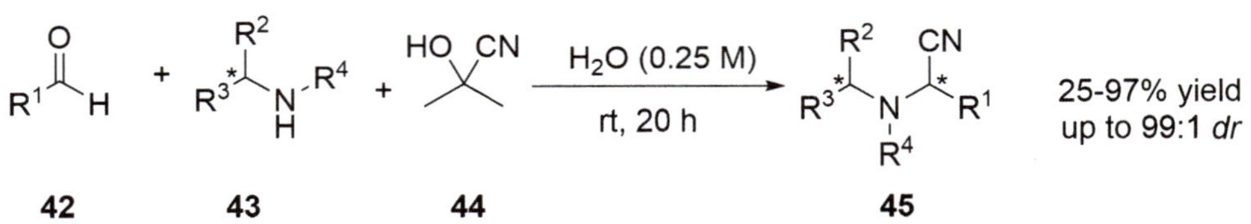

Scheme 10 Diastereoselective Strecker reaction

90% and diastereoselectivies up to 99% were reported. The configuration of the product results from an octahedral coordination complex of Sc^{3+}, which is a typical catalyst for the Povarov reaction, with the imine and enol ether substrates.

Among the most challenging reactions in asymmetric MCRs is the Strecker 3CR, owing largely to the fact that the nucleophile attacking the imine is one of the smallest possible, the cyanide anion. Given this, diastereoselectivity levels when employing chiral inputs were found to be modest with the notable exception of bulky prolinol-derived amines for which the product is isolated as single stereoisomer (Scheme 10) [55]. With the use of water as solvent, ambient conditions and a relatively safe cyanide source (acetone cyanohydrin), this protocol conforms to many of the principles of green chemistry.

Transition metal-catalyzed cascade reactions represent an important part of multicomponent chemistry [101], and often exquisite levels of stereocontrol can be achieved together with the

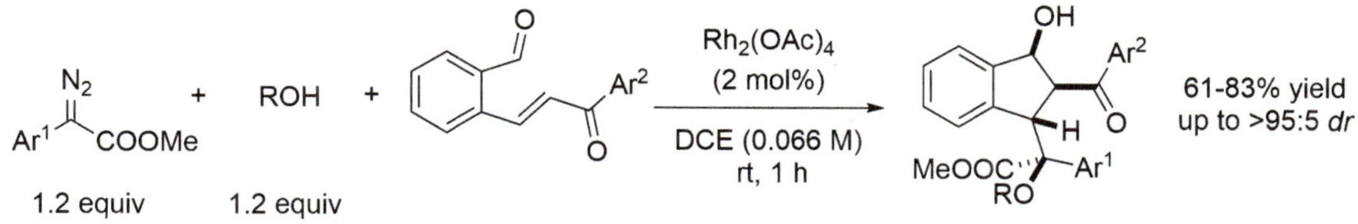

Scheme 11 Diastereoselective Rh-initiated cascade

ability to develop a high degree of molecular complexity in a single step. To highlight one example, Scheme 11 shows the Rh^{II}-catalyzed intramolecular three-component cascade Michael-aldol-type reaction developed by Hu [107]. Interestingly, a delayed proton transfer is observed in this innovative sequence that generates highly functionalized 1-indanol derivatives containing multiple stereocenters with excellent diastereoselectivity.

Much more interesting (and challenging) than diastereoselective approaches are catalytic enantioselective multicomponent processes. Efforts devoted to this area of MCR research have been rewarded with outstanding achievements in the last few years [100], and this chemistry is still rapidly expanding. Most well-known MCRs nowadays have a catalytic asymmetric variant, and numerous other synthetic methodologies have been developed recently for the efficient assembly of complex chiral molecules from simple building blocks via multicomponent condensations [108–119].

2014 was a particularly fruitful year for asymmetric multicomponent synthesis with the first reports of the catalytic enantioselective Orru reaction between ketimines and isocyanoacetates being disclosed by two groups using closely related strategies. The group of Dixon employed a silver salt cinchona-derived aminophosphine (**53**) binary catalytic system (Scheme 12) [120], whereas Nakamura et al. utilized copper catalysis, with again a cinchona alkaloid-based ligand, though with a pyridine nitrogen as donor instead of the phosphine (**52**) [121]. Both groups used the diphenylphosphinoyl substituent as the activating group on the imine, but remarkably despite apparently similar protocols, the diastereoselectivity of the addition is opposite: the silver-catalyzed process gives *anti*-selectivity, whereas with Nakamura's copper catalyst, the major product is *syn*. Both methods have general scope with ee's up to 99% for the valuable imidazoline products. Even though this reaction is not a MCR per se since the phosphinoyl-imines are not prepared in the same pot as the isocyanoacetate addition, these results have a significant degree of relevance to multicomponent chemistry.

A green procedure for the catalytic enantioselective A^3 coupling of aldehydes, amines, and alkynes in water was also reported in 2014 [82]. This protocol employs a *bis*(imidazoline)-CuI catalyst

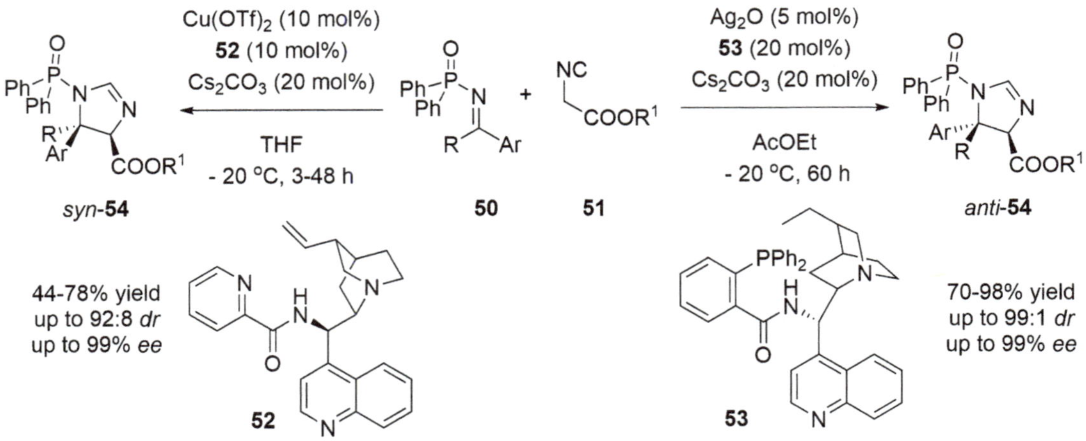

Scheme 12 Catalytic enantioselective Orru reaction

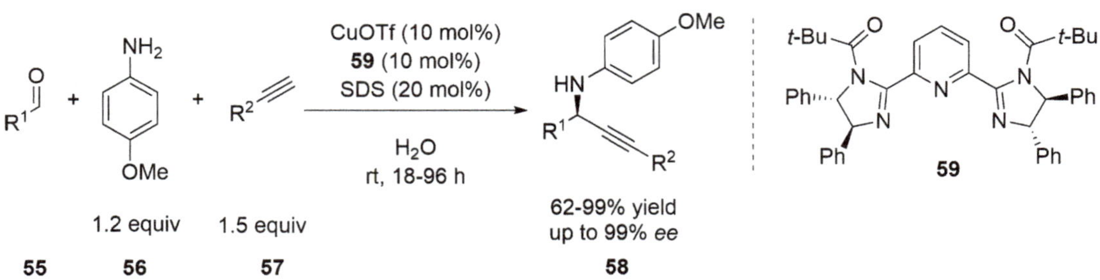

Scheme 13 Catalytic enantioselective A^3 reaction

having a hydrophobic substituent and sodium dodecyl sulfate (SDS) as a surfactant. The enantioselectivity of this process is generally very high (Scheme 13), though the requirement to use *p*-anisidine **56** constitutes a synthetic limitation.

Finally, the development of a catalytic asymmetric variant of the Ugi 4CR represents the holy grail in asymmetric multicomponent chemistry. Despite intensive efforts and promising results in the development of enantioselective Ugi-type reactions [122–124], the race for the first asymmetric variant of the most versatile of MCRs continues. The group of Wulff made an important step forward by developing a highly efficient enantioselective Ugi 3CR in which water plays the role of the acid component [125]. A laborious systematic catalyst screening process concluded with the selection of a boroxinate species as the optimal promotor for the asymmetric multicomponent coupling (Scheme 14). Extending this concept to the long-sought parent four-component condensation remains a future endeavor.

Together with chiral Lewis complexes, organocatalysts also find broad application in enantioselective MCRs. For the pharma industry, this union has great practical value since the green virtues of

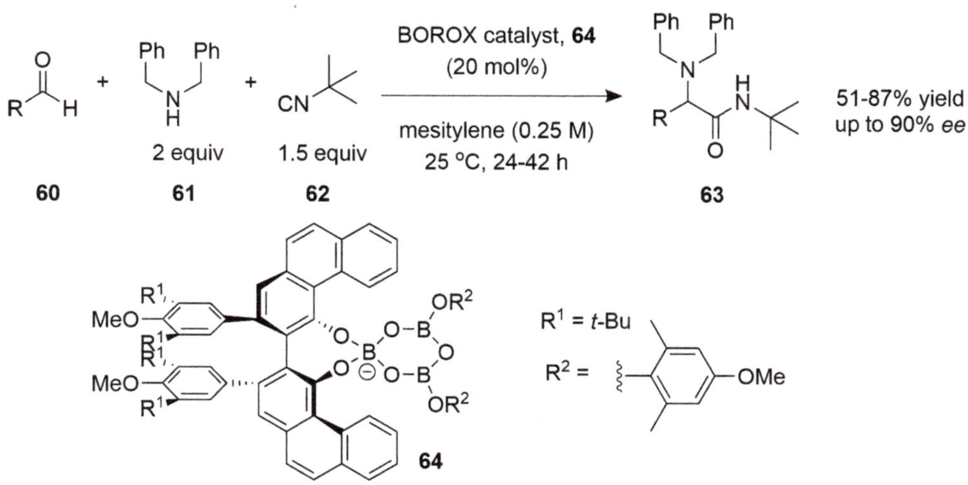

Scheme 14 Catalytic enantioselective Ugi 3CR

organocatalysis are widely accepted [126], and furthermore contamination of APIs (active pharmaceutical ingredient) with metal traces can be avoided. Numerous multicomponent syntheses promoted by various classes of organocatalysts have been reported, and this field is still expanding. For instance, the Biginelli 3CR reaction has received considerable attention, and several organocatalytic enantioselective variants have been disclosed in the last decade. The latest entry to the list was made by Zheng who tackled the notoriously low activity of previous catalytic systems by resorting to bisphosphorylimide catalysis [127]. Thus, double axially chiral bisphosphorylimides were shown to be highly effective in promoting the enantioselective Biginelli reaction in an acceptable time frame. Another application of Brønsted acid catalysis in the multicomponent synthesis of medicinally important scaffolds was reported by Shi [110]. Using aldehydes, 1,2-phenylenediamines, and cyclohexane-1,3-diones, structurally diverse dibenzo[1,4]diazepines were obtained in high yields and with good enantioselectivities using a chiral phosphoric acid catalyst. The formation of these complex products follows a sequence of events very characteristic to multicomponent chemistry, namely, condensation of carbonyls and amines, tautomerization, and nucleophilic addition to the activated imines (Scheme 15).

Similarly, another interesting multicomponent cascade leading to complex chiral molecules via simple elementary steps was developed by Constantieux et al. [109] This process is based on a sequence involving imine formation, Michael addition to an enone, iminium ion generation, and subsequent trapping via a Pictet-Spengler-type cyclization, under prolinol catalysis. This reaction is exquisitely selective for the desired tricyclic product and

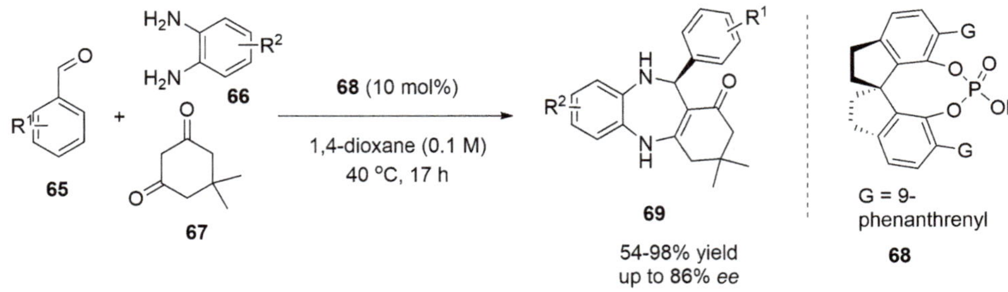

Scheme 15 Shi's Dibenzo[1,4]diazepine synthesis

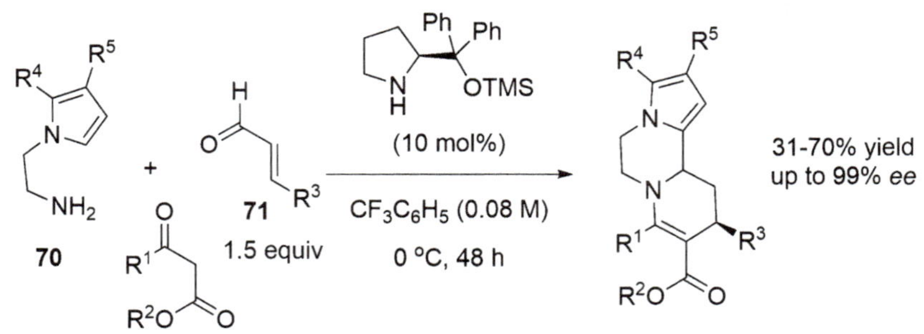

Scheme 16 Multicomponent synthesis of enantioenriched pyrrolopiperazines

further demonstrates the power of the multicomponent approach in assembling complex molecules from simple building blocks in an atom economical manner (Scheme 16).

As demonstrated by the previous example, the versatility of the 100% atom-efficient Michael addition is often exploited in multicomponent cascades. Two other interesting studies conducted in the group of Enders involve the use of nitro-olefins as Michael acceptors and a bifunctional quinine-based squaramide organocatalyst, leading to highly functionalized tetrahydropyrans [128] or tetrahydropyridines [129] with multiple contiguous stereocenters (Scheme 17).

To conclude, asymmetric multicomponent synthesis has reached a high level of sophistication and is now a well-established methodology for the preparation of complex chiral molecules. The further optimization of these protocols toward potentially more sustainable reaction conditions represents a challenge for the future. Indeed, with notable exceptions, most catalytic enantioselective MCRs require cryogenic temperatures, undesirable solvents, and high catalyst loadings, which severely limits their practical application and the transition from academia to industry.

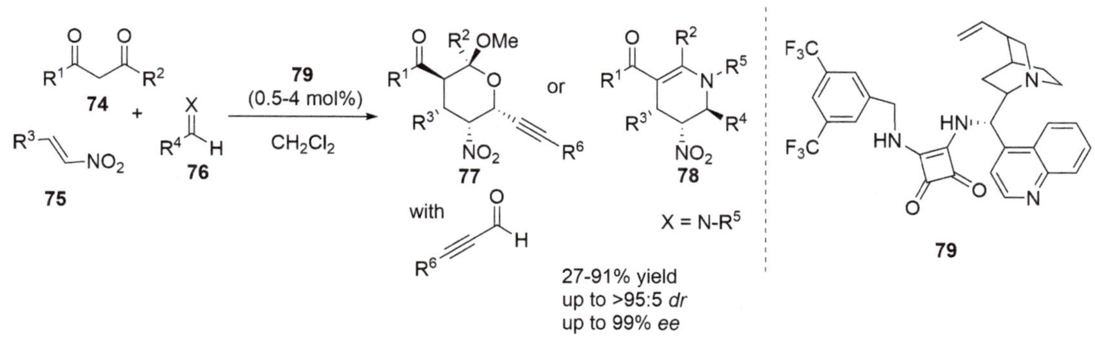

Scheme 17 Enantioselective MCRs catalyzed by a bifunctional quinine squaramide system

6 Biocatalysis

Biocatalysis has emerged as one of the cornerstones for sustainable synthesis [130]. Enzymes operate under mild process conditions and circumvent the necessity of protecting groups due to their excellent chemo- and stereoselectivity. In addition, enzymes are often able to perform transformations that are not possible with conventional reagents and catalysts [131–133], enabling synthetic chemists to redesign entire synthetic routes. In an earlier section, the multicomponent route toward the antiviral drug Telaprevir was discussed [41]. The engineering of the MAO-N enzyme to efficiently produce the novel chiral imine building blocks as starting materials allowed a completely different synthetic strategy (the Ugi-Joullié reaction) to be developed resulting in a significant productivity gain. This example highlights the powerful combination of biocatalysis and multicomponent chemistry toward the sustainable production of pharmaceuticals, and similar approaches are anticipated in the years to come. Since both biocatalysts and multicomponent reactions are advanced attractive tools for green synthesis, their concomitant application in the manufacturing processes of chiral products is highly desirable [134].

If it appears logical to build molecular diversity on a chiral substrate prepared enzymatically, theoretically planning to reverse the order of these operations is more challenging, due to the intrinsic incompatibility between the structural characteristics of MCR products (large molecules heavily decorated with functional groups) and the limited substrate tolerance of biocatalysts. Nevertheless, this strategy proved successful in a number of studies involving the biocatalytic kinetic resolution of racemic mixtures of multicomponent products, particularly dihydropyrimidinones obtained via the Biginelli 3CR [135]. Another interesting example is given by Reinart-Okugbeni et al. who used a lipase-catalyzed kinetic resolution of 3-azabicyclo[3.2.0]heptane derivatives assembled via a multicomponent condensation to deliver enantioenriched products (Scheme 18) [136].

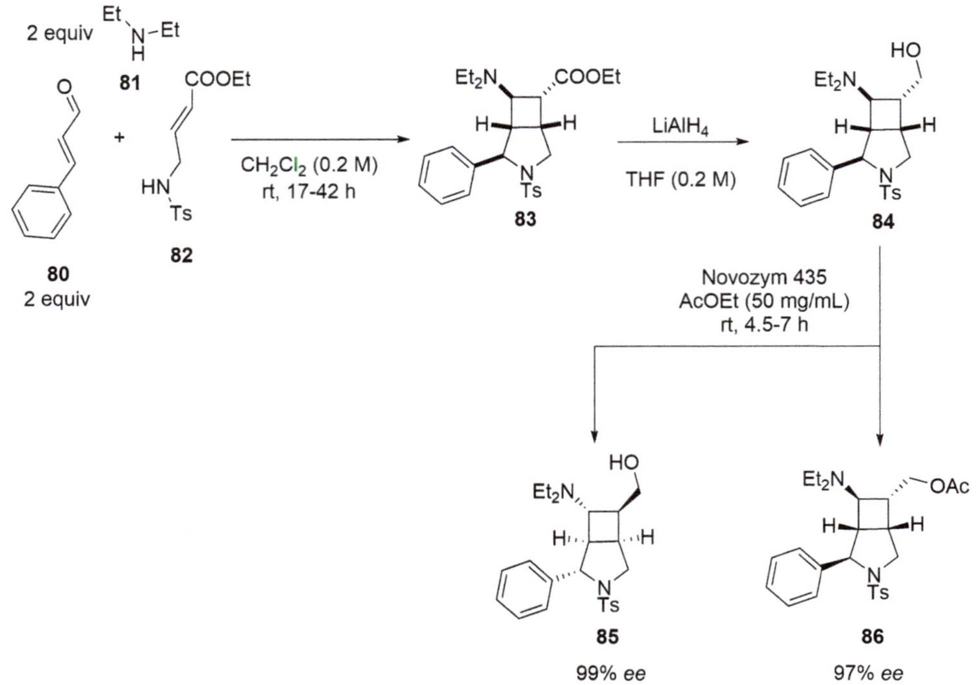

Scheme 18 Multicomponent condensation—biocatalytic resolution sequence

Furthermore biocatalysts play an important direct role in multicomponent chemistry as numerous MCRs were shown to be amenable to enzymatic catalysis. Lipases appear to be the most versatile enzymes as numerous reported MCR applications employ this class of biocatalysts. In addition, proteases are also well represented through trypsin-mediated reactions. In all cases, enzyme promiscuity and not native function is the explanation proposed for their catalytic action. For Mannich- [137, 138], Biginelli- [139, 140], or Hantzsch-type [141] processes, this is perhaps not surprising, as the chemistry involved is well established to be effected by the catalytic machinery of enzymes: hydrolysis, aminolysis, imine formation, nucleophilic addition, dehydration, cyclization, etc. Knoevenagel intermediates are also often encountered in biocatalyzed multicomponent reactions [142–145]. A cyclization step is included in many of these reactions, and this may be the thermodynamic driving force, shifting the sequence of mechanistic events toward the end product [146–148]. Although all the products of these reported condensation processes are chiral, the reactions proceed unfortunately without enantioselectivity (Scheme 19).

Finally, a distinct study investigating the biocatalysis of the Ugi reaction has emerged with promising results [149]. A diverse series of dipeptides has been synthesized under lipase catalysis in good to excellent yields though again, as with the other enzyme-promoted MCRs, no asymmetric induction was observed (Scheme 20).

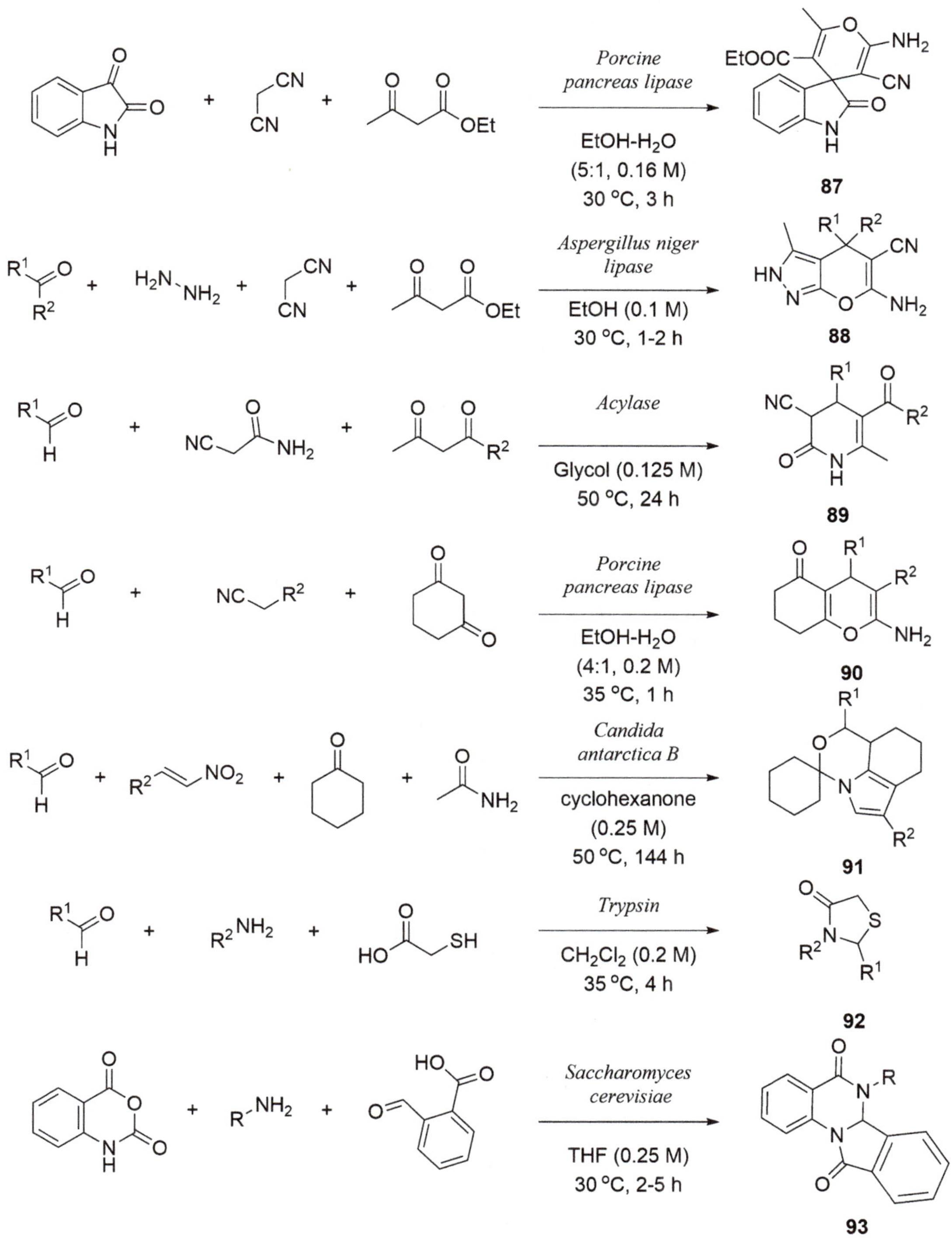

Scheme 19 Enzyme promiscuity in multicomponent chemistry

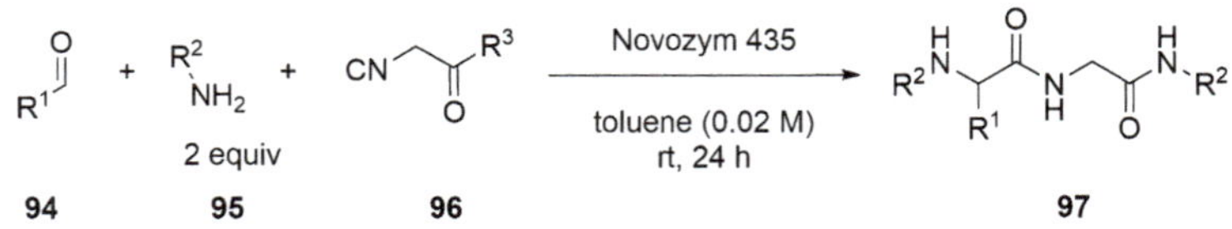

Scheme 20 Enzyme-catalyzed Ugi reaction

7 Flow Chemistry

Flow chemistry has emerged as a key technology for green process development [150, 151], with advantages of adopting to operate a chemical reaction under such a paradigm including time minimization, ease of scale-up, simplified heat transfer, and reduced costs. More importantly for the sustainable dimension of the process, flow chemistry adds value in terms of waste minimization, safer operation, energy efficiency, compatibility with real-time process analysis, etc., and as such this technology intrinsically features many of the 12 green chemistry principles. Since the union of flow chemistry with MCRs supplements this list with characteristics like atom-economy, benign solvents, and reduced use of derivatives, it is evident that the flow operation of multicomponent processes represents a quasi-complete solution for sustainable synthesis. The successful combination of MCRs and flow chemistry has been demonstrated over several examples in recent years [152, 153]. More importantly, multicomponent reactions have been identified as one of the key issues of modern flow chemistry [154], and an expansion of this research area is anticipated for the near future.

The various advantages of flow operation over batch processes have been highlighted for a great number of multicomponent reactions: Strecker [155], Hantzsch dihydropyridine [156] and pyrrole [157] synthesis, Biginelli [158], Mannich [159], Passerini and Ugi [160], Petasis [161], Groebke-Blackburn-Bienaymé [162], and A^3-reaction [163]. For instance, Seeberger et al. showed that with a flow system, both the reaction time and catalyst loading can be dramatically reduced for the Mannich condensation between cyclohexanone and a glyoxylate-derived imine, affording a product with high yield and good optical purity (Scheme 21).

A breakthrough in the field of MCRs in flow was accomplished by the group of Kim who designed a microfluidic system to perform isocyanide chemistry incorporating integrated generation and purification of these key substrates [160]. The continuous four-step in-line process consists of formamide dehydration, extraction, purification, and then direct employment in a multicomponent condensation. In this way, several difficulties in the practical application of isocyanides in synthesis are circumvented, such as the distressing odor, the instability of some derivatives, slow reaction rates, and

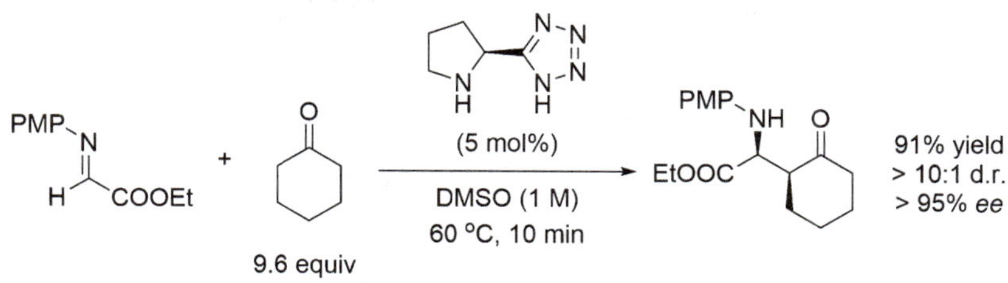

98 **99** **100**

Scheme 21 Organocatalytic Mannich reaction in flow

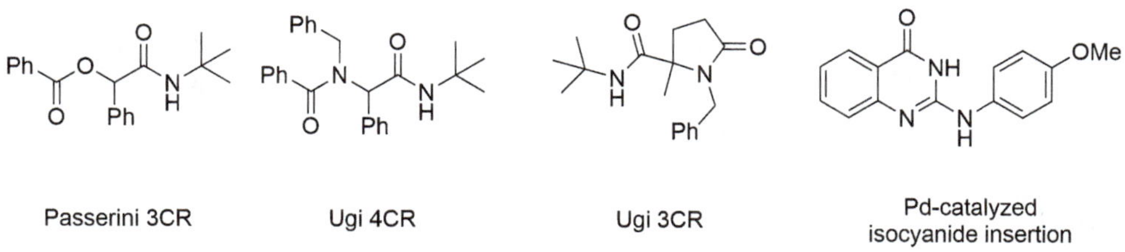

Scheme 22 MCR with in situ generation of isocyanides in microfluidic systems

poor chemoselectivity. Both the classical Ugi and Passerini reactions have been successfully applied in this system, as well as the more challenging palladium-catalyzed isocyanide insertion reactions (Scheme 22) [164].

8 Conclusions

Multicomponent reactions represent an essential strategy to perform sustainable organic synthesis for pharmaceutical applications. Whether it is the discovery of new drug candidates or the industrial production of generic drugs, MCRs represent a technology that provide distinct advantages: high-throughput, ease of automation, large chemical space coverage, mild reaction conditions, operational simplicity, and broad applicability. When it comes to the generation of compound libraries, MCRs are superior to any other class of reactions as they can supply a plethora of scaffolds with large numbers of derivatives in a concise and efficient manner. Considerable effort has been invested in the development of multicomponent chemistry in recent years, particularly in challenging areas like scaffold diversification (explosion of post-MCR functionalization leading to novel pharmacophores) and stereocontrol (catalytic enantioselective MCR variants). On the other hand, MCRs remain relatively underexploited in process chemistry possibly due to limited detailed understanding of the mechanistic pathways involved in many cases. Despite certain benefits, particularly on the sustainability side, the uptake of this technology from academic laboratories in this industrial setting is relatively limited. Retrosynthetically, MCR-based disconnections are less obvious due to the complexity of the transformations, and medicinal chemists are generally less familiar with this class of reactions. However, given the rapidly growing importance of MCRs in applied synthetic chemistry and the increasing awareness of their green virtues, this picture is very likely to change in the future. Multicomponent reactions are perfectly compatible with other green technologies (biocatalysis, flow chemistry, asymmetric synthesis, eco-friendly solvents) and consequently provide the means to cohesively integrate all 12 green chemistry principles in the large-scale preparation of fine chemicals. Multicomponent chemistry plays a leading role in the pharmaceutical industry in the drug discovery phase, and the years to come will tell if they will deservedly grow in importance in generic drug production as well.

References

1. Huang Y, Yazbak A, Dömling A (2012) Multicomponent reactions in green techniques for organic synthesis and medicinal chemistry. John Wiley & Sons, Ltd., Chichester, pp 497–522
2. Strecker A (1850) Ueber die künstliche Bildung der Milchsäure und einen neuen, dem Glycocoll homologen Körper. Justus Liebigs Annalen der Chemie 75(1):27–45
3. Hantzsch A (1881) Condensationsprodukte aus Aldehydammoniak und ketonartigen Verbindungen. Ber Dtsch Chem Ges 14 (2):1637–1638
4. Biginelli P (1891) Ueber Aldehyduramide des Acetessigäthers II. Ber. Dtsch. Chem. Ges 24
5. Wöhler F (1828) Ueber künstliche Bildung des Harnstoffs. Ann Phys 88(2):253–256
6. Dömling A, Wang W, Wang K (2012) Chemistry and biology of multicomponent reactions. Chem Rev 112(6):3083–3135
7. Nguyen LT, De Borggraeve WM, Grellier P, Pham VC, Dehaen W, Nguyen VH (2014) Synthesis of 11-aza-artemisinin derivatives using the Ugi reaction and an evaluation of their antimalarial activity. Tetrahedron Lett 55 (35):4892–4894
8. Reta GF, Chiaramello AI, García C, León LG, Martín VS, Padrón JM, Tonn CE, Donadel OJ (2013) Derivatives of grindelic acid: from a non-active natural diterpene to synthetic antitumor derivatives. Eur J Med Chem 67:28–38
9. Okamoto K, Sakagami M, Feng F, Takahashi F, Uotani K, Togame H, Takemoto H, Ichikawa S, Matsuda A (2012) Synthesis of pacidamycin analogues via an Ugi-multicomponent reaction. Bioorg Med Chem Lett 22(14):4810–4815
10. Peixoto PA, Boulange A, Leleu S, Franck X (2013) Versatile synthesis of acylfuranones by reaction of acylketenes with α-hydroxy ketones: application to the one-step multicomponent synthesis of cadiolide B and its analogues. Eur J Org Chem 16:3316–3327
11. Katayama K, Nakagawa K, Takeda H, Matsuda A, Ichikawa S (2013) Total synthesis of sandramycin and its analogues via a multicomponent assemblage. Org Lett 16 (2):428–431
12. Bourgault JP, Maddirala AR, Andreana PR (2014) A one-pot multicomponent coupling/cyclization for natural product herbicide (±)-thaxtomin A. Org Biomol Chem 12 (41):8125–8127
13. Chiba T, Hosono H, Nakagawa K, Asaka M, Takeda H, Matsuda A, Ichikawa S (2014) Total synthesis of syringolin-A and improvement of its biological activity. Angew Chem 126(19):4936–4939
14. Saito K, Nishimori A, Mimura R, Nakano K, Kotsuki H, Masuda T, Ichikawa Y (2013) A Biomimetic approach to the synthesis of the core structure of the marine sponge terpene halichonadin G. Eur J Org Chem 2013 (31):7041–7043
15. Akritopoulou-Zanze I (2008) Isocyanide-based multicomponent reactions in drug discovery. Curr Opin Chem Biol 12(3):324–331
16. Kalinski C, Umkehrer M, Weber L, Kolb J, Burdack C, Ross G (2010) On the industrial

applications of MCRs: molecular diversity in drug discovery and generic drug synthesis. Mol Divers 14(3):513–522

17. Magedov IV, Kornienko A (2012) Multicomponent reactions in alkaloid-based drug discovery. Chem Heterocycl Compd 48 (1):33–38
18. Slobbe P, Ruijter E, Orru RVA (2012) Recent applications of multicomponent reactions in medicinal chemistry. MedChemComm 3 (10):1189–1218
19. Ruijter E, Orru RVA (2013) Multicomponent reactions – opportunities for the pharmaceutical industry. Drug Discov Today Technol 10 (1):e15–e20
20. Cioc RC, Ruijter E, Orru RVA (2014) Multicomponent reactions: advanced tools for sustainable organic synthesis. Green Chem 16 (6):2958–2975
21. Elders N, Ruijter E, de Kanter FJJ, Groen MB, Orru RVA (2008) Selective formation of 2-imidazolines and 2-substituted oxazoles by using a three-component reaction. Chemistry 14(16):4961–4973
22. Ivachtchenko AV, Ivanenkov YA, Kysil VM, Krasavin MY, Ilyin AP (2010) Multicomponent reactions of isocyanides in the synthesis of heterocycles. Russ Chem Rev 79 (9):787–817
23. Zhu J, Wang Q, Wang M-X (2010) Development and application of isocyanide-based multicomponent reactions in handbook of green chemistry. Wiley-VCH Verlag GmbH & Co. KGaA, Weinheim
24. van Berkel SS, Bogels BGM, Wijdeven MA, Westermann B, Rutjes F (2012) Recent advances in asymmetric isocyanide-based multicomponent reactions. Eur J Org Chem 19:3543–3559
25. Azuaje J, El Maatougui A, García-Mera X, Sotelo E (2014) Ugi-based approaches to quinoxaline libraries. ACS Comb Sci 16 (8):403–411
26. Barlow TMA, Jida M, Tourwe D, Ballet S (2014) Efficient synthesis of conformationally constrained, amino-triazoloazepinone-containing di- and tripeptides via a one-pot Ugi–-Huisgen tandem reaction. Org Biomol Chem 12(36):6986–6989
27. Che C, Yang B, Jiang X, Shao T, Yu Z, Tao C, Li S, Lin S (2014) Syntheses of fused tetracyclic quinolines via Ugi-variant MCR and Pd-catalyzed bis-annulation. J Org Chem 79 (1):436–440
28. Zeng X-H, Wang H-M, Yan Y-M, Wu L, Ding M-W (2014) One-pot regioselective synthesis of β-lactams by a tandem Ugi 4CC/S_N cyclization. Tetrahedron 70(23):3647–3652
29. Polindara-Garcia LA, Vazquez A (2014) Combinatorial synthesis of nicotine analogs using an Ugi 4-CR/cyclization-reduction strategy. Org Biomol Chem 12 (36):7068–7082
30. Sharma N, Li Z, Sharma UK, Van der Eycken EV (2014) Facile access to functionalized spiro[indoline-3,2′-pyrrole]-2,5′-diones via post-Ugi Domino Buchwald–Hartwig/Michael reaction. Org Lett 16 (15):3884–3887
31. Ghabraie E, Balalaie S, Mehrparvar S, Rominger F (2014) Synthesis of functionalized β-lactams and pyrrolidine-2,5-diones through a metal-free sequential Ugi-4CR/cyclization reaction. J Org Chem 79(17):7926–7934
32. Treder AP, Tremblay M-C, Yudin AK, Marsault E (2014) Solid-phase synthesis of piperazinones via disrupted Ugi condensation. Org Lett 16(17):4674–4677
33. Ugi I, Meyr R, Isonitrile I (1960) Darstellung von Isonitrilen aus monosubstituierten Formamiden durch Wasserabspaltung. Chemische Berichte-Recueil 93(1):239–248
34. Hofmann AW (1867) Ueber eine neue Reihe von Homologen der Cyanwasserstoffsäure. Justus Liebigs Annalen der Chemie 144 (1):114–120
35. El Kaim L, Grimaud L, Schiltz A (2009) "Isocyanide-free" Ugi reactions. Org Biomol Chem 7(15):3024–3026
36. El Kaim L, Grimaud L, Schiltz A (2009) Isocyanide-based multicomponent reaction 'without' isocyanides. Synlett 9:1401–1404
37. Pronin SV, Reiher CA, Shenvi RA (2013) Stereoinversion of tertiary alcohols to tertiary-alkyl isonitriles and amines. Nature 501(7466):195–199
38. Koopmanschap G, Ruijter E, Orru RVA (2014) Isocyanide-based multicomponent reactions towards cyclic constrained peptidomimetics. Beilstein J Org Chem 10:544–598
39. Khoury K, Sinha MK, Nagashima T, Herdtweck E, Domling A (2012) Efficient assembly of iminodicarboxamides by a "truly" four-component reaction. Angew Chem Int Ed 51(41):10280–10283
40. Yip Y, Victor F, Lamar J, Johnson R, Wang QM, Barket D, Glass J, Jin L, Liu L, Venable D, Wakulchik M, Xie C, Heinz B, Villarreal E, Colacino J, Yumibe N, Tebbe M, Munroe J, Chen S-H (2004) Discovery of a novel bicycloproline P2 bearing peptidyl α-ketoamide LY514962 as HCV

protease inhibitor. Bioorg Med Chem Lett 14 (1):251–256

41. Znabet A, Polak MM, Janssen E, de Kanter FJJ, Turner NJ, Orru RVA, Ruijter E (2010) A highly efficient synthesis of telaprevir by strategic use of biocatalysis and multicomponent reactions. Chem Commun 46 (42):7918–7920
42. Constable DJC, Curzons AD, Cunningham VL (2002) Metrics to 'green' chemistry—which are the best? Green Chem 4 (6):521–527
43. Curzons AD, Constable DJC, Mortimer DN, Cunningham VL (2001) So you think your process is green, how do you know?—Using principles of sustainability to determine what is green–a corporate perspective. Green Chem 3(1):1–6
44. Znabet A, Ruijter E, de Kanter FJJ, Köhler V, Helliwell M, Turner NJ, Orru RVA (2010) Highly stereoselective synthesis of substituted prolyl peptides using a combination of biocatalytic desymmetrization and multicomponent reactions. Angew Chem Int Ed 49 (31):5289–5292
45. Xia L, Li S, Chen R, Liu K, Chen X (2013) Catalytic Ugi-type condensation of α-isocyanoacetamide and chiral cyclic imine: access to asymmetric construction of several heterocycles. J Org Chem 78(7):3120–3131
46. Thompson MJ, Chen B (2009) Ugi reactions with ammonia offer rapid access to a wide range of 5-aminothiazole and oxazole derivatives. J Org Chem 74(18):7084–7093
47. Nanda KK, Wesley Trotter B (2005) Diastereoselective petasis Mannich reactions accelerated by hexafluoroisopropanol: a pyrrolidine-derived arylglycine synthesis. Tetrahedron Lett 46(12):2025–2028
48. Ghosal P, Shaw AK (2012) A Chiron approach to aminocytitols by petasis-borono-mannich reaction: formal synthesis of (+)-conduramine E and (−)-conduramine E. J Org Chem 77(17):7627–7632
49. Petasis NA, Akritopoulou I (1993) The boronic acid mannich reaction: a new method for the synthesis of geometrically pure allylamines. Tetrahedron Lett 34(4):583–586
50. Wang Y, Saha B, Li F, Frett B, Li HY (2014) An expeditious approach to access 2-arylimidazo[1,2-a]pyridin-3-ol from 2-amino pyridine through a novel Petasis based cascade reaction. Tetrahedron Lett 55(7):1281–1284
51. Gu Y (2012) Multicomponent reactions in unconventional solvents: state of the art. Green Chem 14(8):2091–2128
52. Simon M-O, Li C-J (2012) Green chemistry oriented organic synthesis in water. Chem Soc Rev 41(4):1415–1427
53. Isambert N, Duque MM, Plaquevent J-C, Genisson Y, Rodriguez J, Constantieux T (2011) Multicomponent reactions and ionic liquids: a perfect synergy for eco-compatible heterocyclic synthesis. Chem Soc Rev 40 (3):1347–1357
54. Singh MS, Chowdhury S (2012) Recent developments in solvent-free multicomponent reactions: a perfect synergy for eco-compatible organic synthesis. RSC Adv 2(11):4547–4592
55. Pori M, Galletti P, Soldati R, Giacomini D (2013) Asymmetric strecker reaction with chiral amines: a catalyst-free protocol using acetone cyanohydrin in water. Eur J Org Chem 2013(9):1683–1695
56. Mojtahedi MM, Abaee MS, Abbasi H (2006) Environmentally friendly room temperature Strecker reaction: one-pot synthesis of α-aminonitriles in ionic liquid. J Iran Chem Soc 3(1):93–97
57. Hu XC, Ma YH, Li Z (2012) Eco-friendly synthesis of α-aminonitriles from ketones in PEG-400 medium using potassium Hexacyanoferrate(II) as cyanide source. J Organomet Chem 705:70–74
58. Chaturvedi D, Chaturvedi AK, Dwivedi PK, Mishra N (2013) A novel approach to the synthesis of α-aminonitriles using triphenylphosphine dibromide under solvent-free conditions. Synlett 24(1):33–36
59. Avalani JR, Patel DS, Raval DK (2012) 1-Methylimidazolium trifluoroacetate [Hmim]Tfa: mild and efficient Brønsted acidic ionic liquid for Hantzsch reaction under microwave irradiation. J Chem Sci 124(5):1091–1096
60. Fan XS, Li YZ, Zhang XY, Qu GR, Wang JJ, Hu XY (2006) An efficient and green synthesis of 1,4-dihydropyridine derivatives through multi-component reaction in ionic liquid. Heteroat Chem 17(5):382–388
61. Wang XC, Gong HP, Quan ZJ, Li L, Ye HL (2011) One-pot, three-component synthesis of 1,4-dihydropyridines in PEG-400. Synth Commun 41(21):3251–3258
62. Heravi MM, Zakeri M, Pooremamy S, Oskooie HA (2010) Clean and efficient synthesis of polyhydroquinoline derivatives under solvent-free conditions catalyzed by morpholine. Synth Commun 41(1):113–120
63. Rao GBD, Anjaneyulu B, Kaushik MP (2014) A facile one-pot five-component synthesis of

glycoside annulated dihydropyrimidinone derivatives with 1,2,3-triazol linkage via trans-esterification/Biginelli/click reactions in aqueous medium. Tetrahedron Lett 55 (1):19–22

64. Chavan SS, Sharma YO, Degani MS (2009) Cost-effective ionic liquid for environmentally friendly synthesis of 3,4-dihydropyrimidin-2(1H)-ones. Green Chem Lett Rev 2 (3):175–179
65. Jain SL, Singhal S, Sain B (2007) PEG-assisted solvent and catalyst free synthesis of 3,4-dihydropyrimidinones under mild reaction conditions. Green Chem 9 (7):740–741
66. Xu Z, Jiang YY, Zou S, Liu Y (2014) Bio-based solvent mediated synthesis of dihydropyrimidinthiones via biginelli reaction. Phosphorus Sulfur Silicon Relat Elem 189 (6):791–795
67. Dondoni A, Massi A (2001) Parallel synthesis of dihydropyrimidinones using Yb(III)-resin and polymer-supported scavengers under solvent-free conditions. A green chemistry approach to the Biginelli reaction. Tetrahedron Lett 42(45):7975–7978
68. Luo S, Mi X, Liu S, Xu H, Cheng J-P (2006) Surfactant-type asymmetric organocatalyst: organocatalytic asymmetric Michael addition to nitrostyrenes in water. Chem Commun 35:3687–3689
69. Alvim HGO, Bataglion GA, Ramos LM, de Oliveira AL, de Oliveira HCB, Eberlin MN, de Macedo JL, da Silva WA, Neto BAD (2014) Task-specific ionic liquid incorporating anionic heteropolyacid-catalyzed Hantzsch and Mannich multicomponent reactions. Ionic liquid effect probed by ESI-MS(/MS). Tetrahedron 70 (20):3306–3313
70. Demirkol O, Akbaslar D, Giray S, Anil BB (2014) One-pot synthesis of mannich bases under solvent-free conditions. Synth Commun 44(9):1279–1285
71. Pirrung MC, Sarma KD (2005) Aqueous medium effects on multi-component reactions. Tetrahedron 61(48):11456–11472
72. Andrade CKZ, Takada SCS, Suarez PAZ, Alves MB (2006) Revisiting the passerini reaction under eco-friendly reaction conditions. Synlett 10:1539–1542
73. Koszelewski D, Szymanski W, Krysiak J, Ostaszewski R (2008) Solvent-free passerini reactions. Synth Commun 38(7):1120–1127
74. Banfi L, Basso A, Chiappe C, De Moliner F, Riva R, Sonaglia L (2012) Development of a stereoselective Ugi reaction starting from an oxanorbornene β-amino acid derivative. Org Biomol Chem 10(19):3819–3829
75. Niu TF, Lu GP, Cai C (2011) The Ugi reaction in a polyethylene glycol medium: a mild, protocol for the production of compound libraries. J Chem Res 35(8):444–447
76. Jida M, Malaquin S, Deprez-Poulain R, Laconde G, Deprez B (2010) Synthesis of five- and six-membered lactams via solvent-free microwave Ugi reaction. Tetrahedron Lett 51(39):5109–5111
77. Candeias NR, Veiros LF, Afonso CAM, Gois PMP (2009) Water: a suitable medium for the petasis Borono-Mannich reaction. Eur J Org Chem 2009(12):1859–1863
78. Yadav JS, Reddy BVS, Lakshmi PN (2007) Ionic liquid accelerated Petasis reaction: a green protocol for the synthesis of alkylaminophenols. J Mol Catal A Chem 274 (1–2):101–104
79. Nun P, Martinez J, Lamaty F (2010) Microwave-assisted neat procedure for the petasis reaction. Synthesis 12:2063–2068
80. Shaabani A, Soleimani E, Maleki A (2006) Ionic liquid promoted one-pot synthesis of 3-aminoimidazo[1,2-a]pyridines. Tetrahedron Lett 47(18):3031–3034
81. Vidyacharan S, Shinde AH, Satpathi B, Sharada DS (2014) A facile protocol for the synthesis of 3-aminoimidazo-fused heterocycles *via* the Groebke–Blackburn–Bienayme reaction under catalyst-free and solvent-free conditions. Green Chem 16(3):1168–1175
82. Ohara M, Hara Y, Ohnuki T, Nakamura S (2014) Direct enantioselective three-component synthesis of optically active propargylamines in water. Chem Eur J 20 (29):8848–8851
83. Li ZG, Wei CM, Chen L, Varma RS, Li CJ (2004) Three-component coupling of aldehyde, alkyne, and amine catalyzed by silver in ionic liquid. Tetrahedron Lett 45 (11):2443–2446
84. Zhang Q, Chen J-X, Gao W-X, Ding J-C, Wu H-Y (2010) Copper-catalyzed one-pot synthesis of propargylamines via C-H activation in PEG. Appl Organomet Chem 24 (11):809–812
85. Kabalka GW, Wang L, Pagni RM (2001) A microwave-enhanced, solventless mannich condensation on CuI-doped alumina. Synlett 2001(05):0676–0678
86. Hügel H (2009) Microwave multicomponent synthesis. Molecules 14(12):4936–4972
87. Kruithof A, Ruijter E, Orru RVA (2012) Microwave-assisted multicomponent

reactions in the synthesis of heterocycles in microwaves in organic synthesis. Wiley-VCH Verlag GmbH & Co. KGaA, Weinheim, pp 1099–1171

88. Moseley JD, Kappe CO (2011) A critical assessment of the greenness and energy efficiency of microwave-assisted organic synthesis. Green Chem 13(4):794–806
89. Gawande MB, Bonifacio VDB, Luque R, Branco PS, Varma RS (2014) Solvent-free and catalysts-free chemistry: a benign pathway to sustainability. ChemSusChem 7(1):24–44
90. Banitaba SH, Safari J, Khalili SD (2013) Ultrasound promoted one-pot synthesis of 2-amino-4,8-dihydropyrano[3,2-b]pyran-3-carbonitrile scaffolds in aqueous media: a complementary 'green chemistry' tool to organic synthesis. Ultrason Sonochem 20 (1):401–407
91. Chandralekha E, Thangamani A, Valliappan R (2013) Ultrasound-promoted regioselective and stereoselective synthesis of novel spiroindanedionepyrrolizidines by multicomponent 1,3-dipolar cycloaddition of azomethine ylides. Res Chem Intermed 39(3):961–972
92. Dandia A, Gupta SL, Parewa V (2014) An efficient ultrasound-assisted one-pot chemoselective synthesis of pyrazolo[3,4-b] pyridine-5-carbonitriles in aqueous medium using NaCl as a catalyst. RSC Adv 4 (14):6908–6915
93. Yu SJ, Zhu C, Bian Q, Cui C, Du XJ, Li ZM, Zhao WG (2014) Novel ultrasound-promoted parallel synthesis of trifluoroatrolactamide library via a one-pot Passerini/hydrolysis reaction sequence and their fungicidal activities. ACS Comb Sci 16(1):17–23
94. Pagadala R, Maddila S, Jonnalagadda SB (2014) Ultrasonic-mediated catalyst-free rapid protocol for the multicomponent synthesis of dihydroquinoline derivatives in aqueous media. Green Chem Lett Rev 7 (2):131–136
95. Siddekha A, Azzam SHS, Pasha MA (2014) Ultrasound-assisted, one-pot, four-component synthesis of 1,4,6,8-tetrahydroquinolines in aqueous medium. Synth Commun 44(3):424–432
96. Nagalapalli R, Jaggavarapu SR, Jalli VP, Kamalakaran AS, Gaddamanugu G (2013) Ultrasound promoted green and facile one-pot multicomponent synthesis of 3,4-dihydropyrano[c]chromene derivatives. J Chem 2013:593803
97. Brahmachari G, Das S (2014) L-Proline catalyzed multicomponent one-pot synthesis of gem-diheteroarylmethane derivatives using facile grinding operation under solvent-free conditions at room temperature. RSC Adv 4 (15):7380–7388
98. Maleki A, Javanshir S, Naimabadi M (2014) Facile synthesis of imidazo[1,2-a]pyridines via a one-pot three-component reaction under solvent-free mechanochemical ball-milling conditions. RSC Adv 4(57):30229–30232
99. Cukalovic A, Monbaliu J-CMR, Stevens CV (2010) Microreactor technology as an efficient tool for multicomponent reactions, vol 23. Springer, Berlin
100. de Graaff C, Ruijter E, Orru RVA (2012) Recent developments in asymmetric multicomponent reactions. Chem Soc Rev 41 (10):3969–4009
101. Clavier H, Pellissier H (2012) Recent developments in enantioselective metal-catalyzed Domino reactions. Adv Synth Catal 354 (18):3347–3403
102. Pellissier H (2012) Stereocontrolled Domino reactions. Chem Rev 113(1):442–524
103. Pellissier H (2013) Recent developments in enantioselective multicatalysed tandem reactions. Tetrahedron 69(35):7171–7210
104. Mehta VP, Modha SG, Ruijter E, Van Hecke K, Van Meervelt L, Pannecouque C, Balzarini J, Orru RVA, Van der Eycken E (2011) A microwave-assisted diastereoselective multicomponent reaction to access dibenzo[c,e]azepinones: synthesis and biological evaluation. J Org Chem 76 (8):2828–2839
105. Li Y, Xu M-H (2012) Lewis acid promoted highly diastereoselective petasis Borono-Mannich reaction: efficient synthesis of optically active β,γ-unsaturated α-amino acids. Org Lett 14(8):2062–2065
106. Wang G, Li B, Lou Q, Li Z, Meng X (2013) Diastereoselective povarov-like reaction involving O-pivaloylated D-galactosylimine. Adv Synth Catal 355(2–3):303–307
107. Jiang J, Guan X, Liu S, Ren B, Ma X, Guo X, Lv F, Wu X, Hu W (2013) Highly diastereoselective multicomponent cascade reactions: efficient synthesis of functionalized 1-indanols. Angew Chem 125(5):1579–1582
108. Meng FK, Haeffner F, Hoveyda AH (2014) Diastereo- and enantioselective reactions of bis(pinacolato)diboron, 1,3-enynes, and aldehydes catalyzed by an easily accessible bisphosphine–Cu complex. J Am Chem Soc 136 (32):11304–11307
109. Du H, Rodriguez J, Bugaut X, Constantieux T (2014) Organocatalytic enantioselective multicomponent synthesis of pyrrolopiperazines. Adv Synth Catal 356(4):851–856

110. Wang Y, Tu MS, Shi F, Tu SJ (2014) Enantioselective construction of the biologically significant dibenzo[1,4]diazepine scaffold via organocatalytic asymmetric three-component reactions. Adv Synth Catal 356 (9):2009–2019
111. Calleja J, Gonzalez-Perez AB, de Lera AR, Alvarez R, Fananas FJ, Rodriguez F (2014) Enantioselective synthesis of hexahydrofuro [3,2-c] quinolines through a multicatalytic and multicomponent process. A new "aromatic sandwich" model for BINOL-phosphoric acid catalyzed reactions. Chem Sci 5(3):996–1007
112. Salahi F, Taghizadeh MJ, Arvinnezhad H, Moemeni M, Jadidi K, Notash B (2014) An efficient, one-pot, three-component procedure for the synthesis of chiral spirooxindolopyrrolizidines via catalytic highly enantioselective 1,3-dipolar cycloaddition. Tetrahedron Lett 55(9):1515–1518
113. McGrath KP, Hoveyda AH (2014) A multicomponent Ni-, Zr-, and Cu-catalyzed strategy for enantioselective synthesis of alkenyl-substituted quaternary carbons. Angew Chem Int Ed 53(7):1910–1914
114. Jiang J, Ma XC, Ji CG, Guo ZQ, Shi TD, Liu SY, Hu WH (2014) Ruthenium(II)/chiral Brønsted acid co-catalyzed enantioselective four-component reaction/cascade Aza-Michael addition for efficient construction of 1,3,4-tetrasubstituted tetrahydroisoquinolines. Chem Eur J 20(6):1505–1509
115. Hashimoto T, Takiguchi Y, Maruoka K (2013) Catalytic asymmetric three-component 1,3-dipolar cycloaddition of aldehydes, hydrazides, and alkynes. J Am Chem Soc 135(31):11473–11476
116. Shi F, Tan W, Zhu RY, Xing GJ, Tu SJ (2013) Catalytic asymmetric five-component tandem reaction: diastereo- and enantioselective synthesis of densely functionalized tetrahydropyridines with biological importance. Adv Synth Catal 355(8):1605–1622
117. Qian Y, Jing CC, Liu SY, Hu WH (2013) A highly enantioselective four-component reaction for the efficient construction of chiral β-hydroxy-α-amino acid derivatives. Chem Commun 49(26):2700–2702
118. Gandhi S, List B (2013) Catalytic asymmetric three-component synthesis of homoallylic amines. Angew Chem Int Ed 52 (9):2573–2576
119. Chen ZL, Wang BL, Wang ZB, Zhu GY, Sun JW (2013) Complex bioactive alkaloid-type polycycles through efficient catalytic asymmetric multicomponent Aza-Diels–Alder reaction of indoles with oxetane as directing group. Angew Chem Int Ed 52 (7):2027–2031
120. Ortín I, Dixon DJ (2014) Direct catalytic enantio- and diastereoselective mannich reaction of isocyanoacetates and ketimines. Angew Chem Int Ed 53(13):3462–3465
121. Hayashi M, Iwanaga M, Shiomi N, Nakane D, Masuda H, Nakamura S (2014) Direct asymmetric mannich-type reaction of α-isocyanoacetates with ketimines using cinchona alkaloid/copper(II) catalysts. Angew Chem Int Ed 53(32):8411–8415
122. Hashimoto T, Kimura H, Kawamata Y, Maruoka K (2012) A catalytic asymmetric ugi-type reaction with acyclic azomethine imines. Angew Chem Int Ed 51 (29):7279–7281
123. Su YP, Bouma MJ, Alcaraz L, Stocks M, Furber M, Masson G, Zhu JP (2012) Organocatalytic enantioselective one-pot four-component Ugi-type multicomponent reaction for the synthesis of epoxy-tetrahydropyrrolo[3,4-b]pyridin-5-ones. Chem Eur J 18 (40):12624–12627
124. Yue T, Wang MX, Wang DX, Masson G, Zhu JP (2009) Brønsted acid catalyzed enantioselective three-component reaction involving the α addition of isocyanides to imines. Angew Chem Int Ed 48(36):6717–6721
125. Zhao W, Huang L, Guan Y, Wulff WD (2014) Three-component asymmetric catalytic Ugi reaction—concinnity from diversity by substrate-mediated catalyst assembly. Angew Chem Int Ed 53(13):3436–3441
126. Wende RC, Schreiner PR (2012) Evolution of asymmetric organocatalysis: multi- and retrocatalysis. Green Chem 14(7):1821–1849
127. An D, Fan Y-S, Gao Y, Zhu Z-Q, Zheng L-Y, Zhang S-Q (2014) Highly enantioselective biginelli reaction catalyzed by double axially chiral bisphosphorylimides. Eur J Org Chem 2014(2):301–306
128. Hahn R, Raabe G, Enders D (2014) Asymmetric synthesis of highly functionalized tetrahydropyrans via a one-pot organocatalytic Michael/Henry/ketalization sequence. Org Lett 16(14):3636–3639
129. Blümel M, Chauhan P, Hahn R, Raabe G, Enders D (2014) Asymmetric synthesis of tetrahydropyridines via an organocatalytic one-pot multicomponent Michael/Aza-Henry/cyclization triple domino reaction. Org Lett 16(22):6012–6015
130. Wenda S, Illner S, Mell A, Kragl U (2011) Industrial biotechnology—the future of green chemistry? Green Chem 13 (11):3007–3047

131. Ghislieri D, Green AP, Pontini M, Willies SC, Rowles I, Frank A, Grogan G, Turner NJ (2013) Engineering an enantioselective amine oxidase for the synthesis of pharmaceutical building blocks and alkaloid natural products. J Am Chem Soc 135 (29):10863–10869
132. Yang Y, Liu J, Li Z (2014) Engineering of P450pyr hydroxylase for the highly regio- and enantioselective subterminal hydroxylation of alkanes. Angew Chem Int Ed 53 (12):3120–3124
133. Shoji O, Kunimatsu T, Kawakami N, Watanabe Y (2013) Highly selective hydroxylation of benzene to phenol by wild-type cytochrome P450BM3 assisted by decoy molecules. Angew Chem Int Ed 52 (26):6606–6610
134. Banfi L, Basso A, Moni L, Riva R (2014) The alternative route to enantiopure multicomponent reaction products: biocatalytic or organocatalytic enantioselective production of inputs for multicomponent reactions. Eur J Org Chem 10:2005–2015
135. Barrow JC, Nantermet PG, Selnick HG, Glass KL, Rittle KE, Gilbert KF, Steele TG, Homnick CF, Freidinger RM, Ransom RW, Kling P, Reiss D, Broten TP, Schorn TW, Chang RSL, O'Malley SS, Olah TV, Ellis JD, Barrish A, Kassahun K, Leppert P, Nagarathnam D, Forray C (2000) In vitro and in vivo evaluation of dihydropyrimidinone C-5 amides as potent and selective α_{1A} receptor antagonists for the treatment of benign prostatic hyperplasia. J Med Chem 43(14):2703–2718
136. Reinart-Okugbeni R, Ausmees K, Kriis K, Werner F, Rinken A, Kanger T (2012) Chemoenzymatic synthesis and evaluation of 3-azabicyclo[3.2.0]heptane derivatives as dopaminergic ligands. Eur J Med Chem 55:255–261
137. Li K, He T, Li C, Feng XW, Wang N, Yu XQ (2009) Lipase-catalysed direct Mannich reaction in water: utilization of biocatalytic promiscuity for C–C bond formation in a "one-pot" synthesis. Green Chem 11(6):777–779
138. Chai S-J, Lai Y-F, Zheng H, Zhang P-F (2010) A novel trypsin-catalyzed three-component *Mannich* reaction. Helv Chim Acta 93 (11):2231–2236
139. Li W, Zhou G, Zhang P, Lai Y, Xu S (2011) One-pot synthesis of dihydropyrimidiones via environmentally friendly enzyme-catalyzed biginelli reaction. Heterocycles 83 (9):2067–2077
140. Sharma UK, Sharma N, Kumar R, Sinha AK (2013) Biocatalysts for multicomponent Biginelli reaction: bovine serum albumin triggered waste-free synthesis of 3,4-dihydropyrimidin-2-(1H)-ones. Amino Acids 44 (3):1031–1037
141. Wang JL, Liu BK, Yin C, Wu Q, Lin XF (2011) *Candida antarctica* lipase B-catalyzed the unprecedented three-component Hantzsch-type reaction of aldehyde with acetamide and 1,3-dicarbonyl compounds in non-aqueous solvent. Tetrahedron 67(14):2689–2692
142. Xu JC, Li WM, Zheng H, Lai YF, Zhang PF (2011) One-pot synthesis of tetrahydrochromene derivatives catalyzed by lipase. Tetrahedron 67(49):9582–9587
143. Chai S-J, Lai Y-F, Xu J-C, Zheng H, Zhu Q, Zhang P-F (2011) One-pot synthesis of spirooxindole derivatives catalyzed by lipase in the presence of water. Adv Synth Catal 353 (2–3):371–375
144. Liu Z-Q, Liu B-K, Wu Q, Lin X-F (2011) Diastereoselective enzymatic synthesis of highly substituted 3,4-dihydropyridin-2-ones via domino Knoevenagel condensation–Michael addition–intramolecular cyclization. Tetrahedron 67(50):9736–9740
145. Bora PP, Bihani M, Bez G (2013) Multicomponent synthesis of dihydropyrano[2,3-c] pyrazoles catalyzed by lipase from *Aspergillus niger*. J Mol Catal B Enzym 92:24–33
146. Wang J-L, Chen X-Y, Wu Q, Lin X-F (2014) One-pot synthesis of spirooxazino derivatives via enzyme-initiated multicomponent reactions. Adv Synth Catal 356(5):999–1005
147. Zheng H, Mei YJ, Du K, Shi QY, Zhang PF (2013) Trypsin-catalyzed one-pot multicomponent synthesis of 4-thiazolidinones. Catal Lett 143(3):298–301
148. Avalani JR, Patel DS, Raval DK (2013) *Saccharomyces cerevisiae* catalyzed one pot synthesis of isoindolo[2,1-a]quinazoline performed under ultrasonication. J Mol Catal B Enzym 90:70–75
149. Kłossowski S, Wiraszka B, Berłożecki S, Ostaszewski R (2013) Model studies on the first enzyme-catalyzed Ugi reaction. Org Lett 15(3):566–569
150. Ley SV (2012) On being green: can flow chemistry help? Chem Rec 12(4):378–390
151. Vaccaro L, Lanari D, Marrocchi A, Strappaveccia G (2014) Flow approaches towards sustainability. Green Chem 16(8):3680–3704
152. Wegner J, Ceylan S, Kirschning A (2012) Flow chemistry – a key enabling technology for (multistep) organic synthesis. Adv Synth Catal 354(1):17–57

153. Bremner WS, Organ MG (2007) Multicomponent reactions to form heterocycles by microwave-assisted continuous flow organic synthesis. ACS Comb Sci 9(1):14–16
154. Wegner J, Ceylan S, Kirschning A (2011) Ten key issues in modern flow chemistry. Chem Commun 47(16):4583–4592
155. Wiles C, Watts P (2008) An integrated microreactor for the multicomponent synthesis of α-aminonitriles. Org Process Res Dev 12 (5):1001–1006
156. Bagley MC, Fusillo V, Jenkins RL, Lubinu MC, Mason C (2013) One-step synthesis of pyridines and dihydropyridines in a continuous flow microwave reactor. Beilstein J Org Chem 9:1957–1968
157. Herath A, Cosford NDP (2010) One-step continuous flow synthesis of highly substituted pyrrole-3-carboxylic acid derivatives via in situ hydrolysis of *tert*-butyl esters. Org Lett 12(22):5182–5185
158. Hamlin TA, Leadbeater NE (2013) Raman spectroscopy as a tool for monitoring mesoscale continuous-flow organic synthesis: equipment interface and assessment in four medicinally-relevant reactions. Beilstein J Org Chem 9:1843–1852
159. Odedra A, Seeberger PH (2009) 5-(Pyrrolidin-2-yl)tetrazole-catalyzed Aldol and Mannich reactions: acceleration and lower catalyst loading in a continuous-flow reactor. Angew Chem Int Ed 48(15):2699–2702
160. Sharma S, Maurya RA, Min K-I, Jeong G-Y, Kim D-P (2013) Odorless isocyanide chemistry: an integrated microfluidic system for a multistep reaction sequence. Angew Chem Int Ed 52(29):7564–7568
161. Ceylan S, Coutable L, Wegner J, Kirschning A (2011) Inductive heating with magnetic materials inside flow reactors. Chemistry 17 (6):1884–1893
162. Butler AJE, Thompson MJ, Maydom PJ, Newby JA, Guo K, Adams H, Chen B (2014) Regioselective synthesis of 3-aminoimidazo[1,2-*a*]-pyrimidines under continuous flow conditions. J Org Chem 79 (21):10196–10202
163. Abahmane L, Kohler JM, Gross GA (2011) Gold-nanoparticle-catalyzed synthesis of propargylamines: the traditional A^3-multicomponent reaction performed as a two-step flow process. Chemistry 17(10):3005–3010
164. Vlaar T, Ruijter E, Maes BUW, Orru RVA (2013) Palladium-catalyzed migratory insertion of isocyanides: an emerging platform in cross-coupling chemistry. Angew Chem Int Ed 52(28):7084–7097

Chapter 9

Direct C–H Functionalization Approaches to Pharmaceutically Relevant Molecules

James J. Mousseau and Antonia F. Stepan

Abstract

The development of direct C–H functionalization reactions to construct C–C and C–X bonds has garnered considerable interest over the past 10–15 years. This is due to the prevalence of both aromatic moieties and heteroatoms in pharmaceutical, agrochemical, material, and natural products. More traditional and established cross-coupling approaches rely on tactically installed metal-activating groups, and as such, the transformation of these functional groups limits conventional synthetic strategy. Through the direct functionalization of C–H bonds, the inefficiencies of conventional methods can be mitigated, providing an attractive tool for organic chemists, expediting the synthesis of complex molecules through new "topologically obvious" disconnections. While an attractive concept, significant synthetic challenges are presented by the ubiquitous, and often unreactive, nature of C–H bonds in organic molecules. Recent advances in the field have focused on the development of new catalytic systems that are both highly reactive and predictably selective. These recently developed C–H functionalization methods have the potential to not only streamline synthesis but also allow access to novel heterocycles, thereby enabling late-stage diversification of biologically active entities. As such, they can both accelerate the speed with which structure-activity relationships are generated and the efficiency by which drug targets can be produced. This chapter provides a brief overview of the C–H functionalization strategies and some examples of their application to the synthesis of pharmaceutical targets.

Key words Catalysis, Direct functionalization, Heterocycles, Pharmaceuticals

1 Introduction

Over the past two decades, the field of C–H functionalization has undergone tremendous growth. With the application of various catalytic and structural systems, synthetic chemists now have increased site and reactivity control toward the construction of elaborate molecular architectures. Through the judicious choice of reaction conditions, the field is progressing toward being able to functionalize any targeted sp^2 and sp^3 C–H bond. There is now increasing access to chemical space that was previously reliant on the ability to pre-functionalize the bond of interest. As a result, C–H processes are often more atom-economical, and as the

Paul F. Richardson (ed.), *Green Chemistry in Drug Discovery: From Academia to Industry*, Methods in Pharmacology and Toxicology, https://doi.org/10.1007/978-1-0716-1579-9_9,

transformations are "direct," time and costs can be saved by avoiding the need for substrate pre-functionalization. These green practices are of important interest to the pharmaceutical industry, and combined with the fact that many methods are applicable in late-stage diversification, continual development of this area remains highly desirable.

An increasing toolbox of direct C–H bond functionalization reactions has been reported in the last two decades [1–5]. Transformations include among others arylations, alkylations, alkenylations, aminations, oxidations, halogenations, and borylations. More recently, the emphasis of method development in C–H functionalization processes has been centered on achieving bond activation in a selective yet predictable manner. The major approaches that are used to achieve regioselectivity in C–H functionalization make use of a combination of both substrate- and catalyst-based control. For example:

1. *Intramolecular reactions*: tethered reacting groups are employed to limit the degree of freedom in the system.
2. *Directing groups*: auxiliary groups usually containing Lewis basic heteroatoms to coordinate and bring the metal center into close proximity with a specific C–H bond.
3. *Steric properties*: certain C–H bonds may be more accessible to the catalyst than others.
4. *Electronic properties*: activated C–H bonds targeted, e.g., nucleophilic positions on electron rich heterocycles, or acidic C–H bonds.
5. *Catalyst control*: catalyst systems that favor different mechanisms of activation, e.g., electrophilic Pd(II) systems for targeting "activated" nucleophilic positions, careful choice of base to target acidic C–H bonds via a concerted-metalation deprotonation pathway, or highly coordinating metal catalysts suitable for metallacycle formation in combination with directing groups.

Herein, we will focus our discussion in two parts. In the first part, the discussion will highlight key advances in direct arylation, alkenylation, and alkylation processes. This section is by no means intended to be exhaustive and will only provide the reader with a flavor of the transformations possible. The subsequent section will focus on the application of direct functionalization processes in the synthesis and diversification of pharmaceutically relevant entities as well as tool compounds. As with the first section, the work covered in this section will not be comprehensive and serve to provide the reader with a notion of the application of this cutting-edge chemistry in an industrial setting.

2 C–H Functionalization to Access New Chemical Space for Medicinal Chemistry and the Preparation of Bio-Active Compounds

2.1 Direct Arylation Processes

The construction of aryl C–H bonds is a key method in organic synthesis due to the importance of aryl and heteroaryl motifs in the pharmaceutical industry. Cross-coupling reactions developed since the 1970s provide powerful tools for the construction of aryl C–C bonds and as such, have dominated the way aryl motifs have been constructed. Despite the effectiveness of cross-coupling processes, they suffer from several key drawbacks. Notably, they rely on the synthesis of *pseudo*-electrophiles (**1**) and nucleophile starting materials (**2**), and they generate stoichiometric amounts of sometimes toxic metal salts as reaction by-products (Scheme 1). Consequently, these reactions are neither atom- nor time-economical and may contribute negatively to the environment. Given these facts, more recent focus has revolved around the development of *direct* arylation reactions which minimize the pre-functionalization of the reacting starting materials (Scheme 1). Therefore, these transformations are considered "green" alternatives while simultaneously improving the overall economy of the reaction. These qualities are of importance to the pharmaceutical industry. In addition, direct arylation reactions offer facile access into desirable design space, notably the ability to readily functionalize α to heterocyclic nitrogen atoms.

The Fagnou group played a pivotal role in the development of direct arylation processes. Of particular interest to the pharmaceutical industry was the development of the direct arylation of pyridine *N*-oxides with aryl bromides under palladium catalysis (Scheme 2) [6]. A key advantage of this process over traditional cross-coupling methods is the ability to readily access 2-aryl pyridines. The *N*-oxide offers the advantage of preventing

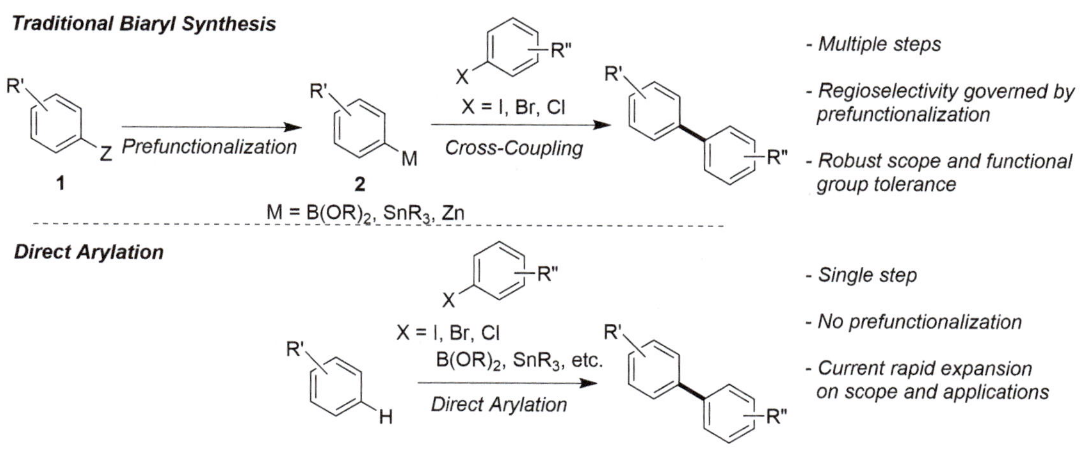

Scheme 1 Comparison of traditional and direct coupling methods

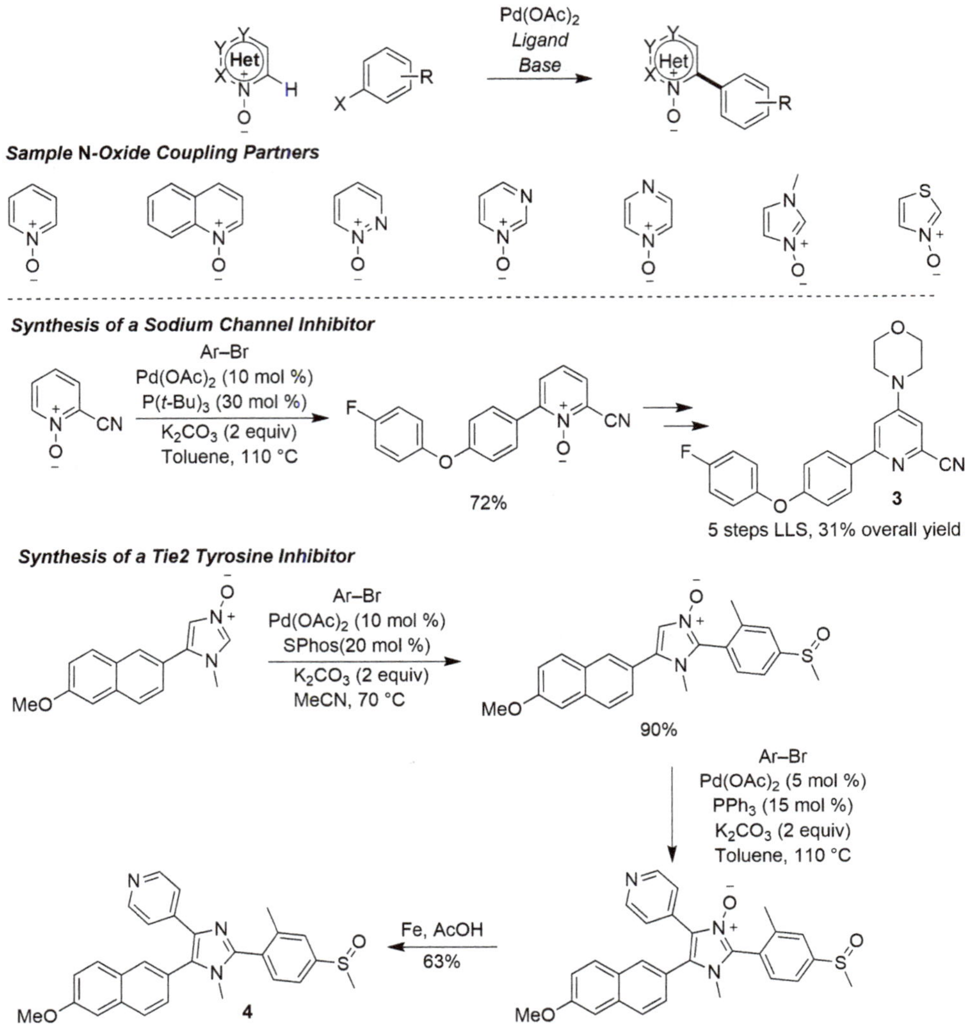

Scheme 2 Some applications of pyridine *N*-oxide direct arylation

nonproductive binding between the transition metal catalyst and the nitrogen lone pair, favoring productive π-binding interactions with the arene ring [7]. Additionally, the *N*-oxide group also helped to increase the electron density of the pyridine ring while simultaneously increasing the Brønsted acidity of the C–H bond at the 2-position. The high stability, wide commercial availability, and ease of synthesis make them an attractive alternative to 2-metallopyridines, which in themselves are often subject to facile protodemetallation. The methodology was expanded to include diazine *N*-oxides as well as various azole *N*-oxides (Scheme 2) [8, 9]. The synthetic utility of this method was demonstrated in the synthesis of a sodium channel inhibitor **3**, as well as a tyrosine kinase inhibitor **4** (Scheme 2).

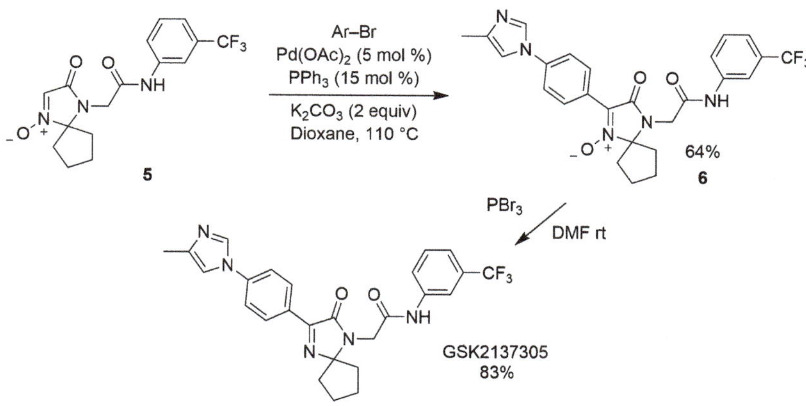

Scheme 3 Application of *N*-oxide arylation *en route* to GSK2137305

Fagnou-like *N*-oxide arylations have been applied in the synthesis of clinical candidates, such as GSK2137305, a glycine transporter inhibitor (Scheme 3) [10]. Here the arylation was applied to the imidazolone *N*-oxide **5** in good yield to afford the arylated product **6** (64% yield), which was then readily deoxygenated to furnish the target compound. The authors noted that the arylation was general with regard to both the aryl bromide coupling partner and the imidazoline *N*-oxide, thus opening up additional avenues to use this chemistry in parallel medicinal chemistry applications.

An elegant example of direct arylation was applied by Itami and co-workers in the synthesis of dragmacidin D, an alkaloid known to exhibit the inhibition of serine-threonine protein phosphatases, which has also been of interest in the treatment of various neurological disorders [11]. The sequence involved the regioselective arylation of indole 7 with thiophene **8** (Scheme 4). The presence of the TIPS protecting group was instrumental in directing the arylation to the C4 position over the C2. Opening of the thiophene to the methyl ketone provided indole **9** that was subjected to oxidative C–H/C–H coupling with pyrazine *N*-oxide to give adduct **10**. Oxygen migration then set up a second oxidative C–H/C–H coupling to give the *bis*-indole **11**, which after two more steps provided dragmacidin D.

Imidazoles and thiazoles are prevalent scaffolds in a range of pharmaceutically active agents and natural structures. Additionally, the varying electronic properties of the carbon centers in these heterocycles should allow for the regioselective functionalization of these rings. Murai recently delineated a sequence taking advantage of these properties to selectively build *tri*-arylated imidazoles [12]. The most nucleophile C5 carbon atom of indole **12** is selectively arylated via a 1,10-phenanthroline-containing palladium catalyst (Scheme 5). This catalyst system was reported to be necessary to mitigate C2 arylation as that position is also highly reactive due to its acidic hydrogen atom. Interestingly, Jain reported a versatile

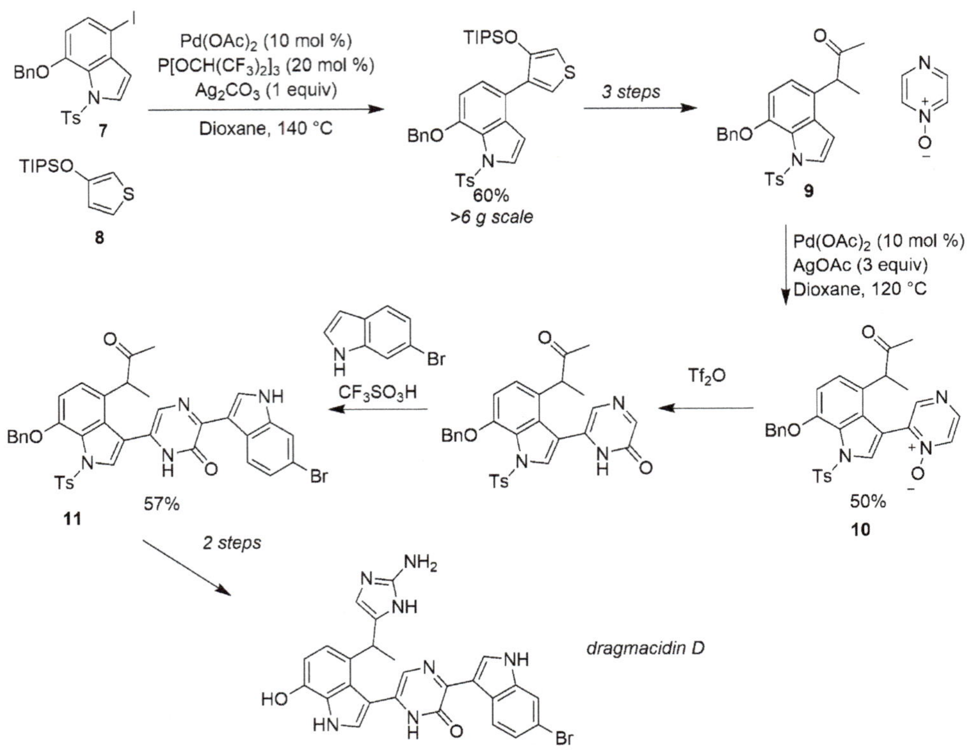

Scheme 4 Preparation of dragmacidin D via a multiple direct arylation sequence

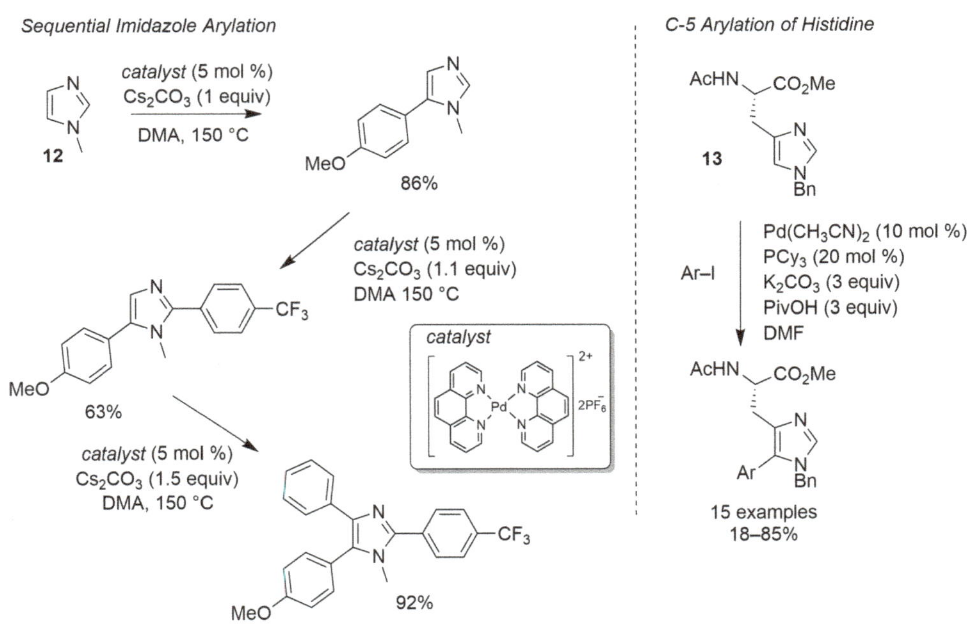

Scheme 5 Various imidazole arylations

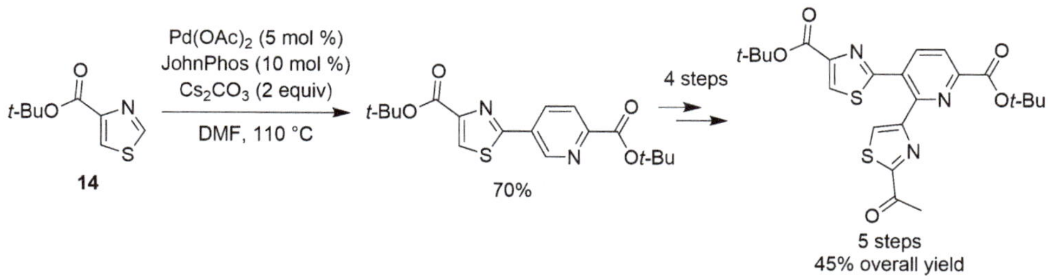

Scheme 6 Application of direct arylation in the synthesis of antibiotics

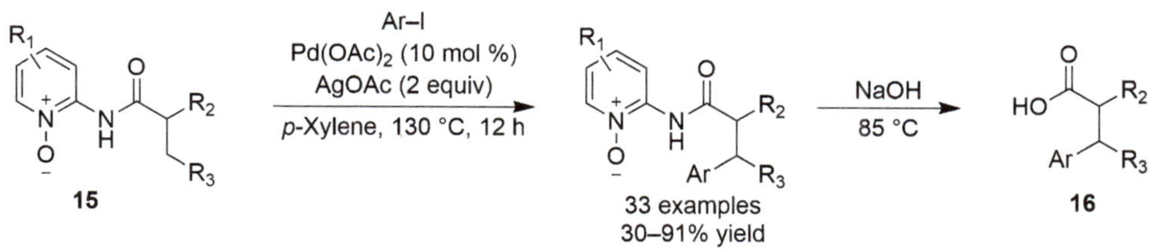

Scheme 7 Use of pyridine *N*-oxide as a directing group in sp^3 arylation processes

selective C5-arylation of L-histidine **13** under Fagnou-type conditions using microwave irradiation in the presence of pivalic acid to facilitate a concerted metalation deprotonation (CMD) mechanism [13]. Following C5 arylation Murai demonstrated that it is then possible to selectively arylate the C2 position followed by position C3, all with the same catalyst system. This approach represents a highly modular means to elaborate these scaffolds.

The arylation of 4-thiazolecarboxylate **14** was applied in the synthesis of the core of thiopeptide antibiotics (Scheme 6) [14]. The use of a $Pd(OAc)_2$/JohnPhos catalyst system permitted the selective arylation of the 2-position. The judicious choice of the ligand was key to achieve sufficient site-selectivity and avoid overarylation. Aryl iodides, bromides, and chlorides were viable coupling partners, and the developed direct arylation reaction was used in the first step of the synthesis of an antibiotic analogue.

While the cases described *vide infra* utilized the *N*-oxide motif solely to activate the arene and not direct carbopalladation, Lu and co-workers described the use of a pyridine *N*-oxide to direct an oxidative arylation of the sp^3 β positions of 2-amidopyridine *N*-oxides (**15**) (Scheme 7) [15]. The scope was general with regard to the pyridinium, amide, and the aryl coupling partners. The pyridine *N*-oxide can be readily cleaved under basic conditions to liberate the corresponding substituted alcohol **16**. Preliminary mechanistic studies have indicated the formation of a pallada-bicyclic intermediate, and work is reported to be underway to apply this unique pathway to other direct functionalization reactions.

While less common, there have been a few accounts of the direct arylation of sp^3 centers. Perhaps the most common examples of such functionalizations involve the application of amino acids or

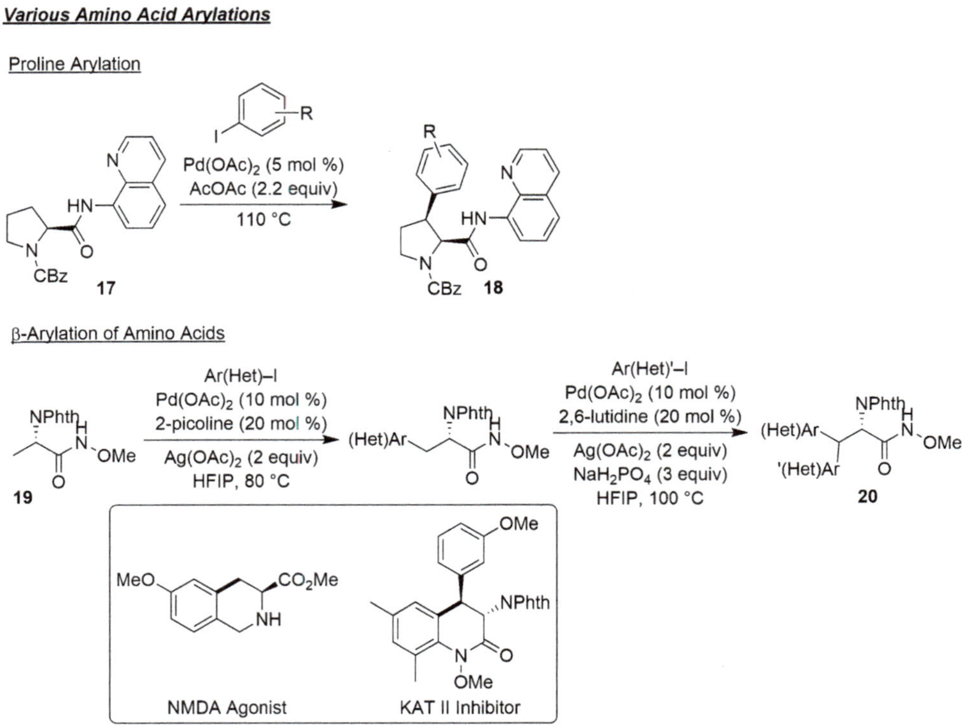

Scheme 8 Direct sp^3 arylation of amino acid derivatives

amino acid derivatives. The Bull group described the palladium-catalyzed oxidative arylation of the 3-position of proline **17** to give functionalized pyrrolidine derivatives **18** (Scheme 8) [16]. The described method converts the carboxylic moiety of the proline to a quinolinamide directing group that stereoselectively guides arylation to the 3-position. A single enantiomer is observed, and the reaction could be performed on multi-gram scale. More recently Yu in collaboration with BMS developed the β-arylation of amino acids [17]. Key to the success of the reaction was the use of a readily removable *N*-methoxylamide auxiliary **19** in addition to a 2-picoline ligand to modulate monoarylation. Diarylated products **20** can be obtained with 2,6-lutidine. The scope of these reactions is very general with a range of aryl, heteroaryl, and intramolecular products reported. Additionally, the method was employed to prepare several biologically active compounds, including a NMDA agonist and KAT II inhibitors (Scheme 8).

Unactivated THF and 1,4-benzodioxane systems **21** have been readily arylated under oxidative conditions [18]. These compounds are of interest due to their prevalence in natural products and other biologically active compounds. As with the other oxidative arylation processes described, the scope is general and employs a quinoline directing group (Scheme 9).

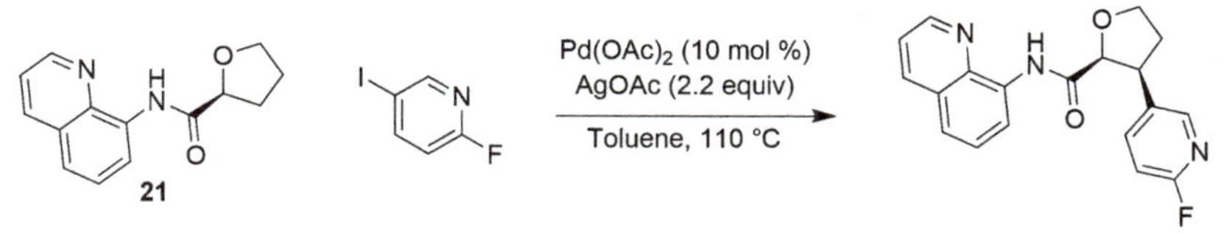

Scheme 9 An example of THF arylation

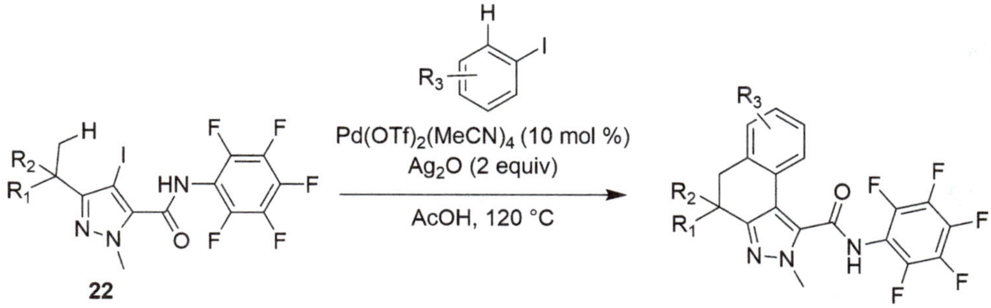

Scheme 10 Rapid synthesis of highly substituted pyrazoles via double direct arylations

Recently Yu and co-workers with chemists from Vertex have reported on a "triple" C–H activation process to enable the rapid assembly of highly substituted pyrazoles (Scheme 10) [19]. Using a pentafluorophenylamido directing group as in analog **22**, the reaction was determined to first pass through the direct arylation of the methyl C–H bond followed by an oxidative C–H/C–H coupling to yield the target products.

2.2 Direct Alkenylation Processes

The Mizoroki-Heck reaction is perhaps the most common means to construct aryl-olefin bonds. However, like with the cross-couplings mentioned *vide infra*, this method requires the pre-activation of the coupling substrates and, in the case of heteroaryl substrates, can suffer from poor reactivity depending on the activation of the electrophile. Curiously, relative to direct arylation processes, the direct C–H alkenylation has received less attention, particularly in the context of the synthesis of biologically relevant molecules. An elegant solution to this problem was presented by Yu and co-workers through the alkenylation of synthetically useful phenylacetic and 3-phenylpropionic acids **23** using a palladium catalyst in the presence of a mono-N-protected amino acid (MPAA) ligand such as Boc-lle-OH (Scheme 11) [20, 21]. In these cases, O_2 acts as the terminal oxidant, and the selectivity of the reaction is governed by the coordination of palladium to the Lewis basic carboxyl group, the use of protecting groups, and the MPAA itself. The ligand is key, accelerating the rate of the C–H

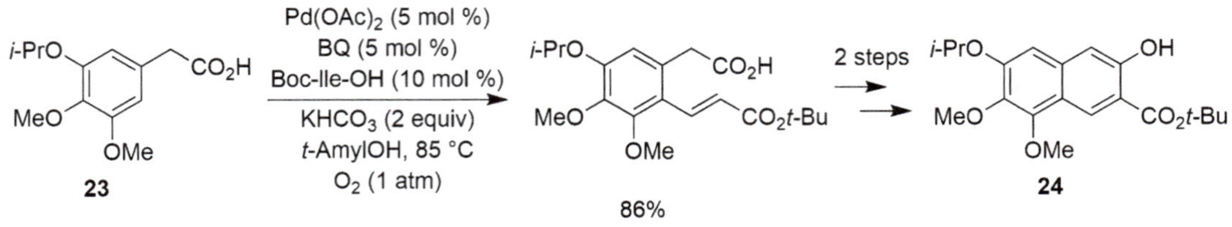

Scheme 11 Application of MPAAs in the direct alkenylation of arenes

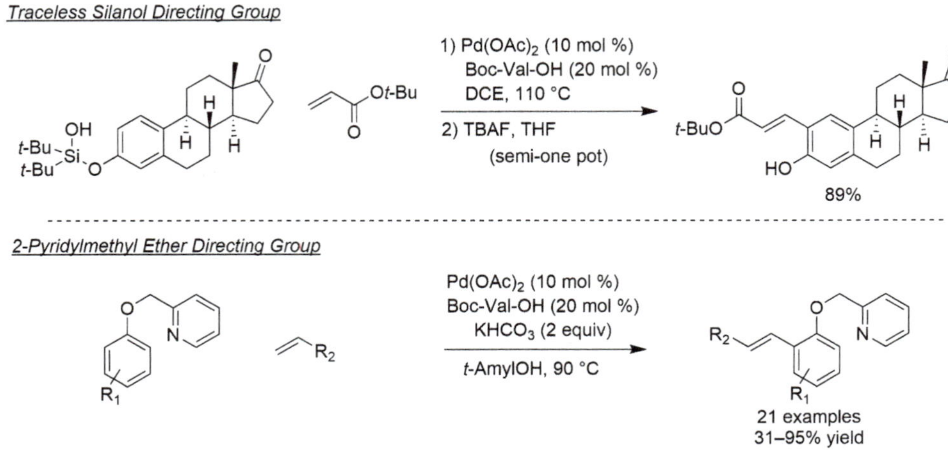

Scheme 12 Various MPAA accelerated direct alkenylation reactions

cleavage step, and thus the desired reaction. In Yu's seminal account, the methodology was applied in the synthesis of the naphthoic acid component of kedarcidin (**24**, Scheme 11).

Phenol derivatives in concert with MPAA systems were also demonstrated to be viable substrates for the alkenylation of arenes. Gevorgyan reported the use of silanols as a "traceless" directing group (**25**) for their *o*-alkenylation in the presence of (+)-menthyl (O_2C)-Leu-OH [22]. A range of substrates were viable, including estrone (Scheme 12), demonstrating the applicability of this methodology in late-stage diversification. Later work by Lan demonstrated that other Lewis basic groups could be applied [23]. Using similar conditions to Yu, with Boc-Val-OH as an MPAA to accelerate the reaction, it was demonstrated that 2-pyridylmethyl esters (**26**) could be used to direct the olefination of arenes. This directing group may be readily removed to reveal the corresponding phenol (Scheme 12).

Yu and co-workers have been able to demonstrate the use of directing groups to selectively arylate the remote *meta* position of aryl bonds [24]. Using a directing group bearing a Lewis basic nitrile at the terminal position to coordinate palladium, in combination with an *N*-acetyl glycine as an MPAA ligand accelerator, the authors were able to demonstrate the selective *meta* arylation of

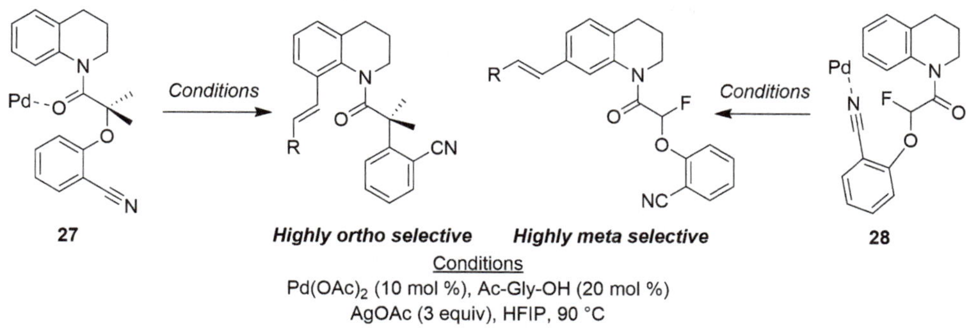

Scheme 13 Yu's site selective MPAA promoted direct alkenylation

tetrahydroquinolines (**27**), 2-phenylpyrrolidine, 2-phenylpiperidines, and other aniline-based substrates. The key to the *meta*-selective arylation lies in the directing group (Scheme 13). When a methylene or dimethyl linker was incorporated, *ortho*-arylation was observed, as expected should the carbonyl oxygen atom direct the palladium C–H insertion. The *meta*-selective C–H arylation arose when a fluorine was incorporated into the methylene chain (**28**). In these substrates, it has been postulated that palladium favors coordination with the nitrile.

Direct alkenylation has been applied in the synthesis of 2-quinolones. Liu and co-workers described a process that proceeds through a C–H activation/C–C bond formation/cyclization cascade reaction [25]. The acetate directing group is generated in situ from the reaction of the aniline starting material **29** and acetic anhydride. The reaction requires tosic acid as an additive to promote cyclization as well as to remove the directing group (**30**) and utilized $Na_2S_2O_8$ as an oxidant to regenerate the active Pd^{II} catalyst. The scope was shown to be general, and the method was applied in the formal synthesis tipifarnib (Scheme 14).

Heterocycles are also suitable substrates for direct alkenylation processes. A recent example by Shi demonstrated the viability of the alkenylation of thiazole derivatives (**31**) that could act as the core of PPAR agonists (**32**) [26]. The functionalization occurs exclusively α to the sulfur atom giving the target products in moderate to very good yields (Scheme 15).

The 3-position of pyridine has been successfully alkenylated through a Rh^{III}-catalyzed oxidative process [27]. This process is limited to 2-amino pyridines that bear a pivalate protecting group that also functions as a directing moiety (**33**). Traditional Heck acceptors appear to be optimal, and the reaction is viable in pyrazoles and indoles, suggesting the versatility of the reaction. The method was demonstrated to be scalable and was applied in the multigram synthesis of napthyridones en route to the synthesis of an anti-malarial reagent **34** (Scheme 16).

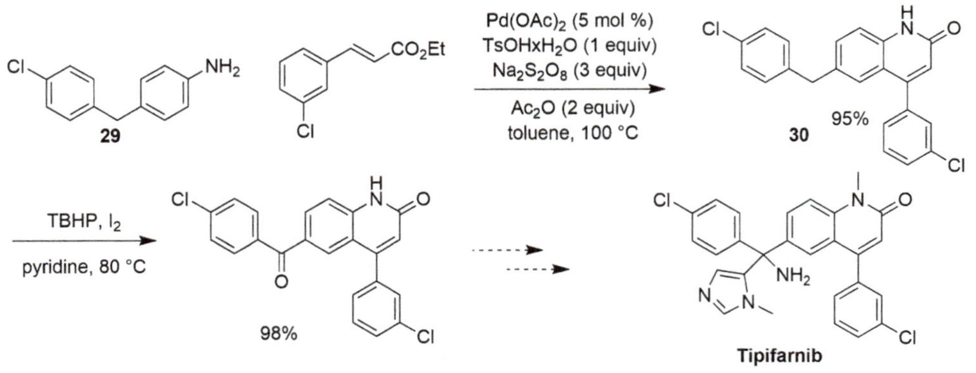

Scheme 14 Tandem alkenylation/cyclization sequence *en route* to tipifarnib

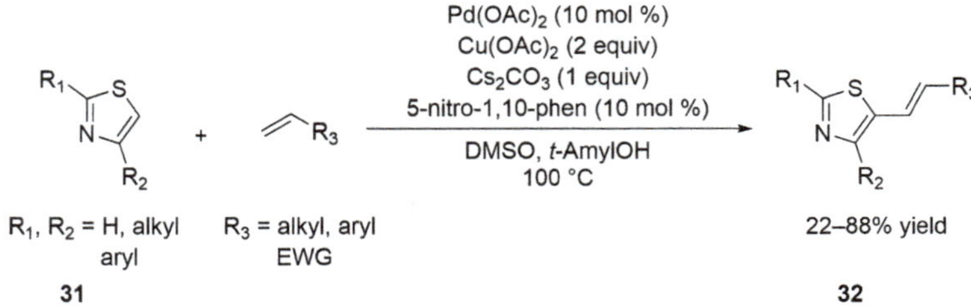

Scheme 15 Shi's alkenylation of thiazole derivatives

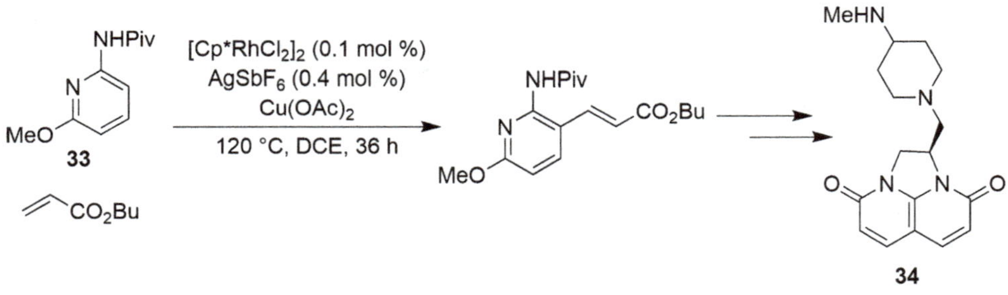

Scheme 16 Rh-catalyzed pyridine alkenylation *en route* to the anti-malarial agent **34**

2.3 Direct Alkylation Processes

While direct arylation and alkenylation reactions are well precedented in the literature, direct alkylation processes remain scarce, and in many cases, rely on activated sp^3 centers. Pyridine *N*-oxides are the most common substrates for direct alkylation processes with one account describing the alkylation of 2-picoline *N*-oxides with 2° alkyl bromides in the presence of $Pd(OAc)_2$dppf and Cs_2CO_3 [28]. The scope is general with regard to the alkyl bromide, affording the targeted products in moderate to good yields (Scheme 17). A near 1:1 kinetic isotope effect (KIE) in addition to observed reaction stereochemistry led to the authors to postulate a hybrid organometallic radical mechanism.

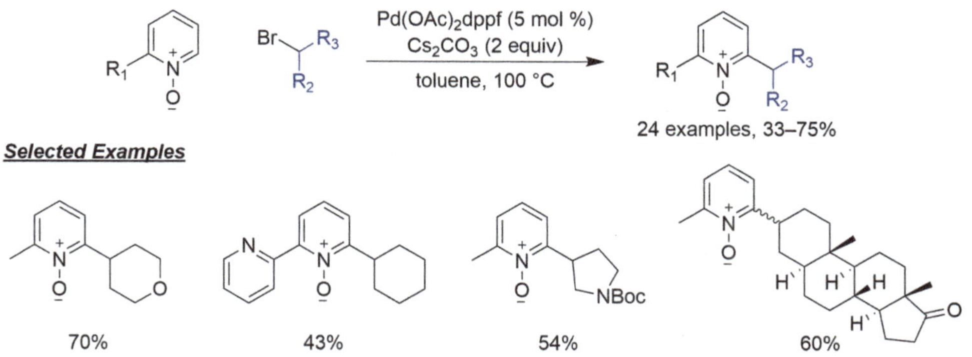

Scheme 17 Direct alkylation of pyridine *N*-oxides with alkyl bromides

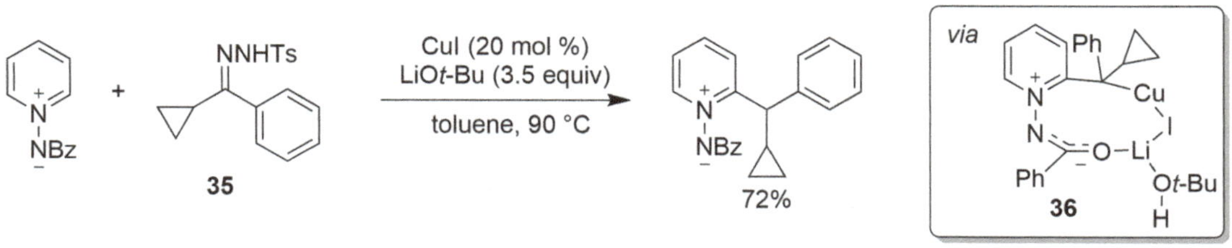

Scheme 18 Direct alkylation of *N*-iminopyridinium ylides with *N*-tosylhydrazones

Wang and co-workers found a protocol to expand the scope beyond 2-substituted *N*-oxides, using *N*-iminopyridinium ylides in a copper-catalyzed process with *N*-tosylhydrazones (Scheme 18) [29]. A density functional (DFT) computational study suggests that the reaction proceeds through Cu-catalyzed decomposition of the *N*-tosylhydrazone **35** to the carbene that then undergoes directed migratory insertion into the pyridinium ylide **36**.

Cramer and co-workers disclosed an elegant intramolecular methine arylation to synthesize spiroindolines in high yield (Scheme 19) [30]. The interest in this structural motif was due its presence in natural products as well as both HIV transcriptase inhibitors and PLK4 antagonists. Under Fagnou-like conditions, 2-bromoanilines with an N-tethered cyclopropane undergo methine arylation at the cyclopropane in good to excellent yields. The reaction presumably occurs due to the acidity of the cyclopropyl C–H bond engendered by its increased *s* character. Interestingly, in cases of 2,6-dibromo substrates, one-pot C–H arylation/Suzuki coupling is possible, enabling the rapid synthesis of a range of highly functionalized spiroindoline products **37** (Scheme 19).

Isoindolinones have been prepared though a copper-catalyzed oxidative coupling of aromatic C–H bonds with malonates. Yu described using an amide-oxazoline directing group **38** in the presence of $Cu(OAc)_2$, Li_2CO_3, and Ag_2CO_3 [31]. The reaction is general with regard to the arene, even tolerating aryl iodides and bromides, leaving a versatile scaffold for future elaboration

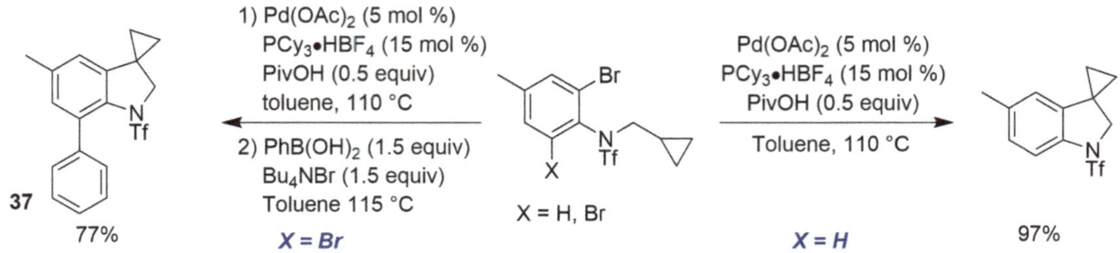

Scheme 19 Synthesis of spiroindolines via direct alkylation

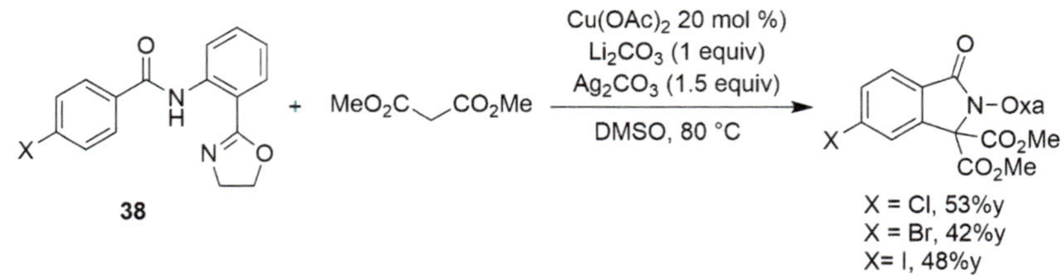

Scheme 20 Cu-catalyzed oxidative alkylation to yield isoinolinones

(Scheme 20). Control experiments suggest that the reaction proceeds first through the arylation of the malonate, followed by arylation of the N–H of the amide.

Yu and workers have described a covalent template strategy for the remote *meta*-alkylation of rings [32]. Using norbornene **39** as the mediator, the team effected alkylations through a Catellani-type process in high yield (Scheme 21) [33]. Any competing *ortho*-alkylation was mitigated through the application of pyridine-type ligands in order to enable 1,2-migratory insertion. A range of benzyl, alkyl, and acetate halides were viable coupling partners; in addition aryl iodides could be utilized to achieve *meta*-arylation.

2.4 Overview

It is possible to see that a wide range of methodologies for the direct C–H functionalization processes have been discovered, many of which cannot be reviewed in the context of this chapter. In addition, there are C–H borylations, carboxylations, halogenations, aminations, orygenation, and others that exist, which are beyond the scope of this chapter. The next section will describe the application of several of the methods described herein in the synthesis of pharmaceutically relevant compounds.

3 Applications of C–H Functionalization to Functionalize Pharmacologically Relevant Entities

As highlighted in the first section of this chapter, advances in transition metal-catalyzed C–H bond functionalization reactions are occurring at a rapid pace and are driven by the desire for atom efficient and green processes. Applied to target-orientated

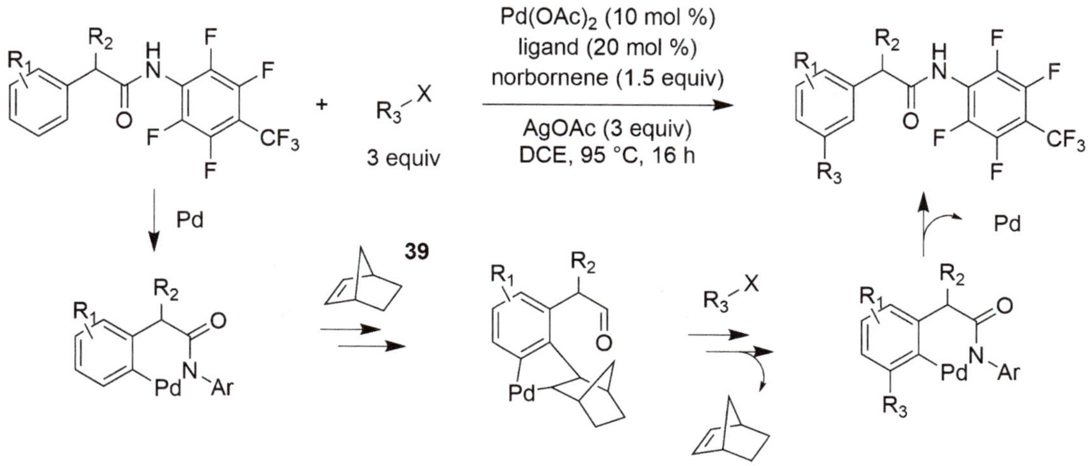

Scheme 21 *Meta*-selective direct alkylation

syntheses, these methods offer routes with alternative disconnection strategies, a possibly improved overall yield and atom efficiency. An additional powerful potential application of C–H functionalization chemistry is found in the late-stage diversification of drug molecules and natural products. Late-stage diversification strategies were first utilized more than 20 years ago to structurally modify steroids through reactions catalyzed by oxidative enzymes [34]. However, the recent advancements in transition metal (TM)-catalyzed chemistry add a substantial amount of additional transformations to the medicinal chemist's diversification toolbox.

Many drug molecules lend themselves to the late-stage diversification using TM, since Lewis basic functionalities are an integral part of their pharmacophore, and as such can function as the directing groups (DG) in a TM-mediated *ortho*-C–H functionalization process. Examples of drug molecules containing such Lewis basic groups are shown in Fig. 1 and include fentanyl (**40**, amide as a potential DG), eszopiclone (**41**, lactam as a potential DG), and valsartan (**42**, acid as a potential DG). Specifically, an analysis of the brand-name drugs has shown that 60% of the 2008 top selling drug molecules contain the Lewis basic DGs required for TM-mediated C–H functionalization [35]. As discussed in the previous section, Yu applied the use of weak coordination to direct C–H functionalization, a concept that has expanded the range of functional groups suitable for this type of chemistry. A collaboration between the Yu group and Pfizer aimed at further expanding this methodology to highly prevalent groups in medicinal chemistry recognized that sulfonamides are present in many drug molecules with commercial examples being the antimigraine drug sumatriptan (**43**), the antibacterial sulfamethoxazole (**44**), and the diuretic azosemide (**45**) (Scheme 22). In order to apply C–H functionalization technologies to these drug scaffolds, it was necessary to establish a protocol that

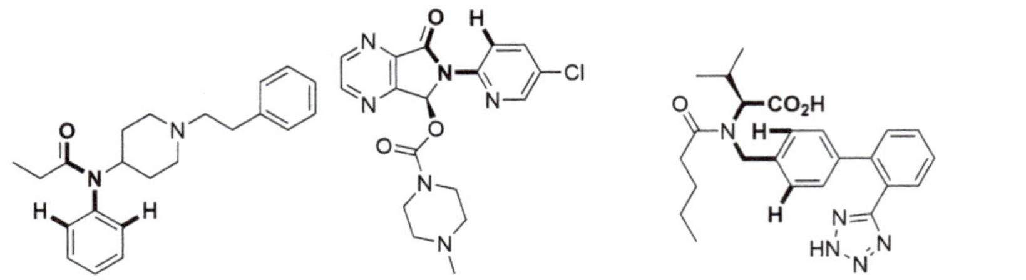

fentanyl (**40**), an opioid analgesic — eszopiclone (**41**), a non-benzodiazepine hypnotic — valsartan (**42**), an angiotensin II receptor antagonist

Fig. 1 Examples of drug molecules with Lewis basic functionalities as potential directing groups for the TM-mediated C–H functionalization

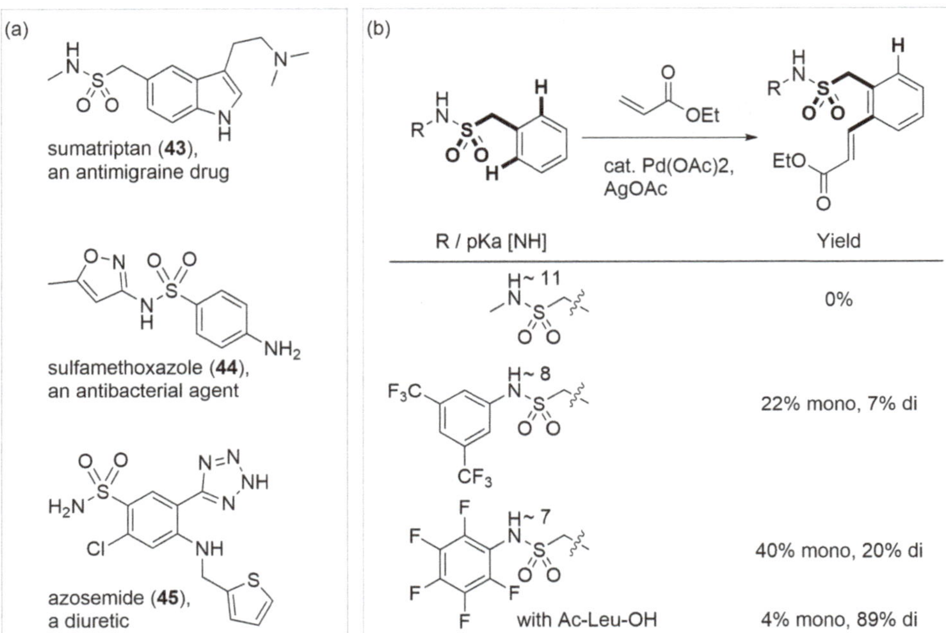

Scheme 22 (**a**) Examples of sulfonamide-containing drug molecules and (**b**) C–H functionalization molecules on benzylsulfonamides

relied on the use of sulfonamides as Lewis basic directing groups, which, despite the pioneering Rh-catalyzed alkoxysulfonamide-directed nitrene insertion reactions reported by DuBois [36], have not been shown to direct the palladation of C–H bonds. Gratifyingly, the early results demonstrated that the *ortho*-olefination of benzylsulfonamides was feasible if the *N*-sulfonamide is sufficiently acidic; in addition, MPAAs such as Ac-Leu-OH enhance the reactivity toward *ortho*-functionalization and increase the overall yield, with mainly di-olefinated product formed in 89% yield (Scheme 22) [37]. The utilization of mono-N-protected amino acids (MPAA) as reaction accelerating ligands had been previously established by Yu and again was outlined in the previous section of this chapter [20, 21].

This reaction could be expanded to the corresponding benzene-sulfonamides and allowed the formation of a broad range of carbon-carbon and carbon-heteroatom bonds. Application of this new set of reactions to the celecoxib analog **46**, readily available from the Pfizer sample bank, then led to six distinct products with exclusive site-selectivity in the presence of multiple reactive C–H bonds, demonstrating the suitability of this sulfonamide-based protocol to diversify a privileged skeleton in a manner that was previously inaccessible (Scheme 23). Of note is the relative ease of protecting group removal, which is reliant on the use of aqueous trifluoroacetic acid to give the unsubstituted sulfonamide derivatives in high yield.

Yu with Pfizer also developed a similar late-stage diversification strategy for α-phenoxyacetic acids, the common pharmacophore in the fibrate class of lipid-lowering agents [38]. Although C–H activation processes through the formation of 5- and 6-membered palladacycles have been developed into robust protocols applicable to a range of systems (see sulfonamide chemistry *vide infra*), C–H diversification processes reliant on 7-membered, or larger, transition states is rare. As seen in Scheme 24, the mono-N-protected amino acids Ac-Gly-OH and Boc-Val-OH were used to access both **47** and **48** from **49** and **50**, respectively, and were key to facilitating the desired regioselective functionalizations. Notably, the weakly coordinating nitrile group leads to the formation of an intermediate 11-membered palladacycle and overrides the intrinsic preference of the substrate for reactivity in the *ortho*- and *para*-position.

Similar late-stage diversification strategies were also applied to the elaboration of biologically active pyridazinones, a motif widely used in medicinal chemistry with examples ranging from PFKB3 inhibitors, COX-2 inhibitors, to m-38 MAP kinase inhibitors (Fig. 2).

In 2015, Zhang reported the development of seven *ortho*-C–H functionalization transformations (arylation to **51**, carboxylation to **52**, olefination to **53**, thiolation, acetoxylation to **54**, halogenation to **55**) facilitated by the pyridazinone pharmacophore element as a directing group (Scheme 25) [39]. Five of the diversification reactions are shown in Scheme 25 and highlight the robustness of the protocol as reflected by the consistently high product yields (>60% for all reactions). The intermediates obtained from these diversification reactions were then treated with NaSH in DMF at room temperature to give the dimeric pyridazinones that were reported as inhibitory analogs of bacterial cysteine protease-transpeptidase sortase (SrTA) [40].

Late-stage diversification can not only be effected through directed *ortho*-C–H functionalization using transition metals but also through a mechanism that relies on the stabilization of a reactive nitrenoid and/or carbenoid species. Although much of the early methodology relied on forging the desired bonds in an

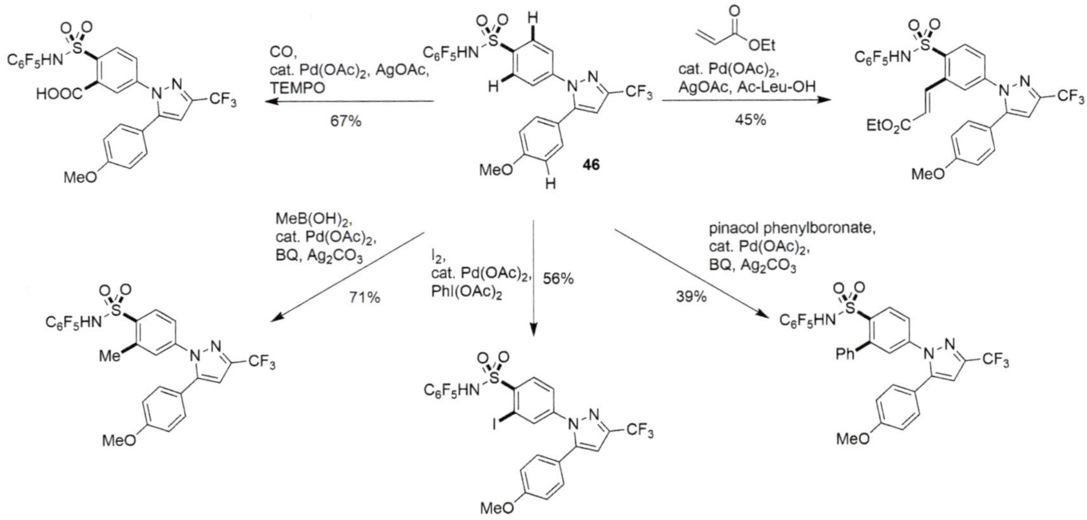

Scheme 23 Late-stage diversification of a celecoxib analog through a TM-mediated, sulfonamide-directed C–H functionalization reaction

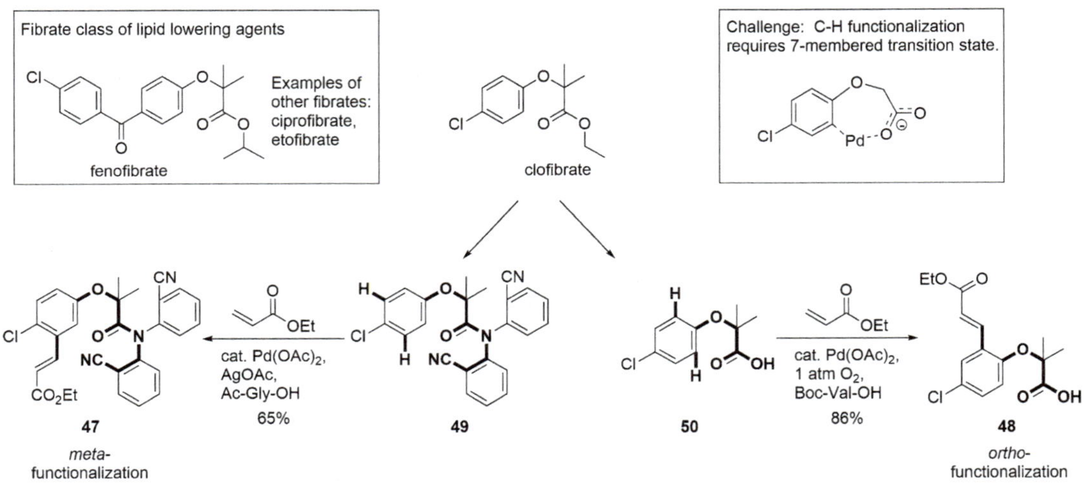

Scheme 24 Late-stage diversification of analogs of clofibrate, a member of the fibrate class of lipid-lowering agents

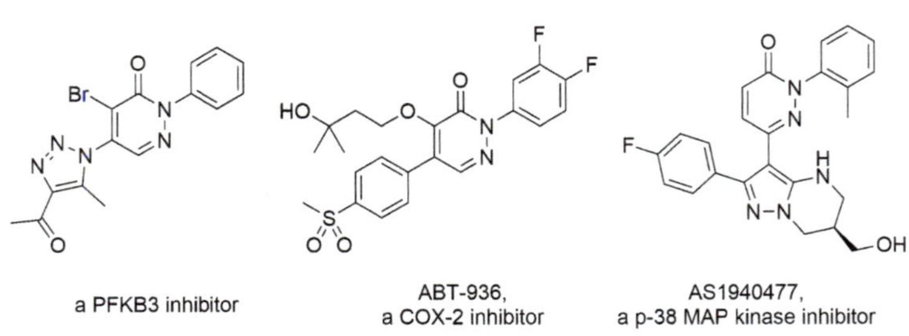

Fig. 2 Examples of drug molecules containing the pyridazinone motif

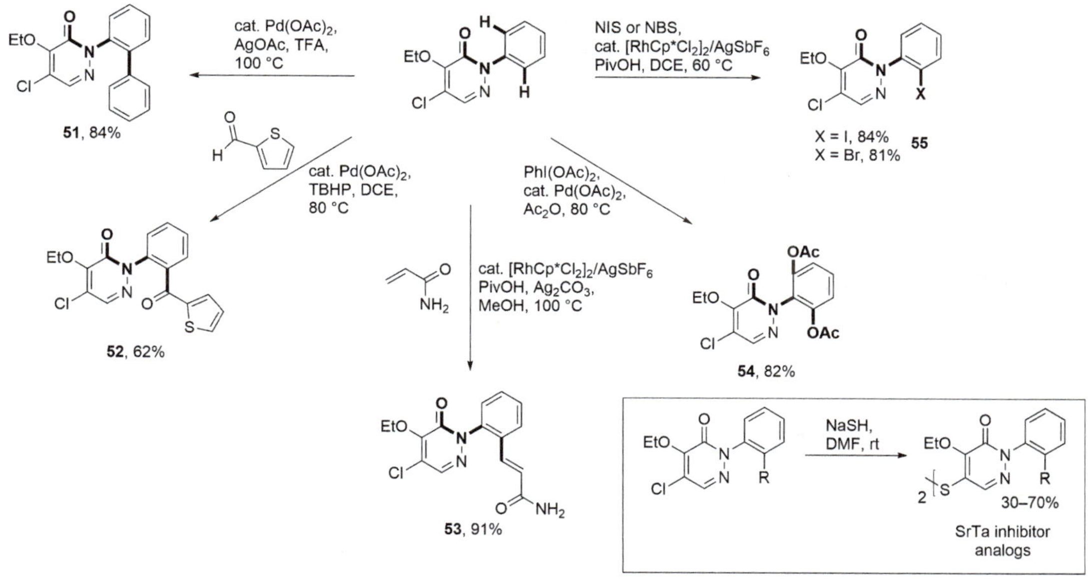

Scheme 25 Pyridazinone-directed C–H functionalizations toward inhibitors of cysteine protease-transpeptidase sortase (SrTA)

intramolecular fashion, catalyst development and mechanistic studies have led to the development of an intermolecular C–H variant that relies on aryloxysulfonamides as the nitrogen source [41]. Romo applied this methodology to the late-stage diversification of natural products with alkyne-containing sulfonamides. These products were subsequently used as cellular probes for mechanism-of-action studies (Scheme 26) [42]. For example, the marine-derived anticancer diterpene eupalmerin acetate (EuPA, **56**) was selectively functionalized at the allylic position with an alkyne-containing sulfonamide to give EuPAyne **57**. EuPAyne was used in gel proteomic profiling of HL-60 cell lines, which led to the identification of several cellular targets that had previously been unknown.

Rhodium-stabilized donor/acceptor carbenoids were used by Davies in collaboration with scientists at Novartis for the late-stage functionalization of alkaloid natural products and pharmaceuticals characterized by the presence of a tertiary amine [43]. Using brucine as a model system, a di-rhodium catalyst screen revealed three reaction systems that functionalized three distinct sites on the complex natural product (Scheme 27). When using $Rh_2(Oct)_4$, the Stevens rearrangement to give compound **58** was the predominant pathway, whereas $Rh_2(TPA)_4$ led to C–H insertion into the methylene group next to the nucleophilic amine and formed product **59** in 50% yield as a single diastereomer. Unexpectedly, the bulky catalyst $Rh_2(BTPCP)_4$ yielded compound **60** through insertion into the most sterically encumbered tertiary C–H bond

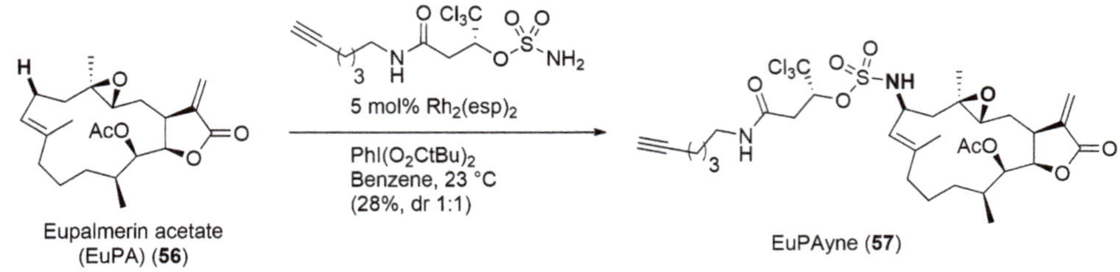

Scheme 26 Romo's late-stage C–H functionalization of the natural product EuPA

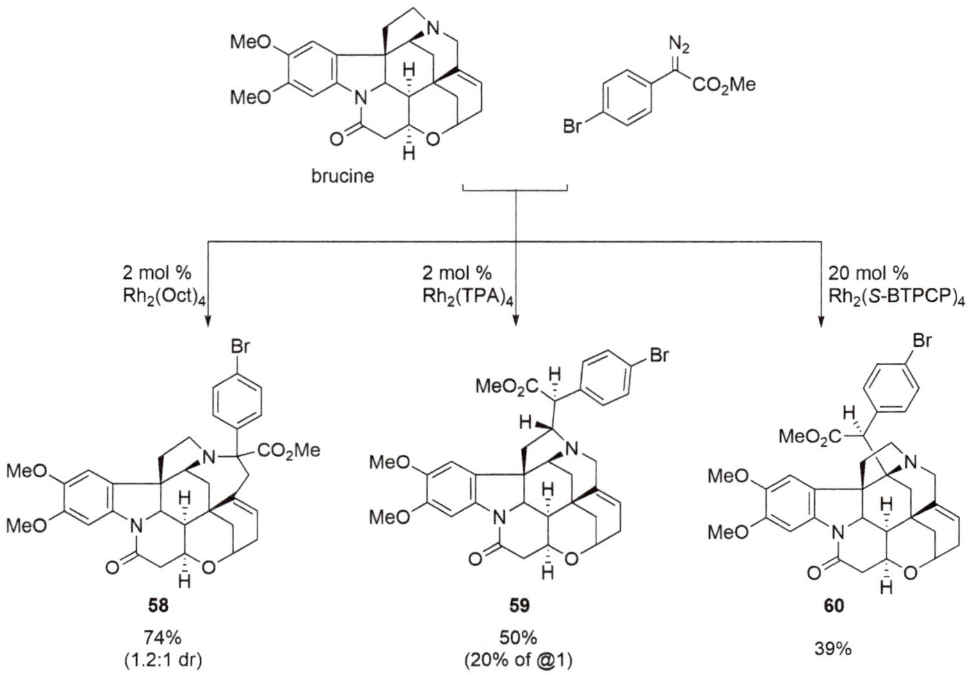

Scheme 27 A carbenoid-based late-stage diversification of the natural product brucine

adjacent to the nitrogen though 20% catalyst loading was necessary to effect sufficient conversion (39% yield, a single diastereomer is formed).

Using similar catalyst systems, the Davies group demonstrated site-specific C–H insertion consistently at the *N*-methyl functionality of various drug molecules (Scheme 28). For example, C–H insertion was affected in >50% yield in the case of sercloremine **61**, noscapine **62**, and dextromethorphan **63** despite the presence of other potentially reactive sites including a number of benzylic positions. The highly chemoselective and regioselective nature of this process suggests potential applications in drug discovery and could include the design of chemical probes for target identification of complex natural products.

Late-stage oxidation methodologies applied to drug molecules are an important aspect of medicinal chemistry as it allows access to possible metabolites for further biopharmaceutical

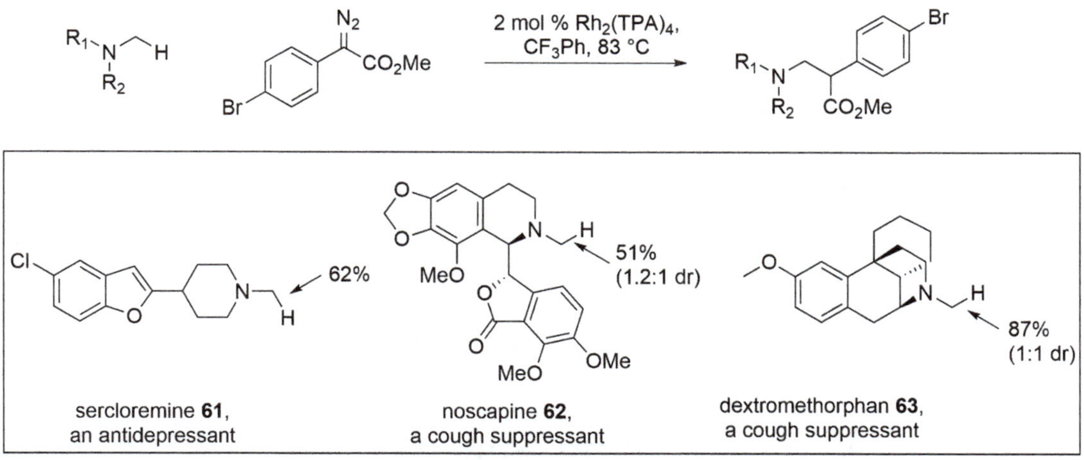

Scheme 28 Selective carbenoid insertion at the *N*-methyl functionality of various drug molecules

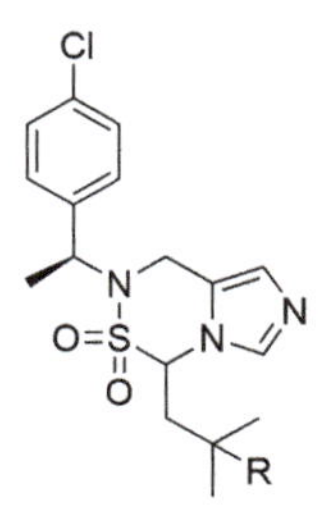

R	*hCYP11B2 IC_{50} (nM)*	*HLM $T_{1/2}$ (min)*	*$clogD_{7.4}$*
H	9.4	6	3.3
OH	11.4	>300	1.7

Fig. 3 CYP11B2 potency, microsomal stability, and solubility of selected aldosterone synthase inhibitors

characterization. These oxidation strategies can also lead to analogs with improved lipophilicity characteristics, since highly lipophilic compounds are often characterized by a suboptimal absorption, distribution, metabolism, and excretion (ADME) profile [44] and suffer from high target promiscuity [45]. An example of a metabolite becoming a lead compound with improved properties is shown in Fig. 3: the observed alcohol, the most likely oxidative metabolite of the parent compound, maintained the original compound affinity for aldosterone synthase (CYP11B2), while considerably improving its metabolic stability in human liver microsomes. The reduced added polarity is also reflected in a higher aqueous solubility (500 μM at pH 6.8) relative to earlier leads within this series [44, 46].

Late-stage oxidation can be achieved through a variety of methods, including biocatalysis with rCYP enzymes and microbial preparations, electrochemistry, as well as biomimetic catalysis [47]. Microbial biocatalysis is often utilized for preparative scale synthesis of drug metabolites for the profiling of their biopharmaceutical and toxicity profile. Recent advances in biomimetic C–H oxidation chemistry provides an alternative way of accessing the

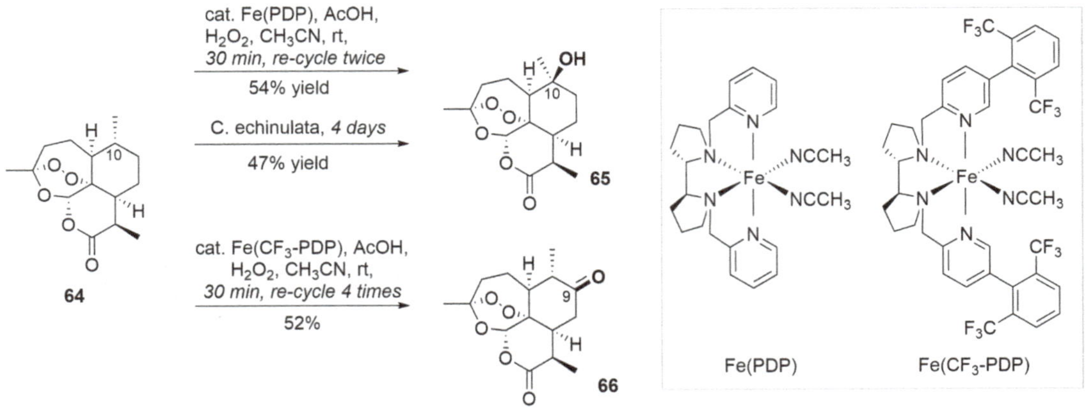

Scheme 29 Oxidation of (+)-artemisinin with White's electrophilic non-heme iron catalysts and *C. echinulata*

desired metabolites and often provides the desired hydroxylated scaffolds in a shorter reaction time and in a higher yield relative to the biocatalytic method. An example is the electrophilic iron-based catalyst Fe(PDP) developed by the White group, which selectively oxidizes tertiary aliphatic C–H bonds using hydrogen peroxide as the oxidant (Scheme 29). This biomimetic non-heme system leads to the selective oxidation of C-10 in (+)-artemisinin (**64**) in 54% yield after twice recycling the starting material with a 30-min reaction time per setup. For comparison, using the microbial culture *Cunninghamella echinulata*, the same hydroxylated analog **65** is obtained in a similar yield (47%), however, after a substantially longer reaction time (4 days), and a lower throughput relative to Fe (PDP) [Fe(PDP): 0.033 M vs. *C. echinulate*: 0.0035 M] [48–50]. The White group also developed a modified Fe(PDP) catalyst, referred to as Fe(CF_3-PDP), that facilitates catalyst controlled oxidation of orthogonal sp^3-hybridised carbons and yields the corresponding C-10 and C-9 oxidized (+)-artemisinin analogs **65** and **66** in 52% combined yield [48].

Other systems for the late-stage oxidation of aliphatics have also been studied and include Crabtree's Cp* iridium precatalyst, which uses $NaIO_4$ as the oxidant. When applied to (+)-artemisinin, this catalytic system did not yield any oxidative products with only 89% of the starting material being recovered [51]. However, in the cases of (−)-ambroxide (**67**) and (+)-sclareolide (**68**), both the White and Crabtree systems yielded a similar set of oxidative analogs, with the Cp* Ir/$NaIO_4$ system giving access to the 2- and 3-oxo-sclareolide analogs in 17% and 5% yield, respectively (Scheme 30).

A limitation of the above oxidation catalysts is that electron-rich aromatics and amines, which are widely present in drug molecules, are not tolerated. In 2013, Touré at Novartis reported a ligand-free copper/molecular oxygen (air) catalyst system that

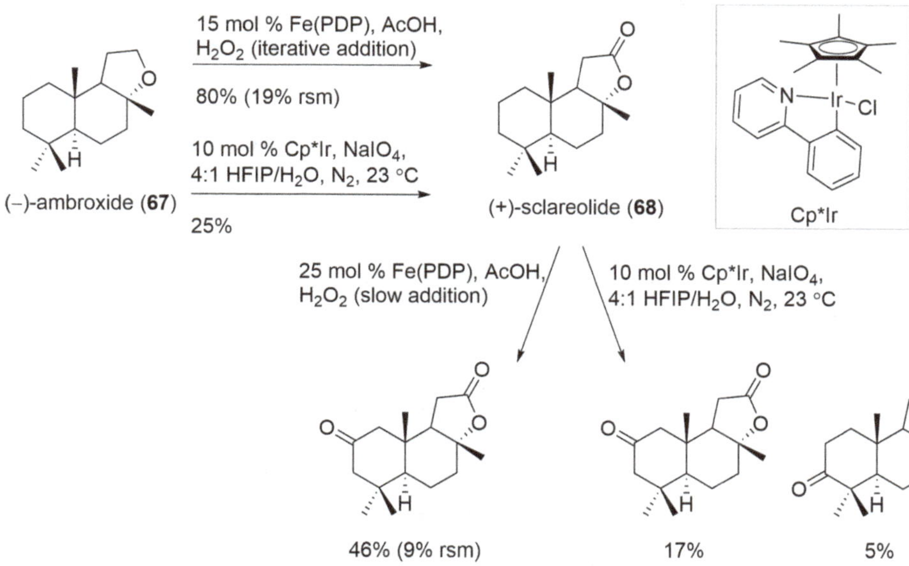

Scheme 30 Comparison of the product profile from the White and Crabtree catalysts using (–)-ambroxide (**67**) and (+)-sclareolide (**68**) as model systems

oxidizes sp^3-hybridized carbons in drug molecules containing these common pharmacophoric elements [52]. The oxidation of imatinib **69**, thenalidine **70**, and levofloxacin **71** indicates the chemoselective nature of this process, with oxidation occurring either at benzylic positions or next to tertiary amines, positions which are commonly CYP oxidation sites (Scheme 31). The authors of this study demonstrate that this functionalization process is viable on scale by subjecting 1 g of imatinib to the oxidative conditions. The four products **72**, **73**, **74**, and **75** were isolated in a combined yield of 39%, thus providing sufficient material for further evaluation of the compounds in vivo. As indicated by the clean conversion of thenalidine to **76** and **77**, electron-rich heterocycles such as thiophenes are tolerated under these conditions. Levofloxacin was oxidized to the corresponding aldehyde **78** under the reaction conditions.

The development of late-stage C–H functionalization methods to introduce fluorine is of importance to drug discovery due to their potential applications to PET (positron emission tomography) imaging. Several isotopes are suitable for PET imaging, e.g., ^{11}C, ^{13}N, ^{15}O, and ^{64}Cu; however, ^{18}F is the most commonly used "radiotag" due to its relatively short half-life (110 min), which allows dosing and studying of ^{18}F-labelled drug molecules in humans without any concerns for residual radioactivity in either human subjects or patients. Historically, PET imaging with ^{18}F has been limited by chemistry, since the attractive short half-life of ^{18}F puts stringent requirements on the available time of the reaction that is being used to prepare the imaging agent as this needs to be

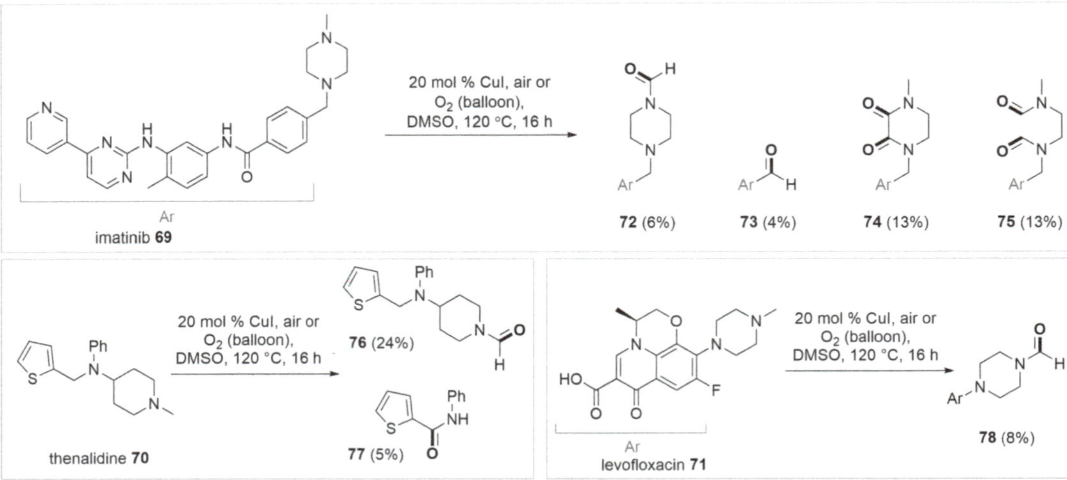

Scheme 31 Oxidation of imatinib, thenalidine, and levofloxacin with Touré copper-based catalyst

prepared very efficiently prior to each single use. Additional requirements is the feasibility of efficient product purification to ensure that the radiofluorinated analog can be isolated in pure form from a mixture of starting material, by-products, and reagents, as well as overall operational simplicity [53]. An example of a C–H fluorination reaction that fits these reaction criteria is that developed by Hooker and Groves which uses a Mn(salen)OTs complex as a fluorine transfer catalyst to radiolabel benzylic positions in a broad range of pharmaceuticals (Scheme 32). Using no-carrier-added [^{18}F]fluoride, labeling is achieved within a short reaction time of 10 min with sufficient radiochemical yields (20–72%, ten examples with pharmaceutical agents) being obtained. Importantly, a laborious dry-down prior to the reaction is not required, since the Mn(salen)OTs catalyst can elute [^{18}F]fluoride from an anion exchange cartridge with 90% recovery of the radiolabel. As shown in Scheme 32, this methodology enables late-stage ^{18}F radio labelling of TFA-rasagiline (**79**), nabumetone (**80**), celecoxib (**81**), and papaverine (**82**), thus indicating reaction compatibility with a range of heterocycles and functional groups including amides, ketones, and alkynes. Since no pre-functionalization is needed, these novel C–H functionalization methods are changing the landscape of molecules viable for PET imaging studies and will likely increase the throughput of evaluating a range of PET ligands [54].

4 Conclusions and Outlook

Direct functionalization strategies have offered the unique opportunity to take advantage of the ubiquity of C–H bonds *en route* to the preparation of small and, in some cases, complex molecules. Although only a small subset of developed reactions was described

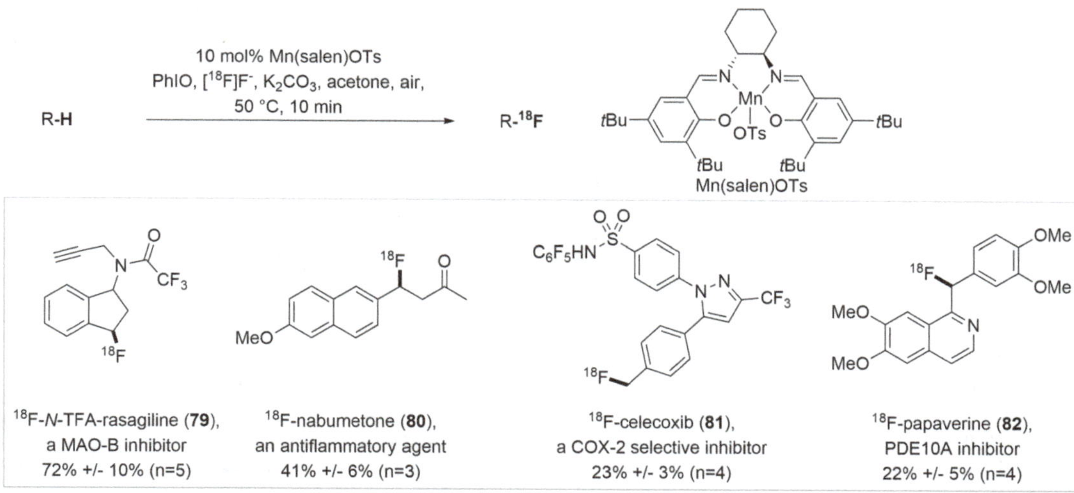

Scheme 32 Groves and Hooker's late-stage fluorination for PET ligand design

herein, the examples highlighted showcase how C–H functionalization reactions can be utilized to accelerate exploration of structure-activity relationships and are therefore of high value for medicinal chemistry applications. C–H functionalization reactions are often more efficient than more traditional transformations, typically utilize catalysis, and employ less toxic reagents, thus offering a "greener" solution to the construction of bioactive compounds.

As seen in the examples presented in this overview, the utilization of a directed C–H functionalization strategy can be particularly powerful when the directing group is part of the molecules intrinsic pharmacophore, or it exploits a functional group that also acts as a handle for further synthetic manipulation. The potential to diversify at a late stage, in order to generate a set of structurally related analogues, enables rapid evaluation of structure-activity relationships. In combination with improved purification techniques, incomplete regioselectivity can be extremely advantageous for the generation of several close-in analogs from a single reaction. The greater hurdle for wider uptake of C–H functionalization approaches comes from the highly specific nature of many published methods to particular substrates. In many cases, translation of reaction conditions to an "analogous" substrate (e.g., an alternative heterocycle or bioisostere) with subtle steric or electronic differences can require extensive re-optimization to enable a related C–H manipulation. Similarly, while C–H functionalization of late-stage intermediates can generate diverse and valuable SAR, efficient re-synthesis and scale-up, where higher yielding reactions and greater selectivity are desired, can be difficult to achieve in a short timescale. Encouragingly, an ever advancing understanding of mechanism, reactivity, and substrate scope offers significant potential to improve this.

To summarize, the recent rapid advancement of direct C–H functionalization technologies means it is becoming increasingly possible to propose a "direct" and topologically obvious disconnection toward specific targets, as well as selectively functionalize or diversify late-stage intermediates to rapidly probe SAR and pharmacophores. There are also an ever increasing number of cutting-edge methods reported that have not yet been demonstrated in complex scenarios that will no doubt find utility in the future.

References

1. Seregin IV, Gevorgyan V (2007) Direct transition metal-catalyzed functionalization of heteroaromatic compounds. Chem Soc Rev 36:1173–1193
2. Alberico D, Scott ME, Lautens M (2007) Aryl-aryl bond formation by transition-metal-catalyzed direct arylation. Chem Rev 107:174–238
3. Lyons TW, Sanford MS (2010) Palladium-catalyzed ligand-directed C-H functionalization reactions. Chem Rev 110:1147–1169
4. Mousseau JJ, Charette AB (2013) Direct functionalization processes: a journey from palladium to copper to iron to nickel to metal-free coupling reactions. Acc Chem Res 46:412–424
5. Cernak T, Dykstra KD, Tyagarajan S, Vachal P, Krska SW (2016) The medicinal chemist's toolbox for late stage functionalization of drug-like molecules. Chem Soc Rev 45:546. https://doi.org/10.1039/C5CS00628G
6. Campeau L-C, Rousseaux S, Fagnou K (2005) A solution to the 2-pyridyl organometallic cross-coupling problem: regioselective catalytic direct arylation of pyridine N-oxides. J Am Chem Soc 127:18020–18021
7. Campeau L-C, Stuart DR, Leclerc J-P, Bertrand-Laperle M, Villemure E, Sun H-Y, Lasserre S, Guimond N, Lecavalier M, Fagnou K (2009) Palladium-catalyzed direct arylation of azine and azole N-oxides: reaction development, scope and applications in synthesis. J Am Chem Soc 131:3291–3306
8. Leclerc J-P, Fagnou K (2006) Palladium-catalyzed cross-coupling reactions of diazine *N*-oxides with aryl chlorides, bromides, and iodides. Angew Chem Int Ed 45:7781–7786
9. Campeau L-C, Bertrand-Laperle M, Leclerc J-P, Villemure E, Gorelsky S, Fagnou K (2008) C2, C5, and C4 azole N-oxide direct arylation including room-temperature reactions. J Am Chem Soc 130:3276–3277
10. Zhao H, Wang R, Chen P, Gregg BT, Hsai MM, Zhang W (2012) Palladium-catalyzed direct arylation of imidazolone N-oxides with aryl bromides and its application in the synthesis of GSK2137305. Org Lett 14:1872–1875
11. Mandal D, Yamaguchi AD, Yamaguchi J, Itami K (2011) Synthesis of dragmacidin D via direct C–H couplings. J Am Chem Soc 133:19660–19663
12. Shibahara F, Yamaguchi E, Murai T (2011) Direct arylation of simple azoles catalyzed by 1,10-phenanthroline containing palladium complexes: an investigation of C4 arylation of azoles and the synthesis of triarylated azoles by sequential arylation. J. Org. Chem. 76:2680–2693
13. Mahindra A, Bagra N, Jain R (2013) Palladium-catalyzed regioselective C-5 arylation of protected l-histidine: microwave-assisted C–H activation adjacent to donor arm. J. Org. Chem. 78:10954–10959
14. Martin T, Verrier C, Hoarau C, Marsais F (2008) Direct C-2 arylation of alkyl 4-thiazolecarboxylates: new insights in synthesis of heterocyclic core of thiopeptide antibiotics. Org Lett 10:2909–2912
15. Liu J, Xie Y, Zeng W, Lin D, Deng Y, Lu X (2015) Pd(II)-catalyzed pyridine N-oxides directed arylation of unactivated Csp3–H bonds. J. Org. Chem. 80:4618–4626
16. Affron DP, Davis OA, Bull JA (2014) Regio- and stereospecific synthesis of C-3 functionalized proline derivatives by palladium catalyzed directed C(sp^3)–H arylation. Org Lett 16:4956–4959
17. Chen G, Shigenari T, Jain P, Zhang Z, Jin Z, He J, Li S, Mapelli C, Miller MM, Poss MA, Scola PM, Yeung K-S, Yu J-Q (2015) Ligand-enabled β-C-H arylation of α-amino acids using a simple and practical auxiliary. J Am Chem Soc 137:3338–3351
18. Parella R, Babu A (2015) Regio- and stereoselective Pd-catalyzed direct arylation of unactivated sp^3 C(3)–H bonds of tetrahydrofuran and 1,4-benzodioxane systems. J. Org. Chem. 80:2339–2355

19. Yang W, Ye S, Fanning D, Coon T, Schmidt Y, Krenitsky P, Stamos D, Yu J-Q (2015) Orchestrated triple C[BOND]H activation reactions using two directing groups: rapid assembly of complex pyrazoles. Angew Chem Int Ed 54:2501–2504
20. Wang D-H, Engle KM, Shi B-F, Yu J-Q (2010) Ligand-enabled reactivity and selectivity in a synthetically versatile aryl C-H olefination. Science 327:315–319
21. Engle KM, Wang D-H, Yu J-Q (2010) Ligand-accelerated C–H activation reactions: evidence for a switch of mechanism. J Am Chem Soc 132:14137–14151
22. Huang C, Chattopadhyay B, Gevorgyan V (2011) Silanol: a traceless directing group for Pd-catalyzed o-alkenylation of phenols. J Am Chem Soc 133:12406–12409
23. Cong X, You J, Gao G, Lan J (2013) 2-Pyridylmethyl ether: a readily removable and efficient directing group for amino acid ligand accelerated *ortho*-C–H olefination of phenols. Chem Commun 49:662–664
24. Tang R-I, Li G, Yu J-Q (2014) Conformation-induced remote *meta*-C–H activation of amines. Nature 507:215–220
25. Wu J, Xiang S, Zeng J, Leow M, Liu X-W (2015) Practical route to 2-quinolinones via a Pd-catalyzed C-H bond activation/C-C bond formation/cyclization cascade reaction. Org Lett 17:222–225
26. Liu X-W, Shi J-L, Wei J-B, Yang C, Yan J-X, Peng K, Dai L, Li C-G, Wang B-Q, Shi Z-J (2015) Diversified syntheses of multifunctionalized thiazole derivatives via regioselective and programmed C-H activation. Chem Commun 51:4599–4602
27. Zhou J, Li B, Hu F, Shi B-F (2013) Rhodium (III)-catalyzed oxidative olefination of pyridines and quinolines: multigram-scale synthesis of naphthyridinones. Org Lett 15:3460–3463
28. Xiao B, Liu Z-J, Liu L, Fu Y (2013) Palladium-catalyzed C-H activation/cross-coupling of pyridine N-oxides with nonactivated secondary alkyl bromides. J Am Chem Soc 135:616–619
29. Xiao Q, Ling L, Ye F, Tan R, Tian L, Li Y, Wang J (2013) Copper-catalyzed direct ortho-alkylation of *N*-iminopyridinium ylides with *N*-tosylhydrazones. J. Org. Chem. 78:3879–3885
30. Saget T, Perez D, Cramer N (2013) Synthesis of functionalized spiroindolines via palladium-catalyzed methine C–H arylation. Org Lett 15:1354–1357
31. Wang H-L, Shang M, Sun S-Z, Zhou Z-L, Laforteza BN, Dai H-X, Yu J-Q (2015) Cu (II)-catalyzed coupling of aromatic C–H bonds with malonates. Org Lett 17:1228–1231
32. Wang X-C, Gong W, Fang L-Z, Zhu R-Y, Li S, Engle KM, Yu J-Q (2015) Ligand-enabled meta-C-H activation using a transient mediator. Nature 519:334–338
33. Catellani M, Ferioli L (1996) An improved synthesis of 1,4-*cis,exo*-hexa- or tetrahydro-methano- or -ethanobiphenylene derivatives catalyzed by palladium complexes. Synthesis:769–772
34. Hogg JA (1992) Steroids, the steroid community, and Upjohn in perspective: a profile of innovation. Steroids 57:593–616
35. Stepan AF (2015) Abstracts of papers, 249th ACS National Meeting & Exposition, Denver, CO, United States, March 22–26, 2015, PagesORGN-37, CODEN:69TQWW
36. Espino CG, Fiori KW, Kim M, Du Bois J (2004) Expanding the scope of C-H amination through catalyst design. J Am Chem Soc 126:15378–15379
37. Dai H-X, Stepan AF, Plummer MS, Zhang Y-H, Yu J-Q (2011) Divergent C-H functionalizations directed by sulfonamide pharmacophores: late-stage diversification as a tool for drug discovery. J Am Chem Soc 133:7222–7228
38. Dai H-X, Gang L, Guo X-G, Stepan AF, Yu J-Q (2013) Pd(II)-Catalyzed *ortho*- or *meta*-C–H olefination of phenol derivatives. J Am Chem Soc 135:7567–7571
39. Li W, Fan Z, Geng K, Xu Y, Zhang A (2015) Late-stage diversification of biologically active pyridazinones via a direct C–H functionalization strategy. Org Biomol Chem 13:539–548
40. Suree N, Yi SW, Thieu W, Marohn M, Damoiseaux R, Chan A, Jung ME, Clubb RT (2009) Discovery and structure-activity relationship analysis of Staphylococcus aureus sortase A inhibitors. Bioorg Med Chem 17:7174–7185
41. Roizen JL, Zalatan DN, Du Bois J (2013) Selective intermolecular amination of C-H bonds at tertiary carbon centers. Angew Chem Int Ed 52:11343–11346
42. Li J, Cisar JS, Zhou C-Y, Vera B, Williams H, Rodríguez AD, Cravatt BF, Romo D (2013) Simultaneous structure–activity studies and arming of natural products by C–H amination reveal cellular targets of eupalmerin acetate. Nat Chem 5:510–517
43. He J, Hamann LG, Davies HML, Beckwith REJ (2015) Late-stage C-H functionalization of complex alkaloids and drug molecules via intermolecular rhodium-carbenoid insertion. Nat Commun 6:5943–5952

44. Stepan AF, Mascitti V, Beaumont K, Kalgutkar AS (2013) Metabolism-guided drug design. Med Chem Commun 4:631–652
45. Hann MM (2011) Molecular obesity, potency and other addictions in drug discovery. Med Chem Commun 2:349–355
46. Adams CM, Hub C-W, Jeng AY, Karki R, Ksander G, LaSala D, Leung-Chu J, Liang G, Liu Q, Meredith E, Rao C, Rigel DF, Shi J, Smith S, Springer C, Zhang C (2010) The discovery of potent inhibitors of aldosterone synthase that exhibit selectivity over 11-β-hydroxylase. Bioorg Med Chem Lett 20:4324–4327
47. Cusack KP, Koolman HF, Lange UEW, Peltier HM, Piel I, Vasudevan A (2013) Emerging technologies for metabolite generation and structural diversification. Bioorg Med Chem Lett 23:5471–5483
48. Gormisky PE, White MC (2013) Catalyst-controlled aliphatic C–H oxidations with a predictive model for site-selectivity. J Am Chem Soc 135:14052–14055
49. Vermeulen NA, Chen MS, White MC (2009) The Fe(PDP)-catalyzed aliphatic C-H oxidation: a slow addition protocol. Tetrahedron 65:3078–3084
50. Chen MS, White MC (2007) A predictably selective aliphatic C-H oxidation reaction for complex molecule synthesis. Science 318:783–787
51. Zhou M, Hintermair U, Hashiguchi BG, Parent AR, Hashmi SM, Elimelech M, Periana RA, Brudvig GW, Crabtree RH (2013) Cp* iridium precatalysts for selective C-h oxidation with sodium periodate as the terminal oxidant. Organometallics 32:957–965
52. Genovino J, Lütz S, Sames D, Touré BB (2013) Complementation of biotransformations with chemical C–H oxidation: copper-catalyzed oxidation of tertiary amines in complex pharmaceuticals. J Am Chem Soc 135:12346–12352
53. Neumann CN, Ritter T (2015) Late-stage fluorination: fancy novelty or useful tool? Angew Chem Int Ed 54:3216–3221
54. Huang X, Liu W, Ren H, Neelamegam R, Hooker JM, Groves JT (2014) Late stage benzylic C-H fluorination with [^{18}F]fluoride for PET imaging. J Am Chem Soc 136:6842–6845

Chapter 10

C–H Activation with Photoredox Catalysis

Joel W. Beatty and Corey R. J. Stephenson

Abstract

In recent years, the chemical community has become increasingly interested in novel modes of reactivity for the production of biologically relevant compounds. While C–H bonds have traditionally been considered inert, methods for activation of C–H bonds with transition metals have become an effective means to selectively functionalize organic substrates based upon careful reaction design. Recently, the use of visible light photoredox catalysis has been shown to be effective in C–H activation in its own right, providing access to a complementary set of reactions to those exhibited by other metal complexes. Here, we summarize some of the key advancements in the use of photoredox catalysis for the C–H activation of organic substrates.

Key words Photoredox catalysis, C–H activation, Visible light, Amine oxidation, Photochemistry, Green chemistry, Solar energy, Electron transfer

1 Introduction

The selective modification of organic substrates is an ongoing pursuit in modern synthetic chemistry. Analyses of a myriad of molecular traits, such as pK_a, coordination sites, electrophilicity and nucleophilicity, electron density, and steric interactions have contributed to orthogonal strategies for complex molecule production. Visible light photocatalysis has been shown in recent years to be extremely effective in mediating radical organic transformations with high efficiency and selectivity [1]. While many investigations have focused on the utilization of polypyridyl metal complexes (Fig. 1) [2–17], a number of reports have detailed the competence of organic dyes as metal-free alternatives [18–21]. Chief among the advantages which these photocatalytic reactions present is a distinct reactivity profile when compared to polar chemical methods, as well as significant environmental benefits over traditional radical chemistry [22]. While most unactivated C–H bonds are viewed as rela-

Paul F. Richardson (ed.), *Green Chemistry in Drug Discovery: From Academia to Industry*, Methods in Pharmacology and Toxicology, https://doi.org/10.1007/978-1-0716-1579-9_10,

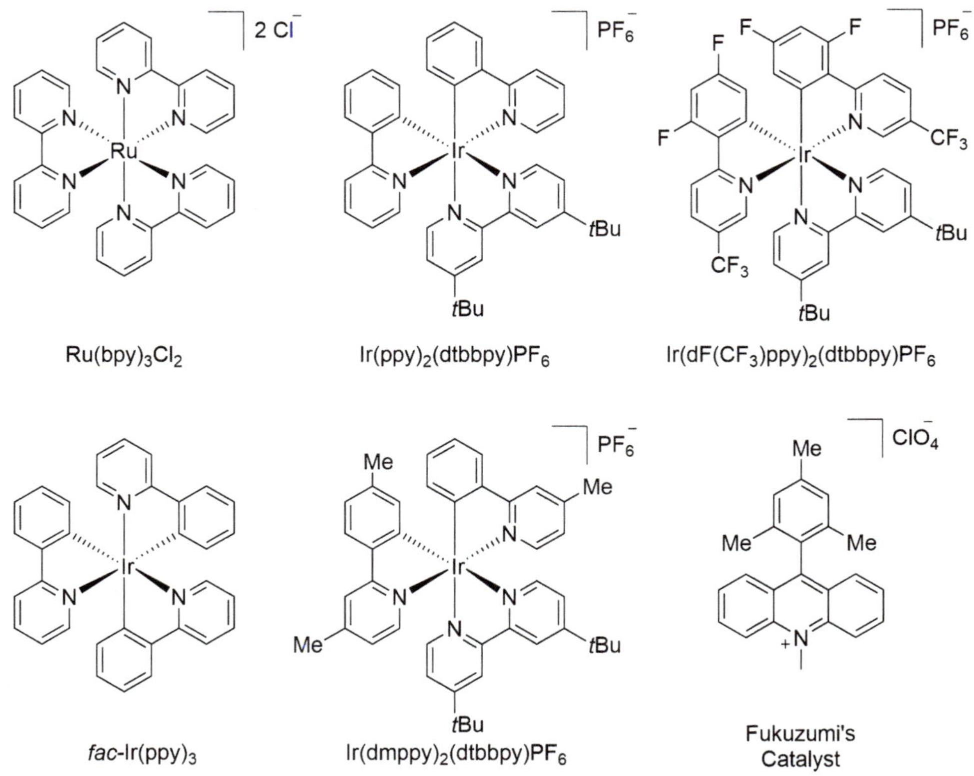

Fig. 1 Prototypical photoredox catalysts

tively inert to polar chemical methods due to their high pK_a, radical reactivity is guided by bond dissociation energies (BDE), which allows for an altered hierarchy of selectivity.

Typically, polypyridyl metal complexes such as $Ru(bpy)_3Cl_2$, Ir $(ppy)_2(dtbbpy)PF_6$, and $Ir(ppy)_3$, among others, are used to promote visible light-mediated redox reactions, but a number of examples have been published demonstrating the competence of metal-free catalysts in these reactions as well (Fig. 1). Upon irradiation with light, these visible light catalysts can absorb a photon to reach an electronically excited singlet state (Fig. 2). In this resulting excited state (denoted with an asterisk), the molecular orbital picture is such that the complex can act as either a reductant or an oxidant, depending upon the reaction conditions. As such, photoredox catalysis offers significant versatility in the design of reaction manifolds, and this versatility allows net reductive, redox neutral, or net oxidative reactivity.

There are two general mechanistic categories into which the following examples of radical sp^3 C–H activation fall [8, 23]. First, and most general, is the activation of a C–H bond through the single-electron oxidation of an adjacent functional group. Second, and less common, consists of direct C–H abstraction from the substrate of interest via a radical propagation mechanism. While

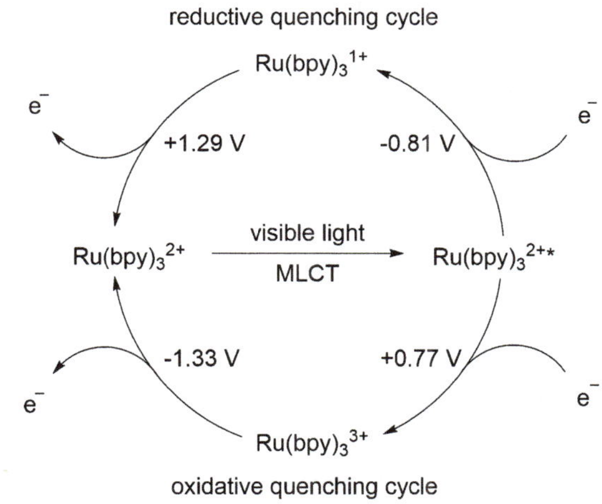

Fig. 2 The photocatalytic cycle of $Ru(bpy)_3^{2+a}$
[a]Potentials are given vs. the saturated calomel electrode (SCE). A full survey of photocatalyst redox potentials can be found in Reference [1]

there are numerous examples of C–H functionalization of sp^2-hybridized centers (e.g., radical-mediated functionalization of an arene), these mechanisms are distinct in that the key mechanistic step is not the C–H bond cleavage itself but a prior C–C bond formation. While the semantics of these distinctions could be discussed at length, examples included in this chapter will consist exclusively of sp^3-hybridized C–H functionalization.

2 α-Amine C–H Activation

The majority of examples of C_{sp}^3–H activation with visible light photoredox catalysis consist of α-amine oxidation, either under aerobic conditions or in the presence of a stoichiometric terminal oxidant. Single electron amine oxidation by the photocatalyst results in the formation of an *N*-centered radical cation (Fig. 3), which can undergo multiple reaction pathways depending upon the conditions employed. Deprotonation can occur easily, as the pK_a of the α-C–H bond drops precipitously upon amine oxidation [24–27], resulting in an α-amino radical, which can either be further oxidized to the corresponding iminium ion or be utilized in new bond-forming steps. The alternative pathway of C–H activation from the amine radical cation is direct C–H abstraction from the α-methylene. These pathways have been utilized to great effect in the literature and will each be discussed in depth.

Amines such as *N*,*N*-diisopropylethylamine and Hantzsch ester (Fig. 3) have been used as additives for several reductive and redox-neutral transformations [28–31]. In several cases the use of amines or tertiary amine salts has provided a synthetically useful source of

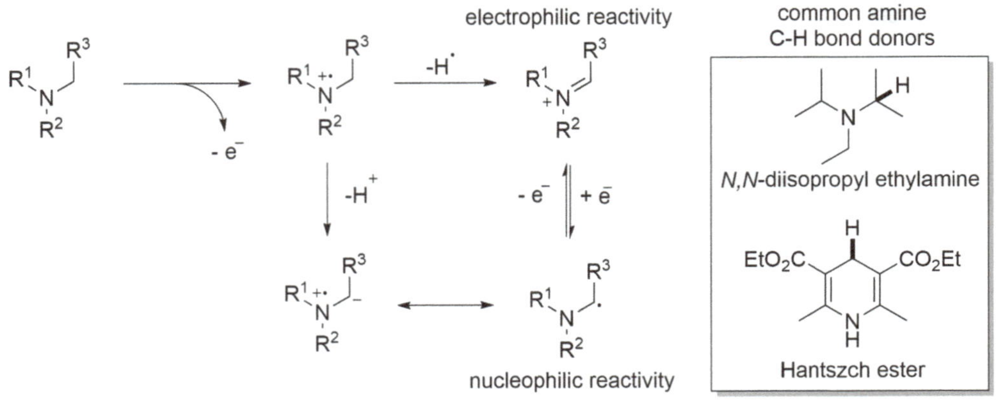

Fig. 3 Amine reactivity upon single-electron oxidation

hydrogen atoms for C–X bond reduction. The proposed formation of iminium by-products in early experimental efforts led to the investigation of iminium formation from a synthetically viable amine substrate [31, 32].

2.1 Iminium Formation Through Amine Oxidation

N-aryl tetrahydroisoquinolines have been shown to be excellent substrates for photochemical oxidation, as their functionalization is highly regioselective (Fig. 4). Early work in our lab demonstrated the ability of visible light photoredox catalysis to regioselectively oxidize the tetrahydroisoquinoline scaffold to the corresponding iminium ion, which is a competent electrophile for aza-Henry reactions [33]. It was found that a variety of tetrahydroisoquinolines are efficiently functionalized by both nitromethane and nitroethane under optimized aerobic conditions. It is worthy of note that *N*-phenyl pyrrolidine was inefficiently functionalized, proceeding to 40% conversion and 27% yield after 72 h.

The oxidative functionalization of **1** is postulated to occur in the following manner (Fig. 5a). Reductive quenching of the excited state of $Ir(ppy)_2(dtbbpy)PF_6$ by **1** provides the radical cation **2** and an [Ir^{2+}] catalytic intermediate, which is itself capable of reducing molecular oxygen. The superoxide thus generated performs a C–H abstraction from **2** to directly produce the iminium ion **3**, priming the tetrahydroisoquinoline for nucleophilic elaboration. A significant background reaction was observed for this transformation in the absence of oxygen (76% conversion, 20 h), and it was postulated that nitromethane may act as a terminal oxidant. Later studies revealed that this oxidation chemistry can be accelerated through careful choice of oxidant, although reactivity is dependent upon a variety of factors [34]. The use of three equivalents of diethyl bromomalonate provided high yields of Strecker product in the presence of NaCN with 1 mol% of $Ru(bpy)_3Cl_2$ in DMF (Fig. 5b). While this oxidant provided an excellent yield, its use is undesirable as the resultant malonyl radical forms by-products in

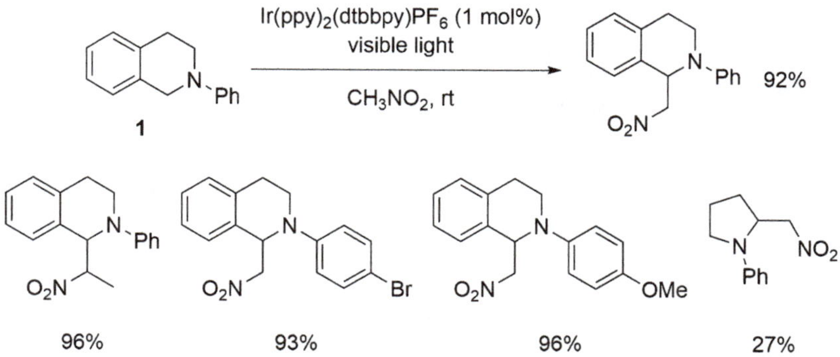

Fig. 4 Photochemical C–H activation of tetrahydroisoquinolines

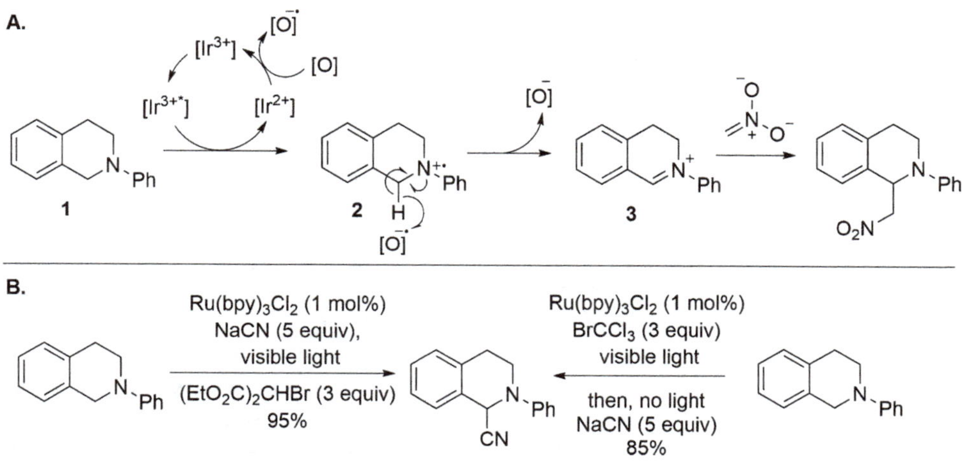

Fig. 5 Proposed mechanism for the photochemical oxidation of tetrahydroisoquinoline

the presence of more electron-rich alkenes and arenes [35]. The use of bromotrichloromethane ($BrCCl_3$) and $Ru(bpy)_3Cl_2$ decreased the reaction time to 3 h as compared to the previous time of 16 h when using molecular oxygen and $Ir(ppy)_2(dtbbpy)PF_6$.

There are many reports detailing the photochemical oxidation of tetrahydroisoquinoline derivatives using photoredox catalysis. Rueping and coworkers have demonstrated the efficiency of organocatalysis in the presence of the photochemical oxidation chemistry, using proline as a catalyst to promote enamine formation in the addition of ketones to the tetrahydroisoquinoline iminium ion [36]. Xia and coworkers have demonstrated the similar addition of silyl enol ethers [37]. Copper-acetylide additions to tetrahydroisoquinoline iminium ions have been shown to be efficient reactions, both in the presence of light [38] and in a one-pot process in which the light source is removed after the oxidation is complete [34]. The tetrahydroisoquinoline oxidation can be achieved with organic dyes as well [39–41]. Difluoroenolates [42] and dialkylphosphites [39, 43] are also competent nucleophiles for the addition reaction.

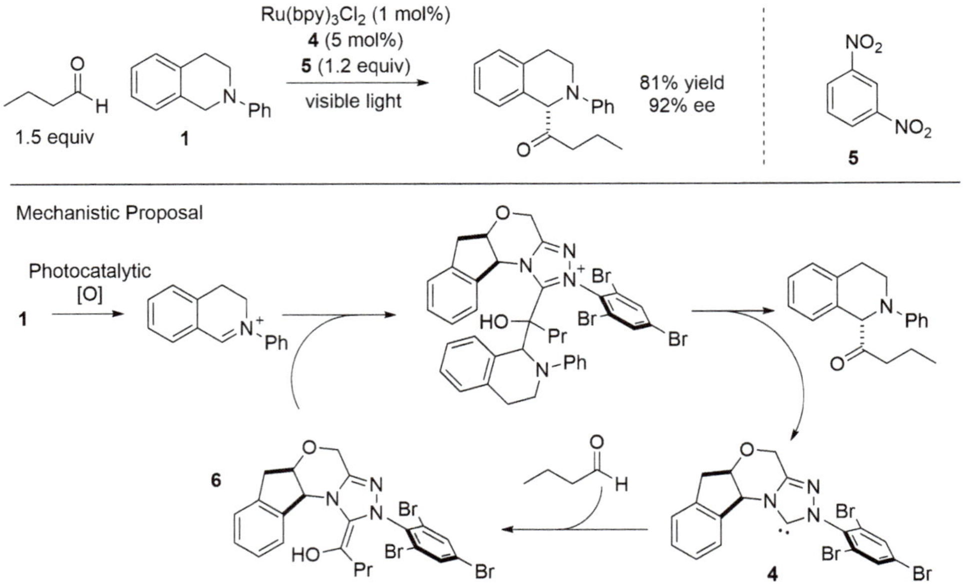

Fig. 6 The merger of photoredox and NHC catalysis

There are a number of asymmetric versions of nucleophilic additions in this context, with an auspicious first entry reported by Rovis and coworkers combining photoredox and *N*-heterocyclic carbene (NHC) catalysis (Fig. 6) [44]. This methodology includes the prototypical tetrahydroisoquinoline oxidation methodology to provide the endocyclic iminium ion. The use of NHC catalysis in this setting allows for the formation of a nucleophilic Breslow intermediate **6** which then adds to the iminium ion in an asymmetric fashion. This work exhibits good scope in terms of the aldehyde coupling partner as long as the steric bulk near the site of reaction is moderate, and overall the enantioselectivity is high. The tetrahydroisoquinoline must be electron rich to undergo iminium formation efficiently, as an electron-poor trifluoromethylated substrate did not form significant amounts of product. One promising aspect of this work is the use of 1,3-dinitrobenzene **5** as a terminal oxidant in the transformation. When **5** was excluded from the reaction conditions, only 13% yield was obtained as opposed to 81% when it is included. No products of dinitrobenzene reduction were observed in the reactions—a rare occasion in which an oxidative quencher remains innocent upon reduction.

A further example of asymmetric additions to the tetrahydroisoquinoline iminium ion was published by Stephenson and Jacobsen, in which they utilized chiral thioureas as anion-binding catalysts (Fig. 7) [45]. A major challenge for this work was the identification of a suitable solvent system, as photoredox catalysis is generally performed in polar media, which was found to be detrimental to the enantioinduction of the thiourea catalysts employed.

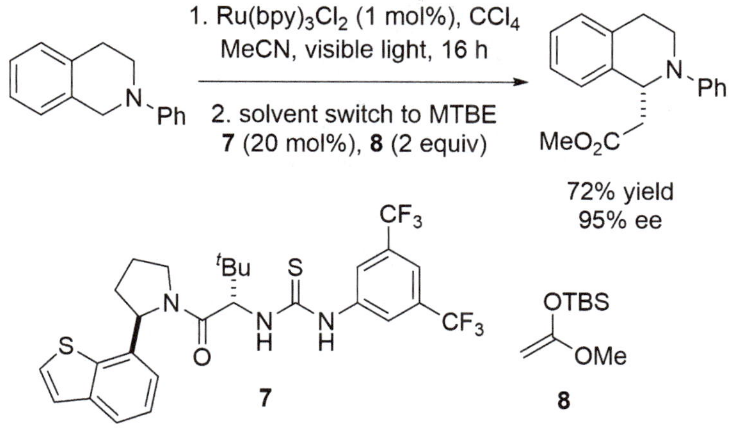

Fig. 7 The merger of photoredox and anion binding catalysis

The solution to this dilemma was to run the photooxidation in acetonitrile before concentrating the reaction mixture and redissolving in methyl *tert*-butyl ether for the nucleophilic addition step. A range of β-amino esters were prepared in this way with high enantiomeric excess and moderate yields.

Xiao has also reported the asymmetric addition of an enolate to the tetrahydroisoquinoline scaffold through nucleophilic catalysis, although only two asymmetric examples are reported [46]. The use of DABCO as a stoichiometric nucleophilic catalyst in the presence of acrolein promotes a Morita-Baylis-Hillman-type reaction in which a transiently generated zwitterion **9** adds to the iminium ion before subsequent elimination to provide the enone product (Fig. 8). The use of the chiral catalyst β-isocupreidine **10** in 20 mol % provides up to an 83:17 enantiomeric ratio of the acroleinated product in 82% yield, serving as a proof of principal for the use of the two catalytic systems in tandem.

In an example of further innovation for the tetrahydroisoquinoline substrate, Xiao and coworkers have also demonstrated that ester-functionalized tetrahydroisoquinoline **11** could undergo facile [3 + 2] cycloadditions with a number of electron-deficient alkenes and alkynes in a cascade process (Fig. 9) [47]. Formation of the endocyclic iminium ion **12** proceeds via an identical initiation pathway as addressed previously; however, the inclusion of the acidic α-amino ester moiety into the substrate provides a proton with a pK_a low enough to be readily removed by the peroxy anion. The so-formed 1,3-dipolar azomethine ylide can then undergo a number of facile [3 + 2] cycloadditions with both symmetrical and unsymmetrical dipolarophiles with complete regioselectivity. Early investigations yielded both pyrrole and pyrrolidine products, and so the authors optimized the procedure by adding 1.1 equivalents of *N*-bromosuccinimide (NBS) upon consumption of starting

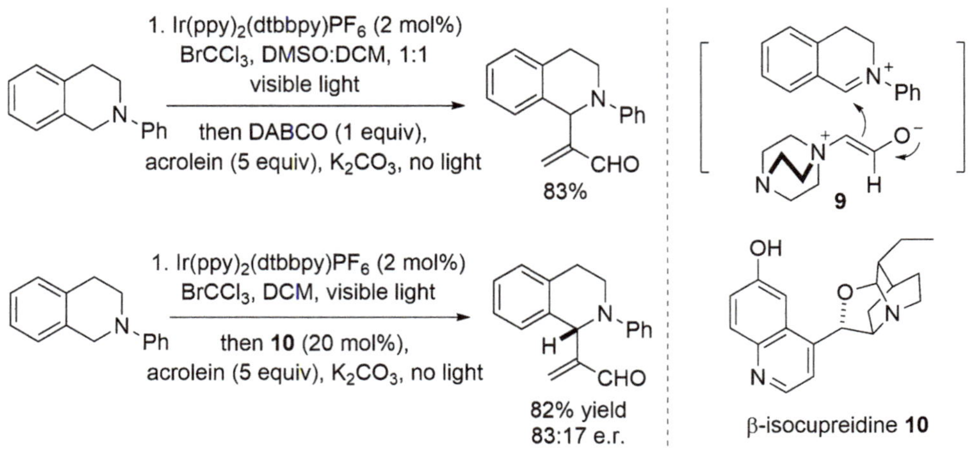

Fig. 8 Asymmetric nucleophilic catalysis for iminium ion addition

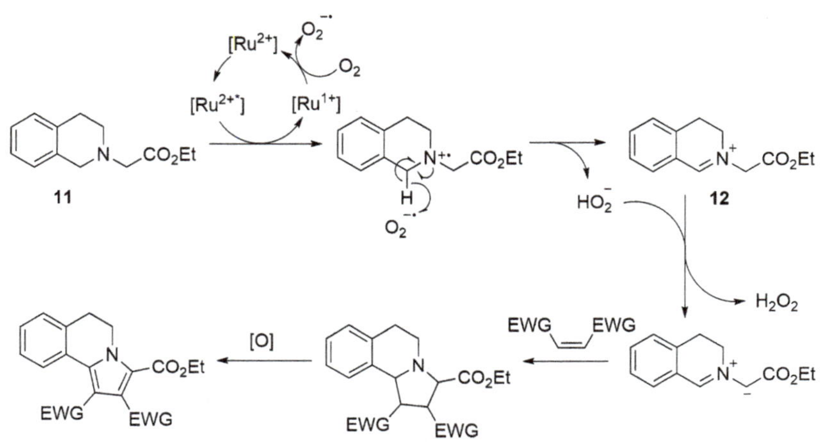

Fig. 9 [3 + 2] Cycloadditions with an isoquinoline azomethine ylide

material to funnel all material to the desired pyrrole. Rueping published a similar report shortly afterward focused on the production of the pyrrolidine products [48].

Another cascade process designed around iminium formation of the tetrahydroisoquinoline scaffold was reported by Zhu and coworkers (Fig. 10) [49]. When α-ketoesters such as **13** are used as nucleophiles in the aerobic photochemical oxidation of **1**, isoxazolidine esters of type **14** are isolated as single diastereomers from the reaction. While the yields are moderate, an impressive number of mechanistic steps are required in order to produce the observed rearrangement products. After nucleophilic addition of the enolate to the electrophile, a retro-aza conjugate addition results in the phenethylamine **15**. Further sequential aerobic oxidations result in the intermediacy of nitrone **16**, which then undergoes a diastereoselective [3 + 2] cycloaddition to afford the isoxazolidine product.

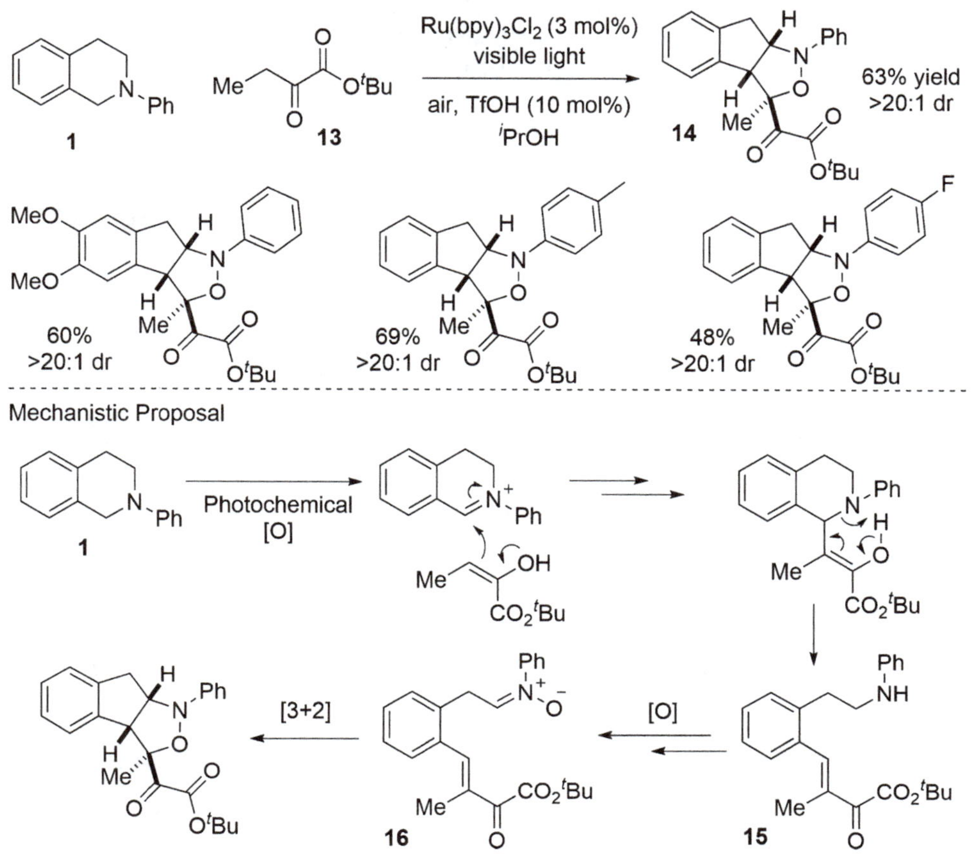

Fig. 10 Aerobic cascade reactivity upon C–H activation

The use of highly activated tetrahydroisoquinoline substrates has been ubiquitous for the development of α-amino C–H activation, but a number of examples of oxidative iminium formation in alternative systems have been reported as well. Xiao and coworkers have reported the diastereoselective synthesis of tetrahydroimidazoles using photoredox catalysis through amine oxidation to an iminium ion followed by trapping of a pendant amine to form the product cyclic aminal (Fig. 11) [50]. Diastereoselectivity in these reactions was achieved through thermodynamic equilibration of the diastereomers produced by increasing the reaction time. While the reported substrates are not tetrahydroisoquinoline derivatives, high regioselectivity in the iminium-forming step is achieved through the benzylic activation of the observed reaction site. The substrate does still contain the *N*-phenyl benzylamine motif found in many of the tetrahydroisoquinoline methodologies reported.

Recently, our group published the semi-synthesis of a number of alkaloids using photoredox catalysis for key bond forming and cleaving reactions (Fig. 12) [51]. In the synthesis of pseudovincadifformine **19**, photochemical amine oxidation of the intermediate

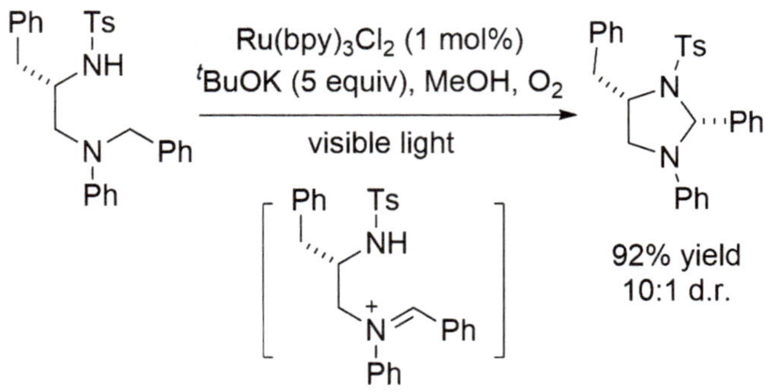

Fig. 11 Photocatalytic tetrahydroimidazole synthesis

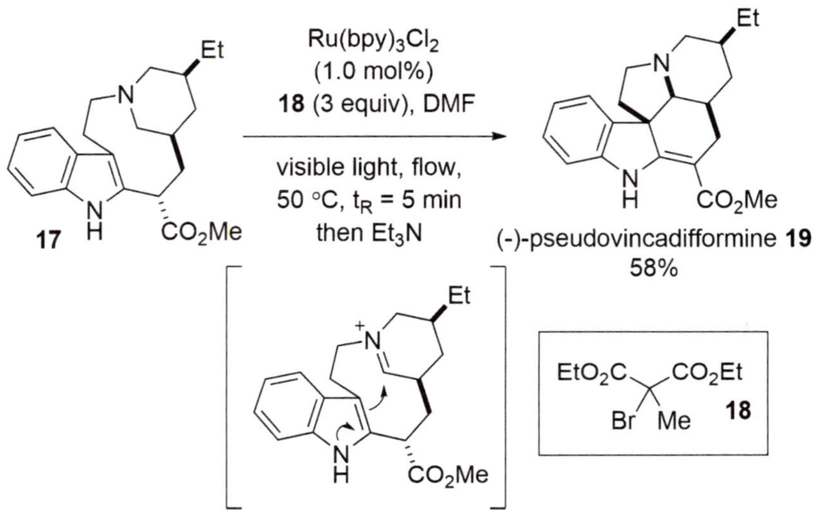

Fig. 12 C–H functionalization to form (−)-pseudovincadifformine

17 in flow to the corresponding iminium ion provided the natural product via a transannular Pictet-Spengler cyclization in a moderate yield of 58%. This represents one of the few examples of the selective oxidation of asymmetrically substituted amines reported to date. Effective examples of iminium formation from symmetrically dialkylated anilines have been reported by Rueping [43] and Tan [41], but the generality of substrate scope for amine oxidation remains a significant challenge for the field.

2.2 α-Amino Radical Formation and Reactivity

Beyond iminium formation, there are numerous examples of α-amine C–H activation which utilize the corresponding α-amino radical [52, 53]. The formation of α-amino radicals from the amine radical cation generally requires the α-deprotonation of an amine radical cation, or alternatively single-electron reduction of an iminium ion (Fig. 3). Provided further oxidation to the iminium ion does not occur, the radical can take part in reactions with electron-poor alkenes.

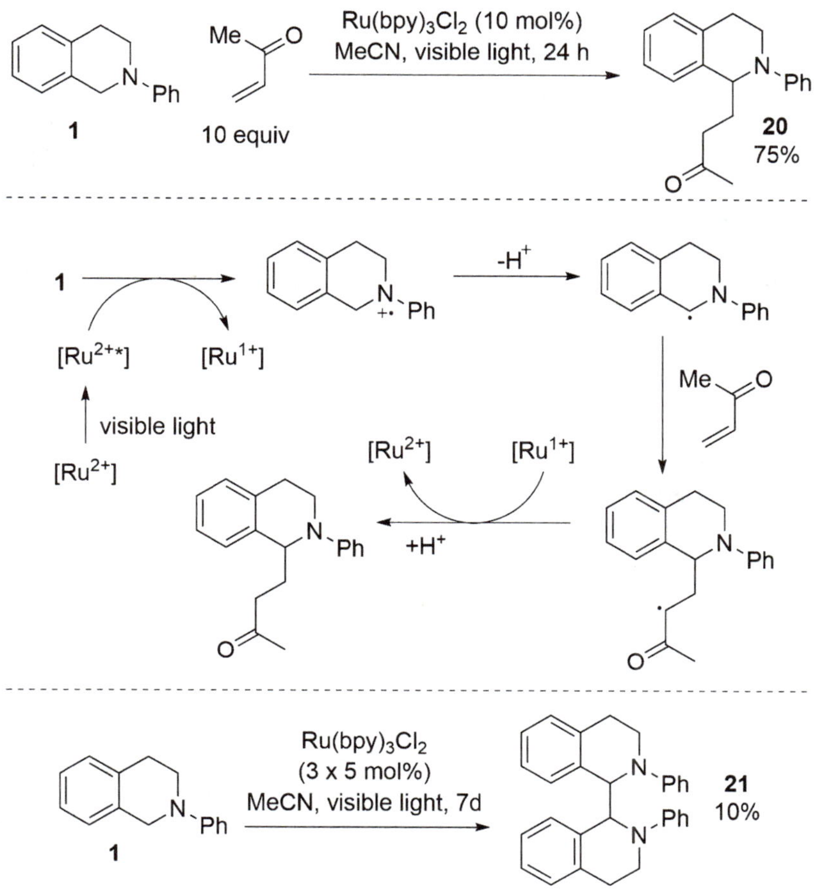

Fig. 13 Intermolecular coupling of a photochemically generated α-amino radical

Reiser and Pandey have successfully utilized the formation of benzylic α-amino radicals from tetrahydroisoquinolines for conjugate additions to a variety of enones (Fig. 13) [54]. Exposure of tetrahydroisoquinoline **1**, $Ru(bpy)_3Cl_2$ (10 mol%), and methyl vinyl ketone to visible light in acetonitrile for 24 h resulted in the formation of coupled product **20** in 75% yield. Catalyst efficacy was dependent on substrate, and in several cases $Ir(ppy)_2(dtbbpy)PF_6$ was found to give higher yields of product. While the formation of the observed product strongly suggests the intermediacy of an α-amino radical, further evidence of this species was obtained through the exclusion of the methyl vinyl ketone. After 7 days and multiple portions of catalyst addition, the dimer **21** was isolated in 10% yield. Interestingly, the authors note that substrate efficiency varied widely with electron density, revealing that a delicate balance of redox potentials may be required to access the α-amino radical reactivity.

A further study by Pandey and coworkers has demonstrated that this reactivity can be achieved at a location distal from the amine in biased systems such as 10-methyl-9,10-dihydroacridine

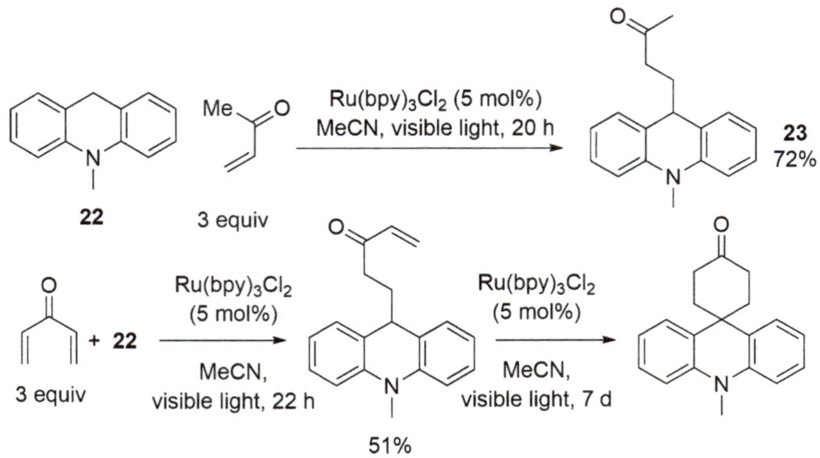

Fig. 14 Alkylation of 9,10-dihydroacridine

22 (Fig. 14) [55]. Merely irradiating the starting material in the presence of three equivalents of methyl vinyl ketone and the photocatalyst Ir(ppy)$_2$(dtbbpy)PF$_6$ (5 mol%) provided the alkylated product **23** after 20 h. Interestingly, while the majority of tetrahydroisoquinoline oxidations result in the mono-functionalization of the benzylic position, the authors were able to di-functionalize the benzylic position of the dihydroacridine through an intramolecular cyclization step. The slow nature of even this intramolecular reaction further cements the previous observations of product stability under various tetrahydroisoquinoline oxidation conditions.

N,N-dimethylaniline has also been demonstrated to be a competent precursor to α-amino radicals for conjugate addition [56]. Xu and Li have shown that moderate yields of conjugate addition products such as **24** may be formed despite some oligomerization of the methacrylate coupling partner (Fig. 15a). The use of Morita-Baylis-Hillman adducts such as **25** also provided a range of elimination products. Works by Bian and Yu [57], Curran and Rueping [58], and Pandey [55] have demonstrated the ability of dimethylanilines to undergo cyclization reactions with electron-poor alkenes. For example, Bian and Yu have shown that *N*-phenyl maleimide and dimethylaniline react under oxidative photocatalysis to provide the annulated product **27**, presumably via cyclization of the proposed radical **28** onto the aniline ring (Fig. 15B) [57]. Nishibayashi and coworkers have obtained excellent yields in the addition of α-amino radicals to ethylidene malonates (Fig. 15C) [59]. Mechanistically, these reactions rely upon the addition of the photocatalytically generated α-amino radical to the unsaturated system, forming the α-carbonyl radical. Subsequent reduction and protonation provides the desired linear product. Curran and Rueping have shown that the product distribution in these cases can be controlled through the exclusion of

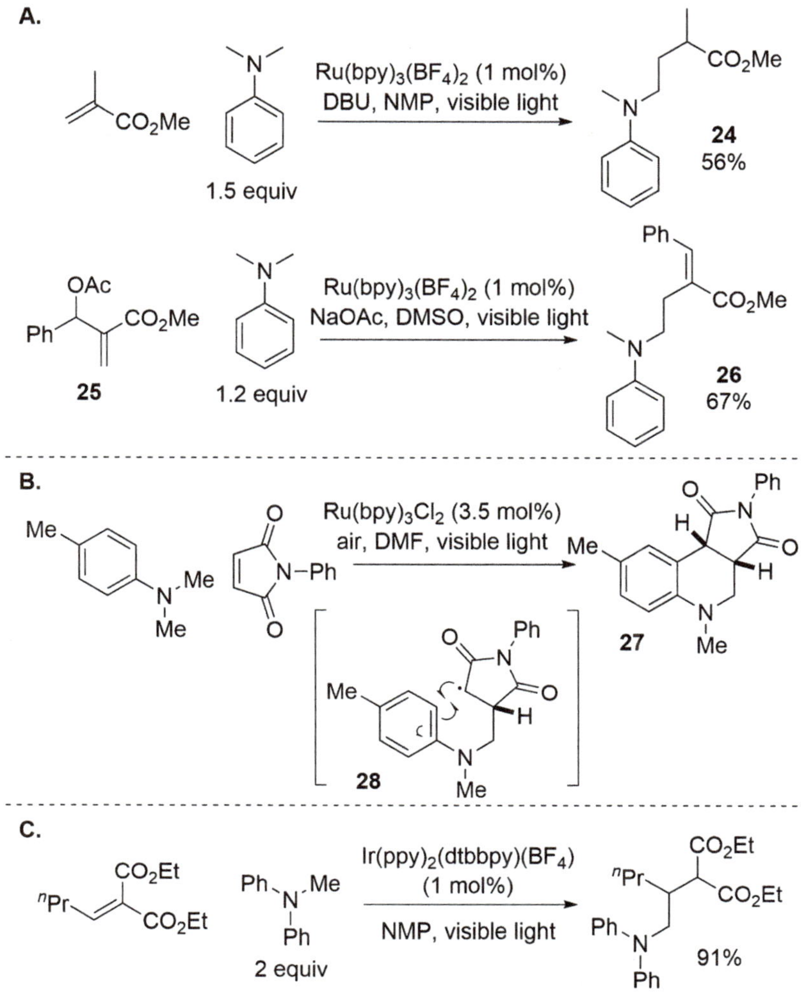

Fig. 15 C–H activation and coupling of *N*-methyl anilines

oxygen from the reaction—in the presence of oxygen, cyclization is observed, but under anaerobic conditions the linear addition products dominate [58].

There are also reports of intramolecular cyclizations of photochemically generated α-amino radicals to pendant unsaturated systems. Curran and Rueping have shown that dibenzylaniline **29** can undergo a C–H activation/cyclization cascade to provide indole-3-carboxaldehydes in moderate yields (Fig. 16) [58]. This transformation involves the oxidative cleavage of a C–C bond, which presumably results from the combination of oxygen with the benzylic radical **30** resulting from benzylic C–H activation of the intermediate indole. Control experiments using tetraphenylporphyrin and Rose Bengal did not produce observable reactivity, which suggests that singlet oxygen is not an active intermediate.

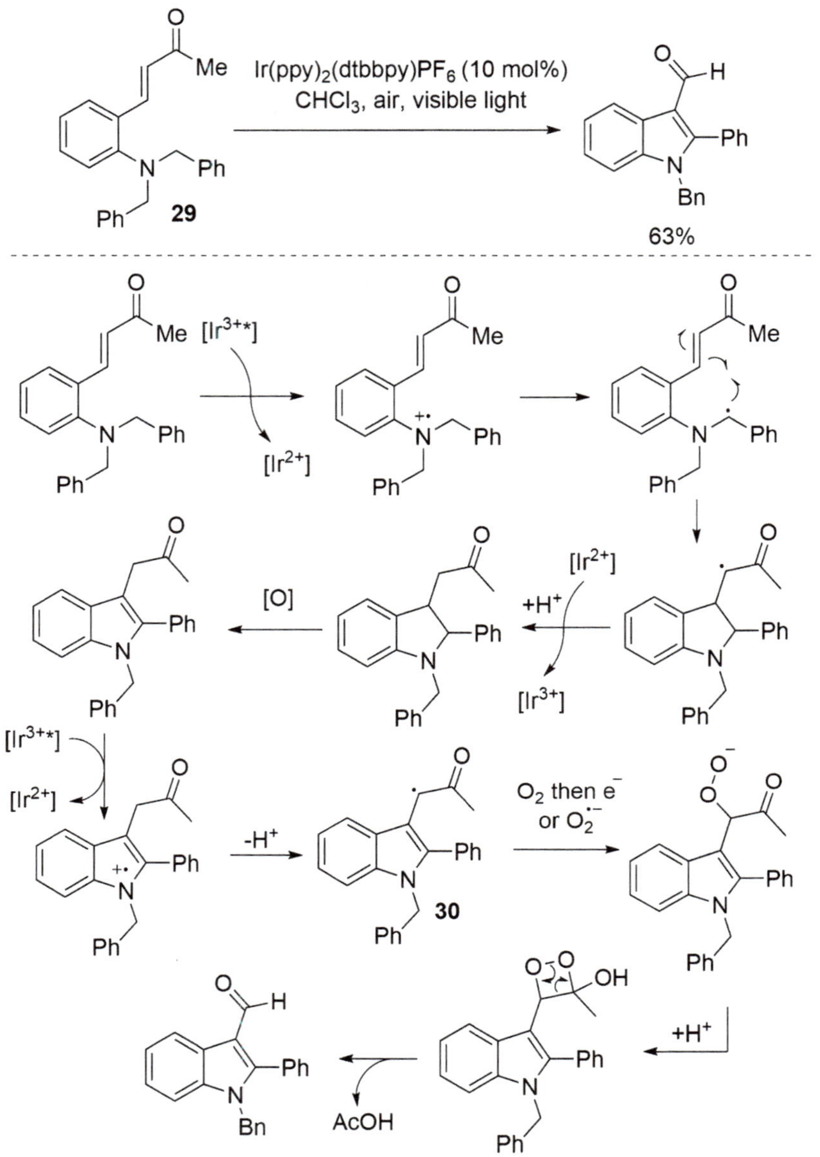

Fig. 16 C–H activation/cyclization onto a pendant enone

A similar cyclization reaction is possible in the presence of a pendant alkyne (Fig. 17) [60]. Xhou and coworkers have shown that dialkylanilines are also competent in this reaction, providing similar yields to the *bis*-benzylic substrates. Radical cyclization onto the alkyne produces an sp^2-hybridized radical, which then combines with oxygen or superoxide to eventually provide the intermediate **31**. Rearomatization and cleavage of the O-O bond results in indole products with carbonyl substituents at the three position. Both cyclization reports used ^{18}O-labeled water to discover no ^{18}O incorporation into the product carbonyl, suggesting that molecular oxygen is the source of benzylic oxygenation.

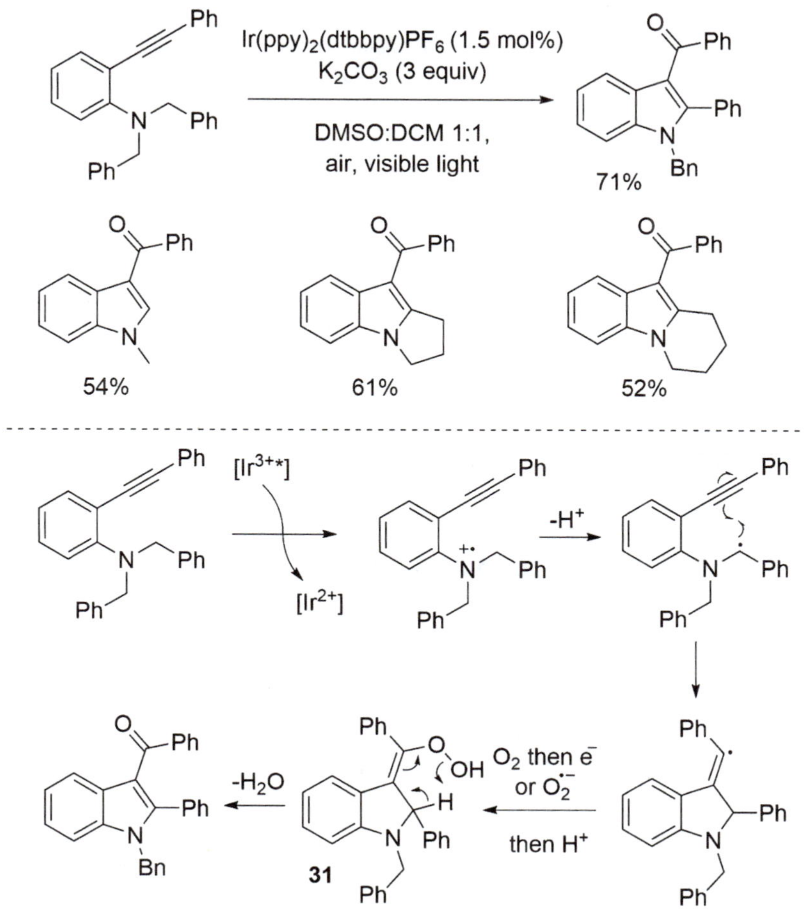

Fig. 17 C–H activation/cyclization onto a pendant alkyne

An interesting sp^3 C–H amination reaction was reported by Nishibayashi and coworkers wherein a tetrahydroquinoline substrate is functionalized at the 2-position (Fig. 18) [61]. Photochemical activation of the tetrahydroquinoline substrate in the presence of di-*tert*-butyl azodicarboxylate **32** results in the formation of aminated products of type **33**. Again, acyclic examples such as **34** resulted in lower reaction efficiencies.

One of the more significant and effective examples of α-amino radical formation with photoredox catalysis is the coupling of dialkylanilines with 1,4-dicyanobenzene developed by MacMillan and coworkers (Fig. 19) [62]. Discovered through high-throughput screening of reaction conditions, this α-arylation reaction is capable of producing coupled products in up to 98% yield with high regioselectivity. A variety of cyanoarenes are capable of participating in the reaction, which is postulated to proceed *via* oxidative quenching of Ir(ppy)$_3$ by the arene. The oxidized [Ir^{4+}] species then performs a single-electron oxidation of the aniline substrate, which is postulated to undergo deprotonation by sodium acetate to provide the

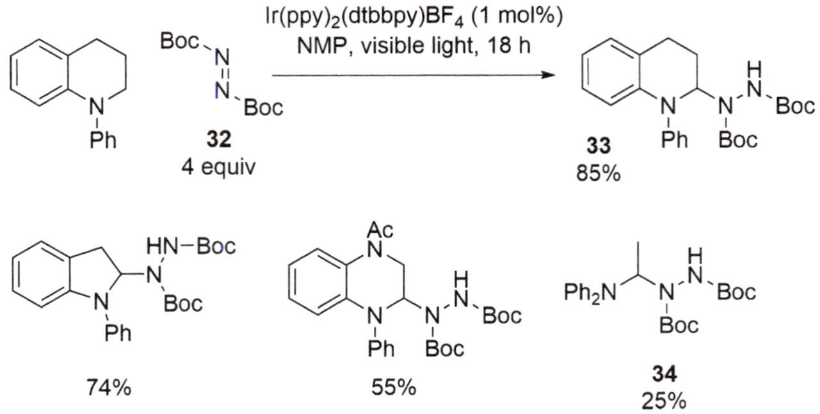

Fig. 18 Radical amination of tetrahydroquinolines

α-amino radical **34**. Subsequent radical-radical coupling then produces the anion **35** which rapidly rearomatizes to form the coupled product. This method has particular relevance to pharmaceutical synthesis, as is exhibited by the derivatization of the antibiotic Zyvox to **36** (Fig. 19). The authors point out that 8 of the top 100 selling pharmaceuticals of 2008 contain the α-aryl amine motif. This coupling methodology has been extended to a number of chloroheteroarenes [63].

MacMillan and coworkers have expanded their utilization of α-amino radicals to include the α-vinylation of *N*-aryl amines [64]. Optimized conditions consist of exposure of a mixture of *N*-phenylpyrrolidine, vinyl sulfone **37**, cesium acetate, and Ir(dF(CF_3)ppy$)_2$(dtbbpy)PF_6 to visible light irradiation for 24 h, providing the allylic amine **38** in 91% yield after 24 h (Fig. 20). Mechanistically, this transformation is hypothesized to proceed through the photocatalyst's reductive quenching cycle. Upon single-electron oxidation of the amine, the resultant radical cation can be deprotonated. The α-amino radical is then postulated to add to the electron-deficient vinyl sulfone, and subsequent β-scission of the resultant radical intermediate would expel the sulfonyl radical. The [Ir^{2+}] catalytic intermediate is presumably oxidized by this radical to regenerate the [Ir^{3+}] species. In reference to the amine substrate scope, the methodology can accommodate differences in ring size and alkyl chain length, as well as electronic alterations to the aromatic ring. Perhaps most fascinating, however, is the complete selectivity in the formation of product **39** for the endocyclic methylene over the exocyclic benzylic group. Only examples of styrenylsulfones are examined as the alkene coupling partner.

In a further co-catalytic advancement for the field, MacMillan and Doyle have reported the combination of photoredox and nickel catalysis for C–H activation cross-coupling reactions (Fig. 21) [65]. Due to the ease with which organic radicals add to Ni

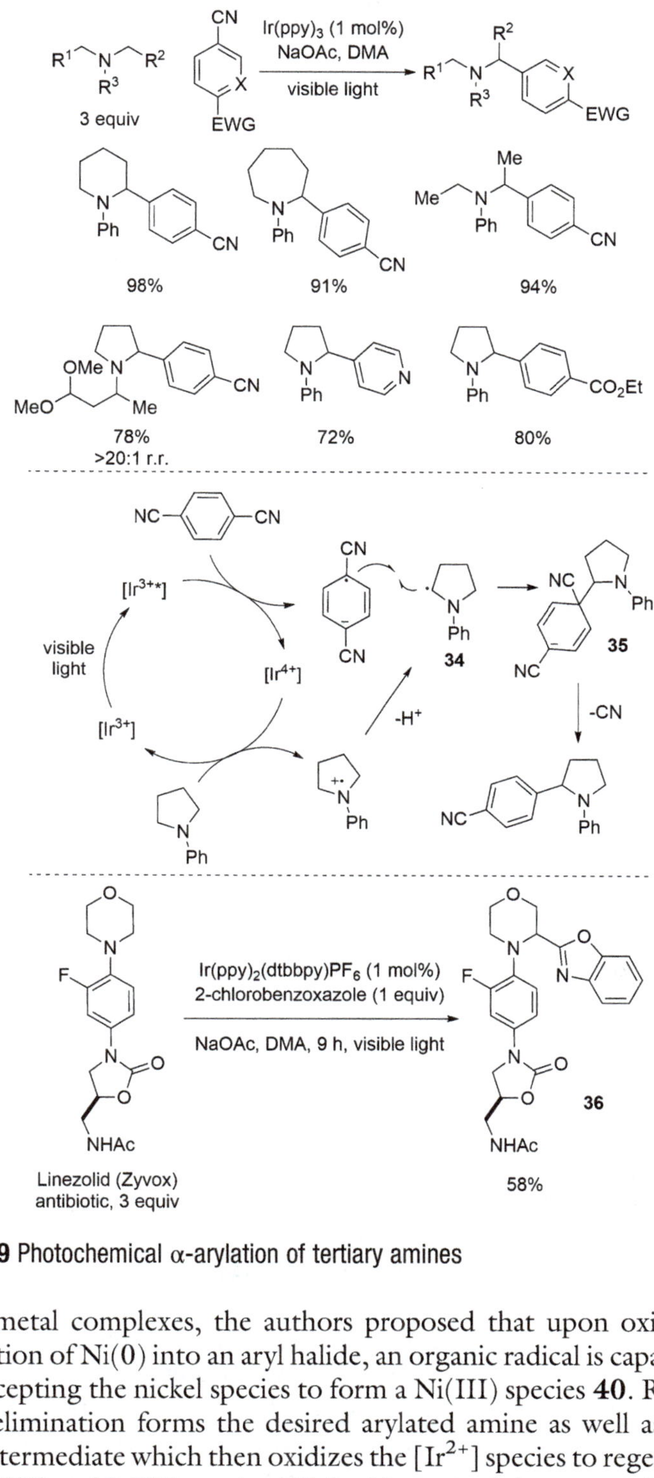

Fig. 19 Photochemical α-arylation of tertiary amines

(II) metal complexes, the authors proposed that upon oxidative insertion of Ni(0) into an aryl halide, an organic radical is capable of intercepting the nickel species to form a Ni(III) species **40**. Reductive elimination forms the desired arylated amine as well as a Ni (I) intermediate which then oxidizes the [Ir^{2+}] species to regenerate the Ni(0) and Ir(III) species. While this methodology was primarily used with *N*-Boc amino acids, dimethylaniline was shown to undergo efficient coupling reactions with both electron-rich and electron-poor arenes. Significantly, while the previous α-arylation

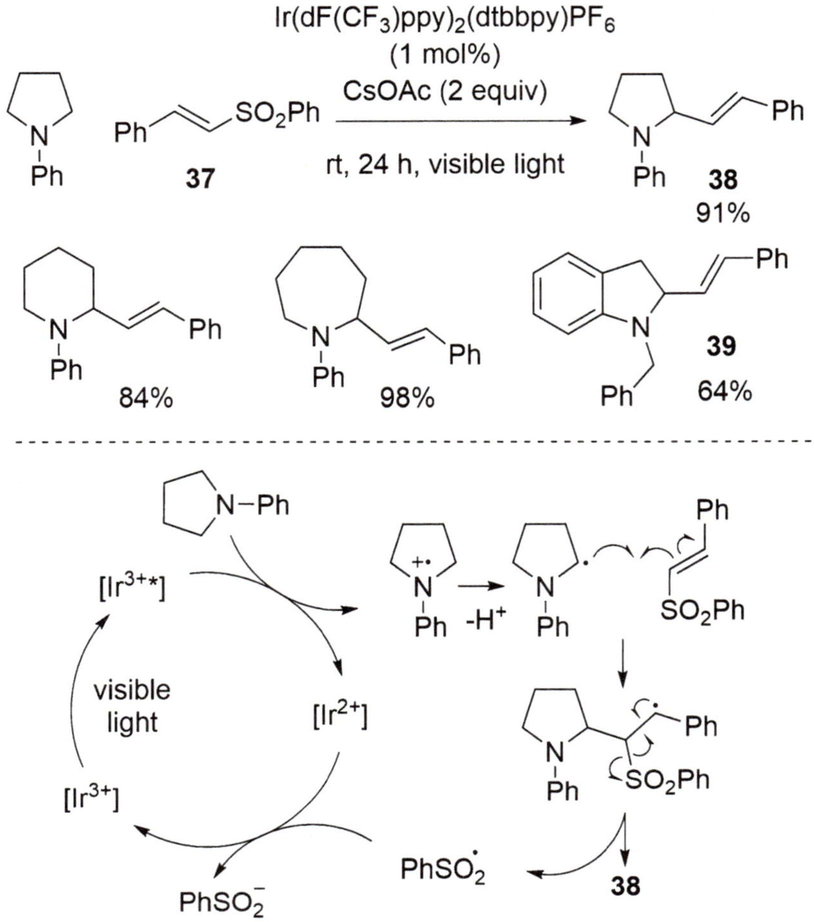

Fig. 20 Photochemical α-vinylation of tertiary amines

methodology reported by MacMillan was electronically limited to electron-poor arenes, the use of the co-catalytic nickel significantly expands the electronic scope of the products that can be formed.

3 Beta-Functionalization of Ketones and Aldehydes

In a significant conceptual advancement, MacMillan and coworkers have reported the direct β-C–H functionalization of aldehydes and ketones [66]. Similar in theory to the idea that single-electron amine oxidation results in a significant drop in the pK_a of the α-amino C–H bonds, MacMillan and coworkers proposed that the pK_a of the C–H bond β to an aldehyde or ketone could be lowered through condensation of an amine organocatalyst and subsequent oxidation of the resulting enamine (Fig. 22). The use of 1,4-dicyanobenzene as an oxidative quencher for the selected Ir (ppy)$_3$ catalyst was hypothesized to be crucial, as the electronic

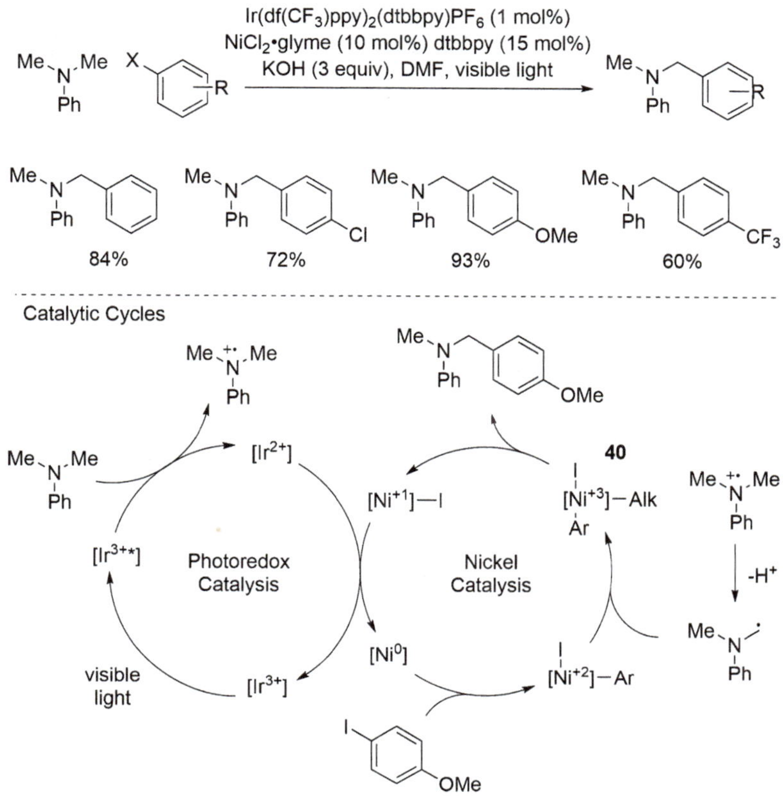

Fig. 21 Photoredox and nickel co-catalysis

mismatch between the radical anion **42** and the enamine intermediate **43** was expected to prevent the formation of α-arylated carbonyl products. This orthogonal electronic configuration allows the enamine to be oxidized by the transient [Ir^{4+}] species produced by oxidative quenching. Upon enamine oxidation and deprotonation, the now electrophilic enaminyl radical can combine with the radical anion **42** to provide a coupled intermediate, whereupon cyanide elimination and enamine hydrolysis provides the β-arylated aldehyde or ketone. The scope of the aldehyde coupling partner is good and is facilitated by the presence of electron-donating groups at the β-position. Significantly, the use of 3-(*p*-cyanophenyl)-propionaldehyde **41** as a starting material resulted in no reaction and thus provided a reasonable explanation as to why only mono-arylated products were observed. A variety of cyano-substituted electron-deficient arenes were found to be effective coupling partners for the reaction, and the use of cinchona alkaloid **44** as the organocatalyst demonstrated the potential for chiral induction in the transformation.

In addition to the seminal β-functionalization work, further effort has been reported by MacMillan in the development of a β-coupling reaction of cyclic ketones with aryl ketones (Fig. 23) [67]. This reactivity proceeds through a similar catalytic cycle,

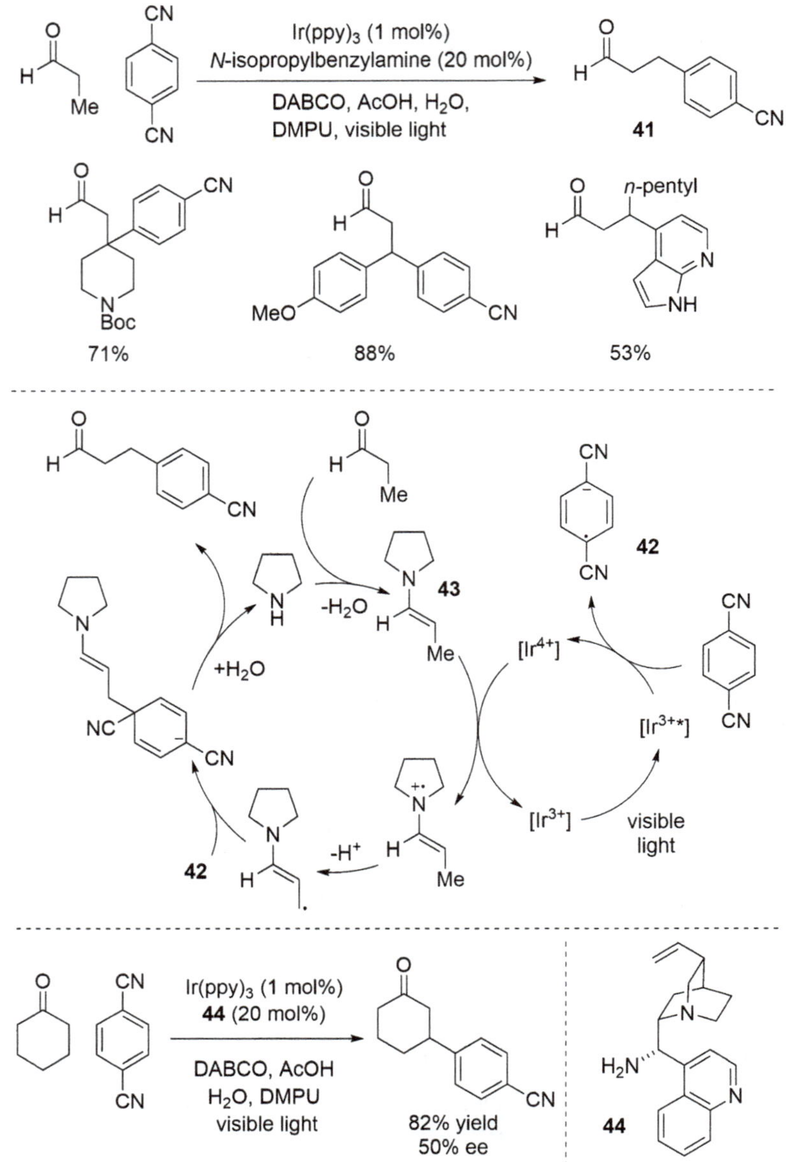

Fig. 22 β-arylation of aldehydes and ketones

where the use of $Ir(ppy)_3$ allows for effective use of the oxidative quenching cycle. Single-electron reduction of the aryl ketone to the ketyl radical is facilitated by Lewis acid activation, and subsequent radical-radical combination with the β-enaminyl $5\pi e^-$ species forms the key carbon-carbon bond.

Further work in the β-functionalization of aldehydes resulted in the discovery of an intermolecular β-coupling of aldehydes with enones (Fig. 24) [68]. This redox-neutral reaction is postulated to proceed through the reductive quenching cycle of the methylated iridium catalyst $Ir(dmppy)_2(dtbbpy)PF_6$ via a single-electron

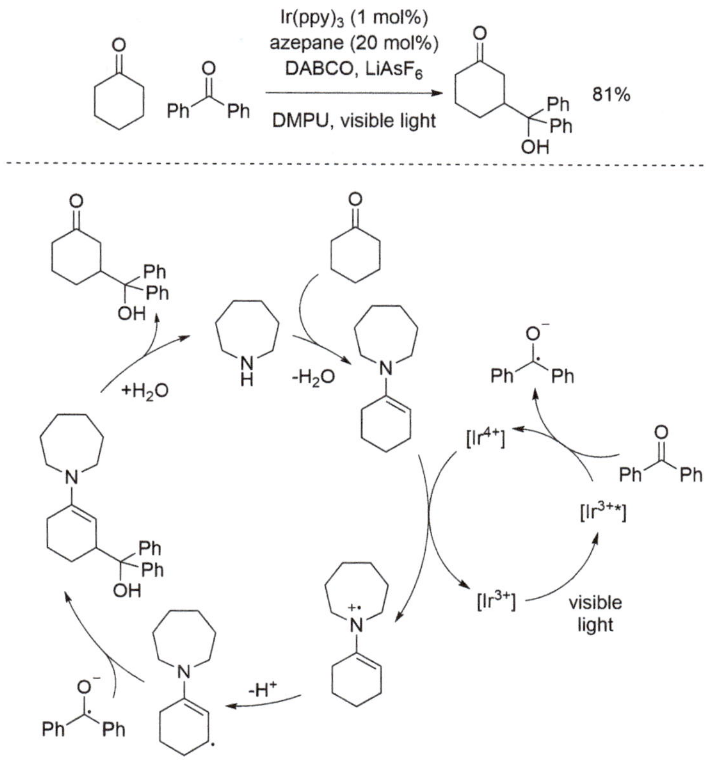

Fig. 23 β-coupling of cyclic ketones with ketyl radicals

oxidation of the enamine intermediate **45**. β-deprotonation of the enamine provides the 5πe$^-$ species **46**, which is electronically matched to add in a conjugate fashion to an enone. Catalyst turnover is accomplished through the reduction of the resultant radical. Organocatalyst selection was found to be crucial, as nucleophilic addition to the acrylate coupling partner was observed as a significant side reaction. The use of bulky dicyclohexylamine satisfactorily solved this issue and allowed for a wide array of products to be synthesized in moderate to good yields.

4 Oxidation of Functional Groups Other than Amines

One of the earliest examples of visible light photocatalysis by Cano-Yelo and Deronzier succeeded in the activation of benzylic C–H bonds of benzyl alcohols [69]. As opposed to the previous examples of C–H activation which are primarily limited to amine oxidation, the authors showed that benzylic C–H bonds can be activated through single-electron arene oxidation (Fig. 25). Oxidative quenching of the catalyst-excited state by the aryl diazonium salt **46** results in the formation of a [Ru^{3+}] species as well as aryl radical **47**. The [Ru^{3+}] intermediate is capable of arene oxidation, which

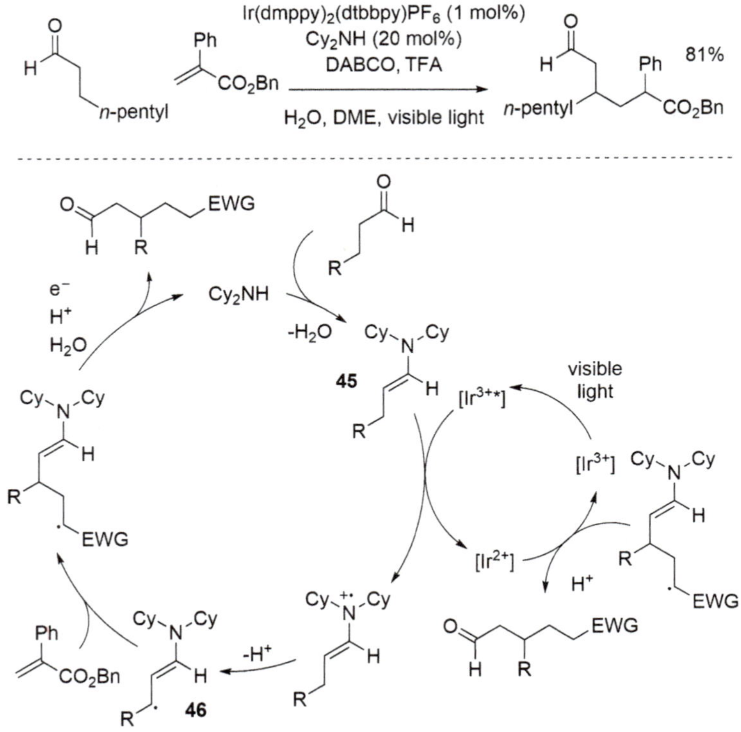

Fig. 24 Photocatalytic β-alkylation of aldehydes

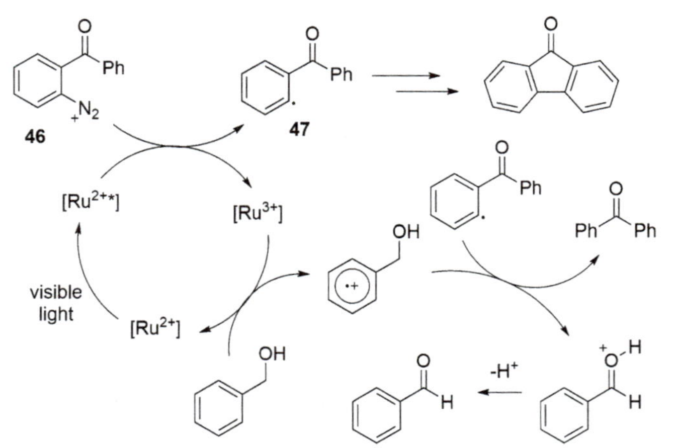

Fig. 25 Oxidation of benzyl alcohols with photoredox catalysis

results in the weakening of the benzylic C–H bonds. Abstraction of a hydrogen atom results in the formation of the desired benzaldehyde, although the yields vary greatly depending upon the substrate used.

In a mechanistic extension of this work, photoredox catalysis can be used to efficiently cleave *p*-methoxy-benzyl (PMB) ethers

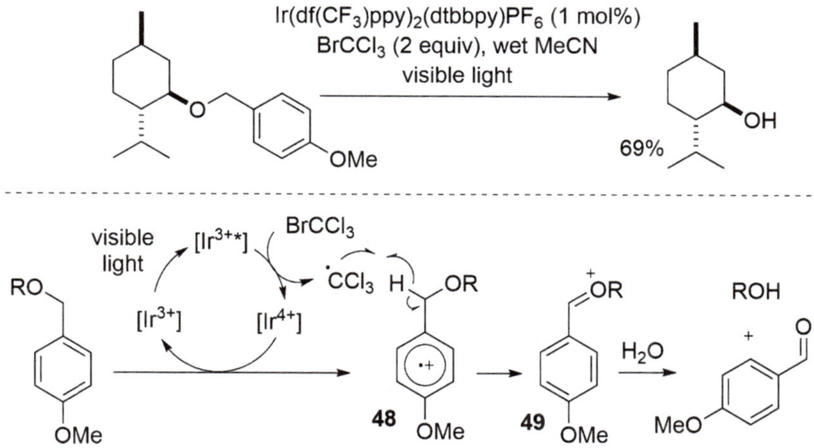

Fig. 26 Photochemical deprotection of PMB-ethers

through the oxidative C–H activation of the benzylic position [70]. Using $BrCCl_3$ as terminal oxidant and oxidative quencher, the $[Ir^{4+}]$ oxidation state of $Ir[dF(CF_3)ppy]_2(dtbbpy)PF_6$ can be accessed (E° $[Ir^{4+}/Ir^{3+}] = 1.69$ V) [71]. This iridium IV species is capable of oxidizing electron-rich arenes (Fig. 26) [72]. Single-electron oxidation of the PMB arene results in radical cation **48**, which then is presumed to disproportionate with the CCl_3 radical to provide an intermediate oxocarbenium ion **49** along with chloroform as a stoichiometric by-product. Hydrolysis of the oxocarbenium ion results in the free alcohol product in good to excellent yields.

A recent development in this chemistry was published by Laha and coworkers, who found that benzylic C–H bonds could be selectively functionalized in the *para* position of a variety of anisoles (Fig. 27a) [73]. Using dicyanonaphthalene as the photocatalyst, the authors were able to intramolecularly trap intermediate *p*-quinone methides with pendant alcohols. The regioselectivity of the reaction is displayed in Fig. 27b and is rationalized both through the estimated pK_a difference between the benzylic positions and the calculated energies of each radical. When the pendant alcohol was replaced by an alkene, the corresponding Prins cyclization was unsuccessful; however, the authors discovered that further oxidation of some of the cyclization products was possible and hypothesized that water could be used as an oxygen source for the direct formation of aryl alkyl ketones from the aryl alkane starting materials (Fig. 27c). While aryl acylation can be achieved through a number of other methods, the benzylic oxidation reaction is mild and chemoselective while avoiding air and moisture sensitive reagents.

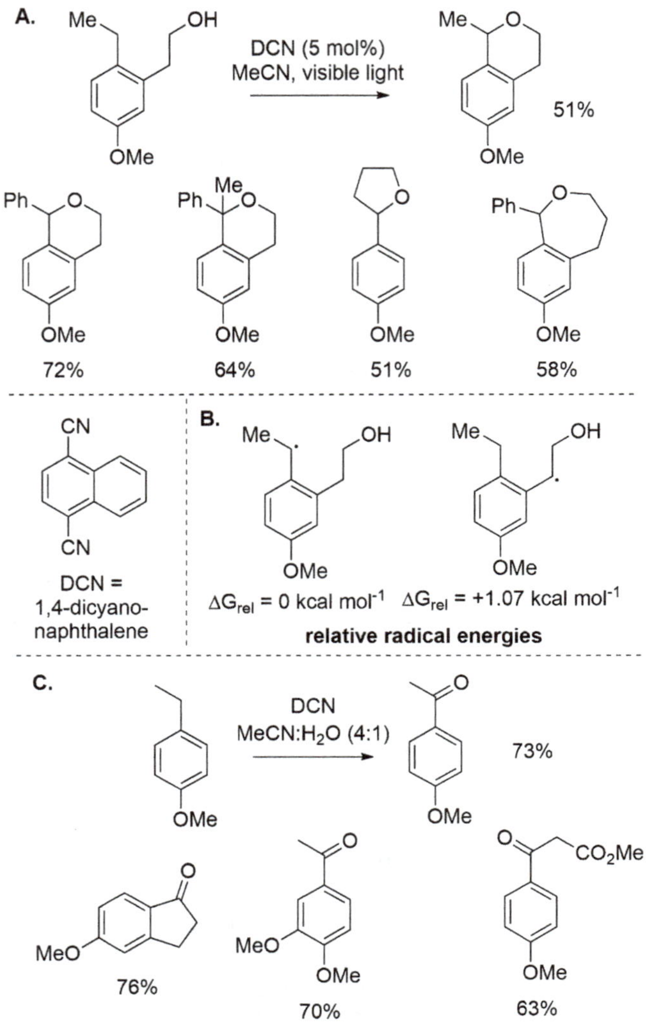

Fig. 27 Photochemical benzylic oxidations

5 Direct C–H Abstraction

While the majority of C–H activation reactions discussed in this chapter involve substrate oxidation which precedes C–H abstraction or deprotonation, there are a few examples of direct C–H abstraction reactions which rely upon the inherent reactivity of the substrate in question. This type of reactivity is most useful in cases where the substrate is not easily oxidized, as with the C–H functionalization of C–H bonds α to amide nitrogens. From DMF, a methyl hydrogen atom abstraction can be achieved by a sulfate radical anion, generated by reduction or thermolysis of ammonium persulfate **50** (Fig. 28) [74]. The α-amino radical **51** so formed is easily oxidized, and *N*-acyliminium ions can be easily formed in this manner.

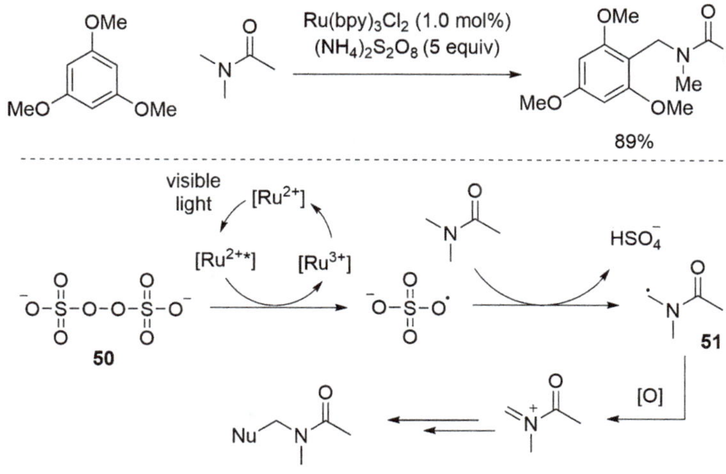

Fig. 28 Photochemical Friedel-Crafts amidoalkylation

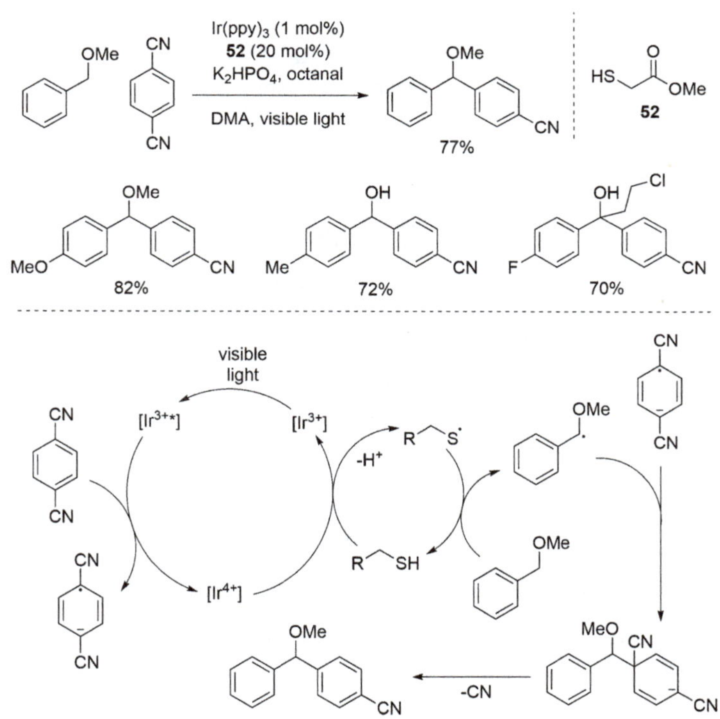

Fig. 29 Photochemical benzylic C–H abstraction

MacMillan and coworkers have also developed an effective C–H abstraction protocol for the functionalization of ethereal benzylic C–H bonds (Fig. 29) [75]. Using 1,4-dicyanobenzene as oxidative quencher for $Ir(ppy)_3$ as well as coupling partner, the $[Ir^{4+}]$ intermediate is formed and is postulated to promote thiyl radical formation via proton-coupled electron transfer (PCET). It is proposed that the thiyl radical thus formed is capable of benzylic C–

H bond abstraction due to favorable comparison between the thiol S-H BDE (86.8–87.2 kcal/mol) [76] and the benzylic C–H bond of benzyl methyl ether (85.8 kcal/mol) [77]. The benzyl radical **53** can then combine with the radical anion to form the desired diaryl alkyl ethers.

6 Conclusion

In conclusion, the ability of visible light photocatalysis to activate C–H bonds is an important field of research with many reports of recent progress. The ability to harness the organic radicals that result from C–H activation is highly dependent upon the conditions used, and there is much possibility for further developments. The ability to perform these catalytic reactions in an environmentally benign manner utilizing solar energy is highly desirable, as it minimizes the potential for the formation of stoichiometric waste by-products. The utility of photoredox catalysis in the modification of pharmaceutically relevant molecules is clear, and it is exciting to witness this technology bloom into a viable option for synthesis.

References

1. Prier CK, Rankic DA, MacMillan DWC (2013) Visible light photoredox catalysis with transition metal complexes: applications in organic synthesis. Chem Rev 113:5322–5363
2. Zeitler K (2009) Photoredox catalysis with visible light. Angew Chem Int Ed 48:9785–9789
3. Teply F (2011) Photoredox catalysis by [Ru(bpy)$_3$]$^{2+}$ to trigger transformations of organic molecules. Organic synthesis using visible-light photocatalysis and its 20th century roots. Collect Czechoslov Chem Commun 76:859–917
4. Narayanam JMR, Stephenson CRJ (2011) Visible light photoredox catalysis: applications in organic synthesis. Chem Soc Rev 40:102–113
5. Xuan J, Xiao W-J (2012) Visible-light photoredox catalysis. Angew Chem Int Ed 51:6828–6838
6. Wallentin C-J, Nguyen JD, Stephenson CRJ (2012) Radical carbon-carbon bond formations enabled by visible light active photocatalysis. Chimia 66:394–398
7. Tucker JW, Stephenson CRJ (2012) Shining light on photoredox catalysis: theory and synthetic applications. J Org Chem 77:1617–1622
8. Hu J, Wang J, Nguyen TH, Zheng N (2013) The chemistry of amine radical cations produced by visible light photoredox catalysis. Beilstein J Org Chem 9:1977–2001
9. Koike T, Akita M (2013) Visible-light-induced photoredox catalysis: an easy access to green radical chemistry. Synlett 19:2492–2505
10. Zou Y-Q, Chen J-R, Xiao W-J (2013) Homogeneous visible-light photoredox catalysis. Angew Chem Int Ed 52:11701–11703
11. Reckenthäler M, Griesbeck AG (2013) Photoredox catalysis for organic synthesis. Adv Synth Catal 355:2727–2744
12. Xuan J, Lu L-Q, Chen J-R, Xiao W-J (2013) Visible-light-driven photoredox catalysis in the construction of carbocyclic and heterocyclic ring systems. Eur J Org Chem 30:6755–6770
13. Xi Y, Yi H, Lei A (2013) Synthetic applications of photoredox catalysis with visible light. Org Biomol Chem 11:2387–2403
14. Hopkinson MN, Sahoo B, Li J-L, Glorius F (2014) Dual catalysis sees the light: combining photoredox with organo-, acid-, and transition-metal catalysis. Chem Eur J 20:3874–3886
15. Koike T, Akita M (2014) Trifluoromethylation by visible-light-driven photoredox catalysis. Top Catal 57:967–974
16. Garlets ZJ, Nguyen JD, Stephenson CRJ (2014) The development of visible-light-driven photoredox catalysis in flow. Isr J Chem 54:351–360

17. Douglas JJ, Nguyen JD, Cole KP, Stephenson CRJ (2014) Enabling novel photoredox reactivity via photocatalyst selection. Aldrichimica Acta 47:15–25
18. Fukuzumi S, Ohkubo K (2013) Selective photocatalytic reactions with organic photocatalysis. Chem Sci 4:561–574
19. Fukuzumi S, Ohkubo K (2014) Organic synthetic transformations using organic dyes as photoredox catalysts. Org Biomol Chem 12:6059–6071
20. Hari DP, König B (2014) Synthetic applications of eosin Y in photoredox catalysis. Chem Commun 50:6688–6699
21. Nicewicz DA, Nguyen TM (2014) Recent applications of organic dyes as photoredox catalysts in organic synthesis. ACS Catal 4:355–360
22. Yoon TP, Ischay MA, Du J (2010) Visible light photocatalysis as a greener approach to photochemical synthesis. Nature Chem 2:527–532
23. Xie J, Jin H, Xu P, Zhu C (2014) When C–H bond functionalization meets visible-light photoredox catalysis. Tetrahedron Lett 55:36–48
24. Nelson SF, Ippoliti JT (1986) The deprotonation of trialkylamine cation radicals by amines. J Am Chem Soc 108:4879–4881
25. Lewis FD (1986) Proton-transfer reactions of photogenerated radical ion pairs. Acc Chem Res 19:401–405
26. Parker VD, Tilset M (1991) Facile proton-transfer reactions of *N,N*-dimethylamine cation radicals. J Am Chem Soc 113:8778–8781
27. Dombrowski GW, Dinnocenzo JP, Zielinski PA, Farid S, Wosinska ZM, Gould IR (2005) Efficient unimolecular deprotonation of aniline radical cations. J Org Chem 70:3791–3800
28. Nguyen JD, D'Amato EM, Narayanam JMR, Stephenson CRJ (2012) Engaging unactivated alkyl, alkenyl and aryl iodides in visible-light-mediated free radical reactions. Nature Chem 4:854–859
29. Nguyen JD, Reiss B, Dai C, Stephenson CRJ (2013) Batch to flow deoxygenation using visible light photoredox catalysis. Chem Commun 49:4352–4354
30. Schnermann MJ, Overman LE (2012) A concise synthesis of (−)-aplyviolene facilitated by a strategic tertiary radical conjugate addition. Angew Chem Int Ed 51:9576–9580
31. Narayanam JMR, Tucker JW, Stephenson CRJ (2009) Electron-transfer photoredox catalysis: development of a tin-free reductive dehalogenation reaction. J Am Chem Soc 131:8756–8757
32. Tucker JW, Narayanam JMR, Krabbe SW, Stephenson CRJ (2010) Electron transfer photoredox catalysis: intramolecular radical addition to indoles and pyrroles. Org Lett 12:368–371
33. Condie AG, González-Gómez JC, Stephenson CRJ (2010) Visible-light photoredox catalysis: aza-henry reactions via C–H functionalization. J Am Chem Soc 132:1464–1465
34. Freeman DB, Furst L, Condie AG, Stephenson CRJ (2012) Functionally diverse nucleophilic trapping of iminium intermediates generated utilizing visible light. Org Lett 14:94–97
35. Furst L, Matsuura BS, Narayanam JMR, Tucker JW, Stephenson CRJ (2010) Visible light-mediated intermolecular C–H functionalization of electron-rich heterocycles with malonates. Org Lett 12:3104–3107
36. Rueping M, Vila C, Koenigs RM, Poscharny K, Fabry DC (2010) Dual catalysis: combining photoredox and Lewis base catalysis for direct Mannich reactions. Chem Commun 47:2360–2362
37. Zhao G, Yang C, Guo L, Sun H, Chen C, Xia W (2012) Visible light-induced oxidative coupling reaction: easy access to Mannich-type products. Chem Commun 48:2337–2339
38. Rueping M, Koenigs RM, Poscharny K, Fabry DC, Leonori D, Vila C (2012) Chem Eur J 18:5170–5174
39. Hari DP, König B (2011) Eosin Y catalyzed visible light oxidative C–C and C–P bond formation. Org Lett 13:3852–3855
40. Liu Q, Li Y-N, Zhang H-H, Chen B, Tung C-H, Wu L-Z (2012) Reactivity and mechanistic insight into visible-light-induced aerobic cross-dehydrogenative coupling reaction by organophotocatalysis. Chem Eur J 18:620–627
41. Pan Y, Wang S, Kee CW, Dubuisson E, Yang Y, Loh KP, Tan C-H (2011) Graphene oxide and Rose Bengal: oxidative C–H functionalization of tertiary amines using visible light. Green Chem 13:3341–3344
42. Li W, Zhu X, Tang Z, Cheng Y, Zhu C (2014) Visible-light-induced direct C(sp3)–H difluoromethylation of tetrahydroisoquinolines with the *in situ* generated difluoroenolates. Chem Commun 50:7521–7523
43. Reuping M, Zhu SQ, Koenigs RM (2011) Visible-light photoredox catalyzed oxidative Strecker reaction. Chem Commun 47:12709–12711
44. DiRocco D, Rovis T (2012) Catalytic asymmetric α-acylation of tertiary amines mediated by dual catalysis mode: N-heterocyclic carbine

and photoredox catalysis. J Am Chem Soc 134:8094–8097

45. Bergonzini G, Schindler CS, Wallentin C-J, Jacobsen EN, Stephenson CRJ (2014) Photoredox activation and anion binding catalysis in the dual catalytic enantioselective synthesis of β-amino esters. Chem Sci 5:112–116
46. Feng Z-J, Xuan J, Xia X-D, Ding W, Guo W, Chen J-R, Zou Y-Q, Lu L-Q, Xiao W-J (2014) Direct sp^3 C–H acroleination of *N*-aryl-tetrahydroisoquinolines by merging photoredox catalysis with nucleophilic catalysis. Org Biomol Chem 12:2037–2040
47. Zou Y-Q, Lu L-Q, Fu L, Chang N-J, Rong J, Chen J-R, Xiao W-J (2011) Visible-light-induced oxidation/[3+2] cycloaddition/oxidative aromatization sequence: a photocatalytic strategy to construct pyrrolo[2,1-*a*]isoquinolines. Angew Chem Int Ed 50:7171–7175
48. Reuping M, Leonori D, Poisson T (2011) Visible light mediated azomethine ylide formation—photoredox catalyzed [3+2] cycloadditions. Chem Commun 47:9615–9617
49. Xie J, Xue Q, Jin H, Li H, Cheng Y, Zhu C (2013) A visible-light-promoted aerobic C–H/C–N cleavage cascade to isoxazolidine skeletons. Chem Sci 4:1281–1286
50. Xuan J, Cheng Y, An J, Lu L-Q, Zhang X-X, Xiao W-J (2011) Visible light-induced intramolecular cyclization reaction of diamines: a new strategy to construct tetrahydroimidazoles. Chem Commun 47:8337–8339
51. Beatty JW, Stephenson CRJ (2014) Synthesis of (−)-pseudotabersonine, (−)-pseudovincadifformine, and (+)-coronaridine enabled by photoredox catalysis in flow. J Am Chem Soc 136:10270–10273
52. Chow YL, Danen WC, Nelsen SF, Rosenblatt DH (1978) Nonaromatic aminiun radicals. Chem Rev 78:243–274
53. Renaud P, Giraud L (1996) 1-Amino and 1-amidoalkyl radicals: generation and stereoselective reactions. Synthesis 8:913–926
54. Kohls P, Jadhav D, Pandey G, Reiser O (2012) Visible light photoredox catalysis: generation and addition of *N*-aryltetrahydroisoquinoline-derived α-amino radicals to Michael acceptors. Org Lett 14:672–675
55. Pandey G, Jadhav D, Tiwari SK, Singh B (2014) Visible light photoredox catalysis: investigation of distal sp^3 C–H functionalization of tertiary amines for alkylation reaction. Adv Synth Catal 356:2813–2818
56. Dai X, Cheng D, Guan B, Mao W, Xu X, Li X (2014) The coupling of tertiary amines with acrylate derivatives via visible-light photoredox catalysis. J Org Chem. 79:7212–7219
57. Ju X, Li D, Li W, Yu W, Bian F (2012) The reaction of tertiary anilines with maleimides under visible light redox catalysis. Adv Synth Catal 354:3561–3567
58. Zhu S, Das A, Bui L, Zhou H, Curran DP, Reuping M (2013) Oxygen switch in visible-light photoredox catalysis: radical additions and cyclizations and unexpected C–C bond cleavage reactions. J Am Chem Soc 135:1823–1829
59. Miyake Y, Nakajima K, Nishibayashi Y (2012) Visible-light-mediated utilization of α-aminoalkyl radicals: addition to electron-deficient alkenes using photoredox catalysts. J Am Chem Soc 134:3338–3341
60. Zhang P, Xiao T, Xiong S, Dong X, Zhou L (2014) Synthesis of 3-acylindoles by visible-light induced intramolecular oxidative cyclization of *o*-alkynylated *N,N*-dialkylanilines. Org Lett 16:3264–3267
61. Miyake Y, Nakajima K, Nishibayashi Y (2012) Direct sp^3 C–H amination of nitrogen-containing benzoheterocycles mediated by visible-light-photoredox catalysis. Chem Eur J 18:16473–16477
62. McNally A, Prier CK, MacMillan DWC (2011) Discovery of an α-amino C–H arylation reaction using the strategy of accelerated serendipity. Science 334:1114–1117
63. Prier CK, MacMillan DWC (2014) Amine α-heteroarylation via photoredox catalysis: a hemolytic aromatic substitution pathway. Chem Sci 5:4173–4178
64. Noble A, MacMillan DWC (2014) Photoredox α-vinylation of α-amino acids and *N*-aryl amines. J Am Chem Soc 136:11602–11605
65. Zuo Z, Ahneman DT, Chu L, Terrett JA, Doyle AG, MacMillan DWC (2014) Merging photoredox with nickel catalysis: coupling of α-carboxyl sp^3-carbons with aryl halides. Science 345:437–440
66. Pirnot MT, Rankic DA, Martin DB, MacMillan DWC (2013) Photoredox activation for the direct β-arylation of ketones and aldehydes. Science 339:1593–1596
67. Petronijević FR, Nappi M, MacMillan DWC (2013) Direct β-functionalization of cyclic ketones with aryl ketones via the merger of photoredox catalysis and organocatalysis. J Am Chem Soc 135:18323–18326
68. Terrett JA, Clift MD, MacMillan DWC (2014) Direct β-alkylation of aldehydes via photoredox catalysis. J Am Chem Soc 136:6858–6861

69. Cano-Yelo H, Deronzier A (1984) Photo-oxidation of some carbinols by the Ru (II) polypyridyl complex-aryl diazonium salt system. Tetrahedron Lett 25:5517–5520
70. Tucker JW, Narayanam JMR, Shah PS, Stephenson CRJ (2011) Oxidative photoredox catalysis: mild and selective deprotection of PMB ethers mediated by visible light. Chem Commun 47:5040–5042
71. Lowry MS, Goldsmith JI, Slinker JD, Rohl R, Pascal RA Jr, Malliaras GG, Bernhard S (2005) Single-layer electroluminescent devices and photoinduced hydrogen production from an ionic iridium(III) complex. Chem Mater 17:5712–5719
72. Lund H (1957) Electroorganic preparations II. Oxidation of carbinols. Acta Chem Scand 11:491–498
73. Pandey G, Pal S, Laha P (2013) Direct benzylic C–H activation for C–O bond formation by photoredox catalysis. Angew Chem Int Ed 52:5146–5149
74. Dai C, Meschini F, Narayanam JMR, Stephenson CRJ (2012) Friedel-Crafts amidoalkylation via thermolysis and oxidative photocatalysis. J Org Chem 77:4425–4431
75. Qvortrup K, Rankic DA, MacMillan DWC (2014) A general strategy for organocatalytic activation of C–H bonds via photoredox catalysis: direct arylation of benzylic ethers. J Am Chem Soc 136:626–629
76. Escoubet S, Gastaldi G, Vanthuyne N, Gil G, Siri D, Bertrand M (2006) Thiyl radical mediated racemization of nonactivated aliphatic amines. J Org Chem 71:7288–7292
77. Ochiai M, Yamane S, Hoque MM, Saito M, Miyamoto K (2012) Metal-free α-CH amination of ethers with hypervalent sulfonylimino-λ^3-bromane that acts as an active nitrenoid. Chem Commun 48:5280–5282

Chapter 11

In Situ Protecting Groups for Chemoselective Transformations

Alan Steven

Abstract

Streamlining the synthesis of synthetic targets of relevance to the pharmaceutical industry has important economic, environmental, and practical drivers. Herein we review in situ protecting group strategies that have been applied, or have the potential to be applied, to the synthesis of small molecule drugs and drug candidates. The discussion will restrict itself to situations where protection has been achieved using a protecting group. In addition, in order to assist with their reduction to practice, we provide example procedures where these methodologies have been used.

Key words α-Amino alkoxide, Boronate ester, *N*-carboxylation, Chemoselectivity, Copper (II) chelate, In situ protection, Lanthanide chloride, *Ortho*-lithiation, LTMDA, Trimethylsilylation

1 Introduction

An ideal synthesis of a drug candidate, or indeed any synthetic target, would only consist of skeleton-constructing reactions or functionalizations. A breathtaking expansion in the methodologies that allow a target to be built using such reactions has taken place since Hendrickson's seminal recognition of this fact [1]. Many of these methodologies have proven to be highly tolerant of functional groups elsewhere in the molecule. More often than not, however, reactions want for chemoselectivity, necessitating the avoidance of the issue altogether through the reordering of synthetic steps or the prior protection of the incompatible functionality. In the case of the latter, protection comes with the need to later deprotect, and both such steps can detract from ideality.

Paul F. Richardson (ed.), *Green Chemistry in Drug Discovery: From Academia to Industry*, Methods in Pharmacology and Toxicology, https://doi.org/10.1007/978-1-0716-1579-9_11,

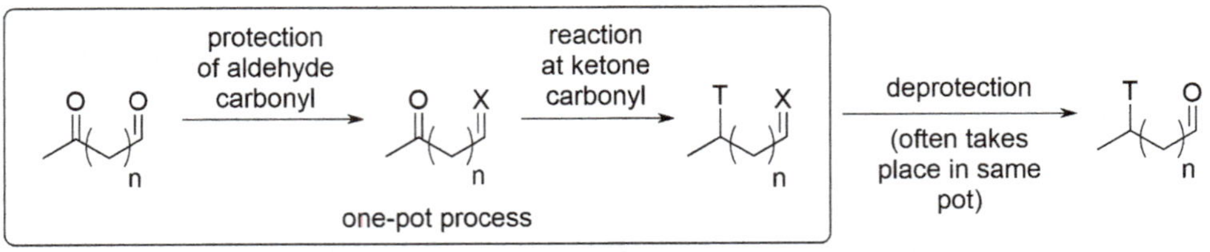

Fig. 1 Generic illustration of a chemical transformation facilitated by in situ protection

An in situ protecting group strategy is one in which the installation of the protecting group and the reaction of interest are performed in the same pot.[1] This is illustrated in Fig. 1 for a dicarbonyl compound, where a ketone carbonyl has been transformed after the protection of an aldehyde carbonyl, in the same pot [2]. An in situ protecting group strategy still necessitates materials and processing time to be spent on introducing the protecting group, though the impact on the yardstick used to measure the green credentials of the process [3] is likely to be second only to an approach that is completely protecting group free. For many of the strategies described herein, the unmasking of the protected functionality readily takes place during the reaction workup, often through the presence of an aqueous medium with a specific pH. Under these circumstances, further efficiencies accrue from the avoidance of a separate process dedicated to the deprotection and chemoselectivity issues that may be associated with the removal of a more securely attached protecting group. The discussion will restrict itself to situations where protection has been achieved using a distinct protecting group. Thus the in situ protection of ketones as enolates [4], for example, ahead of reaction elsewhere in the molecule, will not be discussed further.

2 Control of Solution Acidity

We will start our discussion with the use of the proton, perhaps the simplest of all protecting groups as far as a basic nitrogen is concerned. The attraction of its use stems from its atom economy, as well as our ability to attach or remove it by simply adjusting the acidity of a solution: an operation which aligns itself with in situ protection or deprotection. Armed with knowledge or predictions of solvent-relevant pK_as for the various basic nitrogens present in the starting material and product, the acidity of a solution can be

[1] Processes where the introduction of the protecting group is followed by a concentration step, including a solvent swap, prior to the transformation of interest, are considered in scope for the purpose of this discussion.

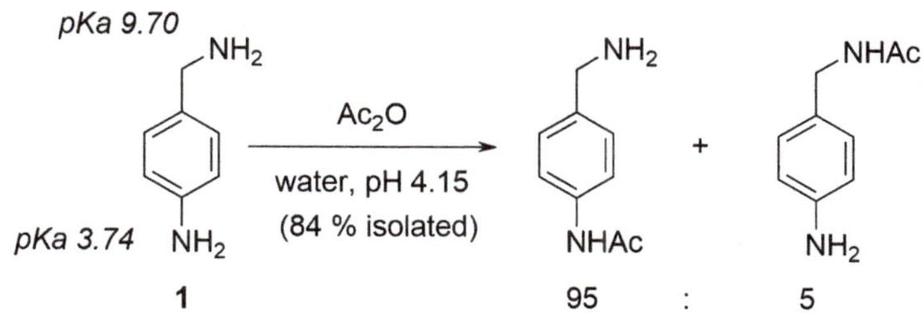

Fig. 2 Regioselective acetylation of 4-(aminomethyl)aniline

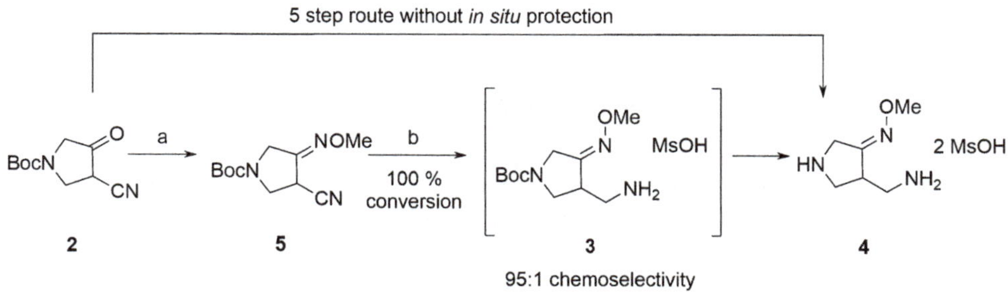

a) $MeONH_2$•HCl, py, MeOH; b) i) H_2, MsOH, 5 % Pd/Al_2O_3, DME, water, ii) evaporation

Fig. 3 In situ protection of an amine as its conjugate acid as part of synthesis of a gemifloxacin intermediate

adjusted in order to protect basic nitrogens in the starting material that may otherwise interfere with the desired reaction at a less basic center [5]. This is illustrated in Fig. 2 where the selective acetylation of the aniline nitrogen was achieved by using the pH of the solution to prevent the acetylation of the benzylamine nitrogen of 4-(aminomethyl)aniline (**1**) [6–8].

The in situ protection of an amine was adeptly deployed as part of a proof of concept synthesis of the dimesylate salt of 4-aminomethyl-3-methoxyiminopyrrolidine **4**, a key intermediate in the manufacture of the quinolone antibacterial gemifloxacin [9]. The synthesis starts from compound **2**, a protected form of 4-oxopyrrolidine-3-carbonitrile, and involves nitrile hydrogenation as shown in Fig. 3. This transformation is potentially susceptible to side reactions, initiated by *O*-methyloxime N-O bond cleavage, that are abetted by the free base form of the newly formed amino group of compound **3** coordinating to the surface of the catalyst. Such coordination also reduces the rate of nitrile reduction. An initial process improvement used di-*tert*-butyl dicarbonate to derivatise in situ the nascent amino group as the corresponding *tert*-butyl carbamate [10]. This procedure was still plagued by a small but significant amount of reaction at the *O*-methyloxime, and it proved difficult to push the desired reduction of the nitrile of **5** to a high level of conversion. Both of these reaction attributes were addressed by performing the reaction under acidic conditions that protonated the amino group, preventing it from coordinating to

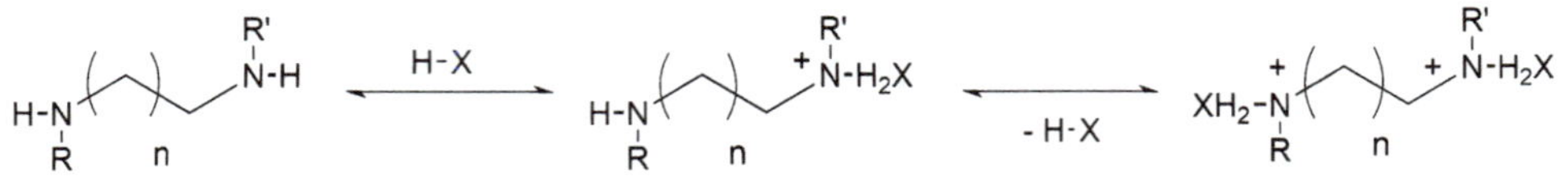

Fig. 4 Selective monoprotection of a diamine

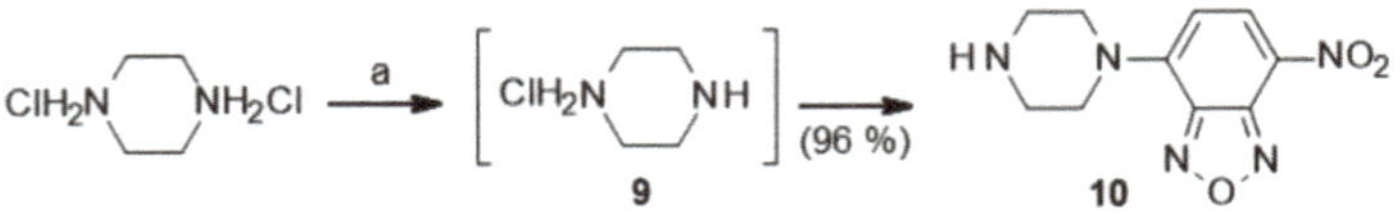

Fig. 5 Selective monofunctionalization of piperazine rings

the catalyst surface [11]. By choosing methanesulfonic acid as the acid, the product of *N*-(*tert*-butyloxycarbonyl) deprotection, 4-aminomethyl-3-methoxyiminopyrrolidine, could be isolated as dimesylate **4**. This dimesylate was the preferred salt form for this material, and thus the process avoided the generation of an environmentally burdensome and salt-rich wastestream as part of the removal of any other acid that could have been used. Additional benefits associated with the replacement of the original route (five steps) with the two-step process shown in Fig. 3 included a decrease in manufacturing time and a reduced use of solvents and reagents.

The selective monoprotection of diamines is an area that is well-suited to exploiting the proton as a protecting group. This is through the 1:1 salt **6** that dominates the equilibrium shown in Fig. 4 when **7** and **8**, the species at its two ends, are treated with an equimolar amount of acid or base, respectively [12]. An example can be found in the protection of the piperazine ring, a privileged structure whose nitrogens typically bear different substituents in natural products or candidate drugs. As illustrated in Fig. 5, the in situ formation of piperazine monohydrochloride **9** has been used to facilitate the protection of the remaining nitrogen with a 7-nitrobenzofurazan (NBD) protecting group as in **10** [14]. This protocol is more efficient and more scalable than the very slow addition of the electrophile, or the use of excess piperazine, in protecting group-free variants of this transformation that achieve similar selectivities [14–17].

3 *N*-Carboxylation

Carbon dioxide is a renewable, inexpensive, and readily available material. Reaction with it can be exploited as a means of temporarily protecting a nucleophilic nitrogen center as a carbamic acid, or a salt thereof [18], in order to shut down side reactions that would otherwise take place.

Practicing synthetic organic chemists who have had to wipe a white precipitate of the carbamate salt from the top of a bottle of the material, or who have struggled to gain acceptable elemental analysis data on their compound, can attest to the facility with which aliphatic amines react with carbon dioxide. For preparative purposes, *N*-carboxylation is achieved by pressurizing the reaction headspace with carbon dioxide gas or by using it in its supercritical form.[2] The unmasking of the amine can often be achieved in the same pot by heating or simply sparging with nitrogen gas. This approach has spawned a number of practical methods that are ripe for exploitation by drug discovery and development programs. These include the suppression of overreaction during the hydrogenation of nitriles and imines [19, 20], or the selective acylation of an alcohol in the presence of a primary amine. For example, the treatment of a mixture of benzylamine and benzyl alcohol with isoprenyl acetate catalyzed by 1,2,4-triazole leads to the quantitative acylation of the latter under an atmosphere of carbon dioxide [21]. Carbon dioxide has also been used to expedite the ring-closing metathesis of substrates bearing functionality that could chelate to the catalyst, hindering turnover [22]. Figure 6 illustrates this in the context of the synthesis of the skeleton of the insect repellent alkaloid epilachnene. The green credentials of this transformation require little burnishing. The isolation of the product **11** eschews the environmental burden associated with the use of

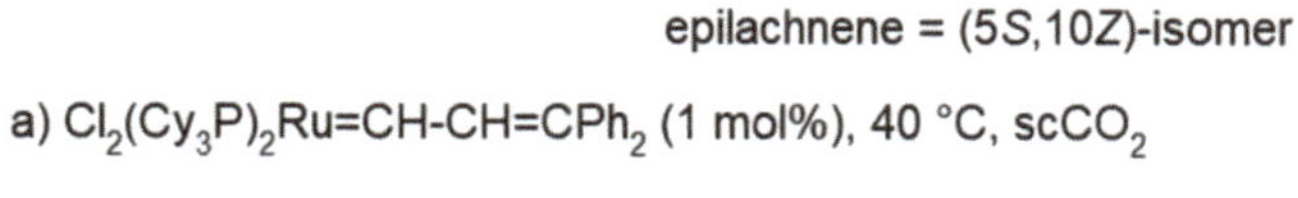

Fig. 6 Temporary *N*-carboxylation of an amino group ahead of a ring-closing metathesis

[2] The protection supercritical carbon dioxide bestows upon basic amines further reinforces the credentials of a reaction medium that is playing an increasing role in drug discovery programmes, in part due to its environmentally and toxicologically benign character and the straightforward isolation of product afforded by its use. Its use in a ring-closing metathesis also allows the competition between cyclization versus oligomerization to be addressed at a constant volume and skewed in favor of the former, by increasing the pressure (and hence density), reducing the molar fraction of the substrate.

liquid-liquid partitioning in favor of the venting of the reactor ahead of a chromatographic purification. The transformation also avoids the use of chlorinated solvents, which are otherwise normally the solvents of choice for a ruthenium-catalyzed olefin metathesis.

The development of in situ *N*-carboxylation as a means of selectively functionalizing NH-bearing aromatic and heteroaromatic compounds has been largely dominated by the Katritzky group [23–30]. The treatment of a deprotonated NH-compound with carbon dioxide forms a carbamate salt, typically a lithium carbamate. In laboratory preparations, in preparation for the next step of the telescope, this is typically isolated by removing volatiles under reduced pressure to afford what can be a stable salt that can be handled in air, prior to its redissolution in fresh tetrahydrofuran. In spite of this operation, these methods have been included within the purview of the wider discussion as their implementation on any sort of scale would seek a distillative solution that would avoid the evaporation to dryness.[3] Once the desired transformation has been achieved elsewhere in the molecule, deprotection can be readily achieved through the exposure of lithium carbamate to heat or mild acid, thus avoiding the side reactions that may plague the removal of a more securely attached protecting group.

The utility of the methodology to drug discovery and development programs is exemplified in Fig. 7 through its application to the functionalization of indoles. Once the indole has been protected as the lithium carbamate **12**, the ensuing functionalization will typically proceed by first lithiating at the 2-position to give **13** [31]. This transformation is expedited by the stabilizing the pres-

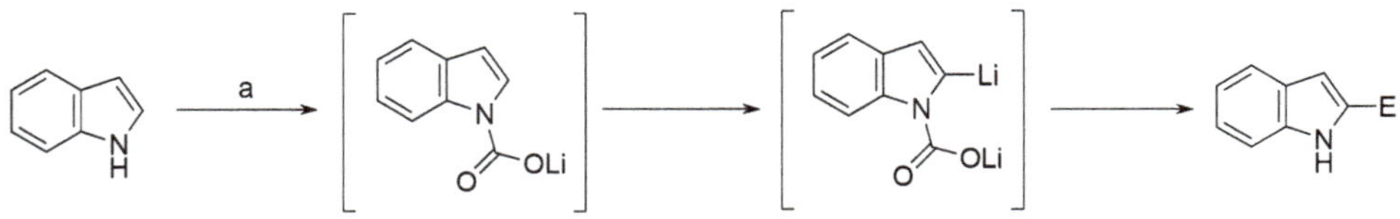

12 **13** **14**

a) i) ^{n}BuLi, THF, *n*-hexane, –70 °C, ii) CO_2, –70 → 20 °C, iii) ^{t}BuLi, –70 °C, iv) E^+, v) NH_4Cl (aq), 20 °C, vi) H_2SO_4 (aq), Δ

Fig. 7 Temporary protection of indole nitrogen as its lithium carbamate salt allows subsequent C2 functionalization via metallation

[3] Indeed, an examination of Henry's Law data for the solubility of carbon dioxide in tetrahydrofuran establishes that simply swing purging the headspace between a vacuum and a nitrogen atmosphere will readily remove carbon dioxide from solution. This has been experimentally verified by the author in an unpublished ReactIR study which also demonstrated that a saturated solution of carbon dioxide in tetrahydrofuran will rapidly lose the former if it is simply left under an overhead sweep of nitrogen gas.

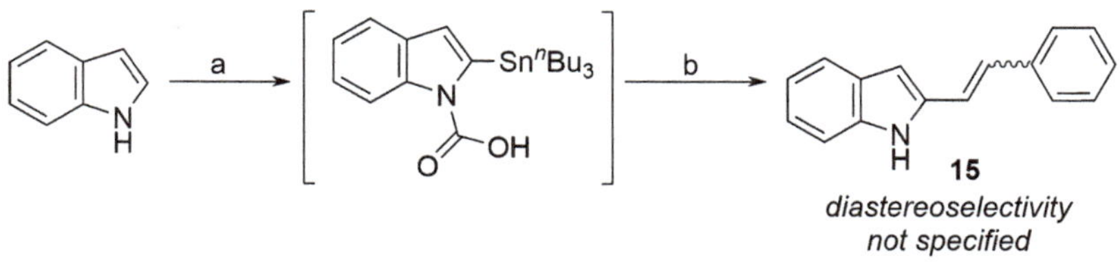

a) i) ^{n}BuLi, THF, –68 °C, ii) CO_2, –68 → 20 °C, iii) ^{t}BuLi, –68 °C, iv) $^nBu_3SnCl$;
b) i) β-bromostyrene, $(Ph_3P)_2PdCl_2$, EtOH, Δ

Fig. 8 Introduction of a tributyltin group at C2 of an indole temporarily protected as its lithium carbamate ahead of a Stille coupling

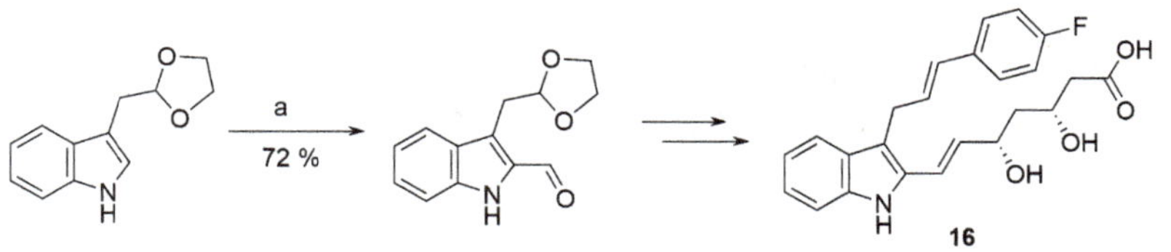

a) i) ^{n}BuLi, THF, –70 °C, ii) CO_2, 25 °C, iii) ^{t}BuLi, THF, –70 → –20 °C, iv) DMF, –70 °C, v) $(NH_4)_2SO_4$ then H_2SO_4 (aq), 0 °C, vi) Δ

Fig. 9 Formylation after temporary protection of an indole as its lithium carbamate

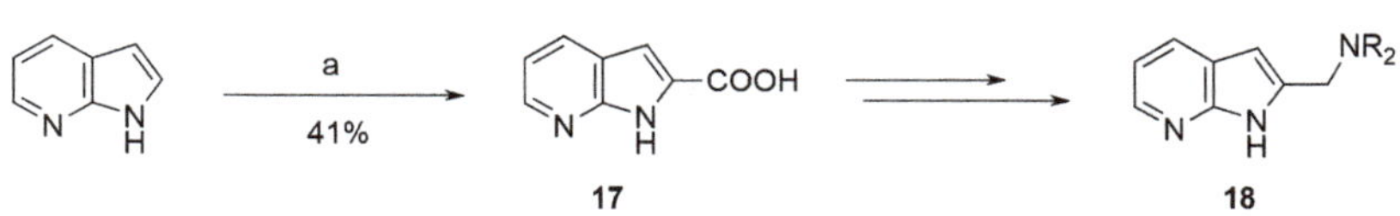

a) i) ^{n}BuLi, THF, –78 °C, ii) CO_2, –78 → 20 °C, iii) ^{t}BuLi, –78 °C, iv) CO_2, –78 → 20 °C, v) H_3O^+

Fig. 10 Carboxylation of 1*H*-pyrrolo[2,3-*b*]pyridine after temporary derivatization of its N–H as a lithium carbamate

ence of the neighboring lithium carbamate. The lithiate can be quenched with a variety of different electrophiles, before the *N*-carboxyl is removed so as to furnish **14**, a 1*H*-indole functionalized at the 2-position. The introduction of a tributylstannyl group, boronic ester, or a halogen atom in this manner [32] and a subsequent cross-coupling offers facile access to 2-aryl- and 2-vinyl-1*H*-indoles [33], motifs that feature heavily in candidate drugs. The sequence is exemplified by the synthesis of 2-(2-styryl) indole (**15**) in Fig. 8 [33]. The formyl group introduced by quenching with *N*,*N*-dimethylformamide [34] played a key role in the synthesis of potential inhibitors of HMG-CoA reductase such as compound **16** (Fig. 9) [35]. Quenching the lithiate itself with carbon dioxide has provided an intermediate **17** in the synthesis of potential dopamine D_4 ligands **18** (Fig. 10) [36]. Alternatively, quenching with an aldehyde or ketone introduces a hydroxyl and

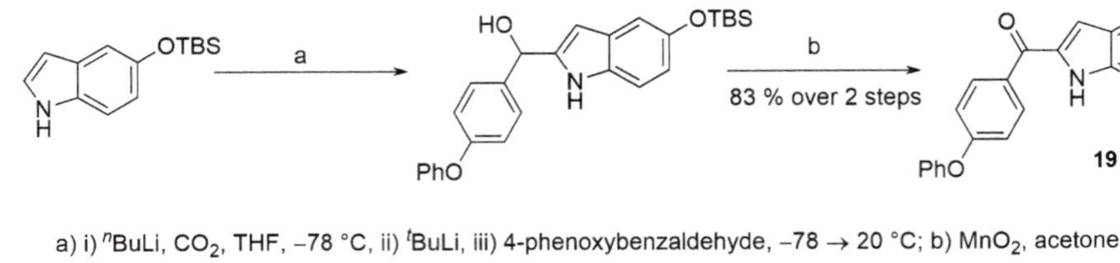

Fig. 11 Two step carbonyl addition-oxidation to acylate an indole after temporary protection of N–H as its lithium carbamate

has found widespread utility [37–42], as was the case with the part of the synthesis shown in Fig. 11 of a CaMKII inhibitor **19** [43].

3.1 Preparation of 2-(styryl)-1H-indole (15) from Indole

n-Butyllithium in hexanes (34 mL of 2.5 M, 85 mmol) was added dropwise over 30 min to a solution of indole (10 g, 85 mmol) in tetrahydrofuran (200 mL) while maintaining a batch temperature of −68 °C. After stirring for 30 min, carbon dioxide was passed through the solution for 10 min. The solution was warmed to 20 °C before it was concentrated in vacuo to ca. 125 mL. Additional tetrahydrofuran (100 mL) was charged, before *t*-butyllithium in pentane (50 mL of 1.7 M, 85 mmol, CARE: pyrophoric) [44] was added while maintaining a batch temperature of −68 °C. After an additional 2 h, tributyltin chloride (29.4 g, 90 mmol) was added over 10 min. The reaction mixture was poured onto ice water (300 g) after 90 min, and the solution acidified with saturated ammonium chloride solution. Extraction with diethyl ether (2 × 200 mL), drying (magnesium sulfate), and concentration afforded 1-carboxy-2-(tributylstannyl)indole as a clear mobile oil (40 g) that was used without further purification. A portion of this material (4.0 g, <8.9 mmol), β-bromostyrene (825 mg, 4.51 mmol), and bis(triphenylphosphine)palladium (II) dichloride (300 mg, 0.427 mmol) in ethanol (40 mL) was refluxed for 48 h. The mixture was then cooled to 20 °C, filtered through a pad of Celite, and concentrated to a solid. Recrystallization from cyclohexane afforded 850 mg (86%) of 2-(styryl)-1*H*-indole (**15**).

Where an *N*-(carboxy)indole bears an alkyl group at the 2-position [45], further functionalization is possible via lateral metallation.[4] This is exploited by the Junjappa-Ila benzannulation shown in Fig. 12, which quenches with α-oxoketene dithioacetal **20** so as to form compound **21** as part of a regioselective means of accessing *N*-unsubstituted carbazoles **22** [46, 47]. Such functionality can be found in a number of drugs that have made it to the marketplace, including the nonsteroidal anti-inflammatory veterinary product carprofen **23** and the beta blocker carvedilol **24**.

[4] The same approach was used for the functionalization of *o*-toluidine, see reference 30.

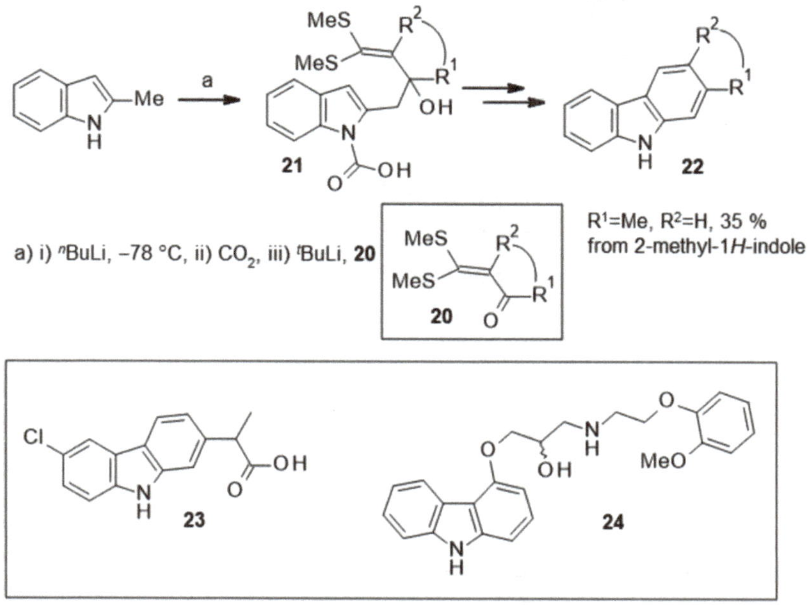

Fig. 12 Use of lateral metallation of 2-methyl-1*H*-indole after temporary protection of N–H bond as its lithium carbamate

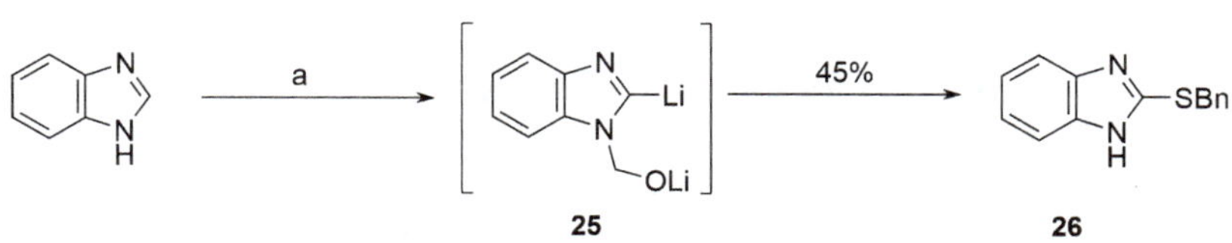

a) i) $HO(CH_2O)_nH$, THF; ii) ^{t}BuLi, −78 → −20 °C; iii) $(BnS)_2$, −78 °C; iv) HCl (aq)

Fig. 13 Functionalization of benzimidazole after temporary protection of N–H bond as a lithiohemiaminal

Whilst Katritzky's carbon dioxide method fails for NH-bearing heterocyclic rings containing more than one heteroatom, such heterocycles can still be analogously functionalized by using *para*-formaldehyde as a formaldehyde equivalent. This sequence has been successfully applied, most notably to benzimidazoles, and proceeds via a dilithiohemiaminal **25** as shown in Fig. 13 [48]. After quenching with an electrophile, exposure to acidic conditions effects the dehydroxymethylation of the protected nitrogen so as to afford a functionalized benzimidazole such as **26**.

4 Boronic Acid, Boric Acid, and Boranes

A variety of boron-containing compounds have found use as in situ protecting groups. The hydroxyl groups of a vicinal diol can be prevented from participating in side reactions in a downstream transformation by using a boronic acid (or boroxine derivative) to

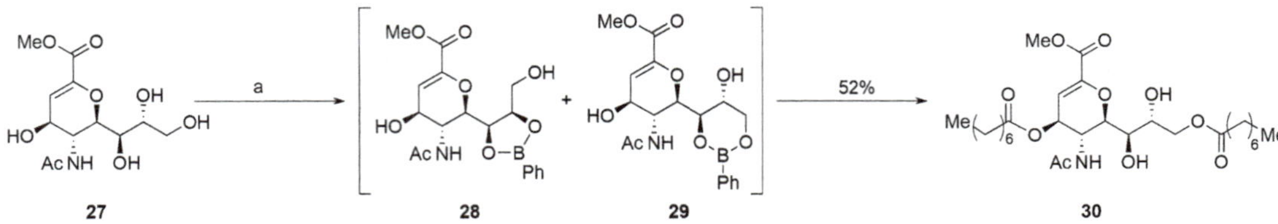

a) i) $PhB(OH)_2$, PhH, pyridine, Δ, ii) $^nC_7H_{15}C(O)Cl$, iii) aq. NaOH

Fig. 14 Selective *bis*-octanoylation of a sialic acid derivative after in situ boronate ester formation involving side chain hydroxyl groups

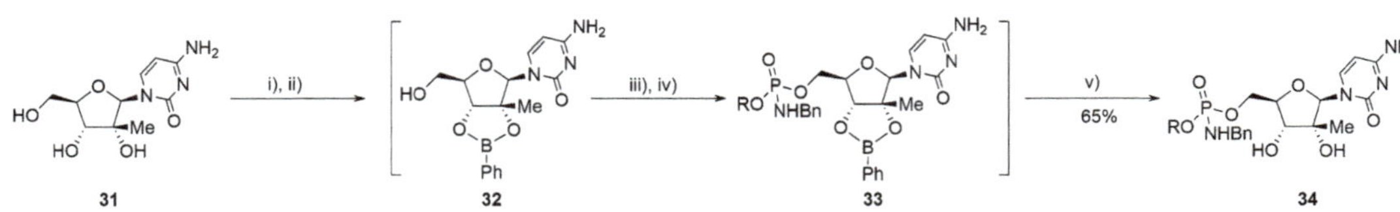

a) i) $PhB(OH)_2$, py, Δ, ii) concentrate, iii) $Et_3NHOPH(O)OR$, EDCI•HCl, MeCN, 41-46 °C, iv) $BnNH_2$, CCl_4, <20 °C, v) citric acid (aq)

R =

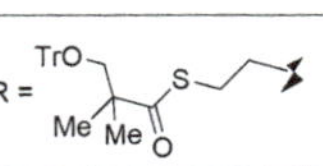

Fig. 15 Selective *H*-phosphonate diester after temporary protection of its vicinal diol as a cyclic boronate

temporarily protect it as a boronate ester [49–51]. Where the water by-product of the protection does not interfere with the ensuing reaction, or judicious choice of reaction solvent allows its distillative removal, it is possible to telescope the protection with the downstream reaction, circumventing the isolation of the boronate ester.[5]

The choice of phenylboronic acid as the boronic acid is especially popular, and it can now be purchased in polymer-bound form. Its use is exemplified in Fig. 14 with the *bis*-octanoylation of a sialic acid derivative **27**. The boronate ester intermediate is actually likely to be a mixture of species (including **28** and **29**) that display Curtin-Hammett-type reactivity with respect to the site of acylation of the triol side chain [52].[6] On repeating the reaction without phenylboronic acid protection, the crude reaction mixture contained multiple materials from which the desired triester **30** was isolated in only 5% yield after purification.

As part of a kilogram-scale manufacture of a nucleoside phosphoramidate prodrug inhibitor of hepatitis C virus NS5B polymerase, phenylboronate protection of its vicinal diol permitted Idenix scientists to form a *H*-phosphonate diester at the primary alcohol of 2′-*C*-methylcytidine (**31**) (Fig. 15) [53]. The 2′-*C*-methylcytidine-

[5] Alternatively, the crystallinity of the boronate ester intermediate often offers the opportunity to isolate it and to use it as a control point in a synthesis.

[6] The same study used a boronate ester formed in situ for the functionalization of mannitol at both the 1- and 6-positions to give analytically pure material without the need for purification using chromatography or recrystallization.

2′,3′-*O*-phenylboronate **32** was formed in refluxing pyridine and the water by-product azeotroped out of the solution. After *H*-phosphonate diester formation to give **33**, the hydrolytic lability of the 1,2-cyclic boronate protecting group enabled its facile removal in the same pot using an aqueous solution of citric acid, in marked contrast to the complex mixture of compounds observed after trying to remove an acetonide protecting group. Indeed, the switch from the use of an acetonide protecting group to the one-pot procedure made a significant contribution to the 82% reduction in process mass intensity that characterized the development performed on the processes for converting **31** into compound **34**. The number of isolations was reduced from five to two, and the overall yield was increased from 20 to 32%. Additionally, per unit mass of **34** produced by each route, there was a 97% reduction in the amount of carbon tetrachloride required for the benzyl phosphoramidate formation on switching from the acetonide to the phenylboronate protecting group.

4.1 Phosphoramidation of 2′-C-Methylcytidine (31)

Phenylboronic acid (78 g, 0.64 mol) was charged to 2′-*C*-methylcytidine (**31**) (150 g, 0.583 mol) in pyridine (2.5 L), and the mixture refluxed for 3 h. The amount of pyridine was then reduced to 250 mL by distillation, and the solution added to a solution of $Et_3NHOPH(O)OCH_2CH_2SC(O)C(Me)_2CH_2OTr$ (615 g, 1.05 mol) in acetonitrile (3 L), followed by 1-ethyl-3-(3-dimethylaminopropyl)carbodiimide hydrochloride (570 g, 2.97 mol). The reaction was stirred at 41–46 °C for 4 h before being cooled to <20 °C. Benzylamine (395 mL, 3.62 mol) was added followed by carbon tetrachloride (260 mL, CARE: evidence of carcinogenicity). After 1 h, ethyl acetate (1 L) was charged to the mixture and acidified to pH 4 with aqueous citric acid (3 L of 20%w/w). The separated aqueous phase was extracted with ethyl acetate (2.5 L). The combined organics were washed with aqueous citric acid solution (3 L of 10%w/w) and then twice with saturated sodium bicarbonate solution (5 L then 2 L). The organic phase was dried (sodium sulfate), evaporated at 25–35 °C, and the residue purified by silica gel chromatography, eluting with methanol (4–7%) in dichloromethane to give trityl phosphoramidate **34** (314 g, 65%).

While the examples given above illustrate the protection of a 1,2-diol as a boronate ester, in the absence of such functionality, a 1,3-diol can also be masked in this manner. As part of a study of the relationship between antiglaucoma activity and the structure of the ω-chain of prostaglandin $F_{2\alpha}$ isopropyl ester analogues, the in situ protection of a 1,3-diol **35** using phenylboronic acid enabled the chemoselective oxidation of an alcohol elsewhere in the substrate (Fig. 16) [54]. The resulting aldehyde **36** was telescoped through a Horner-Wadsworth-Emmons olefination, prior to boronate cleavage to give **37**.

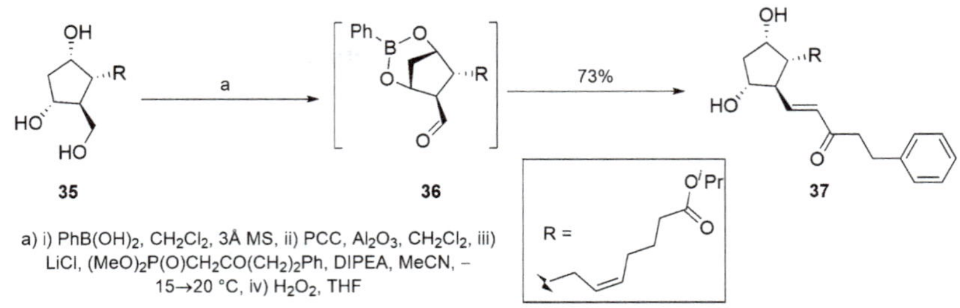

Fig. 16 Protection of a 1,3-diol in 35 as a boronate ester facilitates selective oxidation ahead of Horner-Wadsworth-Emmons olefination

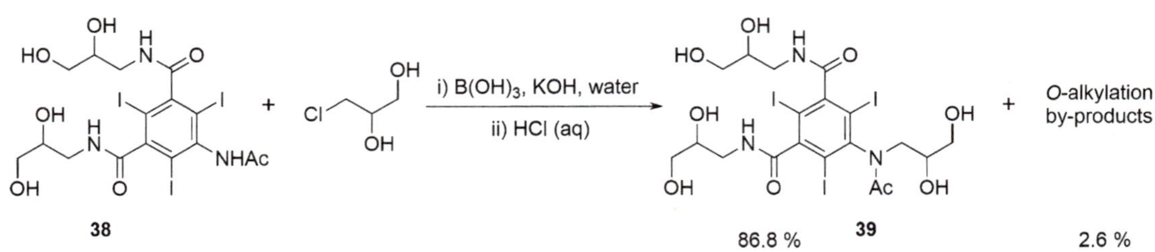

Fig. 17 Use of temporary protection of diols using boric acid

Along the same lines as the use of a boronic acid, boric acid itself has sometimes been used to achieve the same result. A case in point is as part of the synthesis of X-ray contrast agents for medical imaging diagnostics, where protection with boric acid allowed an *N*-alkylation of **38** to **39** while minimizing the formation of by-products arising from alcohol alkylation (Fig. 17) [55].

The in situ protection afforded by boron-containing compounds has been used to streamline the synthesis of compounds other than diols. A case in point is the functionalization of polyamines such as spermine, a polyamine found in all eukaryotic cells. The efficient synthesis of the spermine-containing macrocyclic lactams (±)-verbaskine (**40**) and (±)-verbascenine (**41**) has been made possible through the formation of monocinnamide (±)-verbacine (**42**), as the major product, following the in situ formation of 1,3,2-diazaborinane **43** (Fig. 18) [56].[7] Only dicinnamide formation could be achieved when a protecting group-free strategy was employed. Another example can be found in the chemoselective functionalization of an α-amino acid side chain, which can be expedited after the protection of the α-amino functionality acid as a boroxazolidone derivative such as **44** [57]. Applications of this

[7] 3,5-Bis(trifluoromethyl)phenyl)boronic acid is reported to have produced a superior regioselectivity and chemical yield in this reaction when compared to the use of boronic acid.

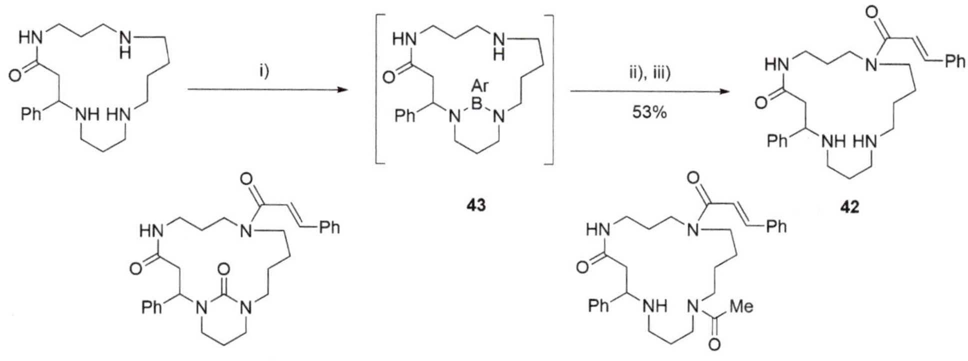

i) 3,5-$(CF_3)_2C_6H_3B(OH)_2$, CH_2Cl_2, −78 → 20 °C, ii) cinnamoyl chloride, CH_2Cl_2, −78 °C, iii) MeOH

Fig. 18 Selective acylation after temporary protection of a polyamine as a 1,3,2-diazaborinane

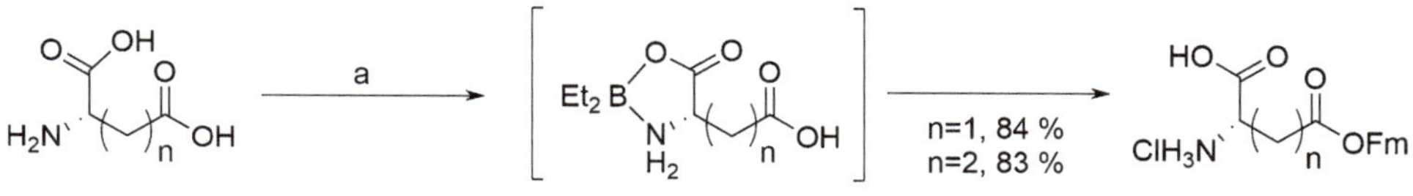

a) i) $B(Et)_3$, THF, Δ, ii) FmOH, DCC, DMAP, THF, 25 °C iii) HCl

Fig. 19 Selective carbamoylations after protection of α-amino acids as boroxazolidones

approach have included a practical and one-pot means of protecting the side chain of L-aspartic acid (**45**) or L-glutamic acid (**46**) with the fluorenylmethyl protecting group (Fig. 19) [58].

4.2 Fluorenylmethyl Protection of the Side Chain of L*-Glutamic Acid (46)*

Triethylborane in tetrahydrofuran (13 mL of 1 M, 13 mmol) was added to a slurry of *L*-glutamic acid (**46**) (1.60 g, 10.9 mmol) in dry tetrahydrofuran (20 mL). The mixture was refluxed until complete dissolution was achieved (2 days), before being filtered. 9-Fluorenemethanol (2.35 g, 12.0 mmol), *N,N*-′-dicyclohexylcarbodiimide (2.47 g, 12.0 mmol), and 4-dimethylaminopyridine (0.15 g, 1.2 mmol) were then added. After 2 h at 25 °C, the slurry was filtered and concentrated to a residue that was subsequently dissolved in ethyl acetate (25 mL). Hydrogen chloride gas was bubbled through the solution for 10 min and (2*S*)-2-amino-5-(9*H*-fluoren-9-ylmethoxy)-5-oxo-pentanoic acid hydrochloride (**47**) collected by filtration before being washed with ethyl acetate (2 × 10 mL). The yield was 3.28 g (83%).

Complexation with borane itself is a useful means of protecting an amine against oxidation [59]. The borane complex can be sufficiently stable to survive multiple transformations and even column chromatography [60, 61], facts that will not surprise any

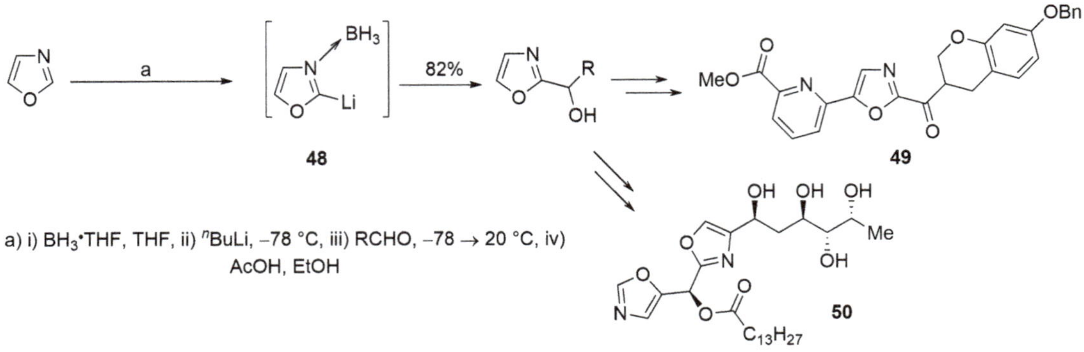

Fig. 20 Use of borane to suppress electrocyclic ring opening of a 2-lithiooxazole

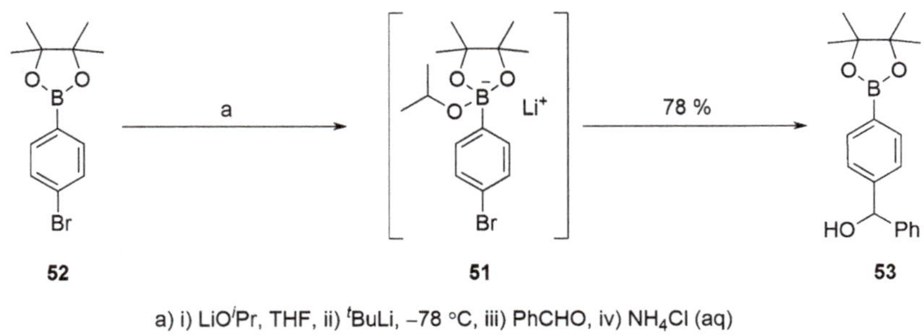

Fig. 21 Temporary protection of a boronic ester *via* an ate complex formation

synthetic organic chemist who has had to resort to vigorous acidic or basic conditions in order to cleave the complex resulting from the borane reduction of an amide. The benefits of this complexation have also been utilized in in situ procedures for the protection of heterocyclic nitrogens, perhaps most notably as part of the functionalization of oxazoles at the 2-position. Precomplexation with borane as in **48** completely suppresses the electrocyclic ring opening of 2-lithiooxazoles, allowing such a species to react smoothly with electrophiles (Fig. 20) [62]. Medicinal chemistry targets to whose synthesis this procedure has been applied include α-ketooxazole inhibitors of fatty acid amide hydrolase [63–66], such as compound **49** [67], and analogues of the antifungal agent bengazole A **50** [68].

We will complete this discussion on how boron-containing functionality can be used as a temporary protecting group, by pointing out that their Lewis acidity dictates that arylboronate esters themselves react with many organometallic reagents during attempts to functionalize another part of the molecule. The tying up of an organometallic reagent in this way can be prevented, however, through the reversible protection of the arylboronate ester itself as an aryltrialkoxyborate species such as **51**. Figure 21 illustrates how this has been achieved in the case of an arylpinacolatoboronate ester **52**, through its prior treatment with lithium isopropoxide, enabling the preparation of compound **53**

[69]. The isopropoxide is selectively removed in the workup, thus preventing the original boronate ester substituents from being scrambled.

5 Trimethylsilylation

The prior trimethylsilylation of alternative sites of reaction provides an operationally simple means of improving the chemoselectivity of certain reactions. Nowhere else is this better illustrated than in peptide chemistry, and we will start by describing its application to the synthesis of protected α-amino acids. Trimethylsilylation, ahead of the protection of an α-amino acid as an N^{α}-carbamate, has been used to significantly reduce the amount of oligomerization that can taint the formation of the most commonly utilized building blocks for both solution and solid phase peptide syntheses [70]. The protection of glycine (**54**) can be particularly problematic, as its lack of steric hindrance can lead to the formation of significant amounts of oligomeric by-products. *N,O-Bis*-trimethylsilylation as in **55**, prior to the introduction of the carbamoyl chloride, eliminates the formation of detectable amounts of oligomer and is illustrated in Fig. 22 [71]. The lability of *O*-TMS and *N*-TMS bonds leads to the ready removal of the protecting group in standard aqueous reaction workups.

Oligomerization aside, the *N*-carbamoylation or *N*-sulfonylation of hydroxyl-bearing amino acids and their derivatives, without prior protection of the hydroxyl functionality, can often lead to a substantial amount of another by-product, arising from *N,O*-biscarbamoylation or *N,O*-bissulfonylation, respectively. Prior trimethylsilylation circumvents this side reaction [72]. An example of its application can be found in the high-yielding isolation of *N*-carbobenzyloxy-L-proline, and substituted variants thereof, such as **56**, as part of the evaluation of the histone deacetylase (HDAC) activity of β-lactam analogues like **57** (Fig. 23) [73].[8] Parallels can

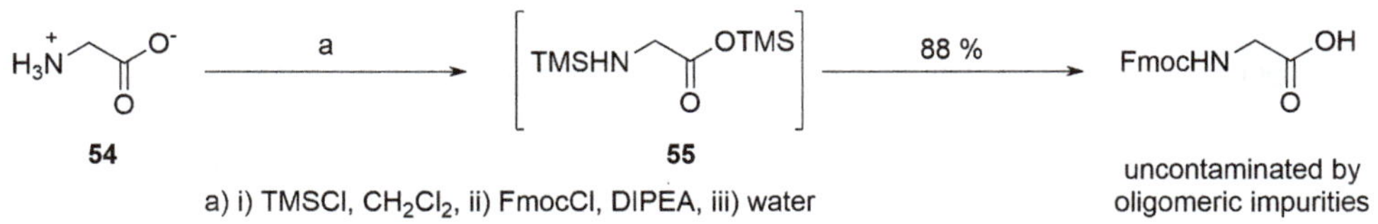

Fig. 22 Use of *N,O-bis*-trimethylsilylation in the *N*-carbamoylation of glycine

[8] It should be noted that there are numerous reports of an essentially quantitative yield being obtained for this transformation when conventional Schotten-Baumann conditions are employed.

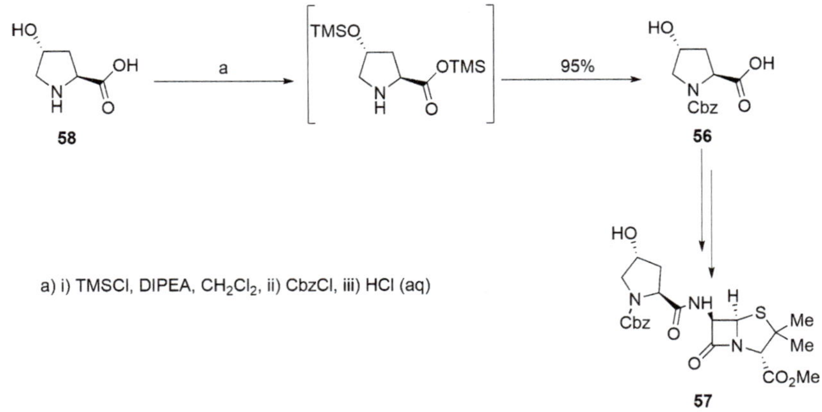

Fig. 23 Temporary *O*-trimethylsilylation as part of the *N*-carbamoylation of **58**

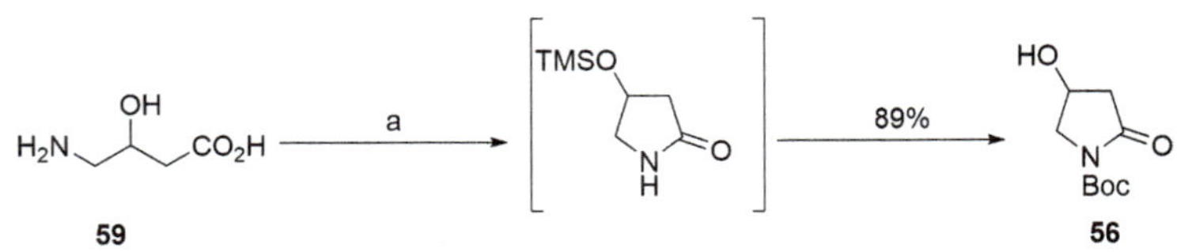

a) i) HMDS, pyridine, reflux, ii) $(Boc)_2O$, DMAP, iii) $Et_3N.HF$

Fig. 24 Temporary *O*-trimethylsilylation as part of the lactamization of (±)-4-amino-3-hydroxybutyric acid (**59**)

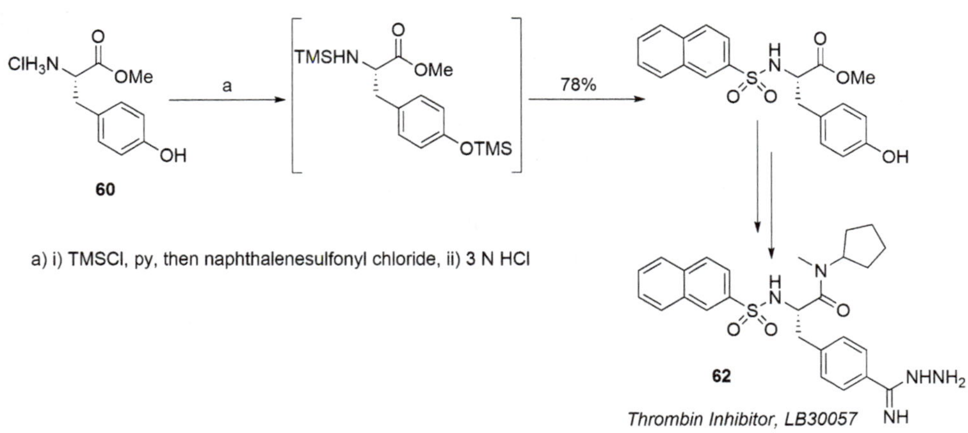

Fig. 25 Temporary *O*-trimethylsilylation as part of *N*-sulfonylation of methyl tyrosinate (**60**)

be drawn here with the increase in the efficiency of the dehydrative intramolecular lactamization of (±)-4-amino-3-hydroxybutyric acid (**59**) that was achieved by in situ trimethylsilylation, as part of the scale-up of an Abbott clinical candidate (Fig. 24) [74]. The same tactic allowed the *N*-selective introduction of a naphthalenesulfonyl group to methyl tyrosinate (**60**) and an *n*-butanesulfonyl to (*S*)-tyrosine (**61**), as part of eminently scalable syntheses of a

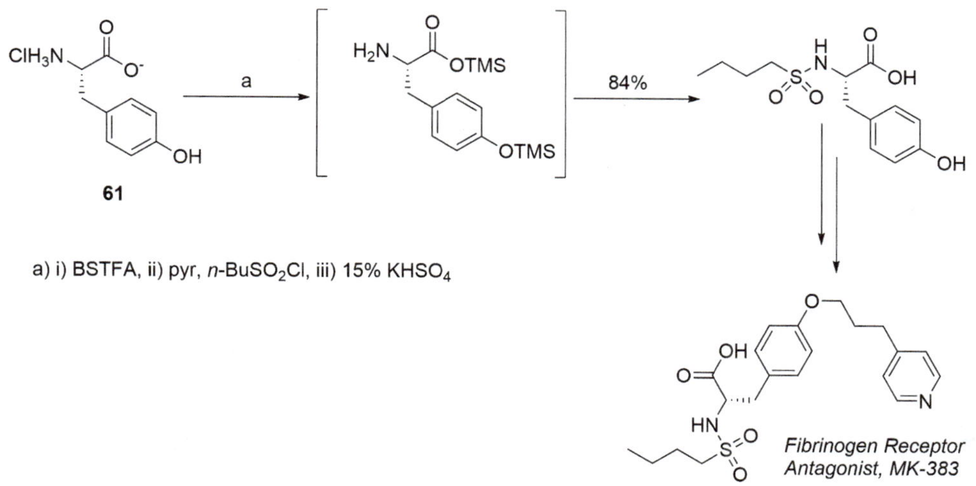

Fig. 26 Temporary *O*-trimethylsilylation as part of *N*-sulfonylation of (*S*)-tyrosine (**61**)

thrombin inhibitor **62** (Fig. 25) [75] and the kilogram-scale synthesis of a fibrogen receptor antagonist (Fig. 26), respectively [76].

5.1 N-Carboxybenzyl Protection of (2S,4R)-4-Hydroxypyrrolidine-2-Carboxylic Acid (58)

N,*N*-Diisopropylethylamine (24.3 g, 188 mmol) and trimethylsilyl chloride (27.9 g, 257 mmol) were sequentially added to a solution of (2*S*,4*R*)-4-hydroxypyrrolidine-2-carboxylic acid (**58**) (7.5 g, 57 mmol) in dry dichloromethane (120 mL), and the mixture refluxed for 2 h. Cooling to 0 °C was followed by the dropwise addition of benzyl chloroformate to the mixture (9.3 g, 55 mmol). After being stirred at 20 °C for 16 h, the reaction mixture was concentrated under reduced pressure, and the residue dissolved in sodium bicarbonate solution (120 mL of 3%). The aqueous layer was separated off and acidified to pH 2 using hydrochloric acid solution (5%). Extraction with ethyl acetate (3 × 100 mL) was followed by the washing of the combined extracts with ammonium chloride solution (200 mL), water (200 mL), and then brine (150 mL). Drying (magnesium sulfate) and concentration to a residue was followed by purification by chromatography on silica gel, eluting with *n*-hexane:ethyl acetate:methanol (70:25:5 v/v) so as to give (2*S*,4*R*)-1-benzyloxycarbonyl-4-hydroxypyrrolidine-2-carboxylic acid (**56**) as a beige oil (14.5 g, 96%).

α-Amino acids aside, prior trimethylsilylation has also expedited the *N*-carbamoylation of 1-aminoalkylphosphinic and 1-phosphonic acids [77–81].[9] In the case of the former, the reaction is thought to proceed through an equilibrating mixture of the

[9] It could be anticipated that this silylation strategy would work for the *N*-protection of related species, including amino sulfonates and amino sulfates.

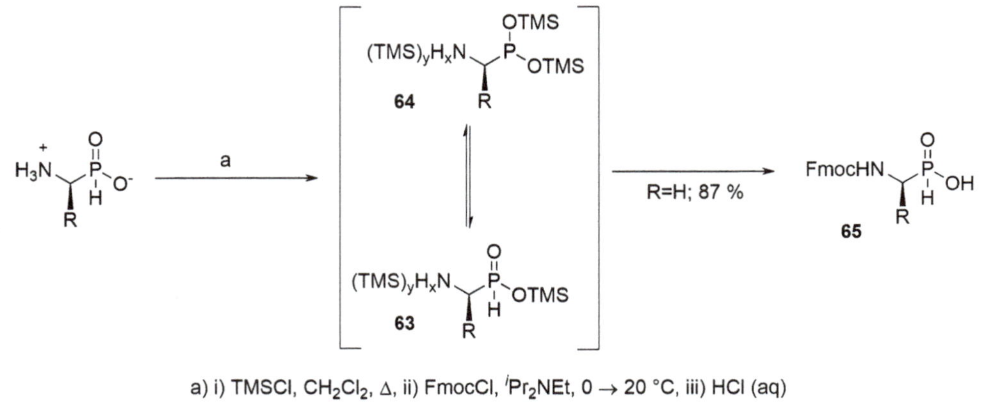

Fig. 27 Temporary trimethylsilylation in the *N*-carbamoylation of 1-aminoalkylphosphinic acids

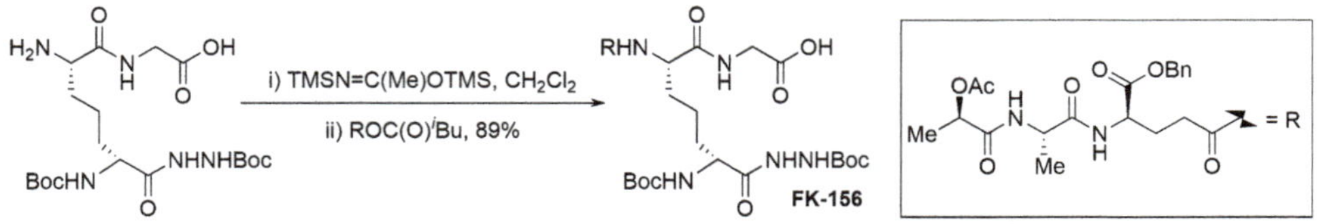

Fig. 28 Use of temporary trimethylsilylation to prevent oligomerization during coupling of peptide fragments

O-silyl *H*-phosphinate ester **63** and *O,O*-bistrimethylsilyl phosphonite **64** (Fig. 27). The carbamate-protected products **65** are important starting materials for the synthesis of phosphino and phosphono peptides that may serve as enzyme inhibitors or haptens for the construction of catalytic antibodies with protease-like specificity.

Moving on from the derivatization of the individual amino acid monomers, the risk of formation of oligomeric by-products during the functionalization of the amino terminus of a peptide bearing an unprotected carboxyl-terminus can be eliminated by the trimethylsilylation of the latter. An example can be found in the part of the synthesis of the immunostimulating peptide FK-156 shown in Fig. 28 [82, 83].

In a twist to the previously mentioned use of trimethylsilylation in the *N*-sulfonylation of amino acids, where it is the amino acid coupling partner that bears more than one potential site of reaction, the method has also been applied where a sulfonyl-bearing coupling partner is the one presenting the interfering functionality. Thus, using *N,O*-bis(trimethylsilyl)acetamide to silylate 4-hydroxybenzenesulfonyl chloride (**66**) prior to the introduction of an amine substrate allows the highly practical synthesis of 4-hydroxybenzenesulfonamides such as **67** (Fig. 29) [84]. This class of compound has been used for the exploration of the structure-activity relationships (SAR) of selective pipecolic acid-

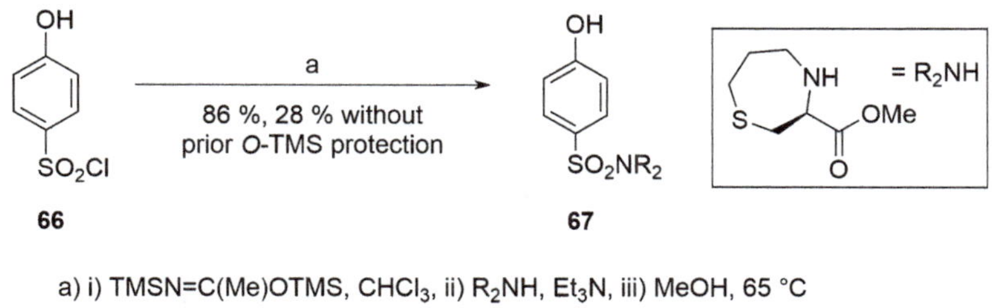

Fig. 29 Use of trimethylsilylation in the sulfonamidation of 4-hydroxybenzenesulfonyl chloride

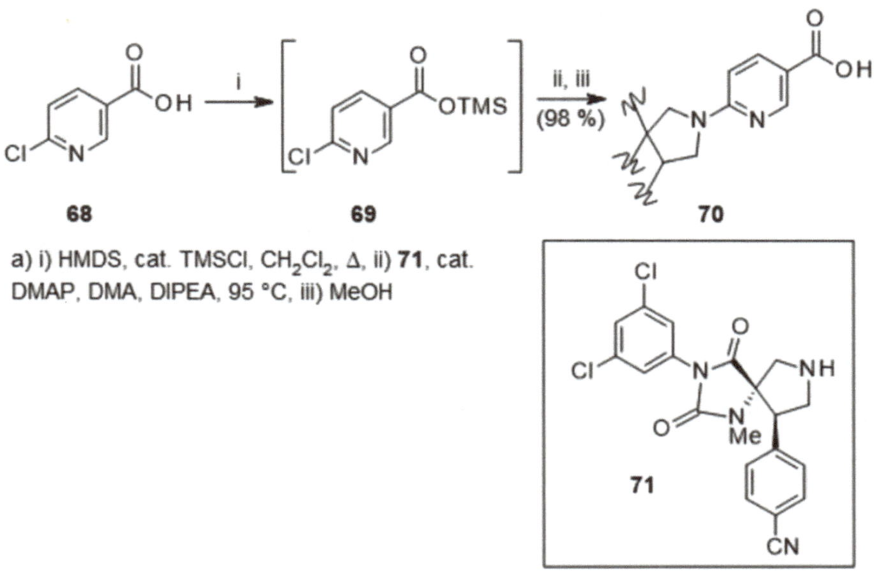

Fig. 30 Temporary protection of a carboxylic acid to improve the efficiency of an S_NAr reaction

based inhibitors of tumor necrosis factor-α-converting enzyme (TACE) [85].

The derivatization of acidic functionalities using trimethylsilylation to improve chemoselectivity has found applications outside the arena of amino acid, peptide, and related chemistry. The efficiency with which a spirocyclic hydantoin core engaged in an S_NAr with 6-chloronicotinic acid (**68**) was improved by the trimethylsilylation of the latter to give **69**, as part of the synthesis of a small molecule inhibitor of the LFA-1/ICAM interaction **70** (Fig. 30) [86]. With its carboxylic acid unprotected, compound **68** proved to be less reactive (presumably due to the deprotonation of this functional group) in the S_NAr, with the result that the hydantoin coupling partner **71** was slowly acetylated by the *N,N*-dimethylacetamide solvent.

O^4-Trimethylsilylation has improved the efficiency of the reaction at C5 of 5-iodouracil, 5-iodouridine, 5-iodo-2′-deoxyuridine, or substituted variants thereof on a number of occasions

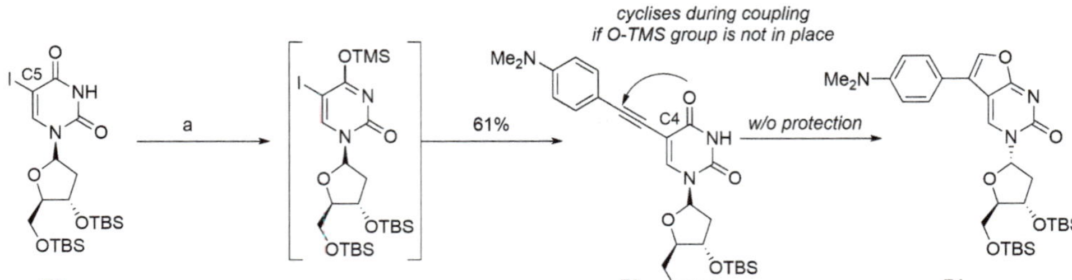

a) i) HMDS, TMSCl, 80 °C, ii) PhMe then concentrate, iii) Pd(PPh$_3$)$_4$, CuI, Et$_3$N, 4-ethynyl-*N,N*-dimethylaniline, THF, 60 °C, iv) chromatography

Fig. 31 Use of in situ trimethylsilylation as part of the functionalization of a 5-iodouridine derivative

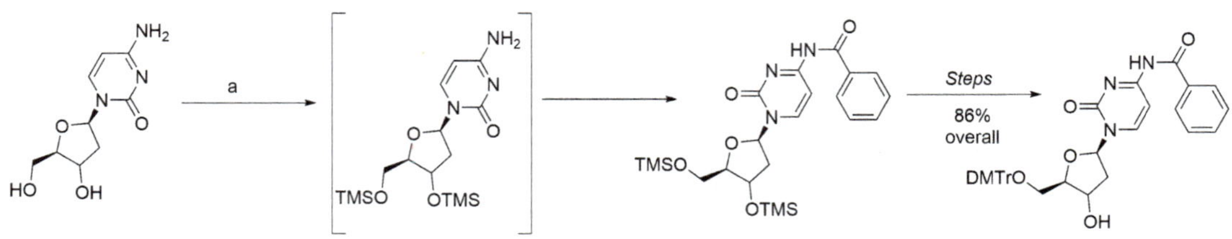

a) i) Et$_3$N, TMSCl, rt, ii) benzoyl chloride, DMAP, - 50 °C, then two steps

Fig. 32 Use of in situ trimethylsilylation for the synthesis of 5'-*O*-dimethoxytrityl-*N*-benzoyl-2′-deoxycytidine (**75**) from 2′-deoxycytidine (**76**)

[87–90]. An example can be found in the introduction of the linker of a *N,N*-dialkylaniline-2′-deoxyuridine conjugate used for DNA-mediated electron transfer studies (Fig. 31) [91]. The transient protection of the C4-carbonyl of **72** prevented the onward cyclization of the Sonogashira coupling adduct **73** into furopyrimidinone **74**. Sticking with deoxynucleosides, the synthesis of 5′-*O*-dimethoxytrityl-*N*-acyl-2′-deoxynucleosides, which are key building blocks for the synthesis of oligodeoxynucleotides has been aided by the in situ *O*-silylation of the 3′ and 5′ positions. The original conditions of Ti et al. [92], which used trimethylsilyl chloride and triethylamine, have been improved by either replacing pyridine by triethylamine in order to prevent *N*-glycosidic bond cleavage by the conjugate acid of the former [93] or by silylating with hexamethyldisilazane [94]. The exploitation of this method for an efficient synthesis of a 5′-*O*-dimethoxytrityl-*N*-acyl-2-′-deoxycytidine derivative **75** from 2′-deoxycytidine (**76**) is shown in Fig. 32.

Finally, in a piece of methodology that has enormous potential for the construction of polysubstituted phenols, such as **77**, in drug substance synthetic targets, an *N*-monoalkylcarbamate group can direct *ortho*-metallation once it has been *N*-silylated as in **78** (Fig. 33) [95]. Such a group can be removed under milder

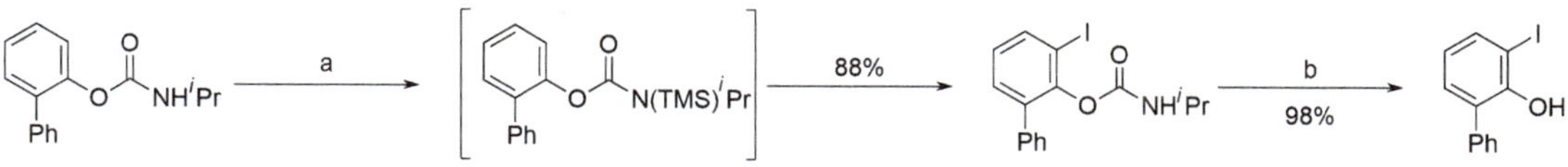

78

79

77

a) i) TMSOTf, TMEDA, Et_2O, 0 °C, ii) ^nBuLi, TMEDA, –78 °C, iii) I_2, iv) 2M HCl, MeOH; b) 2M NaOH, EtOH

Fig. 33 Use of in situ trimethylsilylation as part of functionalization of a phenol derivatized as an *N*-monoalkylcarbamate

conditions compared to the tertiary carbamoyloxy group, a widely recognized group for directed metallation (DMG), lending the *O*-aryl *N*-monoalkylcarbamate potentially greater utility over its more substituted sibling as a DMG for phenol functionalization. In the absence of the in situ protection, the *N*-lithiated carbamate of compound **79** would eliminate a phenoxide, prior to its *ortho*-lithiation, prematurely terminating events.

6 Lanthanide Chlorides

The discussion will now move from the protection of amines, alcohols, and carboxylic acids to the carbonyls of aldehydes and ketones. The efficient differentiation between the carbonyls of an aldehyde and a ketone present in the same molecule, ahead of the onward reaction of only the ketone carbonyl, requires recourse to the derivatization of the aldehyde carbonyl, frequently as an acetal. An acetalization is unlikely to be completely chemoselective, attenuating the yield and introducing the need for a purification step. A separate step would also be required to liberate the aldehyde carbonyl after the intervening reaction of the ketone carbonyl.

Luche and Gemal's use of a lanthanide chloride (sometimes written as lanthanoid chloride) to achieve in situ protection of an aldehyde carbonyl has proven to be a seminal development for such transformations [96].[10] Cerium chloride hexahydrate, the most commonly used lanthanide chloride in this context, temporarily masks the aldehyde carbonyl as a cerium(III)-stabilized geminal diol such as **80**, in the case of the reduction of **81** (Fig. 34).

A closely related strategy exploits the regioselective or chemoselective bias displayed by the lanthanide chloride-catalyzed acetalization of carbonyls. The in situ acetalization of the cyclohexanone of 5α-androstane-3,17-dione (**82**) allowed 17β-hydroxy-5-α-androstan-3-one (**83**) to be regioselectively formed through reduction (Fig. 35) [97]. This constitutes a reversal of the regioselectivity that is otherwise observed when **82** is reduced with sodium borohydride in methanol [98].

[10] The method relies on its success on the transformation of the aldehyde carbonyl into a hydrate. It thus fails with conjugated aldehyde carbonyls where hydrate formation is not facile.

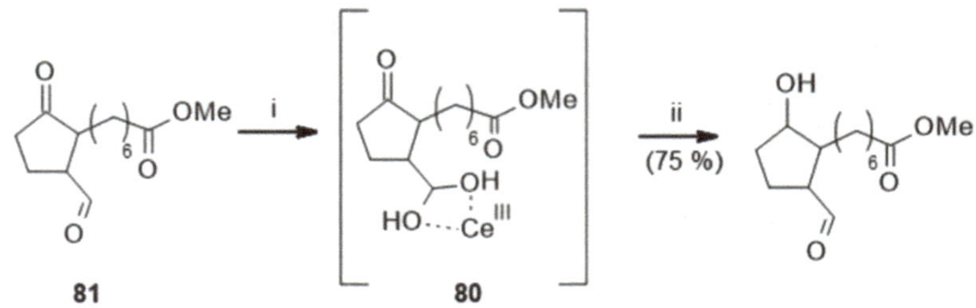

81 80

a) i) $CeCl_3$•$6H_2O$, EtOH-water (6:1 v/v), ii) $NaBH_4$, −15 °C

Fig. 34 Use of cerium(III) chloride to facilitate regioselective reduction of a ketone carbonyl in presence of an aldehyde carbonyl

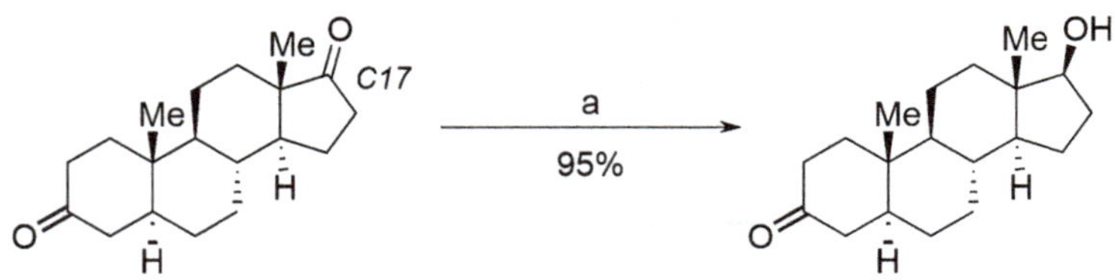

82 83

a) i) $ErCl_3$•$6H_2O$, $HC(OMe)_3$, MeOH, ii) $NaBH_4$, iii) HCl (aq)

Fig. 35 Selective functionalization of C17 carbonyl of 5α-androstane-3,17-dione after C3 carbonyl has been acetalized in situ

6.1 Reduction of 5α-Androstane-3,17-dione (82) to 17β-Hydroxy-5α-androstan-3-one (83)

Trimethyl orthoformate (750 mg, 7.07 mmol) was added to a solution of 5α-androstane-3,17-dione (**82**) (288 mg, 1.00 mmol) and erbium(III) chloride hexahydrate (382 mg, 1.00 mmol) in methanol (5 mL). After being stirred for 15 min at 20 °C, sodium borohydride (80 mg, 2.1 mmol) was added in a single portion. After 5 min, the pH was adjusted to 2 with hydrochloric acid solution (1 M). After a further 20 min, the reaction mixture was partitioned between water and diethyl ether. The organic phase was washed with sodium chloride solution (saturated) and dried (sodium sulfate) before being concentrated to a solid which was recrystallized from a mixture of dichloromethane and hexane so as to afford 17β-hydroxy-5α-androstan-3-one (**83**) (275 mg, 95%).

7 Bulky Lewis Acids

While the complexation of a Lewis acid to a carbonyl will typically enhance the susceptibility of the latter toward nucleophilic attack, such reactivity can be shut down with a sufficiently bulky Lewis acid on the grounds of steric hindrance. Ketone carbonyls in differing steric environments differ in the extent to which they complex to such a Lewis acid, in that the less encumbered one can be

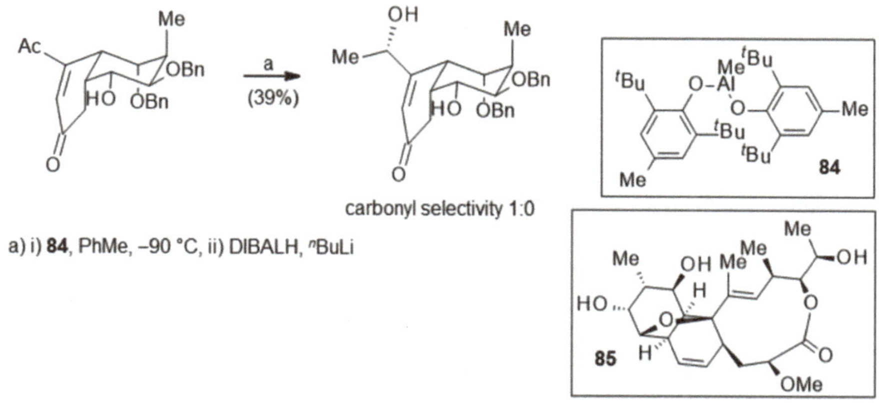

Fig. 36 Use of MAD (**84**) to differentiate between two carbonyl groups such that only one is reduced

preferentially masked from a subsequent reaction in the same pot. Methyl aluminum *bis*(2,6-di-*tert*-butyl-4-methylphenoxide) (MAD, **84**) is an example of a bulky Lewis acid whose use has been exploited in this regard [99], as in studies toward a synthesis of the macrolide antibiotic nodusmicin (**85**) (Fig. 36) [100]. The use of MAD does not extend to aldehydes, in general, where the high reactivity of an aldehyde-MAD complex will still result in reaction at the aldehyde carbonyl. The development of methyl aluminum *bis*(2,6-diphenylphenoxide) (MAPH) overcomes this limitation, allowing a ketone carbonyl to be transformed in the presence of an aldehyde carbonyl or the differentiation of two aldehyde carbonyls in the same molecule [101].

8 Lithium α-Amino Alkoxides

The Comins group developed an alternative means of achieving the differentiated reactivity of aldehyde and ketone carbonyl groups using the knowledge that lithium diisopropylamide will add to certain non-enolizable ketones [102]. This led them to study the addition of a lithium dialkylamide to the carbonyl of a non-enolizable aldehyde [103, 104], effectively masking it as the intermediate of the venerable Bouveault reaction such as **86**, the lithium salt of an *N*,*O*-hemiaminal [105–108]. The formation of this α-amino alkoxide intermediate does not readily reverse on exposure to organometallic reagents. Thus, such reagents can be deployed in the chemoselective functionalization of another part of the molecule, typically a less reactive carbonyl group or an aromatic ring. The remarkable chemoselectivity that can be achieved can be glimpsed through an examination of part of the Danishefsky group's total synthesis of the antitumor agent calicheamicin [109]. This work used lithium *N*-methylanilide to selectively

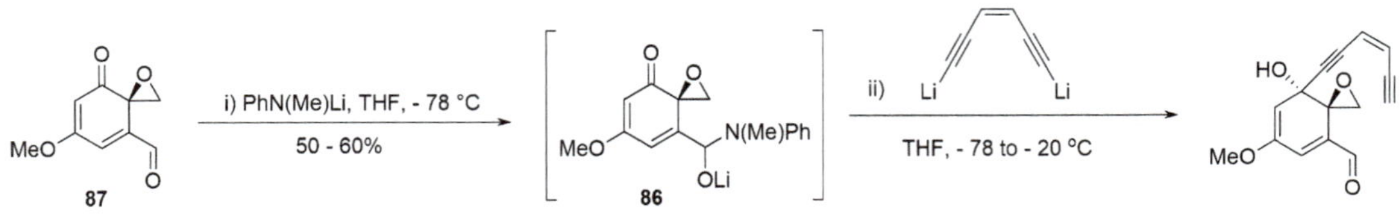

Fig. 37 Protecting an aldehyde carbonyl as an *N,O*-hemiaminal as part of a synthesis of calicheamicin

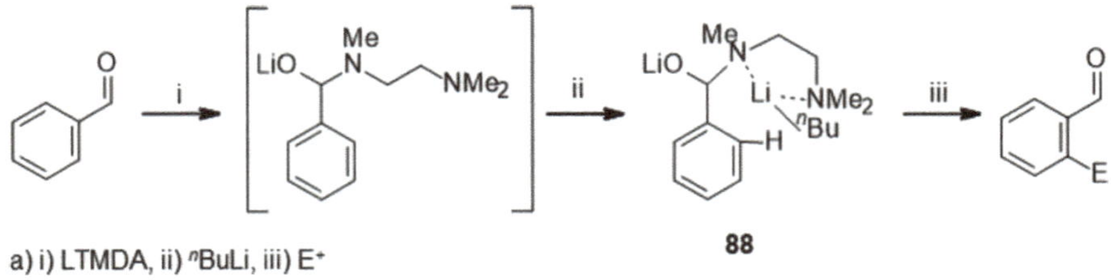

Fig. 38 Demonstration of the facilitation of *ortho*-lithation when α-amino alkoxide is formed from *N,N,N'*-trimethylethylenediamine

protect the aldehyde carbonyl of the ketoaldehyde **87** shown in Fig. 37, in the presence of an epoxide moiety. The ketone carbonyl of the resulting α-amino alkoxide **86** was then engaged with a lithium acetylide.

The organolithiums typically used to metallate an aromatic ring are also strong nucleophiles and would undergo clean 1,2-addition to the unprotected carbonyl of an aromatic or heteroaromatic aldehyde. The prior addition of a dialkylamide, as well as preventing the organolithium from attacking such an aldehyde carbonyl, produces a moiety which can direct a subsequent *ortho*-metallation [110]. Lithium dialkylamides that have been employed as part of such a tactic include the aforementioned lithium *N*-methylanilide, as well as lithium morpholide, lithium *N*-methyl-*N*-(2-pyridyl) amide (LMPA), lithium 1-methylpiperazide (LNMP), lithium *N,O*-dimethylhydroxylamide, and the lithium amide of *N,N,N'*-trimethylethylenediamine (LTMDA) [111]. The α-amino alkoxide formed from the latter is the most powerful *ortho*-director, presumably due to the intramolecular TMEDA-like complexation of the metal component shown for **88** (Fig. 38). The fact that different α-amino alkoxides vary in their ability to direct an *ortho*-lithiation [112] allows the regioselectivity of the lithiation of an aromatic or heteroaromatic aldehyde to be dictated, and indeed tuned, by the competition between the α-amino alkoxide chosen and any other substituent on the ring [113, 114]. An example is shown in Fig. 39 where the *N*-methyl piperazine of compound **89** is unable to outcompete the methoxy group for mesityllithium such that it is the 5-, and not the 2-, position of the ring that is metallated. This results in the formation of compound **90** after quenching with iodomethane.

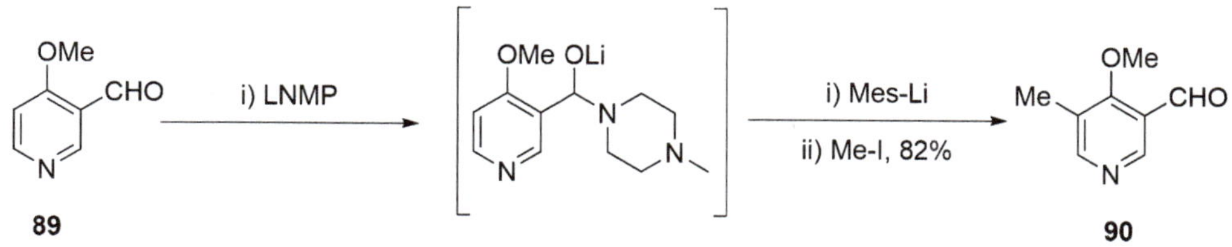

Fig. 39 Protection of an aldehyde as a comparatively weakly *ortho*-directing α-amino alkoxide

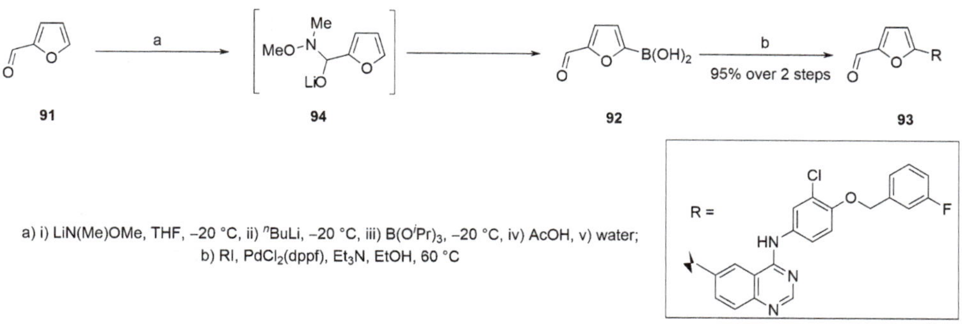

Fig. 40 Protection of aldehyde carbonyl ahead of furan metallation as part of a synthesis of a family of GSK kinase inhibitors

The Comins methodology has been heavily exploited in drug discovery and development programs. In particular, it has been used for the introduction of functionality that opens up myriad options for the modular assembly of a candidate drug target through cross-coupling. The introduction of a tributyltin group has been used as part of a synthesis of the antibiotic WS 5995B [115], while trapping with an electrophilic iodine source has been part of the syntheses of the benzo[*b*]fluorene core of the kinamycin antibiotics and ^{13}C-labeled polyprenylhydroxybenzoic acids [116, 117]. Figure 40 provides an example of the introduction of boronic acid functionality to furfural (**91**). It was elaborated to compound **92** as part of a one-pot synthesis of the pentacyclic core **93** of GlaxoSmithKline (GSK) inhibitors of the ErbB family of protein tyrosine kinases [112]. Here the lithium α-amino alkoxide **94** arising from the use of lithium *N,O*-dimethylhydroxylamide offers minimal *ortho*-directing behavior. Figure 41 demonstrates another use of the Comins methodology by GSK scientists to form a boronic acid, in this case **95**, as part of the synthesis of an insulin secretion modulator. In this instance, the regioselectivity arises from lithium-halogen exchange on the aryl ring of the α-amino alkoxide intermediate, rather than *ortho*-metallation [112].

8.1 Borylation of 2-Furaldehyde

n-Butyllithium (5.00 mL of 2.5 M in hexane, 12.5 mmol) was added to a slurry of *N,O*-dimethylhydroxylamine hydrochloride (623 mg, 6.39 mmol) in tetrahydrofuran (30 mL) while keeping the internal temperature ≤10 °C. The reaction mixture was then

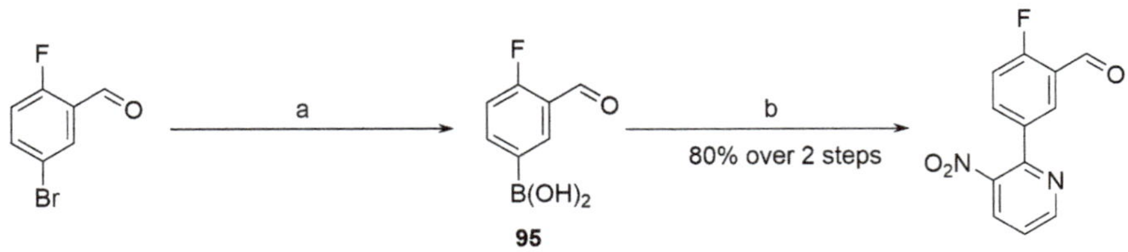

95

a) i) LiN(Me)OMe, −20 °C, ii) B(O^{i}Pr)$_3$, −40 °C, iii) ^{t}BuLi, −40 °C, iv) AcOH, 20 °C, v) water, 20 °C, b) 2-chloro-3-nitro-pyridine, $PdCl_2$(dppf), Na_2CO_3, water, ^{i}PrOH, toluene, 73 °C

Fig. 41 Temporary protection of an aryl aldehyde carbonyl as an α-amino alkoxide ahead of lithium-halogen exchange elsewhere on the ring

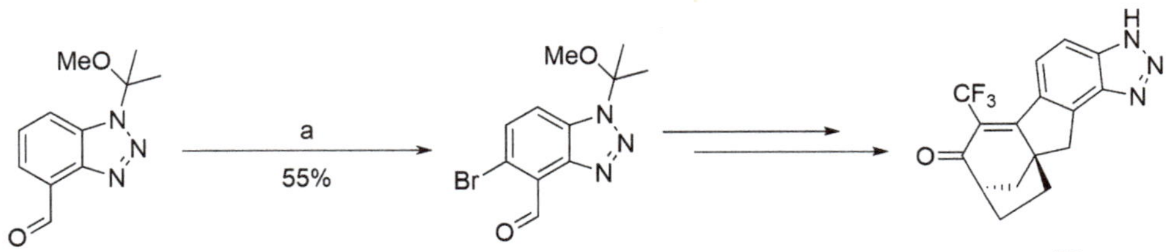

96 97

a) i) $MeNLiCH_2CH_2NMe_2$, ^{n}BuLi, −20 °C, ii) CBr_4, −50 °C

Fig. 42 Temporary protection of an aldehyde carbonyl as an α-amino alkoxide as part of synthesis of an estrogen receptor beta-selective agonist

stirred at −20 °C for 30 min before 2-furaldehyde (0.43 mL, 5.19 mmol) and then *n*-butyllithium (2.70 mL of 2.5 M in hexane, 6.75 mmol) were sequentially added, each at a rate which kept the internal temperature below −15 °C. The pale yellow solution was allowed to stir at −20 °C for 75 min, before triisopropyl borate (1.84 mL, 7.97 mmol) was added at a rate which kept the internal temperature below −15 °C. The reaction mixture was stirred for 15 min before acetic acid (0.75 mL, 13 mmol) was added, and the reaction mixture warmed to 15 °C. Deionized water (470 μL, 26.1 mmol) was then added, and the resulting pale yellow slurry stirred for 10 min. Approximately 10–15 mL of volatiles were solvent swapped with tetrahydrofuran, so as to afford a crude slurry of (5-formyl-2-furyl)boronic acid (**92**) which was carried forward to the Suzuki coupling reaction shown in Fig. 40.

While parlayed as part of an intramolecular acylation, rather than a cross-coupling, the bromination of **96** shown in Fig. 42 was investigated as part of the manufacture of a Merck estrogen receptor β-selective agonist **97** [118]. Batch mode studies were characterized by modest reproducibility and yields and a highly exothermic *ortho*-lithiation. These factors prevented its adoption for the supply of drug substance for early clinical studies; however, there remains the tantalizing possibility of the methodology graduating to the process chemist's armamentarium if accommodated

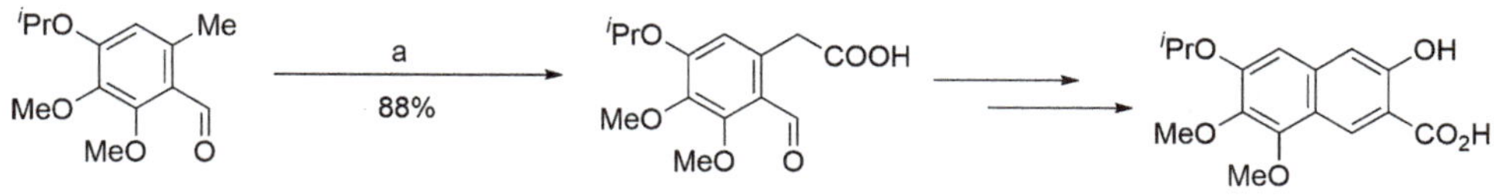

98 99

a) i) $Me_2NCH_2CH_2N(Me)Li$, THF, –20 °C, ii) ^nBuLi, –20 °C, iii) CO_2, –78 → 23 °C

Fig. 43 Temporary protection of an aldehyde carbonyl as an α-amino alkoxide ahead of benzylic functionalization

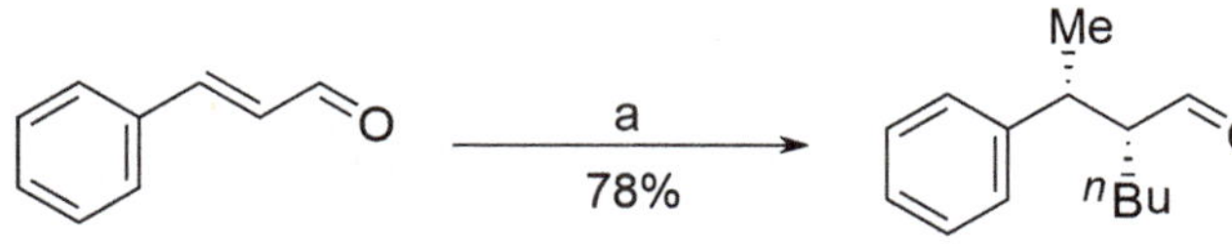

100 101

a) i) $Me_2NCH_2CH_2N(Li)Me$, Et_2O, –40 °C, ii) ^nBuLi, 0 °C, iii) MeI, THF, –80 °C, iv) phosphate buffer

Fig. 44 Temporary protection of carbonyl of cinnamaldehyde as an α-amino alkoxide ahead of carbolithiation of the carbon-carbon double bond

in a flow setting. There are many other examples where the Comins method has been used [119–122], or has the potential to be used [123], to streamline the synthesis of pharmaceutically relevant small molecules through the chemoselective functionalization of aromatic or heteroaromatic aldehydes.

While the α-amino alkoxide intermediate arising from the Comins method is typically formed so as to permit the decoration of an aryl ring, there are also examples where lateral lithiation has been used for the functionalization of a benzylic position [134]. One such example is the carboxylation of compound **98** shown in Fig. 43, which was used to synthesize compound **99**, the naphthoic acid component of the chromophore of kedarcidin, a chromoprotein antibiotic with anticancer activity [135]. The versatility of the Comins method is further showcased with its application to the protection of the carbonyl of cinnamaldehyde (**100**), ahead of the regioselective addition of an organolithium across the olefin [136]. The resulting carbolithiated intermediate can then be further functionalized through alkylation, prior to the unmasking of the aldehyde carbonyl so as to afford products exemplified by **101** (Fig. 44) [137].

While the discussion has hitherto focused on accessing the α-amino alkoxide intermediate by attacking the carbonyl of an aldehyde with an amide base, it can also be accessed by attacking a carboxamide starting material with an organolithium. The chelating effect of an *N*-methoxy that characterizes Weinreb amide

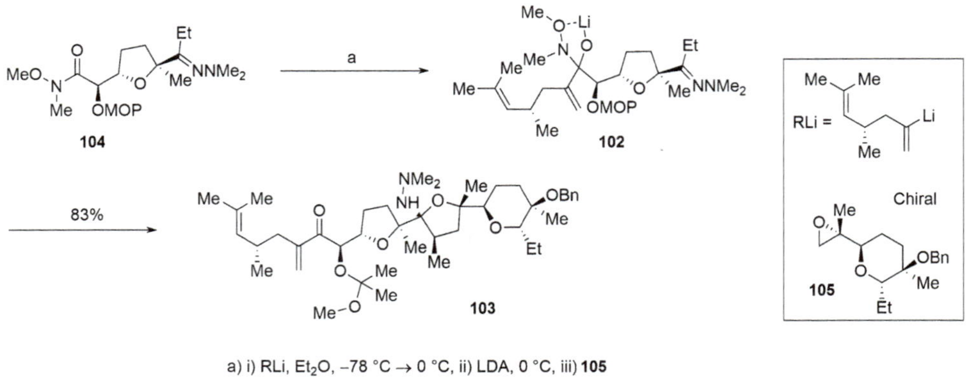

Fig. 45 Use of α-amino alkoxide formed from alkyllithium addition to a Weinreb amide to temporarily protect a ketone carbonyl

functionality in the starting material stabilizes the α-amino alkoxide intermediate (e.g., **102**), once formed, toward collapse. When it does collapse, on working up the reaction, it is the carbonyl of a ketone that is unmasked. Thus this is not a simple case of protecting the original carbonyl functionality in the substrate as this carbonyl is irreversibly transformed by the sequence. Nevertheless, it seems appropriate to discuss this variant of the technology briefly. It was exploited as part of an efficient assemblage of the "right-hand side" of the polyether antibiotic X-206 (**103**), whereby a lithium α-amino alkoxide **102** was formed, through the addition to the Weinreb amide of **104**. The lithium α-amino alkoxide functionality protects the ketone that would otherwise arise at this position were the reaction to have been worked up, allowing a hydrazone elsewhere in the molecule to be efficiently enolized and reacted with epoxide **105** (Fig. 45) [138–140].

In general, the Comins method has the disadvantage that the lithium amide reagents used to access the α-amino alkoxide intermediate are strongly basic, as well as being nucleophilic, making the method most suitable for use on substrates where acidic protons are not present or where their removal does not lead to undesired side reactions [141]. An only modestly basic reagent that has been used to mask aldehyde carbonyls is lithium pyrrolate [142]. The *O*-lithiated pyrrole carbinol is stable to lithium borohydride, Grignard reagents, and *n*-butyllithium at low temperatures. The potential of telescoping these steps into a one-pot procedure is illustrated with the deuteration shown in Fig. 46 of the arene ring of 3-bromobenzaldehyde (**106**). In this case, the aldehyde carbonyl only fleetingly reappears in the midst of a telescoped deprotection-HWE (Horner-Wadsworth-Emmons) sequence which ultimately results in the isolation of enoate **107**.

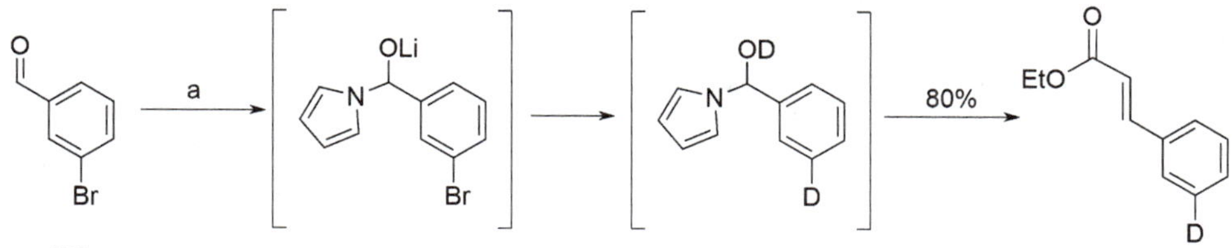

106 107

a) i) lithium pyrrolate, –78 °C, THF, ii) ^{n}BuLi then CD_3OD, iii) $(EtO)_2P(=O)CHLiCO_2Et$, THF, EtOH, –78 → 20 °C

Fig. 46 Masking of an aldehyde carbonyl using lithium pyrrolate

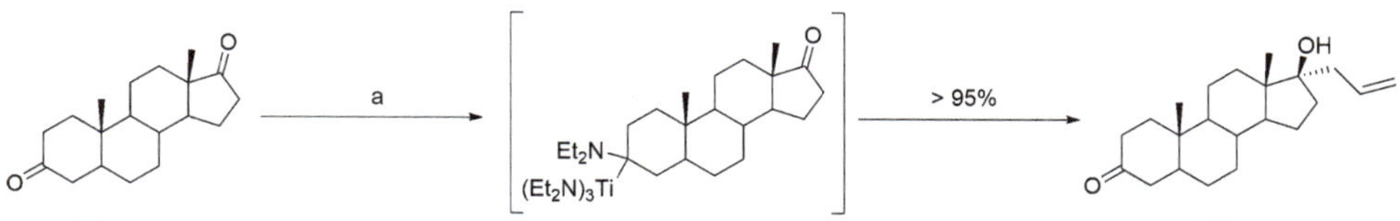

82

108
carbonyl regioselectivity:
>99:1

a) i) $Ti(NEt_2)_4$, THF, ii) CH_2=$CHCH_2MgCl$

Fig. 47 Use of titanium tetrakis(diethylamide) to differentiate between two carbonyl groups

9 α-Amino Alkoxides with Other Metals

Alternative means of avoiding the use of a basic lithium amide reagent include the use of a tetrakis(dialkylamide) of titanium or manganese [143, 144]. These enable the in situ use of Grignards, organolithiums, or enolates as nucleophiles elsewhere in the molecule. The adducts readily unmask the carbonyl on exposure to water during an aqueous reaction workup. The method is demonstrated in Fig. 47 with the allylation of androstane-3,17-dione **82** to give **108**. Here titanium tetrakis(diethylamide) masks the ketone carbonyl that is sterically less encumbered, though it is more commonly used to react a ketone carbonyl in the presence of an aldehyde carbonyl.

Aluminum amides have also proven to be useful means of masking a carbonyl without the limitations imposed by the basicity of the lithium reagents. Aluminum complexes of *N*,*O*-dimethylhydroxylamine (Weinreb amine) have most notably been exploited [145–147]. The corresponding α-amino alkoxide intermediate (e.g., **109**) is reported to be exceptionally resistant to nucleophilic attack, permitting the chemoselective reaction of a ketone or ester carbonyl with an organometallic reagent in the presence of an aldehyde carbonyl and an ester carbonyl to be transformed in the presence of a ketone carbonyl. The disadvantage of the reagent's ability to react with a ketone carbonyl, albeit more slowly than with an aldehyde carbonyl, is that the strict control of the stoichiometry of the reagent is required if useful levels of chemoselectivity are to

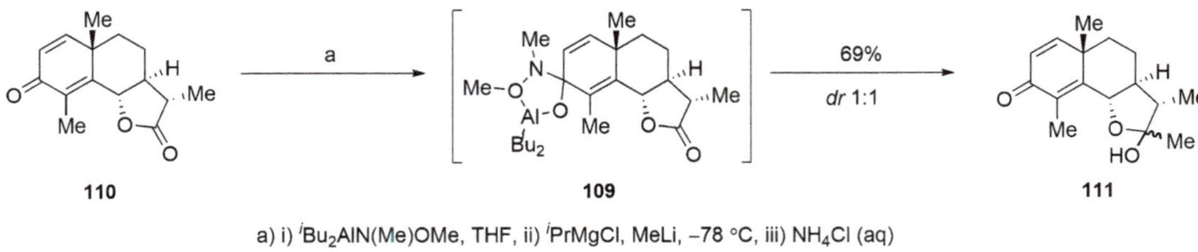

Fig. 48 Temporary protection of a ketone carbonyl using aluminum complex of *N,O*-dimethylhydroxylamine

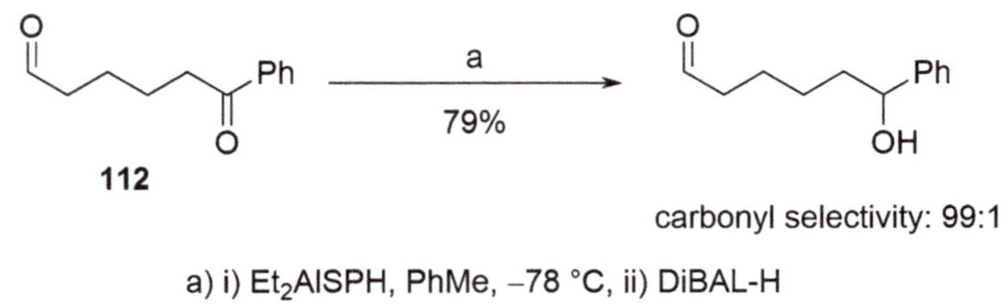

Fig. 49 Use of an *O,S*-acetal to temporarily protect an aldehyde carbonyl ahead of reduction elsewhere in **112**

be achieved in a substrate where both types of carbonyl are present. The utility of this methodology is demonstrated in the reaction of the anthelminthic α-santonin (**110**), where a lactone carbonyl was selectively alkylated to give compound **111** in the presence of a highly reactive dienone moiety (Fig. 48) [145].

10 *O,S*-, *O,P*-, *N,N*- and *O,O*-Acetals

The lithium α-amino alkoxides developed with such success by the Comins group necessarily correspond to the lithium salt of an *N,O*-hemiaminal. A number of groups have shown that success can be achieved with related species that can be classified on account of their heteroatoms. One such group is the *O,S*-acetal formed by the reaction of an aldehyde carbonyl with diethylaluminum benzenethiolate. A ketone carbonyl, as demonstrated in Fig. 49 with **112**, or ester carbonyl, can be chemoselectively reduced in the presence of an aldehyde carbonyl that has been temporarily masked in this way [148].

Through exposure to trimethylsilyl trifluoromethanesulfonate and a phosphine, an aldehyde or ketone carbonyl can be protected as an *O,P*-acetal phosphonium salt, permitting the transformation of another carbonyl or a nitrile elsewhere in the molecule [149, 150]. The use of triphenylphosphine or the more nucleophilic triethylphosphine (CARE: pyrophoric) permits the masking of an aldehyde or ketone carbonyl, respectively. As illustrated in Fig. 50, the former approach was used in conjunction with a Corey-Bakshi-

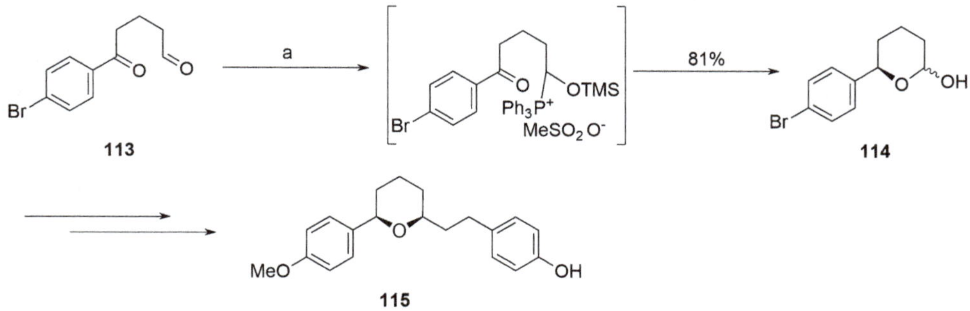

a) i) PPh_3, TMSOTf, CH_2Cl_2, ii) (*S*)-CBS, BH_3•THF, iii) TBAF (THF), $NaHCO_3$, MeOH, 40 °C

Fig. 50 Use of an *O,P*-acetal to temporarily protect an aldehyde carbonyl ahead of reduction elsewhere in **113**

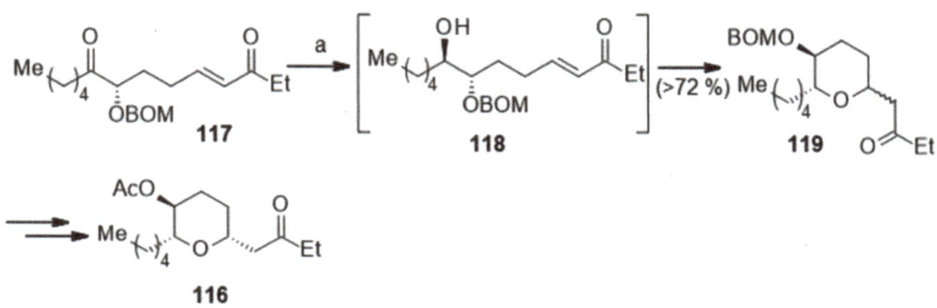

a) i) Ph_3P, TMSOTf, CH_2Cl_2, −78 °C, ii) Red-Al, iii) LiOMe, MeOH

Fig. 51 Use of an *O,P*-acetal to temporarily protect carbonyl of an enone

Shibata reduction for the transformation of **113** to **114** as part of an asymmetric synthesis of the diarylheptanoid (+)-centrolobine (**115**) [149]. A conjugated enone is one class of ketone that can be masked with triphenylphosphine using this methodology, allowing its differentiation from that of a non-conjugated ketone carbonyl [151]. This is exemplified in Fig. 51 by part of a synthesis of decytospolide B (**116**), a compound that has shown in vitro cytotoxic activity toward tumor cell lines A549 and QGY [151]. Keto-enone **117** was converted to hydroxy-enone **118** prior to intramolecular cyclization to pyran **119**.

The fact that the acetal or aminal formation can be driven to completion by simply removing water from a system lends itself to reaction conditions that can be synchronized, albeit typically after a solvent swap in the same pot, with subsequent functionalization elsewhere in the molecule. In the example shown in Fig. 52, the protection of the aldehyde carbonyl of **120** as a 1,3-dimethylimidazolidine **121** permitted the hydroxylation of the aromatic ring to give **122**, which was utilized in a search for differentiation inducers of promonocytic leukaemic cells [152]. As well as assisting with the chemoselectivity, the protection of the aldehyde as a 1,3-dimethylimidazolidine directed the lithiation of

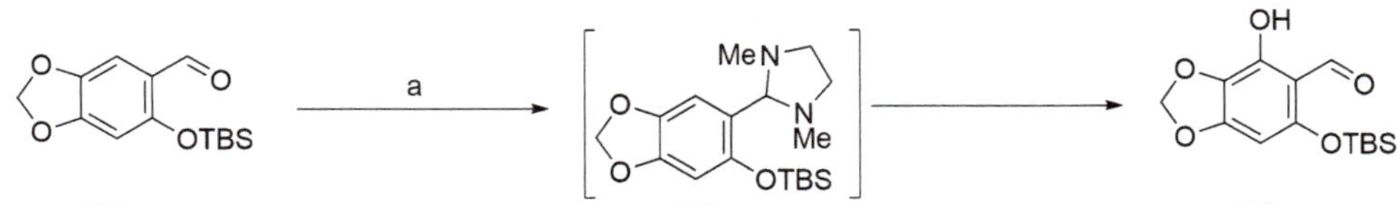

120 **121** **122**

a) i) *N,N'*-dimethylethylene-1,2-diamine, toluene, Δ, ii) concentrate, iii) ^{t}BuLi, Et_2O, −90 °C, iv) $B(OMe)_3$, − 78 → 0 °C, v) H_2O_2, AcOH, vi) HCl (aq)

Fig. 52 Temporary protection of an aldehyde carbonyl as an imidazolidine

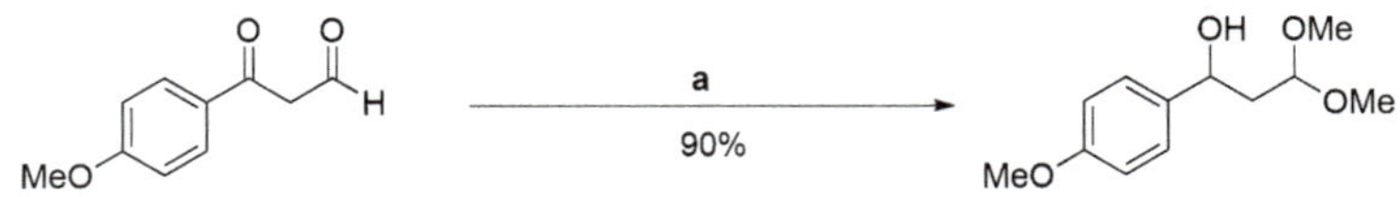

123

a) i) $TiCl_3$, MeOH, Et_3N, ii) NH_3-H_2O, iii) $TiCl_3$, H_2O

Fig. 53 Temporary protection of an aldehyde carbonyl as an acetal ahead of reduction of a ketone carbonyl

the ring to the position *ortho* to the carbonyl group. Mildly acidic conditions have also been used to form an acetal from an aldehyde carbonyl ahead of a transformation, often of a ketone carbonyl, elsewhere in the molecule [153, 154]. Conversely, Porta and co-workers have used mildly basic conditions to protect an aldehyde carbonyl as an acetal, permitting the chemoselective reduction of the carbonyl of an aromatic ketone using titanium(III) chloride [155]. This is exemplified in Fig. 53 with the reduction of the ketone carbonyl of compound **123**.

11 Imines and Formamidines

Certain types of imine, just as with certain types of acetal, also allow a substrate to be chemoselectively manipulated in one pot. Inspired by early work in the steroid area [156], the ability of bulky primary amines like *tert*-butylamine to preferentially mask aldehyde over ketone carbonyls as imines allows the selective reaction of the latter in transformations such as a reduction or an olefination [157]. The unmasking of the aldehyde carbonyl is readily achieved by an aqueous workup. The sequence is exemplified in Fig. 54 where a sterol **124** is synthesized by the reduction of a ketone **125** [158]. The sequence leaves the C20 formyl intact and available for the later introduction of a suitable steroid side chain [159]. Another example can be found in the streamlining by Pfizer scientists of the conversion of 2-chloro-3-aminopyridine (**126**) to 2-phenyl-3-aminopyridine (**127**) [160]. The latter has been a key intermediate in the preparation of 3-amino-2-phenyl-piperidine, a pharmacophoric element of potent non-peptidic NK_1 receptor antagonists such as

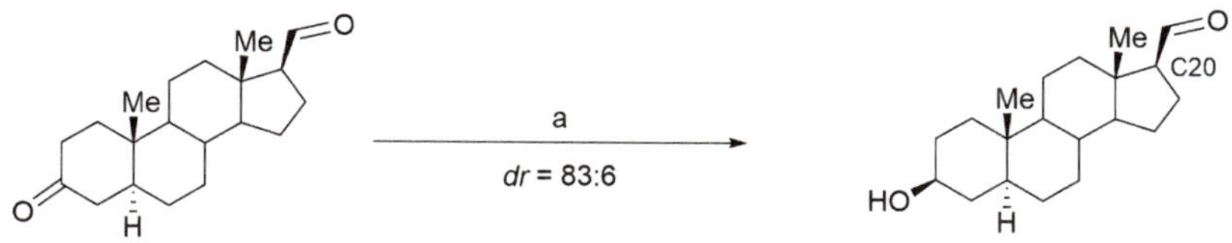

125

124

a) i) $^tBuNH_2$, 4Å MS, THF, ii) $LiAlH(O^tBu)_3$, iii) water, iv) elution on basic alumina

Fig. 54 Temporary protection of an aldehyde carbonyl as its *tert*-butyl imine ahead of reduction of a ketone carbonyl

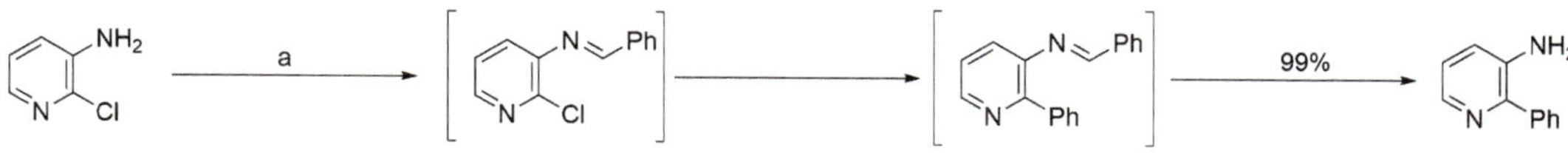

126

128

127

a) i) PhCHO, $PhB(OH)_2$, $Pd(PPh_3)_2Cl_2$ (0.4 mol%), Na_2CO_3 (aq), PhMe, 85 °C, ii) HCl (aq)

Fig. 55 Temporary protection of an amine as an imine ahead of a Suzuki coupling

CP-99,994 (Pfizer) and GR203040 (GSK). An initial three-step procedure, featuring acetamide protection of the amino group, was replaced by the one-pot procedure shown in Fig. 55 [161]. This used benzaldehyde to effect the initial and rapid derivatization of the amino group as imine **128** so that a Suzuki cross-coupling to yield **127** could proceed in quantitative yield in the same pot.

In situ imine formation can also be used to enable the chemoselective derivatization of different amines present in the same molecule. One variant, disclosed by Johnson and Johnson scientists, exploits the temporary protection of primary amines that can be achieved when using methyl isobutylketone (MIBK) as a solvent [162]. High-boiling-free base substrates can be derivatized in high yield by driving off the water by-product of imine formation through azeotropic distillation. Alternatively, a low-boiling-free base can be transformed by gently heating it with a dehydrating agent, and hydrochloride salts can also be used when sodium carbonate is added so as to liberate the free base prior to its masking. With the primary amine temporarily protected, a secondary amine elsewhere in the molecule can be acylated or alkylated. Exposure to acid during the workup or simply heating with water hydrolyses the imine protecting group, unmasking the primary amine, and completing what is often a cost-efficient and scalable means of differentiating between different amino groups. AstraZeneca scientists successfully scaled up the method by converting **129** to **130** as part of a multikilogram delivery of a CCR1 antagonist (Fig. 56) [163–165]. Another example can be found in the temporary protection with MIBK of the amino group of an amino acid ahead of the amidation of its carboxylic acid functionality as part of

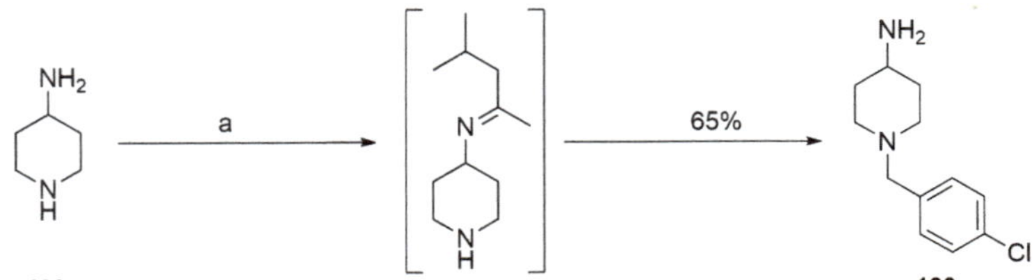

129 130

a) i) MIBK, Δ, ii) 1-chloro-4-(chloromethyl)benzene, NaOH (aq), 20 °C, iii) AcOH

Fig. 56 Temporary protection of a primary amino group as an imine ahead of alkylation of secondary amine

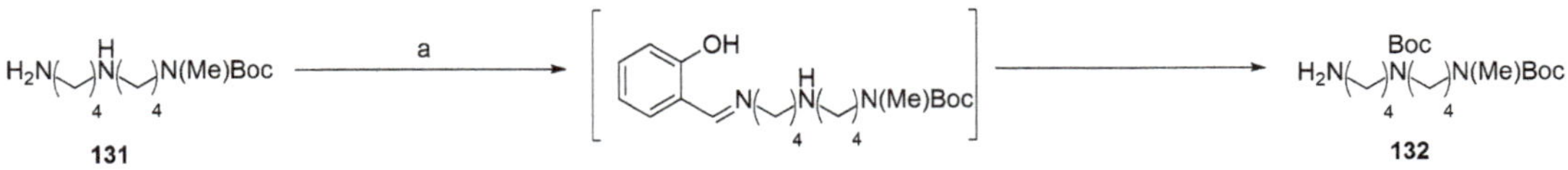

131 132

a) i) salicylaldehyde, Na_2SO_4, MeOH, CH_2Cl_2, ii) Boc_2O, 0 °C, iii) HCl (aq), EtOH

Fig. 57 Temporary protection of a primary amino group as an imine ahead of carbamoylation of a secondary amine

the synthesis of aminoazobenzene-functionalized aluminosilicate nanoparticles [166]. These have been used as a carrier for the chemotherapeutic agent irinotecan.

11.1 Alkylation of Piperidin-4-ylamine (129)

Piperidin-4-ylamine (**129**) (24.7 kg of 94.9%w/w, 234 mol) and MIBK (194 L) were refluxed for up to 24 h. The mixture was cooled to below 30 °C, and a solution of sodium hydroxide (71.5 kg of 20%w/w, 355 mol) and then a solution of 1-chloro-4-(chloromethyl)benzene (37.4 kg, 234 mol) in MIBK (39 L) were sequentially added. The mixture was stirred at 20 °C for up to 24 h, before being washed with water (100 L). Acetic acid (15.5 kg, 257 mol) was added and the resulting precipitate isolated by filtration, before being dried at 50 °C so as to afford 4-amino-1-[(4-chlorophenyl)methyl]piperidinium acetate (**130**) (43.5 kg of 99.5%w/w, 65%).

As with MIBK, the exclusive reaction of salicylaldehyde with a primary amine, in the presence of a secondary amine, has also been exploited in one-pot procedures. This includes the synthesis, part of which is given in Fig. 57 with the conversion of **131** to **132**, of ligands that could be used to treat tumors with high polyamine transport activity [167]. The modest yield is unsurprisingly reported to be due to the accompaniment of the hydrolysis of the imine protecting group by *N*-Boc deprotection under the acidic conditions used. A particularly mild way of removing the *N*-salicylidene protecting group at the end of a sequence, where the use of acid would otherwise cause side reactions, is to use *O*-methyl hydroxylamine [168].

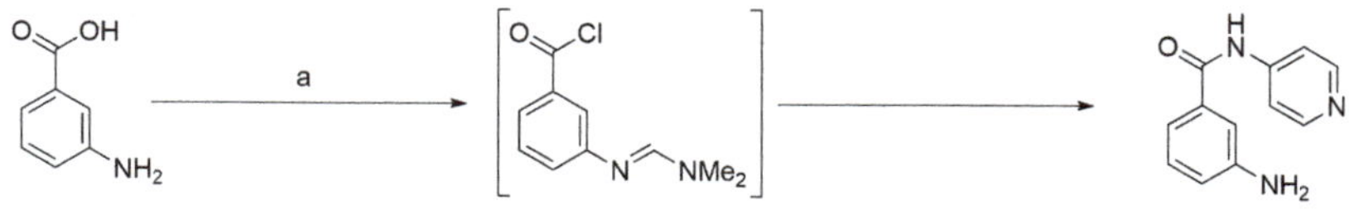

133

134

a) i) DMF, $(COCl)_2$, CH_2Cl_2, 0 → 20 °C, ii) 4-aminopyridine, py, DMAP, iii) $H_2NCH_2CH_2NH_2$, EtOH, Δ

Fig. 58 Temporary protection of an amino group as a formamidine ahead of an amide bond formation

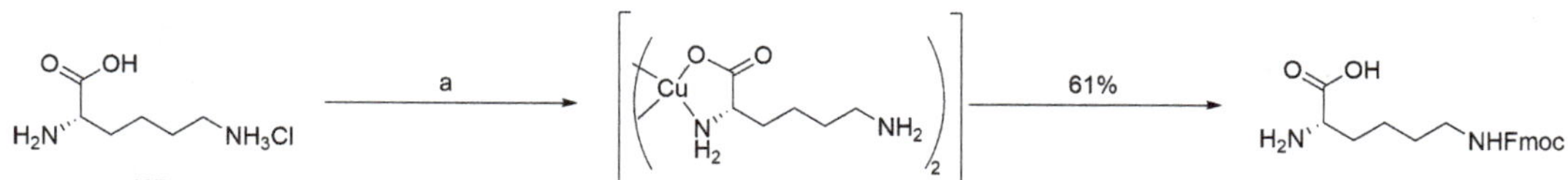

137

135

136

a) i) $CuCO_3 \cdot Cu(OH)_2$, H_2O, Δ, ii) $FmocN_3$, MgO, dioxane, 45 °C, iii) AcOH (aq), iv) Na_2EDTA (aq)

Fig. 59 Temporary protection of α-amino acid of L-lysine ahead of N^{ε}-carbamoylation

The exposure of the primary amino substituent of an aromatic compound to Vilsmeier conditions masks it as a formamidine such as **133** [169]. Through the masking of the amino group and the concomitant activation of the carboxylic acid as an acid chloride, this allows the chemoselective coupling of aminobenzoic acids (and their heteroaromatic analogues) with amine coupling partners, including poorly reactive anilines, using the sequence shown in Fig. 58 [170, 171]. Deprotection can be achieved in the same pot by heating with ethylenediamine derivatives. The amide **134** resulting from the illustration of this sequence in Fig. 58 has featured in the synthesis of a potent non-peptidic bradykinin B_2 receptor antagonist [172]. The method has also been used for the amidation of (*R*)-2-amino-3,3,3-trifluoro-2-methylpropanoic acid to produce compounds that have been screened for PDK inhibitory activity [173].

12 Copper(II) Chelate

The ability of copper(II) to form chelates such as **135** with α-amino acids has been exploited since the time of Fischer [174]. In the case of the dibasic amino acids ornithine and lysine, the chelates are organic-soluble and provide a means of masking the α-amino acid functionality. This allows the amino group of the side chain to be chemoselectively derivatized in a one-pot procedure. This is exemplified by the N^{ε}-urethane **136** formed from *L*-lysine hydrochloride (**137**) in Fig. 59 [58, 175–178]. Decomplexation can be achieved under mild conditions using ethylenediamine tetraacetic acid disodium salt, a chelating ion exchange resin or 8-hydroxyquinoline.

This methodology provides a streamlined means of accessing orthogonally protected diurethane derivatives of dibasic amino acids, compounds that are staple starting materials for the commercial production of a number of peptide drugs.

12.1 N$^{\varepsilon}$-Carbamoylation of L-lysine

A mixture of L-lysine hydrochloride (**137**) (5.00 g, 27.4 mmol) and basic copper(II) carbonate (5.00 g, 22.6 mmol) in water (40 mL) was refluxed for 30 min. The suspension was then filtered while hot and washed with water. After cooling to 25 °C, the solution was basified with magnesium oxide (1.5 g) and a solution of Fmoc-azide (10.6 g, 40 mmol) in dioxane (75 mL) added. After stirring at 45 °C for 16 h, acetic acid solution (25 mL of 2 M) was added to the bulky blue precipitate and the mixture stirred for 60 min before being filtered. The solid was washed with water (3 × 50 mL), dioxane (2 × 25 mL), and chloroform (2 × 25 mL). The amount of N^{ε}-fluorenylmethyloxycarbonyl-L-lysine copper(II) complex after drying and powdering was 6.98 g (64%). A portion of this powder (5.19 g, 6.5 mmol) was added to a supersaturated EDTA disodium salt solution that was freshly prepared by adding EDTA (2.53 g, 8.67 mmol) in portions to a stirred solution of sodium hydrogencarbonate (1.42 g) in water (20 mL). The suspension was agitated until the blue color had been discharged (ca. 1 h). Filtration and washing with water afforded (2*S*)-2-amino-6-(9*H*-fluoren-9-ylmethoxycarbonylamino)hexanoic acid (**136**) as a white solid (4.60 g, 96%).

Where the α-amino group already bears a sulfonamide protecting group, the acidity of this functionality still allows chelate formation, when it is treated at high pH. This was successfully explored by the Merck scientists, in the alkylation of the (*S*)-tyrosine derivative **138** shown in Fig. 60 [76]. In spite of the benefits that such a streamlined approach conferred on the efficiency of the synthesis, it was not adopted due to concerns over the carryover of copper into the drug substance.

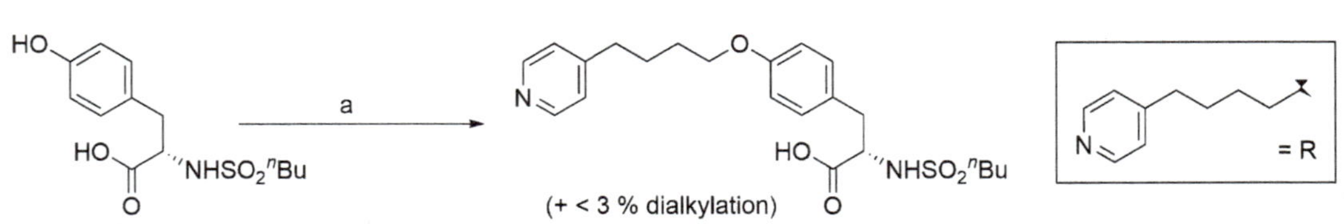

138

a) i) NaOH (aq), $CuSO_4 \cdot H_2O$, 65 °C, ii) RCl•HCl, DMSO or MeOH, NaOH, $^{n}Bu_4NI$ (10 mol %)

Fig. 60 Temporary protection of an α-amino acid derivative of *L*-tyrosine as a copper(II) complex ahead of alkylation of the phenolic functionality

13 Summary

While protecting group chemistry has undoubtedly empowered chemical synthesis, in recent years, the bar has been raised on what can be expected from a protecting group-free synthesis [179–181]. While the chemistries described herein are hardly protecting group free, in as far as a protecting group is attached to (and typically removed from) largely either carbonyl, amine, diol, or amino acid functionality in a one-pot procedure, they constitute a powerful variant of this theme. The examples quoted from drug discovery and development applications illustrate that protecting functionality in the same pot as the subsequent transformation really is a credible strategy for streamlining the synthesis of a compound in a pharmaceutical setting. The simplicity of most of the reagents involved means they are readily available and their use is economically attractive. Such use can also bestow additional practical applications, as is the case where trimethylsilylation solubilizes an amino acid starting material in the organic solvents used for onward processing. The savings in processing that are achieved by protecting a substrate in the same pot as the onward transformation typically also manifest themselves in the unmasking step, which often proceeds under conditions associated with standard workup operations.

In spite of the approaches discussed, there remain myriad opportunities for new methods and new applications of existing methods. For example, to date, once a carbonyl has been temporarily masked, the subsequent transformation has overwhelmingly been the reduction of, or nucleophilic addition to, another carbonyl elsewhere in the molecule. The *N*-carboxylation of an amine is another underexploited approach. That said, the appeal of applying existing methods and developing new ones will surely continue to grow along with the inexorable pressure to demonstrate an improvement in whatever environmental metric is being used to assess performance throughout the lifecycle of a drug project.

References

1. Hendrickson JB (1975) Systematic synthesis design. IV. Numerical codification of construction reactions. J Am Chem Soc 97:5784–5800
2. Yahata K, Fujioka H (2014) *In situ* protection methodology in carbonyl chemistry. Chem Pharm Bull 62:1–11. The *in situ* protection of carbonyl compounds has been reviewed, see this reference
3. Jimenez-Gonzalez C, Ponder CS, Broxterman QB, Manley JB (2011) Using the right green yardstick: why process mass intensity is used in the pharmaceutical industry to drive more sustainable processes. Org Process Res Dev 15:912–917
4. Barton DHR, Hesse RH, Wilshire C, Pechet MM (1977) Enolate anions as protecting groups for ketones during reduction by

hydride. J Chem Soc Perkin Trans:1075–1079

5. Cox BG (2013) Acids and bases: solvent effects on acid-base strength. Oxford University Press, Oxford
6. King JF, Rathore R, Lam JYL, Guo ZR, Klassen DF (1992) pH optimization of nucleophilic reactions in water. J Am Chem Soc 114:3028–3033
7. Tyupalo NF (1977) Catalytic oxidation of heteroaromatic compounds by ozone. Mater. - Vses. Mezhvuz. Konf. Ozonu, 2nd edn. Moskovskii Gos. Univ., Moscow. By protonating the rings of pyrazole and pyridine derivatives using aqueous sulfuric acid as solvent, side chains attached to the rings have been preferentially ozonolysed, see this reference
8. Auth J, Mauleón P, Pfaltz A (2014) Synthesis of functionalized pyridinium salts bearing a free amino group. ARKIVOC:154–169. For the chemoselective quaternisation of a pyridine ring in the presence of a protonated tertiary amine, see this reference
9. Noh HK, Lee JS, Kim Y, Hwang G, Chang JH, Shin H, Nam DH, Lee KH (2004) Synthesis of the intermediate of gemifloxacin by the chemoselective hydrogenation of 4-cyano-3-methoxyimino-1-(*N-tert*-butoxycarbonyl)pyrrolidine. Part 2. The Palladium catalysts in acidic media. Org Process Res Dev 8:788–795
10. Noh HK, Lee JS, Kim Y, Chang JH, Shin H, Nam DH, Lee KH (2004) Synthesis of the intermediate of gemifloxacin by the chemoselective hydrogenation of 4-cyano-3-methoxyimino-1-(*N-tert*-butoxycarbonyl)pyrrolidine. Part 1. Screening of metal catalysts. Org Process Res Dev 8:781–787
11. Baumeister P, Studer M, Roessler F (1997) Handbook of heterogeneous catalysis. Wiley-VCH, New York
12. Lee DW, Ha H, Lee WK (2007) Selective mono-BOC protection of diamines. Synth Commun 37:737–742. For an example of the mono-*N-tert*-butyloxycarbonylation of a diamine using di-*tert*-butyl dicarbonate, see this reference
13. Oganesyan A, Cruz IA, Amador RB, Sorto NA, Lozano J, Godinez CE, Anguiano J, Pace H, Sabih G, Gutierrez CG (2007) High yield selective acylation of polyamines: proton as protecting group. Org Lett 9:4967–4970
14. Krapcho AP, Kuell CS (1990) N-tert-Butoxycarbonyl-α,ω-alkanediamines from α,ω-alkanediamines. Synth Commun 20:2559–2564
15. PCM C, Roon RJ, Koerner JF, Taylor NJ, Honek JF (1995) J Med Chem 38:4433–4438. Whilst the use of an equimolar amount of Boc_2O in methanol is reported to monoprotect a diamine, isolated yields are modest (ca. 35%), popularising methods that rely on a large excess of the diamine. For example see this reference
16. Nudelman R, Ardon O, Hadar Y, Chen Y, Libman J, Shanzer A (1998) Modular fluorescent-labeled siderophore analogues. J Med Chem 41:1671–1678
17. Chhabra SR, Mahajan A, Chan WC (2002) Homochiral 4-azalysine building blocks: syntheses and applications in solid-phase chemistry. J Org Chem 67:4017–4029
18. Dell'Amico DB, Calderazzo F, Labella L, Marchetti F, Pampaloni G (2003) Converting carbon dioxide into carbamato derivatives. Chem Rev 103:3857–3898
19. Xie X, Liotta CL, Eckert CA (2004) CO_2-protected amine formation from nitrile and imine hydrogenation in gas-expanded liquids. Ind Eng Chem Res 43:7907–7911
20. Chatterjee M, Kawanami H, Sato M, Ishizaka T, Yokoyama T, Suzuki T (2010) Hydrogenation of nitrile in supercritical carbon dioxide: a tunable approach to amine selectivity. Green Chem 12:87–93
21. Peeters A, Ameloot R, De Vos DE (2013) Carbon dioxide as a reversible amine-protecting agent in selective Michael additions and acylations. Green Chem 15:1550–1557
22. Fürstner A, Koch D, Langemann K, Leitner W, Six C (1997) Olefin metathesis in compressed carbon dioxide. Angew Chem Int Edit 36:2466–2469
23. Katritzky AR, Akutagawa K (1986) Carbon dioxide: a reagent for the protection of nucleophilic centers and the simultaneous activation of electrophilic attack. Part II. A new synthetic method for the 1-substitution of 1,2,3,4-tetrahydroisoquinolines. Tetrahedron 42:2571–2574
24. Katritzky AR, Fan W, Koziol AE, Palenik GJ (1987) Carbon dioxide: a reagent for the simultaneous protection of nucleophilic centres and the activation of alternative locations to electrophilic attack: Part 81. A novel synthetic route to 4-substituted-2-pyridones. Tetrahedron 43:2343–2348
25. Katritzky AR, Fan WQ, Akutagawa K (1987) Carbon dioxide. A reagent for the protection of nucleophilic centers and the simultaneous activation for electrophilic attack. Part 4. The α-substitution of benzyl alcohol and benzylamine. Synthesis 1987:415–418
26. Katritzky AR, Akutagawawa K (1988) Carbon dioxide. A reagent for protection of the

nucleophilic centers and the simultaneous activation of alternative locations to electrophilic attack. Part XIII. A new synthetic method for the 2-substitution of pyrrole. Org Prep Proced Int 20:585–590
27. Katritzky AR, Fan W (1988) Carbon dioxide: a reagent for the simultaneous protection of nucleophilic centers and the activation of alternative locations to electrophilic attack. XI. A new synthetic route to the 2-substituted thiazolidines. Huaxue Xuebao 46:1134–1138
28. Katritzky AR, Vazquez de Miguel LM, Rewcastle GW(1988) Carbon dioxide: a reagent for the simultaneous protection of nucleophilic centers and the activation of alternative locations to nucleophilic attack. Part 9. Specific synthesis of 1-substituted phenothiazines using carbon dioxide protection of the NH group during lithiation. Synthesis 3:215–217
29. Katritzky AR, Akutagawa K, Fan W (1989) Carbon dioxide: a reagent for the protection of nucleophilic centers and the simultaneous activation of alternative locations to electrophilic attack. 15. A synthetic method for the α-substitution of *N*-phenylbenzylamine. Acta Chim Sin 7:471–476
30. Katritzky AR, Black M, Fan WQ (1991) Carbon dioxide: a reagent for the simultaneous protection of nucleophilic centers and the activation of alternative locations to electrophilic attack. 17. Substitution of *N*-methyl-1- and *N*-methyl-2-naphthylamine and side-chain functionalization of *o*-toluidine. J Org Chem 56:5045–5048
31. Katritzky AR, Akutagawa K (1985) Carbon dioxide: a reagent for the protection of nucleophilic centres and the simultaneous activation of alternative locations to electrophilic attack: Part I. A new synthetic method for the 2-substitution of 1-unsubstituted indoles. Tetrahedron Lett 26:5935–5938
32. Bergman J, Venemalm L (1992) Efficient synthesis of 2-chloro-, 2-bromo-, and 2-iodoindole. J Org Chem 57:2495–2497
33. Hudkins RL, Diebold JL, Marsh FD (1995) Synthesis of 2-aryl- and 2-vinyl-1*H*-indoles via palladium-catalyzed cross-coupling of aryl and vinyl halides with 1-carboxy-2-(tributylstannyl)indole. J Org Chem 60:6218–6220
34. Katritzky AR, Akutagawa K, Jones RA (1988) Carbon dioxide: a reagent for the protection of nucleophilic centers and the simultaneous activation to electrophilic attack. Part XII. One-pot conversion of 3-methylindole into 2-formyl-3-methylindole. Synth Commun 18:1151–1165
35. Sidjimov A (2008) Synthesis of potential indole HMG-CoA reductase inhibitor. God Sofii Univ "Sv Kliment Okhridski" Khim Fak 100:17–28
36. Curtis NR, Kulagowski JJ, Leeson PD, Ridgill MP, Emms F, Freedman SB, Patel S, Patel S (1999) Synthesis and SAR of 2- and 3-substituted 7-azaindoles as potential dopamine D4 ligands. Bioorg Med Chem Lett 9:585–588
37. Hudkins RL, Diebold JL, Angeles TS, Knight E (1997) Fused Pyrrolo[2,3-*c*]carbazol-6-ones: novel immunostimulants that enhance human interferon-γ activity. J Med Chem 40:2994–2996
38. Tholander J, Bergman J (1998) Synthesis of 6-formylindolo[3,2-*b*]carbazole, an extremely potent ligand for the aryl hydrogen (Ah) receptor. Tetrahedron Lett 39:1619–1622
39. Álvarez M, Fernández D, Joule JA (2001) Synthesis of deoxyvariolin B. Tetrahedron Lett 42:315–317
40. Ahaidar A, Fernández D, Danelón G, Cuevas C, Manzanares I, Albericio F, Joule JA, Álvarez M (2003) Total syntheses of Variolin B and Deoxyvariolin B1. J Org Chem 68:10020–10029
41. Wahlström N, Stensland B, Bergman J (2004) Synthesis of 2,3′-diindolylmethanes and substituted indolo[3,2-*b*]carbazoles. Synthesis:1187–1194
42. Guggenheim ER, Ondrus AE, Movassaghi M, Lippard SJ (2008) Poly(ADP-ribose) polymerase-1 activity facilitates the dissociation of nuclear proteins from platinum-modified DNA. Bioorg Med Chem 16:10121–10128
43. Komiya M, Asano S, Koike N, Koga E, Igarashi J, Nakatani S, Isobe Y (2012) Synthesis and structure based optimization of 2-(4-phenoxybenzoyl)-5-hydroxyindole as a novel CaMKII inhibitor. Bioorg Med Chem 20:6840–6847
44. Macias WL (1999) Method for treatment of non-rheumatoid arthritis. WO9909978, 4 Mar 1999. Whilst *tert*-butyllithium is the most common reagent used for the metallation of the lithium carbamate salt, it is worth pointing out that there are scattered examples of the use of the more readily handled *n*-butyllithium, for example see this reference
45. Katritzky AR, Akutagawa K (1986) Carbon dioxide: a reagent for simultaneous protection of nucleophilic centers and the activation of alternative locations to electrophilic attack. V. Activation of the 2-alkyl group of a 2-alkylindole toward proton loss and subsequent

electrophilic substitution. J Am Chem Soc 108:6808–6809

46. Kumar UKS, Patra PK, Ila H, Junjappa H (1998) Benzoannelation of 2-methylindole *via* 1-*N*-carboxy-2-methylindole dianion: a direct regiospecific route to substituted and annelated carbazoles. Tetrahedron Lett 39:2029–2032
47. Kishore K, Reddy KR, Suresh JR, Ila H, Junjappa H (1999) Preparation of lithium 5-lithiomethy-3-methylpyrazole-1-carboxylate and its reaction with α-oxoketene dithioacetals: A new general method for substituted and annelated pyrazolo[1,5-*a*]pyridines. Tetrahedron 55:7645–7652. The same approach has been applied to 5-lithiomethyl-3-methylpyrazole-1-lithiocarboxylate, which leads to the selective formation of a pyrazolo[1,5-*a*] pyridine at the expense of an indazole, see this reference
48. Katritzky AR, Akutagawa K (1989) Formaldehyde: a reagent for the simultaneous protection of nucleophilic centers and the activation and stabilization of alternative locations to electrophilic attack. I. A new synthetic method for the 2-substitution of *N*-unsubstituted benzimidazoles: formaldehyde as a versatile protecting agent for heterocyclic NH. J Org Chem 54:2949–2951
49. Ferrier RJ (1978) Carbohydrate boronates. Adv Carbohydr Chem Biochem 35:31–80
50. Duggan PJ, Tyndall EM (2002) Boron acids as protective agents and catalysts in synthesis. J Chem Soc Perkin Trans 1:1325–1339
51. Mothana S, Grassot J-M, Hall DG (2010) Multistep phase-switch synthesis by using liquid-liquid partitioning of boronic acids: productive tags with an expanded repertoire of compatible reactions. Angew Chem Int Ed 49:2883–2887. Hall and co-workers have developed an elegant means of using the complexation between a polyol and a boronic acid at high pH, not for preparative purposes, but as a means of purifying the latter through liquid-liquid partitioning, see this reference
52. Bhaskar KV, Duggan PJ, Humphrey DG, Krippner GY, Mccarl V, Offermann DA (2001) Phenylboronic acid as a labile protective agent: the selective derivatisation of 1,2,3-triols. J Chem Soc Perkin Trans 1:1098–1102
53. Mayes BA, Arumugasamy J, Baloglu E, Bauer D, Becker A, Chaudhuri N, Latham GM, Li J, Mathieu S, McGarry FP, Rosinovsky E, Stewart A, Trochet C, Wang J, Moussa A (2014) Synthesis of a nucleoside phosphoramidate prodrug inhibitor of HCV NS5B polymerase: phenylboronate as a transient protecting group. Org Process Res Dev 18:717–724
54. Liljebris C, Nilsson BM, Resul B, Hacksell U (1996) Regio- and stereoselective reactions of 17-phenyl-18,19,20-trinorprostaglandin $F_{2\alpha}$ isopropyl ester. J Org Chem 61:4028–4034
55. Priebe H, Cervenka J, Aabye AW, Gulbrandsen T, Bryde AC (2001) A selective process for *N*-alkylation in competition with *O*-alkylation: boric acid, borax, and metaborate as a cheap and effective protecting group applicable for industrial-scale synthetic processes. Org Process Res Dev 5:472–478
56. Ishihara K, Kuroki Y, Hanaki N, Ohara S, Yamamoto H (1996) Antimony-templated macrolactamization of tetraamino esters. Facile synthesis of macrocyclic spermine alkaloids, (±)-Buchnerine, (±)-Verbacine, (±)-Verbaskine, and (±)-Verbascenine. J Am Chem Soc 118:1569–1570
57. Nefkens GHL, Zwanenburg B (1983) Boroxazolidones as simultaneous protection of the amino and carboxyl group in α-amino acids. Tetrahedron 39:2995–2998
58. Albericio F, Nicolas E, Rizo J, Ruiz-Gayo M, Pedroso E, Giralt E (1990) Convenient syntheses of fluorenylmethyl-based side chain derivatives of glutamic and aspartic acids, lysine, and cysteine. Synthesis:119–122
59. Balázs M, Szántay C Jr, Bölcskei H, Szántay C (1993) Borane complexes of vinblastine type diindole derivatives. Tetrahedron Lett 34:4397–4398
60. White JD, Amedio JC, Gut S, Ohira S, Jayasinghe LR (1992) Stereoselective synthesis of the pyrrolizidine alkaloids (−)-integerrimine and (+)-usaramine. J Org Chem 57:2270–2284
61. Picot A, Lusinchi X (1977) Amine-boranes as protecting groups for tertiary amines. Amine-borane derivatives of conanine and dihydroconessine. Bull Soc Chim Fr:1227–1234
62. Vedejs E, Monahan SD (1996) Metalation of oxazole-borane complexes: a practical solution to the problem of electrocyclic ring opening of 2-lithiooxazoles. J Org Chem 61:5192–5193
63. Boger DL, Miyauchi H, Du W, Hardouin C, Fecik RA, Cheng H, Hwang I, Hedrick MP, Leung D, Acevedo O, Guimarães CRW, Jorgensen WL, Cravatt BF (2005) Discovery of a potent, selective, and efficacious class of reversible α-ketoheterocycle inhibitors of fatty acid amide hydrolase effective as analgesics. J Med Chem 48:1849–1856

64. Romero FA, Du W, Hwang I, Rayl TJ, Kimball FS, Leung D, Hoover HS, Apodaca RL, Breitenbucher JG, Cravatt BF, Boger DL (2007) Potent and selective α-ketoheterocycle-based inhibitors of the anandamide and oleamide catabolizing enzyme, fatty acid amide hydrolase. J Med Chem 50:1058–1068
65. Kimball FS, Romero FA, Ezzili C, Garfunkle J, Rayl TJ, Hochstatter DG, Hwang I, Boger DL (2008) Optimization of α-ketooxazole inhibitors of fatty acid amide hydrolase. J Med Chem 51:937–947
66. Ezzili C, Mileni M, McGlinchey N, Long JZ, Kinsey SG, Hochstatter DG, Stevens RC, Lichtman AH, Cravatt BF, Bilsky EJ, Boger DL (2011) Reversible competitive α-ketoheterocycle inhibitors of fatty acid amide hydrolase containing additional conformational constraints in the acyl side chain: orally active, long-acting analgesics. J Med Chem 54:2805–2822
67. Duncan KK, Otrubova K, Boger DL (2014) α-ketoheterocycle inhibitors of fatty acid amide hydrolase: exploration of conformational constraints in the acyl side chain. Bioorg Med Chem 22:2763–2770
68. Mulder RJ, Shafer CM, Dalisay DS, Molinski TF (2009) Synthesis and structure–activity relationships of bengazole A analogs. Bioorg Med Chem Lett 19:2928–2930
69. Jiang Q, Ryan M, Zhichkin P (2007) Use of in situ isopropoxide protection in the metal-halogen exchange of arylboronates. J Org Chem 72:6618–6620
70. Kricheldorf HR (1972) Silylation of amino acids and synthesis of peptides with amino acid trimethylsilyesters. Justus Liebigs Ann Chem 763:17–38
71. Bolin DR, Sytwu II, Humiec F, Meienhofer J (1989) Preparation of oligomer-free N^{α}-Fmoc and N^{α}-urethane amino acids. Int J Pept Protein Res 33:353–359
72. Barlos K, Papaioannou D, Theodoropoulos D (1982) Efficient "one-pot" synthesis of *N*-tritylamino acids. J Org Chem 47:1324–1326. The hydroxyl and carboxylic acid functionalities when disposed as in threonine, for example, can also be protected using a bridged organosilyl through treatment with dichlorodimethylsilane, see this reference
73. Oh S, Jung J, Avery MA (2007) Synthesis of new β-lactam analogs and evaluation of their histone deacetylase (HDAC) activity. Z Naturforsch B Chem Sci 62:1459–1464
74. Tian Z, Rasmussen M, Wittenberger SJ (2002) An efficient scalable process for the synthesis of *N*-Boc-2-*tert*-butyldimethylsiloxypyrrole. Org Process Res Dev 6:416–418
75. Kim BC, Hwang SY, Lee TH, Chang JH, Choi H, Lee KW, Choi BS, Kim YK, Lee JH, Kim WS, Oh YS, Lee HB, Kim KY, Shin H (2006) Development of a scalable synthetic route towards a thrombin inhibitor, LB30057. Org Process Res Dev 10:881–886
76. Chung JYL, Zhao D, Hughes DL, Grabowski EJJ (1993) A practical synthesis of fibrinogen receptor antagonist MK-383. Selective functionalization of (*S*)-tyrosine. Tetrahedron 49:5767–5776
77. Kafarski P, Soroka M, Lejczak B (1988) Synthesis of phosphono peptides from free aminophosphonic acids. Pept Chem:307–310
78. Solodenko V, Kasheva T, Kukhar V (1991) Preparation of *N*-acylated phosphonopeptides with free phosphonic group. Synth Commun 21:1631–1641
79. Mucha A, Kafarski P, Plenat F, Cristau H (1995) Preparation of benzyl *N*-benzyloxycarbonylaminophosphonates and -aminophosphonites. The scope and limitations of *O*-benzyl-*N,N'*-dicyclohexylisourea method. Phosphorus Sulfur Silicon Relat Elem 105:187–193
80. Pudovik AN, Saakyan GM, Khairullin VK, Pudovik MA (1997) Synthesis and properties of *N*-silylated α-amino phosphonates. Russ J Gen Chem 67:1878–1880
81. Li S, Whitehead JK, Hammer RP (2007) Application of in situ silylation for improved, convenient preparation of fluorenylmethoxycarbonyl (Fmoc)-protected phosphinate amino acids. J Org Chem 72:3116–3118
82. Hemmi K, Takeno H, Okada S, Nakaguchi O, Kitaura Y, Hashimoto M (1981) Total synthesis of FK-156 isolated from a *Streptomyces* as an immunostimulating peptide: application of a novel copper chelate amino protection. J Am Chem Soc 103:7026–7028
83. Kitaura Y, Takeno H, Aratani M, Okada S, Yonishi S, Hemmi K, Nakaguchi O, Hashimoto M (1982) Synthesis and RES-stimulating activity of bacterial cell-wall peptidoglycan peptides related to FK-156. Experientia 38:1101–1103. For the application to the synthesis of FK-156 analogues bearing D-alanine instead of glycine, see this reference
84. Levin JI, Du MT, Park K (2004) Practical sulfonylation of amines with 4-hydroxybenzenesulfonyl chloride: *in situ* silylation-desilylation. Synth Commun 34:2773–2781
85. Letavic MA, Axt MZ, Barberia JT, Carty TJ, Danley DE, Geoghegan KF, Halim NS, Hoth LR, Kamath AV, Laird ER, Lopresti-Morrow

LL, McClure KF, Mitchell PG, Natarajan V, Noe MC, Pandit J, Reeves L, Schulte GK, Snow SL, Sweeney FJ, Tan DH, Yu CH (2002) Synthesis and biological activity of selective pipecolic acid-based TNF-α converting enzyme (TACE) inhibitors. Bioorg Med Chem Lett 12:1387–1390

86. Zhang H, Watterson SH, Xiao Z, Dhar TGM, Balasubramanian B, Barrish JC, Chen B (2010) A new and efficient synthesis of 6-[(5*s*,9*r*)-9-(4-cyanophenyl)-3-(3,5-dichlorophenyl)-1-methyl-2,4-dioxo-1,3,7-triazaspiro[4.4]non-7-yl]nicotinic acid, a potent LFA-1/ICAM inhibitor. Org Process Res Dev 14:936–938
87. Bigge CF, Mertes MP (1981) A palladium-catalyzed coupling reaction and a photolytic reaction for the direct synthesis of 5-arylpyrimidine nucleotides. J Org Chem 46:1994–1997
88. Hassan ME (1986) Photochemical synthesis of 5-alkylpyrimidine nucleosides. Recl Trav Chim Pays-Bas 105:30–32
89. Al-Razzak L, Schwepler D, Decedue CJ, Balzarini J, De Clercq E, Mertes MP (1987) 5-Quinone derivatives of 2′-deoxyuridine 5′-phosphate: inhibition and inactivation of thymidylate synthase, antitumor cell, and antiviral studies. J Med Chem 30:409–419
90. Wigerinck P, Pannecouque C, Snoeck R, Claes P, De Clercq E, Herdewijn P (1991) 5-(5-Bromothien-2-yl)-2′-deoxyuridine and 5-(5-chlorothien-2-yl)-2′-deoxyuridine are equipotent to (*E*)-5-(2-bromovinyl)-2′-deoxyuridine in the inhibition of herpes simplex virus type I replication. J Med Chem 34:2383–2389
91. Kerr CE, Eaton BE, Netzel TL (2000) Synthesis of *N,N*-dialkylaniline-2′-deoxyuridine conjugates for DNA-mediated electron transfer studies. Nucleosides Nucleotides Nucleic Acids 19:851–866
92. Ti GS, Gaffney BL, Jones RA (1982) Transient protection: efficient one-flask syntheses of protected deoxynucleosides. J Am Chem Soc 104:1316–1319
93. Kierzek R (1985) The synthesis of 5'-*O*-(dimethoxytrityl)-*N*-acyl-2′-deoxynucleosides. Improved "transient protection" approach. Nucleosides Nucleotides Nucleic Acids 4:641–649
94. McGee DPC, Martin JC, Webb AS (1983) A simple high yield synthesis of N^2-(2-methylpropanoyl)-2′-deoxyguanosine. Synthesis:540–541
95. Kauch M, Hoppe D (2001) Synthesis of substituted phenols by directed *ortho*-lithiation of *in situ* *N*-silyl-protected *O*-aryl *N*-monoalkylcarbamates. Can J Chem 79:1736–1746
96. Luche JL, Gemal AL (1979) Lanthanoids in organic synthesis. 5. Selective reductions of ketones in the presence of aldehydes. J Am Chem Soc 101:5848–5849
97. Gemal AL, Luche JL (1979) Lanthanoids in organic synthesis. 4. Selective ketalization and reduction of carbonyl groups. J Org Chem 44:4187–4189
98. Elisberg E, Vanderhaeghe H, Gallagher TF (1952) Preparation of 3a-hydroxyetiocholane-17-one by a differential reduction with sodium borohydride. J Am Chem Soc 74:2814–2816
99. Maruoka K, Araki Y, Yamamoto H (1988) Methylaluminum bis(2,6-di-*tert*-butyl-4-methylphenoxide) as a protecting group for multifunctional molecules: synthetic utility in selective carbonyl reductions. J Am Chem Soc 110:2650–2652
100. Gössinger E, Schwartz A, Sereinig N (2001) Approach towards an EPC synthesis of nodusmicin. Part 5: Stereoselective introduction of a side chain at the *cis*-decalin part of nodusmicin. Tetrahedron 57:3045–3061
101. Maruoka K, Saito S, Concepcion AB, Yamamoto H (1993) Chemoselective functionalization of more hindered aldehyde carbonyls with the methylaluminum bis(2,6-diphenylphenoxide)/alkyllithium system. J Am Chem Soc 115:1183–1184
102. Kowalski C, Creary X, Rollin AJ, Burke MC (1978) Reductions of α-substituted ketones by lithium diisopropylamide. J Org Chem 43:2601–2608
103. Comins DL, Brown JD (1981) The *in situ* protection of aldehydes *via* α-amino alkoxides. Tetrahedron Lett 22:4213–4216
104. Comins DL (1992) The synthetic utility of α-amino alkoxides. Synlett:615–625
105. Davies JE, Raithby PR, Snaith R, AEH W (1997) Synthesis and novel structure of the first crystallographically characterized chiral α-amino lithium alkoxide. Chem Commun:1721–1722. Crystal structure and conformation data have been determined for the adduct of lithium *N,N,N*′-trimethylethylenediamide and aldehydes, see this reference
106. Clegg W, Liddle ST, Snaith R, Wheatley AEH (1998) The synthesis and solid-state and solution structures of an unprecedented mixed chiral α-amino lithium alkoxide-lithium alkoxide aggregate. New J Chem 22:1323–1326. Crystal structure and conformation data have been determined for the

adduct of lithium *N,N,N'*-trimethylethylenediamide and aldehydes, see this reference

107. Yagi K, Yorimitsu H, Oshima K (2005) Pentamethylcyclopentadienide in organic synthesis: nucleophilic addition of lithium pentamethylcyclopentadienide to aromatic aldehydes and carbon-carbon bond cleavage of the adducts affording the parent aldehydes. Tetrahedron Lett 46:4831–4833. A related method that is still to make an impact in synthetic programmes uses lithium pentamethylcyclopentadienide (C_5Me_5Li, Cp*Li). This chemoselectively reacts with the carbonyl of an aromatic aldehyde, protecting it in preparation for the chemoselective elaboration of another part of the molecule
108. Uemura M, Yagi K, Iwasaki M, Nomura K, Yorimitsu H, Oshima K (2006) Pentamethylcyclopentadienide in organic synthesis: nucleophilic addition of lithium pentamethylcyclopentadienide to carbonyl compounds and carbon-carbon bond cleavage of the adducts yielding the parent carbonyl compounds. Tetrahedron 62:3523–3535. A related method that is still to make an impact in synthetic programmes uses lithium pentamethylcyclopentadienide (C_5Me_5Li, Cp*Li). This chemoselectively reacts with the carbonyl of an aromatic aldehyde, protecting it in preparation for the chemoselective elaboration of another part of the molecule
109. Danishefsky SJ, Mantlo NB, Yamashita DS, Schulte G (1988) Concise route to the calicheamicin-esperamicin series: the crystal structure of an aglycone prototype. J Am Chem Soc 110:6890–6891
110. Comins DL, Brown JD (1984) *Ortho* metalation directed by α-amino alkoxides. J Org Chem 49:1078–1083
111. Plans to use lithium *N,O*-dimethylhydroxylamide in a telescoped sequence should be cognizant of the thermal instability of the *N,O*-dimethylhydroxylamine produced by the sequence if the heating of process solutions is planned
112. Roschangar F, Brown JC, Cooley BE Jr, Sharp MJ, Matsuoka RT (2002) Use of lithium *N,O*-dimethylhydroxylamide as an efficient *in situ* protecting agent for aromatic aldehydes. Tetrahedron 58:1657–1666
113. Comins DL, Killpack MO (1987) Lithiation of heterocycles directed by α-amino alkoxides. J Org Chem 52:104–109
114. Comins DL, Killpack MO (1990) Lithiation of methoxypyridines directed by α-amino alkoxides. J Org Chem 55:69–73
115. de Frutos O, Echavarren AM (1996) Syntheses of fenanthroviridone, gilvocarcin BE-12406X2, and antibiotic WS 5995B based on the palladium and copper catalyzed coupling of organostannanes with bromoquinones. Tetrahedron Lett 37:8953–8956
116. Gonzalez-Cantalapiedra E, de Frutos O, Atienza C, Mateo C, Echavarren AM (2006) Synthesis of the benzo[*b*]fluorene core of the kinamycins by arylalkyne-allene and arylalkyne-alkyne cycloadditions. Eur J Org Chem:1430–1443
117. Lang M, Steglich W (2005) An effective method for the synthesis of ^{13}C-labeled polyprenylhydroxybenzoic acids. Synthesis:1019–1027
118. Maddess ML, Scott JP, Alorati A, Baxter C, Bremeyer N, Brewer S, Campos K, Cleator E, Dieguez-Vazquez A, Gibb A, Gibson A, Howard M, Keen S, Klapars A, Lee J, Li J, Lynch J, Mullens P, Wallace D, Wilson R (2014) Enantioselective synthesis of a highly substituted tetrahydrofluorene derivative as a potent and selective estrogen receptor beta agonist. Org Process Res Dev 18:528–538
119. Tsuyoshi M, Osamu U, Hidenori A, Izumi N (2007) Arylalkanoic acid derivative. EP1829863, 5 Sep 2007
120. Song Y, Shao Z, Dexheimer TS, Scher ES, Pommier Y, Cushman M (2010) Structure-based design, synthesis, and biological studies of new anticancer norindenoisoquinoline topoisomerase I inhibitors. J Med Chem 53:1979–1989
121. Larghi EL, Kaufman TS (2011) Isolation, synthesis and complement inhibiting activity of the naturally occurring K-76, its analogues and derivatives. ARKIVOC:49–102. For the synthesis of structural analogues of the complement inhibitor K-76 by quenching an α-amino alkoxide intermediate with *N,N*-dimethylformamide so as to leave adjacent carbonyls on the aromatic ring post-workup, see this reference
122. Furukawa A, Arita T, Fukuzaki T, Mori M, Honda T, Satoh S, Matsui Y, Wakabayashi K, Hayashi S, Nakamura K, Araki K, Kuroha M, Tanaka J, Wakimoto S, Suzuki O, Ohsumi J (2012) Synthesis and biological evaluation of novel (−)-cercosporamide derivatives as potent selective PPARγ modulators. Eur J Med Chem 54:522–533
123. Sinhababu AK, Borchardt RT (1983) General method for the synthesis of phthalaldehydic acids and phthalides from *o*-bromobenzaldehydes *via ortho*-lithiated aminoalkoxides. J Org Chem 48:2356–2360. The *in situ*

protection of the carbonyl of *o*-bromobenzaldehydes has been exploited as part of an efficient synthesis of phthalides, see this reference

124. Peet NP, McCarthy JR, Sunder S, McCowan J (1986) Convenient, one-step syntheses of 2-fluoroveratrole and 2-fluoroveratraldehyde. Synthetic Commun 16:1551–1556
125. Einhorn J, Luche J, Demerseman P(1988) ortho-Hydroxylation of aromatic aldehydes: a short synthesis of 2-hydroxypyrene-1-carbaldehyde. J Chem Soc Chem Commun:1350–2
126. Gallagher T, Pardoe DA, Porter RA (2000) Synthesis of substituted 2-mercaptobenzaldehydes and 2-substituted benzo[*b*]thiophenes. Tetrahedron Lett 41:5415–5418. Quenching with elemental sulfur has allowed the rapid and one pot synthesis of labile 2-mercaptobenzaldehydes, which have been elaborated through to 2-substituted benzo[*b*]thiophenes, see this reference
127. Bower JF, Szeto P, Gallagher T (2005) Cyclic sulfamidates as lactam precursors. An efficient asymmetric synthesis of (−)-aphanorphine. Chem Commun:5793–5795. Introducing a bromine *ortho* to the aldehyde carbonyl of *p*-anisaldehyde has been used as part of asymmetric total syntheses of (−)-aphanorphine, (−)-variabilin and (−)-glycinol, see this reference
128. Bower JF, Szeto P, Gallagher T (2007) Cyclic sulfamidates as versatile lactam precursors. An evaluation of synthetic strategies towards (−)-aphanorphine. Org Biomol Chem 5:143–150. Introducing a bromine *ortho* to the aldehyde carbonyl of *p*-anisaldehyde has been used as part of asymmetric total syntheses of (−)-aphanorphine, (−)-variabilin and (−)-glycinol, see this reference
129. Calter MA, Li N (2011) Asymmetric total syntheses of (−)-variabilin and (−)-glycinol. Org Lett 13:3686–3689. Introducing a bromine *ortho* to the aldehyde carbonyl of *p*-anisaldehyde has been used as part of asymmetric total syntheses of (−)-aphanorphine, (−)-variabilin and (−)-glycinol, see this reference
130. Grig-Alexa I, Finaru A, Caubere P, Guillaumet G (2006) Lithiation directed by amino alkoxides. Application to the synthesis of new disubstituted dihydrodipyridopyrazines. Org Lett 8:4187–4189. Quenching with dimethyl disulfide, amongst other electrophiles, has been used to regioselectively decorate 5,10-dihydrodipyridopyrazine-4-carbaldehyde at the 6-position, see this reference
131. Bashiardes G, Picard S, Pornet J (2009) Synthesis of nicotine and diverse analogues using intramolecular [3+2] cycloaddition. Synlett:2497–2499
132. Kessler SN, Wegner HA (2012) One-pot synthesis of phthalazines and pyridazino-aromatics: a novel strategy for substituted naphthalenes. Org Lett 14:3268–3271. For the one pot conversion of a substituted benzaldehyde into a phthalazine (or a pyridazino-annulated aromatic), see this reference
133. Dhiman S, SSV R (2013) Indian J Chem Sect A Inorg Bio-Inorg Phys Theor Anal Chem 52A:1103–1112. For the efficient synthesis of 3-formyl-2-furylcarbinols as an enabler for the streamlined synthesis of analogues of cyclohepta[*b*]furans, see this reference
134. Given the choice of either lateral lithiation or *ortho*-lithiation, *N*-lithio-*N*,*N′*,*N′*-trimethylethylenediamine has been shown to promote the former, whereas lithium piperidide promotes the latter
135. Myers AG, Horiguchi Y (1997) Synthesis of the naphthoic acid component of kedarcidin chromophore by routes employing photochemical and thermal electrocyclic ring closure reactions. Tetrahedron Lett 38:4363–4366
136. Bremand N, Normant JF, Mangeney P (2000) Carbolithiation of an α-amino lithio alkoxide derived from cinnamaldehyde. Synlett:532–534
137. Kim S, Park JH, Kim YG, Lee JM (1993) Sulfoniosilylation of α,β-unsaturated carbonyl compounds. Facile nucleophilic substitution of 3-trialkylsilyloxyalk-2-enylenesulfonium salts. J Chem Soc Chem Commun:1188–1189. The same approach can be used to sulfoniosilylate an α,β-unsaturated aldehyde or ketone carbonyl, opening the possibility of their chemoselective reaction with nucleophiles in the presence of an isolated ketone carbonyl, see this reference
138. Evans DA, Bender SL, Morris J (1988) The total synthesis of the polyether antibiotic X-206. J Am Chem Soc 110:2506–2526
139. Taillier C, Bellosta V, Meyer C, Cossy J (2004) Synthesis of ω-hydroxy ketones from ω-benzyloxy Weinreb amides by using a chemoselective nucleophilic addition/Birch reduction process. Org Lett 6:2145–2147. Another example of this approach can be found in the Cossy synthesis of (+)-zoapatanol, see this reference
140. Taillier C, Gille B, Bellosta V, Cossy J (2005) Synthetic approaches and total synthesis of natural zoapatanol. J Org Chem 70:2097–2108. Another example of this approach can be found in the Cossy synthesis of (+)-zoapatanol, see this reference
141. Whilst it can enolise a ketone carbonyl, the use of LMPA has the advantage that it can be

used to mask an aldehyde carbonyl even in the presence of an enolisable ester moiety
142. Dixon DJ, Scott MS, Luckhurst CA (2003) A new chemoselective base-mediated protection/deprotection method for aldehydes. Synlett:2317–2320
143. Reetz MT, Wenderoth B, Peter R (1983) Chemoselective *in situ* protection of aldehydes and ketones using titanium tetrakis(dialkylamides). J Chem Soc Chem Commun:406–408
144. Müller A, Beckert R, Schönecker B, Weiß D (1993) Regioselektive reaktionen an der 17-oxogruppe von androsta-1,4-dien-3,17-dion nach in-situ-schutz mit titantetrakis(dimethylamid). Liebigs Ann Chem:1279–1285
145. Barrios FJ, Zhang X, Colby DA (2010) Dialkylaluminum *N,O*-dimethylhydroxylamine complex as a reagent to mask reactive carbonyl groups *in situ* from nucleophiles. Org Lett 12:5588–5591
146. Barrios FJ, Springer BC, Colby DA (2013) Control of transient aluminum-aminals for masking and unmasking reactive carbonyl groups. Org Lett 15:3082–3085
147. Maruoka K, Araki Y, Yamamoto H (1988) Chemoselective carbonyl alkylation and reduction of aldehydes or ketones. Tetrahedron Lett 29:3101–3104. The complex of *N*-methylaniline, $Me_2AlNMePh$, has also been used, see this reference
148. Bastug G, Dierick S, Lebreux F, Markó IE (2012) Highly chemoselective reduction of carbonyl groups in the presence of aldehydes. Org Lett 14:1306–1309
149. Fujioka H, Yahata K, Kubo O, Sawama Y, Hamada T, Maegawa T (2011) Reversing the reactivity of carbonyl functions with phosphonium salts: enantioselective total synthesis of (+)-centrolobine. Angew Chem Int Ed Engl 50:12232–12235
150. Yahata K, Minami M, Yoshikawa Y, Watanabe K, Fujioka H (2013) Methodology for *in situ* protection of aldehydes and ketones using trimethylsilyl trifluoromethanesulfonate and phosphines: selective alkylation and reduction of ketones, esters, amides, and nitriles. Chem Pharm Bull 61:1298–1307
151. Yahata K, Minami M, Watanabe K, Fujioka H (2014) Selective transformations of carbonyl functions in the presence of α,β-unsaturated ketones: concise asymmetric total synthesis of decytospolides A and B. Org Lett 16:3680–3683
152. Riveiro ME, Maes D, Vázquez R, Vermeulen M, Mangelinckx S, Jacobs J, Debenedetti S, Shayo C, De Kimpe N, Davio C (2009) Toward establishing structure–activity relationships for oxygenated coumarins as differentiation inducers of promonocytic leukemic cells. Bioorg Med Chem 17:6547–6559
153. An JH, Sim TB, Choi J, Yoon NM (1997) Selective reduction of ketones in the presence of aldehydes. Bull Kor Chem Soc 18:111–113
154. Takahashi Y, Mitsudome T, Mizugaki T, Jitsukawa K, Kaneda K (2013) Highly atom-efficient and chemoselective reduction of ketones in the presence of aldehydes using heterogeneous catalysts. Green Chem 15:2695–2698
155. Clerici A, Pastori N, Porta O (2002) Facile reduction of aromatic aldehydes, ketones, diketones and oxo aldehydes to alcohols by an aqueous $TiCl_3/NH_3$ system: selectivity and scope. Eur J Org Chem:3326–3335
156. Herr ME, Heyl FW (1953) "Enamine" derivatives of steroidal carbonyl compounds. III. The synthesis of C11-oxygenated testosterones. J Am Chem Soc 75:5927–5930
157. Paradisi MP, Zecchini GP, Ortar G (1980) A convenient one-pot procedure for the selective reduction of ketones in the presence of aldehydes. Tetrahedron Lett 21:5085–5088
158. Paradisi MP, Zecchini GP (1982) Selective reduction of 3-keto group in steroidal ketoaldehydes. Tetrahedron 38:1827–1829
159. In order to maintain the stereochemical integrity at C20, the aldehyde carbonyl was unmasked using basic alumina
160. Caron S, Massett SS, Bogle DE, Castaldi MJ, Braish TF (2001) An efficient and cost-effective synthesis of 2-phenyl-3-aminopyridine. Org Process Res Dev 5:254–256
161. Whilst conditions now exist for effecting the cross-coupling shown in Figure 55 without even temporary protection of the amino group, such endeavours, whilst attempted, were described as being unproductive by the Pfizer scientists
162. Laduron F, Tamborowski V, Moens L, Horváth A, De Smaele D, Leurs S (2005) Efficient and scalable method for the selective alkylation and acylation of secondary amines in the presence of primary amines. Org Process Res Dev 9:102–104
163. Ainge D, Gnad F, Sinclair R, Vaz L-M, Wells A (2009) Preparation of intermediates ((*R*)-2,2,4-trimethyl-1,3-dioxolane-4-yl) methanol, 3-fluoro-4-nitro-phenol, and 1-(4-chloro-benzyl)piperidin-4-ylamine. WO2009035407, 19 Mar 2009
164. Ainge D, Booker JEM, Pedge N, Sinclair R, Sleigh C, Štefinović M, Vaz L, Way E (2010)

Development of a multikilogram synthesis of a chiral epoxide precursor to a CCR1 antagonist. Use of in situ monitoring for informed optimisation via fragile intermediates. Org Process Res Dev 14:72–84

165. Cardullo F, Donati D, Fusillo V, Merlo G, Paio A, Salaris M, Solinas A, Taddei M (2006) Parallel protocol for the selective methylation and alkylation of primary amines. J Comb Chem 8:834–840. For a related example, see this reference

166. Mitran R, Berger D, Bajenaru L, Nastase S, Andronescu C, Matei C (2014) Azobenzene functionalized mesoporous AlMCM-41-type support for drug release applications. Cent Eur J Chem 12:788–795

167. Muth A, Kamel J, Kaur N, Shicora AC, Ayene IS, Gilmour SK, Phanstiel O (2013) Development of polyamine transport ligands with improved metabolic stability and selectivity against specific human cancers. J Med Chem 56:5819–5828

168. Khomutov AR, Shvetsov AS, Vepsäläinen JJ, Kritzyn AM (2001) Novel acid-free cleavage of *N*-(2-hydroxyarylidene) protected amines. Tetrahedron Lett 42:2887–2889

169. Whilst it is readily destroyed, those synthesising drug candidates should be cognizant of the risk of the formation of the genotoxin dimethylcarbamoyl chloride when using Vilsmeier conditions

170. Zhichkin PE, Peterson LH, Beer CM, Rennells WM (2008) The use of formamidine protection for the derivatization of aminobenzoic acids. J Org Chem 73:8954–8959

171. Whilst this methodology thus avoids the need for separate dedicated protection and deprotection steps, by permitting the presence of an amino group in the aminobenzoic acid starting material, it alternatively avoids recourse to the use of potentially energetic or expensive nitro-bearing surrogates

172. Sawada Y, Kayakiri H, Abe Y, Imai K, Mizutani T, Inamura N, Asano M, Aramori I, Hatori C, Katayama A, Oku T, Tanaka H (2004) A New series of highly potent nonpeptide bradykinin B_2 receptor antagonists incorporating the 4-heteroarylquinoline framework. Improvement of aqueous solubility and new insights into species difference. J Med Chem 47:1617–1630

173. Li T, Shang P, Cheng C, Zhao Y (2013) A convenient one-pot synthesis of (*R*)-2-amino-3,3,3-trifluoro-2-methyl-*N*-phenylpropanamide derivatives. Tetrahedron Lett 54:134–137

174. Fischer E, Zemplen G (1909) Synthesis of the two optically active prolines. Ber Dtsch Chem Ges 42:2989–2997

175. Yamashiro D, Li CH (1973) Adrenocorticotropins. 44. Total synthesis of the human hormone by the solid-phase method. J Am Chem Soc 95:1310–1315

176. Scott JW, Parker D, Parrish DR (1981) Improved syntheses of N^{ε}-(*tert*-butyloxycarbonyl)-L-lysine and N^{α}-(benzyloxycarbonyl)-N^{ε}-(*tert*-butyloxycarbonyl)-L-lysine. Synth Commun 11:303–314

177. Wiejak S, Masiukiewicz E, Rzeszotarska B (1999) A large scale synthesis of mono- and di-urethane derivatives of lysine. Chem Pharm Bull 47:1489–1490

178. Bacarea P-F (2012) Process for the preparation of protected ω-amino-α-amino acids as urethanes by one-pot selective protection *via* a cupric complex formed *in-situ*. RO126201, 30 Aug 2012

179. Hoffmann RW (2006) Protecting-group-free synthesis. Synthesis:3531–3541

180. Young IS, Baran PS (2009) Protecting-group-free synthesis as an opportunity for invention. Nat Chem 1:193–205

181. Li X, Liu A (2010) Inspiration from protecting-group-free synthesis. Huaxue Jinzhan 22:81–90

Part III

Enabling Technologies

Chapter 12

Expanding the Biocatalysis Toolbox

Rajesh Kumar, Carlos A. Martinez, and John W. Wong

Abstract

Research directed toward the development of biocatalytic platforms has been sparked by successful applications of enzymes for sustainable manufacturing of pharmaceutical compounds. This review describes recent progress in the development of ene-reductases, oxygenases, and enzymes for amide and amine synthesis, with a focus on academic and industrial collaborations.

Key words Biocatalysis, Toolbox, Green chemistry, Pharmaceutical, Transaminase, Ketoreductase, Ene-reductase, P450, Monooxygenase, Amine dehydrogenase, Imine reductase

1 Introduction

In a recent review on Green chemistry in process research and development, Dunn [1] highlighted the value of biocatalysis for synthesis of pharmaceutical compounds using enzymatic routes to pregabalin, sitagliptin, and rosuvastatin as case studies. Biocatalysis uses enzymes, which are biodegradable and often sustainably produced, to catalyze highly selective reactions typically operated in aqueous media under mild conditions. These features of biocatalysis give improved efficiency compared to classical synthetic methods by reducing or eliminating the use of organic solvents, avoiding precious metal-based catalysts, shortening synthetic routes by circumventing protection/deprotection reactions, and reducing energy requirements. The biocatalytic route to pregabalin [2] (Scheme 1), the active pharmaceutical ingredient (API) in Lyrica®, illustrates many of the features of biocatalysis that makes this technology particularly well suited for green and sustainable chemistry. Lipolase, a commercially available detergent enzyme, is used in the first step of the synthesis to selectively hydrolyze racemic cyanodiester **1** to give (*S*)-monoester acid **2** and unreacted (*R*)-cyanodiester [(*R*)-**1**], which is removed in the organic phase. The subsequent steps consisting of thermal decarboxylation, ester hydrolysis, and

Paul F. Richardson (ed.), *Green Chemistry in Drug Discovery: From Academia to Industry*, Methods in Pharmacology and Toxicology, https://doi.org/10.1007/978-1-0716-1579-9_12,

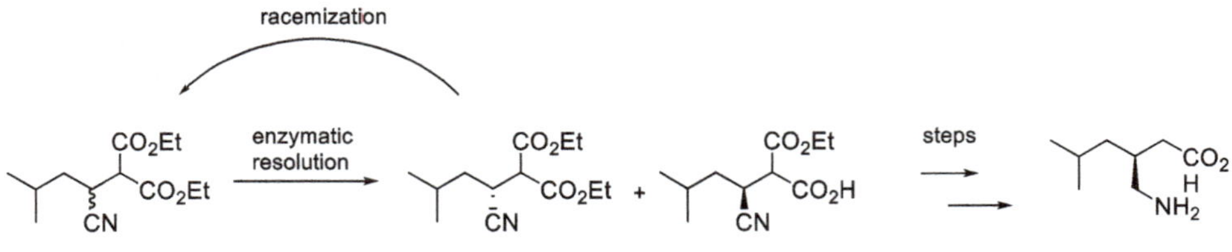

Scheme 1 Biocatalytic route to pregabalin

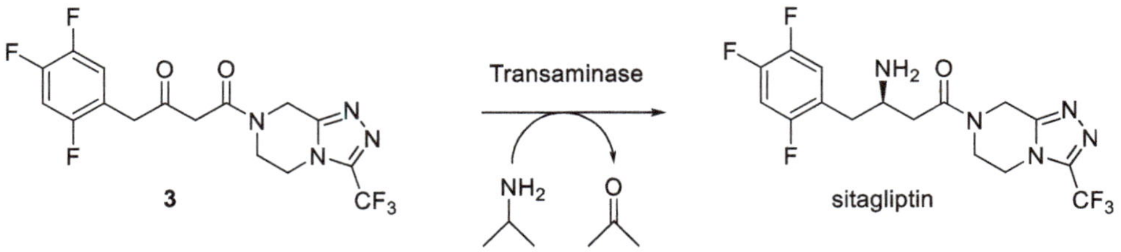

Scheme 2 Biocatalytic route to sitagliptin

catalytic hydrogenation are all carried out in water to convert **2** to pregabalin in high yield and enantiomeric purity. Even without recycle of the undesired (*R*)-**1**, the biocatalytic route yielded a large improvement in process efficiency as indicated by an *E* factor of 17 compared to 86 for the classical resolution route [3]. Recycle of (*R*)-**1** by base-catalyzed racemization further improved the *E* factor of the biocatalytic route to 8, which compares favorably with *E* factors typical of the fine chemicals industry (*E* 5–>50) versus *E* factors of 25–100 for the pharmaceutical industry, as proposed by Sheldon [4]. In addition to waste minimization, the biocatalytic route also reduced energy consumption by 83% compared to the classical resolution route [5].

The synthesis of sitagliptin, the active ingredient in Merck's Januvia®, is another example of the application of biocatalysis for the manufacture of APIs. The biocatalytic route (Scheme 2) utilizes a transaminase to convert ketoamide **3** to sitagliptin (>99.9% ee) in one step, replacing a two-step process consisting of conversion of ketoamide to enamine followed by asymmetric hydrogenation with a rhodium catalyst [6]. Compared to the asymmetric hydrogenation route, the biocatalytic route gave 10–13% improved yield, increased productivity by 53%, eliminated the use of rhodium, and reduced the cost of manufacture. Development of the biocatalytic route required extensive engineering of the transaminase starting from a wild-type enzyme that had no detectable activity for the ketoamide, thus demonstrating the value of enzyme engineering for chemical process development. These achievements were recognized with a Presidential Green Chemistry Challenge award in 2010 to Merck and Codexis as co-recipients in the greener reaction condition category. The value of biocatalysis for green and sustainable manufacturing of APIs was further highlighted since Merck had previously received a Presidential Green Chemistry Challenge award for the development of the asymmetric hydrogenation route to sitagliptin in 2006. The successful implementation of biocatalytic routes to pregabalin and sitagliptin clearly demonstrates the value of biocatalysis for efficient, cost-effective API manufacture, and further application of this technology for pharmaceutical synthesis is expected to increase [7–9] Currently, the number of applications of biocatalysis for pharmaceutical compounds is relatively small, compared to the processes that use standard synthetic methods. Bornscheuer et al. [7] listed 43 biocatalytic processes that have been scaled up for pharmaceutical compounds; however, only a handful of these have been used for commercial manufacture, as this list also included processes scaled up to relatively low scale (grams to low kgs) and processes for compounds no longer in development. Reetz [9] identified the availability of enzymes on suitable scale for practical applications and narrow scope (substrate, selectivity, stability) of wild-type enzymes as two major technical limitations that have contributed to the slow uptake of biocatalysis but have been essentially solved by advances in recombinant DNA technology and directed evolution, respectively. Bornscheuer et al. [7] discussed advances in key technologies (protein engineering, gene synthesis, sequence analysis, bioinformatics, computer modelling) that have driven a resurgence in biocatalysis referred to as "the third wave of biocatalysis." These tools have enabled engineering of industrially useful enzymes with improvements of specific attributes of up to several orders of magnitude compared to the wild-type enzymes. A case in point is the engineering of a halohydrin dehalogenase (HHDH) used in a process for the Lipitor® intermediate, (*R*)-4-cyano-3-hydroxybutanoate. The activity of wild-type HHDH, which suffered from product inhibition and poor stability under process conditions, was improved 2500-fold by changing 35 of the 254 amino acid residues [10]. Equally impressive was the ~40,000-fold improvement of a wild-type transaminase for the sitagliptin process which, as noted previously, had no initial activity for the desired substrate and required changing 27 of 330 amino acid residues to achieve desired targets [11]. These examples clearly illustrate the advantages of using engineered enzymes for industrial synthesis of pharmaceutical compounds but also highlight a need for further advances in enzyme engineering to reduce the time and resources required to engineer enzymes, especially when >20 amino acid residues are changed. As discussed by Bornscheuer et al., engineering an enzyme to give large improvements in properties still requires a large research team and a sufficient return on investment before embarking on such a program. Therefore, with the current state of technology, enzyme engineering for pharmaceutical applications is usually justified for second-generation processes for marketed products or later stage projects where attrition is lower. Further advances are needed to significantly reduce the effort required for enzyme engineering, and this should expand the use of enzyme engineering and result in broader enzyme platforms to fill the biocatalysis toolbox. The biocatalysis toolbox can also be improved by expanding reaction scope. Current biocatalysis applications use a fairly narrow range of enzymes. Of the 43 biocatalytic processes listed by Bornscheuer et al., 33 were carried out with 5 types of enzymes comprised of ketoreductases (KREDs) (17), transaminases (4), hydrolases (6), oxidative enzymes (4), and aldolases (2). The predominance of KRED applications is not surprising as KREDs have been the subject of a great deal of research [12, 13], and enzymes that cover a broad range of substrates are available commercially. More recently, transaminases have been the focus of research [14–16], and many transaminases are commercially available which should lead to an increase in industrial applications. The focus of this chapter will be the expansion of the biocatalysis toolbox with an emphasis on collaborations between academia and industry.

2 Challenges to Further Adoption of Biocatalysis

Biocatalysis is currently impacting the entire drug development cycle, with a larger number of enzyme classes directly applicable to substrates for discovery and early development targets and a smaller group of enzymes applied to late development and manufacturing targets. The majority of applications still center on the preparation of chiral building blocks using ketoreductases (KRED), hydrolases and transaminases (TA), with fewer cases using oxygenases, ene-reductases (EREDs), oxynitrilases, aldolases, epoxide hydrolases, and dehalogenases [2, 7, 12, 13, 17–21]. In today's world, access to enzymes in each of these classes is no longer a bottleneck, and many enzymes are readily available by cloning genomic sequences from databases [22]. The real bottleneck is in the successful expression and characterization of an enzyme against an array of substrates and the demonstration of its true usefulness in chemical synthesis applications. Also, access to useful enzymes is aided by "workhorse" enzymes, which have operational stability and are excellent scaffolds because they are readily expressed and very amenable for the creation of mutant libraries. Examples of workhorse" enzymes are CAL B lipase, *Vibrio fluvialis* aminotransferase and *Lactobacillus kefiri* alcohol dehydrogenase, which have been used in many biocatalysis applications [7, 22, 23]

A key step in designing a synthetic route and determining whether to consider an enzymatic step or a chemical transformation is retrosynthetic analysis, and during this process, it is of prime importance to define the strengths of each approach while making a decision as to how much to study a potential disconnection [24, 25] The existence of gaps in the biocatalysis toolbox becomes clearly visible while evaluating the feasibility of a transformation, particularly with a short timeline (typically 2–4 weeks), and these gaps are defined essentially by one or more limitations:

- One obvious limitation is the lack of enzyme availability, which may be because an enzyme does not exist or has not been discovered for a particular transformation, or because an enzyme is not commercially available. The commercial availability of an enzyme is clearly driven by demand, which in some cases may be for different applications such as enzymes used in the food or detergent industries. However, the commercial availability of enzymes used only for organic synthesis applications would be directly driven by the need for a particular transformation, which in turn is related to the frequency that a particular disconnection appears in retrosynthetic analysis and/or the need for a particular functional group. Alcohols and amines are by far the most common target compounds, and therefore enzymes used for the direct synthesis of these compounds, such as KREDs and TAs, are more commonly offered by commercial suppliers and also cloned in industrial biocatalysis groups. The commercial availabilities of other enzymes such as Baeyer-Villiger monooxygenases, aldolases, dehalogenases, and oxynitrilases are considerably lower than for KREDs and TAs, and this is likely due to a lower need for the transformations catalyzed by these enzymes in pharmaceutical synthesis. Several types of enzymes, such as nitrilases, EREDs, and monooxygenases, are in the middle of the spectrum of availability and are offered by several suppliers. In this discussion, it is reasonable to consider if this is a case of "chicken vs egg problem" in that demand for an enzyme could result from the commercial availability of enzymes, at least initially in quantities sufficient to determine feasibility.
- A second limitation applies to enzymes, available commercially or via in-house cloning efforts, that lack operational stability and are therefore unsuitable for preparative use. This limitation may be overcome by immobilizing an enzyme to improve stability, discovering a better wild-type enzyme, or applying protein engineering techniques to improve the properties of an enzyme.
- A third limitation that is also significant when defining gaps in the tool box is the substrate scope of enzymes. Pharmaceutically relevant substrates are often complex molecules that may be substantially different from an enzyme's natural substrate and therefore pose a challenge when attempting a preparative application. The enzymatic transformation of highly functionalized organic compounds with high molecular weight can be very challenging, and failures are especially common in reactions

involving sugar substrates (using glycosidases and oxygenases) and also natural products as substrates using most enzyme classes. Numerous efforts have been made in these areas with little impact at large scale [10, 26].

- Finally, another gap in the enzyme toolbox is due to enzymatic transformations that are underexploited or require more basic knowledge to develop practical applications. Two enzymatic transformations that may fit in this category of gaps are amide synthesis and oxygenation. In the case of amide formation, the hydrolase-catalyzed coupling of carboxylic acids or esters with amines has been known for several decades and has been exploited industrially for the resolution of chiral amines, but not to any significant extent for the synthesis of pharmaceutically relevant amides. Amide bond formation could potentially be carried out with other non-hydrolytic enzymes, but so far these do not appear to have been exploited for the synthesis of pharmaceutically relevant amides [27–32]. The enzymatic toolbox for the oxygenation of organic substrates is typically limited by low substrate concentrations and conversions. These limitations may be partially due to mass transfer issues resulting from the low solubility of oxygen in aqueous systems [33].

In order to fill these gaps, the future course of biocatalysis should be advanced on multiple fronts including the development of enzymes suitable for preparative applications and also enabling the development of rapid methods for the characterization of novel enzymes against an array of organic substrates. One research area that has received attention over the last few years is cascade reactions, whereby multistep enzyme-catalyzed processes are utilized to achieve enhanced productivity and efficiency by eliminating isolations and displacing reaction equilibrium. Biocatalytic cascade reactions are an intensively investigated research area, as evidenced by more than 100 articles published on this topic in the last 5 years [34, 35]. Another area of active research is the development of new reaction types. Advances include the use of an engineered P450 monooxygenase to perform cyclopropanation reactions, the Michael-type addition of linear aldehydes to β-nitrostyrene by the proline-based enzyme 4-oxalocrotonate tautomerase, the Pictet-Spenglerase activity found in Strictosidine synthase, and continued efforts in the development of Diels-Alderase enzymes [36–40]. The application of process engineering principles is also vital to the development of efficient enzymatic processes, and this includes selecting optimal reactor configurations, ensuring adequate substrate supply, and using reaction monitoring methods to name a few [41–44]. The renaissance of flow systems represents one active area with multiple examples, including the synthesis of chiral amines in organic solvents using transaminase, the biocatalytic production of catechols using a high-pressure tube-in-tube segmented flow microreactor, the use of combined continuous flow reactors for transition-metal-catalyzed heterogeneous hydrogenation/epimerization and Baeyer-Villiger monooxygenation, and the flow-resolution of indole derivatives using immobilized lipases in organic solvent [45–48]. The use of in silico predictive tools to assist enzyme engineering remains an area of active yet challenging research. Promising examples have been reported on the use of these methods on transaminase catalysis. However, the methods are generally applicable to other enzyme systems [49–51].

The remainder of this chapter will discuss the expansion of the biocatalysis toolbox in the areas of amide and chiral amine synthesis, asymmetric reduction of activated CC bonds, and oxygenation, with a focus on research conducted by industrial/academic collaborations.

3 Biocatalytic Amide Synthesis

Amide functionality is one of the most common building blocks in both naturally and synthetically made compounds [52, 53]. Based on some estimates, 25% of all pharmaceuticals on the market and two-third of drug candidates contain amide functionality [54, 55]. In addition to this, amide functionality is also very common in polymers, agrochemicals, and fine chemicals [56, 57].

There are a large number of chemical methods [58] available for amide bond formation starting from the corresponding carboxylic acids and amines. These methods involve thermal activation or acid base catalysis, but are not suitable for large-scale synthesis and are often limited by substrate scope. Most of the industrially relevant chemical syntheses of amide bonds involve activation of carboxylic acids or use of coupling agents [59]. These reactions are useful but not highly efficient, since they typically require stoichiometric amounts of coupling/activation agents which end up as waste and may be difficult to remove, adding to processing costs. Because of these issues, "amide bond formation using environmentally benign methods and with high atom economy" was highlighted and selected as one of the top priorities by the American Chemical Society's Green Chemistry Institute Pharmaceutical round table in 2007 [60].

Enzyme-catalyzed aminolysis [61] of carboxylic acids or esters to form amides is a well-known reaction that offers a potential solution to the issues of low atom economy and environmental challenges, associated with chemical synthesis. However, application of enzymatic amidation technology has been very limited, owing to narrow substrate scope and low activity and stability of the available enzymes. The other major hurdle is the reverse hydrolytic reaction, which not only erodes the selectivity (in the case of chiral amide synthesis), but makes it difficult to achieve good conversions. In the case of enzymatic aminolysis of carboxylic acid esters, the generation of alcohols during the reaction often results in enzyme deactivation, and carboxylic acid esters with bulky alcohols are not very good substrates for aminolysis reactions. In view of these issues, a widely applicable tool box of stable enzymes, with broad substrate scope for amide bond formation, would be very useful [62], especially for the synthesis of structurally diverse compounds for pharmaceutical development. Over the last decade, a significant amount of work has been reported in the literature to address the issues highlighted above as detailed in several reviews [32, 63]. However, enzymatic amide bond formation remains a contemporary challenge. Herein we have reviewed the recent developments in this area with a focus on industrial/academic collaborations, which includes identification and development of novel enzymes, protein engineering, and preparation of more stable enzyme formulations, mainly *via* enzyme immobilization on solid supports.

Based on the mechanism of amidation and carboxylate activation, there are several structurally and functionally distinct protein families that are involved in amide bond formation [32]. In general, based on their mechanistic pathway, they could be classified as ATP-dependent or transacylation enzymes. ATP-dependent enzymes have received much attention for their role in natural system, whereas transacylating enzymes are more established for the synthesis of nonnatural amides. Lipases and proteases such as lipase A and B from *Candida antarctica* (CAL A and CAL B) [64, 65], thermolysin and chymotrypsin, utilize the transacylation strategy for amide bond formation but with limited success. This is mainly due to substrate limitations and enzyme inhibition associated with these enzymes [63, 66]. In general, synthesis of fatty acid amides catalyzed by CAL B is a very efficient process [67], whereas synthesis of peptide amides which is common in nature [68] has been very elusive, except for a few successful examples driven by thermodynamics or in the reverse hydrolysis mode [69, 70]. Several studies are reported in the literature to understand the mechanistic aspect and substrate scope of various hydrolases; however, the number of hydrolases enzymes that could be used successfully, remained unchanged [63]. Recently, CLEA Technologies in collaboration with Pfizer [71] investigated the substrate range of two lipases, CAL B and *Pseudomonas stutzeri* lipase (PSL), and found that PSL showed activity against a range of substrates and, in contrast to CAL B, showed good activity with secondary amines. This was a significant advance in enzymatic amidation technology, as it identified an enzyme (PSL) with complementary selectivity to existing enzymes. The overall conclusion of this study was that enzyme-catalyzed aminolysis was a potential green alternative to existing noncatalytic amide coupling methods and that the application of modern molecular biology tools should lead to enzymes with broader substrate range.

In order to improve the activity and substrate profile for existing hydrolase enzymes, much attention has been focused on improving their protein expression and thermal stability [72]. Several immobilized forms of commonly used enzymes have been developed with improved thermal stability and tolerance to organic solvents. Also, the development of *E. coli*-based expression system for CAL B has been reported to facilitate the development of variants with higher hydrolytic activity *via* protein engineering [73].

A Novozyme sponsored study recently looked at improving the amidase activity of the CAL B via enzyme engineering aided by computational tools [74]. In this study, in silico screening of 393 mutants [75] was compared with experimentally measured activity and showed improved activity for 15 mutants out of 22 designed by computational tools. Their method also identified four of the six most active mutants with >3-fold activity over wild type. In addition to this, several double, triple, and quadruple mutants were generated from the most active single mutant with improved activity. This provided a good initial proof of concept for improving amidase activity via enzyme engineering. A patent application claimed that the enzyme variants could be used to catalyze amidation by ester aminolysis, but this activity was not experimentally tested [76].

4 Biocatalytic Chiral Amine Synthesis

Chiral amines are a key component of both natural and synthetic compounds and are widely present in pharmaceutical, fine chemical, polymer, and agrochemical industries [77–79]. There are numerous chemical methods available for the synthesis of various chiral amines and have been successfully scaled up to industrial scales [80–82]. However, most of the reported methods for chiral amine synthesis require multiple steps including protection and deprotection steps, making the overall synthesis inefficient. In addition to this, several methods for the synthesis of chiral amines use high-energy intermediates such as azides, which pose a safety concern. Enzyme-catalyzed synthesis of chiral amines is highly attractive as this provides a direct route, without the need of protecting or high-energy groups and typically carried out under mild reaction conditions that are environmentally benign [83]. There are various enzyme classes reported in the literature that could be used for the synthesis of chiral amines [84]. Earlier methods employed hydrolases for the resolution of racemic amines via selective *N*-acylation in organic solvents [14, 61]. These methods afforded material with high enantiomeric purity but were limited to primary amines and only gave a theoretical yield of 50%. More recently, there has been a significant growth in the area of asymmetric chiral amine synthesis

Table 1
Enzymatic methods for chiral amine synthesis

Platform	Scheme	Strengths	Gaps
Imine reductase	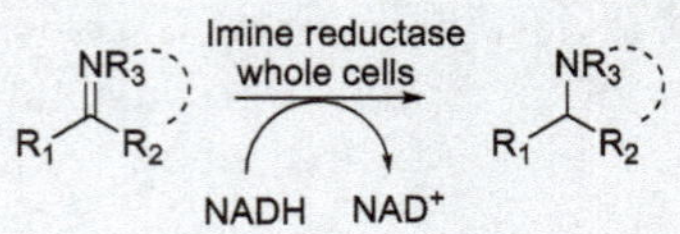	Endo-cyclic imines, Iminium anions, 2°/3° amines	Acyclic N-aliphatic imines (water sensitivity), Substrate scope TBD
Monoamine oxidase	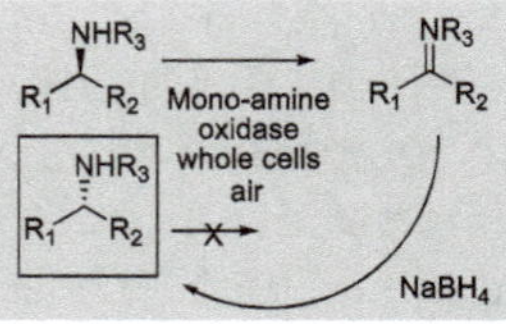	Deracemisation to (*R*)-amines ((*S*)-MAO), In-situ imine formation, 1°/2°/3° amines	Limited for (*R*)-selective oxidation Bulky substrates
ω-Transaminase	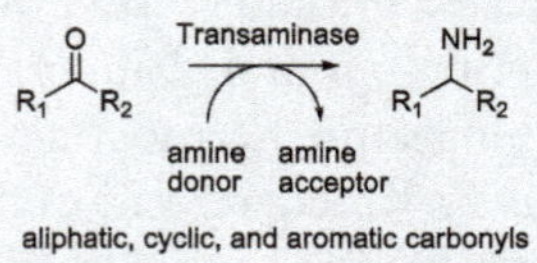	(*S*)- and (*R*)-transaminases, commercially available	Not 2°/3° amines
Amine dehydrogenase	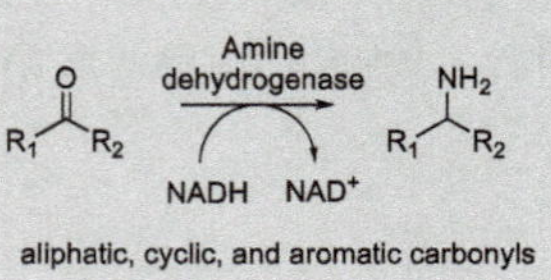	Novel platform	Substrate scope and selectivity TBD, Not 2°/3° amines

using enzymes, driven by a combination of factors [7, 84]. Enzyme classes that have attracted attention for asymmetric chiral amine synthesis include transaminases (also known as amine transferases, TA) [85], lyases [86, 87], imine reductases (IREDs), and amine dehydrogenases (AmDH). Monoamine oxidase enzymes have also been reported for the deracemization or resolution of racemic amines [83]. Table 1 summaries the strengths and gaps of these enzymatic methods.

Among these classes of enzymes, TAs have undergone explosive growth in recent years for the synthesis of chiral amines from corresponding prochiral ketones. As highlighted in Table 1, both *S* and *R* selective TAs are commercially available and have been successfully used for commercial scale manufacturing. TAs are an integral part of the biocatalytic toolbox due to the efforts of many research groups, including several industrial/academic collaborations, that have addressed a variety of problems including substrate scope, enzyme stability, and equilibrium displacement [85, 88–93]. The development of a thermostable TA was carried out by Cambrex in collaboration with Michigan Technological University [94, 95]. This was achieved by the evolution of a mesophilic TA from *Arthrobacter citreus* to a thermostable enzyme using error-prone PCR [95, 96]. Over several rounds of evolution using a high-throughput colorimetric assay, the specific activity of an *S*-selective

TA was improved from 5.9 to 1582.8 IU/g. Since Cambrex developed its TA-based process over a decade ago using its *S*-selective enzyme, there have been several additional *S*-selective TAs identified and improved using bio-informatics, gene sequencing, and protein engineering approaches. A recent review [85] noted generally low activity and selectivity of available TAs and also pointed out the prevalence of *S*-selective TAs compared to only a few *R*-selective TAs.

Recently it was demonstrated that protein engineering could be used to expand the specificity and activity of wild-type TAs [97, 98]. The development of a commercial process for sitagliptin, using an engineered *R*-TA, was reported by Codexis and Merck & Co. and discussed in the introduction to this chapter [11].

Bornscheuer and his group in collaboration with Lonza developed an in silico strategy for a sequence-based prediction of substrate specificity and enantio-preference for TAs [50]. In their approach they used rational protein design to predict key amino acid substitutions for desired activity, followed by a search of protein databases to identify proteins carrying these substitutions. This avoided the laborious work of generating constructs with these mutations and resulted in the identification of naturally evolved proteins. Using this approach they discovered 17 *R*-selective TAs, of which 7 were screened against a variety of substrates and shown to give high conversions and selectivities for *R*-amines [99].

As a part of CoEBIO3 (Center for excellence in Biocatalysis), a multi-university and multi-industry collaboration, development of novel TAs and reactions was one of the priority programs. Bio-informatics tools and enzyme engineering were used to develop a set of engineered *S*-selective TAs that could work on large set of substrates. The CoEBIO3 collaboration also looked at various aspects of TA-catalyzed reactions, such as selection of amine donor and removal of byproduct to drive the reaction in the forward direction [88].

Direct reductive amination of carbonyl compounds with AmDHs could provide another useful method for the synthesis of chiral amines that have a favorable equilibrium compared to transaminations with TAs and do not require the removal of an amine donor byproduct [100]. In this direction a significant milestone was achieved by Bommarius et al. as part of the Center for Pharmaceutical Development, an Industry/University Cooperative Research Program (I/UCRC) sponsored by the NSF. In this program they successfully engineered two (*S*)-selective amino acid dehydrogenases (*L*-leucine and *L*-phenylalanine dehydrogenases) to accept non-amino acid amines as substrates, thus opening up the opportunities for discovering or engineering new novel activities in this area. This was achieved by modifying the carboxylate binding site of wild-type enzymes and by focusing on the mutagenesis of active site residues. In the case of LeuAmDH [101], 11 rounds of enzyme engineering targeting 14 residues were performed to completely change the substrate specificity. A quadruple mutant resulted from this exercise which had little to no activity on its native substrate and showed an activity of 0.69 U/mg with Kcat of 0.46/s for methyl isobutyl ketone. The knowledge gained in this program was applied to engineer a phenylalanine dehydrogenase from *Bacillus badius*, and a double mutant K77S/N276L was shown to have a reductive amination activity for the synthesis of (*R*)-1-(4-fluorophenyl)-propyl-2-amine with high enantioselectivity (>98%) [102]. Further work on expanding the substrate scope and improving the catalytic activity will broaden the potential applicability of this method.

As noted above, enzymatic technology to make chiral primary amines using TAs has been firmly established, and the development of AmDH technology has started. In contrast, the synthesis of secondary or tertiary amines via enzymatic methods is in its early stages. Recently there has been growing interest in the enzymatic asymmetric reduction of imines to make chiral secondary amines [103]. Enzymatic reductions of imines have been scarcely developed, which could be due to their lower stability in aqueous media, restricting it to a few naturally occurring imines [104]. Imine reductases from different sources has been biochemically described and characterized, mainly involving cyclic imines, which are more stable in aqueous media compared to acyclics [105]. Recently imine reductions have been reported using whole cells of *Candida parapsilosis* ATCC 7330, resulting in (*R*)-chiral amines with low yields [106]. This system appears to have very limited substrate scope and low activity even with a high cell loading [107]. More recently imine reductase activity was identified from *Streptomyces* sp. using classical enrichment cultures [107, 108]. In this study two novel NADPH-dependent (*R*)- and (*S*)-selective imine reductases from *Streptomyces* sp. GF3587 and GF 3546 were identified using a cyclic imine, 2-methylpyrroline as a substrate, which is highly stable under enzymatic reaction conditions of neutral aqueous media. Whole-cell-catalyzed reduction of 2-methyl pyrroline produced (2*R*)-methylpyrrolidine (99.2% *ee*, 91–92% yield) and (2*S*)-methylpyrrolidine (92.3% *ee*, 91–92% yield) using *Streptomyces* sp. GF3587 and GF 3546 cells, respectively. The (*R*)-imine reductase (RIR) from *Streptomyces* sp. GF3587 was purified and further characterized and found to be highly specific for 2-methylpyrroline [107]. The gene sequence data analysis of RIR resulted in the identification of a homologue with Uniprot accession number Q1EQE0, in the genome of *Streptomyces kanamyceticus*, which also produced (*R*)-2-methylpyrrolidine from 2-methylpyrroline with 99.6% ee. All three imine reductase were cloned and overexpressed in *E. coli*, and the crystal structure of (*R*)-selective IRED Q1EQE0 from *Streptomyces kanamyceticus* was solved [25]. These enzymes showed activity against a variety of substrates, including

2-methylpyrrolidine, dihydroisoquinolines, and dihydro-B-carbolines. (*S*)-IRED showed good activity on a wide range of substrates, whereas (*R*)-IRED had limited substrate specificity. In addition, these enzymes also showed activity on iminium ions indicating the possibility of tertiary amine synthesis. To further expand the scope of known IREDs, researchers from CoEBIO3 and Chem21 (academic and industrial collaborations) established the Imine Reductase Engineering Database (https://ired.biocatnet.de/ Accessed 14th September 2021) with more than 350 protein entries and used it to find 2 novel *R*-selective and one *S*-selective IREDs by comparing the gene sequences with known IREDs [109]. These novel IREDs showed higher activity compared to previously known IREDs when tested against 2-methylpyrroline. In addition, several key residues were identified by the systematic analysis of potential catalytic residues of *R*- and *S*-selective IREDs, making them amenable to engineering for broader substrate scope and higher activity.

5 Bioreduction of Activated Alkenes with Ene-Reductases

Asymmetric reduction of alkenes is a powerful tool for the synthesis of chiral molecules. The organic chemistry toolbox for alkene reduction is well equipped with tools for *cis*-hydrogenation using precious metal catalysts with chiral ligands [110, 111] and the more recently developed *trans*-hydrogenation with organocatalysts [112, 113]. Biocatalytic tools for alkene reduction, while not as well developed as their counterparts in the chemistry toolbox, are rapidly emerging and are currently comprised of enzymes generally referred to as ene-reductases (EREDs). EREDs belong primarily to the "old yellow enzyme" family of nicotinamide-dependent flavoenzymes [114], which were first discovered more than 80 years ago with the isolation of old yellow enzyme (OYE) from yeast [115]. Recent interest in synthetic applications of EREDs derives from their ability to catalyze the stereoselective reduction of activated alkenes to give products with one or two new chiral centers (Scheme 3). The mechanism of this reduction has been elucidated [116] and consists of a reductive half reaction, in which flavin cofactor is reduced with reducing equivalents from the nicotinamide cofactor, followed by an oxidative half reaction during which hydride is transferred from the reduced flavin to Cβ of the alkene substrate. Transfer of a proton from a tyrosine residue to Cα from the opposite side completes the reaction resulting in the addition of hydrogen across the double bond in trans-specific fashion. The expansion of the ERED platform for biocatalytic alkene reduction will be discussed in the remainder of this section with a focus on research from collaborations between academic and industrial groups.

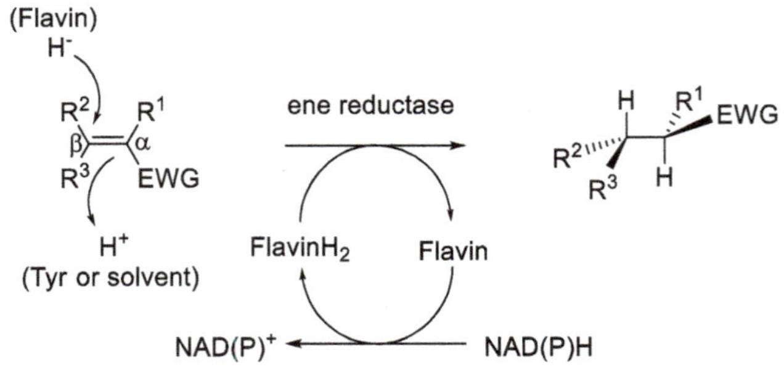

EWG = electron withdrawing group (examples: aldehyde, ketone, nitro, carboxlic acid or ester, imide)

Scheme 3 Bioreduction of activated alkenes with ene reductase. *EWG* electron withdrawing group (examples: aldehyde, ketone, nitro, carboxlic acid or ester, imide)

Asymmetric bioreductions of activated alkenes were initially carried out with whole-cell biocatalysts such as baker's yeast [117]. These whole-cell reactions often gave highly stereoselective reductions of alkene substrates and had the added benefit of not requiring external recycling of the nicotinamide cofactor, but were often plagued by side reactions due to the presence of other enzymes. The bioreduction of citral to citronellal with whole-cell biocatalysts serves as an excellent case in point to illustrate the issues resulting from competing enzyme activities in whole cells [118]. The authors of this study identified yeast, fungi, and bacteria that reduced citral to citronellal with up to 95% *ee* but also observed the formation of nerol, geraniol, and citronellol as major products likely due to the presence of primary alcohol dehydrogenase activities. They concluded that cloning the desired EREDs along with suitable enzymes for cofactor recycling should solve the chemoselectivity issues they encountered with wild-type whole cells. Currently, cloned EREDs are now available as the result of efforts from several groups including industrial/academic collaborations. A recent review [91] lists 30 EREDs from fungi (old yellow enzyme (OYE); old yellow enzyme 1, 2, and 3 (OYE 1-3); old yellow enzyme 2.6 (OYE 2.6); *Kluyveromyces lactis* yellow enzyme (KYE 1); estrogen-binding protein (EBP 1)), bacteria (YqjM, NAD(P)H dependent 2-cyclohexen-1-one reductase (NCR), xenobiotic reductase A and B (XenA, XenB), pentaerythritol tetranitrate reductase (PETNr), TOYE, SYE 1-4, *Gk*OYE, chromate reductase (CrS), morphinone reductase (MR), YersER, *Gluconobacter oxydans* ene-reductase, *N*-ethylmaleimide reductase (NemR), glycerol trinitrate reductase (NerA)), and plants (12-oxophytodienoate reductase 1–3 from *Arabidopsis thaliana* (OPR 1-3) and 12-oxophytodienoate reductase 1–3 from *Solanum lycopersicum* (LeOPR 1-3)) that have been cloned and evaluated for bioreduction of activated alkenes. A collaboration between the University of Graz and BASF profiled the substrate scope of a number of these

EREDs. OPR1, OPR3, and YqjM were evaluated against α,-β-unsaturated dicarboxylic acids (2-methyl maleic acid, 2-methyl fumaric acid, and itaconic acid) and their corresponding dimethyl esters [119]. The results of this study showed that all of the substrates were stereoselectively reduced by OPR 1 and YqjM, except for 2-methyl maleic acid, which was not a substrate for any of the three enzymes tested, and itaconic acid, which was only reduced by YqjM with low conversion but high selectivity. The study also showed that the stereochemical outcome of reactions could be determined by the choice of enzyme or configuration of the double bond in the substrate. In another study, the substrate scope of OPR 1, OPR 3, and YqjM was evaluated against acyclic enals (citronellal, neral), cyclic enones (2-methylcyclopent-2-en-1-one, 3-methylcyclopent-2-en-1-one, 2-methylcyclohex-2-en-1-one, 3-methylcyclohex-2-en-1-one, ketoisophorone,), maleimides (2-methylmaleimide and *N*-phenyl 2-methylmaleimide) and a nitroalkene [58]. Both enals were reduced by all three enzymes, with high conversions and high selectivities for citronellal and moderate selectivities for neral. The 2-substituted enones, 2-methylcyclopent-2-en-1-one and 2-methylcyclohex-2-en-1-one and ketoisophorone, were reduced with high conversions and high selectivities, but both 3-substituted enones were not substrates for OPR1 and very poor substrates for OPR 3 and YqjM. OPR 1, OPR3, and YqjM reduced both maleimides with high conversions and selectivities. The collaboration between Graz and BASF also profiled other EREDs (NCR from *Zymomonas mobilis* and OYE 1–3 from yeasts [58] and PETNr, MR, NemR, and EPB 1 [120]) using essentially the same panel of substrates reported in the studies on OPR 1, OPR 3, and YqjM. The efforts of the Graz and BASF collaboration in profiling the substrate scope of EREDs yielded important results showing that a selection of EREDs could be used to reduce a diverse range of activated alkenes (enals, enones, carboxylic acids, esters, nitriles, cyclic imides, and lactones) and that stereochemical outcome could in many cases be controlled by the choice of enzyme or substrate.

Industrial/academic collaborations were also responsible for several studies on NCR (ERED from *Zymomonas mobilis*). Researchers from Stuttgart University, BASF, and the University of New South Wales collaborated to determine the crystal structure of NCR [121]. In addition to solving the crystal structure of NCR to 1.95 Å, this group also evaluated the effect of mutagenesis on the activity and enantioselectivity of NCR. Eight single-point mutants and four surface loop mutants were evaluated for bioreduction of aliphatic and cyclic α,β-unsaturated aldehydes and ketones. Three of the single point mutants showed increased activity compared to wt NCR. A variant in which tyrosine at position 177, the catalytically active residue, was changed to alanine, surprisingly displayed low to moderate activity, but with markedly

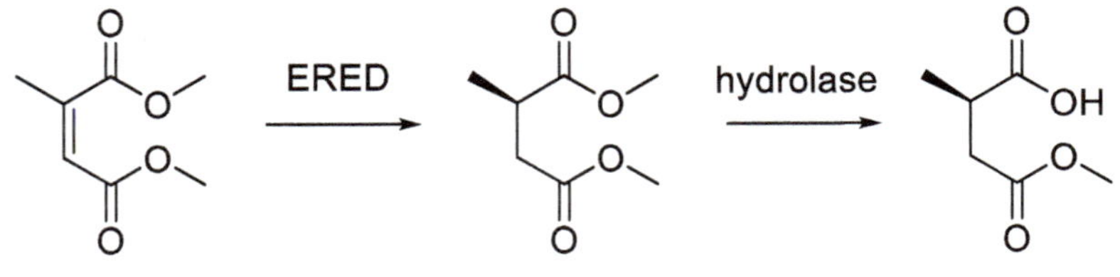

Scheme 4 Biocatalytic synthesis of chiral monoacid

lower stereoselectivities for most of the substrates tested. The four loop variants showed lower activity for all substrates and lower selectivity for all but one of the substrates indicating to the authors that loop regions could be regions for introducing diversity. Further investigation of the loop regions in NCR was carried out by the University of Stuttgart group as part of the Chem21 (chemical manufacturing methods for the twenty-first century pharmaceutical industries) project [122]. This study generated two loop variants, L3_short (loop 3 shortened by seven amino acids) and L4_short (loop 4 shortened by four amino acids), based on a rational protein engineering approach. Evaluation of these loop variants for bioreduction of six substrates (ketoisophorone; 2-methyl-2-pentenal; cinnamaldehydes; 2,4-heptanedienal; neral; and geranial) indicated improved thermostability and organic solvent tolerance (acetone, ethyl acetate, isopropanol, tetrahydrofuran) for L4_short compared to wt NCR and L3_short.

Synthetic applications of ene-reductases for bioreduction of activated alkenes have been recently reviewed [123] and include preparations of chiral building blocks such as (*R*)-3-hydroxy-2-methylpropanoate ('Roche ester') [120], methyl (*S*)-2-bromobutanoate [124], and a chiral monoacid [125]. To date, reported applications of ene-reductases have been carried out at laboratory scale (mg to low grams) with exception of the chiral monoacid [**(*R*)-6**] preparation (Scheme 4) in which bioreduction was carried out on 70-g scale of **4**. Bioreduction of **4** gave **(*R*)-5**, which was then hydrolyzed with a hydrolytic enzyme in a telescoped process to give **(*R*)-6** in 89% overall yield. Another application of EREDs was reported for the synthesis of a pharmaceutical intermediate by a collaboration between scientists at the University of Graz and Pfizer [126]. Asymmetric bioreduction of β-cyanoacrylate esters (Scheme 5) was evaluated as a potential route to pregabalin, a member of the GABA (γ-aminobutyric acid) class of drugs. A similar strategy had previously been reported by Fryszkowska et al. [127] for the synthesis of a precursor to another GABA drug, baclofen. Bioreductions of β-aryl-β-cyano-α,β-unsaturated carboxylic acid salts with crude extracts of *Clostridium sporogenes*, *Ruminococcus productus*, and *Acetobacterium woodii* were reported to give the desired β-aryl-β-cyanopropanoic acid intermediate to baclofen in high yield and >99% ee, while attempted reductions with PETNr and OYE gave no conversions. The Graz/Pfizer researchers systematically investigated the bioreduction of β-cyanoacrylate esters [(*E*)- or (*Z*)-**7a**–**11a**] with various alkyl and ester substituents against a panel of 11 EREDs (OPR1, OPR3, OYE1-3, XenA, XenB, YqjM, EBP1, NerA, and NCR). The results of this study showed that (*E*)-configured ethyl β-cyanoacrylates were good substrates for most of the EREDS except for OPR3, XenA, YqjM, and NerA and always yielded (*S*)-cyanoesters with very high selectivities. The (*Z*)-ethyl β-cyanoacrylates were much poorer substrates compared to the (*E*)-isomers except when R1 was methyl, and typically gave (*R*)-cyanoester products. Stereochemical control of the bioreduction for this series of substrates was almost exclusively controlled by double-bond geometry except for OYE3 which favored (*S*)-**9b** for both (*E*)-**9a** and (*Z*)-**9a**. The (*E*)-configured precursor to pregabalin, (*E*)-**10a**, was reduced in low to moderate conversions by seven of the EREDs with essentially perfect selectivity for the desired (*S*)-**10b**, while (*Z*)-**10a** was not a substrate for any of the enzymes tested. However, the corresponding methyl ester (*Z*)-**11a** was converted by OPR1, OYE3, and NCR to (*S*)-**11b**, which was opposite to the trend observed for the (*Z*)-ethyl cyanoacrylates, which predominantly gave (*R*)-cyanoesters. The potential scalability of the bioreduction was demonstrated by the preparative bioreduction of (*E*)-**10a** at 1-g scale with OPR1 expressed in *E. coli* cells to give the desired (*S*)-**10b** in 69% isolated yield and >99% ee [128].

The Graz/Pfizer collaboration also reported another route to pregabalin based on the bioreduction of β-cyanoacrylic acids (Scheme 6) [129]. In this route, the β-cyanoacrylic acids (*E*)- or (*Z*)-**12a** could be converted to pregabalin in two steps consisting of bioreduction to (*S*)-**13a** followed by hydrogenation, thus eliminating the ester hydrolysis step required for the route starting with β-cyanoacrylate esters (Scheme 5). However, a screen of 14 EREDs showed only marginal conversion of (*E*)-**12a** by OPR1 and a single point mutant of OPR1 and no conversions of (*Z*)-**12a**. Attention was then directed to a diene substrate (*E*)-12b, which was converted by OPR1-wt and several point mutants of OPR1 in high conversion and >99%ee. (*Z*)-**12b** was also screened for bioreduction, but no conversion was observed for any of the EREDs tested.

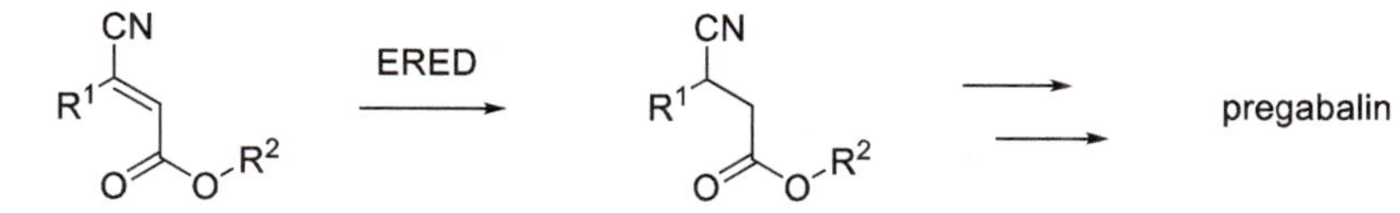

(*E*)- or (*Z*)-**7a - 11a** (*R*)- or (*S*)-**7b - 11b**

7: R^1 = Me, R^2 = Et
8: R^1 = Et, R^2 = Et

Scheme 5 Biocatalytic route to pregabalin via bioreduction of β-cyanoacrylate esters with ERE

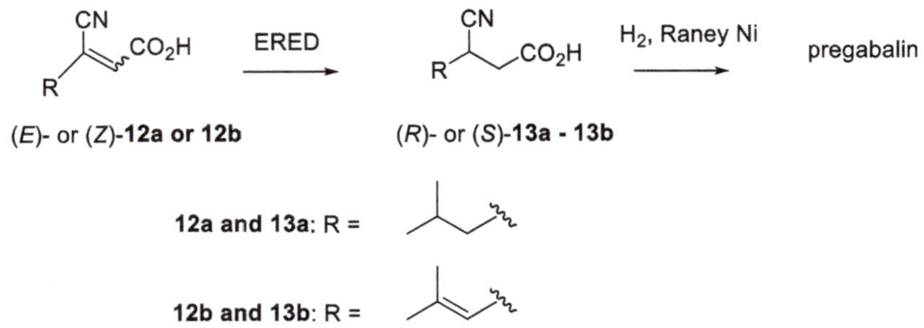

Scheme 6 Biocatalytic route to pregabalin via bioreduction of β-cyanoacrylic acids with EREDs

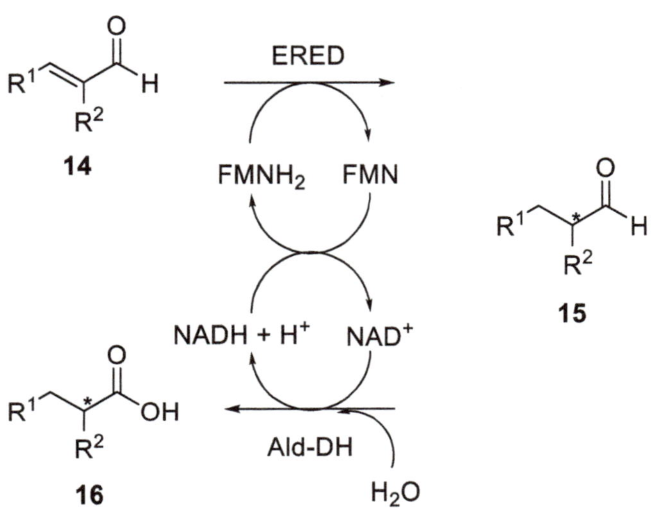

Scheme 7 Biocatalytic hydrogen-borrowing cascade for synthesis of *a*-substituted carboxylic acids

The desired cyanoacid (*R*)-**13b** still contained a C=C bond at the 4,5 position, but this could be reduced along with the nitrile moiety in the final hydrogenation step with Raney nickel. Preparative bioreduction of (*E*)-**12b** at 1.15 g scale with OPR1 gave the desired (*R*)-**13b** in quantitative yield with >99% ee. A deuterium-labelling experiment during the bioreduction of (*E*)-**12b** showed that the nitrile was the activating group.

Researchers at the University of Manchester and GSK recently reported a hydrogen-borrowing cascade involving an ERED and an aldehyde dehydrogenase (Ald-DH) for the preparation of α-substituted carboxylic acids **16** from α-substituted α,-β-unsaturated aldehydes **14** (Scheme 7) [130]. The cascade system was demonstrated at 100-mg scale on (*E*)-2-methyl-3-phenylacrylaldehyde utilizing OYE2 and an *E. coli* Ald-DH to give (*S*)-2-methyl-3-phenylpropanoic acid in 88% isolated yield and >98% *ee*. The self-sufficient redox system eliminated the need for a sacrificial hydride donor to recycle nicotinamide cofactor since oxidized cofactor produced in the bioreduction step was reduced during the oxidation of the aldehyde. Also the cascade system gave access to chiral α-substituted carboxylic acids via bioreduction of enals, which are good substrates for EREDs, versus bioreduction of α,β-unsaturated esters or carboxylic acids, which are only marginal substrates for EREDs.

EREDs are an emerging tool in the biocatalytic toolbox for trans-specific asymmetric bioreduction of activated alkenes. Cloned EREDs from microbial and plant sources and commercial enzymes (Almac, Codexis) from undisclosed sources are available and have been used to reduce diverse substrates for pharmaceutical and fine chemical applications. Large-scale industrial applications of EREDs should follow as research groups continue to identify new enzymes through discovery efforts, improve properties of existing EREDs through protein engineering, and develop new applications. The efforts of industrial/academic collaborations have played a significant role in expanding the ERED platform.

6 Oxygenases

The catalytic oxidation of C–H bonds is one of the most difficult transformations at a laboratory or industrial scale. Over the last decade, advances in the oxidation of unactivated sp^3 C–H bonds have resulted in the discovery of catalysts that are highly reactive and predictably selective [131–134]. Nature uses enzymatic reactions that convert molecular oxygen into highly reactive oxygenated intermediates to perform selective oxidations of organic compounds. Enzymes that use oxygen as the oxidant include oxidases, peroxidases, and oxygenases [135, 136]. The main advantage these biological oxidations have over synthetic organic methods is the use of molecular oxygen, the most environmentally benign, nontoxic, and abundantly available oxidant [137]. Several academic-industrial collaborations have taken place over the last two decades, primarily in the area of monooxygenase research. Monooxygenases catalyze the incorporation of a single atom of molecular oxygen into a substrate (R-H) with the concomitant reduction of the other atom to water, while utilizing NAD(P)H as an external reductant, in the following reaction:

$$R - H + O_2 + NAD(P)H + H^+ \rightarrow R - OH + H_2O + NAD(P)^+$$

Cytochrome P450 (CYP) monooxygenases play a critical role in biochemistry, pharmacology, toxicology, and applied biocatalysis [138–140] and have received most of the attention in these collaborations, directed toward overcoming the challenges facing practical applications. There are five main challenges that are associated with poor productivities in CYP reactions: (1) intrinsic low enzyme activities observed in CYP reactions; (2) the need for efficient expression of not only the CYP gene but also the redox partners;

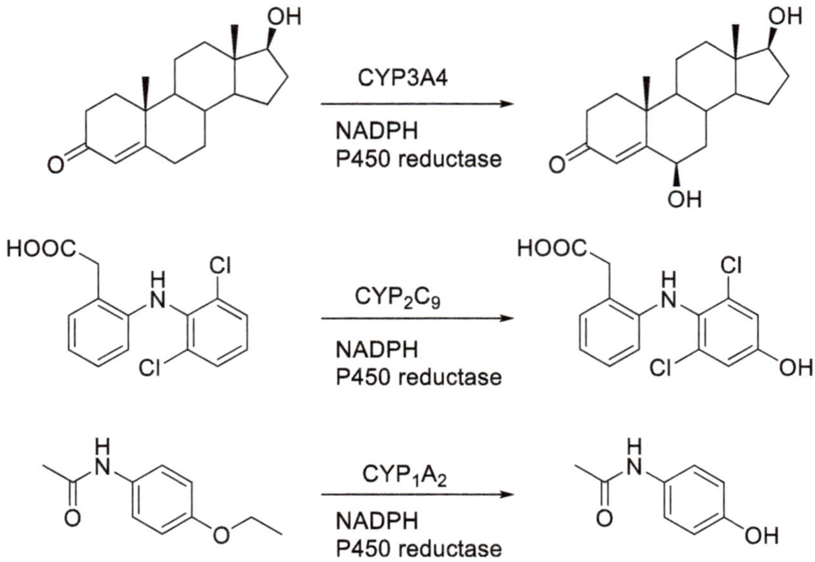

Scheme 8 Metabolites prepared by human P450s expressed in *E. coli*

(3) the need for an efficient NAD(P)H cofactor recycling system; (4) substrate or product inhibition; and (5) degradation of product. The development of a useful mammalian P450 expression systems resulted from a collaboration between the Biomedical Research Centre of the University of Dundee (Scotland, UK) and 14 pharmaceutical companies. During this collaboration a superior cloning strategy was developed where various human CYP cDNAs were fused with a modified ompA leader sequence, directing the emerging enzyme to the bacterial inner membrane [141]. The expression systems derived from that collaboration also co-expressed the oxidoreductase required for catalysis, and the biocatalysts have been used by the industrial partners to express the recombinant human CYP reliably, thus enabling the syntheses of drug metabolites [142]. The same strains can be used for the diversification of organic compounds. In a similar collaboration, a system was developed at the Vanderbilt University School of Medicine and was applied to the synthesis of drug metabolites by scientists at Schering-Plough (Scheme 8) [143].

The low enzyme activities observed with mammalian CYPs can be overcome with the use of engineered bacterial CYPs. A collaboration between Frances Arnold at Caltech and Codexis Inc. enabled the commercialization of several mutants derived from CYP102A1, also known as P450BM3. These mutants were found to be capable of performing metabolite syntheses with relatively high activities and selectivities similar to those seen with mammalian CYPs [144]. Mercian corporation collaborated with the University of Tokyo and the National Institute of Advanced Industrial Sciences and Technologies (AIST) to engineer P450moxA, a CYP that catalyzes the hydroxylation of a broad range of substrates, including fatty acids, steroids, and aromatic substrates. The engineered enzyme had 20 times greater activity compared to the wild-type enzyme [145]. A more recent collaboration in Europe HyPerIn (Integrative Approach to Promote Hydroxylations with Novel P450 Enzymes for Industrial Processes) was set up with the financial support of the European Research Area Industrial Biotechnology (ERA-IB [89]) and involved several academic and industrial partners in Europe. The main goal of the collaboration is to develop P450 biocatalysts for preparative use. The emergence of novel CYP biocatalysts may enable the introduction of alternative oxygen activation pathways, using hydrogen peroxide or other oxygenated precursors. Non-CYP systems with higher productivities have been reported, and these could provide insight that may lead to the development of improved CYP catalyst systems [146–150].

7 Summary

Biocatalysis has enabled greener chemistry in pharmaceutical applications ranging from discovery to manufacturing. Efforts to expand the biocatalysis toolbox will result in new and improved enzymes, which would broaden the application of biocatalysis. Industry and academia have successfully collaborated to expand the biocatalysis toolbox, and continued efforts should reap additional benefits for green chemistry.

References

1. Dunn PJ (2012) The importance of green chemistry in process research and development. Chem Soc Rev 41(4):1452–1461
2. Martinez CA, Hu S, Dumond Y, Tao J, Kelleher P, Tully L (2008) Development of a chemoenzymatic manufacturing process for pregabalin. Org Process Res Dev 12 (3):392–398
3. Hoekstra MS, Sobieray DM, Schwindt MA, Mulhern TA, Grote TM, Huckabee BK, Hendrickson VS, Franklin LC, Granger EJ, Karrick GL (1997) Chemical development of CI-1008, an enantiomerically pure anticonvulsant. Org Process Res Dev 1(1):26–38
4. Sheldon RA (1992) Organic Synthesis—past, present, and future. Chem Ind:903–906
5. Dunn, P.; Hettenbach, K.; Kelleher, P.; Martinez, C. A., The development of a green, energy efficient, chemoenzymatic manufacturing process for pregabalin. In Dunn PJ, Wells AS, Williams MT (Eds) Green chemistry in the pharmaceutical industry. Wiley 2010.pp 161–177
6. Balsells, J.; Hsiao, Y.; Hansen, K. B.; Xu, F.; Ikemoto, N., Synthesis of sitagliptin, the active ingredient in Januvia® and Janumet® 1. In Dunn PJ, Wells AS, Williams MT (Eds) Green chemistry in the pharmaceutical industry. Wiley 2010.
7. Bornscheuer U, Huisman G, Kazlauskas R, Lutz S, Moore J, Robins K (2012) Engineering the third wave of biocatalysis. Nature 485 (7397):185–194
8. Huisman GW, Collier SJ (2013) On the development of new biocatalytic processes for practical pharmaceutical synthesis. Curr Opin Chem Biol 17(2):284–292
9. Reetz MT (2013) Biocatalysis in organic chemistry and biotechnology: past, present, and future. J Am Chem Soc 135 (34):12480–12496
10. Abe I, Piel J (2010) Natural products via enzymatic reactions. Springer
11. Savile CK, Janey JM, Mundorff EC, Moore JC, Tam S, Jarvis WR, Colbeck JC, Krebber A, Fleitz FJ, Brands J (2010)

Biocatalytic asymmetric synthesis of chiral amines from ketones applied to sitagliptin manufacture. Science 329(5989):305–309

12. Huisman GW, Liang J, Krebber A (2010) Practical chiral alcohol manufacture using ketoreductases. Curr Opin Chem Biol 14 (2):122–129
13. Moore JC, Pollard DJ, Kosjek B, Devine PN (2007) Advances in the enzymatic reduction of ketones. Acc Chem Res 40 (12):1412–1419
14. Höhne M, Bornscheuer UT (2009) Biocatalytic routes to optically active amines. ChemCatChem 1(1):42–51
15. Koszelewski D, Lavandera I, Clay D, Guebitz GM, Rozzell D, Kroutil W (2008) Formal asymmetric biocatalytic reductive amination. Angew Chem Int Ed 47(48):9337–9340
16. Nestl BM, Nebel BA, Hauer B (2011) Recent progress in industrial biocatalysis. Curr Opin Chem Biol 15(2):187–193
17. Wolberg M, Dassen BH, Schürmann M, Jennewein S, Wubbolts MG, Schoemaker HE, Mink D (2008) Large-scale synthesis of new pyranoid building blocks based on aldolase-catalysed carbon-carbon bond formation. Adv Synth Catal 350 (11-12):1751–1759
18. Bergeron S, Chaplin DA, Edwards JH, Ellis BS, Hill CL, Holt-Tiffin K, Knight JR, Mahoney T, Osborne AP, Ruecroft G (2006) Nitrilase-catalysed desymmetrisation of 3-hydroxyglutaronitrile: preparation of a statin side-chain intermediate. Org Process Res Dev 10(3):661–665
19. Schmid A, Hollmann F, Park JB, Bühler B (2002) The use of enzymes in the chemical industry in Europe. Curr Opin Biotechnol 13 (4):359–366
20. Griengl H, Schwab H, Fechter M (2000) The synthesis of chiral cyanohydrins by oxynitrilases. Trends Biotechnol 18(6):252–256
21. Kotik M, Archelas A, Wohlgemuth R (2012) Epoxide hydrolases and their application in organic synthesis. Curr Org Chem 16 (4):451–482
22. Davids T, Schmidt M, Böttcher D, Bornscheuer UT (2013) Strategies for the discovery and engineering of enzymes for biocatalysis. Curr Opin Chem Biol 17 (2):215–220
23. Gustafsson C, Minshull J, Govindarajan S, Ness J, Villalobos A, Welch M (2012) Engineering genes for predictable protein expression. Protein Expr Purif 83(1):37–46
24. Clouthier CM, Pelletier JN (2012) Expanding the organic toolbox: a guide to integrating biocatalysis in synthesis. Chem Soc Rev 41 (4):1585–1605
25. Rodríguez-Mata M, Frank A, Wells E, Leipold F, Turner NJ, Hart S, Turkenburg JP, Grogan G (2013) Structure and activity of NADPH-dependent reductase Q1EQE0 from Streptomyces kanamyceticus, which catalyses the R-selective reduction of an imine substrate. ChemBioChem 14 (11):1372–1379
26. Martinez CA, Rupashinghe S (2013) Cytochrome P450 bioreactors in the pharmaceutical industry: challenges and opportunities. Curr Top Med Chem
27. Danger G, Plasson R, Pascal R (2012) Pathways for the formation and evolution of peptides in prebiotic environments. Chem Soc Rev 41(16):5416–5429
28. Dawlaty J, Zhang X, Fischbach MA, Clardy J (2009) Dapdiamides, tripeptide antibiotics formed by unconventional amide ligases†. J Nat Prod 73(3):441–446
29. Hiller DA, Singh V, Zhong M, Strobel SA (2011) A two-step chemical mechanism for ribosome-catalysed peptide bond formation. Nature 476(7359):236–239
30. Hollenhorst MA, Clardy J, Walsh CT (2009) The ATP-dependent amide ligases DdaG and DdaF assemble the fumaramoyl-dipeptide scaffold of the dapdiamide antibiotics. Biochemistry 48(43):10467–10472
31. Marahiel MA, Essen LO (2009) Chapter 13 nonribosomal peptide synthetases: mechanistic and structural aspects of essential domains. In: David AH (ed) Methods in enzymology, vol 458. Academic, pp 337–351
32. Goswami A, Van Lanen SG (2015) Enzymatic strategies and biocatalysts for amide bond formation: tricks of the trade outside of the ribosome. Mol BioSyst 11(2):338–353
33. Geng M, Duan Z (2010) Prediction of oxygen solubility in pure water and brines up to high temperatures and pressures. Geochim Cosmochim Acta 74(19):5631–5640
34. García-Junceda E, Lavandera I, Rother D, Schrittwieser JH (2015) (Chemo) enzymatic cascades—nature's synthetic strategy transferred to the laboratory. J Mol Catal B Enzym 114:1–6
35. Köhler V, Turner NJ (2015) Artificial concurrent catalytic processes involving enzymes. Chem Commun 51(3):450–464
36. Coelho PS, Brustad EM, Kannan A, Arnold FH (2013) Olefin cyclopropanation via carbene transfer catalyzed by engineered cytochrome P450 enzymes. Science 339 (6117):307–310

37. Geertsema EM, Miao Y, Tepper PG, de Haan P, Zandvoort E, Poelarends GJ (2013) Biocatalytic Michael-type additions of acetaldehyde to nitroolefins with the proline-based enzyme 4-oxalocrotonate tautomerase yielding enantioenriched γ-nitroaldehydes. Chem Eur J 19(43):14407–14410
38. Wu F, Zhu H, Sun L, Rajendran C, Wang M, Ren X, Panjikar S, Cherkasov A, Zou H, Stöckigt J (2011) Scaffold tailoring by a newly detected pictet–Spenglerase activity of strictosidine synthase: from the common tryptoline skeleton to the rare piperazino-indole framework. J Am Chem Soc 134 (3):1498–1500
39. Kelly WL (2008) Intramolecular cyclizations of polyketide biosynthesis: mining for a "Diels–Alderase"? Org Biomol Chem 6 (24):4483–4493
40. Eiben CB, Siegel JB, Bale JB, Cooper S, Khatib F, Shen BW, Players F, Stoddard BL, Popovic Z, Baker D (2012) Increased Diels-Alderase activity through backbone remodeling guided by Foldit players. Nat Biotechnol 30(2):190–192
41. Cooney CL (1983) Bioreactors: design and operation. Science 219(4585):728–733
42. Gernaey KV, Cervera-Padrell AE, Woodley JM (2012) A perspective on PSE in pharmaceutical process development and innovation. Comput Chem Eng 42:15–29
43. Kim PY, Pollard DJ, Woodley JM (2007) Substrate supply for effective biocatalysis. Biotechnol Prog 23(1):74–82
44. Rao NN, Lütz S, Seelbach K, Liese A (2006) Basics of bioreaction engineering. Industrial biotransformations. Wiley-VCH, pp 115–145
45. Andrade LH, Kroutil W, Jamison TF (2014) Continuous flow synthesis of chiral amines in organic solvents: immobilization of E. coli cells containing both ω-transaminase and PLP. Org Lett 16(23):6092–6095
46. Tomaszewski B, Schmid A, Buehler K (2014) Biocatalytic production of catechols using a high pressure tube-in-tube segmented flow microreactor. Org Process Res Dev 18 (11):1516–1526
47. Fink MJ, Schön M, Rudroff F, Schnürch M, Mihovilovic MD (2013) Single operation stereoselective synthesis of aerangis lactones: combining continuous flow hydrogenation and biocatalysts in a chemoenzymatic sequence. ChemCatChem 5(3):724–727
48. Falus P, Boros Z, Kovács P, Poppe L, Nagy J (2014) Lipase-catalyzed kinetic resolution of 1-(2-hydroxycyclohexyl) indoles in batch and continuous-flow systems. J Flow Chem 4 (3):125–134
49. Sirin S, Kumar R, Martinez CA, Karmilowicz MJ, Ghosh P, Abramov YA, Martin V, Sherman W (2014) A computational approach to enzyme design: predicting ω-aminotransferase catalytic activity using docking and MM-GBSA scoring. J Chem Inf Model 54(8):2334–2346
50. Höhne M, Schätzle S, Jochens H, Robins K, Bornscheuer UT (2010) Rational assignment of key motifs for function guides in silico enzyme identification. Nat Chem Biol 6 (11):807–813
51. Verma R, Schwaneberg U, Roccatano D (2012) Computer-aided protein directed evolution: a review of web servers, databases and other computational tools for protein engineering. Comput Struct Biotechnol J 2: e201209008
52. Gududuru V, Hurh E, Dalton JT, Miller DD (2004) Synthesis and antiproliferative activity of 2-aryl-4-oxo-thiazolidin-3-yl-amides for prostate cancer. Bioorg Med Chem Lett 14 (21):5289–5293
53. Sun DX, Liu L, Heinz B, Kolykhalov A, Lamar J, Johnson RB, Wang QM, Yip Y, Chen S-H (2004) P4 cap modified tetrapeptidyl α-ketoamides as potent HCV NS3 protease inhibitors. Bioorg Med Chem Lett 14 (16):4333–4338
54. Ghose AK, Viswanadhan VN, Wendoloski JJ (1999) A knowledge-based approach in designing combinatorial or medicinal chemistry libraries for drug discovery. 1. A qualitative and quantitative characterization of known drug databases. J Comb Chem 1(1):55–68
55. Carey JS, Laffan D, Thomson C, Williams MT (2006) Analysis of the reactions used for the preparation of drug candidate molecules. Org Biomol Chem 4(12):2337–2347
56. Dong H, Tang W, Matyjaszewski K (2007) Well-defined high-molecular-weight polyacrylonitrile via activators regenerated by electron transfer ATRP. Macromolecules 40 (9):2974–2977
57. Hosamani KM, Pattanashettar RS (2004) Design and synthesis of novel hydrazides, thiosemicarbazides, oxadiazoles, and triazoles of N, N′-bis (1-carboxy-15-hydroxy-n-pentadec-8-yl) alkyl or-aryl diamides: an approach for their biological evaluation and possible industrial utilization. Ind Eng Chem Res 43 (17):4979–4999
58. Al-Zoubi RM, Marion O, Hall DG (2008) Direct and waste-free amidations and cycloadditions by organocatalytic activation of carboxylic acids at room temperature. Angew Chem 120(15):2918–2921

59. Han S-Y, Kim Y-A (2004) Recent development of peptide coupling reagents in organic synthesis. Tetrahedron 60(11):2447–2467
60. Constable DJ, Dunn PJ, Hayler JD, Humphrey GR, Leazer JL Jr, Linderman RJ, Lorenz K, Manley J, Pearlman BA, Wells A (2007) Key green chemistry research areas—a perspective from pharmaceutical manufacturers. Green Chem 9(5):411–420
61. Gotor-Fernandez V, Gotor V (2006) Enzymatic aminolysis and ammonolysis processes in the preparation of chiral nitrogenated compounds. Curr Org Chem 10(10):1125–1143
62. Pattabiraman VR, Bode JW (2011) Rethinking amide bond synthesis. Nature 480 (7378):471–479
63. Lundberg H, Tinnis F, Selander N, Adolfsson H (2014) Catalytic amide formation from non-activated carboxylic acids and amines. Chem Soc Rev 43(8):2714–2742
64. Gotor-Fernández V, Busto E, Gotor V (2006) Candida antarctica lipase B: an ideal biocatalyst for the preparation of nitrogenated organic compounds. Adv Synth Catal 348 (7-8):797–812
65. Sigmund AE, McNulty KC, Nguyen D, Silverman CE, Ma P, Pesti JA, DiCosimo R (2002) Enantioselective enzymatic aminolysis of a racemic 2-isoxazolylacetate alkyl ester. Can J Chem 80(6):608–612
66. Čeřovský V, Kula MR (2001) Studies on peptide amidase-catalysed C-terminal peptide amidation in organic media with respect to its substrate specificity. Biotechnol Appl Biochem 33(3):183–187
67. Khare SK, Kumar A, Kuo TM (2009) Lipase-catalyzed production of a bioactive fatty amide derivative of 7, 10-dihydroxy-8 (E)-octadecenoic acid. Bioresour Technol 100 (3):1482–1485
68. Kim K-H, Seong BL (2001) Peptide amidation: production of peptide hormonesin vivo andin vitro. Biotechnol Bioprocess Eng 6 (4):244–251
69. Petkov D, Stoineva I (1984) Enzyme peptide synthesis by an iterative procedure in a nucleophile pool. Tetrahedron Lett 25 (34):3751–3754
70. Blum JK, Bommarius AS (2010) Amino ester hydrolase from Xanthomonas campestris pv. campestris, ATCC 33913 for enzymatic synthesis of ampicillin. J Mol Catal B Enzym 67(1):21–28
71. Van Pelt S, Teeuwen R, Janssen M, Sheldon R, Dunn P, Howard R, Kumar R, Martinez I, Wong J (2011) Pseudomonas stutzeri lipase: a useful biocatalyst for aminolysis reactions. Green Chem 13 (7):1791–1798
72. Olaofe OA, Burton SG, Cowan DA, Harrison ST (2010) Improving the production of a thermostable amidase through optimising IPTG induction in a highly dense culture of recombinant Escherichia coli. Biochem Eng J 52(1):19–24
73. Nakagawa Y, Hasegawa A, Hiratake J, Sakata K (2007) Engineering of Pseudomonas aeruginosa lipase by directed evolution for enhanced amidase activity: mechanistic implication for amide hydrolysis by serine hydrolases. Protein Eng Des Sel 20(7):339–346
74. Hediger MR, De Vico L, Svendsen A, Besenmatter W, Jensen JH (2012) A computational methodology to screen activities of enzyme variants. PLoS One 7(12):e49849
75. Hediger MR, De Vico L, Rannes JB, Jäckel C, Besenmatter W, Svendsen A, Jensen JH (2013) In silico screening of 393 mutants facilitates enzyme engineering of amidase activity in CalB. PeerJ 1:e145
76. Besenmatter W, Svendsen A, Rannes JB, Jäckel C 2013 Lipase variants and polynucleotides encoding same. Google Patents.
77. Newman DJ, Cragg GM (2012) Natural products as sources of new drugs over the 30 years from 1981 to 2010. J Nat Prod 75 (3):311–335
78. Welsch ME, Snyder SA, Stockwell BR (2010) Privileged scaffolds for library design and drug discovery. Curr Opin Chem Biol 14 (3):347–361
79. Brooks HW, Guida WC, Daniel KG (2011) The significance of chirality in drug design and development. Curr Top Med Chem 11 (7):760–770
80. Hsieh S-Y, Binanzer M, Kreituss I, Bode JW (2012) Expanded substrate scope and catalyst optimization for the catalytic kinetic resolution of N-heterocycles. Chem Commun 48 (71):8892–8894
81. Zhou S, Fleischer S, Junge K, Das S, Addis D, Beller M (2010) Enantioselective synthesis of amines: general, efficient iron-catalyzed asymmetric transfer hydrogenation of imines. Angew Chem Int Ed 49(44):8121–8125
82. Nugent TC, El-Shazly M (2010) Chiral amine synthesis—recent developments and trends for enamide reduction, reductive amination, and imine reduction. Adv Synth Catal 352(5):753–819
83. Ghislieri D, Turner NJ (2014) Biocatalytic approaches to the synthesis of enantiomerically pure chiral amines. Top Catal 57 (5):284–300
84. Kohls H, Steffen-Munsberg F, Höhne M (2014) Recent achievements in developing the biocatalytic toolbox for chiral amine synthesis. Curr Opin Chem Biol 19:180–192

85. Koszelewski D, Tauber K, Faber K, Kroutil W (2010) ω-Transaminases for the synthesis of non-racemic α-chiral primary amines. Trends Biotechnol 28(6):324–332
86. Heberling MM, Wu B, Bartsch S, Janssen DB (2013) Priming ammonia lyases and aminomutases for industrial and therapeutic applications. Curr Opin Chem Biol 17(2):250–260
87. Lovelock SL, Lloyd RC, Turner NJ (2014) Phenylalanine ammonia lyase catalyzed synthesis of amino acids by an MIO-cofactor independent pathway. Angew Chem 126 (18):4740–4744
88. Green AP, Turner NJ, O'Reilly E (2014) Chiral amine synthesis using ω-transaminases: an amine donor that displaces equilibria and enables high-throughput screening. Angew Chem 126(40):10890–10893
89. https://www.cobiotech.eu/lw_resource/datapool/systemfiles/elements/files/6D54FD7A24704117E0539A695E860CA9/current/document/era_ib-2_final_seminar_berlin_3rd_call_project_hyperin.pdf.
90. Peng Z, Wong JW, Hansen EC, Puchlopek-Dermenci AL, Clarke HJ (2014) Development of a concise, asymmetric synthesis of a smoothened receptor (SMO) inhibitor: enzymatic transamination of a 4-piperidinone with dynamic kinetic resolution. Org Lett 16 (3):860–863
91. Kroutil W, Faber K, Clay D, Hall M, Tasnadi G, Winkler C, Mutti F, Simon R, Fuchs C, Pressnitz D (2012) Surfing the ω-transaminase and ene reductase wave: Biocatalytic asymmetric transformations for preparative organic synthesis. New Biotechnol 29:S236
92. Zhu D, Hua L (2009) Biocatalytic asymmetric amination of carbonyl functional groups—a synthetic biology approach to organic chemistry. Biotechnol J 4(10):1420–1431
93. Tufvesson P, Lima-Ramos J, Jensen JS, Al-Haque N, Neto W, Woodley JM (2011) Process considerations for the asymmetric synthesis of chiral amines using transaminases. Biotechnol Bioeng 108(7):1479–1493
94. Pannuri S, Kamat SV, Martin-Garcia AR (2008) Contacting chiral primary amine, omega-transaminase, and amino acceptor to convert into ketone; stereospecific; gene expression; can be used to enrich enantiomerically a mixture of chiral amines or to synthesize stereoselectively one of a pair of chiral amines. Google Patents
95. Stirling DI, Matcham GW, Zeitlin AL (1994) Using omega-amino acid transaminase. Google Patents
96. Martin AR, DiSanto R, Plotnikov I, Kamat S, Shonnard D, Pannuri S (2007) Improved activity and thermostability of (S)-aminotransferase by error-prone polymerase chain reaction for the production of a chiral amine. Biochem Eng J 37(3):246–255
97. Midelfort KS, Kumar R, Han S, Karmilowicz MJ, McConnell K, Gehlhaar DK, Mistry A, Chang JS, Anderson M, Villalobos A (2013) Redesigning and characterizing the substrate specificity and activity of Vibrio fluvialis aminotransferase for the synthesis of imagabalin. Protein Eng Des Sel 26(1):25–33
98. Narancic T, Davis R, Nikodinovic-Runic J, O'Connor KE (2015) Recent developments in biocatalysis beyond the laboratory. Biotechnol Lett:1–12
99. Schätzle S, Steffen-Munsberg F, Thontowi A, Höhne M, Robins K, Bornscheuer UT (2011) Enzymatic asymmetric synthesis of enantiomerically pure aliphatic, aromatic and arylaliphatic amines with (R)-selective amine transaminases. Adv Synth Catal 353 (13):2439–2445
100. Satake K, Fujita H (1953) Studies on amine dehydrogenases. J Biochem 40(6):547–556
101. Abrahamson MJ, Vázquez-Figueroa E, Woodall NB, Moore JC, Bommarius AS (2012) Development of an amine dehydrogenase for synthesis of chiral amines. Angew Chem Int Ed 51(16):3969–3972
102. Abrahamson MJ, Wong JW, Bommarius AS (2013) The evolution of an amine dehydrogenase biocatalyst for the asymmetric production of chiral amines. Adv Synth Catal 355 (9):1780–1786
103. Gamenara D, de María PD (2014) Enantioselective imine reduction catalyzed by imine reductases and artificial metalloenzymes. Org Biomol Chem 12(19):2989–2992
104. Li H, Williams P, Micklefield J, Gardiner JM, Stephens G (2004) A dynamic combinatorial screen for novel imine reductase activity. Tetrahedron 60(3):753–758
105. Meister A, Buckley SD (1957) Pyridine nucleotide-dependent reduction of the α-keto acid analogue of lysine to L-pipecolic acid. Biochim Biophys Acta 23:202–203
106. Baskar B, Pandian NG, Priya K, Chadha A (2004) Asymmetric reduction of alkyl 2-oxo-4-arylbutanoates and-but-3-enoates by Candida parapsilosis ATCC 7330: assignment of the absolute configuration of ethyl 2-hydroxy-4-(p-methylphenyl) but-3-enoate by 1 H NMR. Tetrahedron Asymmetry 15 (24):3961–3966
107. Mitsukura K, Suzuki M, Shinoda S, Kuramoto T, Yoshida T, Nagasawa T (2011)

Purification and characterization of a novel (R)-imine reductase from Streptomyces sp. GF3587. Biosci Biotechnol Biochem 75 (9):1778–1782
108. Mitsukura K, Suzuki M, Tada K, Yoshida T, Nagasawa T (2010) Asymmetric synthesis of chiral cyclic amine from cyclic imine by bacterial whole-cell catalyst of enantioselective imine reductase. Org Biomol Chem 8 (20):4533–4535
109. Scheller PN, Fademrecht S, Hofelzer S, Pleiss J, Leipold F, Turner NJ, Nestl BM, Hauer B (2014) Enzyme toolbox: novel enantiocomplementary imine reductases. ChemBioChem 15(15):2201–2204
110. Knowles WS (2002) Asymmetric hydrogenations (Nobel Lecture). Angew Chem Int Ed 41(12):1998–2007
111. Noyori R (2002) Asymmetric catalysis: science and opportunities (Nobel Lecture). Angew Chem Int Ed 41(12):2008–2022
112. Yang JW, Hechavarria Fonseca MT, Vignola N, List B (2005) Metal-free, organocatalytic asymmetric transfer hydrogenation of α, β-unsaturated aldehydes. Angew Chem 117(1):110–112
113. List B, Yang JW (2006) The organic approach to asymmetric catalysis. Science 313 (5793):1584–1586
114. Williams RE, Bruce NC (2002) 'New uses for an Old Enzyme'–the old yellow enzyme family of flavoenzymes. Microbiology 148 (6):1607–1614
115. Warburg O, Christian W (1932) Naturtvissenschaften, 20,688,980. Biochem Z 1932 (264):438
116. Kohli RM, Massey V (1998) The oxidative half-reaction of old yellow enzyme the role of tyrosine 196. J Biol Chem 273 (49):32763–32770
117. Servi S (1990) Baker's yeast as a reagent in organic synthesis. Synthesis 1(01):1–25
118. Hall M, Hauer B, Stuermer R, Kroutil W, Faber K (2006) Asymmetric whole-cell bioreduction of an α,β-unsaturated aldehyde (citral): competing prim-alcohol dehydrogenase and C–C lyase activities. Tetrahedron Asymmetry 17(21):3058–3062
119. Stueckler C, Hall M, Ehammer H, Pointner E, Kroutil W, Macheroux P, Faber K (2007) Stereocomplementary bioreduction of α, β-unsaturated dicarboxylic acids and dimethyl esters using enoate reductases: enzyme-and substrate-based stereocontrol. Org Lett 9(26):5409–5411
120. Mueller NJ, Stueckler C, Hauer B, Baudendistel N, Housden H, Bruce NC, Faber K (2010) The substrate spectra of pentaerythritol tetranitrate reductase, morphinone reductase, n-ethylmaleimide reductase and estrogen-binding protein in the asymmetric bioreduction of activated alkenes. Adv Synth Catal 352(2-3):387–394
121. Reich S, Hoeffken HW, Rosche B, Nestl BM, Hauer B (2012) Crystal structure determination and mutagenesis analysis of the ene reductase NCR. ChemBioChem 13 (16):2400–2407
122. Reich S, Kress N, Nestl BM, Hauer B (2014) Variations in the stability of NCR ene reductase by rational enzyme loop modulation. J Struct Biol 185(2):228–233
123. Nestl BM, Hammer SC, Nebel BA, Hauer B (2014) New generation of biocatalysts for organic synthesis. Angew Chem Int Ed 53 (12):3070–3095
124. Brenna E, Gatti FG, Manfredi A, Monti D, Parmeggiani F (2011) Enoate reductase-mediated preparation of methyl (S)-2-bromobutanoate, a useful key intermediate for the synthesis of chiral active pharmaceutical ingredients. Org Process Res Dev 16 (2):262–268
125. Mangan D, Miskelly I, Moody TS (2012) A three-enzyme system involving an ene-reductase for generating valuable chiral building blocks. Adv Synth Catal 354 (11-12):2185–2190
126. Winkler CK, Clay D, Davies S, O'Neill P, McDaid P, Debarge S, Steflik J, Karmilowicz M, Wong JW, Faber K (2013) Chemoenzymatic asymmetric synthesis of pregabalin precursors via asymmetric bioreduction of β-cyanoacrylate esters using ene-reductases. J Organic Chem 78 (4):1525–1533
127. Fryszkowska A, Fisher K, Gardiner JM, Stephens GM (2010) A short, chemoenzymatic route to chiral β-aryl-γ-amino acids using reductases from anaerobic bacteria. Org Biomol Chem 8(3):533–535
128. Debarge S, McDaid P, O'Neill P, Frahill J, Wong JW, Carr D, Burrell A, Davies S, Karmilowicz M, Steflik J (2014) evaluation of several routes to advanced pregabalin intermediates: synthesis and enantioselective enzymatic reduction using ene-reductases. Org Process Res Dev 18(1):109–121
129. Winkler CK, Clay D, Turrini NG, Lechner H, Kroutil W, Davies S, Debarge S, O'Neill P, Steflik J, Karmilowicz M (2014) Nitrile as activating group in the asymmetric bioreduction of β-cyanoacrylic acids catalyzed by ene-reductases. Adv Synth Catal 356 (8):1878–1882

130. Knaus T, Mutti FG, Humphreys LD, Turner NJ, Scrutton NS (2015) Systematic methodology for the development of biocatalytic hydrogen-borrowing cascades: application to the synthesis of chiral α-substituted carboxylic acids from α-substituted α, β-unsaturated aldehydes. Org Biomol Chem 13 (1):223–233
131. Labinger JA, Bercaw JE (2002) Understanding and exploiting C–H bond activation. Nature 417(6888):507–514
132. Wender PA, Hilinski MK, Mayweg AV (2005) Late-stage intermolecular CH activation for lead diversification: a highly chemoselective oxyfunctionalization of the C-9 position of potent bryostatin analogues. Org Lett 7 (1):79–82
133. Das S, Incarvito CD, Crabtree RH, Brudvig GW (2006) Molecular recognition in the selective oxygenation of saturated CH bonds by a dimanganese catalyst. Science 312 (5782):1941–1943
134. Chen MS, White MC (2007) A predictably selective aliphatic C–H oxidation reaction for complex molecule synthesis. Science 318 (5851):783–787
135. Burton SG (2003) Oxidizing enzymes as biocatalysts. Trends Biotechnol 21(12):543–549
136. Hollmann F, Arends IWCE, Buehler K, Schallmey A, Bühler B (2011) Enzyme-mediated oxidations for the chemist. Green Chem 13(2):226–265
137. Roduner E, Kaim W, Sarkar B, Urlacher VB, Pleiss J, Gläser R, Einicke WD, Sprenger GA, Beifuß U, Klemm E (2013) Selective catalytic oxidation of C–H bonds with molecular oxygen. ChemCatChem 5(1):82–112
138. Brown CM, Reisfeld B, Mayeno AN (2008) Cytochromes P450: a structure-based summary of biotransformations using representative substrates. Drug Metab Rev 40(1):1–100
139. Sakaki T (2012) Practical application of cytochrome P450. Biol Pharm Bull 35 (6):844–849
140. Omura T (1999) Forty years of cytochrome P450. Biochem Biophys Res Commun 266 (3):690–698
141. Hanlon SP, Friedberg T, Wolf CR, Ghisalba O, Kittelmann M (2007) Recombinant yeast and bacteria that express human P450s: bioreactors for drug discovery, development, and biotechnology. Modern Biooxidation:233–252
142. Schroer K, Kittelmann M, Lütz S (2010) Recombinant human cytochrome P450 monooxygenases for drug metabolite synthesis. Biotechnol Bioeng 106(5):699–706
143. Vail RB, Homann MJ, Hanna I, Zaks A (2005) Preparative synthesis of drug metabolites using human cytochrome P450s 3A4, 2C9 and 1A2 with NADPH-P450 reductase expressed in Escherichia coli. J Ind Microbiol Biotechnol 32(2):67–74
144. Otey CR, Bandara G, Lalonde J, Takahashi K, Arnold FH (2006) Preparation of human metabolites of propranolol using laboratory-evolved bacterial cytochromes P450. Biotechnol Bioeng 93(3):494–499
145. Kabumoto H, Miyazaki K, Arisawa A (2009) Directed evolution of the actinomycete cytochrome P450 MoxA (CYP105) for enhanced activity. Biosci Biotechnol Biochem 73 (9):1922–1927
146. Bae J-W, Doo E-H, Shin S-H, Lee S-G, Jeong Y-J, Park J-B, Park S (2010) Development of a recombinant Escherichia coli-based biocatalyst to enable high styrene epoxidation activity with high product yield on energy source. Process Biochem 45(2):147–152
147. Baldwin CV, Wohlgemuth R, Woodley JM (2008) The first 200-L scale asymmetric Baeyer–Villiger oxidation using a whole-cell biocatalyst. Org Process Res Dev 12 (4):660–665
148. Collins AM, Woodley JM, Liddell JM (1995) Determination of reactor operation for the microbial hydroxylation of toluene in a two-liquid phase process. J Ind Microbiol 14 (5):382–388
149. Shibasaki T, Mori H, Ozaki A (2000) Enzymatic production of trans-4-hydroxy-L-proline by regio-and stereospecific hydroxylation of L-proline. Biosci Biotechnol Biochem 64(4):746–750
150. Han GH, Shin H-J, Kim SW (2008) Optimization of bio-indigo production by recombinant E. coli harboring fmo gene. Enzym Microb Technol 42(7):617–623

Chapter 13

New Directions in Coupling Chemistry

Gary M. Gallego, Rebecca A. Gallego, and Paul F. Richardson

Abstract

Palladium-mediated cross-couplings are established as a core synthetic technology within the pharmaceutical industry for the assembly of biologically active molecules. Despite both their effectiveness and prevalence, the use of precious metals presents challenges from an environmental perspective. The ability of base metals such as Ni, Fe, and Cu to mediate carbon-carbon bond formations predates the discovery of the Pd-catalyzed methodologies though are far less established in terms of their broad scope and utility. The current chapter focuses on emerging developments in the use of non-precious metals in catalysis with potential applications in both research and development.

Key words Nickel catalysis, Iron catalysis, Copper catalysis, Cross-coupling, Suzuki-Miyaura reaction, Green chemistry, Renewable, Carbon-carbon bond formation, Transition-metal free

1 Introduction

The emergence of palladium-mediated cross-couplings as a core synthetic transformation within the pharmaceutical industry is well established with the importance of these processes being recognized by the award of the 2010 Nobel Prize in Chemistry to Professors Heck, Negishi, and Suzuki [1]. Analysis of medicinal chemistry portfolios on all the reactions utilized for the assembly of new molecular entities highlights the prevalence of these methodologies accounting for 17% of the transformations used [2, 3]. Furthermore, examination of the examples of these reactions carried out on >100 mmol scale highlights the reliance on Pd as the key catalyst with over 90% of reported cases for C-C bond formation utilizing this metal with only isolated examples reporting the use of nickel (primarily Kumada couplings) as an alternative [4].

There are however several compelling reasons to look for alternative metals to Pd such as lowered cost, decreased toxicity, stable long-term supply, as well as metals with a reduced environmental impact as summarized in Table 1 [5].

Paul F. Richardson (ed.), *Green Chemistry in Drug Discovery: From Academia to Industry*, Methods in Pharmacology and Toxicology, https://doi.org/10.1007/978-1-0716-1579-9_13,

Table 1
Comparison of Pd, Ni, Cu, and Fe

Metal	Cost ($/oz)	Annual production (tons)	Oral exposure limits (ppm)	Natural abundance (ppm)	Carbon footprint (kg CO_2/e)	Water withdrawal to extract 1 kg^3 (L)
Pd	976	24	10	0.015	3880	508,000
Ni	0.33	1,350,000	25	90	6.5	390
Cu	0.195	15,000,000	250	68	2.8	79
Fe	0.02	1,200,000,000	1300	56,300	1.5	23

The current chapter examines the use of nickel, iron, and copper catalysis as an alternative to palladium for C-C bond-forming processes. It is important to note that this is an extensive area of research, and as such exhaustive coverage of this topic is impossible, though the focus herein is placed on examples in which emerging methodologies can either seamlessly replace palladium or present advantages in terms of the chemistry scope to the current state-of-the-art technology.

2 Nickel-Mediated Cross-Couplings

2.1 Introduction

Over the past 30 years nickel has enjoyed a period of intensified research interest in the area of cross-coupling chemistry for a variety of reasons [6, 7]. It is considered an abundant, low-cost, and "green" alternative to rarer, more expensive metals such as palladium and platinum. While in elemental form is, mole to mole, roughly 3000 and 10,000 times cheaper than palladium and platinum, respectively [8], it is also capable of undergoing the same elementary reactions. Additionally, it possesses characteristics that make it more advantageous than palladium and platinum in many contexts. It is less electronegative than palladium and platinum, so oxidative addition occurs more readily. This potentiates lower reaction temperatures, allowing for the possibility of fewer side reactions [9]. In fact in 1976 when comparing the nickel- with the palladium-catalyzed cross-coupling between an aryl bromide and alkenyl-aluminum species, Negishi observed that "...the palladium-catalyzed reaction is much slower than the corresponding nickel-catalyzed reaction, and does not offer any advantage over the latter" [10]. Additionally, substrates such as vinyl carbamates [11] and aryl ethers [12] that would be significantly less reactive with palladium and platinum become suitable reactants for nickel-mediated cross-coupling. The decreased electronegativity of nickel relative to palladium and platinum also means that reductive elimination is more difficult. This, along with the relatively higher energy barrier for nickel-carbon bond rotation, is yet another benefit because undesirable processes such as β-hydride elimination are slower [13]. This opens the opportunity for the use of nickel in challenging transformations that are plagued by β-hydride elimination, such as alkyl-alkyl cross-coupling. Finally, nickel can access a variety of oxidation states, allowing for more complex mechanisms and catalytic pathways since it can act in both polar and radical-like modalities. Access to various oxidation states allows nickel to perform transformations otherwise inoperable by palladium and platinum.

2.2 Alkyl-Alkyl Coupling

A major area of research where nickel far exceeds the ability of other metals is in alkyl-alkyl cross-coupling [14, 15]. As mentioned above, the inherent quality of nickel to be less susceptible to β-hydride elimination enables the success of this type of transformation. The field of alkyl-alkyl cross-coupling is considered to have been brought into the modern era by Fu and co-workers in 2003 with the publication of a method for nickel-mediated Negishi coupling of secondary alkyl halides possessing β-hydrogens with primary organozinc reagents [14]. Up until this point, methods for alkyl-alkyl bond formation had been limited to primary electrophiles, so Fu's work represented an impressive advance. His seminal work was further expanded upon to include nickel-mediated alkyl-alkyl Suzuki [15], Hiyama [16], and Stille [17] cross-couplings. This work is particularly impressive considering that there are only a few examples where palladium is capable of performing cross-coupling with secondary sp^3-electrophiles [18, 19].

In recent years, this work has been expanded to a wide range of applications that includes enantioselective transformations. Particularly important in this area are Fu's stereoconvergent alkyl-alkyl Suzuki coupling reactions [20–23] (Fig. 1a–d). Fu's group is able to perform Suzuki cross-couplings on a variety of substrates with and without directing groups. A number of these transformations are particularly novel because the electrophilic components are unactivated (**4**, **7**, **9**).

Apart from his work on alkyl-alkyl Suzuki cross-couplings, Fu and co-workers have developed a stereoconvergent method for access to secondary nitriles (**14**) from α-bromonitriles (**12**) via Negishi cross-coupling [24] (Fig. 2). This can be exploited to form both Csp^3-Csp^3 and Csp^3-Csp^2 bonds. This transformation is useful because the enantioenriched product of this reaction can be further manipulated into a variety of functional groups including heterocycles, carboxylic acids, ketones, esters, amides, and amines.

Impressively, Fu's work has also been extended to achieve nickel-mediated cross-coupling of tertiary electrophiles to form quaternary centers (**17**, **20**) [25] (Fig. 3a, b).

This transformation highlights the special nature of nickel. Since palladium typically oxidatively inserts via an S_N2 or direct

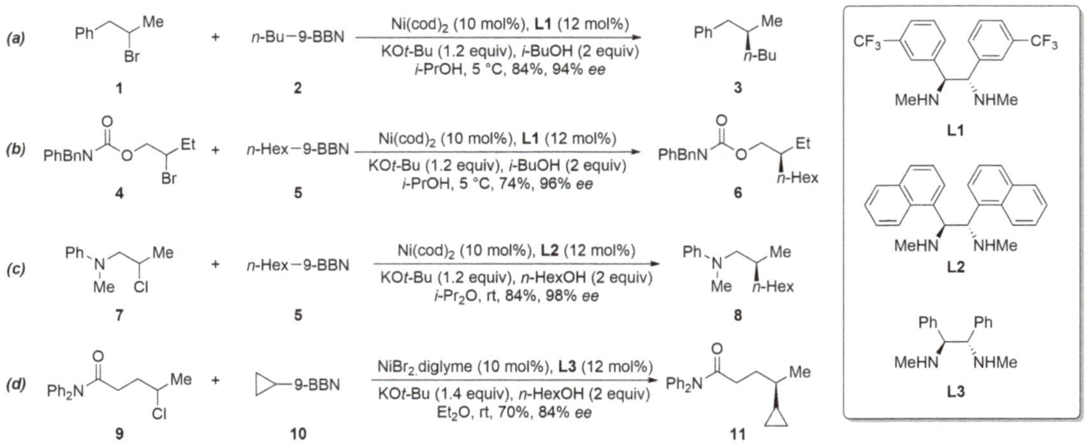

Fig. 1 Fu's stereoconvergent alkyl-alkyl Suzuki cross-couplings

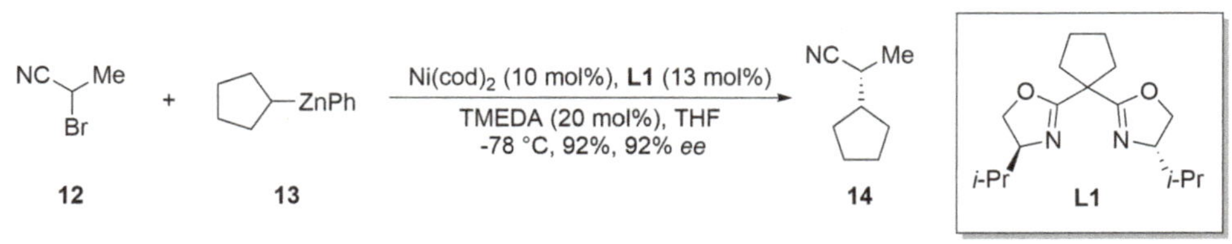

Fig. 2 Negishi cross-coupling for the synthesis of enantioenriched secondary nitriles

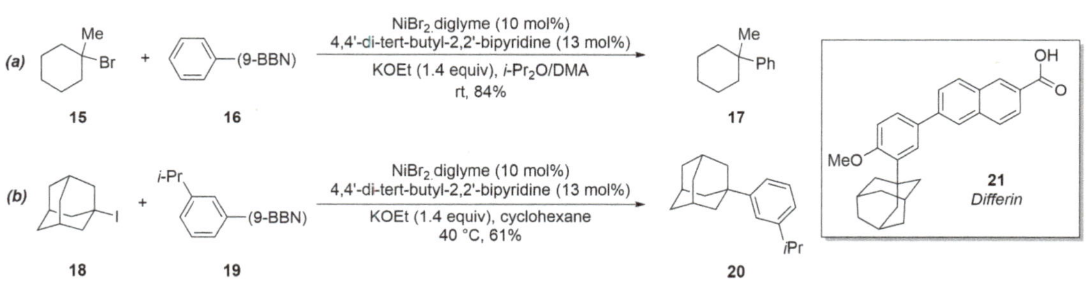

Fig. 3 Nickel-mediated cross-coupling to access quaternary centers

insertion pathway, tertiary electrophiles have been traditionally unusable in palladium-mediated cross-couplings. It is thought that this nickel-mediated reaction proceeds via an inner sphere electron transfer mechanism, allowing the metal to engage hindered alkyl-bromide substrates. Even highly hindered substrates such as 1-iodoadamantane (**18**) readily undergo coupling to give arylated products (**20**) (Fig. 3b). This substrate is particularly notable as adamantanes are found in a number of pharmaceuticals, such as differin (**21**) [26] (Fig. 3).

2.2.1 General Procedure for Suzuki Cross-Coupling with a Tertiary Alkyl Halide

In a nitrogen-filled glove box, $NiBr_2$• diglyme (35.5 mg, 0.10 mmol, 0.10 equiv) and 4,4′-di-*t*-butyl-2,2′-bipyridine (30.0 mg, 0.11 mmol, 0.11 equiv) were added to a 30 mL vial equipped with a magnetic stir bar. Solvent (anhydrous; 24 mL) was added, the vial was sealed with a PTFE-lined cap, and the resulting mixture was stirred vigorously for 2 h (a light-green slurry formed). The solution of the activated aryl-(9-BBN) reagent was then added to the slurry, and the reaction vial was sealed with a PTFE-lined cap and stirred for 20 min (the reaction mixture turned dark-green). The tertiary alkyl halide (neat; 1.0 mmol, 1.0 equiv) was then added to the slurry via microsyringe. The resulting mixture was capped and stirred vigorously at 40 °C for 24 h (outside of the glove box). Next, the reaction mixture was filtered through a plug of silica gel, which was rinsed with diethyl ether, and the filtrate was concentrated using rotary evaporation. The product was purified by chromatography.

2.3 C-O Insertion

Nickel-mediated chemistry is highly useful for the functionalization of carbon-oxygen bonds. Oxidative insertion into carbon-oxygen bonds is more readily achieved by nickel relative to other transition metals due to its decreased electronegativity. This makes nickel a good choice for reaction with these difficult substrates. The functionalization of carbon-oxygen bonds is highly sought after because they are cross-coupling handles that can be carried through many steps as inert functional groups. This enables the use of synthetic strategies such as sequential cross-coupling and late-stage diversification. The use of oxygen-containing handles in the place of halides additionally makes cross-coupling a more environmentally benign process with a less toxic waste stream [27].

A class of carbon-oxygen bonds that is useful in nickel-mediated cross-coupling is phenolic derivatives, such as carbamates. Aryl carbamates are useful because the carbamate group is also an attractive directing group for *ortho*-metalation [28, 29]. There has been much work in this area in recent years [30]. Research done in the Garg group has produced a method for the amination of aryl carbamates under mild nickel-mediated conditions [31] (Fig. 4). Carbamate (**22**) readily undergoes coupling with morpholine (**23**) to give the desired aminated product (**24**) in 71% yield. Related work on the amination of aryl pivalates was previously published by the Chatani group and others [32].

2.3.1 Conversion of Carbamate 22 to Amine 24

A 4 mL reaction vial was charged with a magnetic stir bar, flame-dried under reduced pressure, and allowed to cool under N2. The vial was then charged with Ph–B(pin) (35.71 mg, 0.175 mmol, 35 mol%), anhydrous powdered NaO*t*Bu (108.1 mg, 1.125 mmol, 2.25 equiv), $NiCl_2$(DME) (5.5 mg, 0.025 mmol, 5 mol%), and SIPr•HCl (21.3 mg, 0.05 mmol, 10 mol%). Subsequently, dioxane (2.5 ml), carbamate (**22**) (95.5 mg, 0.50 mmol, 1.0 equiv), and morpholine (**23**) (87.1 μL, 0.9 mmol, 1.8 equiv) were added, sequentially. The resulting heterogenous mixture was stirred for

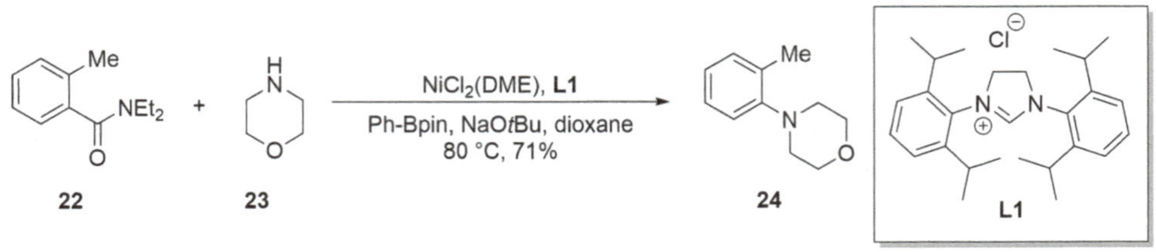

Fig. 4 Nickel-mediated amination of aryl carbamates

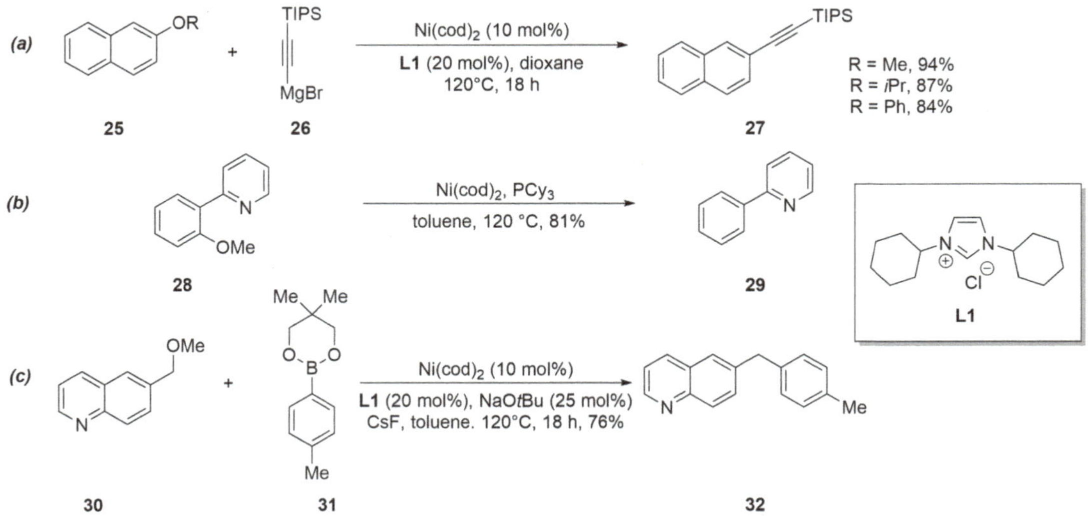

Fig. 5 Nickel-catalyzed ether C-O bond cleavage

1 min while purging with N2, and the vial was sealed with a Teflon-lined screw cap. The mixture was stirred at 23 °C for 1 h and then at 80 °C for 3 h in a preheated aluminum heating block. After cooling the reaction vessel to 23 °C and concentrating the mixture under reduced pressure, the crude residue was purified by flash chromatography (20:1 hexanes/EtOAc) which yielded aminated product (**24**) (71% yield) as a yellow oil (R_f 0.30) (20:1 hexanes/EtOAc).

Another class of carbon-oxygen bonds engaged in nickel catalysis is ethers. Recently, the Chatani, Tobisu, and Martin groups have demonstrated this in a variety of contexts, such as the alkynylation of anisoles [33], the reductive cleavage of aryl ethers [34, 35], and in the context of Suzuki cross-coupling of aryl (**25**, **28**) and benzyl ethers (**30**) [36] (Fig. 5a–c).

Typically, electron-rich carbon-oxygen bonds in ethers are very difficult to engage in oxidative additions because of their electron-rich nature. This has resulted in limited application to cross-coupling chemistry. Recent advances in nickel-mediated cross-coupling have made these substrates valuable reactants for cross-coupling.

2.4 Nickel-Catalyzed Suzuki-Miyaura Reactions

The use of nickel catalyst systems in Suzuki-Miyaura cross-couplings presents several advantages over the well-established palladium-based systems particularly in that it is believed that due to the higher reactivity of Ni (0) toward chloroarenes, more challenging couplings could be achieved with simpler catalyst systems. Despite this, it was 15 years between the original report of the palladium version of the reaction [37] and the disclosure of nickel variants in the coupling of boronic acids with aryl mesylates [38] and chlorides [39], respectively, and over a decade elapsed until this became a topic of significant research interest.

Several drawbacks are evident from much of the initial reports using nickel catalysts, such as the need for high catalyst loadings (3–10%), and limitations in terms of scope, particularly with reactions leading to hetero-biaryl products, which are of greater interest from a pharmaceutical standpoint. Ge and Hartwig reported on a single-component catalyst system, which enables many couplings under mild conditions using low loadings of the ligand (0.5 mol%) [9]. The [(dppf)Ni(cinnamyl)Cl] complex is easy to prepare and stable indefinitely when stored under nitrogen at room temperature, and its demonstrated low reactivity to both moisture and air makes reactions practical to conduct on large or small scale. Couplings are carried out between 2-heteroaryl (sulfur and oxygen containing) boronic acids (**33**) and heteroaryl halides (**34**) either with $K_2CO_3(H_2O)_{1.5}$ as the base in acetonitrile at 50 °C or with K_3PO_4 as base in 1,4-dioxane at 70 °C (Fig. 6). The scope of the reaction is good, and the system tolerates numerous sensitive functional groups such as an unprotected NH_2, aldehydes, ketones with enolizable hydrogens, and nitriles. The catalyst is also stable to high concentrations of pyridines presumably due to the chelating property of the dppf ligand. For the corresponding 3-heteroaryl boronic acids, similar conditions were employed with a slightly elevated temperature (80 °C). For nitrogen containing boronic acids, 1-Boc-pyrrole-2-boronic acid was demonstrated to be a successful coupling partner though pyridyl boronic acids failed to react. The robustness of the chemistry was demonstrated by the ability to obtain practically identical yields when reactions were performed outside the dry box, with the requisite catalyst being weighed out in air.

While this work represents a breakthrough in terms of expanding the scope of the Ni-mediated Suzuki-Miyaura reaction, limitations still exist with this being exemplified by Table 2 comparing four established nickel- and palladium-mediated cross-coupling methodologies with a range of five commonly used heterocyclic electrophiles (**E1**–**E5**) within medicinal chemistry [9, 40, 41] (Fig. 7). As can be seen, the results are somewhat variable, and although conditions can be identified for most of the compounds evaluated utilizing palladium catalysis, the methods based on nickel

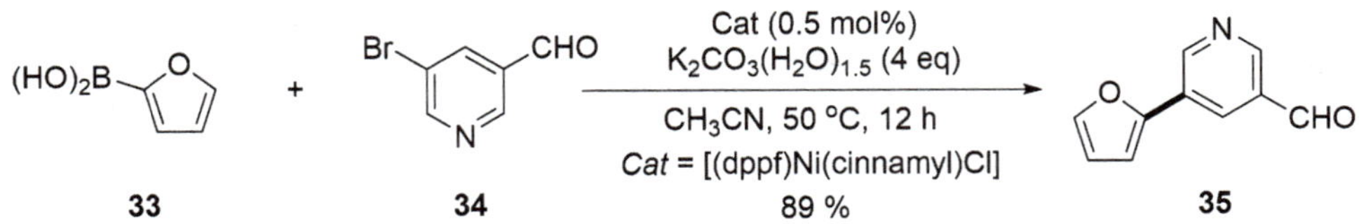

Fig. 6 Hartwig nickel catalyst for Suzuki-Miyaura coupling

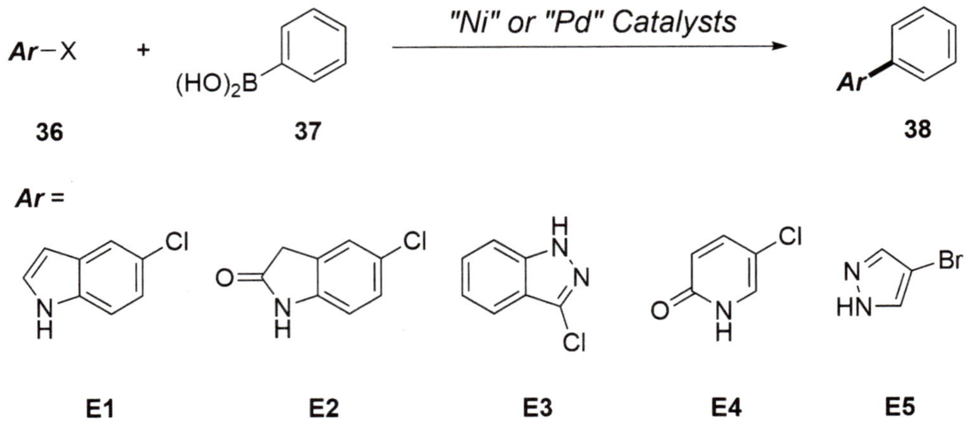

Fig. 7 Comparison of nickel- and palladium-catalyzed methods for heterocyclic electrophiles

Table 2
Comparison of nickel- and palladium-catalyzed methods for heterocyclic electrophiles

Electrophile	(dppf)Ni(cin)Cl	Ni(COD)$_2$/PCy$_3$	Pd(OAc)$_2$/XPhos	X-Phos palladacycle
E1	<1%	100%	100%	100%
E2	<1%	<1%	100%	100%
E3	<1%	3%	84%	100%
E4	<1%	<1%	41%	61%
E5	Decomp.	<1%	<1%	Mostly Ph-Ph

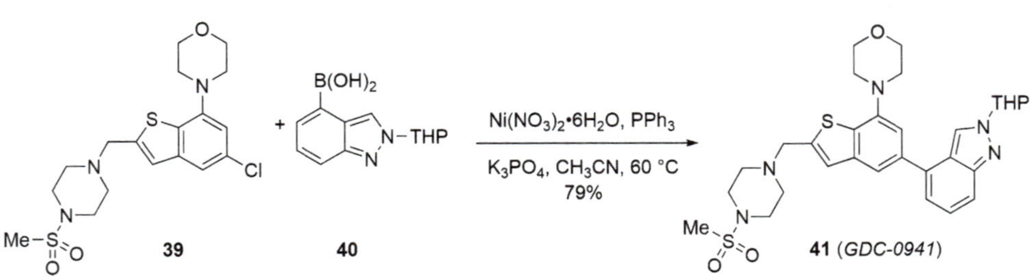

Fig. 8 Scale-up of nickel-catalyzed Suzuki-Miyaura cross-coupling

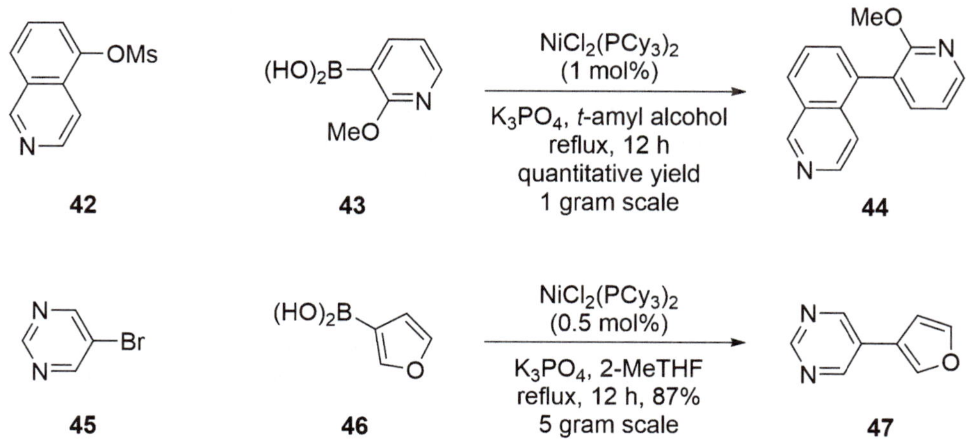

Fig. 9 Garg's use of greener solvents for nickel-mediated cross-couplings

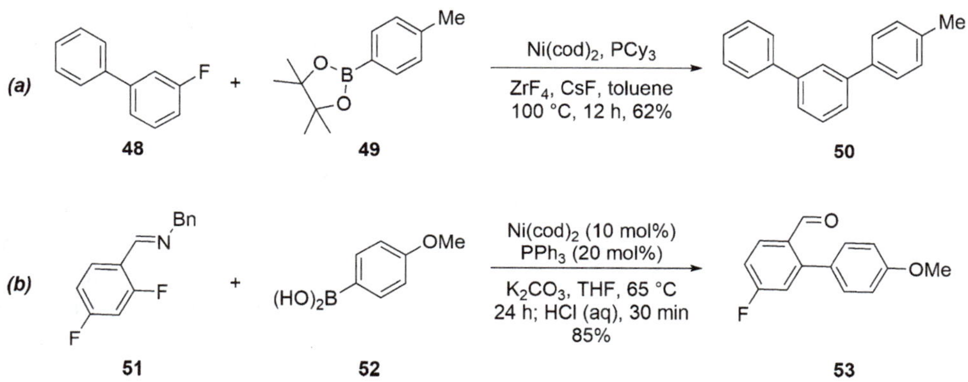

Fig. 10 Nickel-catalyzed Suzuki-Miyaura reaction of aryl fluorides

perform poorly for this range of substrates. One important factor though to consider here is the amount of research that has gone into the development of the palladium-based reaction not only in terms of a mechanistic understanding of the process but also with advances realized in the nature of the ligands as well as the use of preformed catalyst complexes, which are exemplified here. Assessing new ligands for effective nickel catalysis has to some degree been hindered primarily due to the lack of availability of suitable sources of Ni(0) to assess reaction performance. Prior to 2015, Ni $(cod)_2$ was utilized though this complex presents problems through both its air and thermal instability as well as the fact that the cod ligand does not act as an innocent bystander in downstream chemistries and can complicate observed catalytic outcomes. However, this is the precursor utilized in the work of Hartwig and Ge wherein the dppf ligand facilitates a very efficient reduction to (dppf)Ni (0) though this protocol is highly ligand-specific [9]. An alternative approach would be through in situ formation of Ni(II) species such as $(R_3P)_2NiCl_2$, though such a process is ligand-specific and involves strong reductants that can lead to inefficient catalyst activation.

Given this, the identification of nickel-based pre-catalysts that are air-stable and cheap is extremely attractive for both high-throughput experimentation approaches for reaction discovery and large-scale synthesis, and as noted these tend to either not be compatible with a variety of ligand classes, require reducing conditions for activation, or do not display general reactivity. Doyle and co-workers have described the synthesis and applicability of a practical, air-stable Ni(II) pre-catalyst [42]. Using TMEDA and a hindered aryl substituent to shield nickel from associative substitution, the group found that a (TMEDA)Ni(*o*-tolyl)Cl catalyst is stable to air and ambient temperature for 3 months on the benchtop with no detectable decomposition. The catalyst crashes out of the reaction conditions as a crystalline, square planar complex and was used as a substitute for $Ni(cod)_2$ or other Ni(II) catalysts in several reactions, including Suzuki-Miyaura, Buchwald-Hartwig, and oxidative and dehydrogenative coupling reactions. In most cases, with the addition of K_3PO_4, the pre-catalyst was found to be equivalent or superior to the literature reports. Reaction observations suggest that either nickel-boron or nickel-nickel transmetalation is a possible mechanism for pre-catalyst activation, but the reaction must be heated to at least 60 °C for this to occur. Workers at Pfizer have also reported on this catalyst describing the synthesis on multigram scale under neat conditions using $Ni(cod)_2$ as the precursor [43]. They also comment on the stability of the catalyst and note its applicability for use in HTE (high-throughput experimentation) campaigns based on its solubility, stability in solution over 24 h, and the ease of ligand exchange. Again, a comparison is carried with $Ni(cod)_2$ and $NiCl_2$(dme) over a range of reactions including cross-electrophile and Sonogashira couplings, and from these results the authors reinforce the importance of utilizing the appropriate Ni source to maximize the learnings from HTE screening studies.

The advantages and practicality of Ni-based Suzuki-Miyaura couplings on scale are provided by Tian and co-workers at Genentech in the reported synthesis of the PI3K inhibitor, GDC-0941 (**41**) [44] (Fig. 8). The reported conditions utilized cheap $Ni(NO_3)_2/6H_2O$ as the catalyst with simple PPh_3 as the ligand. The reaction is reported on a 54 kg scale with removal of the metal being achieved through an aqueous ammonia wash. The Ni-based reaction provides a 19% improvement in yield over the analogous Pd-mediated reaction, which uses $Pd(PPh_3)_2Cl_2$ as the catalyst and requires an additional treatment with thio-silica to remove residual palladium.

The previous section highlighted the ability of Ni species to undergo C-O insertion to mediate catalysis, and in addition to achieving the amination of aryl carbamates, these conditions are also translatable to heteroaryl carbamates such as pyridines. An example of this is provided by reports on both the Suzuki coupling and amination of heteroaryl carbamates (in addition to sulfamates, chlorides, and bromides) in solvents such as 2-methyl-THF and *t*-amyl alcohol [45, 46] thus enabling more cost-effective nickel-based catalysts to take place under more environmentally friendly conditions. Garg and co-workers have reported on the use of the commercially available $Ni(Cl)_2(PCy_3)_2$ as a pre-catalyst to mediate this transformation [45]. The model reaction between a napthyl sulfamate and phenylboronic acid using potassium phosphate as the base with 5 mol% catalyst was examined in >30 solvents to potentially identify greener alternatives. From these reactions over half the solvents provided a quantitative yield with *tert*-amyl alcohol and 2-MeTHF being selected for further studies. Several other electrophilic partners were evaluated and shown to be effective under the optimal reaction conditions. The scope of the reaction was also examined with numerous heterocyclic halides being shown to couple in high yield. In addition, a range of *bis*-heterocyclic scaffolds (**44**, **47**) were synthesized exploiting this new methodology. Finally, two examples are provided on gram scale with a lowering of the catalyst loading to 0.5 mol% (Fig. 9). The ability to run this reaction in green solvents makes it useful for process scale, which is of great significance to the pharmaceutical industry.

Another highly unique trait of nickel, as stated in the above section on carbon-oxygen bond insertion, is nickel's ability to oxidatively add into otherwise inert bonds. In addition to being able to insert into carbon-oxygen bonds, nickel has been shown to functionalize carbon-fluorine bonds. Organofluorines are useful in cross-coupling chemistry because they are robust to a wide range of conditions, including palladium-catalyzed cross-couplings because of the strength of the carbon-fluorine bond [47]. In 2011 the Chatani group reported a method for nickel-mediated Suzuki cross-coupling of aryl fluorides (**48**) (Fig. 10a). Previous methods achieve Kumada and Negishi reactions with aryl fluorides, but the Suzuki reaction has remained a challenge as previously published examples using aryl fluorides have a very limited substrate scope [48].

The benefit of the Chatani method is that it is applicable to a wide range of substrates, such as electron-rich, electron-deficient, and electron-neutral arenes as well as heteroaromatic fluorides. The reaction is facilitated by the addition of zirconium tetrafluoride. Alternatively, the reaction can be run in the absence of zirconium tetrafluoride if oxidative addition is facilitated through the use of chelation assistance by way of a heteroaromatic directing group. Concurrent with this report, the Love group published a similar chelation-assisted nickel-mediated Suzuki cross-coupling for electronically diverse aryl fluorides, with an imine directing group (**51**) (Fig. 10b). These methods are highly useful for the functionalization of fluorinated building blocks and are enabled by nickel's ability to engage otherwise inert carbon-fluorine bonds.

2.5 Reductive Couplings

Reductive cross-coupling is a valuable method that has gained increasing notoriety in recent years. Modern nickel-mediated reductive cross-coupling technology can achieve coupling without the generation of stoichiometric amounts of organometallic reagents. This has the advantage of greatly increased functional group tolerance. The shift from using organometallic alkanes to alkyl halides additionally provides the advantage of more ready commercial availability of the starting materials. In 2010, the Weix group published a method for the coupling of aryl halides (**54**) with alkyl halides (**55**) with a bench-stable nickel catalyst [49] (Fig. 11).

One benefit of this method over other reductive couplings is that it does not require a large excess of one reagent to avoid homocoupling [50]. Organobromides can be used in this reaction, but, interestingly, selectivity for the homocoupled product is good as long as at least one of the reactants is an organoiodide. This work was later expanded to include the reductive coupling of aryl bromides with alkyl bromides [51]. The substrate scope for both of these reactions includes a wide range of functional groups, including arenes with fluorine substituents, which is of note to the pharmaceutical industry. Additionally, aryl chlorides were shown to be good substrates under slightly modified conditions, although aryl chlorides are significantly less reactive than aryl bromides. The introduction of electron-withdrawing substituents onto the aryl substrate greatly increases yields.

As mentioned above, these methods for reductive cross-coupling exhibit exceptional functional group tolerance. Substrates with acidic protons, such as phenols and sulfonamides, are well tolerated. Additionally, bromoarenes containing pseudohalide functionalization—such as triflates, tosylates, or acetates—are exclusively reactive at the bromo-substituted carbon. Similarly, boronic esters, silanes, and stannanes are stable and unreactive (apart from sometimes undergoing very minor decomposition owing to protonolysis) under the conditions required for reductive coupling. This chemoselectivity is highly useful, as it is possible to selectively functionalize one part of the substrate and leave a handle for later manipulation. This allows for the rapid synthesis of polysubstituted arenes without need for the use of protecting groups.

An interesting aspect to reductive cross-coupling reactions is the role of the reducing agent, which is often Zn(0) or Mn(0). The tolerance of substrates with acidic protons and small amounts of water in the catalyst provides significant evidence that the metal is not forming an organometallic intermediate. Because stoichiometric metal(0) is present in the reaction, Ni(II) can be used as a catalyst instead of air-sensitive Ni(0). In fact, additional mechanistic

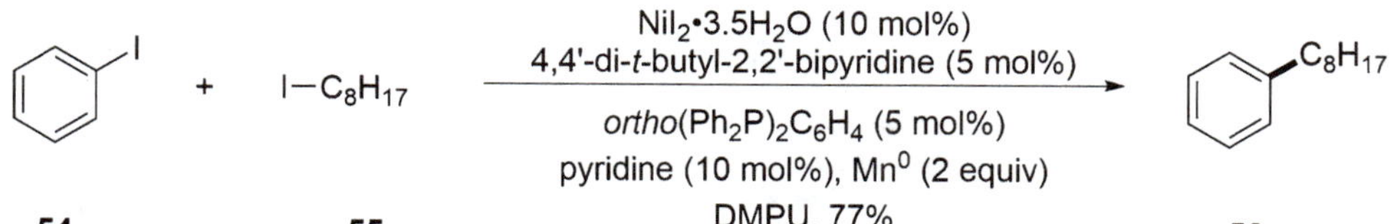

54 55 56

Fig. 11 Weix reductive cross-coupling to form aryl-alkyl bonds

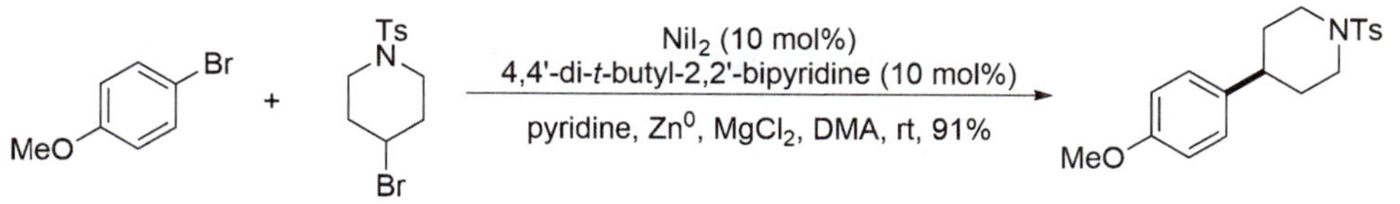

57 58 59

Fig. 12 Synthesis of arylated saturated heterocycles under reductive cross-coupling conditions

investigations have been completed, and they verify that the reaction proceeds via a radical chain mechanism where the stoichiometric Zn(0) or Mn(0) metal serves to reduce Ni(II) to Ni (0) [52]. Thus, the reaction is tolerant of air with any oxidized catalyst being reduced to the catalytically active Ni(0), although the reaction does proceed faster under inert atmosphere. This resiliency of this reaction to small amounts of air and water makes nickel-mediated reductive coupling readily scalable (>25 mmol scale reported) [53].

An area where the Weix reductive coupling is limited is in the identity of the alkyl halide. While they do include a few substrates that are secondary, most are primary alkyl halides. The secondary alkyl halides that are presented typically undergo reaction in under 70% yield. This limitation was improved upon by the Gong group in 2012 [54]. Gong et al. found that the addition of magnesium chloride and increased amounts of pyridine relative to the Weix conditions very effectively promoted the reaction of secondary alkyl bromides (**58**) with aryl bromides (**57**) to provide the desired arylated saturated heterocycles (**59**) in excellent yield (Fig. 12). This work has been highly relevant to synthesis done in the pharmaceutical industry as it enables incorporation of saturated heterocycles onto aromatic rings. They have also been able to perform this reaction using *bis*-pinacolatodiboron as a reductant instead of Zn (0) or Mn(0) [55]. Additionally, these modifications worked well with more electron-deficient aryl chlorides. Finally, this chemistry was also shown to promote the allylation of aryl bromides with allyl acetate.

To a flame-dried Schlenk tube equipped with a magnetic stir bar was loaded alkyl bromide (0.15 mmol, 100 mol%), followed by addition of 4,4′-di-*tert*-butyl-2,2′-bipyridine (4.0 mg,

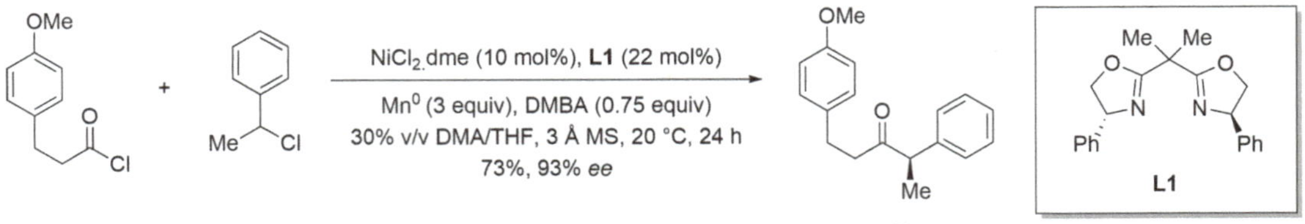

Fig. 13 Reisman enantioselective nickel-mediated reductive cross-coupling

2.5.1 General Procedure for the Coupling of Aryl Bromides and Alkyl Bromides

0.015 mmol, 10 mol%), aryl halide (0.15 mmol, 100 mol%), and zinc powder (19.6 mg, 0.3 mmol, 200 mol%). The tube was moved into a dry glove box, at which point NiI_2 (4.7 mg, 0.015 mmol, 10 mol%) and $MgCl_2$ (14.3 mg, 0.15 mmol, 100 mol%) were added. The tube was capped with a rubber septum, and it was moved out of the glove box. Pyridine (11.9 mg, 0.15 mmol, 100 mol%) and DMA (1.0 mL) were then added via syringe. After the reaction mixture was allowed to stir for 12 h under N_2 atmosphere at 25 °C, it was directly loaded onto a silica column without work-up. The residue in the reaction vessel was rinsed with small amount of DCM. Flash column chromatography (SiO_2/ethyl acetate in hexanes) provided the coupling product.

The Molander group has worked in the area of reductive coupling with secondary electrophiles in collaboration with Pfizer. Their research has shown that a variety of secondary bromides can be coupled to aryl and heteroaryl bromides in moderate to good yield [56]. Further, they have shown that this method can be extended to a variety of secondary alkyl tosylates as substrates [57].

An area where all of the abovementioned reductive cross-coupling methods are limited is in reactivity with heteroaryl halides as substrates. While Weix has been able to use protected indole substrates effectively, other heterocycles such as pyridines have not lead to high yields under the reaction conditions. Similarly, the Gong and Molander groups have shown pyridines and other heterocycles to be challenging substrates. Recently Weix et al. have made advances in this area to expand the utility of this reaction to include 2-chloropyridines [58]. Although this is a valuable advance, much remains to be done in terms of the scope and yields of the reductive cross-coupling of heteroaryl substrates. This is sure to continue to be a hot research topic.

Recent research in the Reisman group has produced the first examples of enantioselective nickel-mediated reductive cross-coupling [59] (Fig. 13). This method relies on direct ketone formation from acyl chlorides (**60**) and secondary alkyl halides (**61**) to produce enantioenriched α,α-disubstituted ketones (**62**). This is an improvement over traditional α-alkylation as selective enolate generation is often challenging. Reisman's conditions constitute a mild method that shows no product epimerization and high functional group tolerance (e.g., esters and alkyl bromides). The substrate scope includes a wide range of electron-donating and electron-withdrawing functionality appended to the benzyl chloride starting material. One note is that *ortho*-substituted benzyl chlorides react to provide the desired product in low yields and enantioselectivity. The Reisman group has further expanded the above work to achieve the enantioselective reductive coupling of benzyl chlorides and vinyl bromides [60].

While the advancement to achieve enantioselective reductive coupling is highly useful, much work remains to be done in this area. Other areas where reductive nickel chemistry requires further expansion include growing the pool of electrophiles used as substrates and the development of better ligands to improve selectivity [61].

2.6 Additional Applications of Ni-Catalyzed Cross-Couplings

A highly useful application of the nickel-mediated coupling is the stereoconvergent coupling of α-halo carbonyls with Grignard reagents (**64**) [62] (Fig. 14a), organozirconium reagents (**67**) [63] (Fig. 14b), and organoboranes (**16**) [64] (Fig. 14c). This method is useful as it provides enantioselective access to α-functionalized ketones and is complementary to the Reisman method in the above section.

Fu et al. have further expanded their work to access enantioenriched tertiary alkyl fluorides (**73**) from racemic α, α-dihaloketones (**71**) in a stereoconvergent Negishi reaction [65] (Fig. 15). Their method is notable because access to stereodefined fluorides is an area of high interest in the pharmaceutical industry.

The reaction has a wide substrate scope, with a variety of aryl groups being tolerated on the ketone. Limitations arise with increasing the steric bulk adjacent to the chlorine or bromine center. For example, going from ethyl to isopropyl at that position decreases the yield of the reaction from 62% to 55% and the *ee* drops from 97% to 82% (see Fig. 15).

2.6.1 General Procedure for Asymmetric Negishi Cross-Coupling

In the air, $NiCl_2$·glyme (33.0 mg, 0.150 mmol) and (4*R*,5*S*)-**L1** were added to an oven-dried 40 mL vial equipped with a stir bar. The vial was closed with a PTFE septum cap and then evacuated and back-filled with nitrogen (three cycles). Diglyme (2.5 mL) was added to the vial, and the mixture was vigorously stirred at room temperature for 45 min. In the air, an oven-dried 4 mL vial was charged with the alpha-bromo-alpha-fluoroketone (1.00 mmol), and then the vial was sparged with nitrogen for 10 min. Diglyme (0.9 mL) was added, and the resulting solution was transferred via syringe to the 40 mL reaction vial. The 4 mL vial was rinsed with diglyme (0.8 mL), and the washings were transferred to the reaction vial. The reaction vial was wrapped with electrical tape and then cooled to −25 °C for 15 min; at the same time, an oven-dried 40 mL vial that contained the arylzinc solution (0.30 M) was also

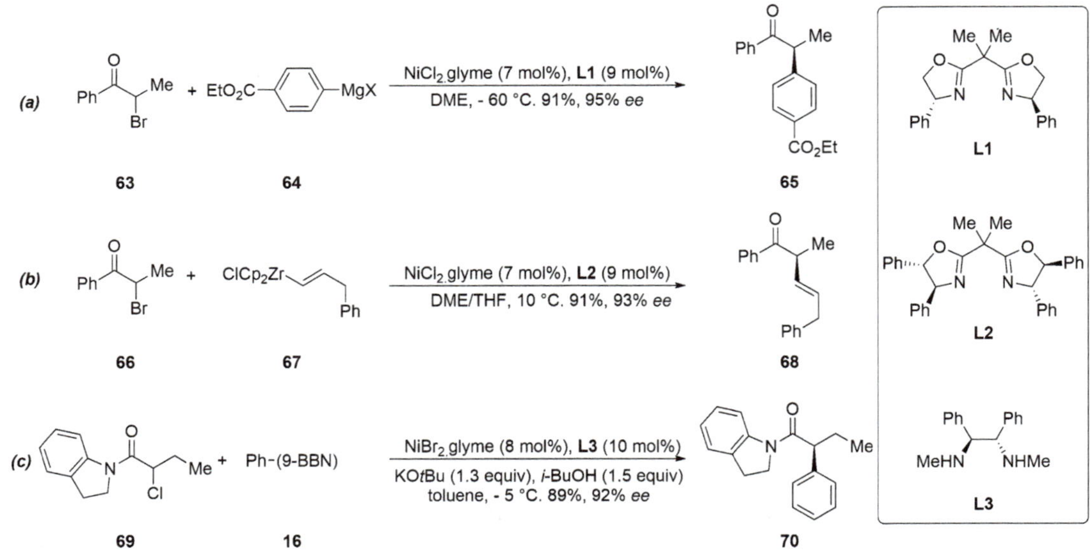

Fig. 14 Coupling of α-carbonyl compounds to form functionalized ketones

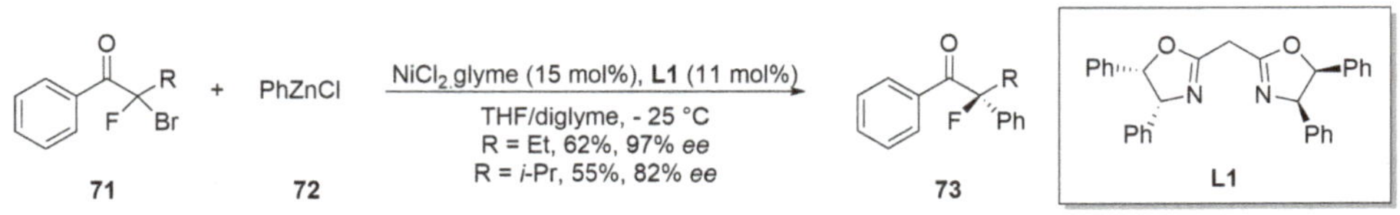

Fig. 15 Synthesis of enantioenriched tertiary alkyl fluorides

cooled to −25 °C for 15 min (a nitrogen-filled balloon was attached to each vial). To the vigorously stirred solution of catalyst and electrophile was added the solution of the arylzinc reagent (0.30 M; 6.67 mL, 2.00 mmol; 2.00 equiv) over 3 min, during which the reaction mixture turned to orange. The balloon was removed, and the septum cap was sealed with grease. The mixture was stirred vigorously at −25 °C for 36 h. Then, the reaction was quenched by the addition of EtOH (2 mL), and the mixture was allowed to warm to r.t. and then diluted with Et_2O (100 mL) and washed with deionized water (4 × 25 mL). The organic layer was dried over Na_2SO_4 and then concentrated, and the residue was purified by flash chromatography.

3 Iron-Mediated Cross-Couplings

3.1 Introduction

The introduction of iron as a competent coupling catalyst dates back to the mid-1940s with the coupling of alkyl Grignard reagents to benzyl bromide [66]. This study was followed by the iron-catalyzed coupling to acetyl chloride in 1953 [67] and, perhaps most significantly, vinyl halides in 1971 [68]. Despite this early start, iron has received considerably less attention than other transition metals (e.g., Pd, Ni, Ir, etc.) capable of performing coupling reactions. This may be partially attributed to the challenge in elucidating reaction mechanism and identification of intermediates [69, 70]. Within the past few decades, however, coupling chemistry has emerged as a mainstay in organic synthesis and the pharmaceutical industry. While metals such as Pd and Ni offer the capability of performing an array of couplings, their associated cost in combination with environmental and health concerns has led to a resurgence in research into iron catalysis [71, 72]. Iron boasts many green and economical properties. The favorable environmental profile of iron along with its high abundance, low cost, and toxicity makes it one of the more ideal metals for coupling, especially on large scale [73, 74]. Additionally, the majority of pre-catalysts are air and moisture stable and the use of exogenous ligands is rarely needed, improving atom economy. Finally, the extraordinary rates of these reactions and subtleties between reaction conditions allow for exquisite chemoselectivity and functional group tolerance.

3.2 Acylation

Construction of unsymmetrical ketones is of fundamental importance in organic synthesis. In this regard, a preliminary report from Cook in 1953 demonstrated that $FeCl_3$ catalyzes the selective monoaddition of *n*-butylmagnesium bromide to acetyl chloride [67]. Notably, 2-hexanone was obtained in 72% yield demonstrating the advantage over the un-catalyzed process, which returned the product in 31% yield. Following reports from Marchese [75, 76], Fürstner made a strong contribution to the generalization of these additions (Fig. 16) [77]. Specifically, he highlighted several features including their high functional group tolerance (including selectivity over aryl bromides), rapid reactivity at low temperature, and maintenance of stereochemical integrity at the α-position of the acid chloride. Also noteworthy is the report demonstrating organozinc [78] compounds can be coupled in good yield.

Beyond traditional acid chlorides, electrophiles, thioesters [79], acyl nitriles [80], and imidoyl chlorides [81] also couple with Grignard reagents to produce the ketone or imine products.

3.3 Vinylation

Vinylation of organometallic dates back to the seminal report by Kochi [68] where he demonstrated that vinyl halides could be successfully coupled with alkyl Grignard reagents in the presence of catalytic amounts of $FeCl_3$ (Fig. 17). An important feature of this reaction is the maintenance of *E,Z* stereochemistry with regard to the vinyl halide; thus **78** and **80** are coupled to produce **79** and **81**, respectively.

Since this original report, a few notable advances have made these reactions more efficient and more general. Iron-1,3-diketonate complexes have generally been shown to produce higher overall yields and, importantly, are easier to handle than the hygroscopic

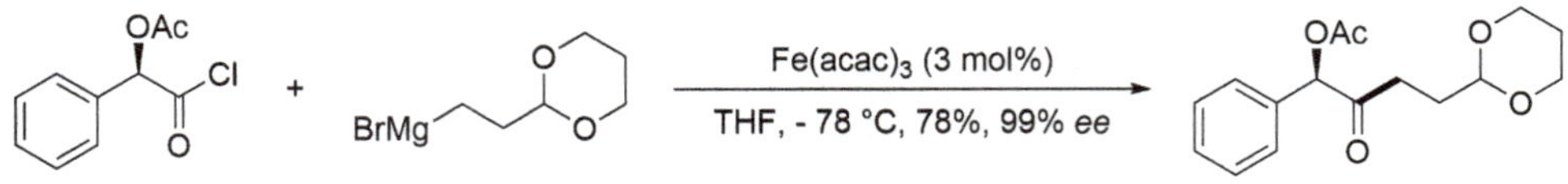

74 75 76

Fig. 16 Iron-mediated addition of Grignard reagents to acyl chlorides

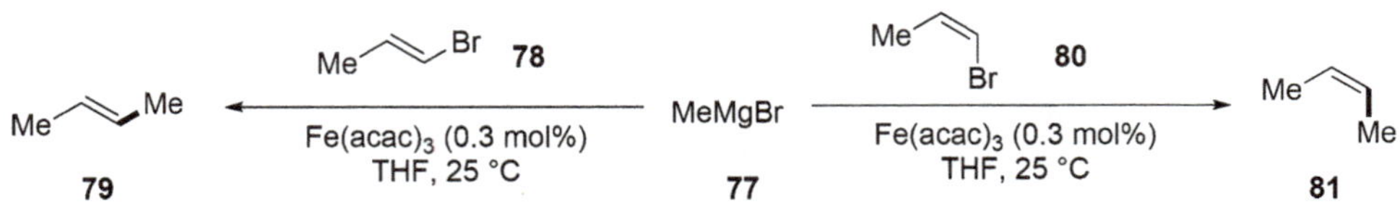

79 77 81

Fig. 17 Maintenance of *E,Z*-stereochemistry in vinylation of MeMgBr

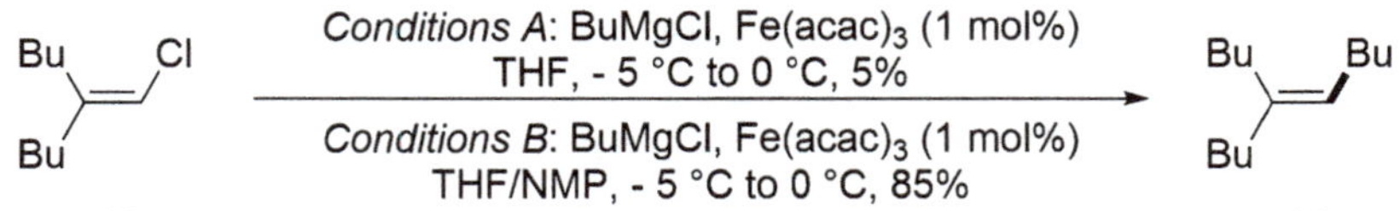

82 83

Fig. 18 Influence of NMP as an additive in Fe-mediated Grignard additions

$FeCl_3$ [82]. An extensive study by the Cahiez group [83] showed that NMP can have a dramatic effect on reactivity (Fig. 18) even allowing for the use of a range of soluble iron pre-catalysts with similar results. These conditions boast green, pharmaceutically friendly attributes including low catalyst loadings (as low as 0.1 mol%), short reaction times, low equivalency of Grignard reagent, and functional group tolerance. Additionally, vinyl iodides, bromides, and chlorides undergo coupling with only slight modification of reaction conditions. Prevention of β-hydride elimination with *tert*-butylmagnesium bromide could also be suppressed with a bulky Fe(dpm)3 pre-catalyst. While Grignard reagents have been the vast majority of nucleophiles employed for this chemistry, organomanganese nucleophiles have also been explored [84–86]. Consistent with Cahiez's improvements, NMP proves to be a crucial cosolvent in these couplings.

3.3.1 Preparation of 5-Butyldec-5-ene from 5-(Chloromethylene)nonane

To a solution of 5-(chloromethylene)nonane (**82**) (25 mmol, 4.37 g) and Fe(acac)[3] (0.25 mmol, 0.088 g) in a mixture of THF (30 mL) and NMP (25 mL) was added dropwise (10 min), between −5 and 0 °C, a 1.2 M solution of *n*-butylmagnesium chloride in THF (27.5 mmol, 22.9 mL). Stirring was continued for 15 min, and then the reaction mixture was hydrolyzed at −10 °C with aq 1 M HCl (80 mL). After decantation, the aqueous layer was extracted with Et_2O, and the combined organic phases were washed with satd aq $NaHCO_3$ solution and H_2O and dried

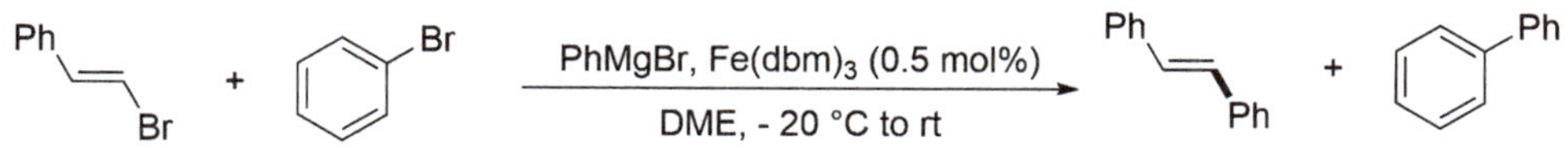

Fig. 19 DME as a superior solvent for Fe-catalyzed vinylation reactions

($MgSO_4$). Solvents were removed under vacuum, and 5-butyldec-5-ene (**83**) was isolated by distillation in 85% yield (4.17 g).

Expansion of the vinyl electrophile profile has been extended beyond halides to vinyl sulfones [87–89], sulfides [90], sulfonyl chlorides [91], tosylates [92, 93], phosphonates [83], triflates [76], and pivalates [94]. With this wide scope in electrophilic components, the opportunity for installation of coupling precursors is large, and further, these partners can be quite stable and thus carried through multiple synthetic manipulations before coupling.

Alkyl nucleophiles are more traditionally employed in these couplings, but aryl and vinyl nucleophiles have also been successfully coupled. The Molander group made a significant contribution in reporting that DME was a superior solvent for vinylation of aryl organomagnesium compounds with vinyl halides and furnished coupled products in high yields and under mild conditions and with equimolar mixtures of halide and Grignard reagent (Fig. 15) [95]. Notably, this reaction was found to be selective for vinyl halides in the presence of aryl halides. Thus, vinyl bromide (**84)** is stereoselectively coupled to form **86** in the presence of bromobenzene (Fig. 19). The yield of biphenyl was only 3% higher than that which results from homocoupling of the Grignard reagent. While a marked improvement was noted in comparison to Kochi's report, this chemistry was shown to be somewhat limited to *E*-alkenyl halides as *Z*-alkenyl halides were shown to produce mixtures when employing sterically encumbered Grignard reagents. Cooperative catalysis (Fe/Cu) has also proven beneficial in the synthesis of 1,1-diarylethenes [96].

3.4 Arylation

One of the major limitations in the use of Grignard reagents in Pd-mediated coupling reactions with aryl electrophiles is their reducing ability leading to the formation of palladium black and the subsequent shutdown of the catalytic cycle. In a seminal study on the analogous iron-catalyzed cross-coupling, Fürstner and co-workers observed that it was this feature that proved critical to the success of the overall process. In this study, they exploited the established observation that $FeCl_2$ reacts with four equivalents of R-MgX to give a new species of the formal composition [Fe(MgX) 2] (**A**), which is termed as an "inorganic Grignard reagent" and is readily soluble in ethereal solvents such as THF [97, 98]. Based on the stoichiometry of this reduction, it suggests that the oxidation

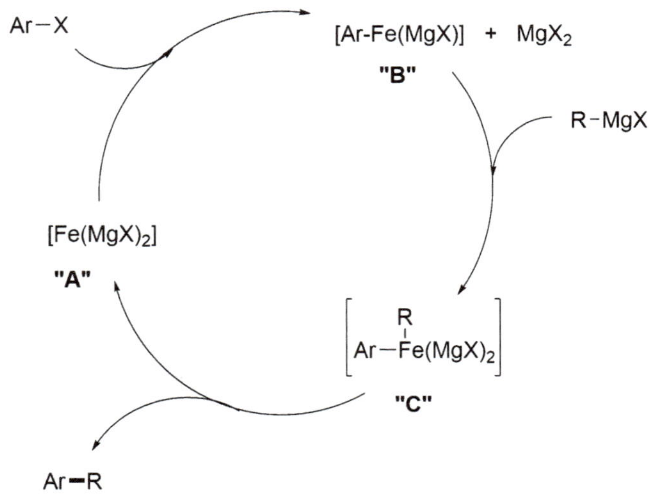

Fig. 20 Mechanism of iron-catalyzed couplings. Intermediacy of an "inorganic Grignard reagent" **A**

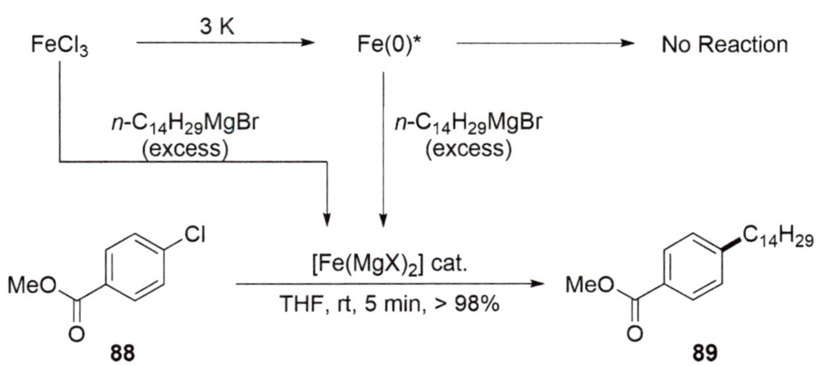

Fig. 21 Effect of iron source on alkyl-aryl cross-coupling

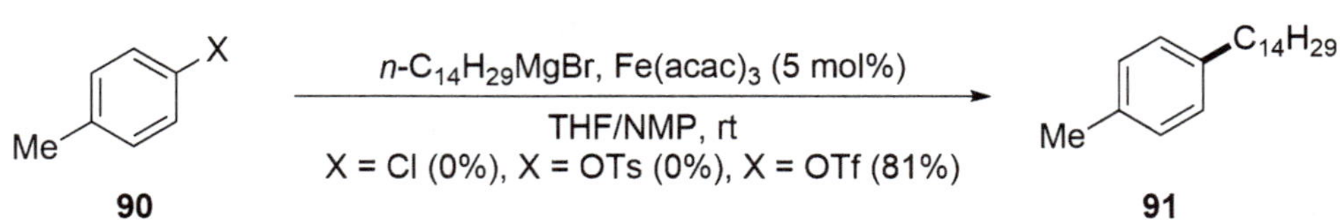

Fig. 22 Influence of the nature of the electrophile on the Fe-mediated coupling

state change does not stop at Fe(0) (complexes of which have often been speculated as the active catalysts in Fe-mediated couplings) but instead leads to a species with a formal negative charge at iron. The Fürstner group hypothesized that such a highly nucleophilic species in the absence of any stabilizing ligands should be able to oxidatively add to aryl halides with the formal Fe(0)-based organometallic intermediate (**B**) formed undergoing further alkylation with excess Grignard reagent to give (**C**). Reductive elimination

Optimal Conditions

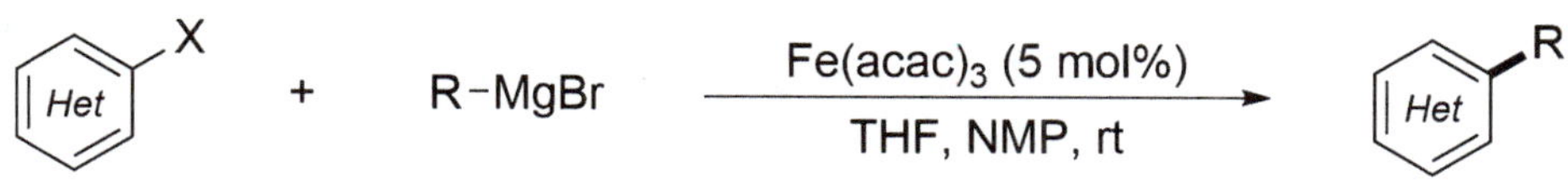

92 **93** **94**

Examples

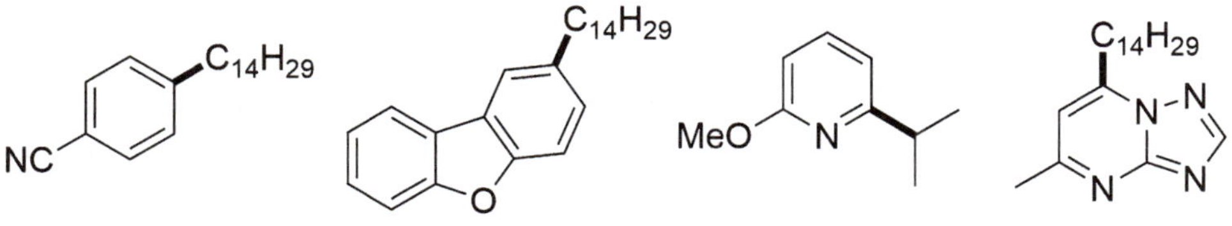

95 91% (X = Cl)

96 81% (X = OTf)

97 56% (X = Cl)

98 67% (X = Cl)

Fig. 23 Optimal conditions/scope of the Fürstner-based conditions

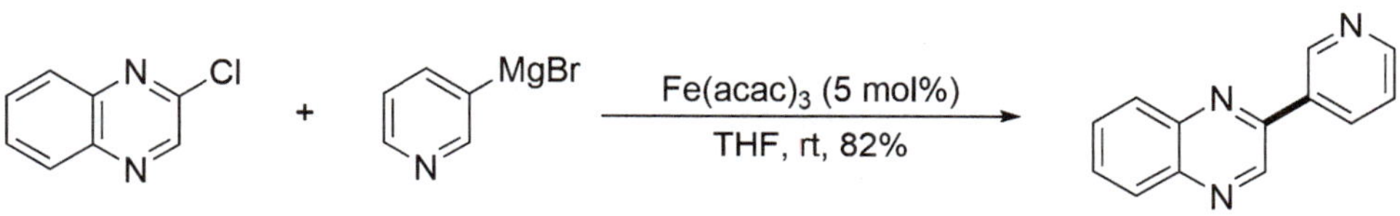

99 **100** **101**

Fig. 24 Fe-mediated coupling for heteroaryl-heteroaryl bond formation

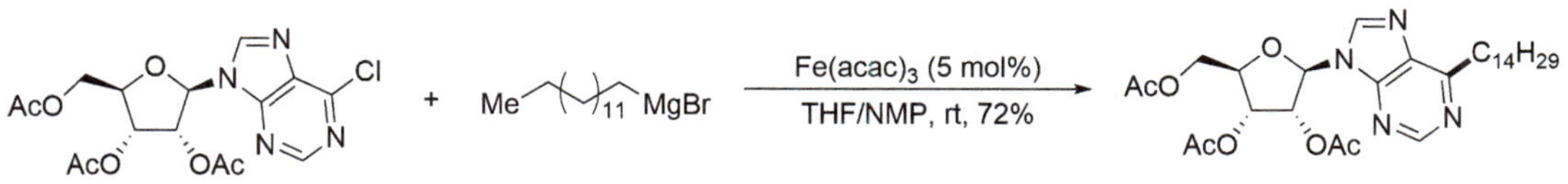

102 **103** **104**

Fig. 25 Fe coupling for the synthesis of ribofuranoside derivative **104**

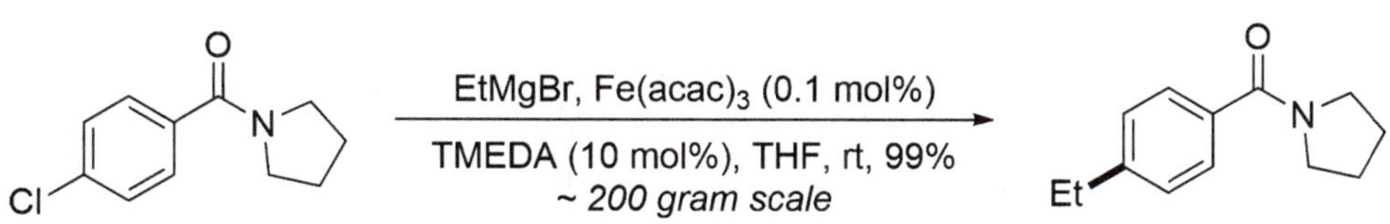

105 **106**

Fig. 26 Scale-up of an Fe-catalyzed aryl-alkyl bond formation

would then yield the Csp^2-Csp^3-coupled product and regenerate the catalytic Fe(-II) species (**A**) (Fig. 20).

To illustrate this, the cross-coupling of 4-chlorobenzoic acid methyl ester with an excess of *n*-tetradecylmagnesium bromide was performed in the presence of 5 mol% $FeCl_x$ ($x = 2, 3$) as pre-catalyst and provided quantitative conversion to the desired cross-coupled product (with no addition of the Grignard reagent into the methyl ester) within 5 min at room temperature (Fig. 21). The reaction even works at –60 °C though only proceeds slowly under forcing conditions when mediated by non-passivated Fe(0)* powder (prepared by the reduction of $FeCl_3$ by potassium). However interestingly, these particles slowly dissolve in THF upon treatment with excess Grignard reagent to give a brown-black solution, which is an effective catalyst for the cross-coupling process.

Given that the catalytically active iron complex is likely to be a highly reduced species, one can envision not only classical oxidative addition via two-electron processes taking place but also single-electron transfer steps inducing radical pathways. This is supported by studies on the nature of the aryl electrophile, and while both bromide and iodide lead to high levels of hydrodehalogenation in the reaction, the chloride and both the triflate and tosylate lead rapidly to high levels of the cross-coupled product (Fig. 22). Electron-deficient chloroarenes reacted similarly to the corresponding triflate and tosylate, and the triflate group was the optimal for electron-rich arenes. A range of iron sources effectively mediate the reaction with $Fe(acac)_3$ being chosen from a practical standpoint being both cheap and non-hygroscopic. *n*-Alkyl and *sec*-alkyl Grignard reagents are successful nucleophiles with the reaction scope being also extended to organozinc and organomanganese reagents though both *n*-BuLi and Et_3Al failed to react purportedly due to being unable to form a significantly covalent Fe-M (M = Mg, Zn, Mn) bond in the active species. Vinyl, allyl, and aryl Grignard reagents are all poor nucleophiles though the latter does lead to a synthetically useful yield in the coupling with 2-chloropyridine.

From a substrate perspective, the reported iron-catalyzed alkyl-aryl cross-coupling shows a broad scope tolerating not only a host of electron-deficient aryl chlorides and tosylates (featuring esters, nitriles, CF_3 groups, etc) but also numerous heterocyclic systems including pyridines, pyrazines, purines, and isoquinolines (Fig. 23). The chemoselectivity is also highlighted with the efficacy of the iron-catalyzed activation of the C-Cl bond allowing the reaction to be performed in the presence of esters, unprotected acidic N-H bonds (though an extra equivalent of Grignard is required for deprotonation for these substrates), and thioethers. Polysubstitution and one-pot consecutive cross-couplings are also successfully demonstrated.

As noted above, the iron-mediated coupling of heteroaryl chlorides (**99**) with (hetero)aryl Grignard reagents (**100**) is also successful with notably these processes being carried out in THF in the absence of NMP as a cosolvent (Fig. 24).

This chemistry further demonstrates the advantage of iron-catalyzed couplings from a pharmaceutical and practical standpoint. For example, per-*O*-acetylated purine-B-D-ribofuranoside (**102**) undergoes smooth alkylation leaving the ester functionalities intact (Fig. 25) which clearly exhibits the beneficial nature of iron catalysis over the un-catalyzed Grignard reaction.

Later studies by Perry expanded the substrate scope to non-activated and electron-rich aryl chlorides [99]. In a limited study, Wang also explored pyrimidine tosylates as electrophiles in couplings with aryl and alkyl Grignard reagents [100].

With an eye toward industrial application, an impressive demonstration by Fox and Hulcoop showed that the coupling of amide (**105**) with ethylmagnesium chloride could be performed on a 210 g scale with low catalyst loading (0.1 mol%) and a crude yield of 99% with >98% purity (Fig. 26). This reaction was performed at ambient temperature, in high concentration (0.625 M) and in the absence of NMP [101].

3.4.1 Preparation of (4-Ethylphenyl)(Pyrrolidin-1-yl)methanone from (4-Chlorophenyl)(pyrrolidin-1-yl)methanone

To a solution of *N*-(4-chlorobenzoyl)pyrrolidine (209.7 g, 1.00 mol), $Fe(acac)_3$ (0.353 g, 1 mmol), and TMEDA (15.0 mL, 0.1 mol) in dry THF (800 mL) was added EtMgCl (600 mL, 2 M solution in THF, 1.2 mol) via cannula at a rate that kept the internal temperature of the reaction mixture below 30 °C (roughly 1 h). Immediately following the completion of the addition, the reaction mixture was added via cannula to a pH 2 buffer solution (1.6 L) (over 45 min) and stirred at room temperature for 16 h. The organics were collected. The aqueous was extracted with EtOAc, and organics were combined, washed with a saturated aqueous NaCl solution, dried (Na_2SO_4), and concentrated in vacuo to give the crude product as a yellow solid (200.5 g, 99%). The purity (by NMR) was >98%. A sample was recrystallized for characterization: mp 85–86 °C.

An important area demonstrating the utility of iron-catalyzed Kumada coupling in a pharmaceutical context is chemoselectivity when multiple electrophilic sites are present, especially on heteroaromatic systems. This area has largely been exploited to perform selective substitutions as well as sequential one-pot couplings and thus shows itself to be quite relevant from a cost and efficiency standpoint [97, 102, 103]. For example, symmetric and pseudo-symmetric pyridines have been sequentially coupled to form the natural products (−)-isooncinotine (**109**) [104] (Fig. 27) and (+)-muscopyridine (**112**) [105] (Fig. 28).

Expansion of the aromatic coupling partner beyond aryl chlorides, tosylates, and triflates has made this reaction even more attractive. The groups of Shi [94], Garg [106], and Cook [93, 107] have explored various phenol derivatives and their reactivity in iron-

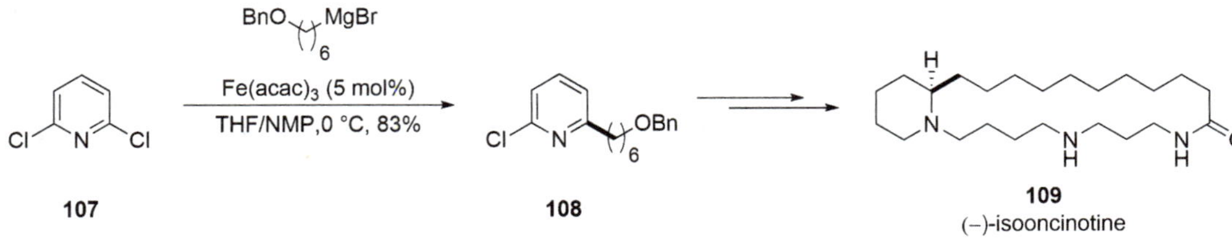

Fig. 27 Application of Fe coupling in the synthesis of (−)-isooncinotine **109**

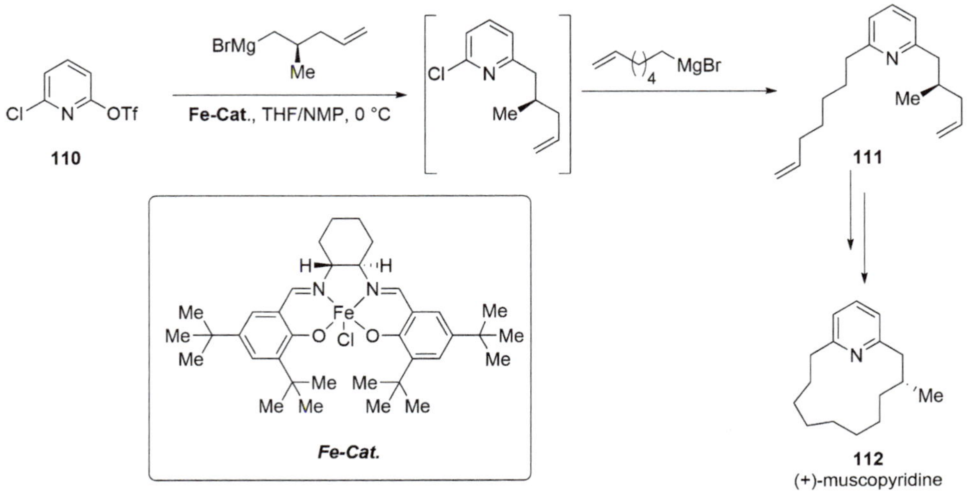

Fig. 28 Application of Fe coupling in the synthesis of (+)-muscopyridine **112**

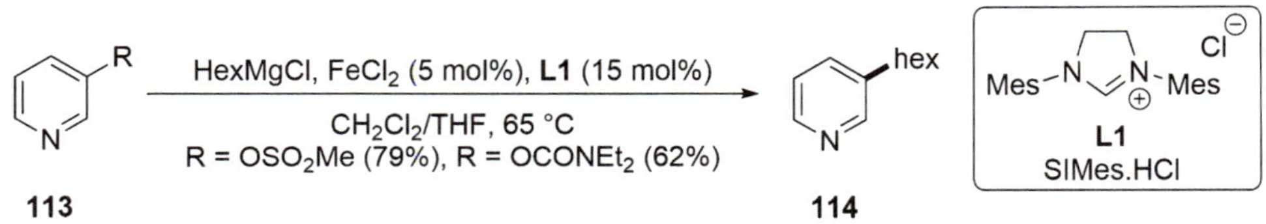

Fig. 29 Iron-mediated coupling of aryl sulfamates and carbamates

catalyzed cross-coupling. Sulfamates and carbamates have been shown to undergo coupling in the presence of an iron catalyst and an NHC ligand (Fig. 29). These handles are useful as they are also useful in directed metalation reactions. Figure 30 nicely demonstrates the principle of sequential and orthogonal reactivity, in the poly-functionalization of arene **115** to deliver arene **118**. Relevant to pharmaceutical interests, the conditions described by Garg are shown to be tolerant of heteroarenes and steric congestion.

Also, noteworthy is the challenge of aryl-aryl bond formation via iron catalysis due to competitive homocoupling. Nakamura demonstrated a remarkable effect in the employment of FeF_3 as the pre-catalyst which successfully suppressed homocoupling of the Grignard reagent [108], and Cook has further extended this

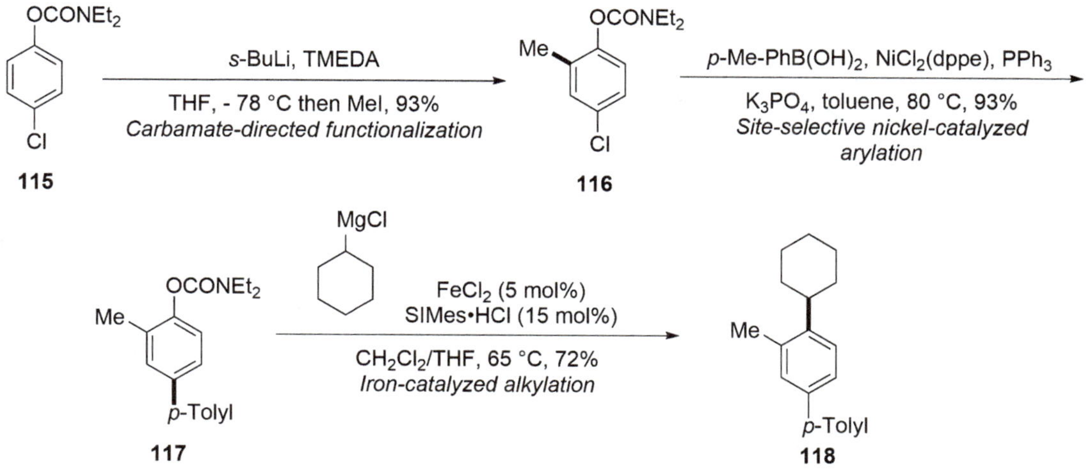

Fig. 30 Controlled substrate functionalization through orthogonal reactivity to Ni/Fe-mediated conditions

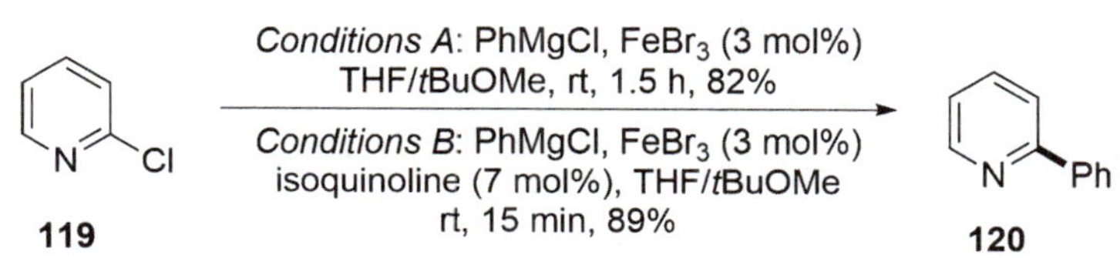

Fig. 31 Addition of isoquinoline as an additive to Fe-catalyzed reactions of N-heterocycles

chemistry from aryl halides to aryl tosylates [93]. Chlorostyrenes have also shown unique reactivity in biaryl couplings [109]. Of note to pharma, Knochel has reported a general method for the coupling of *N*-heterocycles with aryl Grignard reagents [110]. Interestingly, he later found that the addition of isoquinoline is quite beneficial in these couplings, dramatically increasing the reaction rate [111, 112] (Fig. 31).

3.5 Iron-Catalyzed Suzuki-Miyaura Reactions

While the use of iron to mediate Kumada-type couplings is relatively well established (with numerous examples applied on large scale), initial reports utilizing this metal to catalyze Suzuki-Miyaura couplings are limited, and all feature the addition of either pyrophoric alkyllithium reagents to activate the boronic ester [113] and/or magnesium salts purportedly to facilitate the transmetalation [114]. Comparisons to the well-established Pd-catalyzed systems in which basic additives lead to the formation of palladium hydroxides, which are critical for transmetalation, led Byers and co-workers to hypothesize that application of analogous conditions to the corresponding iron-based systems would lead to the formation of iron hydroxides or alkoxides, which would be subsequent to aggregation and subsequent deactivation of the catalytic cycle [115].

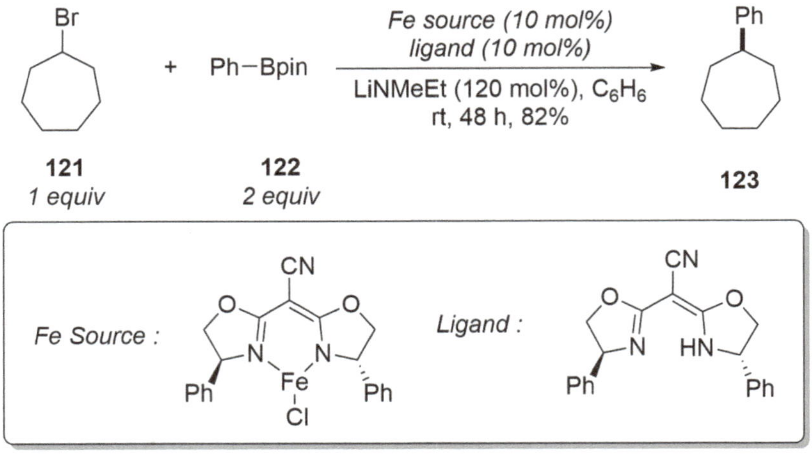

Fig. 32 Byer's iron-catalyzed Suzuki-Miyaura reaction

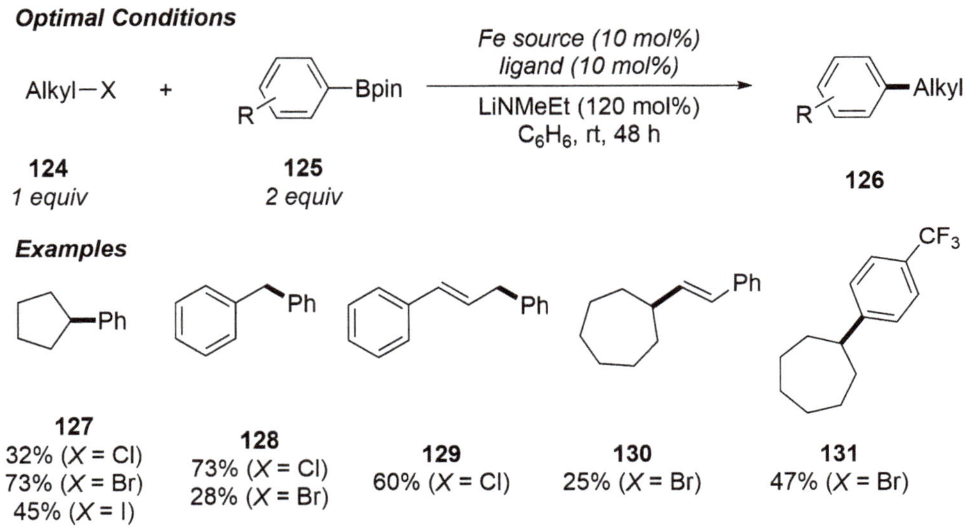

Fig. 33 Optimal conditions/scope of the Byers conditions

An alternative suggestion might be that unfavorable thermodynamics are a barrier to transmetalation, and while computational studies indicated that this was indeed likely to be an uphill process, they also suggested that the energy barrier for transmetalation of iron alkoxides was modest and could be overcome at room temperature. However, experimental studies to explore this were thwarted when attempts to synthesize a mononuclear iron alkoxide complex either through salt metathesis or protonolysis of suitable iron-containing precursors led to a green insoluble material that showed no activity in cross-coupling reactions.

Re-examination of the computational model suggested that switching to an iron-amide species as a possible intermediate for

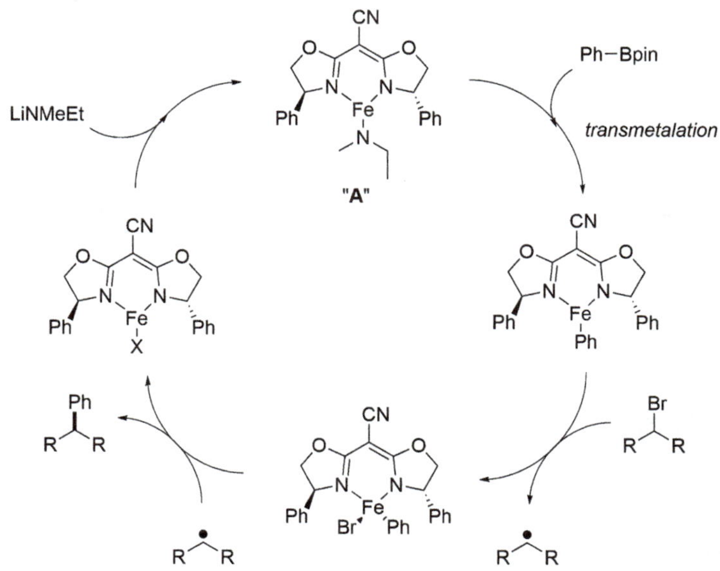

Fig. 34 Proposed mechanism of the iron-mediated Suzuki coupling

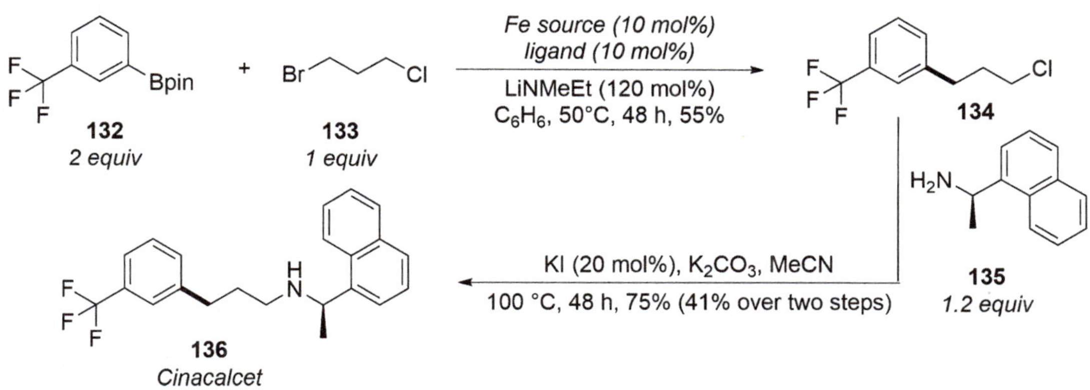

Fig. 35 Synthesis of Cinacalcet using Fe-mediated Suzuki-Miyaura coupling

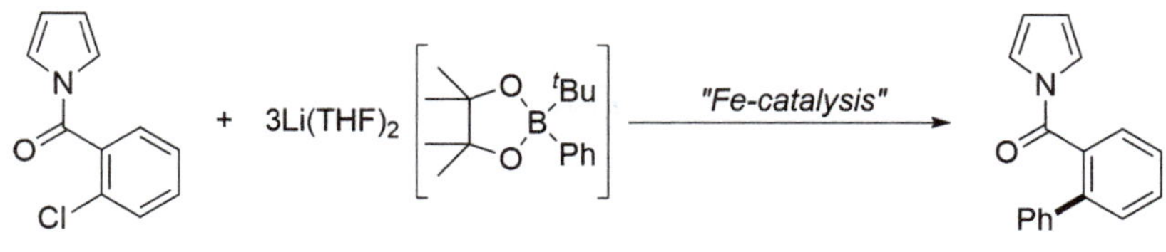

Fig. 36 *N*-Pyrrole amide as a directing group in the iron-mediated Suzuki coupling

transmetalation represents a favorable situation from a thermodynamics perspective and affords a wide scope in terms of potentially tuning both the electronics and sterics of the intermediate. Judicious manipulation of these factors particularly the steric

Optimal Conditions

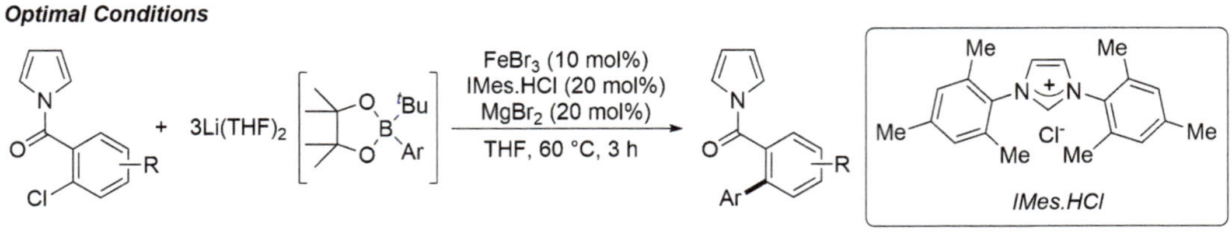

140 **141** **142**

Examples (isolated yields in brackets)

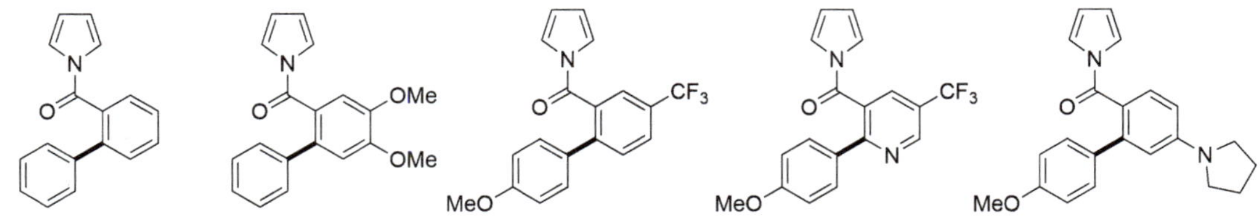

143	**144**	**145**	**146**	**147**
82% (70%)	89% (64%)	48% (42%)	46% (27%)	67% (64%)

Fig. 37 Optimal conditions/scope of the Bedford conditions

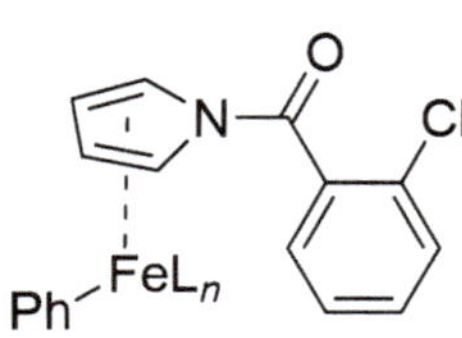

148

Fig. 38 π-Coordination of the ligand tethering the substrate facilitating C-Cl bond activation

environment of the iron amide proved critical to success with LiNMeEt being demonstrated to be optimal for the iron-catalyzed coupling between cycloheptyl bromide and PhB(pin), while both more sterically demanding/electron-deficient and smaller amides were ineffective. Extensive screening of ligands demonstrated that the BOX family was particularly effective with substitution at the bridging carbon shown to be a key factor in obtaining the optimal reaction outcomes. In addition, working with a pre-prepared monoanionic iron-ligand complex proved superior to the in situ formation of the catalyst system with loadings as low as 1% still leading to synthetically useful yields (Fig. 32).

Exploration of the reaction scope showed that both primary and secondary bromide were tolerated, while tertiary alkyl chlorides also provided some level of the desired products (Fig. 33). The alkyl bromides proved to be better coupling partners than the iodides, which in turn were superior to the chloride derivatives (**127**), though the latter proved superior in the coupling of benzylic substrates (**128**) owing to a lesser propensity to undergo homocoupling via radical recombination. The intermediacy of carbon-based radical intermediates was further supported in experiments involving radical clock-based intermediates. A degree of functional group tolerance was tolerated on the alkyl halides. For the boronate esters, electron-rich systems demonstrated reduced efficiency due to slower transmetalation rates, and while alkenyl boronic esters underwent the desired reaction, both heteroaromatic and substrates featuring enolizable protons failed to react. It is important to note that observed modest yields are often due to catalyst deactivation with minimal side-product formation. Given this, filtration and re-subjecting the mixture to the reaction conditions was shown to lead to a 10–25% improvement in yield.

Mechanistically although it is likely that the reaction proceeds through Fe(II)/Fe(III)-based intermediates, previous hypotheses involving an Fe(I)/Fe(II)/Fe(III) pathway cannot be ruled out [116] (Fig. 34). The current reaction exhibits two key features that differentiate it from previously reported iron-catalyzed Suzuki-Miyaura couplings utilizing alkyllithium reagents and magnesium additives in the nature of the ligand employed as well as the involvement of the amide base. The latter is proposed to play a number of roles including being involved in the formation of an anionic cyanoBOX ligand, preventing aggregation of iron species as well as in situ formation of an iron-amide species (**A**) to facilitate transmetalation. The authors however do note that it is unlikely that the current reaction proceeds through borate-based species given that these are observed to play an inhibitory role in reaction progress in this system.

To illustrate the practicality of the new methodology, a two-step synthesis of cinacalcet (a calcimetic for the treatment of secondary hyperparathyroidism) was developed featuring the iron-mediated Suzuki-Miyaura reaction as the initial step in which the difference in reactivity between alkyl bromides and chlorides in **133** is exploited. The reaction proceeds in 55% yield with only 10% of the bis-arylated product being observed. A high-yielding alkylation of a commercially available amine **135** completes the synthesis of the API in 41% overall yield (Fig. 35). This route is shorter than any of the previously reported approaches to this molecule and notably avoids the use of either noble metal catalysts [117] or pyrophoric organometallic reagents [118].

While there are several reports on the iron-catalyzed Suzuki-Miyaura coupling for the synthesis of pharmaceutically relevant biaryl motifs, the reaction remains problematic, and there are significant concerns specifically around reproducibility [119]. Of the cases described, the coupling of 2-halobenzyl halides with activated boronate esters appears to be robust leading to the desired cross-coupling product along with minor amounts of the bi-aryl-coupled side product [120, 121]. Bedford and co-workers hypothesized that the success of this reaction was based on the ability of the substrate to "direct" the activation of the aryl-X bond at the iron center, though noted that reports showed that other classic ortho-directing groups based on tertiary amine, ether, ester, carbamate, or imine functions all failed to give the desired cross-coupling product. Inspired by these results, they highlighted aryl chloride substrates with an *N*-pyrrole amide (**137**) (or related) directing group as being able to successfully participate in the iron-catalyzed Suzuki-Miyaura reaction with activated organoborates (**138**) [122] (Fig. 36).

Optimization studies demonstrated that judicious choice of the ligand was critical and while both amine- and phosphine-based ligands were ineffective, *N*-heterocyclic carbene (NHC) ligands proved beneficial. Herein, the nature of the substitution pattern of the *N*-aryl groups proved important with modest steric bulk at the 2,6-positions providing the best results. Both $FeBr_2$ and $FeBr_3$ were able to catalyze the reaction with no conversion being observed in the absence of iron with the boronate nucleophile acting as a reducing agent for the iron. Chloro proved to be the best halide substituent, and while the 2-bromo analogue also gave a modest yield of the desired coupling product, increased amounts of the hydrodehalogenated side product were observed. This became the major product with the 2-iodo species, while the 2-fluoro proved resistant to the reaction conditions. None of the desired coupling product was observed for either the 3- or 4-chloro derivatives supporting the directing role of the *N*-pyrrole amide function. A range of functional groups could be successfully accommodated on the aryl chloride with the reaction also able to tolerate a modest degree of steric hindrance (Fig. 37). Use of a 2-chloropyridine derivative also furnished the desired product though the other heterocycles examined failed to react.

$MgBr_2$ as an additive was critical for the success of the reaction. Other bromide salts also provide product with varying degrees of success as does $MgCl_2$ though the theory that the additives may form intermediates facilitating the transmetalation to the iron is to be a large degree discounted by the observation that both Zn- and Al-based nucleophiles fail to participate in the reaction when replacing the B-based species. The loading of $MgBr_2$ is also important to control as although the bromide is clearly necessary for activity, above a certain point, it plays an inhibitory role on the reaction. Mechanistic studies based on kinetic analyses (including the delineation of a clear catalyst induction period) and kinetic isotope effects (KIE) enable a simple model for a catalytic cycle to be developed, a critical feature of which is the π-coordination of the pyrrole ligand that is proposed to tether the substrate in the coordination sphere and facilitate activation of the C-Cl bond (**148**) (Fig. 38). Cleavage of this bond is hypothesized to occur via single-electron transfer (SET) rather than classical oxidative addition with this theory supported by addition of TEMPO as a radical trap being able to inhibit the reaction.

3.6 Alkyl-Alkyl Coupling

Iron-catalyzed C(sp^3)-C(sp^2) bond formation has been extensively explored allowing for coupling of alkyl iodides [123, 124], bromides [123–132], chlorides [123, 133–135] thioethers, and sulfones [136]. These reactions have also been shown to tolerate a wider range of nucleophiles beyond Grignard reagents into zinc reagents [137], organolithium reagents [128], and various metallates [134, 135, 138] (e.g., borates, aluminates, and indates).

Generalities of these reactions are noted through the literature. These couplings have been shown to proceed via a radical oxidative addition of the alkyl halide [124, 130, 132–134, 138] to iron, and thus stereochemical integrity of the starting halide is not maintained [123, 128, 129]. The radical nature of these reactions has been taken advantage of in tandem ring-closure/coupling reactions allowing for the synthesis of substituted heterocycles [134]. Analogous to the benefit of NMP to vinylation and arylation discussed above, TMEDA and HMTA have been shown to be advantageous additives in the majority of reports [123, 125, 129, 130, 137]. When reacting with an alkenyl nucleophile (**80**), the olefin geometry within the organometallic can also be maintained using organozinc [137], borate nucleophiles [134, 136], or carefully controlled Grignard formation [125] (Fig. 39).

From a selectivity perspective, conditions which allow for the alkylation of aryl and vinyl nucleophiles have been shown to be orthogonal to those needed for vinylation or arylation [126, 130, 132]; thus cycloheptyl bromide is coupled in high yield to deliver **152;** however iodobenzene (**54**) remains intact under identical conditions (Fig. 40). Fürstner has also shown couplings catalyzed by [Li(TMEDA)$_2$[Fe(C_2H_4)$_4$] demonstrate an impressive functional group tolerance (e.g., isonitriles, esters, TMS-alkynes, nitriles, and more) in the coupling of alkyl iodides and bromides [128]. [Li(TMEDA)$_2$[Fe(C_2H_4)$_4$] can be prepared on large scale but is air sensitive. As an alternative, preparation of [($FeCl_3$)$_2$(TMEDA)$_3$)] as a bench-stable solid allows for a highly practical route toward cross-coupled products with low catalyst loadings and on large scale [129] (Fig. 41).

These reactions also display their usefulness in substrates (**155**) that would be inert to typical S_N2 reactivity that will undergo coupling in good yield [132, 133] (Fig. 42).

Perhaps the most challenging iron-catalyzed coupling reaction is that of C(sp^3)-C(sp^3) bond formation due to competing homocoupling, β-elimination, and disproportionation which hinder productive reactivity of unactivated alkyl electrophiles. The Chai group made a seminal contribution and demonstrated that in the presence of Fe(OAc)$_2$ and the bulky, bidentate bisphosphine Xantphos, unactivated alkyl bromides (**157**) could be successfully coupled to

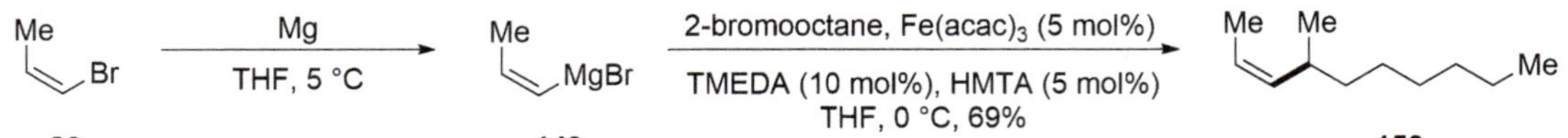

80
Z = 99%
149
Z/E = 97/3
150
Z/E = 97/3

Fig. 39 Beneficial effects of TMEDA/HMTA as additives in Fe-mediated couplings

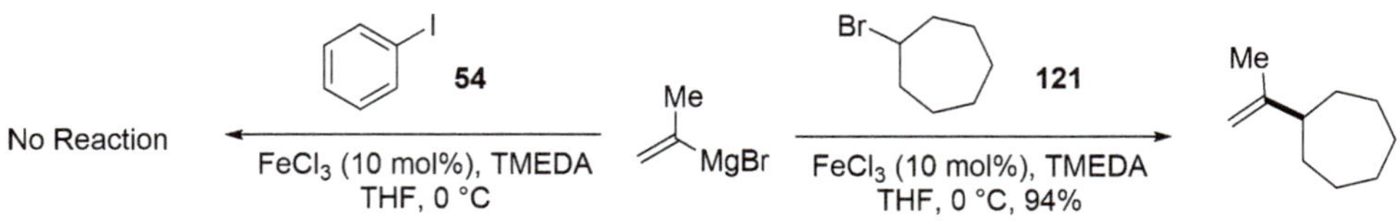

151
152

Fig. 40 Orthogonal reactivity in the Fe-catalyzed alkylation of aryl and vinyl halides

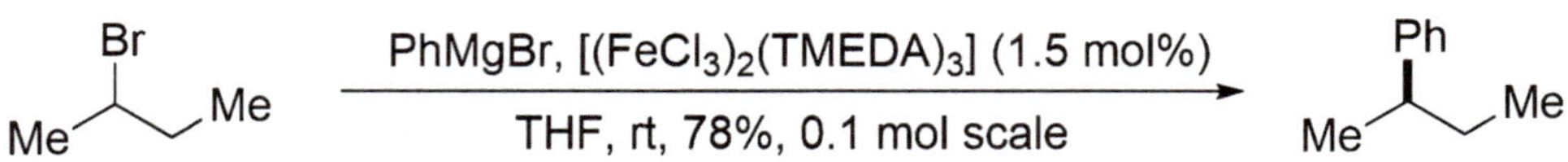

153
154

Fig. 41 [$(FeCl_3)_2(TMEDA)_3$)], a bench-stable solid as a cross-coupling catalyst

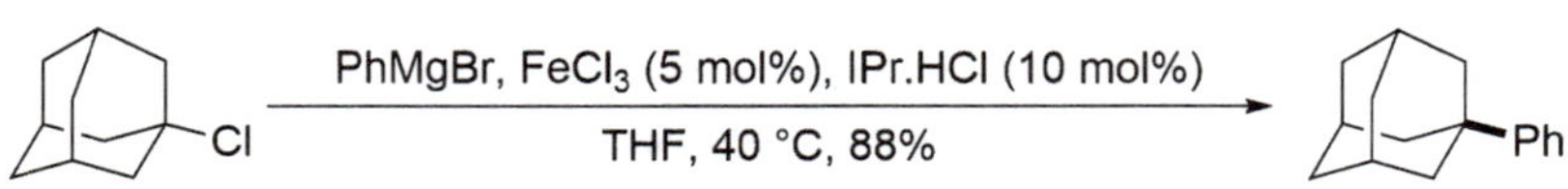

155
156

Fig. 42 Fe-mediated cross-coupling of tertiary alkyl chlorides

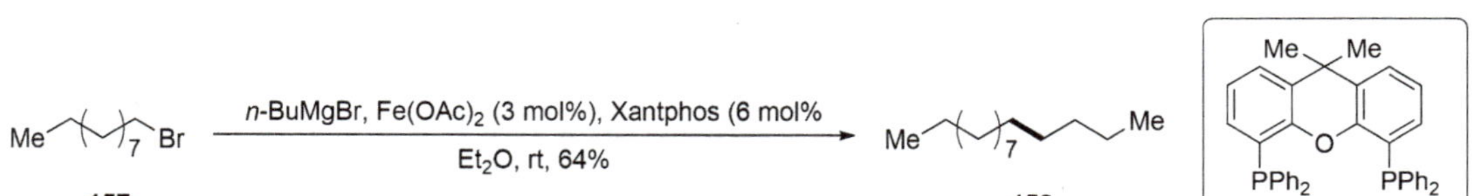

Fig. 43 Xantphos-mediated coupling of unactivated alkyl bromides

alkyl Grignard reagents in moderate yields (Fig. 43) [139]. Expansion of this chemistry by Nakamura has resulted in a tandem hydroboration/iron-catalyzed Suzuki-Miyuara coupling of the resulting tetraalkylborates [140].

From the standpoint of an enantioselective cross-coupling catalyzed by iron, the literature is much sparser. Only one report of a catalytic, enantioselective coupling of an organometallic compound has been published [141]. This reaction generates α-aryl esters from α-chloroesters with moderate enantioselectivities and is complimentary to the couplings presented by nickel and cobalt.

4 Copper-Mediated Cross-Couplings

4.1 Introduction

Copper represents a highly attractive option for cross-coupling chemistries due to its low cost, high natural abundance, and low toxicity, and its use in organic synthesis predates the disclosure of the Mizoroki-Heck reaction by almost 100 years. The Glaser homocoupling of alkynes was first reported in 1869 with a catalytic version reported by Hay in 1962 [142]. The unsymmetrical coupling of alkynes (Castro-Stephens reaction) was reported in 1963 [143]. Further classical examples of copper-mediated transformations include the Rosenmund-von Braun cyanation of aryl bromides [144] and Hurtley's C-arylation of CH-acids with *o*-bromobenzoic acid in the presence of sodium ethoxide [145]. In addition, the Ullmann coupling primarily for the formation of C-N and C-O bonds enjoyed extensive use until the development of Pd-based methods, though has also undergone a significant renaissance in recent years with catalytic methods enabling milder conditions being reported as well as the development of the Chan-Lam coupling [146]. Typically, these pioneering copper-based reactions required stoichiometric amounts of the Cu source, as well as forcing reaction conditions, and thus became somewhat neglected as Pd-based paradigms emerged as a highly versatile source of bond-forming catalysis.

The use of copper has also been widely recognized as advantageous in tandem with several Pd-based methodologies. For example, the addition of Cu(I) salts has long been known to accelerate the Stille reaction purportedly through the scavenging of Pd-based ligands thus facilitating the rate-determining transmetalation [147]. To develop a purely Cu-based version of this reaction, highly polar solvents have been shown to favor Sn/Cu transmetalation, while both use of excess Cu and addition of reagents to scavenge excess Sn reagents have been explored as strategies to successfully drive the reaction to the desired cross-coupled product [148]. Several examples of Cu-based Stille couplings have been reported in total syntheses [149, 150]. Similarly, the intermediacy of Cu acetylides has long been implicated in the venerable Pd-catalyzed Sonogashira coupling with the polymeric, insoluble nature of these species often being cited as a barrier to the development of a protocol for this transformation based on copper catalysis alone [151]. Recognizing this, numerous reports of palladium-free Sonogashira reactions have been reported often featuring the addition of ancillary ligands to solubilize and break up copper acetylides, thus enabling a Cu(I)/Cu(III)-based catalytic cycle to operate [152, 153].

Despite the wealth of reports of copper-based coupling chemistries [154, 155], this section will focus on examples of the couplings of Grignard and organoboron-based reagents. The coupling of the former was first reported by Kharasch using Co salts [156] with Tamura and Kochi describing the first examples of copper-catalyzed cross-couplings of Grignard reagents with alkyl halides in 1971 [157].

4.2 Alkyl-Alkyl Bond Formation with Grignard Reagents

Early studies on the formation of carbon-carbon bonds utilizing copper-based reagents focused on the use of either "lower-order" cuprates ($R_2CuLi.LiX$ commonly referred to as Gilman reagents) or "high-order" cuprates ($R_2CuCNLi_2$) wherein R can be saturated alkyl, allylic, vinyl or aryl groups [158, 159]. Whereas Gilman reagents only reliably couple with primary alkyl halides and sulfonates, the use of the higher-order cuprates increases the scope to couplings with secondary alkyl halides. However, drawbacks do exist in utilizing these reagents based mainly on typically excess cuprate being required to drive the reaction even though the reagents themselves contain two potentially transferable R groups. Furthermore, the greater accessibility of Grignard reagents compared to the corresponding organolithiums led to the natural progression to investigate the copper-catalyzed cross-coupling of these reagents with various electrophilic species. In connection with synthetic studies around the design of meta-cyclophanes (**159**) as components for molecular devices, Burns and co-workers studied the synthesis of a versatile *bis*-anisole fragment (**160**) based on the coupling of a *bis*-electrophilic alkyl component (**161**) of varying chain lengths to link two molecules of 2-bromoanisole (**162**) [160] (Fig. 44).

Problems were rapidly encountered attempted to utilize 2-lithiated anisoles in the coupling processes, and this approach would likely be further complicated by the need for downstream elaboration to access the target cyclophanes. Given this, several systems utilizing Grignard reagents in tandem with copper catalysis were evaluated to mediate the desired coupling. Initial studies evaluated the coupling between two equivalents of (2-methoxyphenyl)magnesium bromide with 1,3-dibromopropane to form the desired *bis*-anisole (**165**) (Fig. 45). The commonly utilized catalysts Li_2CuCl_4 and CuBr (with and without HPMA as an additive) were initially assessed in the coupling though led only to poor yields of the bis-coupled product (~ 15–20%) even when excess Grignard reagent was used though these pilot studies did indicate a correlation between catalyst solubility and reaction efficiency with high yields of the mono-coupled anisole (> 80%) being obtained with the more soluble copper-based species.

With this hypothesis in mind, attempts were made to prepare a more soluble Cu(II) catalyst through exchanging Cl ligands with Br, I, and/or SPh ligands, which are known to stabilize the active Cu(I)-coupling catalyst. This is further supported by NMR studies on the reaction using Li_2CuCl_4, which demonstrated two distinct rates of reaction based on the nature of the Cu species in the

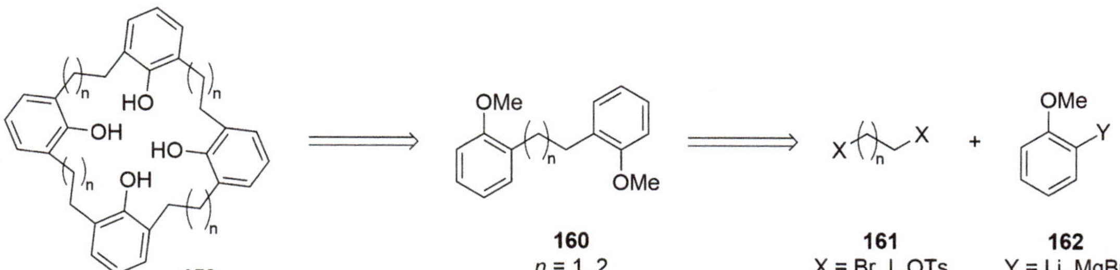

159

160
$n = 1, 2$

161
X = Br, I, OTs

162
Y = Li, MgBr

Fig. 44 Retrosynthetic analysis of metacyclophanes **159**

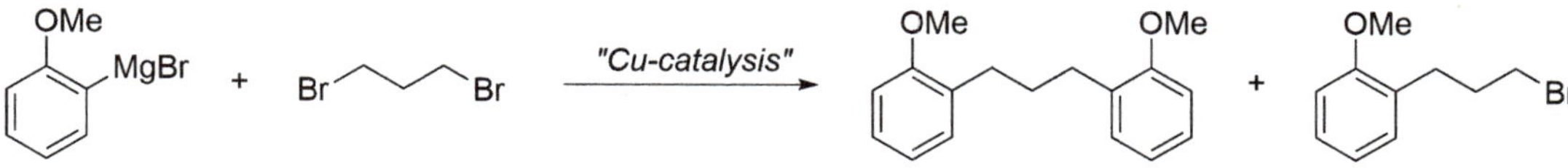

163 164 165 166

Fig. 45 Model reaction for the formation of the *bis*-anisole **165**

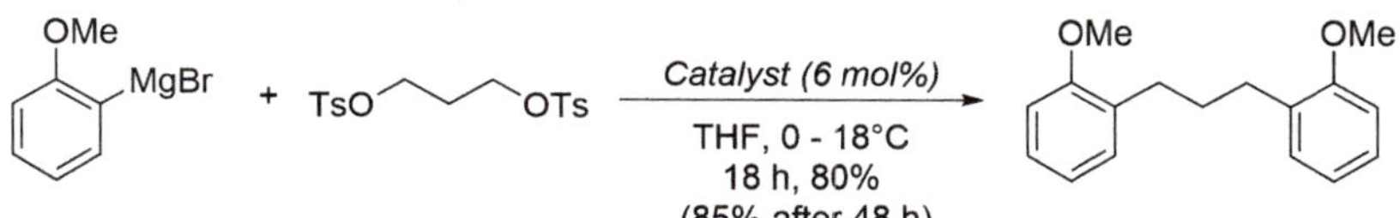

163 167 165

Catalyst = CuBr-SMe_2/LiBr/LiSPh

Fig. 46 Initial coupling with the Burns catalyst system

Table 3
Variation of electrophile 169 in catalyst screenings

Entry	X =	Catalyst	Conc. (M)	Yield (%)
1	Br	Li_2CuCl_4	0.5	54
2	I	Li_2CuCl_4	0.5	57
3	OTs	Li_2CuCl_4	0.5	70
4	Br	CuBr-SMe_2/LiBr/LiSPh	0.5	57
5	I	CuBr-SMe_2/LiBr/LiSPh	0.5	59
6	OTs	CuBr-SMe_2/LiBr/LiSPh	0.5	77
7	OTs	CuBr-SMe_2/LiBr/LiSPh	1.0	60
8	OTs	CuBr-SMe_2/LiBr/LiSPh	0.1	75
9	OTs	CuBr-SMe_2/LiBr/LiSPh	0.05	31

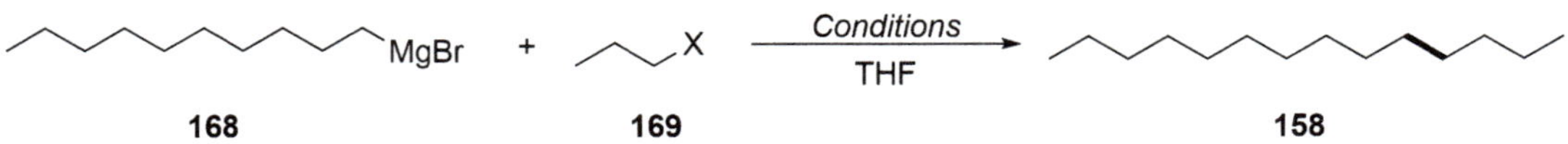

Fig. 47 Screening Cu-based catalysts against various electrophiles

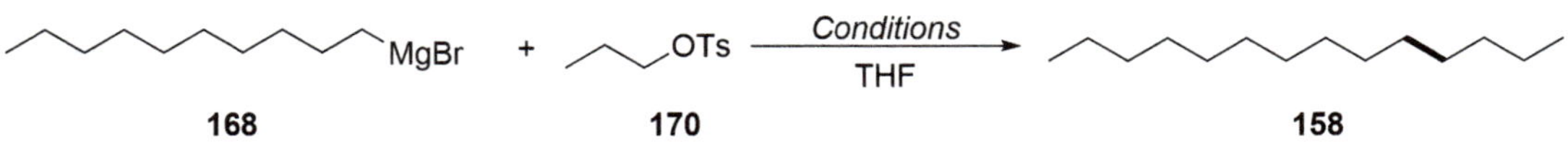

Fig. 48 Temperature/concentration and additive effects

Table 4
Temperature/concentration and additive effects

Entry	Catalyst	Temp. (°C)	Additive	Yield (%)
1	Li_2CuCl_4	25	0.5	52
2	$CuBr\text{-}SMe_2/LiBr/LiSPh$	25	0.5	55
3	Li_2CuCl_4	45	0.5	21
4	$CuBr\text{-}SMe_2/LiBr/LiSPh$	45	0.5	33
5	$CuBr\text{-}SMe_2/LiBr/LiSPh$	0	6% HMPA	57
6	$CuBr\text{-}SMe_2/LiBr/LiSPh$	25	6% HMPA	78
7	$CuBr\text{-}SMe_2/LiBr/LiSPh$	25	6% HMPA	87
8	Li_2CuCl_4	67	6% HMPA	80
9	$CuBr\text{-}SMe_2/LiBr/LiSPh$	67	6% HMPA	80
10	CuBr/HMPA	67	6% HMPA	44

reaction. Addition of Li salts (either the iodide or bromide) to $CuBr_2$ led to discernible color changes to the solutions with also enhanced yields of the desired bis-coupling products being obtained (~ 50–60%) when the bis-tosylate was employed as the electrophile though only after relatively long reaction times (3–4 days) and with high catalyst loadings (12–15%).

Mixing $CuBr_2$ with LiSPh in THF at room temperature led to a yellow solid containing a Cu(I) species, which is presumably formed through reduction of thiophenol leading to concomitant formation of the corresponding disulfide. Given that the thiophenol acts both as a ligand and the reducing agent, an alternative preparation was devised to avoid issues with control of the stoichiometry. To achieve this, one equivalent of $CuBr\text{-}SMe_2$ was mixed with one equivalent of each LiBr and LiSPh in THF forming a yellow suspension after 15–30 min. Mixing the reactants at 0 °C led to a homogeneous solution, which was maintained for several hours prior to any signs of precipitation. This solution (or suspension)

Optimal Conditions

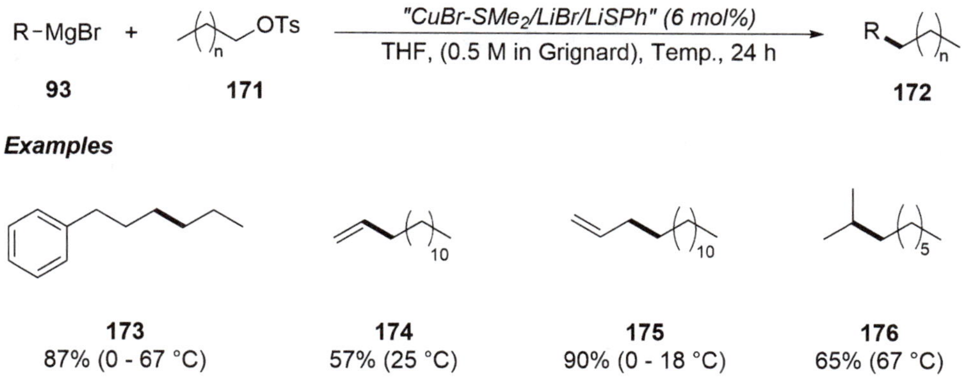

Fig. 49 Optimal conditions/scope of the Burns conditions

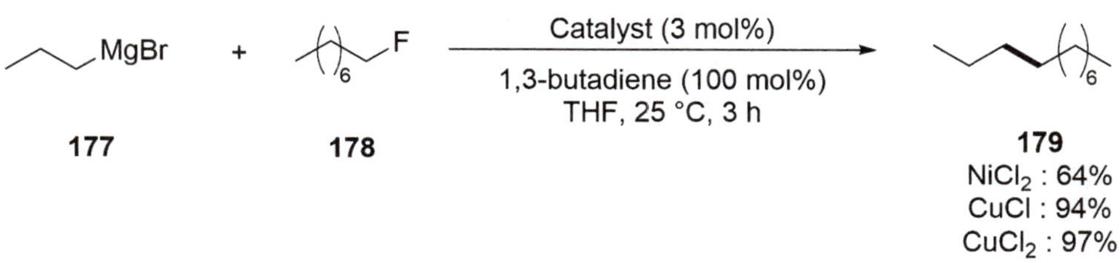

Fig. 50 Cu-mediated cross-coupling of alkyl fluorides

Optimal Conditions (yields in brackets are without additive)

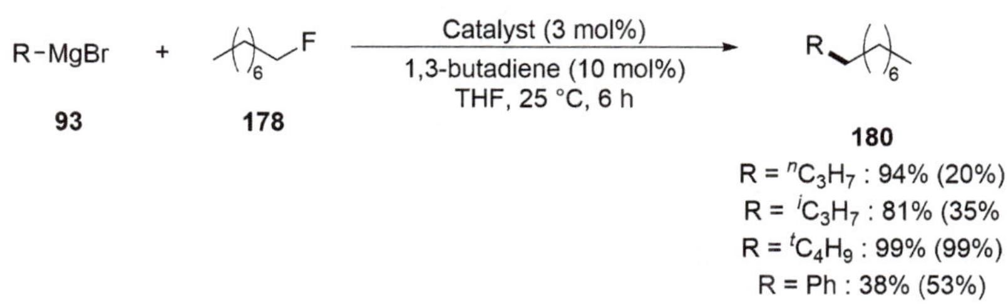

Fig. 51 Variation of the Grignard reagent **93** in the coupling of alkyl fluorides

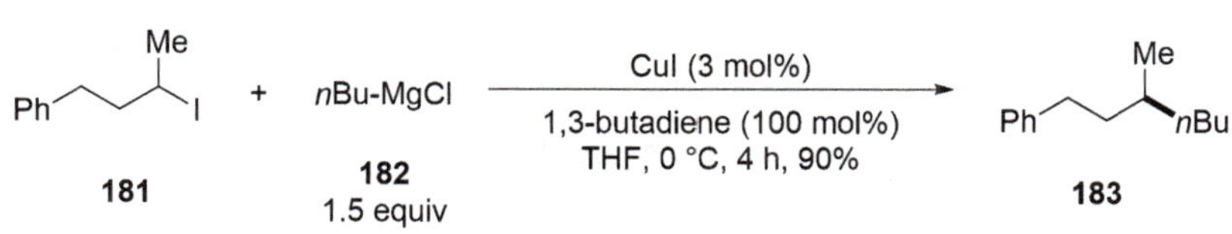

Fig. 52 1,3-Butadiene as an additive for the coupling of unactivated secondary alkyl iodides

proved to be more effective (80% yield) at coupling the anisole-based Grignard reagent (**163**) with the *bis*-tosylate (**167**) (derived from the 1,3-disubstituted propane) than any of the previously studied catalyst systems (Fig. 46).

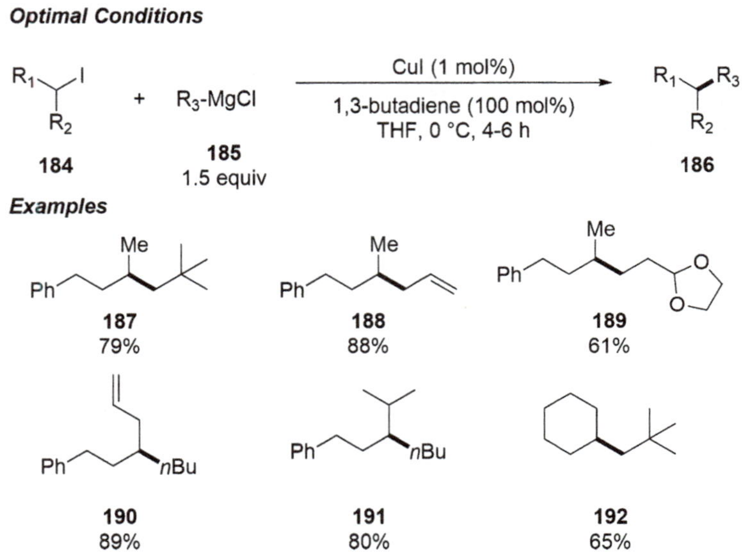

Fig. 53 Optimal conditions/scope of the 1,3-butadiene assisted coupling of secondary alkyl iodides

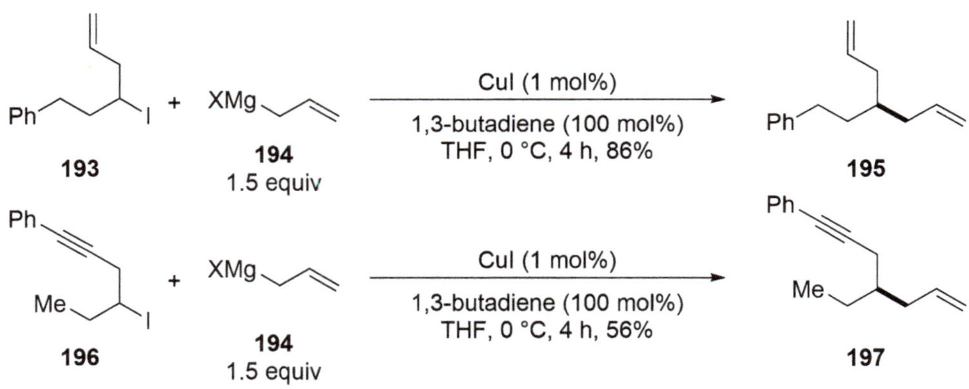

Fig. 54 Copper-catalyzed couplings for the synthesis of dienes and enynes

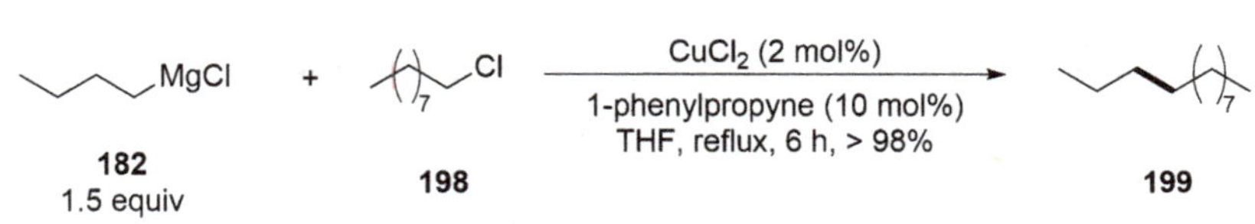

Fig. 55 Use of 1-phenylpropyne as an additive in the coupling of alkyl chlorides/tosylates

Based on this promising result, a more comprehensive study of the newly developed catalyst system was undertaken with the specific aim to benchmark its performance against both Li_2CuCl_4 and CuBr (in THF/HMPA). Model studies on the reaction between decylmagnesium bromide (**168**) and electrophiles based on 1-substituted propane (importantly using one equivalent of each

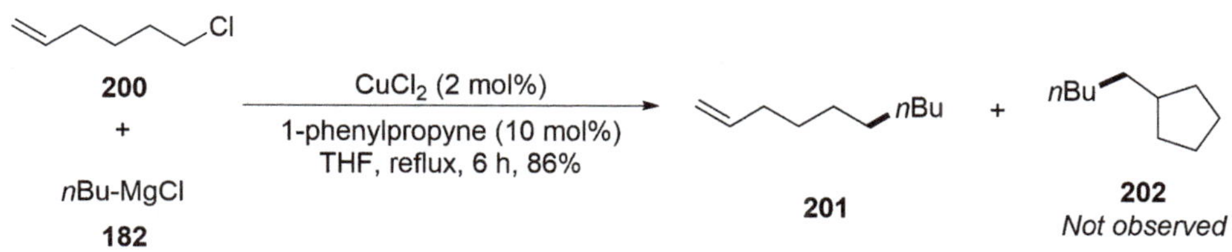

Fig. 56 "Radical-clock" experiment for the copper-mediated cross-coupling

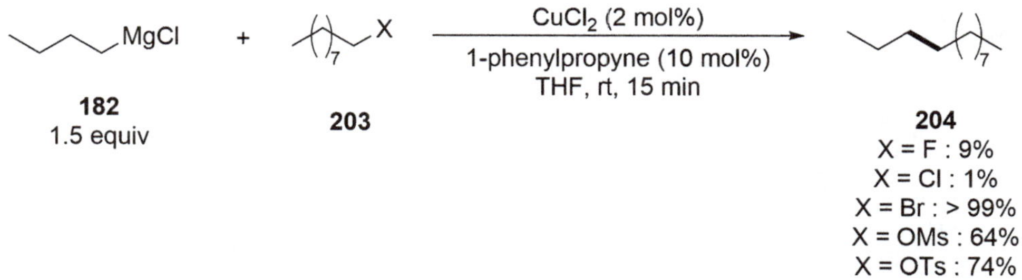

Fig. 57 Comparative performance of electrophiles **203** in the Cu-mediated coupling

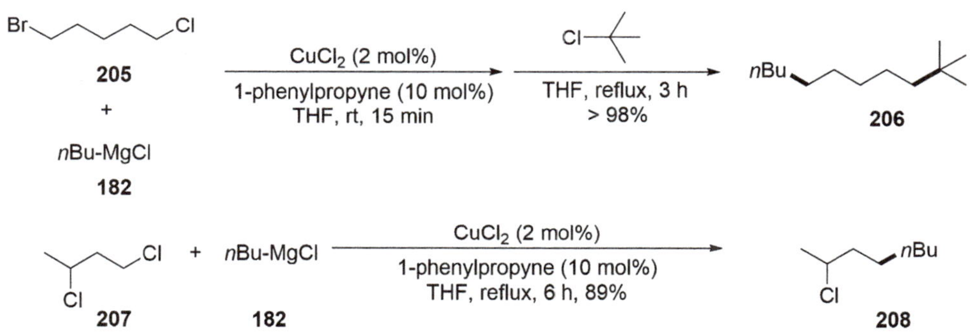

Fig. 58 Reactivity differences enable selective Cu-mediated couplings

coupling partner) were carried out to identify the optimal conditions for establishing the scope of the coupling chemistry (Fig. 47) (Table 3). The nature of the electrophile (**169**) was shown to be important, and while both the bromide and iodide coupled in modest yields, the tosylate was shown to be superior (compare entries 4, 5, and 6). In equivalent cases, the new catalyst system gave ~5–10% higher yields than Li_2CuCl_4 with the typical by-products observed being Wurtz-type dimers (5–10%), thiophenol-substituted electrophiles (0–3%), and quenched Grignard reagent. Control of the Grignard reagent concentration was also shown to be a key factor, and although synthetically useful yields could be obtained at 1.0 M, the highest yields were achieved at concentrations between 0.1 and 0.5 M. As noted previously, the nature of preparation of the new catalyst did not impact the outcome of the reaction.

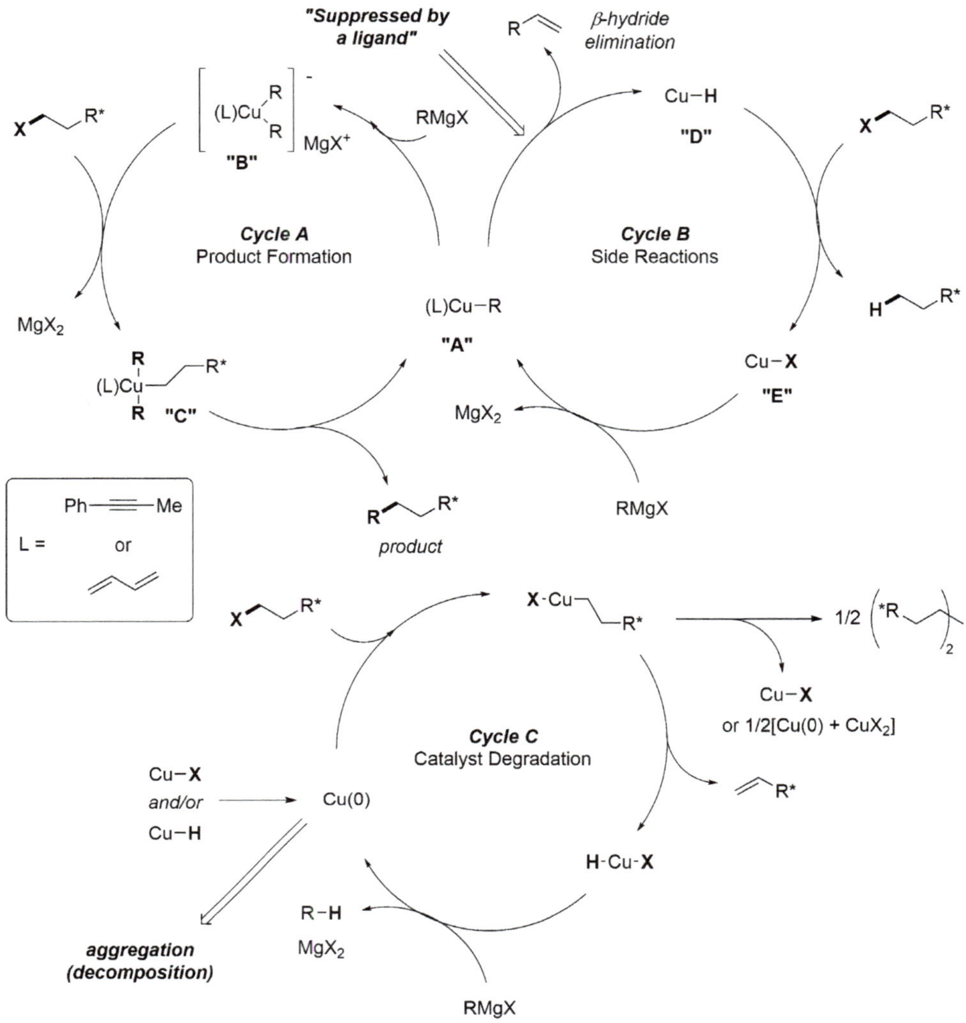

Fig. 59 Mechanistic studies on additive effects in copper-catalyzed alkyl-alkyl cross-couplings

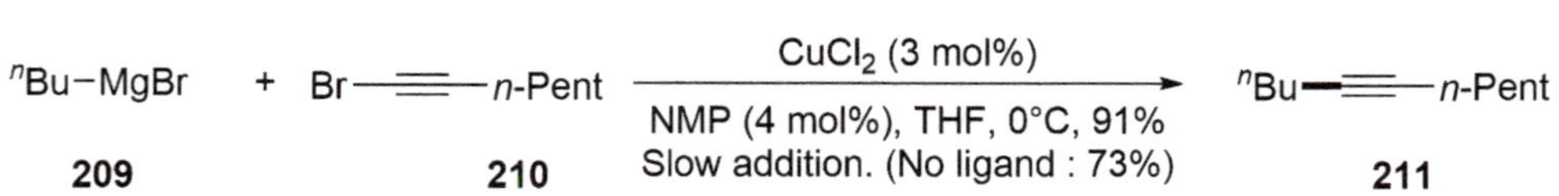

Fig. 60 NMP as an additive in the coupling of alkynyl halides

Both the Li_2CuCl_4 and the $CuBr.SMe_2$/LiBr/LiSPh systems show a significant sensitivity to temperature with yields dropping using the later in the model coupling from 77% (0 to 18 °C—warm-up protocol) to 33% (45 °C reaction temperature) (Fig. 48, Table 4) (compare entry 2 and 4). Temperatures below 0 °C proved not to be practical due to a slow reaction rate. Addition of HMPA (6% by volume) as a cosolvent proved to not only be able to increase the coupling yields at ambient temperature (87% at 25 °C) but also

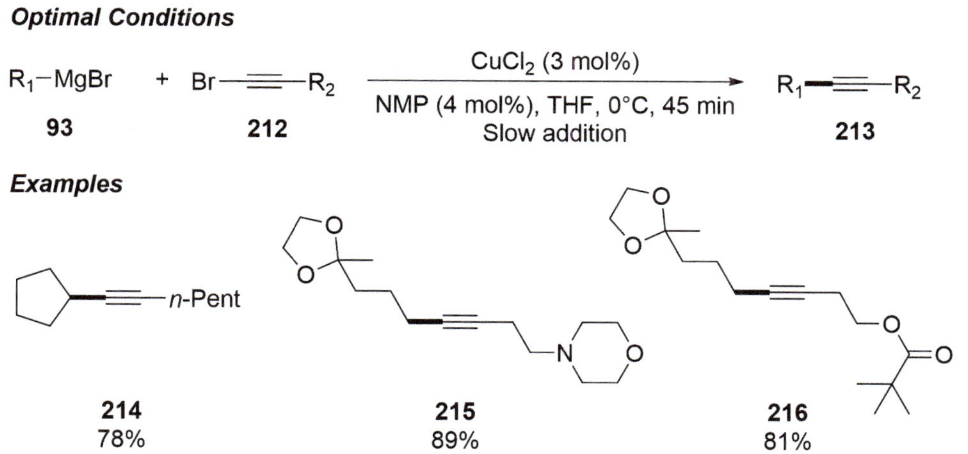

Fig. 61 Optimal conditions/scope of the alkyl/alkynyl coupling reaction

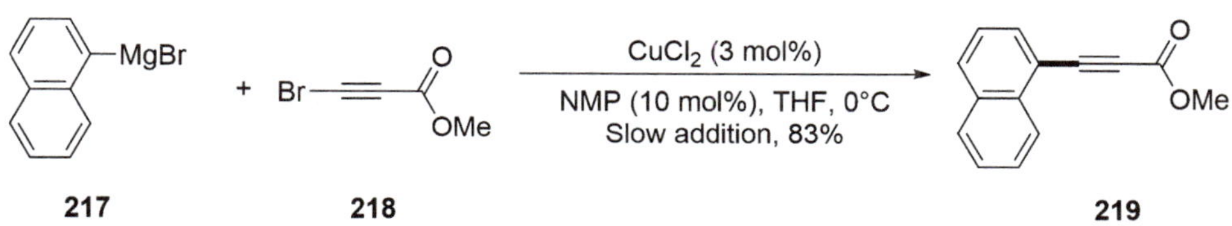

Fig. 62 Copper-mediated synthesis of arylacetylenes

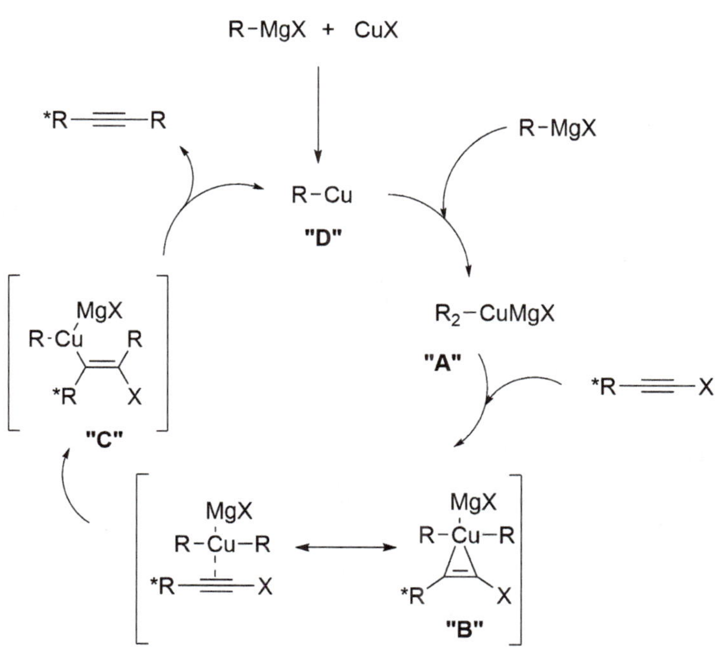

Fig. 63 Mechanistic hypotheses for the coupling of haloalkynes

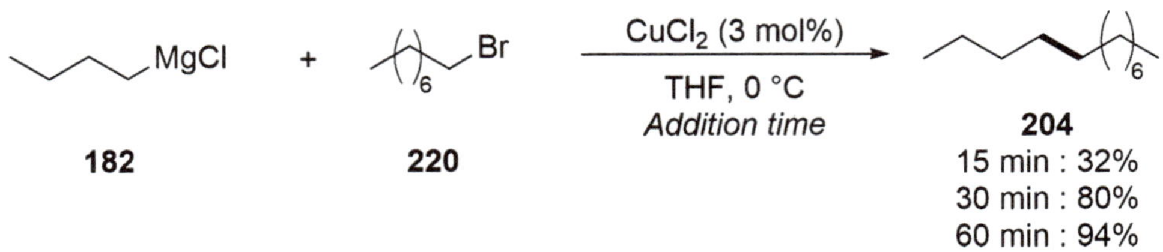

Fig. 64 Slow addition to facilitate "additive-free" alkyl-alkyl cross-couplings

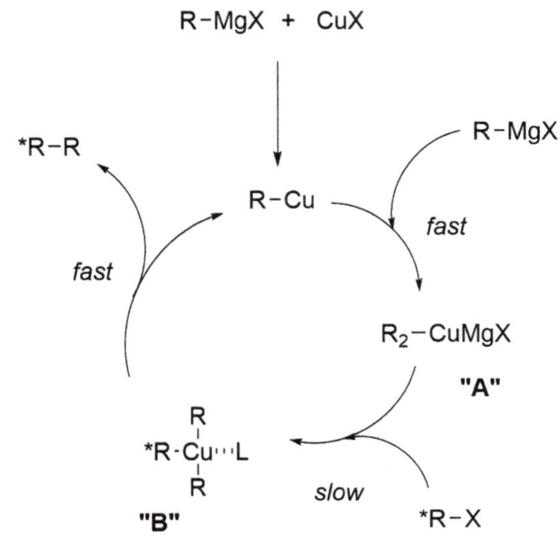

Fig. 65 Proposed mechanism to explain the benefits of slow Grignard addition

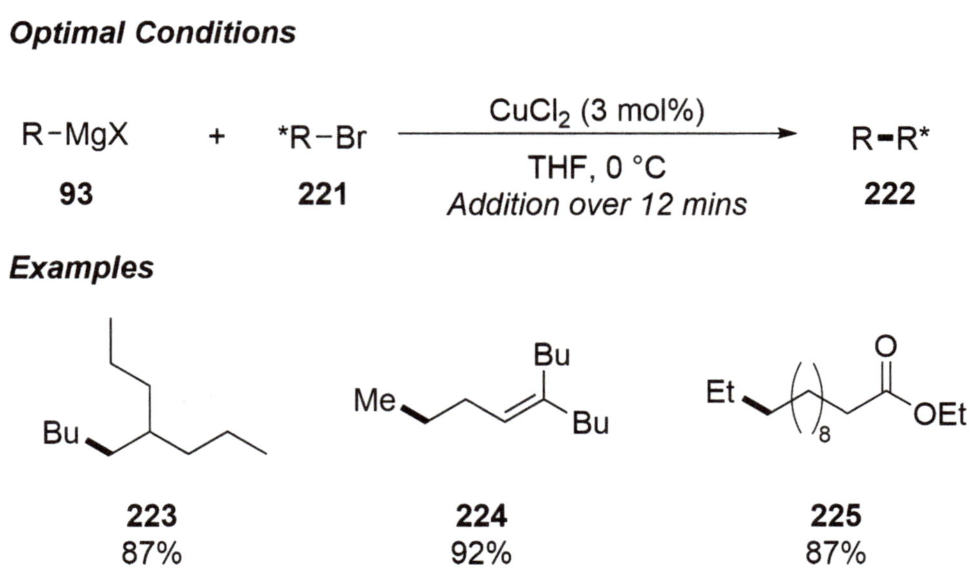

Fig. 66 Optimal conditions/scope of the "slow addition" protocol

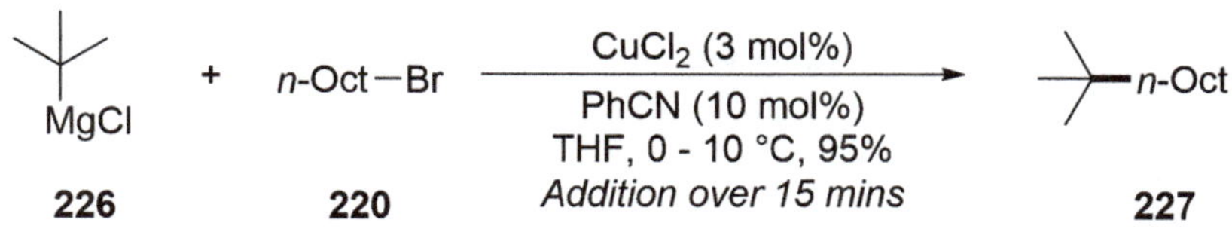

Fig. 67 Benzonitrile as an additive for the coupling of tertiary Grignard reagents

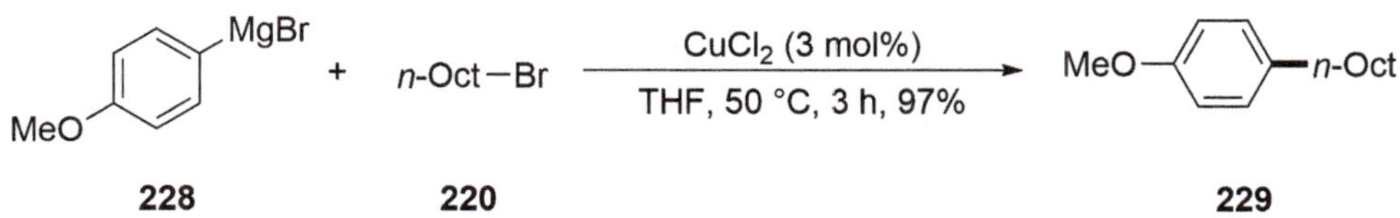

Fig. 68 Ligand-free copper-catalyzed alkyl-aryl cross-coupling

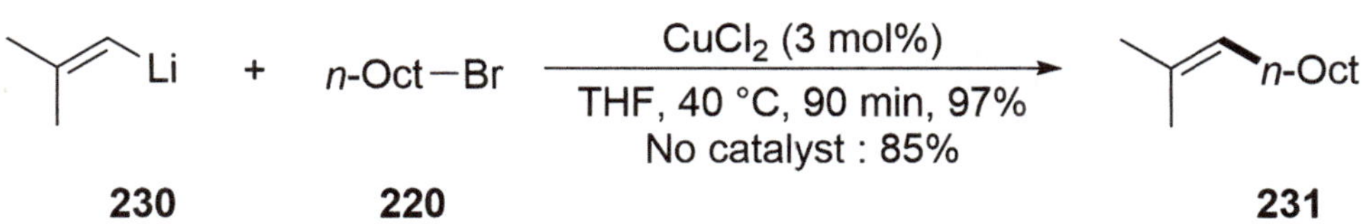

Fig. 69 Copper-mediated coupling of alkenyllithium reagents

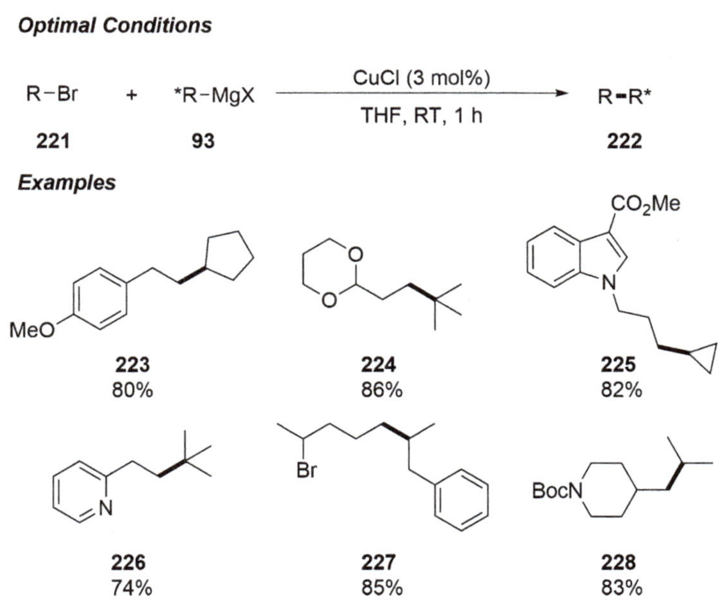

Fig. 70 Optimal conditions/scope of coupling of functionalized alkyl electrophiles **221**

enhance the catalyst's stability/efficiency at elevated temperatures (80% at 67 °C) with a faster reaction rate (3–4 h as opposed to 24 h) (entry 9). Interestingly, addition of the HMPA cosolvent led to a slightly lower yield for the reaction at 0 °C (entry 5).

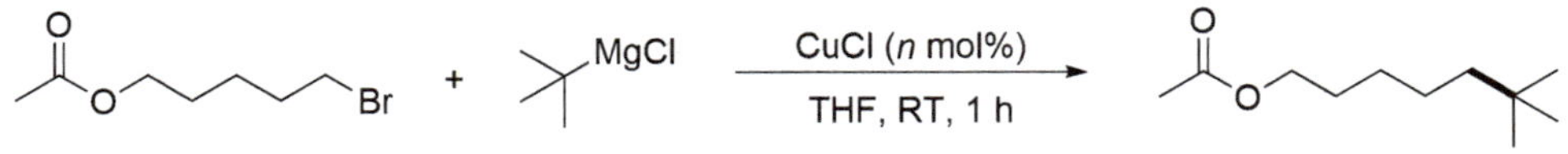

229 **226** **230**

0% (*n* = 0)
94% (*n* = 0.5)
94% (*n* = 3)

Fig. 71 Functional group tolerance in the CuCl-mediated cross-coupling

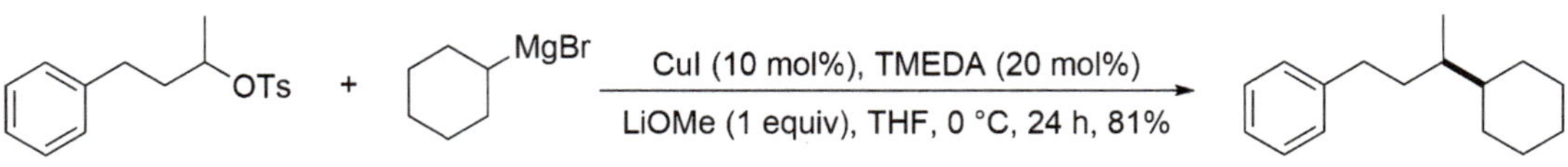

231 **232** **233**

2 equiv.

Fig. 72 Liu conditions for the coupling of sterically congested electrophiles

Optimal Conditions

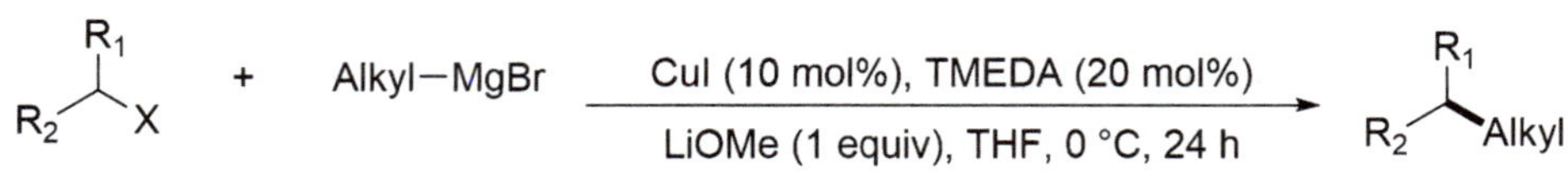

234 X = Br, OTs

93 2 equiv.

235

Examples

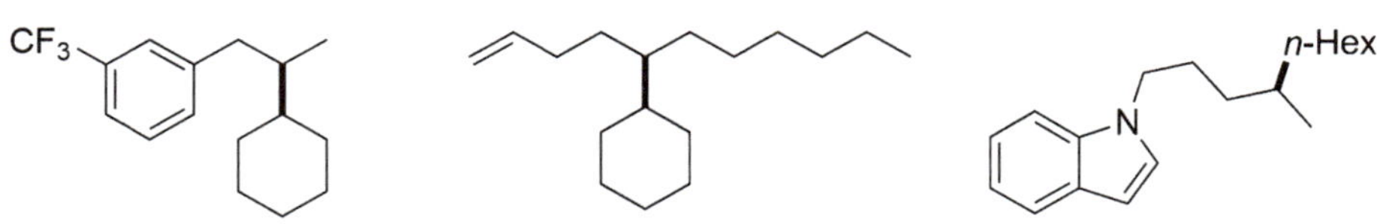

236 80% (X = OTs)

237 56% (X = OTs)

238 64% (X = Br)

Fig. 73 Optimal conditions/scope of the Liu conditions

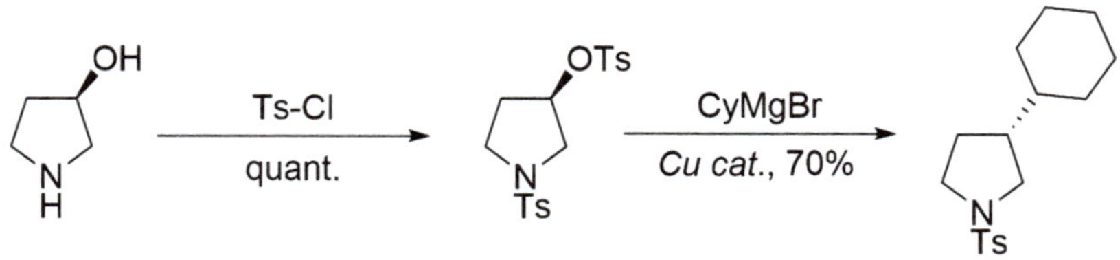

239 **240** **241**

Fig. 74 Cross-coupling proceeds with clean inversion of configuration

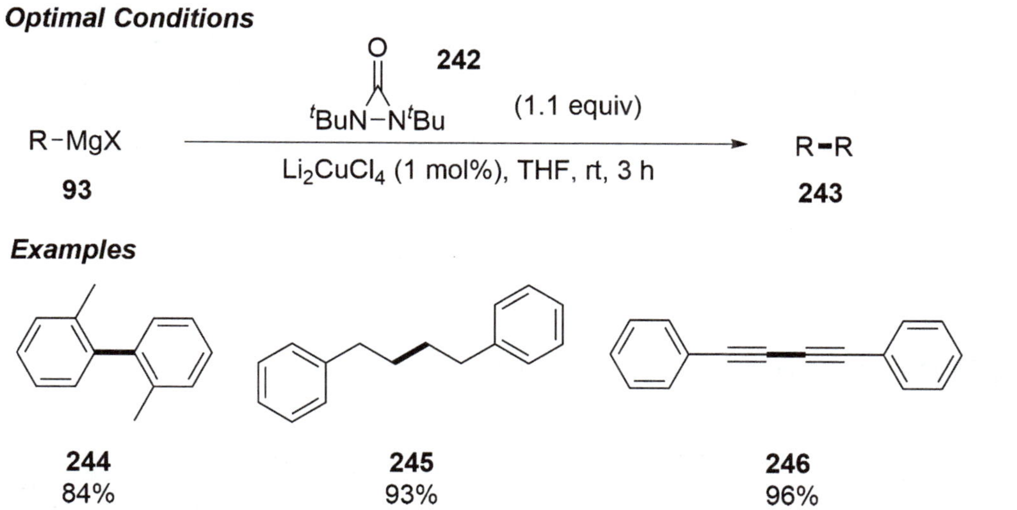

Fig. 75 Oxidative homocoupling of Grignard reagents mediated by **242**

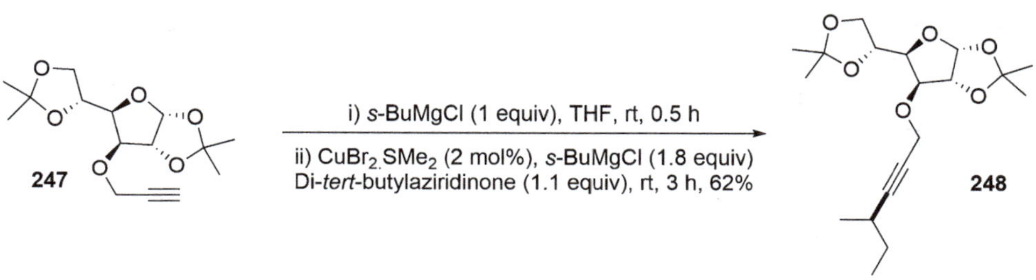

Fig. 76 Oxidative cross-coupling of densely functionalized systems

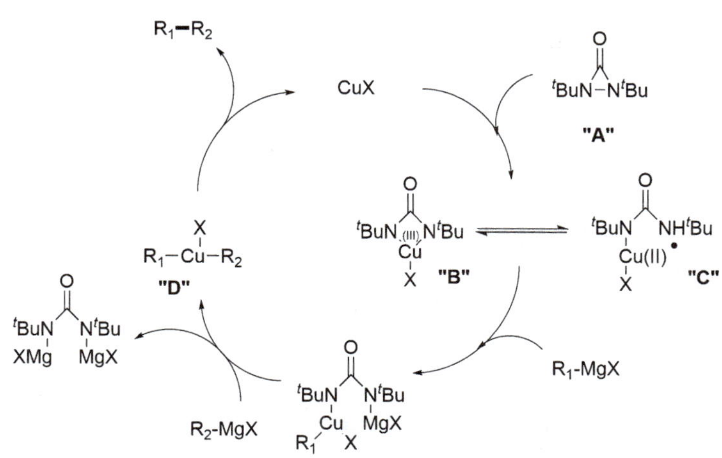

Fig. 77 Mechanistic proposal for the oxidative cross-coupling

Scope-wise the newly developed catalyst system showed a definite edge in reactivity in the coupling of one equivalent of primary, aromatic, vinyl, and allylic Grignard reagents with one equivalent of

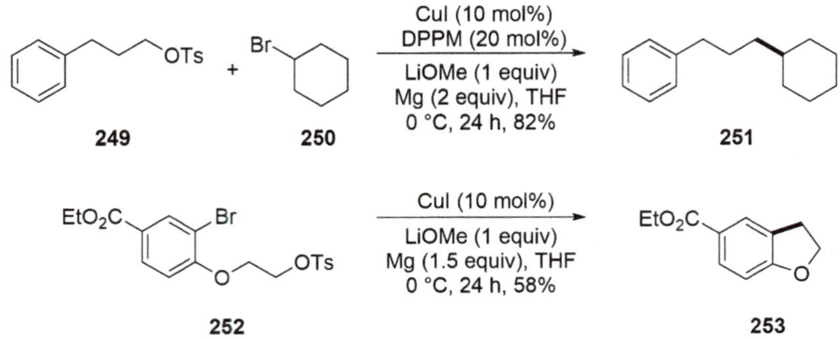

Fig. 78 Inter- and intramolecular copper-catalyzed reductive cross-coupling

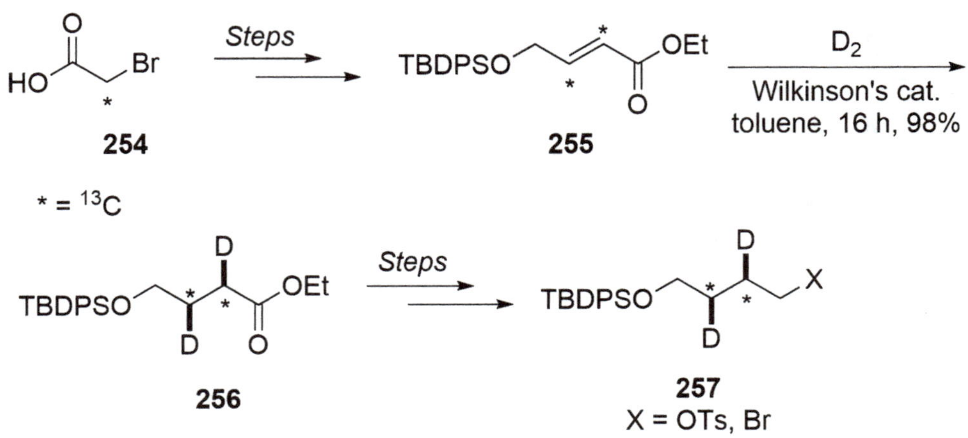

Fig. 79 Synthesis of key synthon **257**

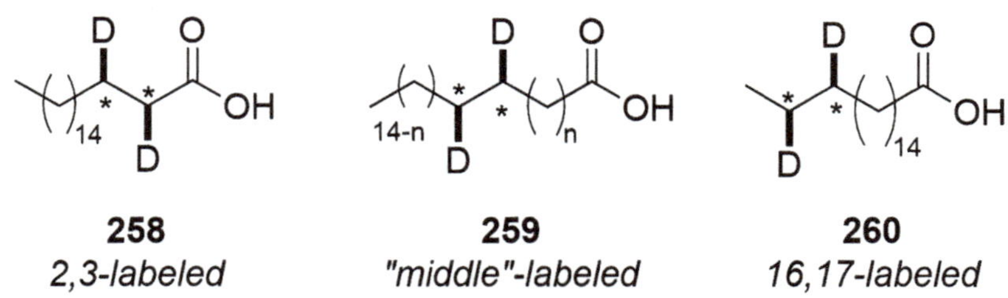

Fig. 80 Labeled stearic acid targets

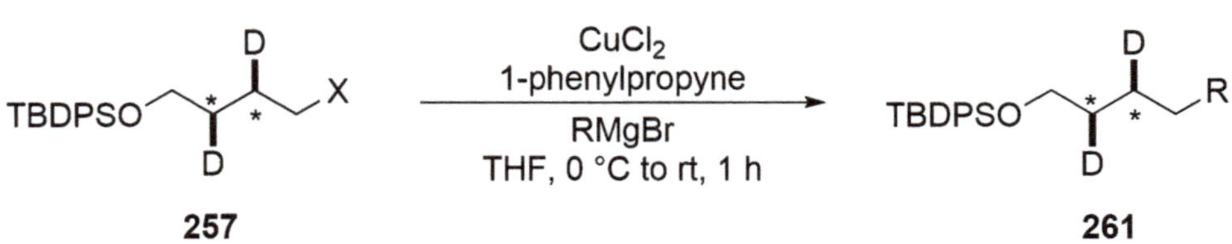

Fig. 81 Effect of electrophile leaving group in cross-coupling

Table 5
Leaving group/chain length and additive loadings in Cu-mediated coupling

Entry	X =	$CuCl_2$ (mol%)	Acetylene (mol%)	R =	RMgBr (equiv)	Yield (%)
1	Br	2	10	C_6H_{13}	1.5	50
2	Br	5	20	C_6H_{13}	2	80
3	OTs	5	20	C_6H_{13}	2	96
4	OTs	5	20	$C_{14}H_{29}$	2	98

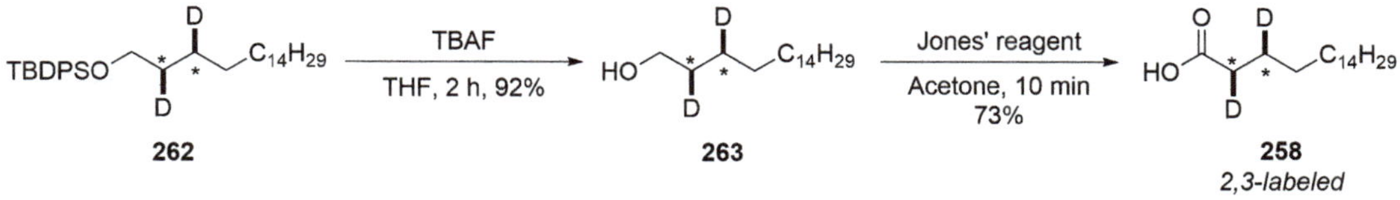

Fig. 82 Synthesis of 2,3-labeled stearic acid derivative **258**

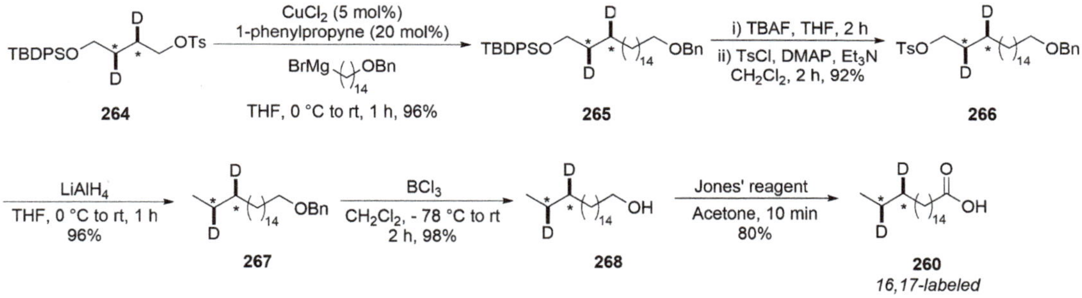

Fig. 83 Synthesis of 16,17-labeled stearic acid derivative **260**

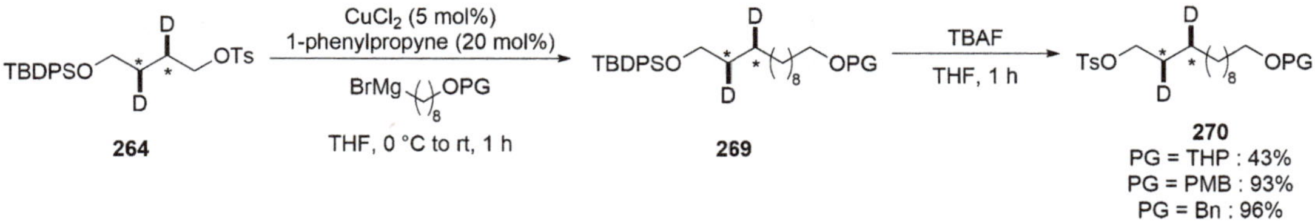

Fig. 84 Effect of protecting group in the coupling of **264**

primary alkyl tosylates with more consistent yields also being obtained compared to previously studied systems (Fig. 49). In addition, the coupling of secondary and tertiary Grignard reagents was shown to be possible in equivalent yields to that realized by the Gilman reagents with the benefit herein that only a stoichiometric amount of the organometallic reagent is used. Further extension of the methodology to the coupling of secondary electrophiles with primary Grignard reagents led to only modest yields of the desired products, though a dramatic improvement could be realized

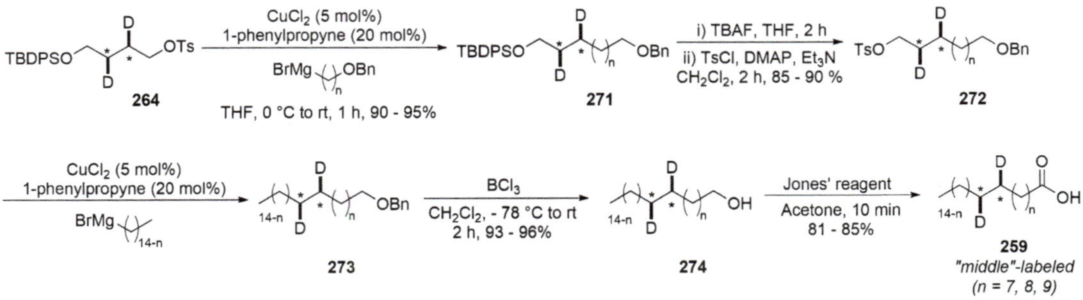

Fig. 85 Synthesis of "middle"-labeled stearic acid derivatives **259**

switching to the less sterically encumbered mesylates in a similar manner to that previously described for "higher-order" cuprates.

Incorporation of an ester into the electrophile highlighted an aspect of chemoselectivity of the CuBr.SMe2/LiBr/LiSPh catalyst with smooth coupling being realized (in a similar manner to a cuprate addition) and no addition into the carbonyl moiety in contrast to the same reaction conducted in the absence of the catalyst. Extensive ^{1}H NMR and UV/Vis spectrophotometric studies indicate that in solution there is a complicated mixture of Cu species featuring ligation with thiophenol, THF, and LiBr in various aggregated forms.

Carbon-fluorine bonds are the strongest single bonds in organic compounds, and given this, examples of sp^3 C-F bond fission mediated by transition metal complexes are rare. Kambe and co-workers have developed a nickel- and copper-mediated coupling of alkyl fluorides (**178**) with Grignard reagents (**177**) [161]. Key to the success of these reactions was the addition of 1,3-butadiene as an additive with the copper salts (CuCl or $CuCl_2$) showing the highest activities among the catalysts evaluated while also completely suppressing by-product formation (Fig. 50).

n-Octyl fluoride (**178**) was shown to react with a range of primary and secondary Grignard reagents in the presence of 1,3-butadiene though for the reactions of tertiary alkyl and phenyl Grignard reagents, the additive had little effect with good yields of the desired coupling product being obtained in its absence (Fig. 51). The corresponding alkyl bromides were also shown to be successful substrates under the developed conditions, while alkyl chlorides gave only poor yields of the coupling products.

The use of 1,3-butadiene as an additive was also extended to the copper-mediated coupling reaction between unactivated secondary alkyl iodides and alkyl Grignard reagents [162]. This reaction leads to the formation of tertiary, stereogenic carbon centers and is typically more challenging to achieve in a selective manner. Model studies on the reaction between 4-phenylbutan-2-yl iodide (**181**) and *n*-BuMgCl (**182**) in the presence of a stoichiometric amount of 1,3-butadiene catalyzed by CuI led to the desired product in 90% yield with only trace amounts of the side products resulting from elimination and reduction (<2% total) (Fig. 52). Repeating the reaction in the absence of the additive led to a significant decrease in yield (56%) with a corresponding increase in by-product formation (~23% total). Other additives were also shown to be successful in mediating the reaction though were not as effective as 1,3-butadiene, while the amount of this additive could be reduced with synthetically useful yields still being obtained from the reaction.

A range of secondary iodides and Grignard reagents were successfully evaluated in the process with the reaction shown to be tolerant of a moderate degree of steric hindrance around the reacting centers as well as several functional groups (alkene, ketal) (Fig. 53). The corresponding secondary alkyl bromides were not successful substrates for the reaction.

Preliminary studies on the mechanism of the transformation suggest that it proceeds through an ionic mechanism through the formation of magnesiocuprates from the Grignard reagent and the copper catalyst with the additive hypothesized to stabilize the cuprate intermediate and thus prevent β-elimination. Experiments involving a radical clock-based substrate indicate that free alkyl radicals are not formed, and this observation enables the chemistry to be extended to the preparation of dienes (**195**) and enynes (**197**) through incorporation of unsaturation in both reacting substrates (Fig. 54).

One gap in the initial copper-mediated cross-coupling methodology for the reaction of alkyl halides and sulfonates with Grignard reagents was the lack of applicability to the cheap, readily available alkyl chlorides owing probably to the strength of the C-Cl bond relative to the C-Br and C-I congeners. Screening additives in the reaction between *n*-nonyl chloride (**198**) and *n*-BuMgCl (**182**), Kambe and co-workers reported that high yields of the desired coupling product could be obtained in the presence of the additive 1-phenylpropyne in THF at reflux [163] (Fig. 55).

The reaction was successful with either CuCl or $CuCl_2$ as the catalyst and proceeded at room temperature though more slowly. Several other additives were also evaluated with 1,3-butadiene and styrene shown to be far less effective, while steric hindrance around the triple bond of the alkyne was also found to have an impact on the reaction. It is hypothesized that the role of the 1-phenylpropyne is through coordination to the Cu(I) ion preventing the decomposition of the thermally unstable alkylcopper (I) intermediates in the catalytic cycle. Increased concentrations of the alkyne additive were shown to lower the initial rate of the coupling reaction, and this is proposed to be due to a shift in the equilibrium toward *bis*(alkyne)copper (I) complexes that represent potential resting states of the catalyst (*vide supra*). Substrate studies indicate that a radical mechanism is not involved with inversion of configuration observed during the cross-coupling process suggesting primarily an S_N2 mechanism (Fig. 56).

Examination of independent reactions of the alkyl electrophiles (**203**) indicated that the reactivities increased in the order: chloride < fluoride < mesylate < tosylate < bromide with the slightly enhanced reactivity of the alkyl fluoride compared to the corresponding chloride being attributed to the formation of a strong Mg-F bond during the reaction (Fig. 57).

These reactivity differences enabled site-selective sequential cross-coupling reactions to be carried out on dihaloalkanes (**205**) while selectivity could also be realized between primary and secondary halides in the same substrate (**207**) (Fig. 58).

Kambe and co-workers have carried out extensive mechanistic studies on the effects that additives such as 1-phenylpropyne and 1,3-butadiene have on the copper-catalyzed alkyl-alkyl coupling reaction as well as evaluating the influence of catalyst loadings on both the rate and reaction profile [164, 165]. These demonstrate that very high TONs of up to 10^6 and TOFs of $10^4\ h^{-1}$ are attainable for the transformation. The loading of the additive was shown to be a key parameter as although this suppresses not only catalyst degradation but also the side reactions though increasing the loading of 1-phenylpropyne also leads to a decrease in reaction rate suggesting that higher concentrations generate Cu complexes coordinated to two acetylene molecules that represent a resting state of the catalytic system. Prolonging the reaction time though leads to >80% conversion to the desired product. Initial formation of an alkyl copper species (**A**) through reaction of a copper salt and Grignard reagent is followed by further reaction with the Grignard to form an intermediate dialkyl cuprate (**B**) that undergoes nucleophilic substitution with the alkyl halide through an ionic mechanism with the desired coupling product formed via reductive elimination of two cis-oriented alkyl groups from **C** with the concomitant regeneration of the alkyl copper species (*Cycle A*) (Fig. 59). Alternatively owing to the thermal instability of the alkyl copper species, this can undergo β-hydrogen elimination to give the olefin and a copper hydride (**D**) that can also react with the alkyl halide to give the reduced alkane by-product (*Cycle B*). Catalyst degradation is triggered through the disproportionation or homolysis of either Cu-H (**D**) or Cu-X (**E**) forming Cu(0) that leads to deactivation through aggregation as well as promoting further side-product formation through dehydrohalogenation and homocoupling processes (*Cycle C*). Sequestering of the Cu-H (**D**) by the additive through formation of a vinyl- or allyl-copper intermediate that can regenerate the alkyl copper species (**A**) through transmetalation was investigated and shown to be feasible though is not thought to be the main mechanism by which an additive suppresses catalyst degradation. The additives are believed to coordinate to Cu thus preventing β-hydrogen elimination (*Cycle B*) while also potentially accelerating formation of the dialkyl cuprate complex (**B**) (*Cycle A*). Additional observations demonstrate that catalyst deactivation is accelerated at higher catalyst loadings though while running the reaction with low Cu loadings alleviated this issue, selectivity control was an issue in the absence of an additive. While these studies provide some guidance as to obtaining optimal yields for the coupling process, the precise mechanistic details remain elusive owing to the ease of one-electron redox, aggregation, and the disproportionation of copper species.

Couplings between alkyl and alkynyl reagents can be achieved through two distinct pathways. Whereas alkylation of a metal acetylide with a primary alkyl halide can be used, drawbacks of this method include poor functional group compatibility as well as limited substrate scope particularly due to competing elimination reactions. Reversing the polarity of the coupling partners and reacting an alkynyl halide with an alkyl-metal reagent represents an alternative approach, and while potentially offering a wider scope in terms of substrate compatibility, reports of this coupling paradigm are somewhat sparse.

Cahiez and co-workers sought to develop the copper-catalyzed variant of this coupling through model studies on the reaction between *n*-BuMgCl (**209**) and heptynyl bromide (**210**) [166]. Previous reports had indicated that only low yields could be obtained with halogen/magnesium exchange being the predominant reaction. However, a 73% yield could be obtained using $CuCl_2$ as the catalyst simply by controlled syringe-pump addition of the Grignard reagent into the reaction mixture. The influence of ligands/additives (10 mol%) was also investigated, and while the previously described 1-phenylpropyne led to a slight increase in yield, σ-donor ligands such as TMEDA, $(EtO)_3PO$, Me_2S, and NMP (for which loadings as low as 4 mol% were similarly effective) led to excellent yields. Use of the bidentate phosphine ligand dppe shut down the reaction completely, while a monophosphine PPh_3 led to an 84% yield of the desired coupling product (Fig. 60).

A range of aliphatic internal alkynes were prepared in good to excellent yields under the optimal conditions (3 mol% $CuCl_2$, 4 mol % NMP) from the reactions between primary and secondary (acyclic/cyclic) Grignard reagents with alkynyl bromides with good functional group compatibility observed with respect to both reaction components (Fig. 61). The reaction failed for alkynyl bromides derived from arylacetylenes with Br/Mg exchange being observed though this issue could be resolved using the corresponding alkynyl chlorides in the reaction with herein the additive NMP no longer required. Primary, secondary, and even tertiary Grignard reagents all participated effectively in this reaction, which offers an alternative to the classical Sonogashira coupling. The versatility of this chemistry was further demonstrated with the preparation of a range of functionalized arylacetylenes starting initially from 2-bromobenzonitrile.

A related approach to arylacetylenes (**219**) was also developed through the arylation of alkynyl bromides (**218**) (Fig. 62). The conditions developed for the analogous alkylation reaction proved to be effective with a slightly higher loading of the NMP additive used. Grignard reagents featuring both electron-donating and electron-withdrawing groups were successful substrates with Br/Mg exchange rarely encountered as a problem, while a range of functional groups were tolerated on the alkynyl bromide.

A mechanistic hypothesis is provided suggesting that initially a cuprate derivative (**A**) is formed from the Grignard reagent, which then undergoes carbocupration with the haloalkyne to form an unstable vinylcopper-base species (**C**) through the proposed intermediacy of a metallacyclopropene (**B**) (Fig. 63). β-Halogen elimination then generates the product and an organocopper species (**D**) that can react with further Grignard reagent to regenerate the cuprate. The ability of σ-ligands to promote the transformation is attributed to their ability to favor complexation of the Cu(I) species to the triple bond by increasing the electron density of the copper atom.

While numerous additives have been disclosed to promote the Cu-catalyzed alkylation of Grignard reagents, Cahiez and co-workers have reported on a ligand-free methodology, which focuses on the slow, controlled addition of the organometallic reagent to the reaction to achieve robust, consistent yields [167]. Investigating the coupling reaction between *n*-BuMgCl (**182**) and 1-bromooctane (**220**) in the presence of 3 mol% $CuCl_2$ demonstrated that while a 94% yield of dodecane could be realized through syringe-pump addition of the Grignard reagent over 1 hour, this yield dropped precipitously to 32% on addition from a dropping funnel over 15 min (Fig. 64).

Consideration of the reaction mechanism enables an explanation for the dramatic effect that the rate of addition of the Grignard reagent has on the reaction outcome. Initial reaction of the Grignard leads to a cuprate (**A**) through transmetalation, which is followed by reaction with the alkyl halide to generate a transient Cu (III) species (**B**) that undergoes reductive elimination to generate the product (Fig. 65). Studies indicate that the reaction of the alkyl halide with the intermediate cuprate species is likely the slow step (and this is purportedly accelerated when additives are utilized in the reaction thus negating the need for slow addition) and therefore for efficient reaction to occur control of the amount of the Grignard reagent present is critical to prevent any excess reacting with the intermediate cuprate (**A**) in unproductive side reactions (elimination, Mg/X exchange, formation of higher-order cuprates).

A range of copper sources were also effective catalysts, and the modified reaction protocol was demonstrated with several primary Grignard reagents and primary alkyl bromides as substrates (Fig. 66). For the latter, several common functional groups were well tolerated, and steric hindrance at the β-position did not affect the coupling efficiency. Alkyl chloride though was not a suitable coupling partner.

The coupling of *s*-BuMgCl with octyl bromide (**220**) was also shown to be successful, and this reaction was used to highlight the critical influence of temperature (5–10 °C was optimal) on the reaction as well as the ability of benzonitrile to function as a simple and inexpensive ligand to promote the reaction. Use of this ligand was also shown to be effective in the coupling of tertiary Grignard reagents (**226**) (Fig. 67).

Direct coupling of aryl Grignard reagents with primary alkyl bromides under ligand-free conditions was also demonstrated with now the reactions being conducted at 50 °C due to the enhanced stability of the organometallic reagents and without the need for slow addition as the Grignard does not react directly with the alkyl halide leading to side products (Fig. 68).

With alkenyl Grignard reagents, only modest yields could be obtained (~40%) owing to the poor reactivity of the intermediate vinylic copper species as well as the constraints on the temperature owing to the slow reaction rate in contrast to the rate of decomposition of the intermediate. However, the analogous alkenyllithium reagents (**230**) proved to be effective coupling partners with excellent yields being obtained with and without copper catalysis with relatively fast reaction times (Fig. 69).

Alkyl- and aryl-lithium reagents were also demonstrated to be suitable coupling partners with either primary alkyl bromides or chlorides. Again, reactions occurred with or without $CuCl_2$ as the catalyst though higher yields (with the effect much more pronounced for the alkyl chlorides) were obtained uniformly in the presence of the catalyst.

One of the drawbacks of a number of protocols developed for copper-catalyzed alkyl-alkyl cross-couplings is the limited synthetic applicability in cases involving functionalized alkyl electrophiles. Reporting on model studies between a series of functionalized alkyl bromides (**221**) and a tertiary alkyl Grignard reagent, Hu and co-workers demonstrated that excellent yields of the desired coupling products could be obtained for substrates featuring a range of functional groups (ester, acid, nitrile, alcohol, etc.) using a catalytic amount of CuCl at room temperature with no additive and THF as the solvent [168] (Fig. 70).

Several heterocyclic motifs were also well tolerated, and while ketone-containing substrates proved to be more challenging, addition of NMP as an additive enabled synthetically useful yields of these coupling products to be obtained. Aldehyde, nitro, and succinimide groups were not compatible with this method. Alkyl iodides and tosylates were also successful substrates though alkyl chlorides and secondary alkyl halides could not be coupled. This as well as the lack of reactivity of aryl bromides allows selective and sequential functionalization to be achieved to rapidly build up molecular complexity (Fig. 71).

Primary, secondary, and tertiary acyclic and cyclic Grignard reagents were all shown to couple successfully though preliminary attempts to extend the reaction to an enantioselective variant using a chiral ligand failed. Control studies show that the reaction most likely proceeds through an S_N2 mechanism.

While several methodologies have been disclosed for the copper-mediated cross-coupling of alkyl halides with Grignard reagents, challenges still exist specifically in the reactions of the more sterically congested secondary alkyl electrophiles with secondary alkyl nucleophiles, thus somewhat limiting the overall utility of the methodology in the assembly of complex carbon skeletons. The combination of "CuBr/LiSPh/LiBr/THF" [160] has been shown to mediate the cross-coupling of secondary alkyl sulfonates with primary Grignard reagents with the NMR-driven hypothesis being that the active species consisted of Cu ligated with LiBr in aggregated forms. Inspired by these studies, Liu and co-workers investigated the cross-coupling between 4-phenylbutan-2-yl tosylate (**231**) and cyclohexylmagnesium bromide (**232**) with the intention of developing a general catalyst system to mediate this reaction in a stereo-controlled fashion [169] (Fig. 72). Initial studies using CuI as the catalyst in a mixed solvent system of THF/HMPA with TMEDA as an additive to suppress side reactions lead only to low yields of the desired product. Further optimization revealed the beneficial effect of Li-based additives, and while LiCl and LiBr had little effect on the reaction, LiI, LiO*t*Bu and LiOMe all gave an increase in yield, while HMPA was shown not be essential for the reaction. CuI was demonstrated to be the optimal Cu source, and while other additives (PBu3, 1-phenylpropyne, NMP) previously shown to be successful in Cu-mediated cross-couplings were also shown to promote the reaction, TMEDA was the best. Switching the secondary alkyl electrophile to the corresponding bromide was also successful, but the reaction failed for the iodide, chloride, and mesylate.

Scope studies indicate that the new protocol tolerates a range of synthetically useful functional groups including olefins, alcohols, and heterocycles on the alkyl bromide or tosylate and that acyclic and cyclic Grignard reagents are successful substrates (Fig. 73). In addition, the conditions can also be applied to the coupling of tertiary alkyl Grignard reagents. Several interesting chemoselectivity aspects of the reaction are also presented with both alkyl tosylates and bromides reacting in the presence of an aryl bromide while a secondary tosylate undergoes selective cross-coupling in the presence of a primary alkyl chloride (note the catalysts is active enough to promote cross-couplings of primary alkyl chlorides with secondary Grignard reagents) enabling site-selective sequential cross-coupling reactions to be performed.

Evaluation of the reaction on chiral secondary alkyl tosylates derived from the corresponding easily accessible chiral alcohols showed that the cross-coupling occurs with inversion of configuration through an S_N2 process thus enabling the stereo-controlled formation of C-C bonds (Fig. 74).

Oxidative coupling reactions of various organometallic reagents represent an alternative strategy for C-C bond formation though while numerous examples of oxidative homocouplings of Grignard reagents have been developed, examples of the corresponding cross-couplings remain scarce. Shi and co-workers have developed a mild methodology for the Cu-mediated cross-coupling of Grignard reagents employing di-*tert*-butyldiaziridinone (**242**) as the oxidant [170].

Initial model studies focused on the homocoupling of phenylmagnesium bromide and demonstrated that a wide range of transition metal catalysts (including several copper sources) were able to successfully mediate the reaction. As expected, only trace amounts of the homocoupled product were observed in the absence of either the oxidant or the metal catalyst with the transformation extended to a range of (hetero)aryl-, benzylic-, alkyl-, and alkynyl Grignard reagents with excellent yields being uniformly obtained (Fig. 75).

For the challenging cross-coupling, CuBr.SMe2 was shown to be the catalyst of choice with the reaction successfully demonstrated for the formation of C(sp2)-C(sp2), C(sp)-C(sp2), and C (sp)-C(sp3) bonds with good yields being obtained. Alkyl Grignard reagents possessing β-hydrogens were well tolerated with the desired cross-coupling products being formed with high selectivity. The reaction was demonstrated on gram scale and further utilized for the derivatization of densely functionalized substrates (**247**) (Fig. 76).

Mechanistically, the reaction is proposed to proceed through insertion of CuX into the N-N bond of the diaziridinone (**A**) to form a four-membered Cu(III) species (**B**) (and/or Cu (II) nitrogen radical, **C**) that undergoes sequential transmetalations with the Grignard reagents followed by reductive elimination from **D** to yield the cross-coupled product (Fig. 77).

In contrast with the methodologies presented so far, direct reductive cross-coupling between two organic halides presents practical advantages in avoiding the prior preparation and handling of sensitive or expensive organometallic reagents. Several metals have been shown to successfully mediate these transformations though the majority tend to be aryl-alkyl coupling, thus avoiding the challenges of the homocoupling of both alkyl halides in the reductive alkyl-alkyl cross-coupling.

Liu and co-workers have developed a copper-catalyzed reductive cross-coupling between non-activated alkyl tosylates and mesylates (**249**) with alkyl or aryl bromides (**250**), which can be extended to an intramolecular variant for the construction of oxygen-containing heterocycles (**253**) [171] (Fig. 78). Initial studies on the reaction between (3-bromopropyl)benzene and bromocyclohexane employing previously developed conditions (CuI/TMEDA/LiOMe/THF) for the cross-coupling of an alkyl halide with an alkyl Grignard reagent with the addition of Mg powder as a reductant lead encouragingly to a moderate (54%) yield of the desired cross-coupled product though synthetic utility was limited owing to formation of significant quantities of the corresponding homocoupled products. Optimization efforts failed to alleviate this issue though switching one of the reaction components to the corresponding alkyl pseudohalide and addition of a phosphine ligand represented a viable path forward for reaction development with synthetically useful yields being obtained for a range of substrates.

The utility of copper-based couplings for alkyl-alkyl bond formation is exemplified by studies by Murata and co-workers on the syntheses of selectively labeled fatty acid derivatives [172]. Within this work, the goal was to access a range of stearic acids bearing labeled positions with a ^{13}CHD-^{13}CHD moiety incorporated into the aliphatic chain to enable the investigation of the conformations of these molecules and their ensuing impact on protein-lipid interactions. Primarily a modular approach was desired employing a versatile, readily available selectively labeled synthon, which could be diversified in a robust manner.

Construction of a suitable four-carbon synthon was achieved through the use of two molecules of [2-^{13}C]-bromoacetic acid (**254**) to form the desired ^{13}C-^{13}C spin pair, while the *threo*-configuration for C2/C3 of the common intermediate was achieved through *syn*-selective deuteration of the precursor olefin (**255**). Conversion of the unprotected alcohol to either the bromide or tosylate (**257**) enabled the effect of leaving group to be assayed in the subsequent coupling chemistry (Fig. 79).

With the requisite synthons in hand, strategies were developed in order to incorporate into specific positions within the stearic acid molecule to potentially allow discrimination between the gauche+ (G^+) and gauche- (G^-) conformations of the alkyl chain utilizing the labeled butylene unit. Specifically, the plan was to introduce this moiety adjacent to the acid head group (**258**) (2,3-labeled), at the opposite termini (**260**) (16,17-labeled), and to develop a versatile approach to introduce the fragment throughout the chain (**259**) ("middle"-labeled stearic acid) (Fig. 80).

Direct coupling followed by a deprotection/oxidation would enable access to the 2,3-labeled stearic acid. Model studies on the coupling of the bromide with a shortened alkyl chain

($C_6H_{13}MgBr$) demonstrated that judicious modification of the conditions of Kambe in using increased loadings of both the $CuCl_2$ catalyst and 1-phenylpropyne additive allowed high yields (80%) to be obtained (Fig. 81, Table 5, entry 2).

However, in contrast to the original report, the tosylate-based synthon proved to be a superior coupling partner to the bromide providing quantitative yields for both the model short and longer ($C_{14}H_{29}$) alkyl chains (entry 3 and 4). Routine deprotection and oxidation provided the 2,3-labeled stearic acid in 66% overall yield (Fig. 82).

Coupling the labeled butylene-based synthon to the fatty acid head group though proved to be more challenging specifically given the propensity of carboxylic acids to bind to metals. A typical approach to overcome this is to utilize the corresponding ester though herein various methodologies to couple this failed leading to a strategy being adopted in which a protected alcohol is employed as the coupling partners with subsequent oxidation steps providing the desired acid. Reports of the copper-mediated coupling Grignard reagents, which are freshly prepared prior to use, featuring heteroatoms are relatively rare, and in the current study, three common protecting groups (THP, PMB, Bn) were evaluated (Fig. 83). Each of these was subjected to a model reaction to test coupling with the tosylate under the previously identified conditions followed by desilylation, and while the THP derivative led only to modest yields, both the Bn- and PMB-protected groups proved to be sufficiently robust to provide the desired protected products in >90% yield over the two steps. Utilizing the slightly superior Bn-protecting group in the starting material with the appropriate chain length enabled the desired 16,17-labeled stearic acid to be synthesized in 65% overall yield over six steps.

With strategies devised to introduce the labeled fragment in close proximity to both termini of the stearic acid, attention shifted to the regiocontrolled incorporation of this fragment throughout the rest of the molecule. To achieve this, a hybrid approach was developed using two copper-mediated cross-couplings in tandem with a range of standard deprotection/activation and oxidation reactions with the aforementioned four-carbon-labeled synthon (**264**) forming the core of this modular strategy. Operationally the fragment that would eventually become the acyl termini was coupled first to avoid both volatility issues and potential problematic separations if the sequence were to be reversed. Evaluation of a series of protected Grignard reagents identified the Bn group as superior in terms of coupling efficiency (Fig. 84).

Initial coupling with the Bn-protected hydroxyl chain followed by desilylation and tosylation provided the precursor (**272**) for the second coupling step to attach the tail. Coupling and debenzylation led to the stearyl alcohols (**274**) that could be oxidized to the desired acid using Jones reagent with three differentially labeled stearic acids being obtained in *ca.* 65% yield over the six-step sequence demonstrating the overall versatility of the Cu-catalyzed coupling in the assembly of biorelevant molecules (Fig. 85).

4.3 Reactions of Organoboron Reagents

Despite the utility of the copper-mediated cross-coupling reaction developed using Grignard reagents, enabling similar processes using organoboron reagents presents several advantages most notably the greater commercial availability of such compounds as well as the higher-functional group tolerance displayed in their reactions. Liu and co-workers have disclosed a copper-catalyzed cross-coupling of organoboron compounds (**275**) with primary alkyl halides (**276**) and pseudohalides [173] (Fig. 86). Model studies on reaction of *n*-dodecyl-*p*-toluenesulfonate with a range of organoboron species derived from phenylboronic acid demonstrated that CuI was a competent catalyst with the use of LiO*t*Bu as the base crucial for high yields to be obtained. With respect to the nature of the nucleophile, boronic acids, esters (Bpin and neopentyl), and cyclic boroxine derivatives all successfully coupled with a range of functional groups being tolerated on the aromatic nucleus including halogens creating the potential for further downstream cross-coupling processes. A reactivity profile was established for the various leaving groups (I > Br > OTs > OMs > Cl) thus allowing both selective and sequential transition-metal-mediated cross-couplings to be achieved.

Coupling of alkyl-boron derivatives was also shown to be possible but required the use of 9-BBN-derived reagents (**282**) (Fig. 87). Mechanistically, the reaction is believed to proceed via an S_N2-type substitution with the LiO*t*Bu acting in a dual capacity with the Li^+ stabilizing the organocopper intermediates while the tBuO- anion facilitates the transmetalation through coordination to the boron atom.

The transmetalation of an organoboron species to generate an organocopper-based intermediate represents a paradigm shift when envisioning a catalytic cycle for a copper-mediated variant of the classical Suzuki-Miyaura aryl-aryl bond formation. In contrast to the Pd-mediated process, initial transmetalation generates a nucleophilic Ar-Cu(I) species (**A**) that can react with an electrophilic aryl halide to give an intermediate Cu(III)-based species that upon reductive elimination provides the desired product while regenerating the Cu(I) catalyst (Fig. 88).

Brown and co-workers have exploited this reactivity and through a model reaction between the neopentyl-ester of *p*-tolylphenylboronic acid (**275**) and iodobenzene identified copper chloride with the ligand Xantphos as an effective system to mediate the process in an efficient manner [174] (Fig. 89). Several features emerged from studies on the scope of the reaction with aryl bromides being unreactive while boronic acids rapidly decomposed through a copper-mediated protodemetalation process. Alternative

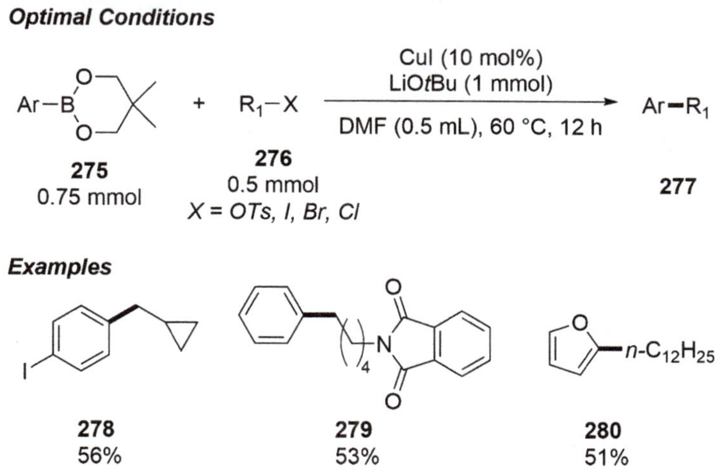

Fig. 86 Optimal conditions/scope of the Liu conditions for the coupling of organoboron compounds **275**

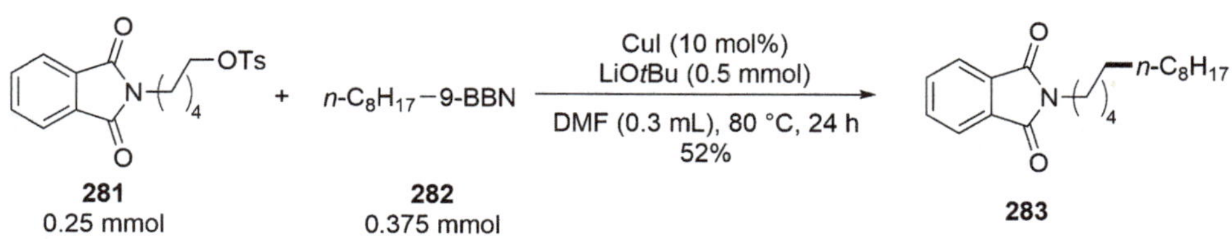

Fig. 87 Liu conditions for the coupling of alkylboron compounds using 9-BBN derivatives

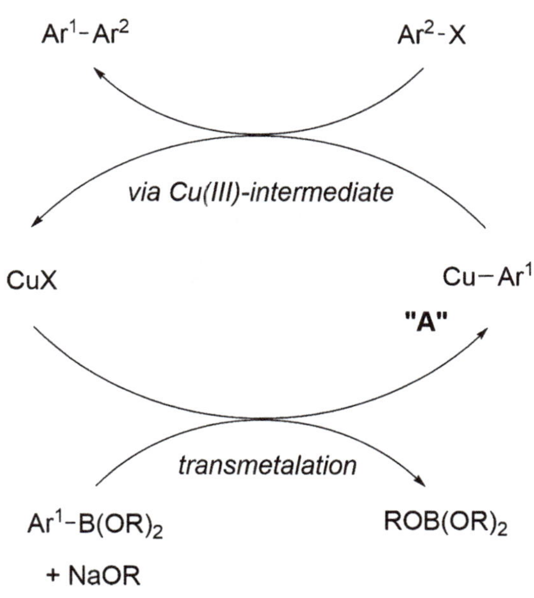

Fig. 88 Mechanistic hypothesis for aryl-aryl bond formation through formation of an organocopper-based intermediate **A**

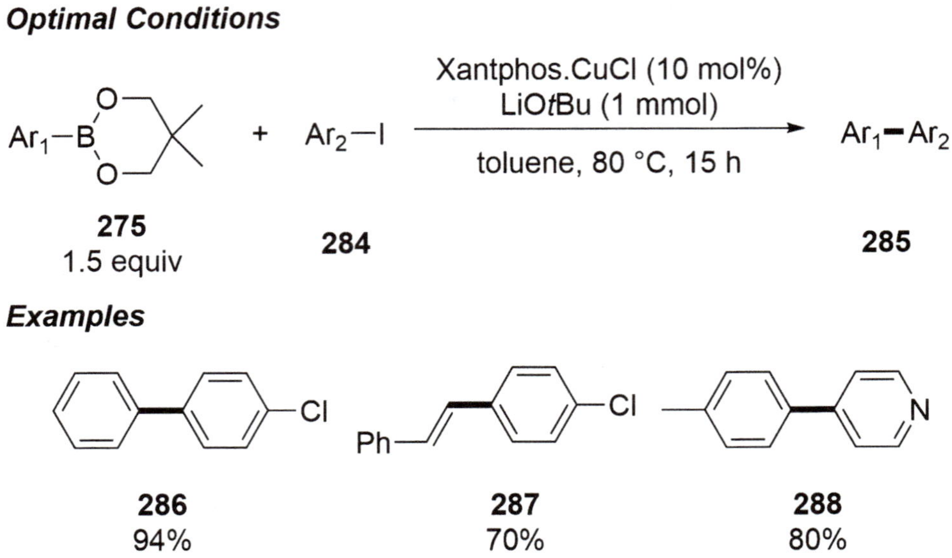

Fig. 89 Optimal conditions/scope of the Brown conditions for aryl-aryl bond formation

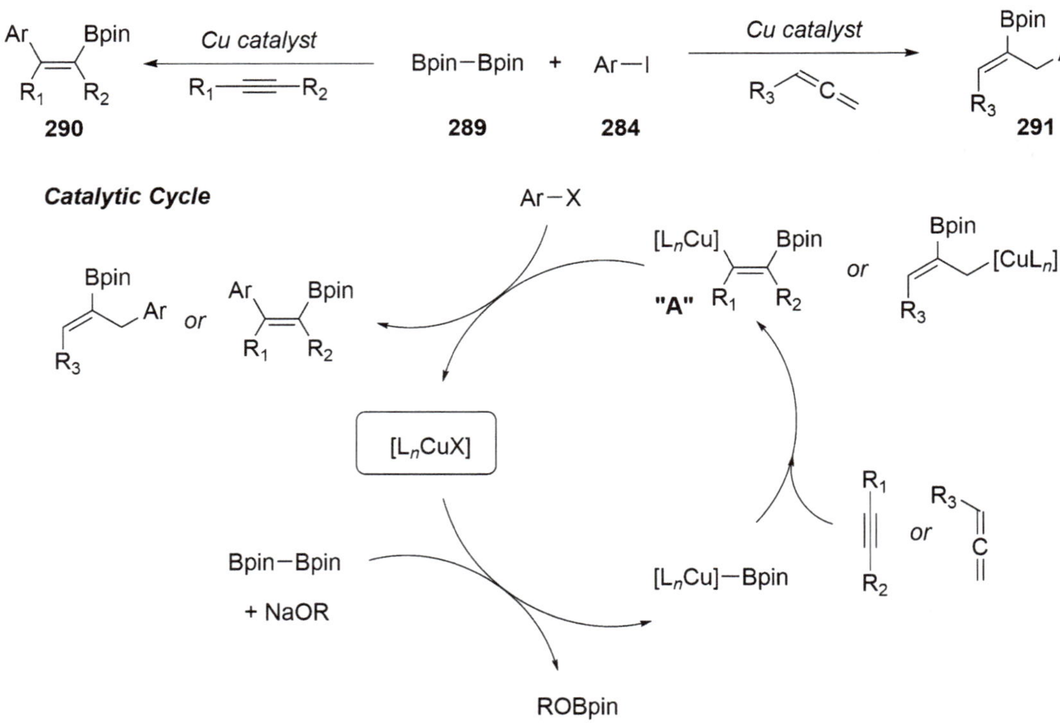

Fig. 90 Mechanistic proposal for the development of a three-component coupling

boronic esters (Bpin) were suitable substrates though led to lower yields. Both electron-donating and electron-withdrawing substituents were tolerated on either of the coupling partners, while

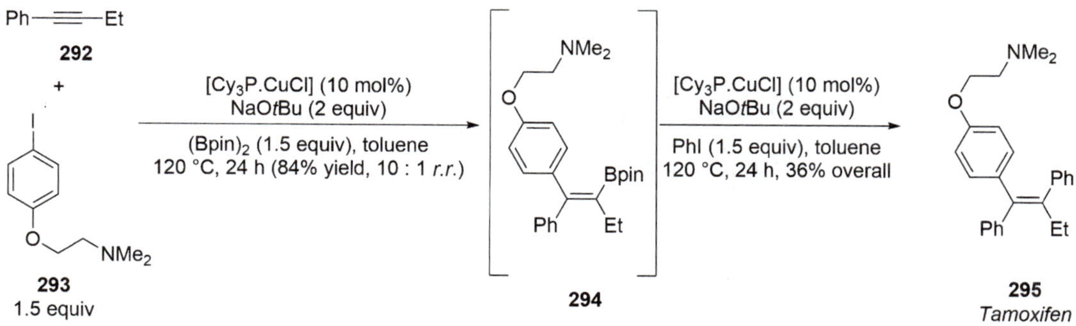

Fig. 91 Three-component coupling in the synthesis of tamoxifen **295**

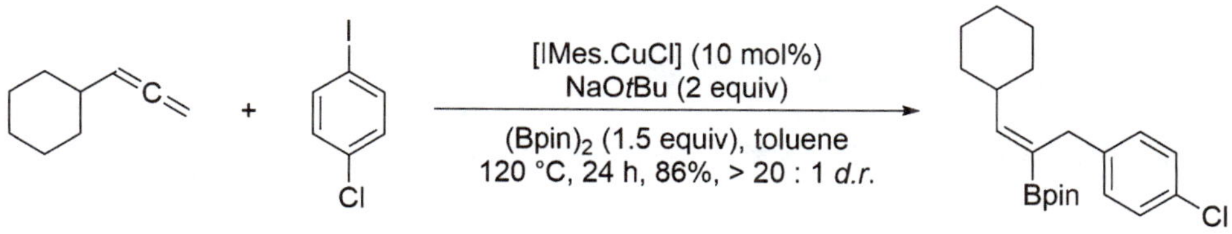

Fig. 92 Allenes in the copper-catalyzed three-component coupling reaction

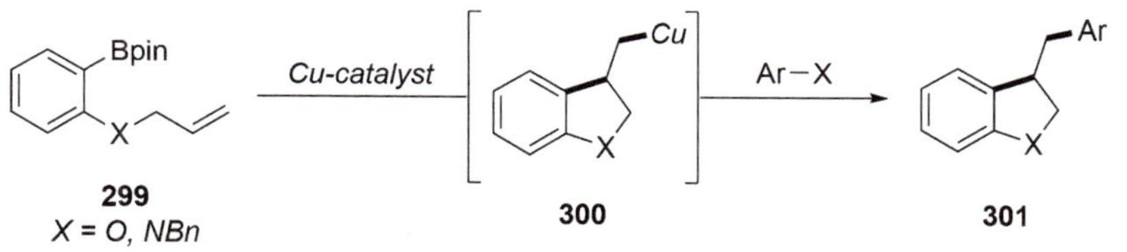

Fig. 93 Copper-mediated intramolecular diarylation of alkenes

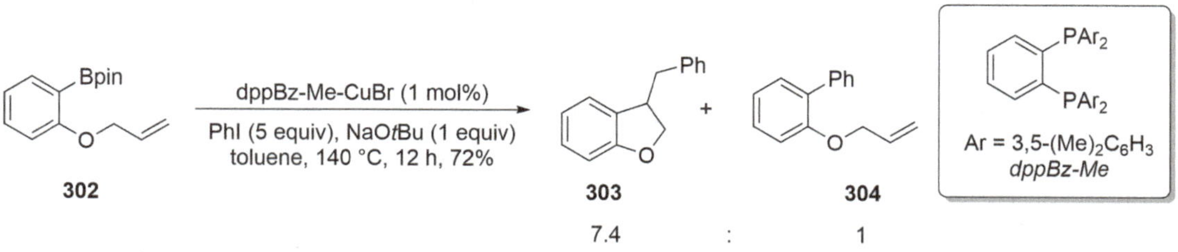

Fig. 94 Ligand screen in the intramolecular diarylation of alkenes

sterically hindered boronic esters also underwent the coupling process albeit under slightly modified conditions.

While initial mechanistic studies suggest that radical intermediates are not involved, the intermediacy of RCu(I) (R = aryl or vinyl) (**A**) species generated by a transmetalation process led to a hypothesis for the development of a new three-component reaction. Specifically, if the RCu(I) species (**A**) could be generated through

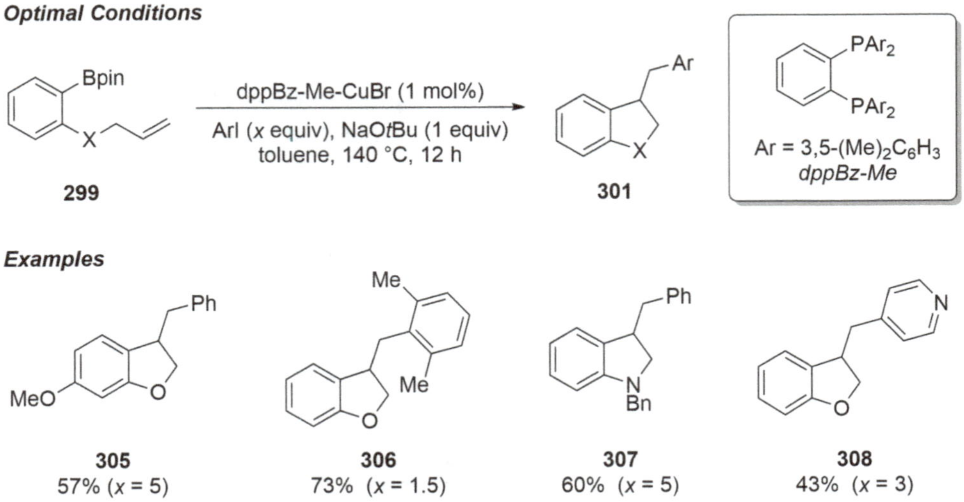

Fig. 95 Optimal conditions/scope of the intramolecular diarylation reaction

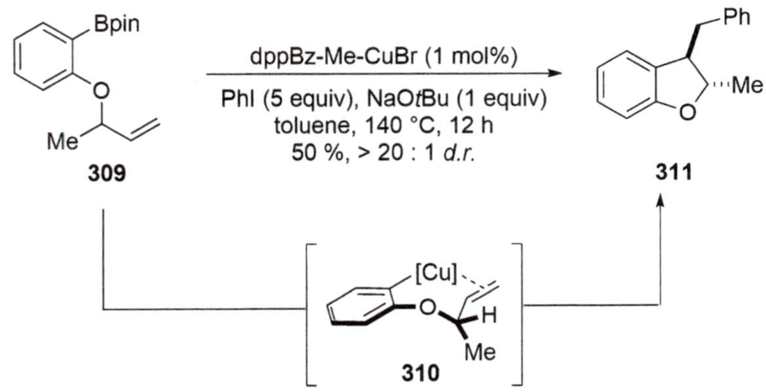

Fig. 96 Linker-substitution enables high diastereoselectivity to be achieved

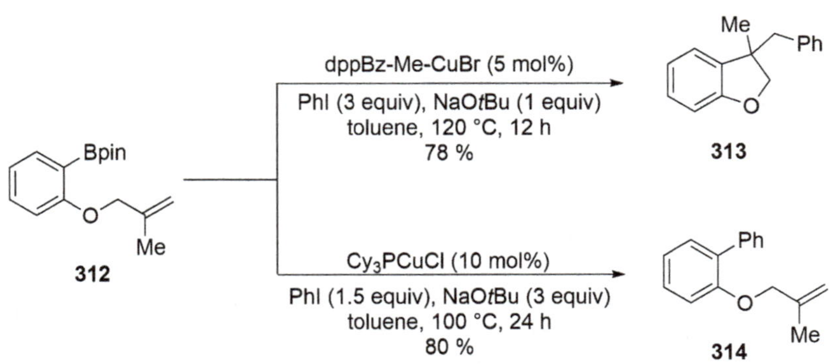

Fig. 97 Influence of the ligand on the reaction of 2-substituted alkene derivatives

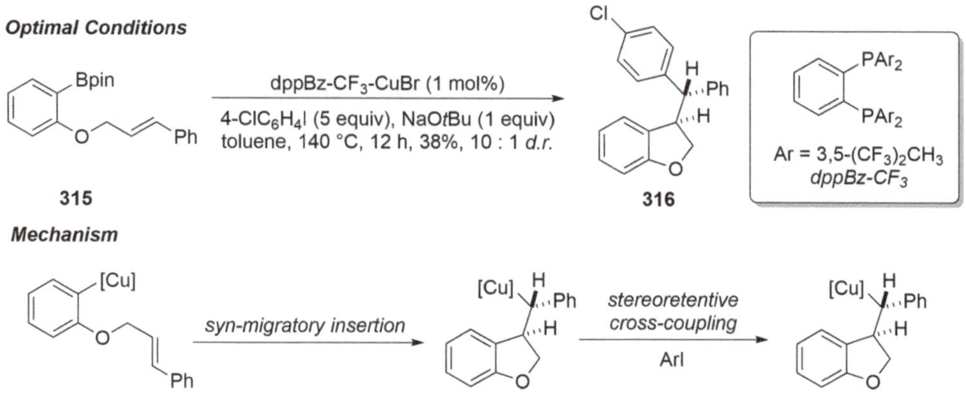

Fig. 98 Optimal conditions and proposed mechanism for the cyclization of styrene-based derivatives

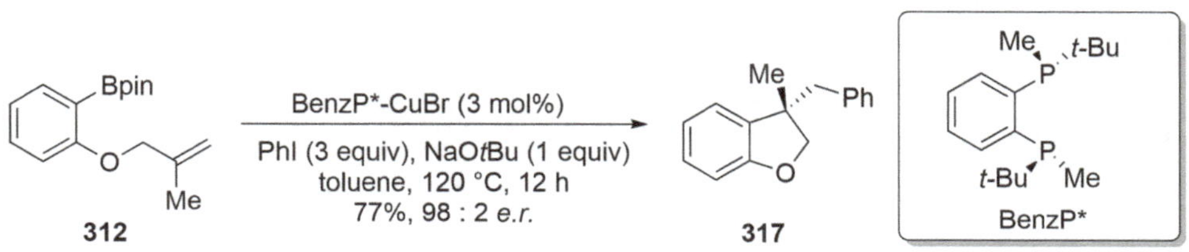

Fig. 99 Enantioselective induction through use of a chiral diphosphine ligand

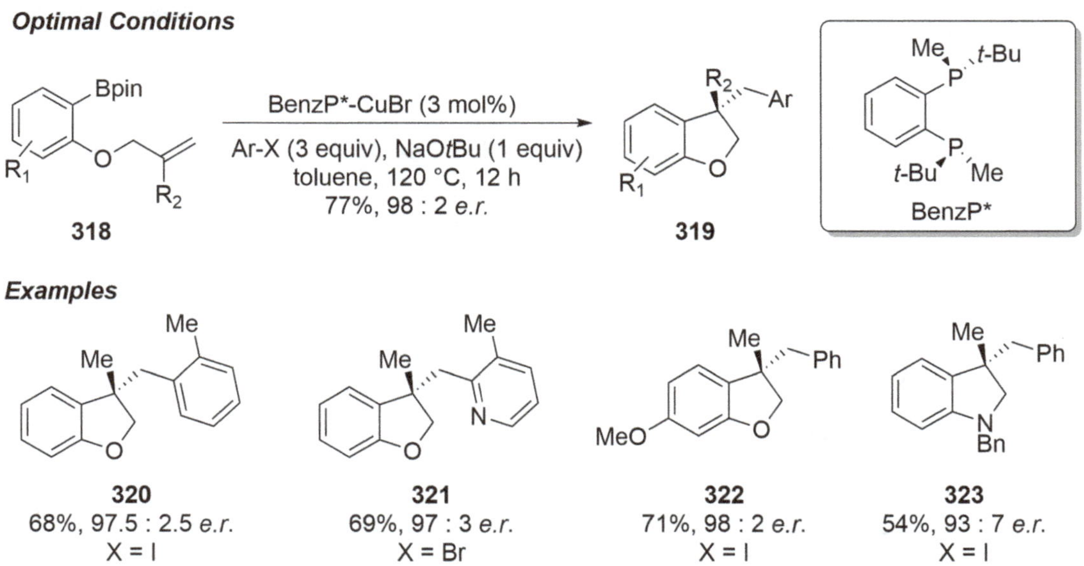

Fig. 100 Optimal conditions/scope of the enantioselective intramolecular diarylation process

migratory insertion across C-C π-bonds and then react with an aryl halide, then an overall carboboration procedure could be achieved (Fig. 90).

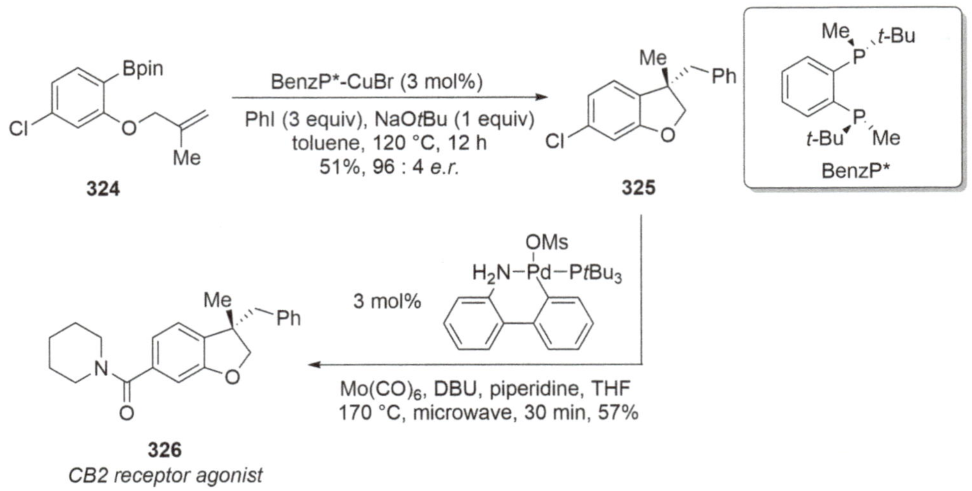

Fig. 101 Enantioselective synthesis of a CB2 receptor agonist **326**

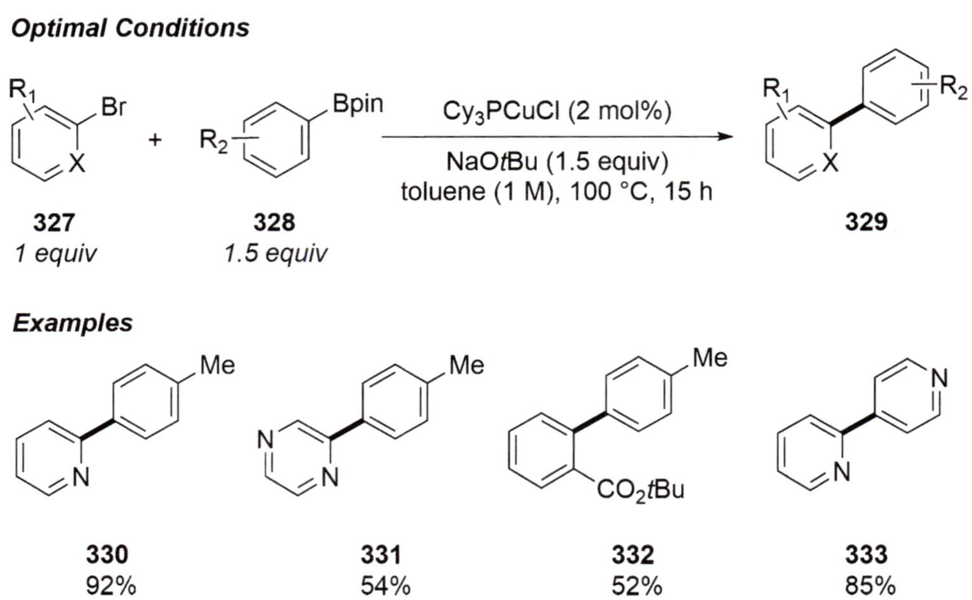

Fig. 102 Coupling of heteroaromatic and electron-deficient aryl bromides: Optimal conditions/scope

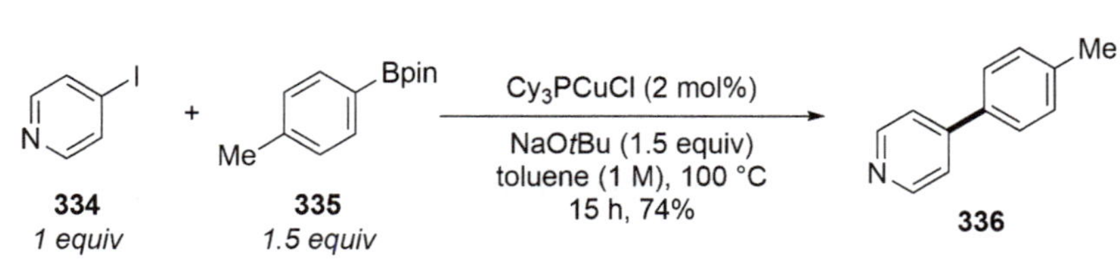

Fig. 103 Cy_3PCuCl-mediated cross-coupling of aryl iodides

Initial studies indicated that the vinyl-copper bond (the intermediacy of such a vinyl-copper species (**A**) is confirmed by

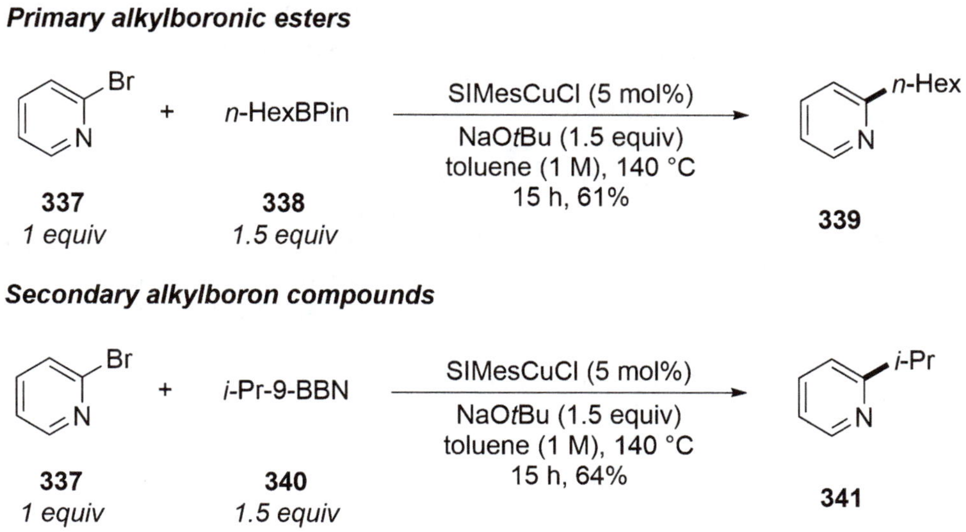

Fig. 104 Cross-coupling of primary and secondary alkylboron reagents with **337**

subsequent mechanistic studies) generated through the migratory insertion of a Cu/Bpin complex could be subsequently coupled with an aryl iodide leading to highly substituted alkene derivatives (**290**). The optimal catalyst was shown to be [Cy_3PCuCl] with the reaction proceeding under relatively mild conditions with a range of aryl iodides. Internal alkynes performed best with symmetrical derivatives giving the product as a single alkene isomer, while unsymmetrical derivatives gave high levels of regioselectivity in the case of aryl/alkyl substitution though with two aryl substituents, the observed regioselectivities were poor. Terminal alkynes reacted badly leading to the disubstituted vinyl borane as the major product due to protonation of the intermediate vinyl copper species. The overall key to the success of the reaction is due to the superior rate of the initial insertion to the alkyne compared to the reaction with the aryl iodide. The synthetic utility of the process is demonstrated through its application to an efficient synthesis of the estrogen receptor antagonist tamoxifen (**295**) (Fig. 91).

Switching to [IMesCuCl] as the catalyst enabled the carboboration of allenes (**296**) to be achieved leading to the formation of stereodefined vinyl boronic esters (**298**) with coupling of the iodide occurring at the least-substituted position (Fig. 92). This as well as the exclusive formation of the (Z)-alkene suggests that migratory insertion of pinB-CuIMes across the allene occurs from the least hindered π-face to generate the least-substituted allylcopper complex.

While the Cu-mediated reactions are successful for both alkynes and allenes, alkenes have proven more challenging to functionalize through an analogous migratory insertion/cross-coupling-type process. Successful application of this strategy would enable direct functionalization of both positions of the π-bond, and this allows the one-step conversion of readily available alkenes into significantly more complex structures.

The first successful reports to achieve this utilized a dual-metal-base system with a Pd catalyst required to facilitate reaction with the aryl halide. Brown and co-workers reported an extension of this approach with a solely copper-mediated diarylation of alkenes for the synthesis of both benzofuran- and indoline-based scaffolds [175] (Fig. 93). Many methods to develop such reactions have relied on the careful design of the substrates to either suppress or take advantage of β-hydride elimination pathways though in the studies of Brown et al., there is not an issue due to the intermediacy of C(sp^3)-Cu complexes.

Optimization studies on the easily prepared borane (**302**) to form the benzofuran framework revealed that the nature of the ligand was critical to the outcome of the reaction. While in the absence of or using a monodentate phosphine ligand led to the direct cross-coupling product (**304**), bidentate phosphine systems gave a predominance of the desired (~10:1) benzofuran product (**303**) (Fig. 94). Both steric bulk of the ligand and the molar excess employed by the Ar-I substrate were shown to be key factors in optimizing the ratio of desired cyclic product formed though it was shown to be important to balance these with respect to overall yield. Excess iodide could be recovered after the reaction, while surprisingly it was shown that yield increased with decreased catalyst loadings.

From a substrate perspective, a range of electron-deficient, electron-rich, and heteroaryl iodides were well tolerated, while variation of the tether demonstrated that both *O*- and *N*-linkages underwent the reaction, but attempts to form either carbocycles or six-membered rings failed (Fig. 95). Substitution on the linker was shown to be feasible with the cyclic products formed with high levels of diastereoselectivity presumably due to a pre-transition state formed with migratory insertion occurring from the less hindered face of the alkene.

Substitution on the linker was shown to be feasible with the cyclic products formed with high levels of diastereoselectivity presumably due to a pre-transition state (**310**) formed with migratory insertion occurring from the less hindered face of the alkene (Fig. 96).

Substitution of the olefin moiety was also investigated with the possibility for the formation of quaternary centers. Alkyl-substitution at the 2-position (**312**) further confirmed the critical influence of the ligand on the reaction pathway with direct cross-coupling (**313**) being obtained in the presence of Cy_3PCuCl, while the 3,3′-disubstituted benzofuran (**314**) was obtained in excellent yield using the bidentate phosphine-copper-based system (Fig. 97).

For 1,2-dialkyl-substituted alkenes, only cross-coupling was observed though for the analogous styrene-based systems, the desired cyclized products were formed with the level of diastereoselectivity achieved increasing with the steric bulk of the aryl iodide used (Fig. 98). The origin of the diastereoselectivity is believed not to be due to equilibration of an intermediate C(sp^3)-Cu complex to a thermodynamic minimum but arises instead from a syn-migratory insertion followed by a stereoretentive coupling with the aryl iodide. These results suggest that radical intermediates are not involved, while mechanistic studies utilizing deuterium labeling indicate that the reaction proceeds via carbometalation of an Ar-Cu complex across and alkene followed by cross-coupling with Ar-I.

A natural extension of this work would be to test the ability of chiral diphosphine ligands to induce enantioselectivity into the transformation [176]. Initial evaluation of DuPhos-based ligands showed that although these were capable of mediating the reaction, only modest levels of enantioselectivity were achieved and while increasing the steric bulk of the ligand system increased the ee, this was accompanied by a decrease in yield. Pleasingly use of both QuinoxP* and BenzP* ligands led to good yields and enantioselectivity with the latter being shown to be superior (Fig. 99). Interestingly these high levels of enantioselectivity could be maintained while lowering the catalyst loadings from 5 to 3 mol% and were achieved despite the high temperature required for reactivity.

From a scope perspective, a variety of aryl iodides were able to furnish the 2,3-dibenzofurans with uniformly good selectivities (Fig. 100). Notably both heterocyclic systems and sterically hindered aryl iodides also performed well while the reaction could also be extended to 2-bromopyridines. Both electron-rich and electron-poor substituents on the arylboronic ester were tolerated though the reaction failed for *ortho*-substituted derivatives.

Again indane- and six-membered ring products could not be formed, and the reaction was also unsuccessful with alkenes substituted at the terminal position. However sterically bulky alkyl substituents were tolerated on the 2-position of the alkene with very little deviation in the enantioselectivities with this hypothesized to be due to the key enantiodetermining event being the migratory insertion with the orientation of the CH_2O substituent toward the smaller methyl group of the catalyst in the pre-transition state driving the selectivity, while the alkyl substituent points away from the steric bulk of the catalyst. The protocol was applied to a concise synthesis of a CB2 receptor agonist (**326**) (Fig. 101).

Given the success of using heteroaromatic bromides (specifically 2-pyridyl) derivative in the migratory insertion/cross-coupling cascade to access benzofurans, Brown and co-workers initiated an investigation to see whether these substrates could serve as competent electrophiles in more general cross-coupling reactions with organoboron compounds [177]. Key within this study looking to expand the scope of the transformation was looking to identify a cheaper ligand alternative to Xantphos, reducing the equivalents of base and running the reaction at increased concentrations.

Model studies on the reaction between 2-bromopyridine and the Bpin ester of *p*-tolylboronic acid demonstrated that Cy_3PCuCl represented a cheaper alternative to XantphosCuCl to effectively mediate the coupling with loadings as low as 0.5 mol% shown to be adequate (though typically 2 mol% was used). From a concentration perspective, the reaction could be run at up to 1 M in toluene though at higher scale and concentrations proved difficult to stir. A range of boronate ester derivatives including cyclic boroxines were well tolerated though neither boronic acids nor potassium trifluoroborates were suitable nucleophiles. From a substrate perspective, electron-rich, electron-poor, heterocyclic, and sterically hindered boronic esters were tolerated though some issues with functional group tolerance were encountered with ketones and nitriles. While 2-bromopyridines, 2-bromoisoquinolines, and 2-bromopyrazines were all successfully coupled, 3- and 4-bromo-substituted systems failed to react under the current conditions. The reaction though could be extended to electron-deficient aryl bromides, while CuCl alone was shown to be an effective catalyst for the coupling of 2-bromopyridine though was inferior to Cy_3PCuCl and was ineffective for aryl bromides (Fig. 102).

Application of the newly developed system to the Xantphos-based conditions previously disclosed for the coupling of aryl iodides demonstrated that for all the substrate classes (electron-rich, electron-deficient, hindered, and heterocycles) moderate to good yields of the desired products could be generated (Fig. 103).

Alkyl-based boron reagents continued to present challenges with the various reports utilizing the more sensitive alkyl-9-BBN derivatives to realize cross-couplings. In the current study, it proved possible to successfully couple alkyl Bpin substrates (**338**) using SIMesCuCl as the catalyst at elevated temperatures with both 2-bromopyridines and aryl iodides. Application of these conditions for secondary organoboron derivatives did however require reverting to the use of alkyl-9-BBN (**340**) derivatives for successful couplings to be realized though notably no *n*-alkyl side products (potentially arising from β-hydride elimination/re-insertion pathways) were observed under these conditions though aryl iodides now failed to react (Fig. 104).

4.4 Copper-Catalyzed Suzuki-Miyaura/Hiyama Reactions

While several reports have emerged describing the Cu-catalyzed Suzuki-Miyaura coupling, these examples are typically limited to the coupling of non-heteroaryl boronic acids with non-heteroaryl iodides [119]. Furthermore, the attractive nature of using earth-abundant copper sources to mediate this transformation is

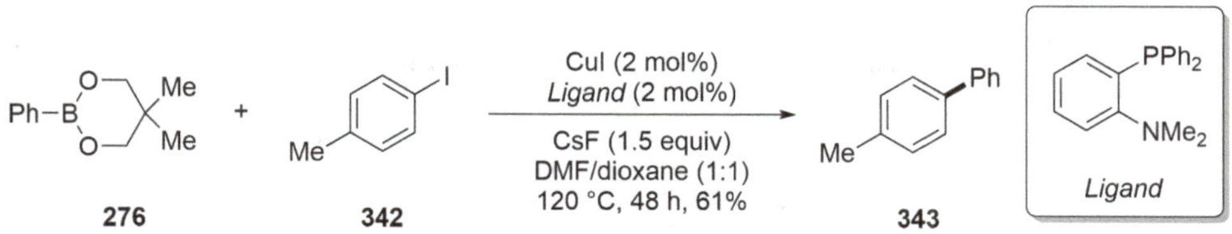

Fig. 105 Initial conditions for the Suzuki-Miyaura reaction

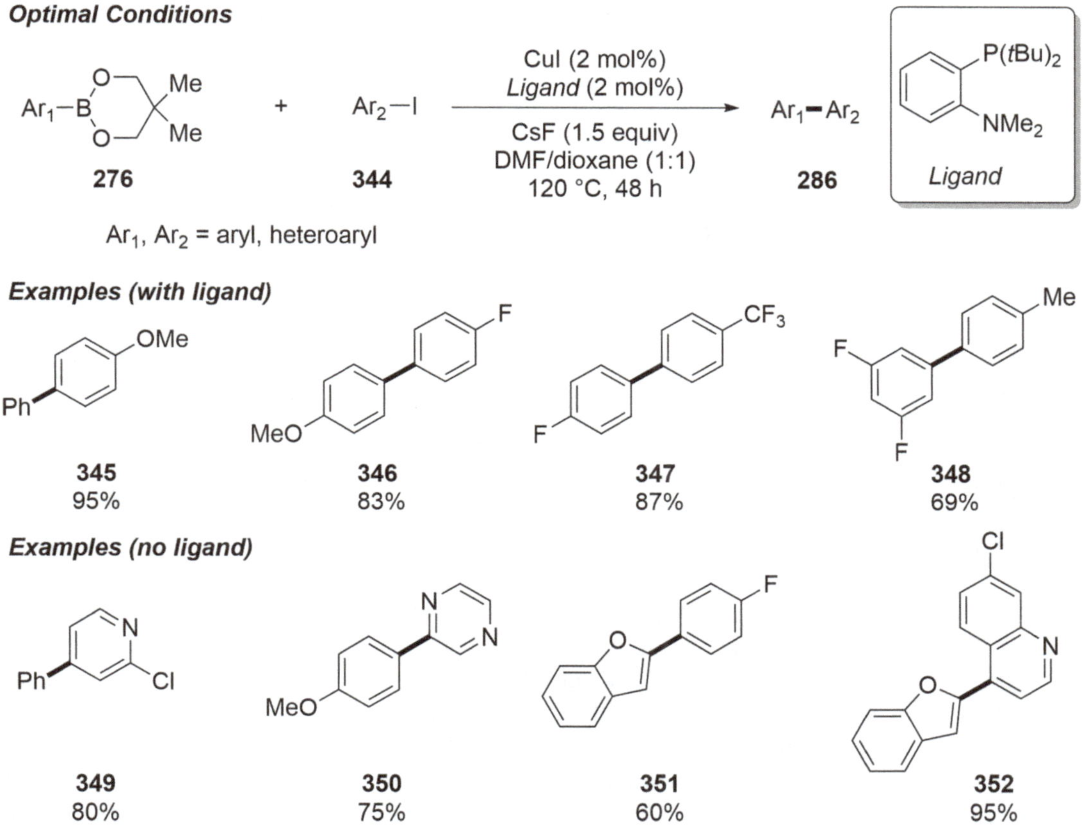

Fig. 106 Optimal conditions/scope of the Giri conditions

somewhat offset by low catalytic turnovers often requiring high loadings of both catalysts and ligands (10–20 mol%), and even stoichiometric amounts in some cases, as well as limited mechanistic insight into these processes thus hindering hypotheses for better catalyst design.

Giri and co-workers have reported on the development of a copper-catalyzed Suzuki-Miyaura coupling, which overcomes many of these limitations in terms of substrate scope and proceeds effectively with low catalyst loadings (2 mol%) [178]. In addition, detailed mechanistic studies including the isolation and characterization of several discrete reaction intermediates were carried out enabling the delineation of a possible catalytic cycle.

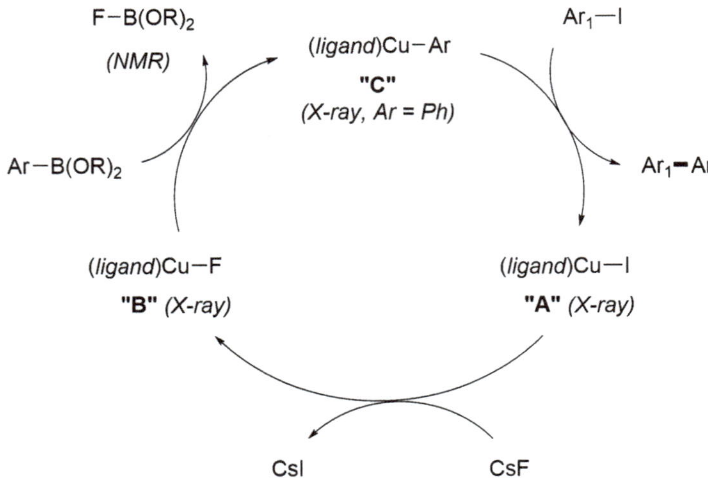

Fig. 107 Mechanistic studies of intermediates in the Cu-mediated Suzuki coupling

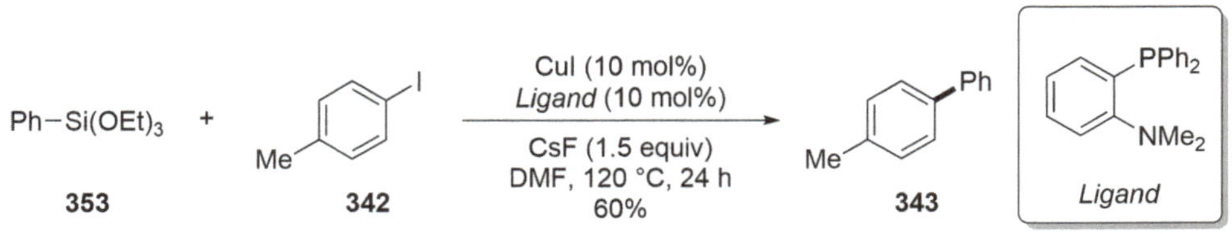

Fig. 108 Copper-mediated Hiyama coupling

Previous work on the copper-mediated Hiyama coupling (*vide infra*) revealed that P,N-based ligands (specifically *o*-(diphenylphosphino)-*N,N*-dimethylaniline) were effective in catalyzing the reactions of $ArSi(OEt)_3$ with aryl iodides in the presence of cesium fluoride (CsF). Direct application of these conditions to the Suzuki-Miyaura coupling of phenylboronic acid neopentyl glycol ester (**276**) with iodotoluene (**342**) leads to the formation of the desired product in 61% yield (Fig. 105).

Optimization studies demonstrated that use of more electron-rich and sterically hindered P,N-based ligands led up to 95% yield being realized with CsF shown to be by far the most effective base for the transformation. CuI was chosen as the copper source though Cu(Ot-Bu) could also be utilized, and while a range of phenyl boron-based reagents could be employed interchangeably (including the parent boronic acid), notably the corresponding MIDA esters failed to react.

The optimal conditions were applied to a series of reactions between non-heterocyclic aryl boronate esters and non-heterocyclic aryl iodides and provided products in good to excellent yields with both EDGs and EWGs being tolerated on either/both coupling partners (Fig. 106). In addition, the

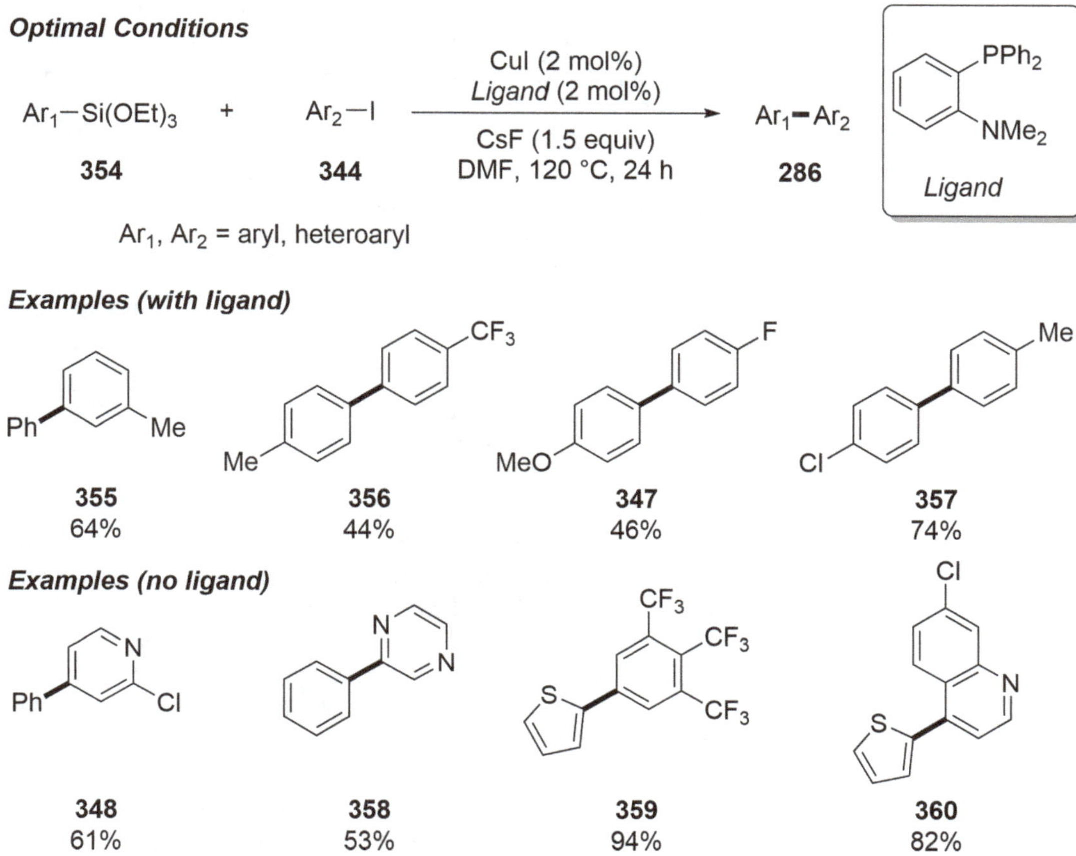

Fig. 109 Optimal conditions/scope of the copper-catalyzed Hiyama coupling

reactions were highly chemoselective for aryl iodides with other halogens being well tolerated providing a functional handle for further product elaboration. Similar functional-group compatibilities were observed for aryl-heteroaryl and heteroaryl-heteoaryl-based couplings though in these cases it was shown that high yields could be obtained even in the absence of the ligand.

Mechanistic studies were carried out involving the synthesis and characterization of a series of ligated Cu complexes, stoichiometric and catalytic reactions, as well as extensive ^{1}H, ^{19}F, ^{11}B, and ^{31}P NMR studies. These led to a proposed catalytic cycle, which features CsF-mediated conversion of a ligated Cu-I complex (**A**) to the corresponding Cu-F-based species (**B**) that undergoes transmetalation with the aryl boronate ester to give **C** prior to coupling with the aryl iodide. Although this is supported by both X-ray and NMR evidence, an alternative scenario featuring a fluoride activation of the boronate ester prior to transmetalation cannot be excluded (Fig. 107).

Despite their low toxicity and ease of both synthesis and handling, organosilicon compounds are far less frequently used in metal-mediated coupling process than the corresponding organoboron reagents, and when employed the so-called Hiyama coupling is typically conducted with Pd catalysts either with or without a fluoride source. While Cu salts have been shown to improve the yield of Pd-based Hiyama couplings, and an isolated example of a Cu salt [CuOC6F6]-promoting reactions between a limited scope of aryl silicon reagents and aryl iodides has been reported, investigations into a Cu-catalyzed Hiyama coupling have been limited.

Giri and co-workers investigated the copper-mediated coupling of arylsilicon reagents with *p*-iodotoluene (**342**) and found that when $PhSi(OEt)_3$ (**353**) was used as the coupling partner, a synthetically useful yield (60%) of the desired biphenyl-based product could be obtained using CuI as the catalyst, a P,N-based bidentate ligand in the presence of CsF [179] (Fig. 108). Interestingly, acceptable levels of the product were still obtained either with structurally similar P,N-based ligands (45–50%) or even in the absence of the ligand (40%). DMF was shown to be the best solvent for the transformation, and of the aryl halides evaluated, only the iodide underwent reaction. While other copper sources were effective in the reaction, replacement of CsF with reagents such as LiF, NaF, or TBAF led to only trace yields of the product. Similarly switching from the aryltrialkoxysilanes to the corresponding aryltrialkylsilanes gave no reaction.

Investigation of the scope of the reaction demonstrated an interesting phenomenon with regard to whether the P,N-based ligand was included or not. Whereas in the case of aryl-aryl couplings when use of the ligand increased the yields by ca. 20–54% relative to it being omitted, both aryl-heteroaryl and heteroaryl-heteroaryl couplings showed an opposite effect with ligand inclusion leading to 14–34% decreases in yield (Fig. 109). A hypothesis for this observation is provided in that in cases featuring the heteroaryl-based substrates, the substrate present in excess acts as secondary ligands binding to the Cu, which is already ligated to the P,N ligand thus inhibiting the necessary approach of the C-I bond of the heteroaryl iodide required for reaction. In all cases, a range of functional groups are well tolerated on both reaction partners including both chlorides and bromides, while varying the electronics of either (hetero)aryl ring had little impact on the reaction efficiency or yield.

Mechanistic studies involving studying the reactivity of an independently synthesized [CuPh] species with iodotoluene demonstrated the key roles of CsF in not only facilitating the transmetalation of the organosilicon reagent with CuI but also in breaking up poorly reactive Cu-based aggregates while the ligand was shown to reduce homocoupling. Radical probe experiments suggest that the reaction does not involve aryl radical intermediates.

5 Transition-Metal-Free Cross-Couplings

The focus of this chapter and indeed much of current cross-coupling research has centered on the study of C-C and C-N/O bond forming reactions employing cheap and abundant metals, e.g., Ni, and Fe, Cu. An undercurrent to the development of these coupling methodologies can be seen in recent research aimed at developing transition metal-free approaches to these valuable disconnections. This emerging area has recently been reviewed [180]. The reader is directed to this review and the references provided within for an extensive overview. The Barluenga coupling, however, is of particular importance to the practicing medicinal chemist due to the ubiquitous nature of commercially available boronic acids/esters and their inherent stability [181–184]. Therefore, the latest developments and potential pharmaceutical applications will be presented here.

Classical methods for the formation of C-C bonds via the union of stabilized α-diazo carbonyl compounds and organoboranes were first demonstrated in pioneering works by Hooz [185] and Brown [186] (Fig. 110a). Related transformations exploiting tosylhydrazones and alkyl lithium reagents (**365**) as coupling partners were developed by Vedejs [187]. This approach was later modified to incorporate silylated tosylhydrazones (**364**) as reported by Myers [188] which ultimately lead to broader substrate scope (Fig. 110b). Contemporaneous studies by Wang et al. demonstrated the use of aryl and vinyl boroxines as competent coupling partners for α-diazo carbonyl compounds [189]. Later in 2009, the Barluenga group reported on the cross-coupling of readily available tosylhydrazones (**367**) and boronic acids (**368**) whereby the reactive diazo intermediates are generated under the reaction conditions employing potassium carbonate as a basic additive [190] (Fig. 110c).

The optimized Barluenga coupling conditions tolerate the use of unhindered aryl, alkenyl, and aliphatic boronic acids delivering the coupled products in good to excellent yields. Additionally, a one-pot protocol, in which the requisite tosylhydrazone is generated in situ, was shown to deliver high yields of cross-coupled products thus setting the stage for broad adoption of this methodology in a parallel medicinal chemistry setting (Fig. 111a). A detailed study of the substituent effects when employing alkenyl boronic acids has also been reported [191]. Lastly, studies demonstrating the in situ generation of diazo species directly via diazotization with sodium nitrite have provided unique approaches to metal-free trifluoroethylation (**373**) [192, 193] and α-arylation of α-amino esters/nitriles (**376**) [194] (Fig. 111b, c).

A mechanistic understanding of this process can be gleaned from early studies by Bamford and Stevens on the thermal decomposition of sulfonylhydrazones to diazomethanes in basic media

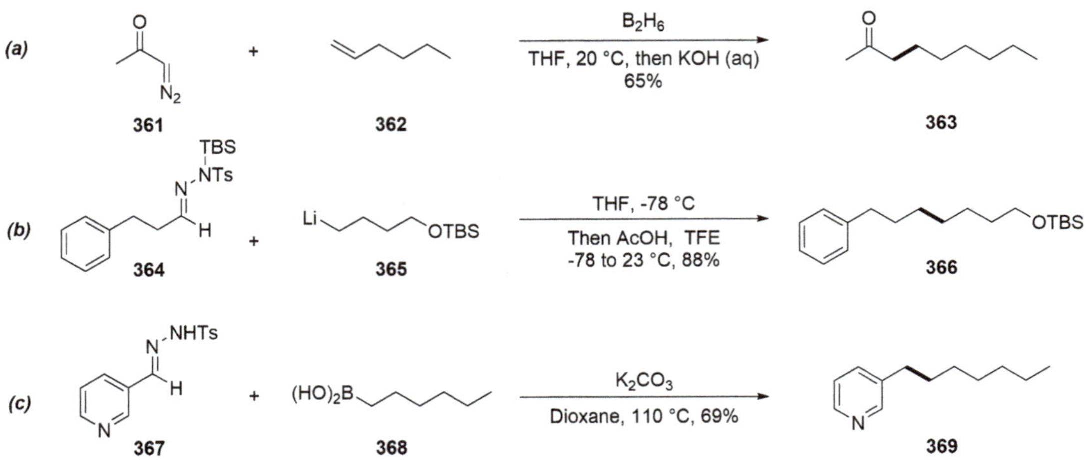

Fig. 110 Transition-metal-free cross-coupling of diazo-based compounds

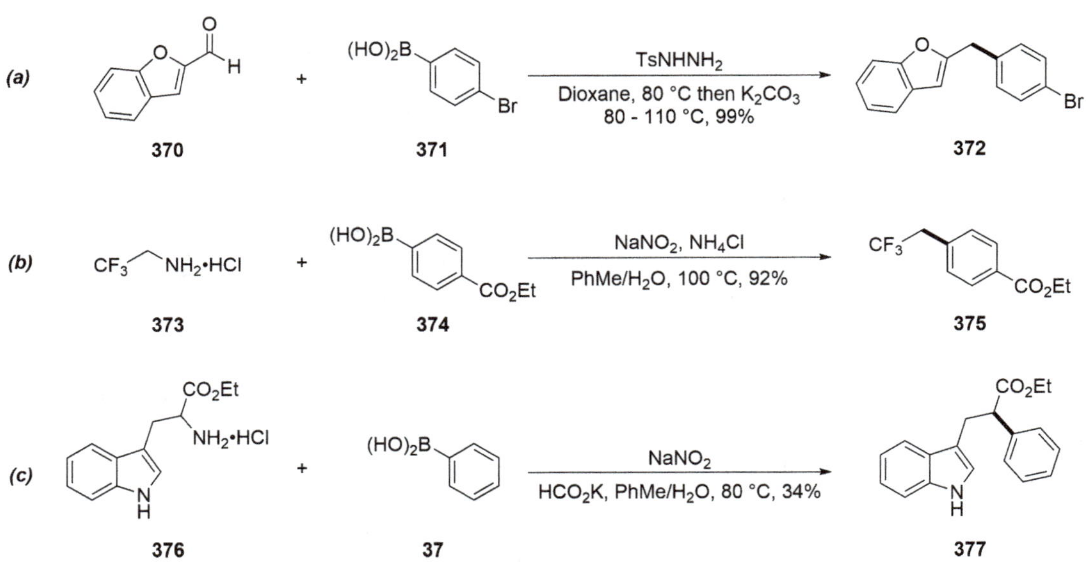

Fig. 111 In situ formation of diazo compounds for cross-coupling

[195]. Once the diazo species has been generated, two pathways have been proposed for the fate of this highly reactive species (Fig. 112). Nucleophilic addition to the aryl boronic acid could be envisaged to provide a diazoboronate complex **A** (*Associative Pathway*). Then 1,2-migration of the aryl group with concomitant extrusion of gaseous nitrogen would deliver a tertiary boronate species **B** with *in situ* protodeborylation furnishing the observed coupling product. Alternatively, α-elimination of nitrogen could give rise to a singlet carbene **C** (*Dissociative Pathway*). Complexation of the aryl boronic acid to give **D** followed by a 1,2-shift (to give **B**) and subsequent deborylation is also reasonable. Mechanistic details have been difficult to obtain due to the high reaction

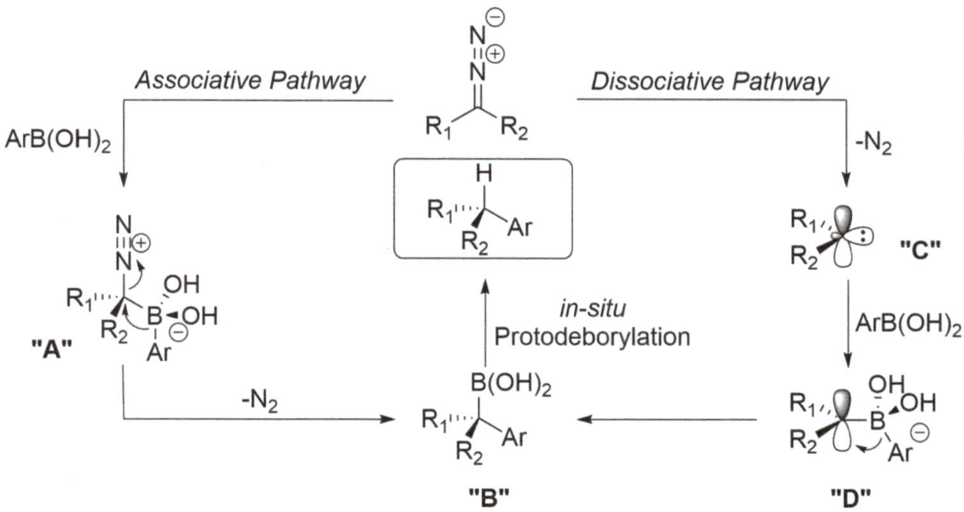

Fig. 112 Possible mechanistic pathways for the cross-coupling of diazo compounds

temperatures and transient nature of the reactive species involved in this process.

The simplicity of the Barluenga coupling and the commercial availability of the coupling partners make this process particularly attractive for library synthesis. In fact, Nakagawa and co-workers at Pfizer set out to investigate this coupling for the preparation of drug-like/drug fragment-like molecules containing hydrogen bond donors/acceptors with high polar surface area in a parallel synthesis format [196] (Fig. 113a). Here they showed successful examples employing sulfonylhydrazones containing azetidine, imidazole, and pyrrolopyridine functional group arrays. In the same vein, the direct coupling of heterocyclic motifs such as pyrrolidines, piperidines, and oxetanes has been demonstrated by the Ley group [197] (Fig. 113b). These efforts have greatly expanded the chemical space accessible using the Barluenga coupling methodology.

As previously stated, one of the proposed mechanistic pathways for the Barluenga coupling invokes the intermediacy of highly reactive carbene species. Naturally, this hypothesis has garnered interest from the synthetic community toward further development of this chemistry in flow. Kirschning and Kupracz have reported a two-step flow synthesis wherein tosylhydrazones are formed in the first flow step and subsequently coupled to boronic acids in the second reactor [198]. Most recently, the Ley group has developed an extremely mild protocol whereby a solution of hydrazone can be oxidized by MnO_2 to the reactive diazo species under flow conditions and coupled directly to boronic acids via translocation at room temperature [199].

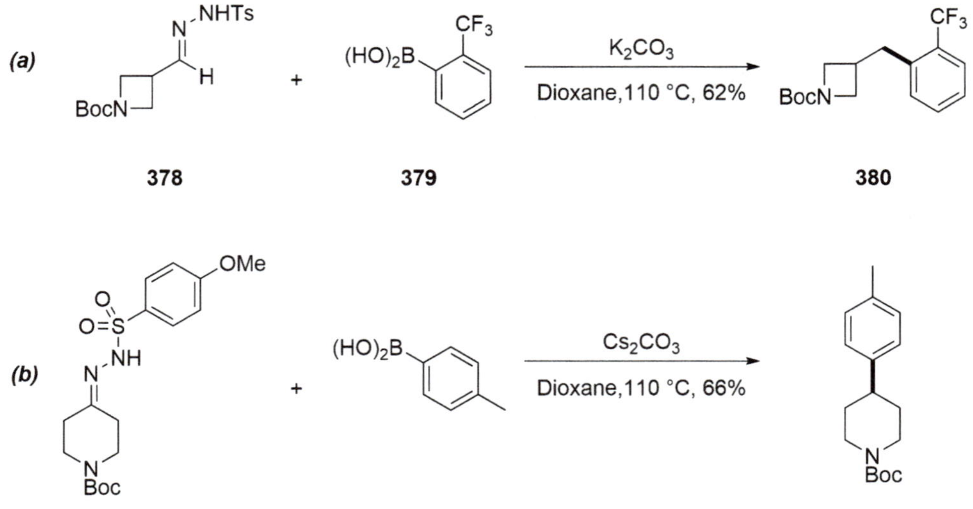

Fig. 113 Synthetic applications of the Barluenga coupling

5.1 General Procedure

A reaction tube was charged with potassium carbonate (1.5 mmol), dioxane (4 mL), boronic acid (1.5 mmol), and tosylhydrazone (1.0 mmol). The reaction was refluxed at 110 °C with stirring. The reaction was monitored by GCMS. When the reaction was complete, the crude mixture was allowed to reach room temperature and the solvent was removed. Dichloromethane and a saturated solution of $NaHCO_3$ were added and the layers were separated. The aqueous phase was extracted three times with dichloromethane. The combined organic layers were washed with two portions of a saturated solution of $NaHCO_3$ and one portion of brine and then dried over $MgSO_4$ and filtered. The solvent was removed under reduced pressure. Where necessary, products were purified by chromatography on silica gel or alumina.

6 Conclusions

Although not comprehensive, the intention of the present chapter has been to illustrate not only examples of the uses of base metals such as Ni, Fe, and Cu to mediate carbon-carbon bond formations but also the challenges that still exist for this methodology to acquire the same widespread utility that palladium-mediated cross-couplings occupy within the pharmaceutical industry today. Pivotal to achieving this aim is to understand and exploit the different mechanistic paradigms involved in these transformations. While palladium-based C-C coupling reactions typically proceed through a three-step Pd(0)/Pd(II) cycle (oxidative addition, transmetallation, and reductive elimination), nickel (0) complexes prefer to react through one-electron redox steps. With regard to ligands, palladium chemistry is dominated by bulky phosphine and N-heterocyclic carbene ligands for the Pd(II)-based precatalysts, while the analogous Ni(II) systems require stronger reductants to be converted to low-valent nickel, thus reducing the substrate scope of many relevant processes.

While base metals represent attractive alternatives to the use of platinum group metals for catalysis owing to lower cost, reduced toxicity, enhanced reactivity, and higher natural abundancies, key to their successful use will be a broader understanding and application of the fundamental differences in electronic structure and bonding most notably the largely varying redox potentials, free reaction energies for both homolytic and heterolytic M-L bond dissociation and the relative stabilities, and accessibility of ground and excited spin states, respectively. The examples presented herein already demonstrate the promise of this methodology and better understandings of the reaction manifolds coupled with innovative ligand design and/or coupling with emerging technologies such as the renaissance of electrochemistry should lead to the development of a plethora of new industrially applicable methodologies in the not too distant future.

References

1. Johansson Seechurn CCC, Kitching MO, Colacot TJ, Snieckus V (2012) Palladium-catalyzed cross-coupling: a contextual historical perspective to the 2010 Nobel prize. Angew Chem Int Ed 51:5062–5085
2. Cooper TWJ, Campbell IB, Macdonald SJF (2010) Factors determining the selection of organic reactions by medicinal chemists and the use of these reactions in arrays (small focused libraries). Angew Chem Int Ed 49:8082–8091
3. Roughley SD, Jordan AM (2011) The medicinal chemist's toolbox: an analysis of reactions used in the pursuit of drug candidates. J Med Chem 54:3451–3479
4. Magano J, Dunetz JR (2011) Large-scale applications of transition metal-catalyzed couplings for the synthesis of pharmaceuticals. Chem Rev 111:2177–2250
5. Bryan MC, Dunn PJ, Entwistle D, Gallou F, Koenig SG, Hayler JD, Hickey MR, Hughes S, Kopach ME, Moine G, Richardson P, Roschangar F, Steven A, Weiberth FJ (2018) Key Green Chemistry research areas from a pharmaceutical manufacturers' perspective revisited. Green Chem 20:5082–5103
6. Wilke G (1988) Contributions to organo-nickel chemistry. Angew Chem Int Ed 27:185–206
7. Jana R, Pathak TP, Sigman MS (2011) Advances in transition metal (Pd,Fe,Ni)-catalyzed cross-coupling reactions using alkyl-organometallics as reaction partners. Chem Rev 111:1417–1492
8. Tasker SZ, Standley EA, Jamison TF (2014) Recent advances in homogeneous nickel catalysis. Nature 509:200–309
9. Ge S, Hartwig JF (2012) Highly reactive, single-component nickel catalyst precursor for Suzuki-Miyaura cross-coupling of heteroaryl boronic acids with heteroaryl halides. Angew Chem Int Ed 51:12837–12841
10. Negishi E-I, Baba S (1976) Novel stereoselective alkenyl–aryl coupling *via* nickel-catalysed reaction of alkenylanes with aryl halides. J Chem Soc Chem Commun:596–597
11. Kocienski P, Dixon NJ (1989) Stereoselective synthesis of homoallylic alcohols by migratory insertion reactions of higher-order cyanocuprates and nickel-catalysed coupling reactions involving enol carbamates. Synlett:52–54
12. Dankwardt JW (2004) Nickel-catalyzed cross-coupling of aryl Grignard reagents with aromatic alkyl ethers: an efficient synthesis of unsymmetrical biaryls. Angew Chem Int Ed 43:2428–2432

13. Lin B-L, Liu L, Fu Y, Lu S-W, Chen Q, Guo Q-X (2004) Comparing nickel- and palladium-catalyzed Heck reactions. Organometallics 23:2114–2123
14. Rudolph A, Lautens M (2009) Secondary alkyl halides in transition-metal-catalyzed cross-coupling reactions. Angew Chem Int Ed 48:2656–2670
15. Ishiyama T, Abe S, Miyaura N, Suzuki A (1992) Palladium-catalyzed alkyl-alkyl cross-coupling reaction of 9-alkyl-9-BBN derivatives with iodoalkanes possessing beta-hydrogens. Chem Lett:691–694
16. Zhao J, Fu GC (2003) Cross-couplings of unactivated secondary alkyl halides: room-temperature nickel-catalyzed Negishi reactions of alkyl bromides and iodides. J Am Chem Soc 125:14726–14727. and references cited therein
17. Zhao J, Fu GC (2004) Suzuki cross-couplings of unactivated secondary alkyl bromides and iodides. J Am Chem Soc 126:1340–1341
18. Powell DA, Fu GC (2004) Nickel-catalyzed cross-couplings of organosilicon reagents with unactivated secondary alkyl bromides. J Am Chem Soc 126:7788–7789
19. Powell DA, Maki T, Fu GC (2005) Stille cross-couplings of unactivated secondary alkyl halides using monoorganotin reagents. J Am Chem Soc 127:510–511
20. Owston N, Fu GC (2010) Asymmetric alkyl—alkyl cross-couplings of unactivated secondary alkyl electrophiles: stereoconvergent Suzuki reactions of racemic acylated halohydrins. J Am Chem Soc 132:11908–11909
21. Lu Z, Wilsily A, Fu GC (2011) Stereoconvergent amine-directed alkyl–alkyl Suzuki reactions of unactivated secondary alkyl chlorides. J Am Chem Soc 133:8154–8157
22. Zultanski SL, Fu GC (2011) Catalytic asymmetric gamma-alkylation of carbonyl compounds via stereoconvergent Suzuki cross-couplings. J Am Chem Soc 132:15362–15364
23. Wilsily A, Tramutola F, Owston NA, Fu GC (2012) New directing groups for metal-catalyzed asymmetric carbon—carbon bond-forming processes: stereoconvergent alkyl-alkyl Suzuki cross-coupling of unactivated electrophiles. J Am Chem Soc 134:5794–5797
24. Choi J, Fu GC (2012) Catalytic asymmetric synthesis of secondary nitriles via stereoconvergent Negishi arylations and alkenylations of racemic alpha-bromonitriles. J Am Chem Soc 134:9102–9105
25. Zultanski SL, Fu GC (2013) Nickel-catalyzed carbon—carbon bond-forming reactions of unactivated tertiary alkyl halides: Suzuki arylations. J Am Chem Soc 135:624–627
26. Lamoureux G, Artavia G (2010) Use of adamantane structure in medicinal chemistry. Curr Med Chem 17:2967–2978
27. Cornella J, Zarate C, Martin R (2014) Metal-catalyzed activation of ethers via C-O bond cleavage: a new strategy for molecular diversity. Chem Soc Rev 43:8081–8097
28. Snieckus V (1990) Directed ortho metalation. Tertiary amide and O-carbamate directors in synthetic strategies for polysubstituted aromatics. Chem Rev 90:879–993
29. Macklin TK, Snieckus V (2005) Directed ortho metalation methodology. The N, N-dialkyl aryl O-sulfamate as a new directed metalation group and cross-coupling partner for Grignard reagents. Org Lett 7:2519–2522
30. Rosen BM, Quasdorf KW, Wilson DA, Zhang N, Resmerita A-M, Garg NK, Percec V (2011) Nickel-catalyzed cross-couplings involving carbon-oxygen bonds. Chem Rev 111:1346–1416
31. Hie L, Ramgren SD, Mesganaw T, Garg NK (2012) Nickel-catalyzed amination of aryl sulfamates using an air-stable precatalyst. Org Lett 14:4182–4185
32. Shumasaki T, Tobisu M, Chatani N (2010) Nickel-catalyzed amination of aryl pivalates by the cleavage of aryl C-O bonds. Angew Chem Int Ed 49:2929–2932. See this reference and references cited therein
33. Mamoru T, Takahira T, Ohtsuki A, Chatani N (2015) Nickel-catalyzed alkynylation of anisoles via C-O bond cleavage. Org Lett 17:680–683
34. Tobisu M, Yamakawa K, Shimasaki T, Chatani N (2011) Nickel-catalyzed reductive cleavage of aryl-oxygen bonds in alkoxy- and pivaloxyarenes using hydrosilanes as a mild reducing agent. Chem Commun 47:2946–2948
35. Álvarez-Bercedo P, Martin R (2010) Ni-catalyzed reduction of inert C-O bonds: a new strategy for using aryl ethers as easily removable directing groups. J Am Chem Soc 132:17352–17353
36. Tobisu M, Yasutome A, Kinuta H, Nakamura K, Chatani N (2014) 1,3-Dicyclohexylimidazol-2-ylidene as a superior ligand for the nickel-catalyzed cross-couplings of aryl and benzyl methyl esters with organoboron reagents. Org Lett 16:5572–5575
37. Miyaura N, Yanagi T, Suzuki A (1981) The Palladium-catalyzed cross-coupling of

phenylboronic acid with haloarenes in the presence of bases. Syn Commun 11:513–519

38. Percec V, Bae J-Y, Zhao M, Hill DH (1995) Aryl mesylates in metal-catalyzed homocoupling and cross-coupling reactions. 1. Functional symmetrical biaryls from phenols via nickel-catalyzed homocoupling of their mesylates. J Org Chem 60:176–185
39. Saito S, Sakai M, Miyaura N (1996) A synthesis of biaryls *via* nickel(0)-catalyzed cross-coupling reaction of chloroarenes with phenylboronic acids. Tetrahedron Lett 37:2993–2996
40. Yang J, Liu S, Zheng JF, Zhou J (2012) Room-temperature Suzuki–Miyaura coupling of heteroaryl chlorides and tosylates. Eur J Org Chem:6248–6259
41. Düfert NA, Billingsley KL, Buchwald SL (2013) Suzuki-Miyaura cross-coupling of unprotected, nitrogen-rich heterocycles: substrate scope and mechanistic investigation. J Am Chem Soc 135:12877–12885
42. Shields JD, Gray EE, Doyle AG (2015) A modular air-stable nickel precatalyst. Org Lett 17:2166–2169
43. Magano J, Monfette S (2015) Development of an air-stable, broadly applicable nickel source for nickel-catalyzed cross-coupling. ACS Catal 5:3120–3123
44. Tian Q, Cheng Z, Yajima HM, Savage SJ, Green KL, Humphries T, Reynolds ME, Babu S, Gosselin F, Askin D, Kurimoto I, Hirata N, Iwasaki M, Shimasaki Y, Miki T (2013) A practical synthesis of a PI3K inhibitor under noncryogenic conditions via functionalization of a lithium triarylmagnesiate intermediate. Org Process Res Dev 17:97–107
45. Ramgren SD, Hie L, Ye Y, Garg NK (2013) Nickel-catalyzed Suzuki-Miyaura couplings in green solvents. Org Lett 15:3950–3953
46. Nathel NFF, Kim J, Hie L, Jiang X, Garg NK (2014) Nickel-catalyzed amination of aryl chlorides and sulfamates in 2-methyl THF. ACS Catal 4:3289–3293
47. Amii H, Uneyama K (2009) C-F bond activation in organic synthesis. Chem Rev 109:2119–2183
48. Tobisu M, Xu T, Shimasaki T, Chatani N (2011) Nickel-catalyzed Suzuki-Miyaura reaction of aryl fluorides. J Am Chem Soc 133:19505–19511. and references cited therein
49. Everson DA, Shrestha R, Weix DJ (2010) Nickel-catalyzed reductive cross-coupling of aryl halides with alkyl halides. J Am Chem Soc 132:920–921
50. Krasovskiy A, Duplais C, Lipshutz BH (2009) Zn-mediated, Pd-catalyzed cross-couplings in water at room temperature without prior formation of organozinc reagents. J Am Chem Soc 131:15592–15593
51. Everson DA, Jones BA, Weix DJ (2012) Replacing conventional carbon nucleophiles with electrophiles: nickel-catalyzed reductive alkylation of aryl bromides and chlorides. J Am Chem Soc 134:6146–6159
52. Biswas S, Weix DJ (2013) Mechanism and selectivity in nickel-catalyzed cross-electrophile coupling of aryl halides with alkyl halides. J Am Chem Soc 135:16192–16197
53. Everson DA, George DT, Weix DJ (2013) Nickel-catalyzed cross-coupling of aryl halides with alkyl halides: ethyl 4-(4-(4-methylphenylsulfonamido)-phenyl)butanoate. Org Synth 90:200–214
54. Wang S, Qian Q, Gong H (2012) Nickel-catalyzed reductive coupling with secondary alkyl bromides and allylic acetate. Org Lett 14:3352–3355
55. Xu H, Zhao C, Qian Q, Deng W, Gong H (2013) Nickel-catalyzed cross-coupling of unactivated alkyl halides using bis (pinacolato)diboron as reductant. Chem Sci 4:4022–4029
56. Molander GA, Traister KM, O'Neill BT (2014) Reductive cross-coupling of nonaromatic, heterocyclic bromides with aryl and heteroaryl bromides. J Org Chem 79:5771–5780
57. Molander GA, Traister KM, O'Neill BT (2015) Engaging nonaromatic, heterocyclic tosylates in reductive cross-coupling with aryl and heteroaryl bromides. J Org Chem 80:2907–2911
58. Everson DA, Buonomo JA, Weix DJ (2014) Nickel-catalyzed cross-electrophile coupling of 2-chloropyridines with alkyl bromides. Synlett 25:233–238
59. Cherney JH, Kadunce NT, Reisman SE (2013) Catalytic asymmetric reductive acyl cross-coupling: synthesis of enantioenriched acyclic alpha,alpha-disubstituted ketones. J Am Chem Soc 135:7442–7445
60. Cherney AH, Reisman SE (2014) Nickel-catalyzed asymmetric reductive cross-coupling between vinyl and benzyl electrophiles. J Am Chem Soc 136:14365–14368
61. Everson DA, Weix DJ (2014) Cross-electrophile coupling: principals of reactivity and selectivity. J Org Chem 79:4793–4798
62. Lou S, Fu GC (2010) Nickel/bis(oxazoline)-catalyzed asymmetric Kumada reactions of alkyl electrophiles: cross-couplings of racemic

alpha-bromoketones. J Am Chem Soc 132:1264–1266
63. Lou S, Fu GC (2010) Enantioselective alkenylation via nickel-catalyzed cross-coupling with organozirconium reagents. J Am Chem Soc 132:5010–5011
64. Lundin PM, Fu GC (2010) Asymmetric Suzuki cross-couplings of activated secondary alkyl electrophiles: arylations of racemic alpha-chloroamides. J Am Chem Soc 132:11027–11029
65. Liang L, Fu GC (2014) Catalytic asymmetric synthesis of tertiary alkyl fluorides: Negishi cross-couplings of racemic alpha, alpha-dihaloketones. J Am Chem Soc 136:5520–5524
66. Vavon G, Chaminade C, Quesnel G (1945) Action du bromure de benzyle sur les magnesiens C R Hebd. Seances Acad Sci 220:850–852
67. Percival WC, Wagner RB, Cook NC (1953) Grignard reactions. XXI. Synthesis of aliphatic ketones. J Am Chem Soc 75:3731–3734
68. Tamura M, Kochi J (1971) Vinylation of Grignard reagents. Catalysis by iron. J Am Chem Soc 93:1487–1489
69. Sears JD, Neate PGN, Neidig ML (2018) Intermediates and mechanism in iron-catalyzed cross-coupling. J Am Chem Soc 140:11872–11883. and references therein
70. Mako TL, Byers JA (2016) Recent advances in iron-catalysed cross coupling reactions and their mechanistic underpinning. Inorg Chem Front 3:766–790
71. Fürstner A (2016) Iron catalysis in organic synthesis: a critical assessment of what it takes to make this base metal a multitasking champion. ACS Cent Sci 2:778–789. and references therein
72. Nakamura E, Hatakeyama T, Ito S, Ishizuka K, Ilies L, Nakamura M (2014) Iron-catalyzed cross-coupling reactions. Org React 83:1–209. and references therein
73. Piontek A, Bisz E, Szostak M (2018) Iron-catalyzed cross-couplings in the synthesis of pharmaceuticals: in pursuit of sustainability. Angew Chem Int Ed 57:11116–11128. and references therein
74. Bolm C (2009) A new iron age. Nat Chem 1:420
75. Cardellicchio C, Fiandanese V, Marchese G, Ronzini L. (1987) Functionalized ketones by iron mediated reaction of Grignard reagents with acyl chlorides. Tetrahedron Lett 28:2053–2056
76. Fiandanese V, Marchese G, Martina V, Ronzini L (1984) Iron catalyzed cross-coupling reactions of acyl chlorides with Grignard reagents. A mild. General, and convenient synthesis of aliphatic and aromatic ketones. Tetrahedron Lett 25:4805–4808
77. Scheiper B, Bonnekessel M, Krause H, Fürstner A (2004) Selective iron-catalyzed cross-coupling reactions of grignard reagents with enol triflates, acid chlorides and dichloroarenes. J Org Chem 69:3943–3949
78. Reddy CK, Knochel P (1996) New cobalt- and iron-catalyzed reactions of organizinc compounds. Angew Chem Int Ed 35:1700–1701
79. Cardellicchio C, Fiandanese V, Marchese G, Ronzini L (1985) a highly efficient synthetic route to ketones through sequential coupling reactions of Grignard reagents with S-phenyl carbonochloridothioate in the presence of nickel or iron catalysts. Tetrahedron Lett 26:3595–3598
80. Duplais C, Bures F, Sapountzis I, Korn TJ, Cahiez G, Knochel P (2004) An efficient synthesis of diaryl ketones by iron-catalyzed arylation of aroyl cyanides. Angew Chem Int Ed 43:2968–2970
81. Ottesen LK, Ek F, Olsson R (2006) Iron-catalyzed cross-coupling of imidoyl chlorides with Grignard reagents. Org Lett 8:1771–1773
82. Neumann SM, Kochi JK (1975) Synthesis of olefins. Cross coupling of alkenyl halides and Grignard reagents catalyzed by iron complexes. J Org Chem 40:599–606
83. Cahiez G, Avedissian H (1998) Highly stereo- and chemoselective iron-catalyzed alkenylation of organomagnesium compounds. Synthesis:1199–1205
84. Cahiez G, Marquais S (1996) Cu-catalyzed alkylation of Fe-catalyzed alkenylation of organomanganese reagents. Pure Appl Chem 68:53–60
85. Cahiez G, Marquais S (1996) Highly chemo– and stereoselective Fe-catalyzed alkenylation of organomanganese reagents. Tetrahedron Lett 37:1173–1776
86. Fürstner A, Brunner H (1996) Preparation of allyl-, alkenyl- and of functionalized arylmanganese reagents by oxidative insertion of manganese-graphite into organic halides. Tetrahedron Lett 37:7009–7012
87. Fabre JL, Julia M, Verpeaux JN (1982) Couplage Mixte Entre Sulfones Vinyliques et Réactifs de Grignard en Presence de sels de Metal de Transition: Synthèse Stéréosélective d'oléfines Trisubstituées. Tetrahedron Lett 23:2469–2472

88. Alvarez E, Cuvigny T, du Penhout CH, Julia M (1988) Syntheses with sulfones XLVIII: stereoselective synthesis of 2-isopropyl 1,4-dienes through the iron-catalysed cross-coupling reactions of 2-benzenesulfonyl 1,4-dienes and isopropylmagnesium chloride. Tetrahedron 44:111–118
89. Alvarez E, Cuvigny T, du Penhout CH, Julia M (1988) Syntheses with sulfones XLIX: stereo- and enantioselective synthesis of (s)-(−)-3,9-dimethyl 6-(1-methylethyl) (e)-5,8-decadien 1-ol acetate, sexual pheromone of yellow scale. Tetrahedron 44:119–126
90. Itami K, Higashi S, Mineno M, Yoshida J (2005) Iron-catalyzed cross-coupling of alkenyl sulfides with Grignard reagents. Org Lett 7:1219–12222
91. Volla CMR, Vogel P (2008) Iron-catalyzed desulfinylative C-C cross-coupling reactions of sulfonyl chlorides with Grignard reagents. Angew Chem Int Ed 47:1305–1307
92. Nishikado H, Nakatsuji H, Ueno K, Nagase R, Tanabe Y (2010) Mild, efficient, and robust method for stereocomplementary iron-catalyzed cross-coupling using (*E*)- and (*Z*)-enol tosylates. Synlett:2087–2092
93. Agrawal T, Cook SP (2014) Iron-catalyzed coupling of aryl sulfamates and aryl/vinyl tosylates with aryl Grignards. Org Lett 16:5080–5083
94. Li BJ, Xu L, Wu ZH, Guan BT, Sun CL, Wang BQ, Shi ZJ (2009) Cross-coupling of alkenyl/aryl carboxylates with grignard reagents via Fe-catalyzed C-O bond activation. J Am Chem Soc 131:14656–14657
95. Molander GA, Rahn BJ, Shubert DC, Bonde SE (1983) Iron catalyzed cross-coupling reactions. Synthesis of arylethenes. Tetrahedron Lett 24:5449–5452
96. Hamze A, Brion JD, Alami M (2012) Synthesis of 1,1-diarylethenes via efficient iron/copper co-catalyzed coupling of 1-arylvinyl halides with Grignard reagents. Org Lett 14:2782–2785
97. Fürstner A, Leitner A, Méndez M, Krause H (2002) Iron-catalyzed cross-coupling reactions. J Am Chem Soc 124:13856–13863
98. Fürstner A, Leitner A (2002) Iron-catalyzed cross-coupling reactions of alkyl-grignard reagents with aryl chlorides, tosylates, and triflates. Angew Chem Int Ed 41:609–612
99. Perry MC, Gillett AN, Law TC (2012) An unprecedented iron-catalyzed cross-coupling of primary and secondary alkyl Grignard reagents with non-activated aryl chlorides. Tetrahedron Lett 53:4436–4439
100. Chen X, Quan ZJ, Wang XC (2015) Iron-catalyzed cross-coupling of heteroaromatic tosylates with alkyl and aryl Grignard reagents. Appl Organometal Chem 29:296–300
101. Rushworth PJ, Hulcoop DG, Fox DJ (2013) Iron/tetramethylethylenediamine-catalyzed ambient-temperature coupling of alkyl Grignard reagents and aryl chlorides. J Org Chem 78:9517–9521
102. Hocek M, Dvořáková H (2003) An efficient synthesis of 2-substituted 6-methylpurine bases and nucleosudes by Fe- or Pd-catalyzed cross-coupling reactions of 2,6-dichloropurines. J Org Chem 68:5773–5776
103. Malhotra S, Seng PS, Koenig SG, Deese AJ, Ford KA (2013) Chemoselective SP2-SP3 cross-couplings: iron-catalyzed alkyl transfer to dihaloaromatics. Org Lett 15:3698–3701
104. Scheiper B, Glorius F, Leitner A, Fürstner A (2004) Catalysis-based enantioselective total synthesis of the macrocyclic spermidine alkaloid isooncintine. Proc Natl Acad Sci U S A 101:11960–11965
105. Fürstner A, Leitner A (2003) A catalytic approach to (*R*)-(+)-muscopyridine with integrated "self-clearance". Angew Chem Int Ed 42:308–311
106. Silberstein AL, Ramgren SD, Garg NK (2012) Iron-catalyzed alkylations of aryl sulfamates and carbamates. Org Lett 14:3796–3799
107. Agrawal T, Cook SP (2013) Iron-catalyzed cross-coupling reactions of alkyl Grignards with aryl sulfamates and tosylates. Org Lett 15:96–99
108. Hatakeyama T, Nakamura M (2007) Iron-catalyzed selective biaryl coupling: remarkable suppression of homocoupling by the fluoride anion. J Am Chem Soc 129:9844–9845
109. Gülak S, Wangelin AJV (2012) Chlorostyrenes in iron-catalyzed biaryl coupling reactions. Angew Chem Int Ed 51:1357–1361
110. Kuzmina OM, Steib AK, Flubacher D, Knochel P (2012) Iron-catalyzed cross-coupling of *n*-heterocyclic chlorides and bromides with arylmagnesium reagents. Org Lett 14:4818–4821
111. Kuzmina OM, Steib AK, Markiewicz JT, Flubacher D, Knochel P (2013) Ligand-accelerated iron- and cobalt-catalyzed cross-coupling reactions between N-heteroaryl halides and aryl magnesium reagents. Angew Chem Int Ed 52:4945–4949

112. Kuzmina OM, Steib AK, Fernandez S, Boudot W, Markiewicz JT, Knochel P (2015) Practical iron- and cobalt-catalyzed cross-coupling reactions between N-heterocyclic halides and aryl or heteroaryl magnesium reagents. Chem Eur J 21:8242–8249
113. Hatakeyama T, Hashimoto T, Kondo Y, Fujiwara Y, Seike H, Takaya H, Tamada Y, Ono T, Nakamura M (2010) Iron-catalyzed Suzuki-Miyaura coupling of alkyl halides. J Am Chem Soc 132:10674–10676
114. Bedford RB, Brenner PB, Carter E, Carvell TW, Cogswell PM, Gallagher T, Harvey JN, Murphy DM, Neeve EC, Nunn J, Pye DR (2014) Expedient iron-catalyzed coupling of alkyl, benzyl and allyl halides with arylboronic esters. Chem Eur J 20:7935–7938
115. Crockett MP, Tyrol CC, Wong AS, Li B, Byers JA (2018) Iron-catalyzed Suzuki-Miyaura cross-coupling reactions between alkyl halides and unactivated arylboronic esters. Org Lett 20:5233–5237
116. Hedström A, Izakian Z, Vreto I, Wallentin C-J, Norrby P-O (2015) On the radical nature of iron-catalyzed cross-coupling reactions. Chem Eur J 21:5945–5953
117. Lei F, Qu B, Li X, Guo L, Guan M, Hai L, Jin H, Wu Y (2014) Efficient synthesis of substances related to cinacalcet hydrochloride via heck coupling. Syn Commun 44:2879–2885
118. Tewari N, Maheshwari N, Medhane R, Nizar H, Prasad M (2012) A novel method for the large scale synthesis of Cinacalcet hydrochloride using iron-catalyzed C-C coupling. Org Process Res Dev 16:1566–1568
119. Tailor SB, Manzotti M, Asghar S, Rowsell BJS, Luckham SLJ, Sparkes HA, Bedford RB (2019) Revisiting claims of the iron-, cobalt-, nickel- and copper-catalyzed Suzuki biaryl cross-coupling of aryl halides with aryl boronic acids
120. Guo Y, Young DJ, Hor TSA (2008) Palladium-free Suzuki-Miyaura cross-coupling at elevated pressures. Tetrahedron Lett 49:5620–5621
121. Bedford RB, Gallagher T, Pye DR, Savage W (2015) Towards iron-catalyzed Suzuki biaryl cross-coupling: unusual reactivity of 2-halobenzyl halides. Synthesis 47:1761–1765
122. O'Brien HM, Manzotti M, Abrams RD, Elorriaga D, Sparkes HA, Davis SA, Bedford RM (2018) Iron-catalysed substrate-directed Suzuki biaryl cross-coupling. Nat Catal 1:429–437
123. Nakamura M, Matsuo K, Ito S, Nakamura E (2004) Iron-catalyzed cross-coupling of primary and secondary alkyl halides with aryl Grignard reagents. J Am Chem Soc 126:3686–3687
124. Bedford RB, Betham M, Bruce DW, Danopoulos AA, Frost RM, Hird M (2006) Iron-phosphine, –phosphite, –arsine, and -carbene catalysts for the coupling of primary and secondary alkyl Halides with aryl Grignard reagents. J Org Chem 71:1104–1110
125. Cahiez G, Duplais C, Moyeux A (2007) Iron-catalyzed alkylation of alkenyl Grignard reagents. Org Lett 9:3253–3254
126. Nagano T, Hayashi T (2004) Iron-catalyzed Grignard cross-coupling with alkyl halides possessing β-hydrogens. Org Lett 6:1297–1299
127. Bedford RB, Bruce DW, Frost RM, Goodby JW, Hird M. (2004) Iron(III) Salen catalysts for the cross-coupling of aryl Grignards with alkyl halides bearing β-hydrogens. Chem Commun:2822–2823
128. Martin R, Fürstner A (2004) Cross-coupling of alkyl halides with aryl Grignard reagents catalyzed by low-valent iron complex. Angew Chem Int Ed 43:3955–3957
129. Cahiez G, Habiak V, Duplais C, Moyeux A (2007) Iron-catalyzed alkylation of aromatic Grignard reagents. Angew Chem Int Ed 46:4364–4366
130. Guérinot A, Reymond S, Cossy J (2007) Iron-catalyzed cross-coupling of alkyl halides with akenyl Grignard reagents. Angew Chem Int Ed 46:6521–6524
131. Fürstner A, Martin R, Krause H, Seidel G, Goddard R, Lehmann CW (2008) Preparation, structure, and reactivity of nonstabilized organoiron compounds. Implications for iron-catalyzed cross coupling reactions. J Am Chem Soc 130:8773–8787
132. Hatakeyama T, F U, Okada Y, Itoh T, Hashimoto T, Kawamura S, Ogata K, Takaya H, Nakamura M (2011) Kumada-tamao-corriu coupling of alkyl halides catalyzed by an iron-bisphosphine complex. Chem Lett:1030–1032
133. Ghorai SK, Jin M, Hatakeyama T, Nakamura M (2012) Cross-coupling of non-activated chloroalkanes with aryl Grignard reagents in the presence of iron/*N*-heterocyclic Carbene catalysts. Org Lett 2012(14):1066–1069
134. Hashimoto T, Hatakeyama T, Nakamura M (2012) Stereospecific cross-coupling between alkenyl boronates and alkyl halides catalyzed

by iron-bisphosphine complexes. J Org Chem 77:1168–1173

135. Hateakeyama T, Hashimoto T, Kondo Y, Fujiwara Y, Seike H, Takaya H, Tamada Y, Ono T, Nakamura M (2010) Iron-catalyzed Suzuki-Miyaura coupling of alkyl halides. J Am Chem Soc 132:10674–10676

136. Denmark SE, Cresswell AJ (2013) Iron-catalyzed cross-coupling of unactivated secondary alkyl thio ethers and sulfones with aryl Grignard reagents. J Org Chem 78:12593–12628

137. Hatakeyama T, Nakagawa N, Nakamura M (2009) Iron-catalyzed Negishi coupling toward an effective olefin synthesis. Org Lett 11:4496–4499

138. Bedford RB, Brenner PB, Carter E, Clifton J, Cogswell PM, Gower NJ, Haddow MF, Harvey JN, Kehl JA, Murphy DM, Neeve EC, Neidig ML, Nunn J, Snyder BER, Taylor J (2014) Iron phosphine catalyzed cross-coupling of tetraorganoborates and related group 13 nucleophiles with alkyl halides. Organometallics 33:5767–5780

139. Dongol KG, Koh H, Sau M, Chai CLL. (2007) Iron-catalysed sp3–sp3 cross-coupling reactions of unactivated alkyl halides with alkyl Grignard reagents. Adv Synth Catal:1015–1018

140. Hatakeyama T, Hashimoto T, Kathriarachchi KKADS, Zenmyo T, Seike H, Nakamura M (2012) Iron-catalyzed akyl-alkyl Suzuki-Miyaura coupling. Angew Chem Int Ed 51:8834–8837

141. Jin M, Adak L, Nakamura M (2015) Iron-catalyzed enantioselective cross-coupling reactions of α-chloroesters with aryl Grignard reagents. J Am Chem Soc 137:7128–7134

142. Sindhu KS, Anilkumar G (2014) Recent advances and applications of Glaser coupling employing greener protocols. RSC Adv 4:27867–27887. and references therein

143. Stephens RD, Castro CE (1963) The substitution of aryl iodides with Cuprous acetylides. A synthesis of tolanes and heterocyclics. J Org Chem 28:3313–3315

144. Koelsch CF, Whitney AG (1941) The Rosenmund von Braun nitrile synthesis. J Org Chem 6:795–803

145. Hurtley WRH (1929) CCXLIV—replacement of halogen in *ortho*bromo-benzoic acid. J Chem Soc:1870–1873

146. Ley SV, Thomas AW (2003) Modern synthetic methods for copper-mediated C(aryl)-O, C(aryl)-N, and C(aryl)-S bond formation. Angew Chem Int Ed 42:5400–5449. and references therein

147. Lee V (2019) Application of copper(I) salt and fluoride promoted Stille coupling reactions in the synthesis of bioactive molecules. Org Biomol Chem:9095–9123 and references therein

148. Allred GD, Liebeskind LS (1996) Copper-mediated cross-coupling of organostannanes with organic iodides at or below room temperature. J Am Chem Soc 118:2748–2749

149. Schuppan J, Wehlan H, Keiper S, Koert U (2001) Synthesis of apoptolidinone. Angew Chem Int Ed 40:2063–2066

150. Maleczka RE, Terrell LR, Geng F, Ward JS (2002) Total synthesis of proposed amphidinolide a via a highly selective ring-closing metathesis. Org Lett 4:2841–2844

151. Chinchilla R, Nájera C (2007) The Sonogashira reaction: a booming methodology in synthetic organic chemistry. Chem Rev 107:874–922

152. Rattan K, Gujadhur RK, Bates CG, Venkataraman D (2001) Formation of aryl-nitrogen, aryl-oxygen, and aryl-carbon bonds using well-defined copper(I)-based catalysts. Org Lett 3:4315–4317

153. Liori AA, Stamatopoulos IK, Papastavrou AT, Pinaka A, Vougioukalakis GC (2018) A sustainable, user-friendly protocol for the Pd-free Sonogashira coupling reaction. Eur J Org Chem:6134–6139 and references therein

154. Maaliki C, Thiery E, Thibonnet (2017) Emergence of copper-mediated formation of C-C bonds. Eur J Org Chem:209–228

155. Thapa S, Shrestha B, Gurung SK, Giri R (2015) Copper-catalysed cross-coupling: an untapped potential. Org Biomol Chem 13:4816–4827

156. Kharasch MS, Fuchs CF (1945) Factors influencing the course and mechanism of Grignard reactions XVII. Interchange of radicals in the reaction of Grignard reagents and organic halides in the presence of metallic halides. J Org Chem 10:292–297

157. Tamura M, Kochi J (1971) Coupling of Grignard reagents with organic halides. Synthesis:303–305

158. Nakamura E, Mori S (2000) Wherefore art thou copper? Structures and reactions mechanism of organocuprates in organic chemistry. Angew Chem Int Ed 39:3751–3770

159. Lipshutz BH, Sengupta S (1992) Organocopper reagents: substitution, conjugate addition, carbo/metallocupration, and other reactions. Org React 41:135–631. and references therein

160. Burns DH, Miller JD, Chan H-K, Delaney MO (1997) Scope and utility of a new soluble

copper catalyst [CuBr-LiSPh-LiBr-THF]: a comparison with other copper catalysts in their ability to couple one equivalent of a grignard reagent with an alkyl sulfonate. J Am Chem Soc 119:2125–2133. and references therein

161. Terao J, Ikumi A, Kuniyasu H, Kambe N (2003) Ni- or Cu-catalyzed cross-coupling reactions of alkyl fluorides with Grignard reagents. J Am Chem Soc 125:5646–5647

162. Shen R, Iwasaki T, Terao J, Kambe N (2012) Copper-catalyzed coupling reaction of unactivated secondary alkyl iodides with alkyl Grignard reagents in the presence of 1,3-butadiene as an effective additive. Chem Commun 48:9313–9315

163. Terao J, Todo H, Begum SA, Kuniyasu H, Kambe N (2007) Copper-catalyzed cross-coupling reaction of Grignard reagents with primary alkyl-halides: remarkable effect of 1-phenylpropyne. Angew Chem Int Ed 46:2086–2089

164. Iwasaki T, Imanishi R, Shimizu R, Kuniyasu H, Terao J, Kambe N (2014) Copper-catalyzed alkyl-alkyl cross-coupling reactions using hydrocarbon additives: efficiency of catalyst and roles of additives. J Org Chem 79:8522–8532

165. Terao J, Kambe N (2008) Cross-coupling reaction of alkyl halides with Grignard reagents catalyzed by Ni, Pd or Cu complexes with p-carbon ligand(s). Acc Chem Res 41:1545–1554

166. Cahiez G, Gager O, Buenidia J (2010) Copper-catalyzed cross-coupling of alkyl and aryl Grignard reagents with alkynyl halides. Angew Chem Int Ed 49:1278–1281

167. Cahiez G, Gager O, Buendida J (2010) New insights into the copper-catalyzed alkylation of Grignard and organolithium reagents. Synlett:299–303

168. Ren P, Stern L-A, Hu X (2012) Copper-catalyzed cross-coupling of functionalized alkyl halides and tosylates with secondary and tertiary alkyl Grignard reagents. Angew Chem Int Ed 51:9110–9113

169. Yang C-T, Zhang Z-Q, Liang J, Liu J-H, Lu X-Y, Chen H-H, Liu L (2012) Copper-catalyzed cross-coupling of nonactivated secondary alkyl halides and tosylates with secondary alkyl Grignard reagents. J Am Chem Soc 134:11124–11127

170. Zhu Y, Xiong T, Han W, Shi Y (2014) Copper-catalyzed oxidative homo- and cross-coupling of Grignard reagents using diaziridinone. Org Lett 16:6144–6147

171. Liu J-H, Yang C-T, Lu X-Y, Zhang Z-Q, Xu L, Cui M, Lu X, Xiao B, Fu Y, Liu L (2014) Copper-catalyzed reductive cross-coupling of nonactivated alkyl tosylates and mesylates with alkyl and aryl bromides. Chem Eur J 20:15334–15338

172. Lethu S, Matsuoka S, Murata M (2014) Highly efficient preparation of selectively isotope cluster-labeled long chain fatty acids via two consecutive C_{sp3}-C_{sp3} cross-coupling reactions. Org Lett 16:844–847

173. Yang C-T, Zhang Z-Q, Liu Y-C, Liu L (2011) Copper-catalyzed cross-coupling of organoboron compounds with primary alkyl halides and pseudohalides. Angew Chem Int Ed 50:3904–3907

174. Zhou W, You W, Smith KB, Brown MK (2014) Copper-catalyzed cross-coupling of boronic esters with aryl iodides and application to the carboboration of alkynes and allenes. Angew Chem Int Ed 53:3475–3479

175. You W, Brown MK (2014) Diarylation of alkenes by a Cu-catalyzed migratory insertion/cross-coupling cascade. J Am Chem Soc 136:14730–14733

176. You W, Brown MK (2015) Catalytic enantioselective diarylation of alkenes. J Am Chem Soc 137:14578–14581

177. Bergmann AM, Oldham AM, You W, Brown KM (2018) Copper-catalyzed cross-coupling of aryl-, primary alkyl-, and secondary alkylboranes with heteroaryl bromides. Chem Commun 54:5381–5384

178. Gurung SK, Thapa S, Kafle A, Dickie DA, Giri R (2014) Copper-catalyzed suzuki-miyaura coupling of arylboronate esters: transmetalation with (PN)CuF and identification of intermediates. Org Lett 16:1264–1267. and references therein

179. Gurung SK, Thapa S, Vangala AS, Giri R (2013) Copper-catalyzed hiyama coupling of (hetero)aryltriethoxysilanes with (hetero)aryl iodides. Org Lett 15:5378–5381. and references therein

180. Sun C-L, Shi Z-J (2014) Transition-metal-free coupling reactions. Chem Rev 114:9219–9280

181. Roscales S, Csákÿ (2014) Transition-metal-free C–C bond forming reactions of aryl, alkenyl and alkynylboronic acids and their derivatives. Chem Soc Rev 43:8215–8225

182. Li H, Zhang Y, Wang J (2013) Reaction of Diazo compounds with organoboron compounds. Synthesis 45:3090–3098

183. Shao Z, Zhang H (2012) *N*-Tosylhydrazones: versatile reagents for metal-catalyzed

and metal-free cross-coupling reactions. Chem Soc Rev 41:560–572
184. Barluenga J, Valdés C (2011) Tosylhydrazones: new uses for classic reagents in palladium-catalyzed cross-coupling and metal-free reactions. Angew Chem Int Ed 50:7486–7500
185. Hooz J, Linke S (1968) The reaction of trialkylboranes with diazoacetone. A new ketone synthesis. J Am Chem Soc 90:5936–5937
186. Brown HC, Midland MM, Levy AB (1972) Reaction of dialkylchloroboranes with ethyl diazoacetate at low temperatures. Facile two-carbon homologation under exceptionally mild conditions. J Am Chem Soc 94:3662–3664
187. Vedejs E, Stolle WT (1977) Reductive alkylation of aldehyde tosylhydrazones with organolithium reagents. Tetrahedron 2:135–138
188. Myers AG, Movassaghi M (1998) Highly efficient methodology for the reductive coupling of aldehyde tosylhydrazones with alkyllithium reagents. J Am Chem Soc 120:8891–8892
189. Peng C, Zhang W, Yan G, Wang J (2009) Arylation and vinylation of α-diazocarbonyl compounds with boroxines. Org Lett 11:1667–1670
190. Barluenga J, Tomás-Gamasa M, Aznar F, Valdés C (2009) Metal-free carbon–carbon bond-forming reductive coupling between boronic acids and tosylhydrazones. Nat Chem 1:494–499
191. Pérez-Aguilar MC, Valdés C (2012) Olefination of carbonyl compounds through reductive coupling of alkenylboronic acids and tosylhydrazones. Angew Chem Int Ed 51:5953–5957
192. Argintaru OA, Ryu D, Aron I, Molander GA (2013) Synthesis and applications of α-trifluoromethylated alkylboron compounds. Angew Chem Int Ed 52:13656–13660
193. Wu G, Deng Y, Wu C, Wang X, Zhang Y, Wang J (2014) Switchable 2,2,2-trifluoroethylation and *gem*-difluorovinylation of organoboronic acids with 2,2,2-trifluorodiazoethane. Eur J Org Chem:4477–4481
194. Wu G, Deng Y, Wu C, Zhang Y, Wang J (2014) Synthesis of α-aryl esters and nitriles: deaminative coupling of α-aminoesters and α-aminoacetonitriles with arylboronic acids. Angew Chem Int Ed 53:10510–10514
195. Bamford WR, Stevens TS (1952) The decomposition of toluene-*p*-sulphonylhydrazones by alkali. J Chem Soc:4735–4740
196. Nakagawa S, Bainbridge KA, Butcher K, Ellis D, Klute W, Ryckmans T (2012) Application of barluenga boronic coupling (BBC) to the parallel synthesis of drug-like and drug fragment-like molecules. ChemMedChem 7:233–236
197. Allwood DM, Blakemore DC, Brown AD, Ley SV (2014) Metal-free coupling of saturated heterocyclic sulfonylhydrazones with boronic acids. J Org Chem 79:328–338
198. Kupracz L, Kirschning A (2013) Two-step flow synthesis of biarylmethanes by reductive arylation of tosylhydrazones. J Flow Chem 3:11–16
199. Tran DN, Battilocchio C, Lou S-B, Hawkins JM, Ley SV (2015) Flow chemistry as a discovery tool to access sp^2–sp^3 cross-coupling reactions *via* diazo compounds. Chem Sci 6:1120–1125

Chapter 14

Flow Chemistry as an Enabling Technology for Synthetic Organic Chemistry

Nicholas E. Leadbeater

Abstract

Continuous-flow processing is proving to be an enabling technology for synthetic organic chemists. After an introduction to the advantages and disadvantages of flow processing and an overview of the equipment currently available, the use of flow chemistry in a range of avenues of organic chemistry is showcased. Attention is focused on areas such as hazardous transformations, multistep synthesis, photochemistry, electrochemistry, and organocatalysis. The scope of the chapter is also broadened to techniques used for monitoring flow processes and the incorporation of flow chemistry into the undergraduate teaching laboratory.

Key words Flow chemistry, Continuous manufacturing, Scale-up, Hazardous chemistry, Oxidation, Organometallic, Organolithium, Oxidation, Organofluorine chemistry, Solid-supported reagents, Heterogeneous catalysis, Photochemistry, Organocatalysis, Gases, In-line spectroscopy, Reaction monitoring, Electrochemistry, Education, Undergraduate laboratory

1 Introduction

Looking back in history, most manufactured products were made individually by hand. Craftsmen would use their skills and tools to create the constituent parts and then would assemble them into the final product. This approach was both time-consuming and labor intensive. The Venetian Arsenal, dating to about 1104, took a different approach. Ships under construction were moved down a canal and outfitted by various expert tradesmen on the way. At its peak, over 16,000 people were employed and they were capable of producing one ship each day. This was an example of a production line in operation. Moving to the twentieth century, the Ford motor company perfected the production line. They put constituent parts of a car together in succession and then arranged the flow of these units at the right time and in the right place to a moving final assembly line from which came the completed car. As a result of

Paul F. Richardson (ed.), *Green Chemistry in Drug Discovery: From Academia to Industry*, Methods in Pharmacology and Toxicology, https://doi.org/10.1007/978-1-0716-1579-9_14,

these optimized production methods, cars came off the line in 3-min intervals. They were able to increase production eightfold while using less manpower. Today, automation features front and center in many aspects of advanced manufacturing.

A similar story is unfolding in chemistry. Historically, all reactions were performed in batch. Reagents would be added to a flask, and this often either heated or cooled for prolonged periods before the reaction was deemed complete. This is time-consuming, labor intensive, and often energy inefficient. In addition, by-products can be formed over time, thus compromising yield. By turning to continuous-flow processing, reactions can often be streamlined, multiple steps can be linked together, and reactions that were otherwise impossible or else highly challenging in batch are now available for chemists to use [1–9]. In addition, there is better control of heating and mixing, allowing for reactions to be performed under very precise and reproducible conditions.

Flow chemistry is not, however, a panacea. Just as it is not possible to transition the manufacture of a Steinway piano (which takes over a year to make by hand from over 12,000 constituent parts) to a fast-moving production line, there are some chemical reactions that are just not amenable to flow chemistry. For example, reactions that involve the manipulation of solids and reactions that are by their very nature slow pose significant challenges although solutions, certainly for the former, are already on the way [10–14]. Another issue is simply unfamiliarity. From the time they first start practical work as a student, chemists are trained to perform their reactions in batch. Moving to flow processing requires them to learn new skills, albeit simple now that modern flow apparatus is available and to overcome the barrier of believing the way that they have always done things is the only way.

All things considered, the inherent advantages of safety, process intensification, and efficiency mean that flow processing can be a great asset when trying to develop greener, more sustainable approaches to preparative chemistry [15–18]. The objective of this chapter is to highlight some of the key advantages and applications of flow processing, as well as showing how the technology can be incorporated into education at the undergraduate level thereby training and broadening the mindset of future chemists.

2 Equipment

Flow chemistry equipment can be as simple as a syringe pump and a length of tubing. Much of the early work using flow processing, and indeed many examples today, involves the use of homemade apparatus. Coils of tubing placed in hot water or oil, or in cryogenic baths, allow for heating and cooling, respectively, and simple plumbing allows for addition and mixing of reagents along the way. With the increasing interest in flow chemistry, a number of companies now produce equipment of various shapes and sizes [19]. With this, parameters such as temperature, residence time, mixing, and reagent stoichiometry can be accurately monitored and controlled. In addition, scientific flow equipment is built with safety in mind. This is important, especially when performing reactions for the first time or indeed for extended times or when potentially hazardous reagents are generated or used. Europe leads the way when it comes to flow chemistry equipment. There is a particularly high concentration of flow chemistry companies (Syrris, Uniqsis, and Vapourtec) based around Cambridge, UK, close to the laboratories of Professor Steven Ley, one of the pioneers of flow chemistry. Two companies are based in the Netherlands (Chemtrix and Future Chemistry) and one in Hungary (ThalesNano). As well as making flow reactors and pumps, many of the companies also offer additional tools such as gas-loading modules, photochemical devices, and supported catalyst cartridges.

3 Flow Chemistry as a Tool for Single Chemical Transformations

Continuous-flow processing is used extensively for performing single-step reactions. Indeed a cornucopia of examples exist in the literature [1–9]. Perhaps the most popular are those reactions that traditionally require prolonged times at elevated temperatures or those performed using microwave heating [20] such as cycloadditions and metal-catalyzed couplings [21]. By being able to operate safely and easily at high temperatures and under pressure using a flow approach, reaction times can be reduced and yields often improved [22–26]. A case in point is the solvent and catalyst-free Huisgen cycloaddition of 2,6-difluorobenzyl azide (**1**) with methyl-3-methoxyacrylate (**2**), followed by the elimination of methanol to yield the 1,4-triazole **3**, a key intermediate in the synthesis of the antiepileptic drug rufinamide (Scheme 1). In batch, this reaction takes 28 h to reach completion and requires heating to 135 °C [27], with the dipolarophile, **2**, being added in batches over the period of the reaction. Moving to flow, the reaction can be performed at 220 °C in a capillary reactor (750μm ID and 4.7 m long) with a residence time of 10 min [28]. The product is obtained in an isolated yield of 84%. However, since the reaction is performed solvent-free and the product is a crystalline solid with a melting point of 136–137 °C, the issue of clogging, a bug-bear of flow chemistry, needs to be addressed. While the product is a liquid in the reaction zone (220 °C), once it exits there is the potential for blocking the tubing. In addition, since the entire flow process is performed under pressure (~70 bar), the material has to exit through a back-pressure regulator (BPR). The use of a BPR, while an integral part of many flow processes, has the effect of

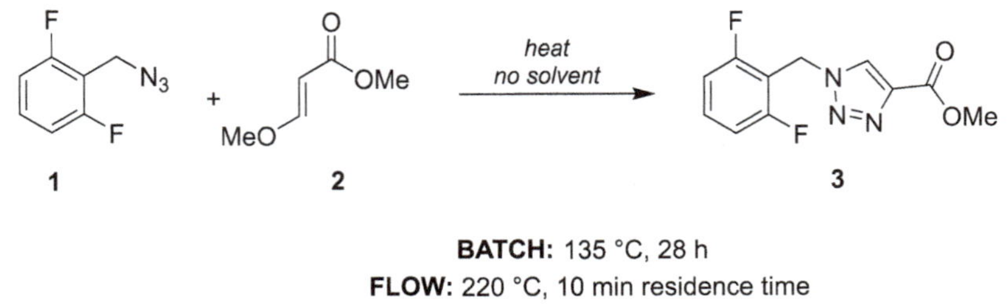

Scheme 1 The preparation of **3**, a key intermediate in the synthesis of rufinamide

increasing the tendency for clogging. The buildup of solids behind the BPR occurs rapidly, rendering the process inoperable. To overcome this problem, one approach is to intercept the product stream upon exiting the heated zone with a flow of a suitable solvent, thus solubilizing the product and allowing it to pass through the BPR unimpeded [29]. In the case of **3**, by introducing a stream of acetonitrile or methanol, the product can be sufficiently solubilized to allow it to pass through the BPR, held at 60 °C, into a collection vessel. Upon cooling, the product crystallizes out in pure form.

While reactions using flow processing can be performed at elevated temperatures, they can also be performed under sub-ambient conditions. Often it is not necessary to cool reactions to the same degree as required in batch because of the differences in heat transfer between a flask and a narrow-gauge tube [30–32]. One example is the Swern oxidation. This reaction converts primary alcohols to the corresponding aldehydes in the presence of an amine base using DMSO activated by an electrophile such as oxalyl chloride or trifluoroacetic anhydride (TFAA). It avoids the use of toxic metals such as chromium, and the aldehyde products do not react further to give carboxylic acids. However, a significant drawback to the reaction is that it is very temperature dependent. In batch, the reaction must be kept below −60 °C when using oxalyl chloride as the activator or −30 °C when using TFAA. This prevents side reactions, in particular the generation of mixed thioacetals. By performing the reaction in flow (Scheme 2), it is possible to operate at −10 °C using oxalyl chloride without compromising the selectivity of the reaction [33]. A solution of the requisite alcohol and DMSO is first mixed with oxalyl chloride in flow, generating the dimethylchlorosulfonium ion, **4**, in situ which then reacts with the alcohol to yield an alkoxysulfonium ion, **5**. The reactant stream is then intercepted with a flow of triethylamine to deprotonate **5** and leads to product formation.

The Swern oxidation can even be performed at room temperature using flow processing [34]. The key to this success is the use of fast flow rates and short pathlengths. This technique has been dubbed "flash chemistry" because from start to finish the reaction is performed on the second or sub-second timescale [35–37]. This

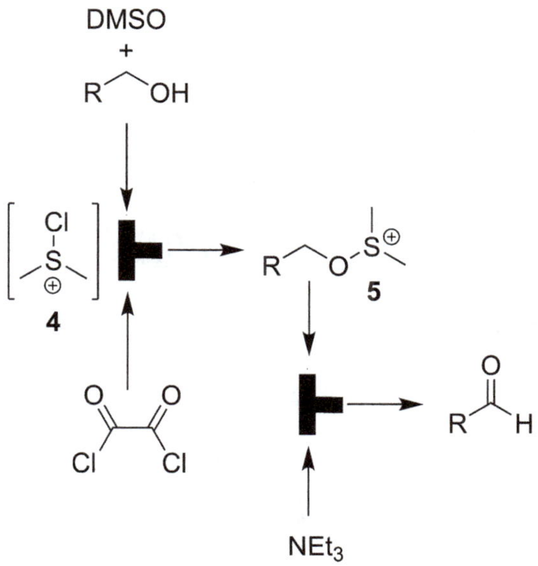

Scheme 2 The Swern oxidation performed in flow at −10 °C

is not as daunting as it may at first appear. For example, the mean residence time in a tube reactor of length 10 mm at a flow rate of 1 m/s is 10 ms. By being able to operate at these flow rates, unstable intermediates can be quickly transferred to subsequent steps, and, as a result, many of the issues associated with by-product formation can be mitigated. Flash chemistry is generally performed using tubing of internal diameter 1 mm or less, and highly efficient mixing, achieved using T-shaped micromixers, is also an important component for its success [38, 39]. In the case of the Swern oxidation (Scheme 3), DMSO is first reacted in flow with TFAA at room temperature for 0.01–2 s before a stream of the alcohol is introduced. Between 1 and 2 s further downstream, triethylamine is introduced, and the reaction mixture flowed through a coil heated to 30 °C for a further 7 s before exiting the reactor.

Flash chemistry can also be used for performing reactions without the need for protecting groups, offering another greener advantage over conventional batch approaches [40]. Reactions involving organolithium reagents exemplify this [41–44]. Ketones react very rapidly with organolithium reagents and so are generally protected prior to reaction at another site in the molecule. However, using flash chemistry aryllithium species bearing ketone functionalities can be generated and then quenched with electrophiles without concomitant reaction at the carbonyl group (Scheme 4) [45]. The flow of aryllithium intermediate, which is produced by lithium halogen exchange from the corresponding aryl iodide by reaction with mesityllithium, is intercepted by a stream of the electrophile within 0.003 s of their generation. The reaction is

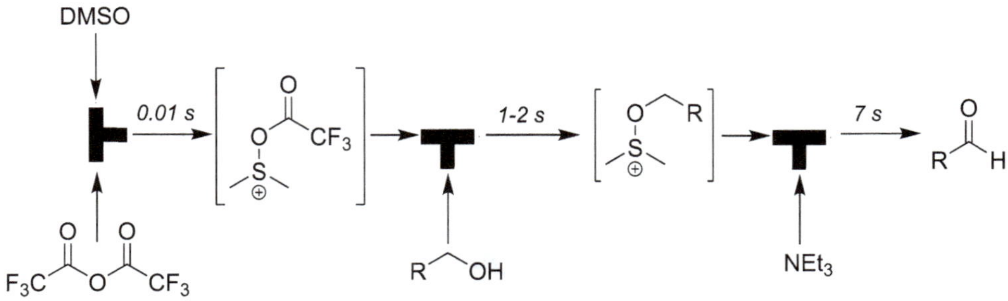

Scheme 3 The Swern oxidation performed at room temperature using "flash chemistry"

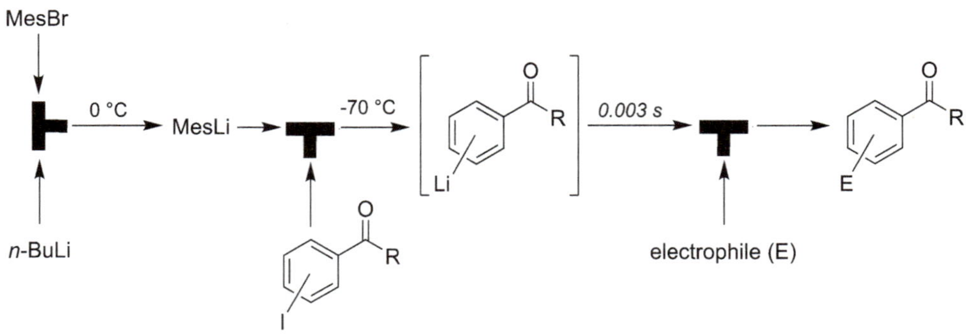

Scheme 4 Generation and quenching of organolithium species using "flash chemistry"

performed at −70 °C. Longer residence times or higher temperatures lead to decomposition and side reactions of the aryllithium species.

The concept is further exemplified in the reaction of difunctional electrophiles with one stoichiometric equivalent of phenyllithium [46]. It is possible to add the organolithium species selectively to the more reactive electrophilic site using flash chemistry. For example, when using 4-benzoylbenzaldehyde (**6**), phenyllithium selectively reacts with the aldehyde functionality. The reaction is not only more selective but also significantly higher yielding than the batch equivalent (Scheme 5).

One final example that shows the power of flash chemistry is the diisobutylaluminum hydride (DIBAL-H) reduction of esters to aldehydes [47]. In batch this reaction is often shunned since over-reduction of the ester to the alcohol is frequently observed. Indeed, it is not unusual for chemists to circumnavigate the problem by reduction of the ester to a primary alcohol and then re-oxidation to the aldehyde. However, using flow processing it is possible to perform the selective reduction of the ester to the aldehyde in one step and in high yield. Key to the success of the methodology is performing the reaction in the sweet spot of reaction time and temperature. A residence time of 60 s or less in a coil cooled to between −40 °C and −78 °C proves optimal (Scheme 6); the exact conditions for each substrate being individually optimized by

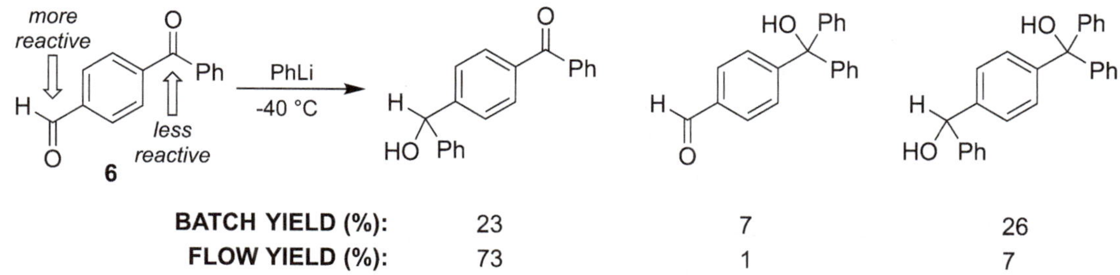

Scheme 5 The selective reaction of phenyllithium with a difunctional electrophile

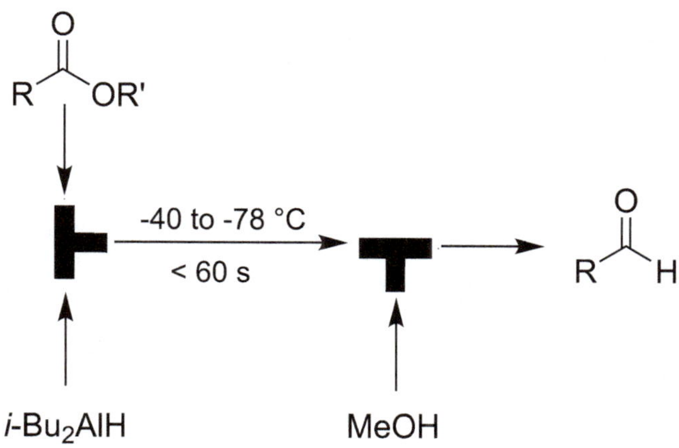

Scheme 6 DIBALH reduction of esters to aldehydes using "flash chemistry"

performing a series of reactions at different retention time/reactor temperature combinations and the data compiled in a contour plot. Overall, the flash chemistry approach brings back to life a useful chemical transformation that has been avoided by chemists for many years.

4 Flow Chemistry as a Tool for Safely Performing Hazardous Chemistry

When considered for batch processing, there are some reactions that are quickly taken off the table due to safety concerns. Only the foolhardy would consider performing reactions on anything more than a small scale using reagents such as hydrazoic acid, phosgene, hydrogen cyanide, diazomethane, or elemental fluorine. However, when using a flow approach, the hazards can be mitigated by the fact that only small volumes of reactants and reagents are exposed to one another at a given time. A particularly valuable aspect of flow chemistry is that hazardous intermediates can be generated, used, and quenched all within one reactor. This on-demand preparation of the reagent means that they do not need to be transported or stockpiled. As illustrated in Scheme 7, all of the reagents above have been safely used in synthetic chemistry by leveraging the inherent

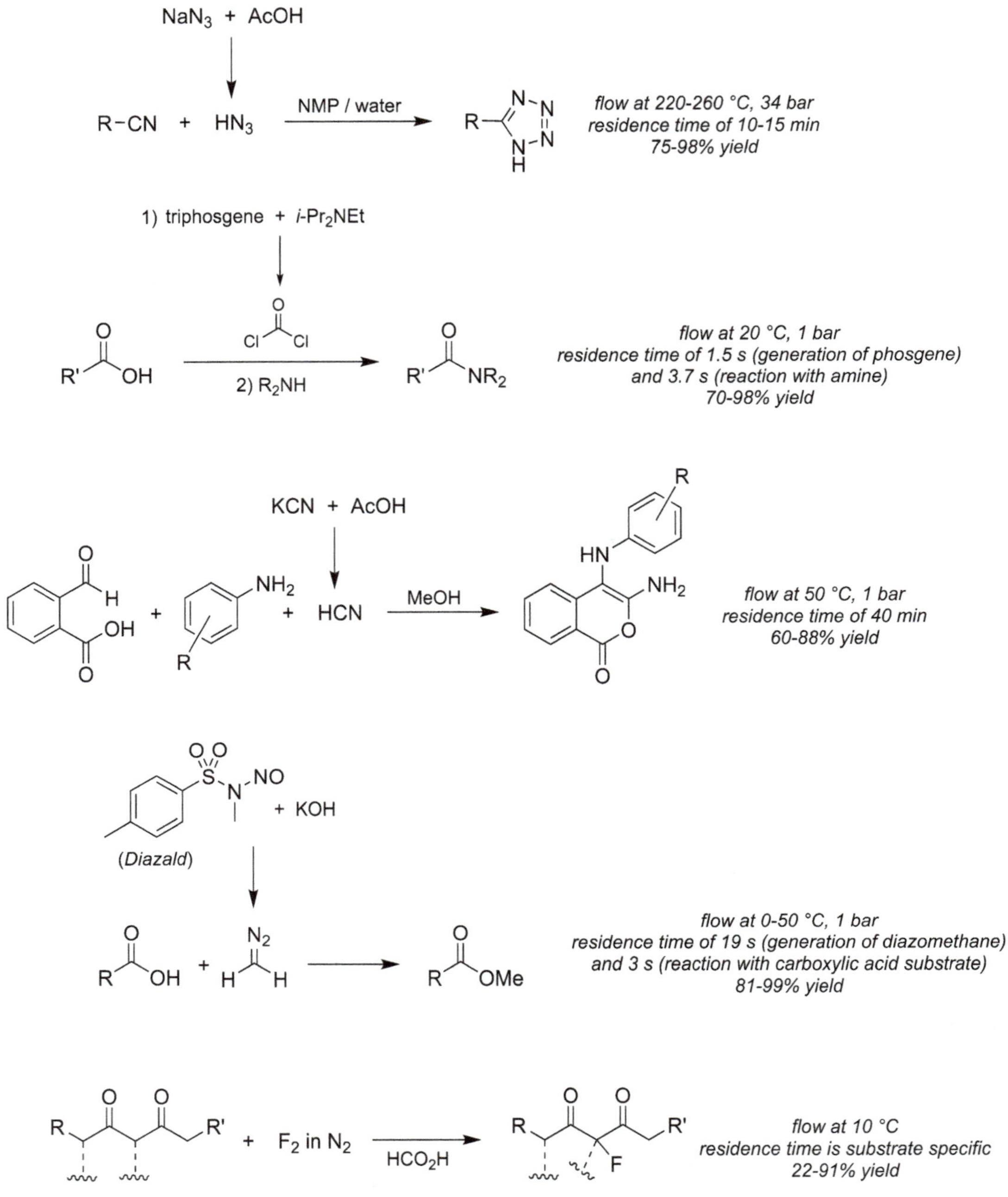

Scheme 7 Illustrating the use of hazardous reagents such as hydrazoic acid [48–50], phosgene [48–50], hydrogen cyanide [48–50], diazomethane [48–50], or elemental fluorine [48–50] using flow chemistry

advantages of flow processing. Employment of a flow approach in nitration reactions has also garnered significant attention because these reactions are dangerous, exothermic, and often non-specific when using traditional batch reactors. Toluene [51–53], phenol [54], benzaldehyde [55], and salicylic acid [56] have all been mono-nitrated in flow, often employing the standard conditions

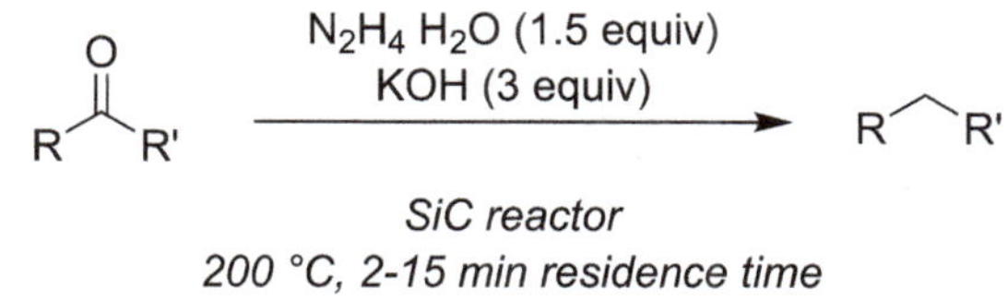

Scheme 8 The Wolff-Kishner reduction of aldehydes and ketones

of nitric acid and sulfuric acid, but the reaction times are shorter and the product purity far superior than when using batch conditions.

Upping the ante even more, flow chemistry has been used to perform reactions using hydrazine, a reagent that is both explosive and challenging to handle. When using it in the Wolff-Kishner reduction of aldehydes and ketones to the corresponding alkanes, a further issue arises. The reaction requires the use of strong base and heating at temperatures in excess of 200 °C. Over time this can etch glass reaction vessels. Compounding the problem, metals are known to catalyze the decomposition of hydrazine, so contact with stainless steel should be avoided. This makes batch processing very challenging. Moving to flow chemistry could mitigate some of the hazards associated with the use of hydrazine, such as buildup of anhydrous hydrazine in the headspace of a reaction vessel. However, the problem of vessel compatibility still remains. Most flow apparatus uses either glass, steel, or thermally sensitive polymer tubing. All of these are incompatible with the reaction conditions required. One solution has been to fabricate and use a microreactor made of silicon carbide [57]. This ceramic material has exceptional chemical compatibility, temperature stability, and thermal conductivity. Using this approach, the Wolff-Kishner reaction can be performed effectively in flow, not only shortening reaction times but also reducing the amount of hydrazine required for the reaction (Scheme 8). Hydrazine has also been used as a precursor to diimide (N_2H_2) for transfer hydrogenation of alkenes to alkanes in flow [58]. Diimide is highly unstable at temperatures greater than −180 °C, making it very hard to handle and use effectively in batch mode. However, by using a flow approach, the diimide can be generated on demand in a reaction with oxygen and immediately consumed by the alkene reagent, thereby offering a valuable, high yielding, and highly selective, route to alkanes.

5 Flow Chemistry as a Tool for Multistep Organic Synthesis

Just as the automobile industry uses production lines for the assembly of cars from multiple components, chemists can use flow processing as a molecular factory. Concatenating synthetic transformations in a streamlined flow has proven to be a very valuable approach to building up molecular complexity [59]. Starting with a

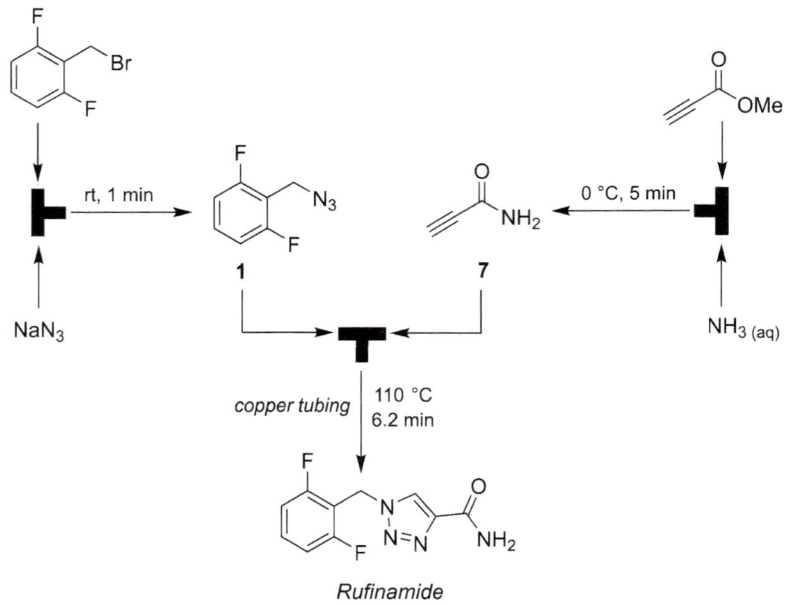

Scheme 9 Preparation of rufinamide

simple case in point, and building on a previous discussion in this chapter, rufinamide has been prepared in three steps (Scheme 9) [60]. Azide **1** is prepared by an S_N2 reaction of the requisite benzyl bromide with sodium azide. The stream of **1** is then mixed with a flow of propiolamide (**7**), itself prepared in the reaction of methyl propiolate with aqueous ammonia. The [3 + 2] cycloaddition of **1** and **7** is performed in a coil of copper tubing. The release of trace amounts of ionic copper complexes into the reaction stream catalyzes the click reaction. Of note is that the desired cycloaddition product is generated with greater than 20:1 regioselectivity. From start to finish, the total residence time of the synthesis is approximately 11 min, and rufinamide is obtained in 92% overall yield. In a similar vein, flow approaches have been taken to the synthesis of (*E*/*Z*) tamoxifen [61] and amitriptyline [62], both of which are important pharmaceuticals.

In batch, multistep synthesis often involves the isolation and purification of intermediates along the way. When considering flow chemistry, a similar approach often has to be taken. One option is to perform a step in flow, isolate the product using conventional techniques, and then continue on to the next step. Indeed, this is what is generally done since many flow reactions require some sort of purification step. But offline product purification is time-consuming and also defeats part of the object of using flow chemistry in the first place, namely, ease and speed of operation. In-line separation and purification offers a much more streamlined approach [63–68]. One example of this in action is in the preparation of 3,3,3-trifluoromethylpropenes from trifluoromethyl ketones by a Peterson olefination approach (Scheme 10). In

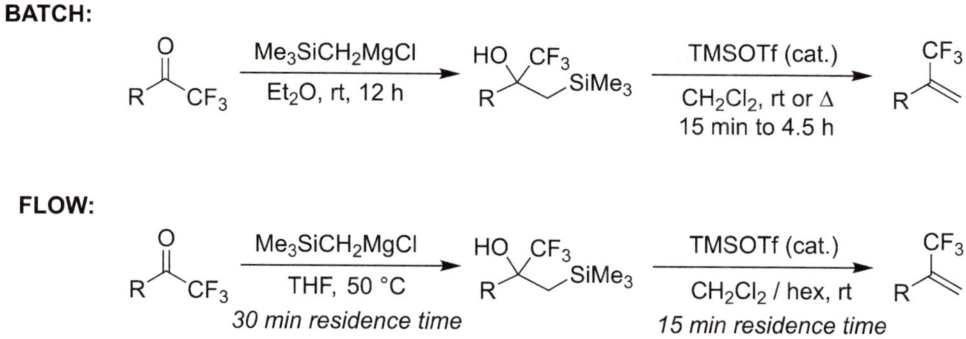

Scheme 10 Preparation of 3,3,3-trifluoromethylpropenes

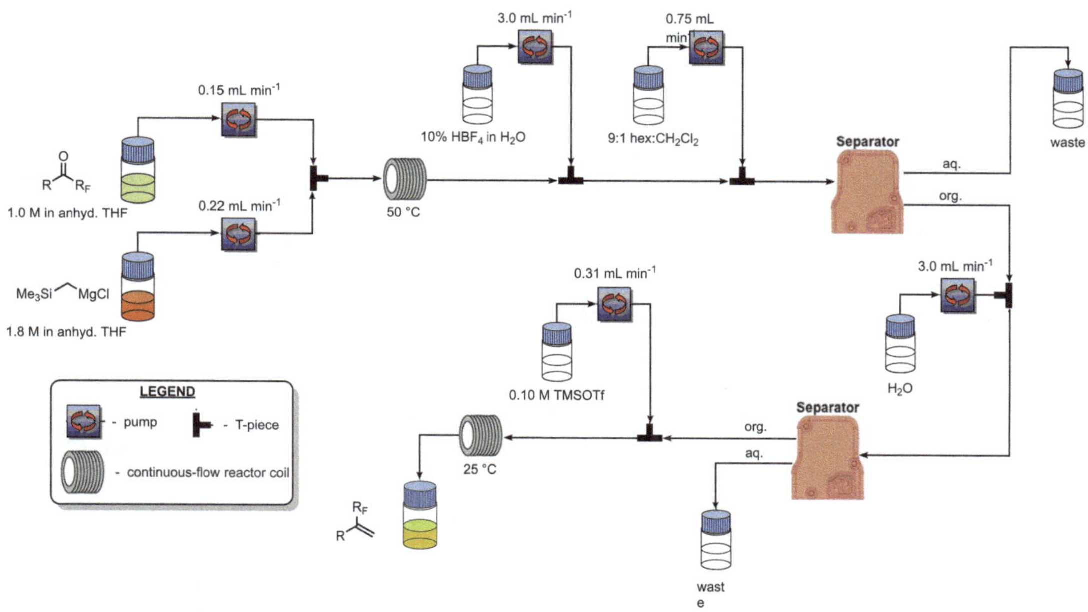

Fig. 1 The flow configuration used for the preparation of 3,3,3-trifluoromethylpropenes, involving in-line separation

batch, the approach involved Grignard addition of (trimethylsilyl)-methylmagnesium chloride to the trifluoromethyl ketone followed by dehydrative desilylation of the α-trifluoromethyl-β-hydroxysilyl alcohol using a catalytic amount of TMSOTf [69]. Moving to flow poses a challenge [70]. The first step of the reaction is performed in a Lewis basic solvent (diethyl ether or THF), but these solvents have a severely detrimental effect on the dehydrative desilylation step. Therefore an in-flow solvent switch is required prior to introducing the TMSOTf, at the same time removing magnesium by-products formed in the Grignard addition step. To achieve this, a membrane-based liquid-liquid separator is used [71]. A schematic for the flow approach is shown in Fig. 1. The strategy taken is to perform the Grignard reaction in THF in one flow coil,

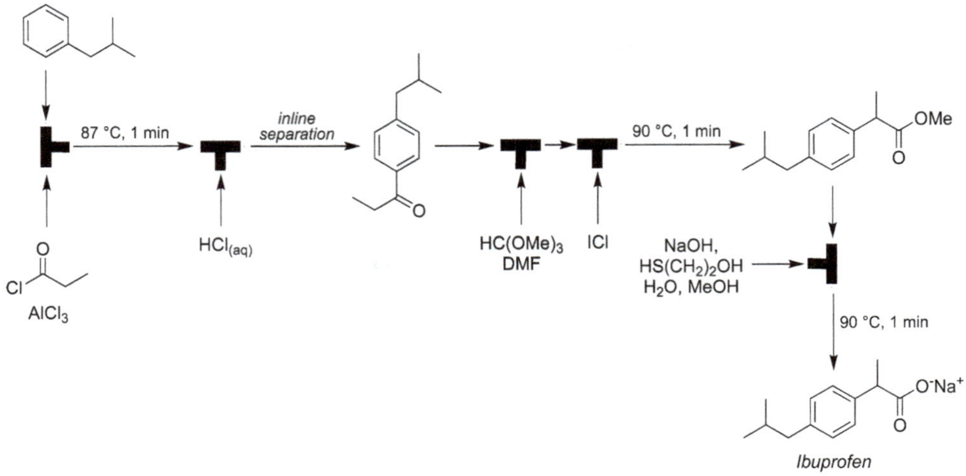

Scheme 11 Rapid preparation of ibuprofen in flow

introduce a stream of dilute tetrafluoroboric acid (to quench the Grignard reaction) followed by 9:1 hexane/dichloromethane, and then send this mixture into the aqueous/organic separator. The aim is to have the intermediate alcohol remain in the organic phase while at the same time partitioning the THF into the aqueous phase, along with the magnesium by-products. The separation is effective with the alcohol remaining entirely in the organic phase. However there is also a significant amount of THF left in the organic phase as the increased ionic strength of the aqueous stream decreased the solubility of THF. To circumvent this issue, a flow of deionized water is introduced into the organic stream leaving the first separator and then the resultant mixture passed through a second separator to remove more of the THF. The organic phase is then flowed into a reactor, together with a stream of TMSOTf and the desilylation successfully performed. Overall, the flow approach is not only faster but also 11–25% higher yielding than the corresponding batch methodology.

The same separator has been used in an expedited synthesis of ibuprofen (Scheme 11) [72]. Three chemical transformations are performed in a total of 3 min (1 min/step), each with a yield in excess of 90%. The in-line separator is used to purify the product from a Friedel-Crafts acylation. The approach also showcases some of the previously discussed advantages of flow chemistry, namely, that the use of solvent can be minimized and aggressive reagents, in this case neat ICl and a solution of $AlCl_3$ in propionyl chloride, can be employed without incident.

Moving beyond liquid-liquid extraction, immobilized scavengers packed in columns and cartridges are also of great benefit to provide in-line purification of a flow stream [73–75]. Using these,

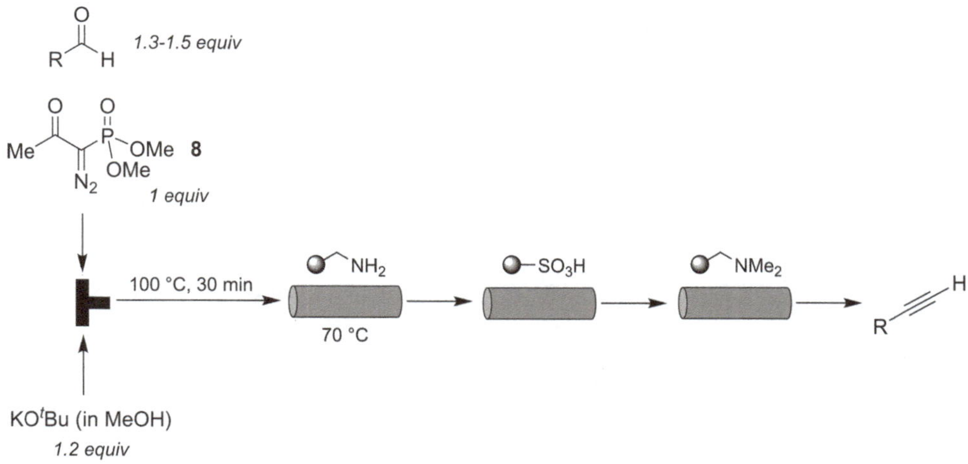

Scheme 12 Synthesis and in-line purification of alkynes from aldehydes

by-products or excess reagents can be sequestered from solution, and it is possible to avoid traditional work-up protocols, thereby allowing progression of the pure product flow stream to the next step of the synthetic process. An example of supported reagents in action is the preparation of alkynes from aldehydes using the Bestmann-Ohira reagent (**8**) [76]. After exiting the reactor coil, the product mixture is passed through a series of three scavenger columns. First, a tube is packed with benzylamine resin at 70 °C to remove any unreacted aldehyde, then supported sulfonic acid removes the base from the mixture and protonates any phosphoric groups arising from **8**, and finally an immobilized amine removes any acid material formed in the prior step (Scheme 12).

In a similar vein, supported catalysts [77–80] and reagents [76, 81–83] prove valuable in this endeavor. In essence, as the reagents pass through a bed of supported material, they "see" an excess of the catalyst or reagent and so reactions can take place very rapidly. In addition, by minimizing the components in solution at any one time, in-line product purification is again facilitated. Bringing all these concepts together, natural products and medicinally important compounds have been prepared using flow chemistry with no purification between steps [84]. These include the alkaloid oxomaritidine (**9**) synthesized in seven steps (five supported reagents, one supported catalyst, and one scavenger; along with a solvent switch) [85], and (*R*)-rolaripram (**10**), an anti-inflammatory drug, prepared in four steps (all using supported catalysts) [86]. The latter is particularly attractive since it employs only supported catalysts. While supported reagents and scavengers streamline a synthetic process, they need to be replaced (or in some cases regenerated) over time by simple virtue of their nature.

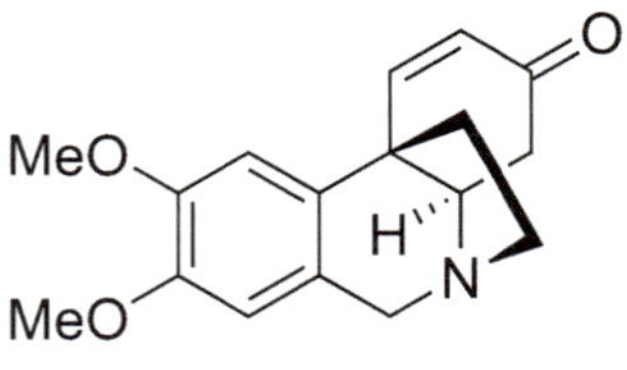

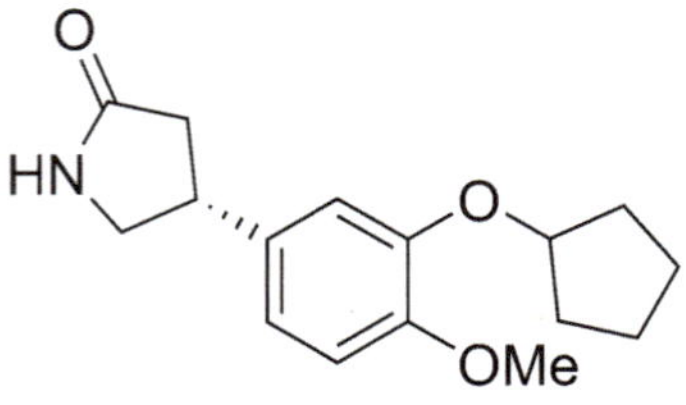

9 **10**

It is also possible to marry flow processing with conventional batch chemistry or purification for targets of high molecular complexity. Take for example (+)-dumetorine (**11**) which can be prepared in a synthetic strategy comprising of five synthetic steps using packed columns of reagents or scavengers, and one off-line chromatographic separation of diastereoisomers [87]. The overall yield for the entire sequence is 30%, compared to just 1% for the batch-only predecessor. Likewise, spirangien A methyl ester (**12**) and spirodienal (**13**) have been prepared in an approach that integrates the best of both worlds: batch and flow processing [88].

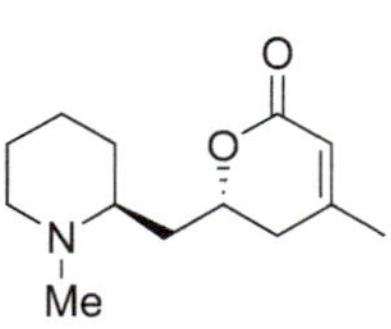

11

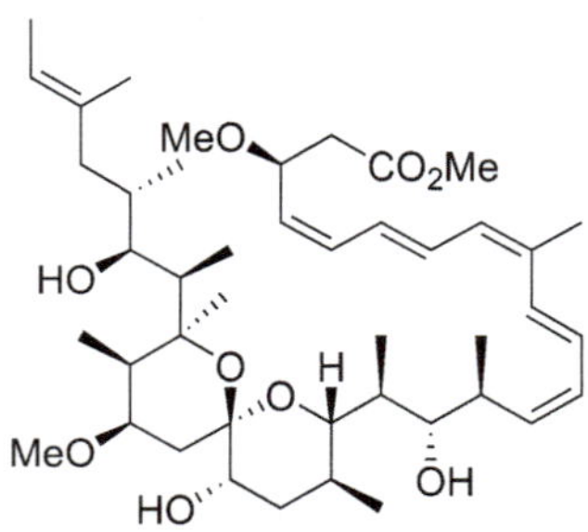

12

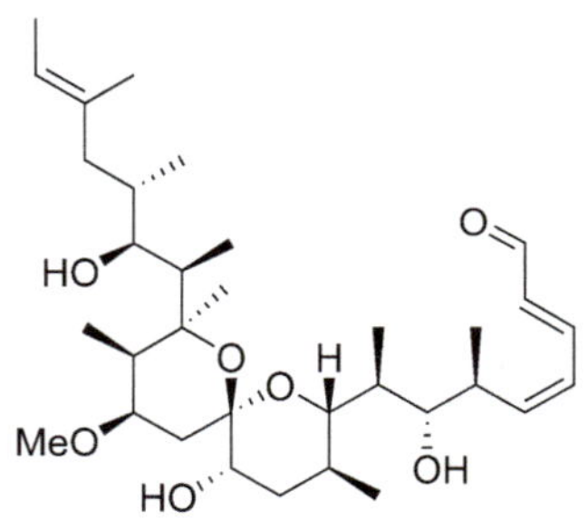

13

Another concern when performing multistep synthesis is monitoring the efficiency of each individual step. Any issues upstream in the process will impact subsequent steps. This is an issue of increasing importance as industry starts to implement flow processes on a manufacturing scale. The Federal Drug Administration (FDA) is on board with continuous processing, but of importance is the analysis of the product stream to demonstrate consistency in terms of performance. To this end, a number of spectroscopic tools have been interfaced with flow chemistry apparatus [89–91]. These include infrared [92–97], UV-visible [98, 99], NMR [100, 101], Raman [102–105], and mass [106–109] spectroscopy. This opens the avenue for fast, reliable assay in comparison with the traditional approach in which performance is evaluated based on off-line product analysis. In addition, being able to monitor reactions in real time opens avenues for rapid experimentation with a feedback loop for "on the fly" optimization of conditions.

6 Emerging Opportunities Using Flow Chemistry

6.1 Reactions Involving Gases

Interfacing gaseous reagents with flow processes is an area of growing interest. The small volumes of flow reactors ameliorate the hazards of high-pressure gas reactions and enable improved mixing with the liquid phase. Historically, reactions involving gases have relied on plug-flow approaches in which gas bubbles are introduced to a liquid stream. This has a number of drawbacks. An excess of gas is required since its dissolution relies on interfacial contact of the gas bubbles with the liquid phase. Also, obtaining consistent bubble flow is challenging since the size and regularity of gas bubbles fluctuate with changes in system pressure, temperature, and solution viscosity. A way to alleviate some of the operational issues is to employ a "tube-in-tube" approach, which comprises an outer PTFE tube and a gas-permeable Teflon inner tube [110]. The reagent stream flows within the inner membrane tubing, while the gas fills the PTFE outer tubing. Diffusion through the inner tube membrane allows for the transfer of gas into the reagent stream. This has been successful for numerous reactions including the ozonolysis of alkenes [111], carboxylation of Grignard reagents with carbon dioxide [62, 112], Heck reactions using ethene [113], direct C-H bond functionalization [114], Wacker oxidation [115] using oxygen, and both homogeneous [116] and heterogeneous [117] hydrogenation reactions (Scheme 13). It has also been employed as a tool for removing gas from solution during the course of a reaction [118] as well as for the generation and subsequent consumption of anhydrous diazomethane [119]. The first generation of tube-in-tube design, while excellent for ambient

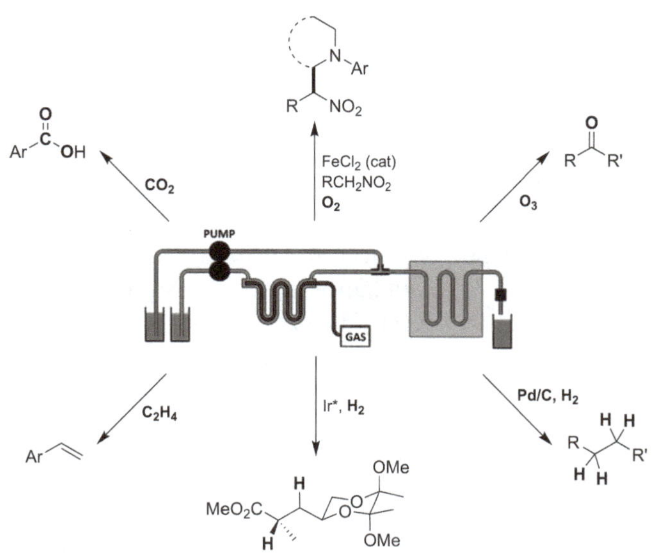

Scheme 13 Selected applications of a tube-in-tube gas reactor

reactions, is not amenable to operation at elevated temperatures. This is firstly because the liquid runs through the inner tube and so is difficult to heat externally. Secondly, the gas-permeable Teflon tube is not particularly robust. The reaction is thereby limited by the quantity of gas that can be loaded into the solution in the tube-in-tube unit prior to entering the heated zone. This can necessitate running reactions at more dilute concentrations to match the concentration of reagent and gas or performing multiple passes through the entire system to ensure complete consumption of starting materials. These issues have been overcome in a variant of the design. A coil of stainless steel tubing is used through which liquid flows on the outside of a gas-permeable inner membrane [120]. Using this, both gas input and heating can be performed simultaneously. Examples using hydrogen [121], carbon monoxide [122], and oxygen [123] have all been reported. Reactions can be performed at significantly higher concentrations than were possible before due to the fact that the supply of gas in solution can be continuously replenished as it is consumed. Hydrogenation reactions are run at 3 M and carbonylations at 1 M. On the flip side, while the inner gas-permeable membrane is more rugged, it is not as porous as in the first-generation design, thus limiting the size of the gas that can be used to small diatomic molecules like hydrogen, carbon monoxide, and oxygen.

6.2 Continuous-Flow Photochemistry

By bringing together flow chemistry and other tools, new vistas in synthetic chemistry are being discovered. One example is at the nexus of photochemistry and flow processing. In theory, photochemistry is an ideal synthetic tool for green chemistry, as was noted almost a century ago [124]. However, performing photochemical reactions in batch is not trivial, especially when working on a scale greater than milligrams. One of the key issues when using an immersion-well (batch) reactor is that light penetration to the surrounding solution is limited by the absorption of the substrate and the solvent and falls off rapidly with distance from the lamp. Generally, 90% of the light is absorbed within 0.01–1 mm of the lamp meaning that the majority of the reaction mixture is not being irradiated at any one time and good stirring is essential with long reaction times often being a consequence. In addition, immersion-well lamps generate a significant amount of heat, meaning that the reaction mixture often has to be cooled. Flow chemistry addresses all of these challenges. Due to the narrow tubing or microchannel chips that are used, light penetrates the entire cross section of the reaction mixture, and reaction times can be reduced from hours to minutes. Heat transfer is also not an issue due to the high surface area to volume ratio, so cooling the reactor is often not necessary. It is therefore no wonder that flow photochemistry is seeing a huge uptick in popularity [125–127]. Some of the pioneering work was undertaken using a homemade assembly of an immersion-well

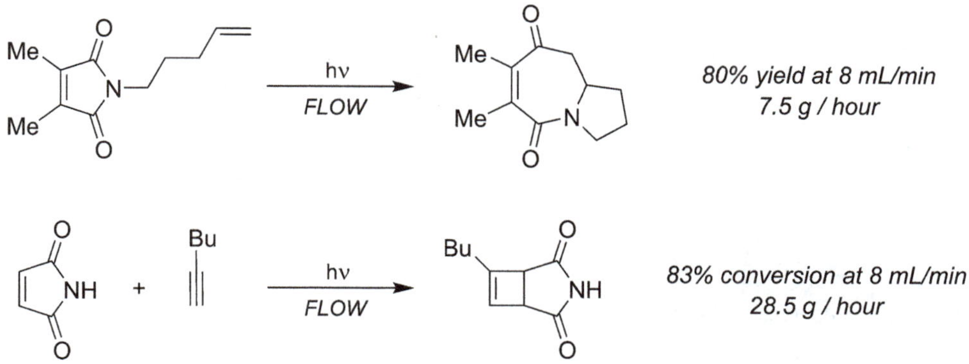

Scheme 14 Cycloadditions performed photochemically in flow

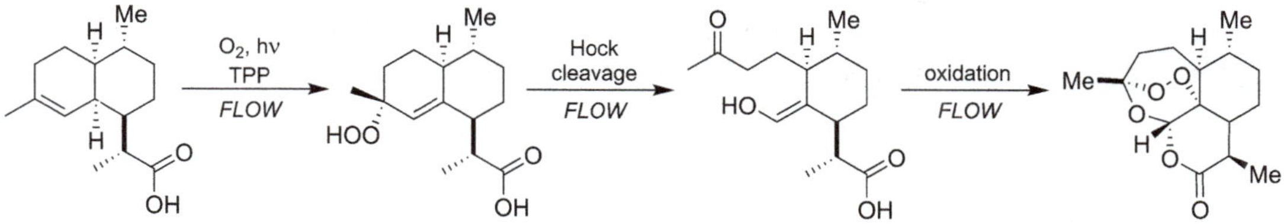

Scheme 15 Preparation of artemisinin using flow chemistry

lamp, around which UV-transparent tubing was wrapped [99, 128, 129]. It is possible to perform photocycloadditions in shorter times and higher yields than the corresponding batch approach (Scheme 14). More recently, photochemical accessories for commercially available flow reactors have come on to the market; both lamps and wavelength filters being available. The field has rapidly developed, and flow photochemistry has been used for transformations as varied as C-H bond activation, radical reactions, cyclizations, rearrangements, glycosylations, and the synthesis of polymer colloids. Perhaps the poster child for modern flow photochemistry is the continuous route to the antimalarial drug artemisinin (Scheme 15) [130, 131]. The route begins with dihydroartemisinic acid (**14**), a starting material that is available on large scale via fermentation. The first step of the process is a photochemical singlet oxygen-based ene reaction to yield a peroxide. The reaction is performed using tetraphenylporphyrin (TPP) as an oxygen sensitizer. The oxygen is introduced into a dichloromethane stream of **14** and sensitizer in a plug-flow approach before the mixture then passes through a tube wrapped around an immersion-well lamp. Under these conditions, 1.5 mmol of the peroxide product is produced per minute, in 91% conversion and 75% yield. The peroxide is then combined with trifluoroacetic acid and undergoes a Hock cleavage followed by further oxidation with oxygen gas to yield artemisinin in 69% yield from **14**. This corresponds to a throughput of 165 g of artemisinin per day. Similar approaches

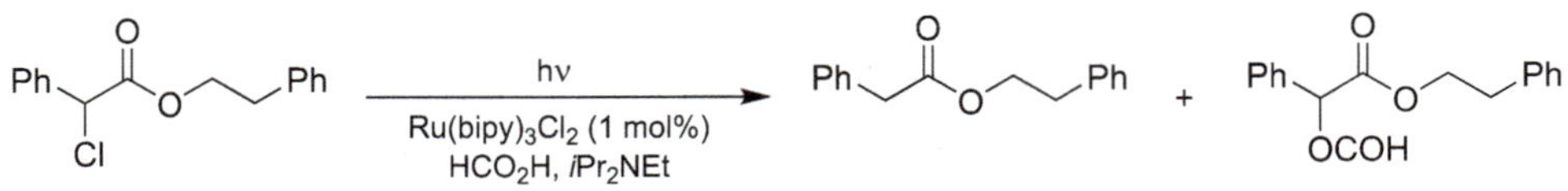

	residence time	ratio of **16** to **17**	conversion (%)
batch	24 h	84:16	98
flow	30 min	100:0	>95

Scheme 16 Cycloadditions performed photochemically in flow

have also been developed using liquid carbon dioxide and aqueous ethanol as the solvent instead of dichloromethane, with an eye to greener processing [132].

Flow photochemistry is not limited to UV irradiation. It is finding application in the field of visible light photoredox catalysis [133]. The paucity of visible light absorption by many organic molecules limits the application of photochemistry in synthesis. This barrier can be overcome by employing visible light absorbing photocatalysts and utilizing their electron/energy transfer processes to sensitize organic molecules to carry out reactions [134, 135]. The most commonly used photocatalyst is $Ru(bipy)_3Cl_2$ (bipy = 2,2-bipyridine). The excited-state species $[Ru(bipy)_3]^{2+*}$ can be easily oxidized or reduced by a quencher in solution. Oxidative quenching of $[Ru(bipy)_3]^{2+*}$ generates $[Ru(bipy)_3]^{3+}$, a strong oxidant, while reductive quenching generates $[Ru(bipy)_3]^{+}$, a strong reducing agent. Therefore, depending on the conditions used, $[Ru(bipy)_3]^{2+*}$ can be employed as either a single-electron oxidant or reductant. Since $Ru(bipy)_3Cl_2$ has a very high extinction coefficient, flow chemistry again shines through as a valuable tool, with light-emitting diodes (LEDs) being the irradiation source of choice. An example of the tool in action is the reductive dehalogenation of **15** to yield **16** (Scheme 16) [136]. Flow processing reduces the reaction time as well as eliminating by-product **17** generated in the corresponding batch process.

6.3 Continuous-Flow Electrochemistry

Just as the use of photochemistry as a synthetic tool is being revolutionized by the application of flow processing, the same is true for electrochemistry, though in this case the transition is still somewhat in its infancy [137, 138]. However, there is a common thread between the two. Electrochemical reactions offer a clean route for synthesis; electrons can be added or removed from substrates under mild conditions without the need for chemical oxidizing or reducing agents [139, 140]. Expensive equipment is not necessary; electrochemical synthesis can even be performed using a simple 6 V battery [141]. Batch electrochemistry does however have its drawbacks. For the technique to work well, supporting electrolytes often have to be used to dissipate charge beyond the very close proximity of the electrodes and hence ameliorate the large current gradient as a function of distance. These electrolytes must be removed after the reaction, often increasing the complexity of product isolation. In addition, conventional electrochemical reactors are operated in a galvanostatic mode (controlling the current). A high current is used but the potential on the electrodes cannot be controlled. Overvoltage and undefined current density distribution along the electrodes can lead to side products being formed, as well as eroding the electrode itself over time. Flow chemistry, specifically that using microreactors, can address these and other issues [142–145]. The distance between the anode and cathode can be very small, meaning that charge can be dissipated across the entire cross section of the reaction mixture, often without the need for supporting electrolytes. Another benefit of the small sample volume in the electrochemical cell is that reactions can be performed in a potentiostatic mode (controlling the potential difference between the electrodes by means of a third, reference, electrode), thus increasing the selectivity of the reaction (less side products formed) and prolonging the lifetime of the electrodes. Flow electrochemistry can be employed for a range of transformations (Scheme 17), examples being di- and trifluoromethylation by means of a decarboxylative dimerization reaction (Kolbe electrolysis) [146], reductive deprotection of the isonicotinyloxycarbonyl (*i*Noc) group [147], the oxidation of cyclic amines [148], and a range of benzylic compounds [149], as well as the synthesis of drug metabolites [150]. It is also possible to use electrochemistry and flow processing for the catalytic 2,2,6,6-tetramethylpiperidine-1-oxyl (TEMPO) oxidation of alcohols, with the oxidant being regenerated electrochemically [151].

Electrochemistry has also been married with flash chemistry to perform rapid, selective synthesis [152]. Just as highly reactive organolithium intermediates are generated and then quickly consumed before they become miscreant, electrochemically generated intermediates can be produced and used in the same way. A case in point is the Friedel-Crafts reaction of *N*-acyliminium ion **19** with 1,3,5-trimethoxybenzene (Scheme 18) [153]. The cation **19** is generated by the anodic oxidation of carbamate **18**. The solution containing **19** is flowed in to a micromixer in which the arene reactant is introduced. After a residence time of 0.03 s, triethylamine is flowed in as a quencher. Using this approach, the very rapid monoalkylation reaction takes place, but there is not enough time for a second, slower alkylation to take place.

6.4 Continuous-Flow Organocatalysis

At first glance, the phrase "organocatalysis in flow" seems a misnomer. Organocatalytic reactions, while highly specific and useful, are notoriously slow [154]. Reactions can take hours, if not days, to reach completion in many cases. Therefore flow chemistry, with

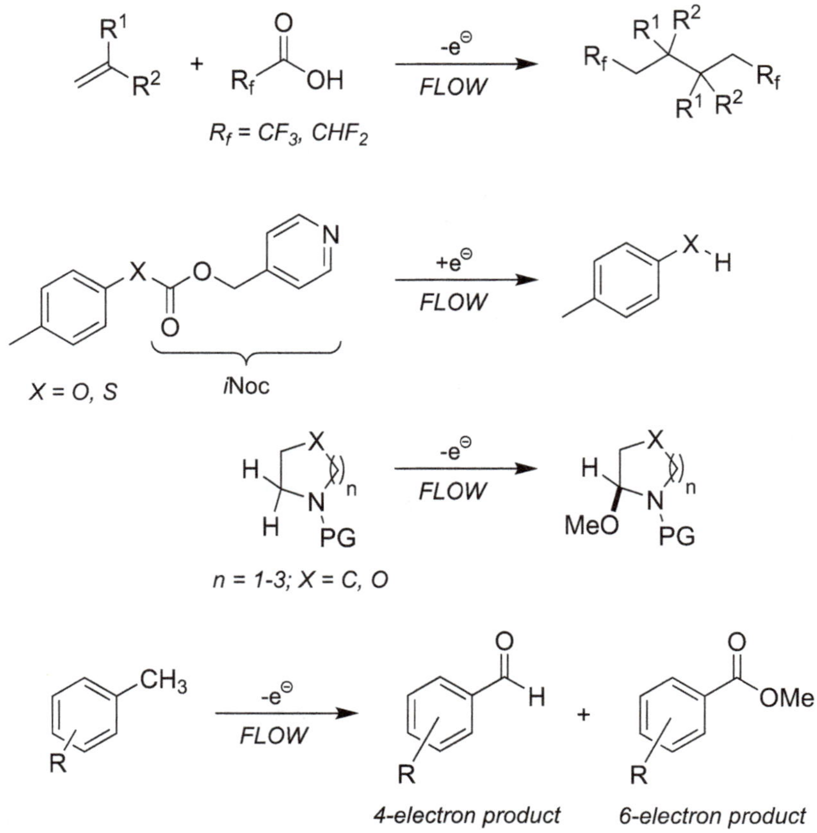

Scheme 17 Reactions performed electrochemically in flow

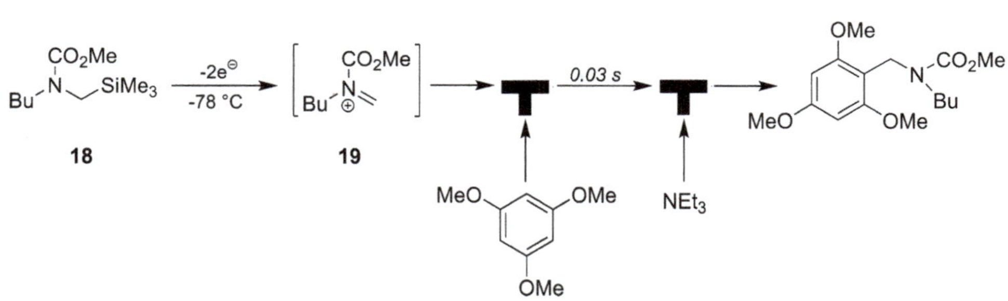

Scheme 18 Combination of electrochemistry and flash chemistry

prospective residence times as long as this, does not seem to be a method of choice. However, flow chemistry *can* be used [155, 156]. By passing through slightly heated reactor coils, times for homogeneous organocatalyzed reactions can be decreased to a level where reasonable reagent throughput can be obtained. Take for example the aldol reaction. Using proline-derived tetrazole catalyst **20**, it is possible to perform the reaction between acetone and 4-nitrobenzaldehyde in flow at 60 °C with a residence

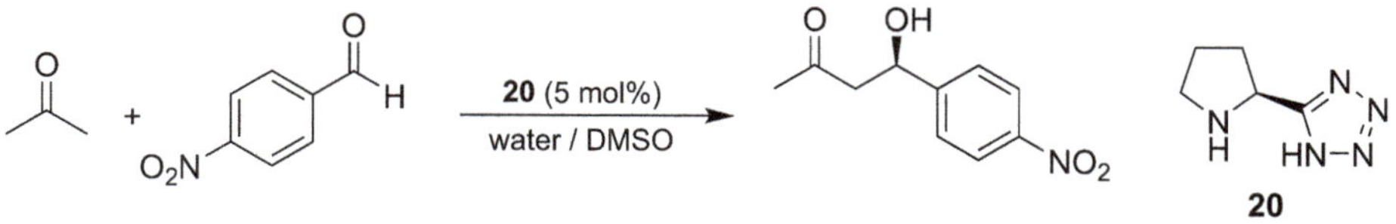

FLOW: 60 °C, 20 min residence time, 79% yield, 75% ee
BATCH: room temperature, 40 h, 77% yield, 74% ee *[60 °C, 20 min, identical to flow results]*

Scheme 19 Homogeneous asymmetric aldol reaction performed in flow

time of 20 min (Scheme 19) [157]. At room temperature in batch, the reaction takes 40 hours and the product is obtained in a similar yield. However, a bone of contention voiced by defenders of batch chemistry is that the reaction would have gone equally as fast in a flask if run at the same temperature [158]. It does—in the same yield and enantioselectivity.

Where flow processing begins to pull ahead of analogous batch reactions in homogeneous organocatalysis is when the synergy with photochemistry is leveraged. Either by concatenating a photochemical transformation, such as a cyclization, together with an organocatalyzed step [159, 160], or by using an organic dye as a photosensitizer in the presence of an organocatalyst [161], novel chemistry can be performed.

Flow chemistry truly comes in to its own when immobilized organocatalysts are used, and it is this area that has seen the most attention [162]. In essence, when passing a stream of reagents through a bed of supported catalyst, the reaction mixture sees a superstoichiometric quantity of the organocatalyst. As a result, issues associated with catalyst loading that are seen in batch do not exist in flow and reaction times can be significantly shortened. Another benefit is that the catalyst can, in principle, be reused multiple times. This is important given the complexity (and hence cost) of many organocatalysts. Again, a case in point is the asymmetric aldol reaction between acetone and 4-nitrobenzaldehyde, this time using a supported tripeptide as the organocatalyst [163]. A residence time of 6 min is enough to give product yields and selectivities on par with those in a batch reaction taking 6 hours (Scheme 20). The field of flow organocatalysis does have a way to go before it will be widely applicable. Deactivation of supported organocatalysts with time is an issue that needs to be addressed. In addition, many of the examples in the literature have been "optimized to order," in that they work well for the particular substrate being showcased, but the conditions are not widely applicable to a range of starting materials. That said, this is often a shortfall in the methodology and not on the applicability of flow to it.

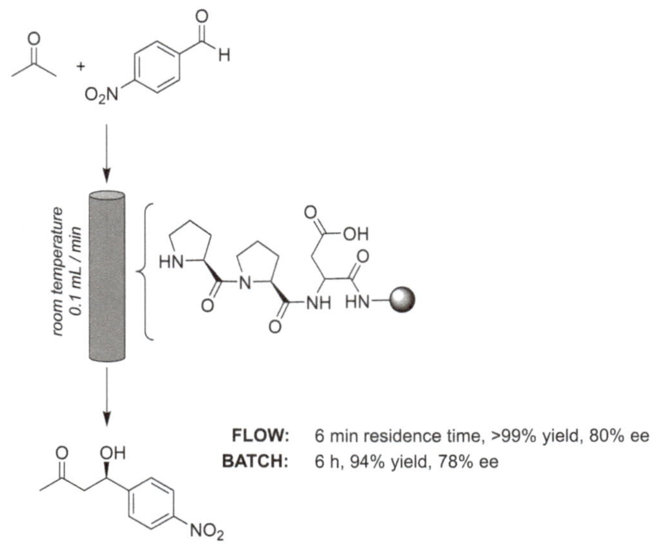

Scheme 20 Asymmetric aldol reaction performed in flow using a supported organocatalyst

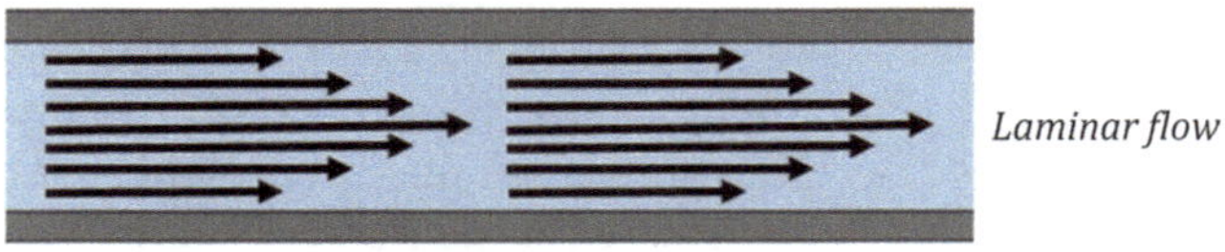

Fig. 2 Laminar flow of a liquid in a tube

6.5 Automation

Real-time monitoring of flow chemistry can lead to the diagnostic capture of information for use in controlling reaction parameters. One issue when performing multistep flow reactions is that of dispersion. As a reaction mixture passes along a length of tubing in a flow reactor, the material in the center of the tube begins to move faster than that near the walls. This is the result of a phenomenon known as "laminar flow" (Fig. 2). This is a problem because it means that some of the fluid takes longer to travel through the reactor than the rest. When a flow reactor is used to process a finite volume of reagents, the leading and trailing ends of the product emerging from the end of the reactor will have mixed to some extent with the fluid (usually solvent) that preceded or followed it. This means that there are zones at the leading and trailing ends of the product stream in which the concentration of product is variable and a steady-state portion between these in which the concentration is constant. When performing multistep processes, metering in reagents at the right time and at the right concentration becomes increasingly hard due to this dispersion. This can lead to compromised yields and product purification issues. However, using in-line IR monitoring interfaced with the appropriate software, it is possible to control pumps to dispense reagents in real

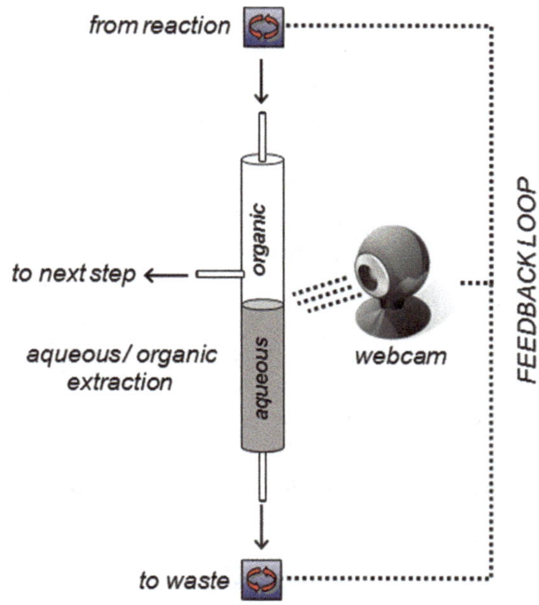

Fig. 3 Monitoring an in-line separation using a webcam

time based upon the concentration of reaction intermediates [164]. This enables precise mixing with perfect timing, thus greatly increasing product quality and enabling chemical flow processing to be used in extended reaction sequences. Gases can also be accurately metered by employing in-line IR monitoring. For example, the uptake of CO in palladium-catalyzed alkoxycarbonylation reactions can be quantified by monitoring the intensity of the CO stretching frequency at 2142 cm^{-1} [165].

Building on this concept, it is possible for flow processes to be self-optimizing [166, 167]. In essence, this involves performing the reaction and then varying parameters such as stoichiometry, temperature, and residence time, until optimal conditions are found. These iterations can be performed automatically by means of feedback loops between in situ reaction monitoring tools and the flow apparatus [95, 168–171]. Reaction monitoring is not limited to the use of spectroscopic tools. Something as simple as a webcam can be employed to observe changes and feedback to the flow equipment. For example, in an aqueous-organic extraction using a gravity separator, a webcam can be used to monitor the level of the phase boundary (Fig. 3) [172, 173]. The extraction involves an upper organic phase that passes on to the next step of the reaction and an aqueous phase that is pumped off to waste. If the level rises too high, the pump removing the aqueous phase can be speeded up or, if the level falls too low, the pump can be slowed down. A similar approach can be used to monitor reservoirs of reagents that are built up during the process of a multistep synthesis [174]. For instance, if an intermediate is generated at one flow rate, and then taken on to the next step of the synthesis at a slower flow rate, a reservoir helps buffer this change. Feedback loops are valuable in monitoring and maintaining the levels of these holding tanks. Turning to another tool, balances can be used to monitor the quantities of feedstocks present, the information then being used to ensure that pump flow rates are accurate and stoichiometric ratios consistent over extended periods [173]. Since all of these tools are computer controlled, they can be networked and overseen by a human operator anywhere in the world [175, 176]. It gives a whole new perspective to preparative chemists who can rightly claim they can do their work from home (or the beach).

6.6 Flow Chemistry in the Undergraduate Teaching Laboratory

A look at a standard organic chemistry laboratory manual shows that while a range of reactions are covered, they represent only a few of the many possibilities. In addition, the equipment used in undergraduate organic chemistry laboratories has not changed significantly in the last 50 years. This is of course very different to industrial chemistry facilities and most academic research laboratories where state-of-the-art equipment is often found. With flow chemistry being an enabling technology of the future, and with undergraduates being the chemists of the future, it is important to develop laboratory-based materials for incorporation of flow chemistry into the undergraduate teaching laboratory. This is beginning to happen, but there are some key metrics that come up in every discussion. The initial cost of the apparatus is often considered to be the major obstacle. The cheapest option may be a "build-your-own" design, but this can be time-consuming and also takes the instructor into areas of uncertainty when it comes to safety and reliability. That said, it does open up interesting opportunities for collaborations between chemistry and chemical engineering students, developing an in-house system together. Many of the commercially available flow reactors are out of the budget of an undergraduate teaching laboratory, especially if multiple units are needed to accommodate the number of students in the laboratory at any one time. Another issue faced with many of the systems is that there is a fairly steep learning curve for their effective use. Priming pumps, avoiding clogging due to particulate matter, and ease of operation of the user interface are all concerns when considering undergraduate students with little to no background with these types of equipment. To address this, two equipment manufacturers in particular have developed "undergraduate-friendly" versions of their equipment. One is based on a microreactor platform [177]. It operates at flow rates of 0.012–2.9 mL/min and at temperatures up to 200 °C. A series of seven experiments have been developed for use in a teaching setting [178]. They range from a simple aldol condensation to a more complex Wittig reaction using either homogeneous or biphasic conditions, as well as a Swern oxidation and the synthesis of silver nanoparticles. The reactions

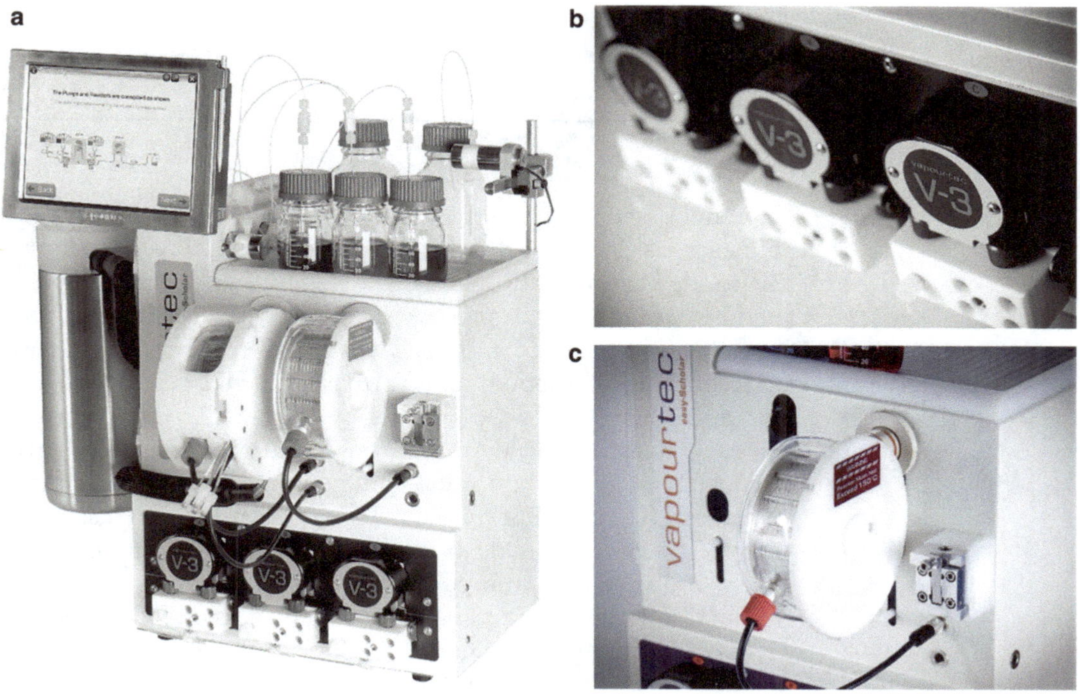

Fig. 4 The Vapourtec E-series system showing (**a**) the whole system, (**b**) the pump modules, and (**c**) a reactor coil in position (Reproduced with permission from Vapourtec Ltd)

take between 4 and 8 h to perform, which in some regard is too long for a regular undergraduate laboratory period. In another example of the use of microreactors in the teaching laboratory, two series of experiments have been developed to show the advantages of flow chemistry as well as, in one case, a direct comparison between batch and flow processing [179, 180]. The other approach is to perform reactions on a larger scale. A flow chemistry manufacturer has developed a unit comprised of two or three self-priming peristaltic pumps (Fig. 4) [181]. Flow rates range from 0.1 to 10 mL/min, and a variety of reactors can be used with the system, the most commonly used for educational purposes being 10 mL PFA coils capable of operation up to 150 °C. An advantage of this from an undergraduate laboratory setting is that throughput can be higher and students can make more material per unit time, shortening reaction times and ameliorating the effects of loss of small quantities of product in transfers or work-up stages. A set of ten experiments has been developed for incorporation into the undergraduate laboratory [182]. As before, each experiment is designed to highlight a particular concept in organic chemistry. Many of the experiments reinforce material that is covered in mainstream organic chemistry courses, such as cycloadditions, rearrangements, and the reaction chemistry of carbonyl compounds. It also extends the scope of experiments undergraduates can perform in a laboratory class to "Nobel prize-winning" examples, such as ring-closing metathesis and metal-catalyzed couplings.

7 The Matter of Scale

Continuous-flow processing has been, and is still, performed on immense scales in the petrochemical arena. Here there are stringent environmental regulations, many competitors, and small profit margins meaning that processes have to be high-performing, cost-effective, safe, and atom efficient in order to be viable. However, the range of reactions performed is focused on a few processes. For flow chemistry to be adopted widely in the fine chemicals industry, the wider array of reactions used on a day-to-day basis need to be performed on scales commensurate with the quantities of material required. But scale-up of flow processes is often an easier prospect than considering the corresponding batch approach. In batch, if a process is transitioned from the discovery lab to the production facility, the reaction conditions often need to be modified to take into account changes in mixing and heating or cooling characteristics. However, in flow, processes can be very effectively scaled without having to go to great lengths re-optimizing reaction conditions. To increase the scale of a microscale flow reactor while still taking advantage of the very efficient mixing and heat transfer, one option is to just run multiple reactors simultaneously, that is, to scale the process *out*. A bank of reactors can be assembled and used as a factory for the production of significant quantities of material. A case in point (and a pause for thought) is the production of nitroglycerin on a scale of 10 kg/h using microreactor technology [183]. This poisonous, explosive compound (which is notably for medical use and not nefarious activity) is prepared by a highly exothermic nitration process that runs efficiently and safely in the microreactor assembly.

The other option is to scale *up* by increasing the reactor length and diameter. This helps overcome some of the technical challenges of scaling out, namely, the cost of the individual reactors and the challenge of pumping liquid evenly through a bank of microreactors. It is also still possible to ensure effective heat transfer and mixing when using wider-bore tubing, with the correct engineering in place. Returning to the theme of organolithium chemistry, one example is in the case of **21** which can be formylated on the kilogram scale using two 6 mL volume stainless steel coil reactors of 0.25 in. diameter immersed in a dry ice/acetone bath (Scheme 21) [184]. A stream of the lithium salt of **21**, formed in batch by the reaction of **21** with MeLi, is first combined with a flow of *n*-butyl lithium to form a dianion in the first flow reactor. After a residence time of 2 s, a stream of *N*,*N*-dimethylformamide (DMF) in THF is introduced and the reaction mixture passed through a

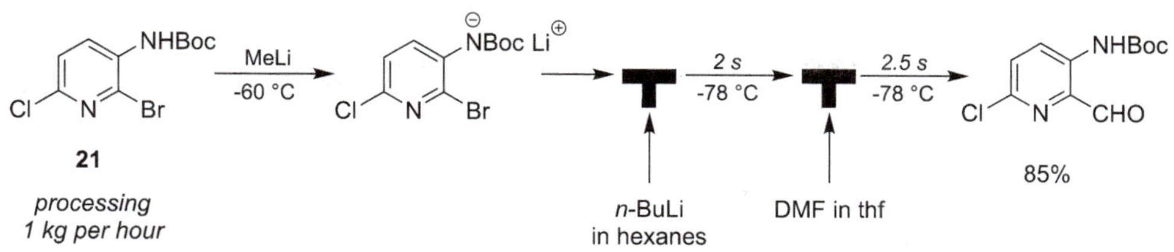

Scheme 21 Scale-up of an organolithium reaction in flow

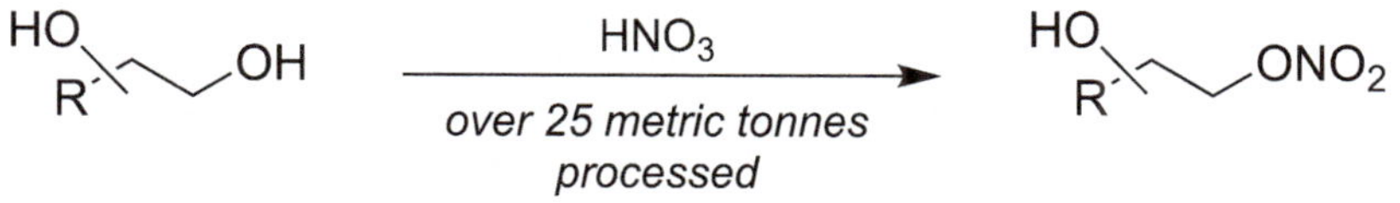

Scheme 22 Scale-up and scale-out of a nitration reaction in flow

second reactor coil for a residence time of 2.5 s before being fed into an aqueous quench solution. Even at a flow rate of 175 mL/min, efficient cooling can be maintained and 1 kg of **18** can be processed per hour. The purity of the product is notably higher than that from a corresponding batch approach; the level of debrominated side product was markedly lower (4% in flow *vs* 7–8% in batch).

Sometimes the best option is to scale up and out. Nitration chemistry provides an example. In the nitration of an undisclosed symmetric terminal diol, a two-phase organic/aqueous reaction mixture is ideal for performing the reaction on a small scale in a microreactor where both heat transfer and efficient mixing are possible (Scheme 22) [185]. When it comes to increasing the throughput, one option is to use a large number of microreactors, but the quantity required is financially prohibitive. The other option is to use a wide-bore tube, but as tube diameter is increased, mixing becomes less efficient as does heat transfer, both of which are essential in a biphasic, highly exothermic reaction like this. The solution is to use a number of medium-bore tube reactors, each of them with a working volume of 150 mL and capable of processing 13 kg of material per hour. Using this approach over 25 metric tonnes of product has been prepared in a single campaign. Another significant benefit of using flow processing is that the mono-nitrated product can be made selectively.

8 Concluding Remarks

Continuous-flow processing is a real game changer in synthetic chemistry. The examples showcased in this chapter are really only the tip of the iceberg. Flow chemistry is opening avenues for safer,

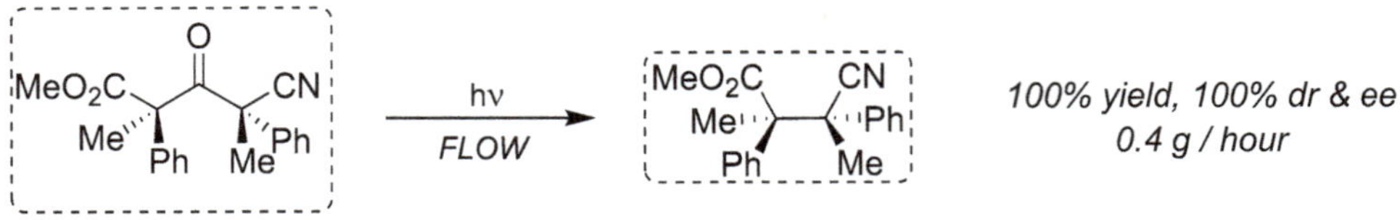

Scheme 23 Photodecarboxylation of nanocrystalline ketones in flow

easier, more automated preparation of a wide array of chemicals. This whole book is centered on the tenets of green chemistry, their application to drug synthesis, and the transition from academia to industry. Flow processing fits very well in this regard. It offers the best, safest, and easiest way to perform a range of transformations that in batch would be prohibited on any reasonable scale or at least very dangerous. Also, the interweaving of work performed in academic laboratories with that undertaken in an industrial setting has been shown. Reactions and technology developed in a university can be readily adopted in industry, and the reverse has been true too. Scale-up, scale-out, or a combination of the two can be used to turn a process making a few milligrams into one making several kilograms or more.

The synergistic combination of flow chemistry and other enabling tools has been highlighted. Both photochemistry and electrochemistry have huge potential as clean, green ways to perform synthetic chemistry in a very selective manner, but when weighed on the scales of efficiency in batch, they are found wanting. By using flow chemistry, a paradigm shift is observed and new vistas for synthetic chemistry become open. This is exemplified by the generation of compounds bearing adjacent quaternary chiral centers by means of a continuous-flow photodecarbonylation reaction of solid ketone precursors (Scheme 23) [186]. Nanocrystals of acyclic, homochiral, hexasubstituted ketones are suspended in water and flowed past two UV lamps where the decarbonylation reaction occurs generating a product bearing two adjacent stereogenic, all-carbon substituted quaternary centers, in quantitative chemical yield and 100% diastereoselectivity and enantiomeric excess.

The drive toward automation has also been discussed. It is possible to perform in-line extractions or separations and even chromatography [187]. The reactions can be monitored in real time using an arsenal of spectroscopic tools, as well as something as simple as a webcam. Flow chemistry can also be a mini-factory, with multiple chemical steps being performed in sequence, opening the avenue for building molecular complexity in a novel manner. Pushing the boundaries of this concept, one recent example is the end-to-end integrated continuous manufacturing plant for aliskiren hemifumarate (**22**) [188–190]. The plant started with an advanced intermediate (**23**), two synthetic steps away from the final active pharmaceutical ingredient, and ended with finished tablets. Flow processing is used for synthesis, purification, formulation, and final tablet production in one controlled sequence.

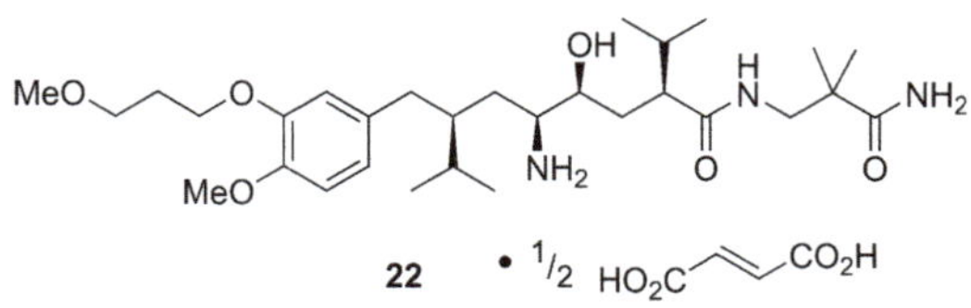

22

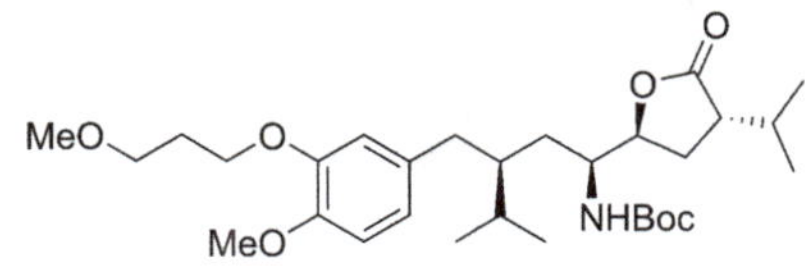

23

A life-cycle analysis of a number of flow processes has been undertaken and the results compared with those from batch approaches for the same reactions [191–195]. On the whole, flow chemistry comes out on top, with energy demand being substantially lower. Economics are also in the favor of flow processing. Once the capital equipment has been purchased, costs are lowered due to the decrease in personnel required to operate a large-scale continuous process compared to running the reactions in batch. That said, flow chemistry is not a magic bullet. There are drawbacks in some cases, as discussed at various junctures of this chapter. Additionally, flow chemistry cannot make a poorly designed synthetic route to a target product better, though can potentially make it safer and easier.

In closing, just as the factory production line has revolutionized the manufacturing of consumer goods, continuous-flow processing is changing the way that chemists in drug discovery and beyond are thinking of performing their reactions. It is opening avenues to new chemical space and doing so in a safe, easy, and reliable way. This, in addition to the increased level of automation that is possible, means that it will not be long before we see compact and efficient systems comprising of flow reactors, in-line purification tools, and reaction monitoring modules that a chemist can walk up to and say "make me my molecule," be they in an academic laboratory, an industrial setting, or even in a remote location where the compound is needed [48–50].

Acknowledgements

The students in our laboratory involved in flow chemistry projects are thanked for their hard work and dedication. Our efforts have been funded by the National Science Foundation (CAREER Award CHE-0847262) and the University of Connecticut. We also thank Vapourtec Ltd. for equipment support as well as input on a number of projects.

References

1. Darvas F, Dormán G, Hessel V (eds) (2014) Flow chemistry: volume 1 (fundamentals) and volume 2 (applications). DeGruyter, For books on flow chemistry see this reference
2. Reschetilowski W (ed) (2013) Microreactors in preparative chemistry: practical aspects in bioprocessing, nanotechnology, catalysis and more. Wiley-VCH, For books on flow chemistry see this reference
3. Wiles C, Watts P (2011) Micro reaction technology in organic synthesis. CRC Press, For books on flow chemistry see this reference
4. Luis SV, Garcia-Verdugo E (eds) (2010) Chemical reactions and processes under flow conditions. RSC Green Chemistry, For books on flow chemistry see this reference
5. Gutmann B, Cantillo D, Kappe CO (2015) Continuous-flow technology—a tool for the safe manufacturing of active pharmaceutical ingredients. Angew Chem Int Ed. https://doi.org/10.1002/anie.201409318, For selected recent reviews, see this reference
6. Jensen KF, Reizman BJ, Newman SG (2014) Tools for chemical synthesis in microsystems. Lab Chip 14:3206–3212, For selected recent reviews, see this reference
7. Baxendale IR (2013) The integration of flow reactors into synthetic organic chemistry. J Chem Technol Biotechnol 88:519–552, For selected recent reviews, see this reference
8. Protasova LN, Bulut M, Ormerod D, Buekenhoudt A, Berton J, Stevens CV (2013) Latest highlights in liquid-phase reactions for organic synthesis in microreactors. Org Process Res Dev 17:760–791, For selected recent reviews, see this reference
9. McQuade DT, Seeberger PH (2013) Applying flow chemistry: methods, materials, and multistep synthesis. J Org Chem 78:6384–6389, For selected recent reviews, see this reference
10. Deadman BJ, Browne DL, Baxendale IR, Ley SV (2015) Back pressure regulation of slurry-forming reactions in continuous flow. Chem Eng Technol 38:259–262. See this reference, for example
11. Hartman RL (2012) Managing solids in microreactors for the upstream continuous processing of fine chemicals. Org Process Res Dev 16:870–887. See this reference, for example
12. Browne DL, Deadman BJ, Ashe R, Baxendale IR, Ley SV (2011) Continuous flow processing of slurries: evaluation of an agitated cell reactor. Org Process Res Dev 15:693–697. See this reference, for example
13. Hartman RL, Naber JR, Zaborenko N, Buchwald SL, Jensen KF (2010) Overcoming the challenges of solid bridging and constriction during Pd-catalyzed C−N bond formation in microreactors. Org Process Res Dev 14:1347–1357. See this reference, for example
14. Roberge DM, Ducry L, Bieler N, Cretton P, Zimmermann B (2005) Chem Eng Technol 28:318–323. See this reference, for example
15. Vaccaro L, Lanari D, Marrocchi A, Strappaveccia G (2014) Flow approaches towards sustainability. Green Chem 16:3680–3704. For reviews see this reference
16. Wiles C, Watts P (2014) Continuous process technology: a tool for sustainable production. Green Chem 16:55–62. For reviews see this reference
17. Wiles C, Watts P (2014) Continuous flow reactors: a perspective. Green Chem 16:38–54. For reviews see this reference
18. Newman SG, Jensen KF (2013) The role of flow in green chemistry and engineering. Green Chem 15:1465–1472. For reviews see this reference
19. Fekete M, Glasnov T (2014) Technology overview /overview of the devices. In: Darvas F, Dormán G, Hessel V (eds) Flow chemistry: volume 1. DeGruyter. For an overview of equipment currently available, see this reference
20. Glasnov TN, Kappe CO (2011) The microwave-to-flow paradigm: translating high-temperature batch microwave chemistry to scalable continuous-flow processes. Chem Eur J 17:11956–11968
21. Noel T, Buchwald SL (2011) Cross-coupling in flow. Chem Soc Rev 40:5010–5029
22. Hessel V, Kralisch D, Kockmann N, Noël T, Wang Q (2013) Novel process windows for enabling, accelerating, and uplifting flow chemistry. ChemSusChem 6:746–789. For a review, see this reference
23. Lengyel LC, Sipos G, Sipőcz T, Vágó T, Dormán G, Gerencsér J, Makara G, Darvas F (2015) Synthesis of condensed heterocycles by the Gould-Jacobs reaction in a novel three-mode pyrolysis reactor. Org Process Res Dev 19:399–409. For examples, see this reference
24. Kobayashi H, Driessen B, van Osch DJGP, Talla A, Ookawara S, Noël T, Hessel V (2013) The impact of novel process windows

on the Claisen rearrangement. Tetrahedron 69:2885–2890. For examples, see this reference

25. Cantillo D, Sheibani H, Kappe CO (2012) Flash flow pyrolysis: mimicking flash vacuum pyrolysis in a high-temperature/high-pressure liquid-phase microreactor environment. J Org Chem 77:2463–2473

26. Lengyel L, Nagy TZ, Sipos G, Jones RV, Dorman G, Urge L, Darvas F (2012) Highly efficient thermal cyclization reactions of alkylidene esters in continuous flow to give aromatic/heteroaromatic derivatives. Tetrahedron Lett 53:738–743

27. Mudd WH, Stevens EP (2010) An efficient synthesis of rufinamide, an antiepileptic drug. Tetrahedron Lett 51:3229–3231

28. Borukhova S, Noël T, Metten B, de Vos E, Hessel V (2013) Solvent- and catalyst-free Huisgen cycloaddition to rufinamide in flow with a greener, less expensive dipolarophile. ChemSusChem 6:2220–2225

29. Kelly CB, Lee CX, Leadbeater NE (2011) An approach for continuous-flow processing of reactions that involve the in situ formation of organic products. Tetrahedron Lett 52:263–265. See this reference for example

30. Browne DL, Baumann M, Harji BH, Baxendale IR, Ley SV (2011) A new enabling technology for convenient laboratory scale continuous flow processing at low temperatures. Org Lett 13:3312–3315. See this reference for example

31. Gustafsson T, Sörensen H, Pontén F (2012) Development of a continuous flow scale-up approach of reflux inhibitor AZD6906. Org Process Res Dev 16:925–929. See this reference for example

32. Hessel V, Hofmann C, Löwe H, Meudt A, Scherer S, Schönfeld F, Werner B (2004) Selectivity gains and energy savings for the industrial phenyl boronic acid process using micromixer/tubular reactors. Org Process Res Dev 8:511–523

33. ThalesNano application note: cryogenic applications in flow chemistry enabled by the IceCube™ flow reactor—1st application: Swern oxidation. http://thalesnano.com/index.php?route=module/application_note/download&download=48. Accessed 2 Oct 2015

34. Kawaguchi T, Miyata H, Ataka K, Mae K, Yoshida J-i (2005) Room-temperature Swern oxidations by using a microscale flow system. Angew Chem Int Ed 44:2413–2416

35. Yoshida J-i (ed) (2008) Flash chemistry: fast organic synthesis in microsystems. Wiley. For an introduction to "flash chemistry", see this reference

36. Yoshida J-i, Takahashia Y, Nagakia A (2013) Flash chemistry: flow chemistry that cannot be done in batch. Chem Commun 49:9896–9904. For reviews, see this reference

37. Yoshida J-i, Nagaki A, Yamada T (2008) Flash chemistry: fast chemical synthesis by using microreactors. Chem Eur J 14:7450–7459. For reviews, see this reference

38. Hessel V, Lowe H, Schönfeld F (2005) Micromixers—a review on passive and active mixing principles. Chem Eng Sci 60:2479–2501. For background, see this reference

39. Kakuta M, Bessoth FG, Manz A (2001) Microfabricated devices for fluid mixing and their application for chemical synthesis. Chem Rec 1:395–405. For background, see this reference

40. Young IS, Baran PS (2009) Protecting-group-free synthesis as an opportunity for invention. Nat Chem 1:193–205

41. Nagaki A, Matsuo C, Kim S, Saito K, Miyazaki A, Yoshida J-i (2012) Lithiation of 1,2-dichloroethene in flow microreactors: versatile synthesis of alkenes and alkynes by precise residence-time control. Angew Chem Int Ed 51:3245–3248. See this reference, for example

42. Nagaki A, Kim H, Yoshida J-i (2009) Nitro-substituted aryl lithium compounds in microreactor synthesis: switch between kinetic and thermodynamic control. Angew Chem Int Ed 48:8063–8065. See this reference, for example

43. Nagaki A, Kim H, Moriwaki Y, Matsuo C, Yoshida J-i (2010) A flow microreactor system enables organolithium reactions without protecting alkoxycarbonyl groups. Chem Eur J 16:11167–11177. See this reference, for example

44. Usutani H, Tomida Y, Nagaki A, Okamoto H, Nokami T, Yoshida J-i (2007) Generation and reactions of o-Bromophenyllithium without benzyne formation using a microreactor. J Am Chem Soc 129:3046–3047. See this reference, for example

45. Kim H, Nagaki A, Yoshida J-i (2011) A flow-microreactor approach to protecting-group-free synthesis using organolithium compounds. Nat Commun 2:264

46. Nagaki A, Imai K, Ishiuchi S, Yoshida J-i (2015) Reactions of difunctional electrophiles with functionalized aryllithium compounds: remarkable chemoselectivity by flash chemistry. Angew Chem Int Ed 54:1914–1918

47. Webb D, Jamison TF (2012) Diisobutylaluminum hydride reductions revitalized: a fast, robust, and selective continuous flow system for aldehyde synthesis. Org Lett 14:568–571
48. Choi EJ, GSF L (2015) Manufacturing battlefield medicine: paradigm shift for pharmaceuticals. PDA J Pharm Sci Tech 68:312–312. The foundations are already being laid for this. See this reference
49. Baxendale IR, Braatz RD, Hodnett BK, Jensen KF, Johnson MD, Sharratt P, Sherlock J-P, Florence AJ (2015) Achieving continuous manufacturing: technologies and approaches for synthesis, workup, and isolation of drug substance. J Pharm Sci 104:781–791. The foundations are already being laid for this. See this reference
50. Lee SL, O'Connor TF, Yang X, Cruz CN, Chatterjee S, Madurawe RD, Moore CMV, Yu LX, Woodcock J (2015) Modernizing pharmaceutical manufacturing: from batch to continuous production. J Pharm Innov. https://doi.org/10.1007/s12247-015-9215-8. The foundations are already being laid for this. See this reference
51. Halder R, Lawal A, Damavarapu R (2007) Nitration of toluene in a microreactor. Catal Today 125:74–80
52. Panke G, Schwalbe T, Stirner W, Taghavi-Moghadam S, Wille G (2003) A practical approach of continuous processing to high energetic nitration reactions in microreactors. Synthesis:2827–2830
53. Antes J, Boskovic D, Krause H, Loebbecke S, Lutz N, Tuercke T, Schweikert W (2003) Analysis and improvement of strong exothermic nitrations in microreactors. Chem Eng Res Des 81:760–765
54. Ducry L, Roberge DM (2005) Controlled autocatalytic nitration of phenol in a microreactor. Angew Chem Int Ed 44:7972–7975
55. Kulkarni AA, Kalyani VS, Joshi RA, Joshi RR (2009) Continuous flow nitration of benzaldehyde. Org Process Res Dev 13:999–1002
56. Kulkarni AA, Nivangune NT, Kalyani VS, Joshi RA, Joshi RR (2008) Continuous flow nitration of salicylic acid. Org Process Res Dev 12:995–1000
57. Newman SG, Gu L, Lesniak C, Victor G, Meschke F, Abahmane L, Jensen KF (2014) Rapid Wolff–Kishner reductions in a silicon carbide microreactor. Green Chem 16:176–180
58. Pieber B, Martinez ST, Cantillo D, Kappe CO (2013) In situ generation of diimide from hydrazine and oxygen: continuous-flow transfer hydrogenation of Olefins. Angew Chem Int Ed 52:10241–10244
59. Webb D, Jamison TF (2010) Continuous flow multi-step organic synthesis. Chem Sci 1:675–680. For an overview, see this reference
60. Zhang P, Russell MG, Jamison TF (2014) Continuous flow total synthesis of rufinamide. Org Process Res Dev 18:1567–1570
61. Murray PRD, Browne DL, Pastre JC, Butters C, Guthrie D, Ley SV (2013) Continuous flow-processing of organometallic reagents using an advanced peristaltic pumping system and the telescoped flow synthesis of (E/Z)-tamoxifen. Org Process Res Dev 17:1192–1208
62. Kupracz L, Kirschning A (2013) Multiple organolithium generation in the continuous flow synthesis of amitriptyline. Adv Synth Catal 355:3375–3380
63. Newby JA, Huck L, Blaylock DW, Witt PM, Ley SV, Browne DL (2014) Investigation of a lithium-halogen exchange flow process for the preparation of boronates by using a Cryo-Flow reactor. Chem Eur J 20:263–271. See this reference, for example
64. Varas AC, Noël T, Wang Q, Hessel V (2012) Copper(I)-catalyzed azide–alkyne cycloadditions in microflow: catalyst activity, high-T operation, and an integrated continuous copper scavenging unit. ChemSusChem 5:1703–1707. See this reference, for example
65. Cervera-Padrell AE, Morthensen ST, Lewandowski DJ, Skovby T, Kiil S, Gernaey KV (2012) Continuous hydrolysis and liquid–liquid phase separation of an active pharmaceutical ingredient intermediate using a miniscale hydrophobic membrane separator. Org Process Res Dev 16:888–900. See this reference, for example
66. Hu DX, O'Brien M, Ley SV (2012) Continuous multiple liquid–liquid separation: diazotization of amino acids in flow. Org Lett 14:4246–4249. See this reference, for example
67. O'Brien M, Koos P, Browne DL, Ley SV (2012) A prototype continuous-flow liquid–liquid extraction system using open-source technology. Org Biomol Chem 10:7031–7036. See this reference, for example
68. Sahoo HR, Kralj JG, Jensen KF (2007) Multistep continuous-flow microchemical synthesis involving multiple reactions and separations. Angew Chem Int Ed 46:5704–5708. See this reference, for example

69. Hamlin TA, Kelly CB, Cywar RM, Leadbeater NE (2014) Methylenation of perfluoroalkyl ketones using a Peterson olefination approach. J Org Chem 79:1145–1155
70. Hamlin TA, Lazarus GML, Kelly CB, Leadbeater NE (2014) A continuous-flow approach to 3,3,3-trifluoromethylpropenes: bringing together Grignard addition, Peterson elimination, inline extraction, and solvent switching. Org Process Res Dev 18:1253–1258
71. Adamo A, Heider PL, Weeranoppanant N, Jensen KF (2013) Membrane-based, liquid–liquid separator with integrated pressure control. Ind Eng Chem Res 52:10802–10808
72. Snead DR, Jamison TF (2015) A three-minute synthesis and purification of ibuprofen: pushing the limits of continuous-flow processing. Angew Chem Int Ed 54:983–987
73. Hodge P (2005) Synthesis of organic compounds using polymer-supported reagents, catalysts, and/or scavengers in benchtop flow systems. Ind Eng Chem Res 44:8542–8553. For background on the use of supported reagents, catalysts, and scavengers in flow chemistry, see this reference
74. Baumann M, Baxendale IR, Ley SV (2011) The flow synthesis of heterocycles for natural product and medicinal chemistry applications. Mol Divers 15:613–630. See this reference, for example
75. Seeberger PH (2009) Organic synthesis: scavengers in full flow. Nature Chem 1:258–260
76. Baxendale IR, Ley SV, Mansfield AC, Smith CD (2009) Multistep synthesis using modular flow reactors: Bestmann–Ohira reagent for the formation of alkynes and triazoles. Angew Chem Int Ed 48:4017–4021
77. Rojo MV, Guetzoyan L, Baxendale IR (2015) A monolith immobilised iridium Cp* catalyst for hydrogen transfer reactions under flow conditions. Org Biomol Chem 13:1768–1777. For examples, see this reference
78. Maestre L, Ozkal E, Ayats C, Beltrán Á, Díaz-Requejo MM, Pérez PJ, Pericàs MA (2015) A fully recyclable heterogenized Cu catalyst for the general carbene transfer reaction in batch and flow. Chem Sci 6:1510–1515. For examples, see this reference
79. Reynolds WR, Plucinski P, Frost CG (2014) Robust and reusable supported palladium catalysts for cross-coupling reactions in flow. Cat Sci Technol 4:948–954. For examples, see this reference
80. Battilocchio C, Hawkins JM, Ley SV (2014) Mild and selective heterogeneous catalytic hydration of nitriles to amides by flowing through manganese dioxide. Org Lett 16:1060–1063. For examples, see this reference
81. Baxendale IR, Ley SV (2006) Solid supported reagents in multi-step flow synthesis. Ernst Schering Found Symp Proc 3:151–185. For background, see this reference
82. Baumann M, Baxendale IR (2013) The rapid generation of isothiocyanates in flow. Beilstein J Org Chem 9:1613–1619. For examples, see this reference
83. Hopkin MD, Baxendale IR, Ley SV (2010) A flow-based synthesis of Imatinib: the API of Gleevec. Chem Commun 46:2450–2452. For examples, see this reference
84. Pastre JC, Browne DL, Ley SV (2013) Flow chemistry syntheses of natural products. Chem Soc Rev 42:8849–8869
85. Baxendale IR, Deeley J, Griffiths-Jones CM, Ley SV, Saaby S, Tranmer GK (2006) A flow process for the multi-step synthesis of the alkaloid natural product oxomaritidine: a new paradigm for molecular assembly. Chem Commun:2566–2568
86. Tsubogo T, Oyamada H, Kobayashi S (2015) Multistep continuous-flow synthesis of (R)- and (S)-rolipram using heterogeneous catalysts. Nature 520:329–332
87. Riva E, Rencurosi A, Gagliardi S, Passarella D, Martinelli M (2011) Synthesis of (+)-dumetorine and congeners by using flow chemistry technologies. Chem Eur J 17:6221–6226
88. Newton S, Carter CF, Pearson CM, de Alves CL, Lange H, Thansandote P, Ley SV (2014) Accelerating spirocyclic polyketide synthesis using flow chemistry. Angew Chem Int Ed 53:4915–4920
89. Holmes N, Bourne RA (2014) Analysis and optimisation of continuous processes. In: Letcher T, Scott J, Patterson DA (eds) Chemical processes for a sustainable future. Royal Society of Chemistry. For a perspective, see this reference
90. Yue J, Schouten JC, Nijhuis TA (2012) Integration of microreactors with spectroscopic detection for online reaction monitoring and catalyst characterization. Ind Eng Chem Res 51:14583–14609. For a perspective, see this reference
91. Ferstl W, Klahn T, Schweikert W, Billeb G, Schwarzer M, Loebbecke S (2007) Inline analysis in microreaction technology: a suitable tool for process screening and optimization. Chem Eng Technol 30:370–378. For a perspective, see this reference

92. STR M, Murat A, Maillos D, Lesimple P, Hellier P, Wirth T (2015) Rapid generation and safe use of carbenes enabled by a novel flow protocol with in-line IR spectroscopy. Chem Eur J 21:7016–7020. For selected examples, see this reference
93. Rydzak JW, White DE, Airiau CY, Sterbenz JT, York BD, Clancy DJ, Dai Q (2015) Real-time process analytical technology assurance for flow synthesis of oligonucleotides. Org Process Res Dev 19:203–214. For selected examples, see this reference
94. Moore JS, Jensen KF (2014) "Batch" kinetics in flow: online IR analysis and continuous control. Angew Chem Int Ed 53:470–473. For selected examples, see this reference
95. Moore JS, Jensen KF (2012) Automated multitrajectory method for reaction optimization in a microfluidic system using online IR analysis. Org Process Res Dev 16:1409–1415. For selected examples, see this reference
96. Brodmann T, Koos P, Metzger A, Knochel P, Ley SV (2012) Continuous preparation of arylmagnesium reagents in flow with inline IR monitoring. Org Process Res Dev 16:1102–1113. For selected examples, see this reference
97. Carter CF, Lange H, Ley SV, Baxendale IR, Wittkamp B, Goode JG, Gaunt NL (2010) ReactIR flow cell: a new analytical tool for continuous flow chemical processing. Org Process Res Dev 14:393–404. For selected examples, see this reference
98. Benito-Lopez F, Verboom W, Kakuta M, Gardeniers JGE, Egberink RJM, Oosterbroek ER, van Den Berg A, Reinhoudt DN (2005) Optical fiber-based on-line UV/Vis spectroscopic monitoring of chemical reaction kinetics under high pressure in a capillary microreactor. Chem Commun:2857–2859
99. Lu H, Schmidt MA, Jensen KF (2001) Photochemical reactions and on-line UV detection in microfabricated reactors. Lab Chip 1:22–28
100. Gökay O, Albert K (2012) From single to multiple microcoil flow probe NMR and related capillary techniques: a review. Anal Bioanal Chem 402:647–669
101. Bart J, Kolkman AJ, Oosthoek-de Vries AJ, Koch K, Nieuwland PJ, Janssen JWG, van Bentum PJM, Ampt KAM, Rutjes FPJT, Wijmenga SS, Gardeniers JGE, Kentgens APM (2009) A microfluidic high-resolution NMR flow probe. J Am Chem Soc 131:5014–5015
102. Hamlin TA, Leadbeater NE (2013) Raman spectroscopy as a tool for monitoring mesoscale continuous-flow organic synthesis: equipment interface and assessment in four medicinally-relevant reactions. Beilstein J Org Chem 9:1843–1852
103. Chaplain G, Haswell SJ, Fletcher PDI, Kelly SM, Mansfield A (2013) Development and evaluation of a Raman flow cell for monitoring continuous flow reactions. Aust J Chem 66:208–212
104. Mozharov S, Nordon A, Littlejohn D, Wiles C, Watts P, Dallin P, Girkin JM (2011) Improved method for kinetic studies in microreactors using flow manipulation and noninvasive Raman spectrometry. J Am Chem Soc 133:3601–3608
105. Rinke G, Ewinger A, Kerschbaum S, Rinke M (2011) In situ Raman spectroscopy to monitor the hydrolysis of acetal in microreactors. Microfluid Nanofluid 10:145–143
106. Bristow TWT, Ray AD, O'Kearney-McMullan A, Lim L, McCullough B, Zammataro A (2014) On-line monitoring of continuous flow chemical synthesis using a portable, small footprint mass spectrometer. J Am Soc Mass Spectrom 2014 (25):1794–1802
107. Hamilton SE, Mattrey F, Bu X, Murray D, McCullough B, Welch CJ (2014) Use of a miniature mass spectrometer to support pharmaceutical process chemistry. Org Process Res Dev 18:103–108
108. Browne DL, Wright S, Deadman BJ, Dunnage S, Baxendale IR, Turner RM, Ley SV (2012) Continuous flow reaction monitoring using an on-line miniature mass spectrometer. Rapid Commun Mass Spectrom 26:1999–2010
109. Koster S, Verpoorte E (2007) A decade of microfluidic analysis coupled with electrospray mass spectrometry: an overview. Lab Chip 7:1394–1412
110. Brzozowski M, O'Brien M, Ley SV, Polyzos A (2015) Flow chemistry: intelligent processing of gas–liquid transformations using a tube-in-tube reactor. Acc Chem Res 48:349–362. For a review, see this reference
111. O'Brien M, Baxendale IR, Ley SV (2010) Flow ozonolysis using a semipermeable Teflon AF-2400 membrane to effect gas/liquid contact. Org Lett 12:1596–1598
112. Polyzos A, O'Brien M, Petersen TP, Baxendale IR, Ley SV (2011) The continuous-flow synthesis of carboxylic acids using CO_2 in a tube-in-tube gas permeable membrane reactor. Angew Chem Int Ed 50:1190–1193
113. Bourne SL, Koos P, O'Brien M, Martin B, Schenkel B, Baxendale IR, Ley SV (2011) The continuous flow synthesis of styrenes

using ethylene in a palladium-catalysed heck cross-coupling reaction. Synlett:2643–2647

114. Brzozowski M, Forni JA, Savage GP, Polyzos A (2015) The direct α-C(sp3)-H functionalisation of N-aryl tetrahydroisoquinolines via an iron-catalysed aerobic nitro-Mannich reaction and continuous flow processing. Chem Commun 51:334–337

115. Bourne SL, Ley SV (2013) A continuous flow solution to achieving efficient aerobic anti-markovnikov wacker oxidation. Adv Synth Catal 355:1905–1910

116. Newton S, Ley SV, Casas Arce E, Grainger DM (2012) Asymmetric homogeneous hydrogenation in flow using a tube-in-tube reactor. Adv Synth Catal 354:1805–1812

117. O'Brien M, Taylor N, Polyzos A, Baxendale IR, Ley SV (2011) Hydrogenation in flow: homogeneous and heterogeneous catalysis using Teflon AF-2400 to effect gas-liquid contact at elevated pressure. Chem Sci 2:1250–1257

118. Skowerski K, Czarnocki SJ, Knapkiewicz P (2014) Tube-in-tube reactor as a useful tool for homo- and heterogeneous olefin metathesis under continuous flow mode. ChemSusChem 7:536–542

119. Mastronardi F, Gutmann B, Kappe CO (2013) Continuous flow generation and reactions of anhydrous diazomethane using a Teflon AF-2400 tube-in-tube reactor. Org Lett 15:5590–5593

120. Mercadante MA, Leadbeater NE (2012) Development of methodologies for reactions involving gases as reagents: microwave heating and conventionally-heated continuous-flow processing as examples. Green Proc Synth 1:499–507. For a review, see this reference

121. Mercadante MA, Kelly CB, Lee C, Leadbeater NE (2012) Continuous flow hydrogenation using an on-demand gas delivery reactor. Org Process Res Dev 16:1064–1068

122. Mercadante MA, Leadbeater NE (2011) Continuous-flow, palladium-catalysed alkoxycarbonylation reactions using a prototype reactor in which it is possible to load gas and heat simultaneously. Org Biomol Chem 9:6575–6578

123. Rudzinski DM, Leadbeater NE (2013) Microwave heating and conventionally-heated continuous-flow processing as tools for performing cleaner palladium-catalyzed decarboxylative couplings using oxygen as the oxidant—a proof of principle study. Green Proc Synth 2:323–328

124. Ciamician G (1912) The photochemistry of the future. Science 36:385–394

125. Gilmore K, Seeberger PH (2014) Continuous flow photochemistry. Chem Rec 14:410–418. For reviews, see this reference

126. Schuster EM, Wipf P (2014) Photochemical flow reactions. Isr J Chem 54:361–370. For reviews, see this reference

127. Knowles JP, Elliott LD, Booker-Milburn KI (2012) Flow photochemistry: old light through new windows. Beilstein J Org Chem 8:2025–2052. For reviews, see this reference

128. Hook BDA, Dohle W, Hirst PR, Pickworth M, Berry MB, Booker-Milburn KI (2005) A practical flow reactor for continuous organic photochemistry. J Org Chem 70:7558–7564

129. Foraboschi FP (1959) Conversion in a photochemical flow reactor. Chim Ind 41:731–737. Some work had preceded this. See this reference, for example

130. Kopetzki D, Lévesque F, Seeberger PH (2013) A continuous-flow process for the synthesis of artemisinin. Chem Eur J 19:5450–5456

131. Lévesque F, Seeberger PH (2012) Continuous-flow synthesis of the anti-malaria drug artemisinin. Angew Chem Int Ed 51:1706–1709

132. Amara Z, Bellamy JFB, Horvath R, Miller SJ, Beeby A, Burgard A, Rossen K, Poliakoff M, George MW (2015) Applying green chemistry to the photochemical route to artemisinin. Nat Chem 7:489–495

133. Garlets ZJ, Nguyen JD, Stephenson CRJ (2014) The development of visible-light photoredox catalysis in flow. Isr J Chem 54:351–360. For a review, see this reference

134. Prier CK, Rankic DA, MacMillan DWC (2013) Visible light photoredox catalysis with transition metal complexes: applications in organic synthesis. Chem Rev 113:5322–5363. For a review, see this reference

135. Xuan J, Xiao W-J (2012) Visible-light photoredox catalysis. Angew Chem Int Ed 51:6828–6838. Narayanam MR, Stephenson CRJ (2011) Visible light photoredox catalysis: applications in organic synthesis. Chem Soc Rev 40:102–113

136. Bou-Hamdan FR, Seeberger PH (2012) Visible-light-mediated photochemistry: accelerating $Ru(bpy)_3^{2+}$-catalyzed reactions in continuous flow. Chem Sci 3:1612–1616

137. Watts K, Baker A, Wirth T (2015) Electrochemical Synthesis in Microreactors. J Flow Chem 4:2–11. For reviews, see this reference
138. Schuelein J, Löwe H (2014) Electrochemistry in flow. In: Darvas F, Dormán G, Hessel V (eds) Flow chemistry: volume 1. DeGruyter. For reviews, see this reference
139. Waldvogel SR, Janza B (2014) Renaissance of electrosynthetic methods for the construction of complex molecules. Angew Chem Int Ed 53:7122–7123. For perspectives, see this reference
140. Frontana-Uribe BA, Little RD, Ibanez JG, Palmad A, Vasquez-Medrano R (2010) Organic electrosynthesis: a promising green methodology in organic chemistry. Green Chem 12:2099–2119
141. Frey DA, Wu N, Moeller KD (1996) Anodic electrochemistry and the use of a 6-volt lantern battery: a simple method for attempting electrochemically based synthetic transformations. Tetrahedron Lett 37:8317–8320
142. Watts K, Gattrell W, Wirth T (2011) A practical microreactor for electrochemistry in flow. Beilstein J Org Chem 7:1108–1114. For designs, see this reference
143. Kuleshova J, Hill-Cousins JT, Birkin PR, Brown RCD, Pletcher D, Underwood TJ (2011) A simple and inexpensive microfluidic electrolysis cell. Electrochim Acta 56:4322–4326
144. Ziogas A, Kolb G, O'Connell M, Attour A, Lapicque F, Matlosz M, Rode S (2009) Electrochemical microstructured reactors: design and application in organic synthesis. J Appl Electrochem 39:2297–2313
145. Küpper M, Hessel V, Löwe H, Stark W, Kinkel J, Michel M, Schmidt-Traub H (2003) Micro reactor for electroorganic synthesis in the simulated moving bed-reaction and separation environment. Electrochim Acta 48:2889–2896
146. Arai K, Watts K, Wirth T (2014) Difluoro- and trifluoromethylation of electron-deficient alkenes in an electrochemical microreactor. ChemistryOpen 3:23–28
147. Arai K, Wirth T (2014) Rapid electrochemical deprotection of the isonicotinyloxycarbonyl group from carbonates and thiocarbonates in a microfluidic reactor. Org Process Res Dev 18:1377–1381
148. Kabeshov MA, Musio B, Murray PRD, Browne DL, Ley SV (2014) Expedient preparation of nazlinine and a small library of indole alkaloids using flow electrochemistry as an enabling technology. Org Lett 16:4618–4621
149. Roth GP, Stalder R, Long T, Sauer D, Djuric S (2013) Continuous-flow microfluidic electrochemical synthesis: investigating a new tool for oxidative chemistry. J Flow Chem 3:34–40
150. Stadler R, Roth GP (2013) Preparative microfluidic electrosynthesis of drug metabolites. ACS Med Chem Lett 4:1119–1123
151. Hill-Cousins JT, Kuleshova J, Green RA, Birkin PR, Pletcher D, Underwood TJ, Leach SG, Brown RCD (2012) TEMPO-mediated electrooxidation of primary and secondary alcohols in a microfluidic electrolytic cell. ChemSusChem 5:326–331
152. Yoshida J-i (2005) Flash chemistry using electrochemical method and microsystems. Chem Commun:4509–4516
153. Nagaki A, Togai M, Suga S, Aoki N, Mae K, Yoshida J-i (2003) Control of extremely fast competitive consecutive reactions using micromixing. Selective Friedel−Crafts aminoalkylation. J Am Chem Soc 127:11666–11675
154. Dalko PI (2013) Comprehensive enantioselective organocatalysis: catalysts, reactions, and applications. Wiley-VCH
155. Atodiresei I, Vila C, Rueping M (2015) Asymmetric organocatalysis in continuous flow: opportunities for impacting industrial catalysis. ACS Catal 5:1972–1985. For general reviews, see this reference
156. Finelli FG, Miranda LSM, de Souza ROMA (2015) Expanding the toolbox of asymmetric organocatalysis by continuous-flow process. Chem Commun 51:3708–3722
157. Odedra A, Seeberger PH (2009) 5-(Pyrrolidin-2-yl)tetrazole-catalyzed aldol and Mannich reactions: acceleration and lower catalyst loading in a continuous-flow reactor. Angew Chem Int Ed 48:2699–2702
158. Valera FE, Quaranta M, Moran A, Blacker J, Armstrong A, Cabral JT, Blackmond DG (2010) The flow's the thing…or is it? Assessing the merits of homogeneous reactions in flask and flow. Angew Chem Int Ed 49:2478–2485
159. Liao H-H, Hsiao C-C, Sugiono E, Rueping M (2013) Shedding light on Brønsted acid catalysis—a photocyclization-reduction reaction for the asymmetric synthesis of tetrahydroquinolines from aminochalcones in batch and flow. Chem Commun 49:7953–7955
160. Sugiono E, Rueping M (2013) A combined continuous microflow photochemistry and asymmetric organocatalysis approach for the enantioselective synthesis of tetrahydroquinolines. Beilstein J Org Chem 9:2457–2462

161. Neumann M, Zeitler K (2012) Application of microflow conditions to visible light photoredox catalysis. Org Lett 14:2658–2661
162. Rodríguez-Escrich C, Pericàs MA (2015) Organocatalysis on tap: enantioselective continuous flow processes mediated by solid-supported chiral organocatalysts. Eur J Org Chem:1173–1188
163. Ötvös SB, Mándity IM, Fülöp F (2012) Asymmetric aldol reaction in a continuous-flow reactor catalyzed by a highly reusable heterogeneous peptide. J Catal 295:179–185
164. Lange H, Carter CF, Hopkin MD, Burke A, Goode JG, Baxendale IR, Ley SV (2011) A breakthrough method for the accurate addition of reagents in multi-step segmented flow processing. Chem Sci 2:765–769
165. Koos P, Gross U, Polyzos A, O'Brien M, Baxendale IR, Ley SV (2011) Teflon AF-2400 mediated gas-liquid contact in continuous flow methoxycarbonylations and in-line FTIR measurement of CO concentration. Org Biomol Chem 9:6903–6908
166. Ley SV, Fitzpatrick DE, Ingham RJ, Myers RM (2015) Organic synthesis: march of the machines. Angew Chem Int Ed 54:3449–3464. For reviews, see this reference
167. Fabry DC, Sugiono E, Rueping M (2014) Self-optimizing reactor systems: algorithms, on-line analytics, setups, and strategies for accelerating continuous flow process optimization. Isr J Chem 54:341–350
168. Sans V, Porwol L, Dragone V, Cronin L (2015) A self optimizing synthetic organic reactor system using real-time in-line NMR spectroscopy. Chem Sci 6:1258–1264. For examples, see this reference
169. Skilton RA, Parrott AJ, George MW, Poliakoff M, Bourne RA (2013) Real-time feedback control using online attenuated total reflection Fourier transform infrared (ATR FT-IR) spectroscopy for continuous flow optimization and process knowledge. Appl Spectrosc 67:1127–1131
170. Parrott AJ, Bourne RA, Akien GR, Irvine DJ, Poliakoff M (2011) Self-optimizing continuous reactions in supercritical carbon dioxide. Angew Chem Int Ed 50:3788–3792
171. McMullen JP, Jensen KF (2011) Rapid determination of reaction kinetics with an automated microfluidic system. Org Process Res Dev 15:398–407
172. Ingham RJ, Battilocchio C, Fitzpatrick DE, Sliwinski S, Hawkins JM, Ley SV (2015) A systems approach towards an intelligent and self-controlling platform for integrated continuous reaction sequences. Angew Chem Int Ed 54:144–148. For examples, see this reference
173. Guetzoyan L, Ingham RJ, Nikbin N, Rossignol J, Wolling M, Baumert M, Burgess-Brown NA, Strain-Damerell CM, Shrestha L, Brennan PE, Fedorov O, Knapp S, Ley SV (2014) Machine-assisted synthesis of modulators of the histone reader BRD9 using flow methods of chemistry and frontal affinity chromatography. MedChemComm 5:540–546. For examples, see this reference
174. Ingham RJ, Battilocchio C, Hawkins JM, Ley SV (2014) Integration of enabling methods for the automated flow preparation of piperazine-2-carboxamide. Beilstein J Org Chem 10:641–652
175. Skilton RA, Bourne RA, Amara Z, Horvath R, Jin J, Scully MJ, Streng E, SLY T, Summers PA, Wang J, Perez E, Asfaw N, GLP A, Dupont J, Comak G, George MW, Poliakoff M (2015) Remote-controlled experiments with cloud chemistry. Nat Chem 7:1–5. For commentaries, see this reference
176. Peplow M (2014) Organic synthesis: the robo-chemist. Nature 512:20–22
177. www.futurechemistry.com. (accessed September 14th, 2021)
178. Flow chemistry course. https://futurechemistry.com/flow-chemistry-education/. Accessed September 14, 2021
179. König B, Kreitmeier P, Hilgers P, Wirth T (2013) Flow chemistry in undergraduate organic chemistry education. J Chem Educ 90:934–936
180. Feng ZV, Edelman KR, Swanson BP (2015) Student-fabricated microfluidic devices as flow reactors for organic and inorganic synthesis. J Chem Educ 92:723–727
181. Vapourtec E-series. http://vapourtec.co.uk/products/eseriessystem). Accessed 30 May 2015
182. Flow chemistry guide. http://vapourtec.co.uk/news/FlowChemistryGuide. Accessed 30 May 2015
183. Thayer AM (2005) Harnessing Microreactions. Chem Eng News 83(22):43–52
184. Grongsaard P, Bulger PG, Wallace DJ, Tan L, Chen Q, Dolman SJ, Nyrop J, Hoerrner RS, Weisel M, Arredondo J, Itoh T, Xie C, Wen X, Zhao D, Muzzio DJ, Bassan EM, Shultz CS (2012) Convergent, kilogram scale synthesis of an Akt kinase Inhibitor. Org Process Res Dev 16:1069–1081
185. Braune S, Pöchlauer P, Reintjens R, Steinhofer S, Winter M, Lobet O, Guidat R,

Woehl P, Guermeur C (2009) Industry perspective—selective nitration in a microreactor for pharmaceutical production under cGMP conditions. Chem Today 27:26–29

186. Hernandez-Linares MG, Guerrero-Luna G, Perez-Estrada S, Ellison M, Ortin M-M, Garcia-Garibay MA (2015) Large-scale green chemical synthesis of adjacent quaternary chiral centers by continuous flow photodecarbonylation of aqueous suspensions of nanocrystalline ketones. J Am Chem Soc 137:1679–1684
187. O'Brien AG, Horváth Z, Lévesque F, Lee JW, Seidel-Morgenstern A, Seeberger PH (2012) Continuous synthesis and purification by direct coupling of a flow reactor with simulated moving-bed chromatography. Angew Chem Int Ed 51:7028–7030
188. Mascia S, Heider PL, Zhang H, Lakerveld R, Benyahia B, Barton PI, Braatz RD, Cooney CL, Evans JMB, Jamison TF, Jensen KF, Myerson AS, Trout BL (2013) End-to-end continuous manufacturing of pharmaceuticals: integrated synthesis, purification, and final dosage formation. Angew Chem Int Ed 52:12359–12363
189. Heider PL, Born SC, Basak S, Benyahia B, Lakerveld R, Zhang H, Hogan R, Buchbinder L, Wolfe A, Mascia S, Evans JMB, Jamison TF, Jensen KF (2014) Development of a multi-step synthesis and workup sequence for an integrated, continuous manufacturing process of a pharmaceutical. Org Process Res Dev 18:402–409
190. Lakerveld R, Benyahia B, Heider PL, Zhang H, Wolfe A, Testa CJ, Ogden S, Hersey DR, Mascia S, Evans JMB, Braatz RD, Barton PI (2014) The application of an automated control strategy for an integrated continuous pharmaceutical pilot plant. Org Process Res Dev. https://doi.org/10.1021/op500104d
191. Hessel V, Wang Q, Kralisch D (2014) From green chemistry principles in flow chemistry towards green flow process design in the holistic viewpoint. In: Darvas F, Dormán G, Hessel V (eds) Flow chemistry: volume 1. DeGruyter. For a review, see this reference
192. Wang Q, Gürsel IV, Shang M, Hessel V (2013) Life cycle assessment for the direct synthesis of adipic acid in microreactors and benchmarking to the commercial process. Chem Eng J 234:300–311. For examples, see this reference
193. Van der Vorst G, Aelterman W, De Witte B, Heirman B, Van Langenhove H, Dewulf J (2013) Reduced resource consumption through three generations of Galantamine·HBr synthesis. Green Chem 15:744–748
194. Kralisch D, Streckmann I, Ott D, Krtschil U, Santacesaria E, Di Serio M, Russo V, De Carlo L, Linhart W, Christian E, Cortese B, de MHJM C, Hessel V (2012) Transfer of the epoxidation of soybean oil from batch to flow chemistry guided by cost and environmental issues. ChemSusChem 5:300–311
195. Kralisch D, Kreisel G (2007) Assessment of the ecological potential of microreaction technology. Chem Eng Sci 62:1094–1100

Chapter 15

Reaction Optimization: A High-Throughput Experimentation Approach

Simon Berritt, Jason R. Schmink, and Ana Inés Bellomo Peraza

Abstract

This chapter reviews the various technologies and applications of high-throughput experimentation (HTE) in developing or optimizing chemical reactions. A brief history of HTE and current technology is presented, followed by an overview of HTE examples from the literature from 2007 to 2015, many of which were developed at the laboratories of Merck and Company and the University of Pennsylvania.

Key words High-throughput Experimentation (HTE), Catalysis, Microscale, Transition metal, Cross-coupling, Palladium, Ligands, Optimization

1 Introduction

For all practical purposes, the 1980s marked the decade in which scientists' long-standing desire to survey broad swaths of the chemical landscape finally became a reasonable goal. While researchers had always ***imagined*** the possibility of screening hundreds, thousands, or even hundreds of thousands of sets of possibilities, the logistics of scientific research in the past made this no more than a pipe dream. However, the end of the twentieth century marked a period of significant improvement of three specific areas of industry, without which scientists would still be running only a handful of reactions at a time. First, and most important, the development and continual improvement of the PC and its associated software was crucial for the development of today's modern chemistry laboratory. The PC is so important that it plays a significant role in each discrete step of modern screening approaches. Next, on the hardware side, the miniaturization and automation of the modern laboratory has played a large role in our ability to screen thousands of sets of conditions at a time. Lastly, it is hard to envision the logistics of high-throughput chemistry and biochemistry without multichannel pipettes and automated HPLCs.

Paul F. Richardson (ed.), *Green Chemistry in Drug Discovery: From Academia to Industry*, Methods in Pharmacology and Toxicology, https://doi.org/10.1007/978-1-0716-1579-9_15,

Over the past decades, "chemical screening" has itself evolved to suit the specific needs and desires of particular areas of research, and one only needs to consider the plethora of catchphrases associated with chemical screening to fully appreciate the evolving utility and application. The earliest chemical screening is most commonly referred to as ***high-throughput screening*** or HTS. Originally carried out in 96-well plates, today's modern HTS lab is fully automated and run in 384-, 1536-, or 3456-well plates. It is no surprise that the first area to develop screening approaches was one in which there are relatively few variables to consider. As a typical example, HTS generally seeks to answer only whether a specific molecule inhibits (or not) an enzyme (or cell line, embryo, etc.). To answer this question, a protein in buffered aqueous solution is dosed across all 96 wells before layering atop each well a different small molecule inhibitor. The one-dimensional nature of the question lends uniformity and hence simplicity!

As years passed, many scientists saw the potential within high-throughput methods and began to apply portions of the technique to slightly different areas, and as time passed, biochemists passed the baton to medicinal chemists. To improve the efficiency of drug discovery, high-throughput methods evolved to include a variety of experimental techniques in which thousands of compounds could be created rapidly using ***parallel synthesis***. However, as the complexity of the system increased (i.e., more variables), additional obstacles had to be overcome. Where the biochemists had the advantage of needing a platform suitable for a single set of conditions (buffered water, room temperature, perhaps some amount of organic co-solvent), medicinal chemists needed to be able to change the reaction environment as they explored different chemistries. That said, the same set of reaction conditions could be used for each well in order to produce a collection of closely related organic compounds [1]. Interestingly, an argument can be made that the biochemists' newly found ability to rapidly screen large libraries of compounds put pressure on medicinal chemistry to make larger, more diversified small molecule libraries at an even faster rate, thus leading to the birth of parallel synthesis. Like medicinal chemists before us, organic chemists are now borrowing from our biochemistry (enzymatic catalysis) [2] and medicinal chemistry (library synthesis) [3] colleagues, working to develop screening approaches to expedite the development of new methodologies. The focus of this chapter will be on the recent developments in ***high-throughput experimentation*** (HTE). HTE was conceived as a valuable tool to accelerate the discovery and implementation of efficient new methodologies. Much of this development was fostered on the process chemistry side within industry. Where parallel synthesis seeks to use a previously developed protocol in order to make new compounds, HTE aims to rapidly develop the best conditions to affect a discrete transformation, often times on a very specific and/or problematic substrate. HTE allows for the rapid refinement of existing protocols, can assist in the development of completely new reaction parameters, and often affords nonobvious lead "hits" that might have been otherwise unexplored using more traditional approaches. With HTE, there is a wide range of variables being examined simultaneously, and as such, the logistical difficulty of implementing this approach increases geometrically when compared to high-throughput screening or even parallel synthesis. When employing HTE, a scientist may simultaneously be examining five to six discrete variables (e.g., solvent, catalyst, ligand, base, substrate, additive/cosolvent, etc.) within the same plate. The logistics to support such a platform were difficult to develop, and it can be argued that continued improvements are still needed. Largely, issues of mass transfer have been the most problematic. With so many variables within a given screen, it is possible (likely even) that a plate will have some wells that are homogeneous, some that are heterogeneous, and some that are biphasic. Similarly, the search for ideal reaction conditions will span across a range of solvents with very different inherent characteristics (substrate solubility, product solubility, by-product solubility, boiling point, etc.).

Indeed, while enzymatic catalysis and library synthesis approaches seem obvious to build upon, the adaptations of these platforms are not as trivial as they first appear. Both of these approaches have one very powerful advantage: the platforms need to accommodate only ***a single set of reaction conditions!*** For instance, enzyme catalysis is run in a buffered aqueous or aqueous with organic co-solvent media at 5–37 °C. Similarly, library syntheses have as their variables only the coupling partners and utilize the identical set of reaction conditions across the plate. In HTE, the array of requirements is inverted. We examine a single transformation across a screen with a diverse set of reaction conditions. This dynamic situation necessitated significant platform development in order to make HTE screening as robust as possible and amenable to the widest range of potential reaction conditions.

This chapter will discuss specifics of the HTE platform, including the approach, necessary equipment, and ultimate execution of large screens. To accompany this discussion of HTE, this chapter will also touch briefly on a range of similar contemporary screening approaches. Next, the impact of HTE on "green chemistry" will be detailed. Finally, a diverse range of recent examples will be outlined, showing the flexibility and continued evolution of modern screening techniques in the laboratory.

2 HTE Screening Platform and Platform Development

As mentioned, HTE presents the opportunity to explore and evaluate novel chemistry in a rapid manner, removing the traditional bottleneck of methodology development: initial reaction optimization. There are three crucial components to HTE: (a) the use of software to design large screens, (b) the robust and reliable physical reaction platform, and (c) rapid reaction analysis.

2.1 Software

While any spreadsheet software could potentially be utilized to assist the scientist in designing large screens, the wide range of commercially available software from vendors designed to specifically aid in the design and execution of large screens make them a more attractive option. It is often difficult to disentangle much of the software and hardware advances over the past decade as they've been developed in tandem by companies/institutions such as Symyx (became Freeslate and now Unchained Laboratories), HTE (BASF) [4], Chemspeed [5], Avantium [6], and Biotage [7], among others. While any of the proprietary software applications could be chosen and discussed, here we will use Library Studio®, a PC-based virtual laboratory suite that facilitates the design of large screens, as the backdrop for further discussion. Library Studio is currently a part of the "Lab Execution and Analysis (LEA)" software available from Freeslate. This software allows a researcher to design complex high-throughput experiment arrays using a graphical user interface (Fig. 1). Details such as mass and moles of reagents, solvent volume(s), catalyst loading and stoichiometry, and reaction volume are calculated by the program, saved, and easily accessed by the researcher in the future. The software generates step-by-step recipes that can either be handed off to the scientist and followed manually, or as is becoming more customary, can be inputted into automated solid and/or liquid handling robotics, which dose the reaction plates.

2.2 Platform

In HTE, we strive to envision the screening platform as a scaled-down version of the round-bottom flask. This allows us to access any conditions that we would conceivably use at the lab bench on a more traditional scale. This approach, in turn, necessitates a ***general*** platform and one that must fulfill all of the following requirements:

1. The HTE platform must be able to accommodate the full range of conceivable reaction conditions employed by the bench chemist.
2. The platform must operate with high fidelity across the screen.
3. The results obtained on the small scale (<1–20 μmol) must scale with high correlation to synthetically useful scales, e.g., 1–100 plus mmol.

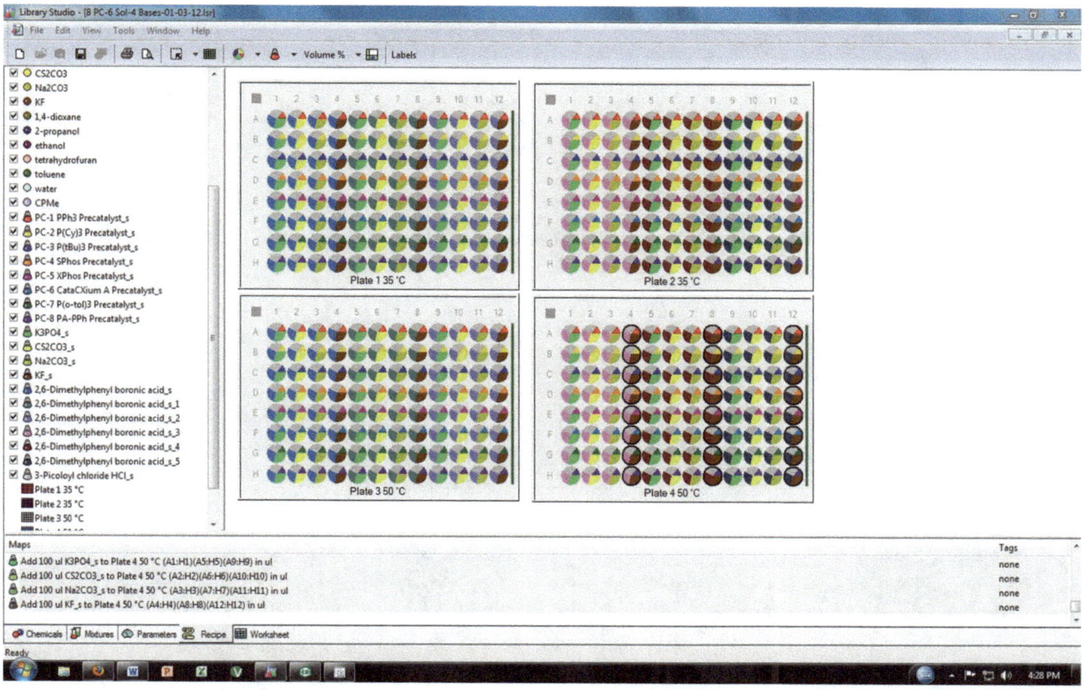

Fig. 1 Screenshot of Library Studio® software program utilized to assist the scientist with experiment design and execution

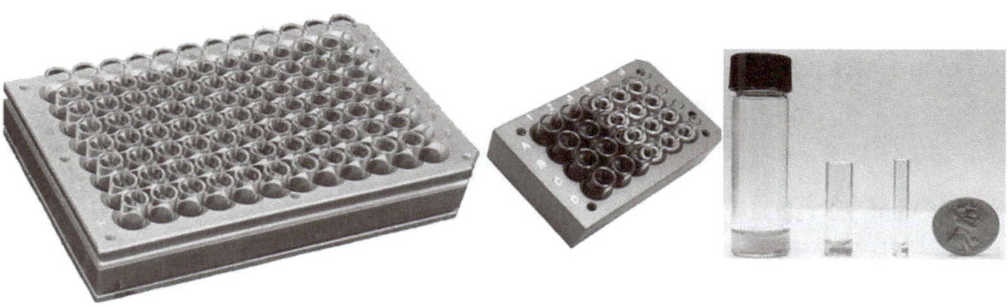

Fig. 2 96- and 24-well reactor blocks and three vial sizes (8 mL, 1 mL, and 250 μL)

The platform must be able to accommodate the full complement of solvents regularly utilized by bench chemists and must operate across a range of temperatures inclusive of both cryogenic conditions through temperatures that well exceed the boiling point of a selected solvent. Because these reactions are typically carried out with between 20 and 100 μL of solvent in 250–8000 μL vials (Fig. 2), even when operating below a solvent's boiling point, the screening platform must guarantee lossless seals on individual reaction vials to prevent evaporation. Next, the platform must effortlessly accommodate homogeneous, heterogeneous, and biphasic reaction conditions, the latter two of which create unique concerns at the microscale. The platform needs to be prepared in such a

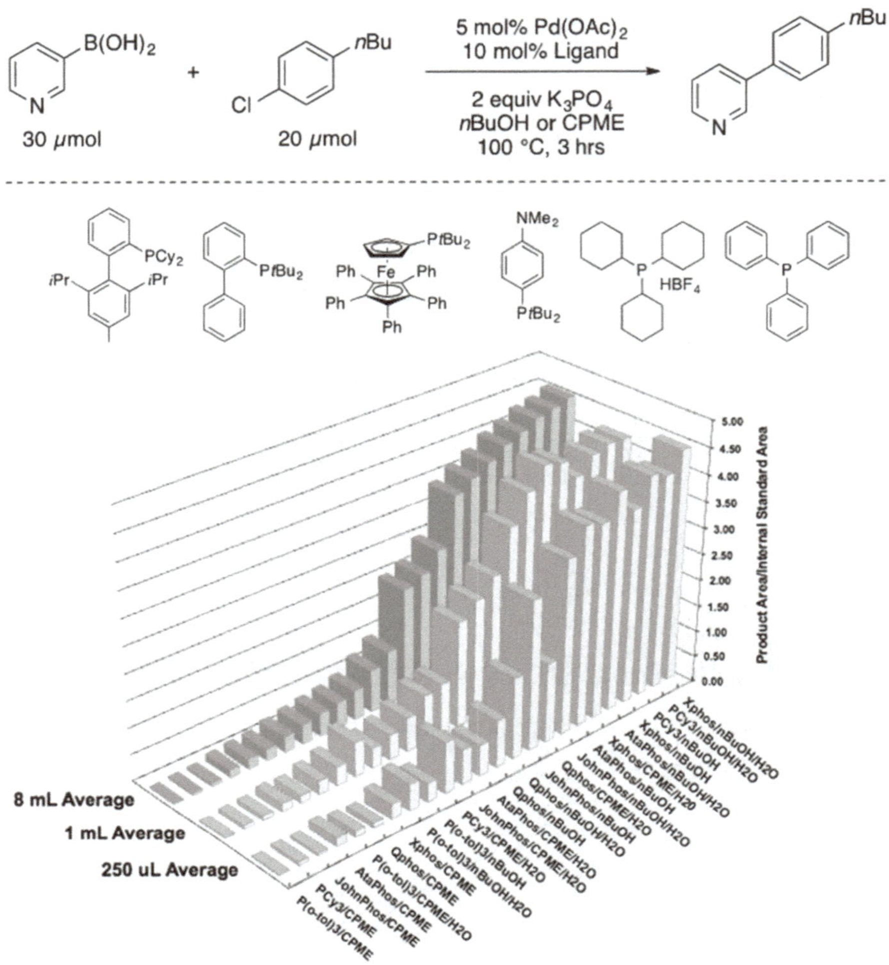

Fig. 3 Validation of HTE reaction scales rank ordered conditions based on 0.16 mmol

manner to allow for anhydrous conditions if needed and to allow reactions to be run under an inert atmosphere or even run under an atmosphere of a reactive gas such as H_2 or CO.

Next, vial-to-vial fidelity must be achieved. The scientist examining hundreds of reactions per day needs to be reasonably confident that when a reaction fails, it did so because the tested reaction conditions are nonideal and not simply because a catalyst was incorrectly dosed, solvent was lost, mass transfer was poor, or the desired temperature on that particular location on the plate was not achieved.

Validation of reaction scale demonstrated that the platform would indeed give identical results upon increasing the scale of reactions. Chemists at Merck & Co. used a relatively nontrivial Suzuki-Miyaura reaction [8, 9] at three different reaction scales: 250 μL vials/5 μmol, 1 mL vials/20 μmol, and 8 mL vials/0.16 mmol. A 6 × 2 × 2 screen with six diverse ligands and two solvents and with and without water was carried out in triplicate. As Fig. 3 demonstrates, good reproducibility is seen across the three different reaction scales.

The final crucial criteria inherent in the development of an HTE platform is that the reaction size must be able to be scaled such that running 24, 96, or even 1536 reactions does not require an exorbitant outlay of reagents. Ideally, the chemistry should be able to be developed using on the order of 0.1–1 mg or less of substrate per reaction. Without this critical aspect, the HTE approach would never be widely adopted. In the case study described above, the excellent fidelity among the different reaction scales allows the chemist to conduct 24 discrete reactions at the 5 μmol scale, using only approximately 32 mg of material,[1] and the scientist will be confident in the ability to scale the screening hits with similar results at a preparative scale. Continued improvements to the platform, robotics, and analytics surrounding HTE have allowed researchers at Merck to now move into the nanomolar scale, running plates of 1536 reactions using 0.02–0.05 mg of material per cell! [10].

3 Brief Summary of Alternative Screening Approaches in Reaction Development

The past decade has seen numerous entries into the field that can be described broadly as "high-throughput reaction optimization," and these reports can be broken into two classes. While recent screening developments can be compared and contrasted at length, a helpful understanding of screening approaches can be summarized as "scientist-driven" vs. "chance-driven." These two approaches evolved due to the very different end goals of the scientists who use them. A recent review of the latter approach by Collins, Gensch, and Glorius does a fantastic job of detailing the many variations of screening approaches in modern synthetic laboratories [11]. Experimental scientists at the cutting edge of research are looking for new chemistry, yet they have no requirement for ***what*** that new chemistry is. The authors rightfully acknowledge the importance of happenstance on reaction discovery, and their review details a range of approaches that seek to amplify serendipity in order to discover new reactions at a faster rate.

The so-called "multidimensional" [12–15] or "accelerated serendipity" [16] approaches will often yield interesting leads, in turn leading to the development of novel, high-impact, and even transformative science. Indeed, when driven by the desire to "discover something new," there may be no more powerful approach than

[1] 1 Based on small molecule with molecular weight of 523.3 at 0.1M concentration.

these full-spectrum pursuits coupled with a well-trained scientist. However, for the chemist faced with a ***specific task***, one that is looking to optimize a ***discrete transformation*** in order to advance a ***particular project***, this approach is not a viable strategy. The scientist looking to improve the outcome for a particular step in the synthesis of kilogram quantities of a small molecule therapeutic relies upon literature precedent and the screening of a directed subset of reaction parameters and far less upon serendipity.

The final consideration when considering the myriad "screening" approaches is that many of these reported approaches tend to be quite reaction specific and/or reaction-type specific. While these works have contributed greatly to scientific advancement, what most lack is the use of broadly applicable techniques and/or approaches that are scientist-driven. The broader chemistry community is slow to adopt those reports that are too reaction specific because it is recognized that the technology has a narrow range of potential applications.

In contrast, the HTE platform was purposefully developed with the notion, indeed the absolute requirement that a wide variety of chemistry conditions could and would need to be screened. The platform was engineered to handle, among others, heterogeneous and homogeneous reactions; high temperatures and cryogenic ones; polar and nonpolar solvents; reactions that run under air, under a blanket of inert gas, or beneath an atmosphere of reactive gas; and accommodating reactions that run at both atmospheric pressure and elevated pressures. The targeted, scientist-driven HTE platform presented here has the advantages of reliability, scalability, and fidelity across the platform and is amenable to the entire range of reaction conditions that a synthetic organic chemist expects to have in their arsenal at any time in a well-equipped fumehood.

4 Importance of Green Chemistry in HTE

Besides the inherent advantages of rapid reaction optimization, HTE is exceptionally well suited to develop new methodologies in a more benign fashion. Green chemistry [17] is becoming a driving force in synthetic organic chemistry at all levels from academia to industry, and pushing the boundaries of science with minimal impact to the environment is paramount. It behooves the serious scientist to be well versed in the tenants of green chemistry as few research settings don't stress enough the importance of judicious solvent selection, toxicity of reagents used, the utility of catalysis, and how our work impacts the world around us.

Because HTE develops conditions while working on the 2.5–10 μmol scale, it inherently fulfills many of the "12 Principles of Green Chemistry," including the minimization of waste (#1), reduction of required energy (#6), and the minimization of potential chemical accidents (#12). Furthermore, it is ideally suited to examine the use of alternative and more benign reagents (#3) and the implementation of catalysis for a transformation (#9).

5 The HTE Approach

The development of new synthetic methodologies is a cornerstone of the twenty-first-century organic chemistry. Arguably, the primary driving force in the development of new synthetic methodologies has been the chemist's ability to take advantage of the flexible and catalytic approaches afforded by transition metals [18, 19]. The reactivity of transition metals toward the specific and desired bond-forming event can be finely tuned by altering the catalyst environment via the ligand type (e.g., N, P, O), the steric and electronic properties of the ligand, and even the peripheral reaction conditions such as solvent, base, temperature, additives, and co-catalysts, among other variables. When one considers the diverse range of potential metals, ligands, solvents, and bases or other additives that are routinely employed in catalytic bond-forming events, it should be no surprise that chemists looking to rapidly optimize transition metal-mediated approaches have embraced the HTE screening platform [20].

After a thorough literature search, a synthetic chemist typically will devise a model reaction on which to develop a new method, though picking the "best" model reaction on which to develop this new chemistry is not always obvious. Ideally, the model reaction should be representative, though should avoid potentially problematic substrates with reactive and/or intolerant functional groups or ones that might be electronically or sterically unfavorable. Judicious selection of the correct model reaction therefore presents somewhat of a dichotomy. If the model reaction is too esoteric and the initial screens yield no results, the scientist is left to wonder: "does the chemistry not work or does the chemistry not work on ***this particular substrate***?" Of course, the opposite approach is no better when benign substrates are selected, which are so bland (or worse yet, manipulated) that the newly developed conditions are narrow in scope, and therefore applicable only to very similar compounds. Finally, we must demand that our newly developed chemistry shows improved reactivity, is more selective toward our desired transformation, and is tolerant of a wider range of functional groups than previous approaches.

Often, HTE reaction optimization utilizes an iterative approach. An initial screen(s) aims to cast a wide net, exploring diverse conditions. Typically, qualitative/semiquantitative analysis will yield preliminary hits, and even though these early conditions may only provide the desired compound in $<10\%$ conversion, these initial results can be used as the basis for the design of a second iteration of screens and therefore be rapidly improved upon.

In order to demonstrate the utility and flexibility of the HTE screening platform, this chapter ends with a diverse range of examples where the screening approach was utilized to help develop and optimize advancements in particular projects. We ask the readers to pay close attention to the publication date of the chemistry presented. Almost without exception, the reports contained on the following pages have been published within 5 years of writing this review. We point this out only to reiterate the value that HTE techniques have added to research groups both in academia and industry and to highlight the rapid uptake of this approach by a wide synthetic community. Stated simply: ***the HTE approach facilitates the discovery of new chemistry, and researchers continue to find significant value in utilizing this technique.***

6 Representative Examples of HTE in New Methodology Development

What follows is a survey of representative examples that highlight results from both industry and academia. This collection is not meant to be comprehensive, but rather is meant to convince the reader of the power and flexibility of the HTE platform as it affords the scientist a green, rapid, reliable, and systematic approach to developing new synthetic methodologies. With few exceptions, we have focused on the most recent examples, and most of the presented work was published between 2012 and 2015. However, the rapid uptake of this enabling technology by the wider synthetic community, especially in the past few years, makes it difficult to keep up with, much less yet compile a comprehensive review.

HTE was utilized by Kozlowski and co-workers to develop two related strategies, first using palladium to catalyze the α-arylation of nitroacetates [21] before expanding the chemistry to include the coupling with the far less reactive nitromethane. The authors turned toward HTE to conduct a comprehensive screen of ligands, suitable bases, and solvents. Twenty-four ligands, four bases (K_3PO_4, Cs_2CO_3, $CsHCO_3$, NaOAc), two solvents (toluene, dimethoxyethane), and two palladium sources (Pd_2dba_3•$CHCl_3$, [PdCl(allyl)$_2$]) were chosen for evaluation. For the α-arylation of nitroacetates, only three ligands gave rise to any product: BrettPhos, Me$_4$*t*BuXPhos, and *t*BuXPhos with *t*BuXPhos providing the cleanest reaction profile. After the initial leads, iterative screening led the scientist to disclose the best conditions as outlined in Fig. 4 (top panel) [22]. The *t*-butyl and methyl ester nitroacetate also participated in this reaction. The group quickly followed up this work, demonstrating that the related arylnitromethane moiety could be accessed by coupling aryl halides directly with nitromethane under palladium-catalyzed conditions (Fig. 4, bottom panel) [23].

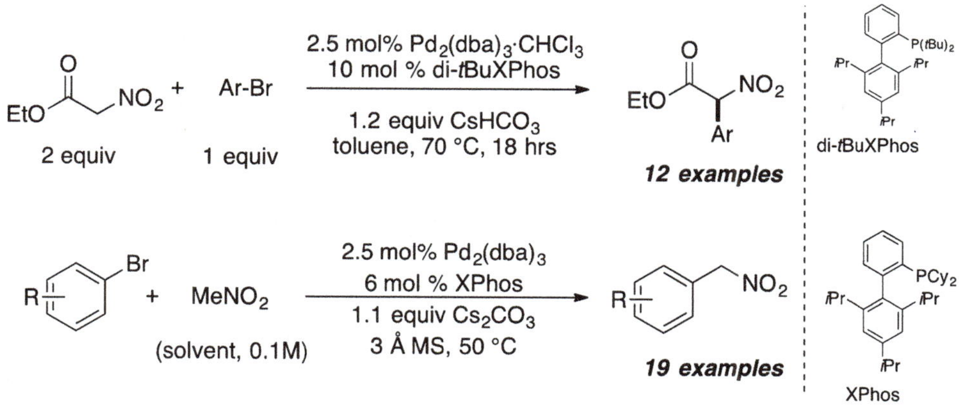

Fig. 4 Palladium-catalyzed α-arylation of ethyl nitroacetate and nitromethane

In 2013, the group of Patrick Walsh disclosed a route to synthesize α-benzyl aldehydes by employing a tandem palladium-catalyzed ring opening of cyclopropanols via a C-C cleavage step followed by arylation of the palladium homoenolate [24]. HTE was used to develop room temperature conditions, which revealed Q-Phos as the best ligand. Overall, 21 examples were disclosed with reported yields ranging from 59 to 93% (Fig. 5).

Further work by Walsh and co-workers focused on the development of an intermolecular arylations of weakly acidic sp^3 C-H bonds using HTE. Building on earlier work detailing α-arylation of (η^6-tolyl)$Cr(CO)_3$ complexes [25], they recognized the potential to affect this catalytic transformation stereospecifically. One hundred ninety-two chiral mono- and bidentate phosphine ligands were screened from which cy-MandyPhos provided the best enantioenrichment. The authors report that using TMEDA to break up lithium aggregates improves reactivity and allows aryl bromides and triflates to be coupled to the chromium-complexed arenes in yields of up to 87% and *ee*'s up to 92%. This strategy was used to report the first catalytic asymmetric arylation of these activated benzylic amines (Fig. 6, top panel) [26]. Building upon this work, the η^6-tolyl chromium scaffold was then shown to participate in the palladium-catalyzed Tsuji-Trost allylic substitution reaction [27]. The authors used HTE to screen conditions, finding that $Pd(COD)Cl_2$, XantPhos, and LiHMDS/Et_3N were the optimum combination of reagents (Fig. 6, bottom panel) [28].

Following this work, Walsh et al. turned their attention to identifying conditions to effect chromium-free direct C(sp^3)-H arylation of *non*-activated diarylmethanes, a reaction that the authors termed a deprotonative cross-coupling process (DCCP). Iterative HTE screens were conducted leading to the identification of the optimum conditions which critically highlighted a single base/ligand pair (KHMDS/NiXantphos) out of a possible 1344 (12 bases × 112 ligands) combinations of the most important

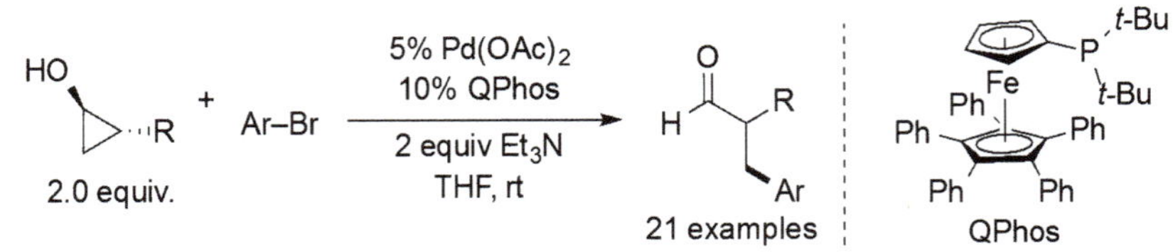

Fig. 5 Palladium-catalyzed cyclopropanol ring opening/arylation yielding aryl homoenolates

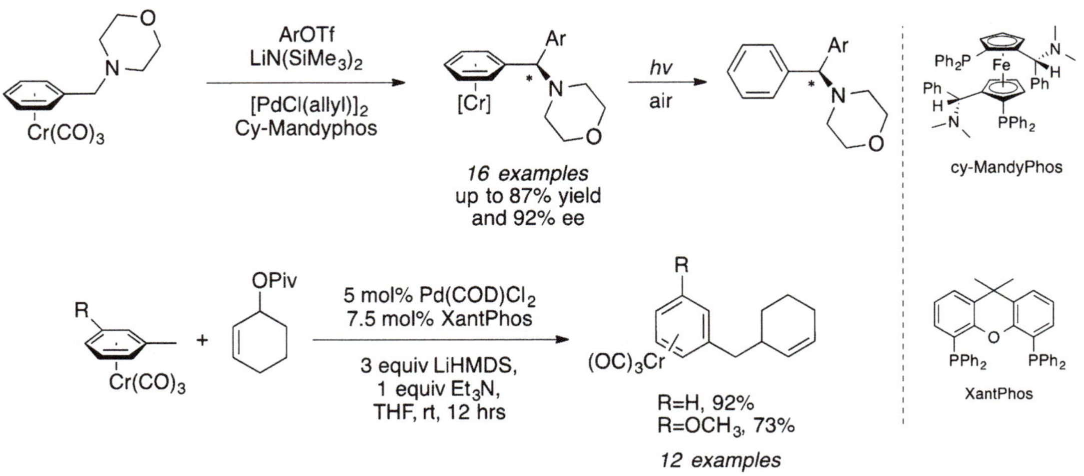

Fig. 6 Asymmetric α-arylation and allylic substitution reactions of η^6-benzyl chromium species

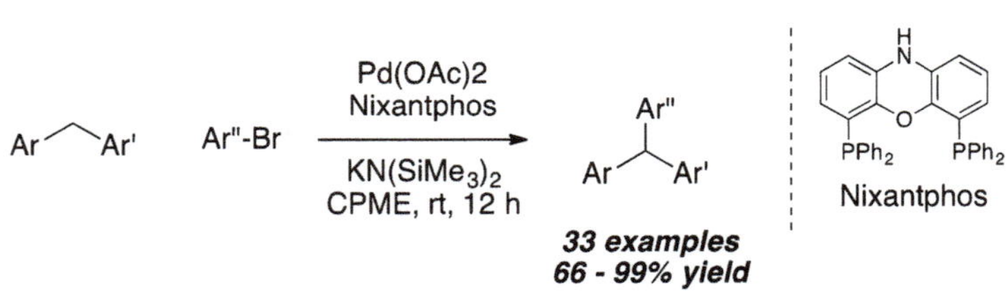

Fig. 7 Arylation of diarylmethanes

variables. The protocol developed enabled the synthesis of a variety of sterically and electronically diverse aryl- and heteroaryl-containing triarylmethanes at room temperature (Fig. 7) [29]. During their investigation the authors noted that LiHMDS and NaHMDS did not give rise to the triaryl product formation. Knowing that addition of amine additives increased reaction performance in their earlier studies, the authors conducted a rapid HTE screen of mono-and polydentate Lewis base additives [30–32] finding that running the reaction with the tetradentate amine, 1,1,4,7,10,10-hexamethyltriethylenetetramine, allowed the chemistry to be expanded to include heteroaryl bromides [33].

Building on their growing body of examples for the coupling of weakly acidic substrates, Walsh and co-workers reported a palladium-catalyzed α-arylation of unactivated sulfoxides (Fig. 8).

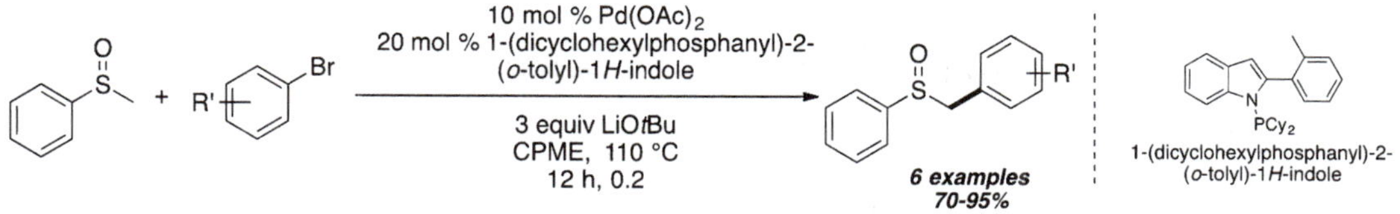

Fig. 8 α-Arylation of arylsulfoxides

HTE explored a range of strong bases, various solvents, and Pd $(OAc)_2$ with various ligands. Of the 112 ligands examined, the indole-based phosphine [34] ligand showed a significant propensity for promoting this transformation. Under the initially optimized conditions, aryl bromides were cross-coupled successfully with sulfoxides. However, aryl chlorides failed to undergo the desired cross-coupling, but a modification that included using a palladium precatalyst, μ-chloro-dimer biphenylamino-palladacyle, subsequently allowed aryl chlorides to be coupled in moderate yields.

In 2014, the groups of Walsh [35] and Schmink [36] simultaneously and independently reported the palladium-catalyzed cross-coupling of aryl bromides with 2-aryl-1,3-dithianes (Fig. 9). Both groups employed HTE screening techniques to examine the ability of a range of ligands, solvents, and bases to effect the desired reaction. Though working independently, the groups developed remarkably similar sets of conditions with both reporting NiXantphos as the optimum ligand for their protocols.

In 2014, researchers at Merck & Co. disclosed a protocol that employed palladium to effect the α-arylation of cyclopropyl nitriles with aryl and heteroaryl bromides [37]. The initial lead was generated from a 384-reaction HTE screen that identified racemic BINAP with $Pd_2(dba)_3$ as a promising catalyst system (Fig. 10). The authors report the cross-coupling leading to 15 diverse heteroaryl- and aryl-cyclopropyl nitriles before demonstrating that the system could also be applied to the corresponding cyclobutyl and cyclopentyl nitriles.

In 2014, Walsh and co-workers were able to overcome the tendency for allyl arenes to engage with aryl halides in Heck-type reactions and instead influence the catalytic system to cross-couple at the sp^3-benzylic position, allowing access to 25 diarylallylmethanes [38]. To optimize the desired coupling of aryl bromides with allyl arenes, the authors utilized HTE screens to investigate 29 different ligands (Fig. 11). Both PCy_3 and BrettPhos showed the highest desired selectivity, though ultimately the less expensive PCy_3 was chosen.

Again seeking to engage weakly acidic C-H bonds in a DCCP, Walsh and co-workers reported that benzylic phosphine oxides were suitable substrates to engage with aryl bromides under palladium-mediated conditions [39]. HTE screens revealed that

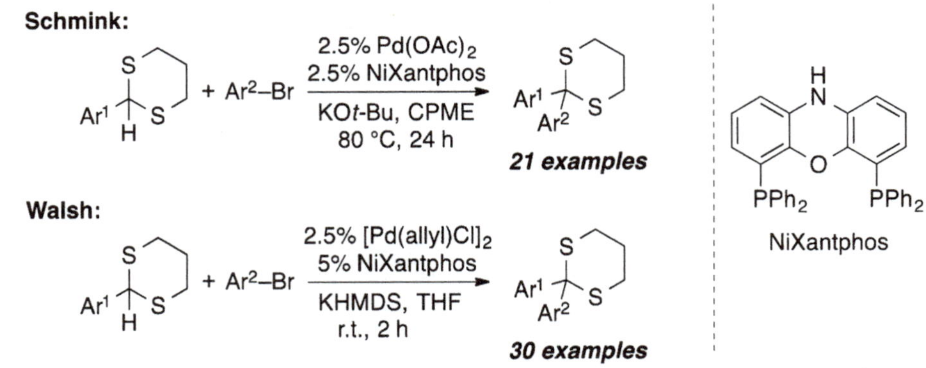

Fig. 9 Reports demonstrating the polarity-reversed cross-coupling of 2-aryl-1,3-dithianes was optimized using HTE

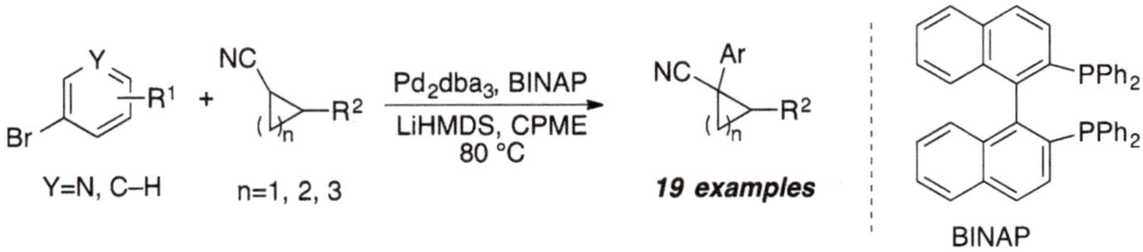

Fig. 10 The palladium-catalyzed α-arylation of cyclopropyl nitriles was rapidly developed using HTE

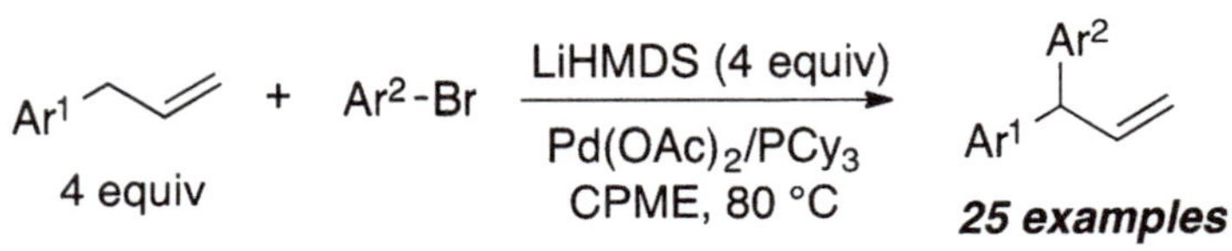

Fig. 11 Arylation of allyl arenes

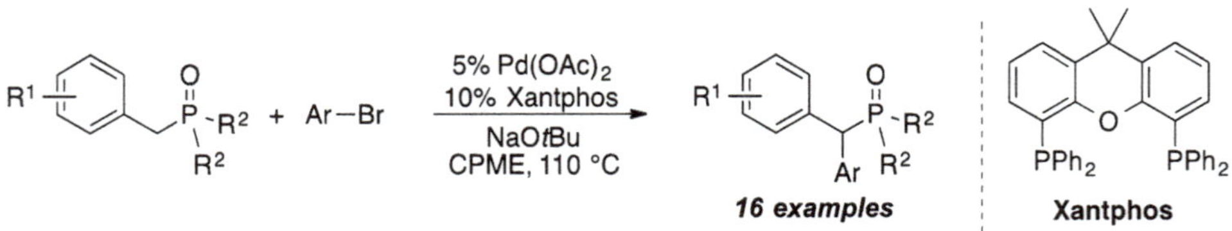

Fig. 12 Arylation of benzylic phosphine oxides

XantPhos and NaO*t*Bu were the optimal ligand and base, respectively, for the coupling (Fig. 12). The authors note that the system is quite sensitive to the cation of the base chosen: while both NaO*t*Bu and NaHMDS smoothly promoted the desired coupling, the potassium and lithium variants of both bases were completely ineffective. This situation is a perfect example of what makes HTE so valuable. When a researcher can run only a handful of reactions per day, and no desired cross-coupling is seen with, for example, LiHMDS, the researcher would likely not think that the next experiment should look for the desired reactivity using NaHMDS. Yet, when screening 384 reactions at a time in the HTE lab, a researcher is free to screen for reactivity leaving no discrete variable alone, often leading to surprising and unpredictable results as well as synergistic interactions between reagents.

Three catalyst systems (Figs. 13–15) capable of coupling secondary organotrifluoroborates with sterically and electronically demanding aryl chlorides and bromides were discovered by the Molander group [40]. This body of work represented the first comprehensive study of alkylboron coupling to aryl chlorides focusing on secondary alkylboron partners. Isomerism of more hindered substrates was problematic and was believed to occur via a ligand-dependent β-hydride elimination/reinsertion mechanism (Fig. 13).

HTE was used to develop two sets of conditions for the Suzuki-Miyaura reaction (SMR) of diversely functionalized primary alkyltrifluoroborates (Figs. 14 and 15) [41]. Aryl bromides, iodides, and triflates were effective coupling partners. Conditions previously used for the cross-coupling of secondary alkyltrifluoroborates (Fig. 13) with aryl chlorides were not optimal for the primary alkyltrifluoroborates (Fig. 14).

Though closely related, the development of cross-coupling conditions for the alkoxyethyl trifluoroborates and their related acetals found AtaPhos to be the best-performing ligand. Although the motifs illustrated in Figs. 14 and 15 appear to be closely related as primary trifluoroborates, the pendant functionality impacted their chemical reactivity to such an extent that a unique set of conditions needed to be developed, demonstrating the value of parallel experimentation to rapidly identify novel, substrate-specific results.

Cross-coupling of a diverse range of aminomethyltrifluoroborates [42–44], with aryl and heteroaryl bromides, was next investigated by the group using HTE to access the biologically relevant methylamine motifs via the SMR [45]. The reported optimum cross-coupling conditions included the inexpensive $P(^{t}Bu)_3$ aminobiphenyl palladium precatalyst to couple the trifluoroborates with aryl bromides (Fig. 16).

In 2010, the authors reported on the stereospecific cross-coupling of enantioenriched *non-benzylic* secondary alkyl boron compounds. The Buchwald ligands, SPhos or XPhos, were found to show high selectivity toward product formation over the undesired β-hydride elimination. Inversion of stereochemistry was observed, which was speculated to occur due to coordination of the ancillary amide carbonyl to the diorganopalladium intermediate (Fig. 17).

In 2012, the Molander group employed HTE techniques to develop a palladium-catalyzed SMR between potassium 1-(alkoxy/acyloxy)alkyltrifluoroborates and aryl/heteroaryl chlorides.

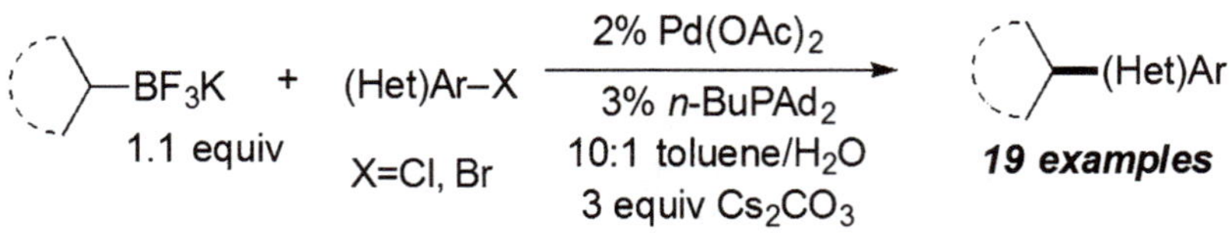

Fig. 13 Cross-coupling of aryl chlorides and bromides with secondary potassium alkyltrifluoroborates

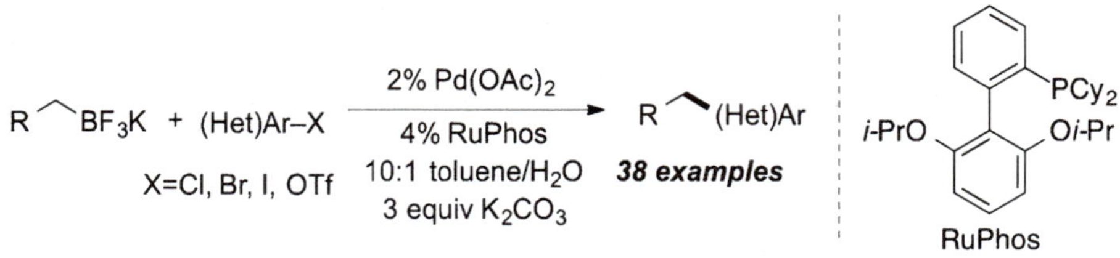

Fig. 14 Cross-coupling of aryl and heteroaryl chlorides, bromides, iodides, and triflates with primary potassium organotrifluoroborates

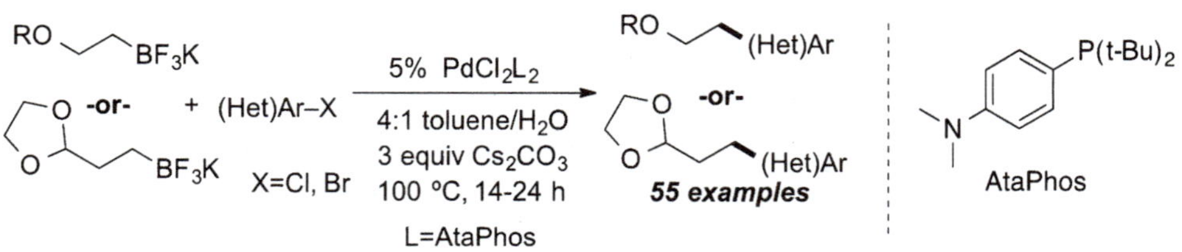

Fig. 15 Cross-coupling of aryl and heteroaryl chlorides and bromides with primary alkoxyethyl potassium organotrifluoroborates and related acetals

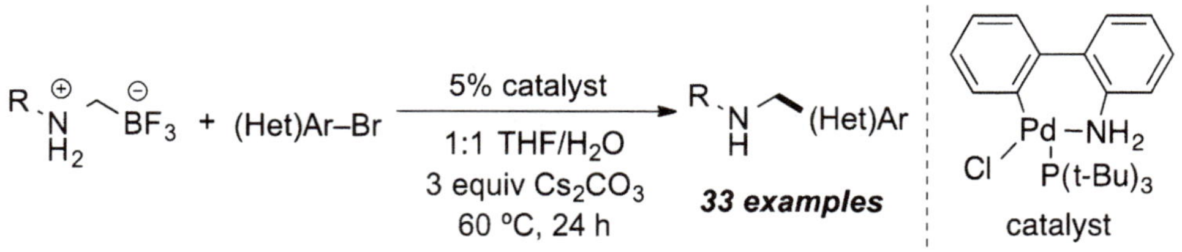

Fig. 16 Cross-coupling of aryl and heteroaryl bromides with aminomethyl trifluoroborates

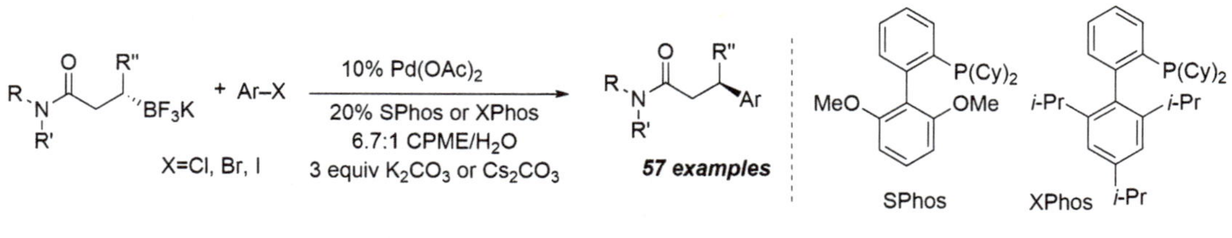

Fig. 17 Cross-coupling of aryl chlorides, bromides, and iodides with secondary potassium organotrifluoroborates containing an ancillary amide, which invokes an unanticipated inversion of stereochemistry at the reactive center

Stabilization of the diorganopalladium intermediate through coordination of the benzyl protecting group to the metal center avoided the non-productive β-hydride elimination pathway. Without the ancillary amide present, inversion of stereochemistry was also avoided with 48 diverse examples being reported (Fig. 18).

Workers at Merck & Co. utilized HTE to report the rapid development of the palladium-catalyzed SMR of nitrogen-bearing heterocyclic chloromethyl derivatives with aryl and heteroaryl boronic acids [46]. The authors opted to couple only the most problematic substrates (heteroaryls featuring basic nitrogens and boronic acids with multiple electron-withdrawing groups) and to attempt to reduce the reaction temperature in order to negate unwanted side reactions. The comprehensive substrate scope and respectable yields highlight the synthetic utility of this reaction (Fig. 19).

Two nickel-catalyzed methods for the cross-coupling of alkyl electrophiles with various potassium aryl- and heteroaryltrifluoroborates were reported by Molander et al. [47] Aryltrifluoroborates and primary or secondary chlorides were effectively coupled using a $NiCl_2$•glyme/*L*-prolinol catalyst system. A $NiBr_2$•glyme/bathophenanthroline catalyst enabled the chemoselective coupling of C(sp^3)–Br bonds in the presence of C(sp^2)–Br bonds (Fig. 20).

A protocol that utilized palladium catalysis to synthesize α-(hetero)aryl esters and amides through a SMR was disclosed by the Molander group [48]. Organotrifluoroborate salts were cross-coupled with α-chloro esters and amides utilizing a Pd-XPhos biaryl precatalyst. Catalytic amounts of Cu_2O improved yields for the cases in which secondary amides were coupled with trifluoroborate salts. A variety of functional groups and heterocyclic compounds were tolerated (Fig. 21).

The Molander group was able to expand considerably the functionalization of azaborines by using HTE techniques to rapidly identify reaction conditions to effect the cross-coupling of aryl- and heteroaryltrifluoroborates, alkenyl trifluoroborates, and acetylenes using different combinations of palladium/ligand catalysis [49]. In total, the synthesis of 42 novel compounds was reported in the 2014 publication demonstrating the value of HTE in rapidly enabling conditions to access new chemical space (Fig. 22).

HTE was utilized to develop conditions to synthesize boronic acids using palladium catalysis. In four publications, Molander et al. report a more atom economical approach using XPhos or CataCXium-A-based precatalyst systems [49–52] (Fig. 23) The authors affect the new C-B bond using either *bis*-boronic acid (BBA) or its synthetic precursor, *tetrakis*(dimethylamino)diboron, leading a significant gain in atom economy compared to B_2pin_2-based approaches. For both ease of isolation and to preserve the often-sensitive C-B bond, the boronic acids were directly converted to their more stable trifluoroborate counterparts (61 diverse

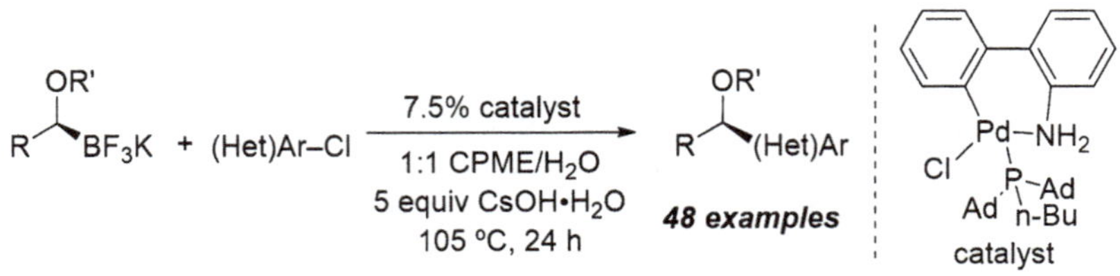

Fig. 18 Cross-coupling of aryl and heteroaryl chlorides with alkoxy/acyloxy potassium organotrifluoroborates

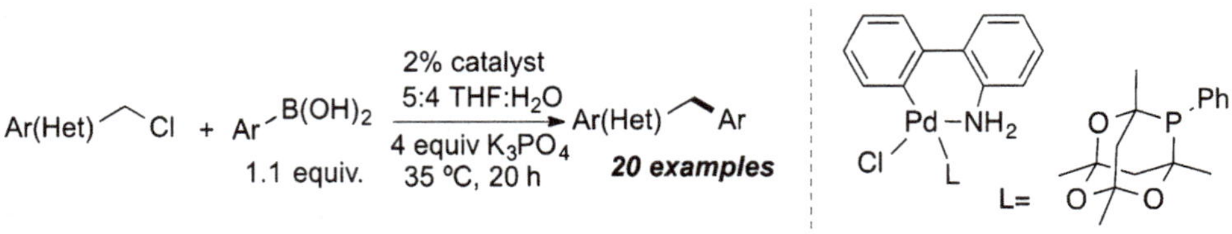

Fig. 19 Cross-coupling of chloromethyl heterocycles with boronic acids

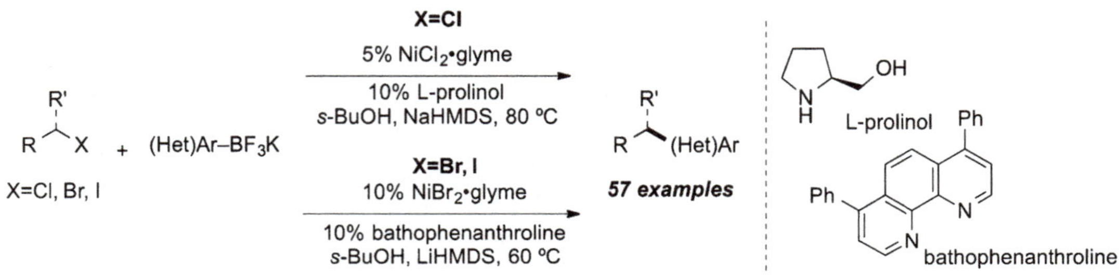

Fig. 20 Nickel-catalyzed cross-coupling of aryl and heteroaryl organotrifluoroborates with primary chlorides, bromides, and iodides

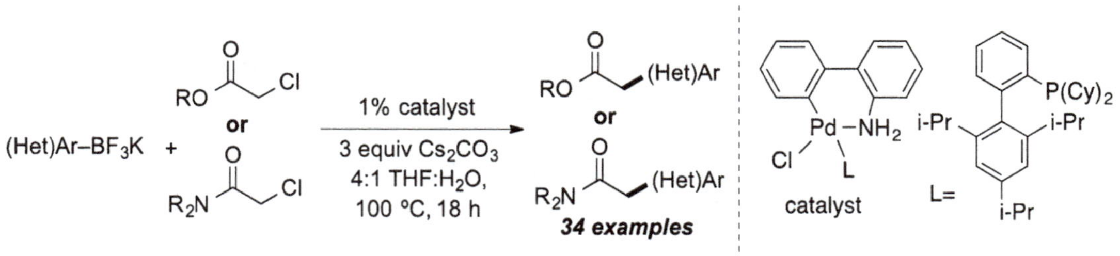

Fig. 21 Cross-coupling of aryl and heteroaryl organotrifluoroborates with α-chloro esters and amides

examples). Sequential palladium-catalyzed borylation followed by SMR with a second aryl or heteroaryl halide to yield the biaryl compounds in a single pot sequence was also reported (26 examples).

A collaboration between researchers at Michigan State University and Merck & Co. disclosed a comprehensive study of iridium-catalyzed borylation via C-H activation of aryl systems [53]. Specifically, the authors sought to investigate any synergy between various components of the reaction such as solvent, ligand,

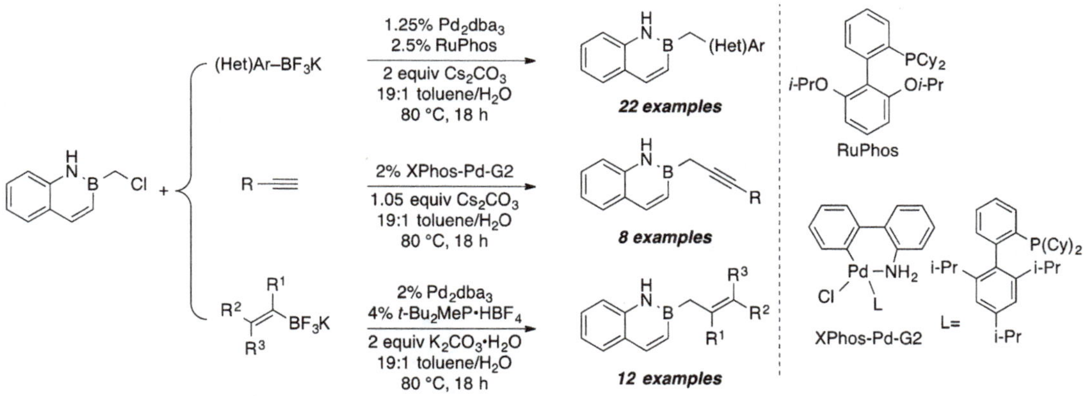

Fig. 22 Syntheses of diversely substituted azaborines

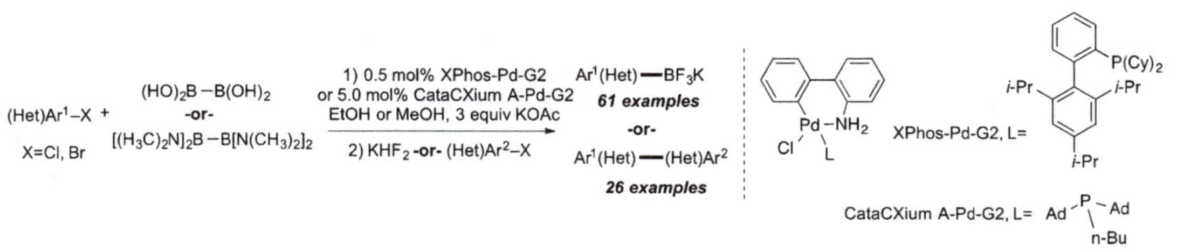

Fig. 23 Palladium-catalyzed borylation of aryl and heteroaryl chlorides and bromides using *bis*boronic acid or *tetrakis*(dimethylamino)diboron

catalyst/precatalyst, temperature, and order of addition. Given the multidimensional nature of this study, the authors logistically reasoned that HTE techniques could be used to more rapidly explore the large reaction space of this important transformation. Through this unbiased approach, the authors made many discoveries with some expected while others were more surprising. Importantly, this thorough investigation improved the overall scope of the reaction, unearthing conditions that would cleanly borylate substrates known to perform poorly under more traditional sets of conditions (Fig. 24).

In 2015, researchers at the University of Pennsylvania disclosed the first selective coupling between a carbon nucleophile and C_1-C_4 alkyl arenes *without* the use of a directing group [54]. Here, the authors use palladium to catalyze a double C-H activation that shows remarkable selectivity for a terminal methyl activation in the presence of electronically similar methylene groups as well as the traditionally more reactive sp^2-hybridized C-H bonds. After the initial discovery of running the reaction in neat toluene, HTE was used to screen conditions, including alternative metal catalysts, external oxidants, and the most ideal co-solvents. Kinetic isotope studies lead the authors to conclude that the reaction proceeds via

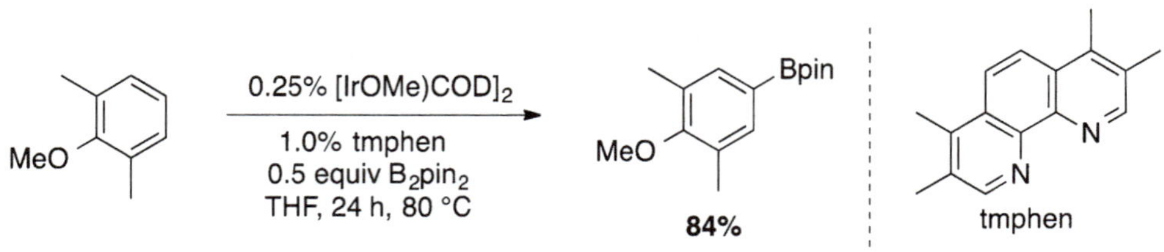

Fig. 24 Iridium-catalyzed C-H activation/borylation

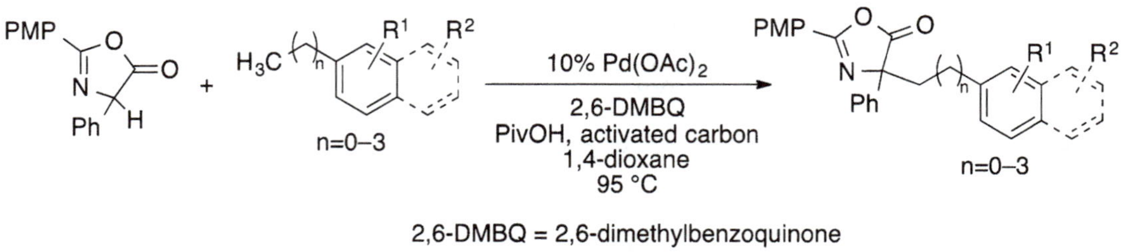

Fig. 25 Oxidative palladium-catalyzed double C-H activation

palladium C-H activation rather than a radical-type process (Fig. 25).

In 2012, researchers used HTE for the optimization of an unexpected palladium-catalyzed oxidative esterification of aldehydes in the presence of alcohols [55]. Aliphatic and aromatic aldehydes were successfully converted into their corresponding esters using $Pd(OAc)_2$ and XPhos. This approach utilized a hydrogen-transfer protocol, employing acetone as a hydrogen acceptor, providing an inexpensive and sustainable approach that avoids the need for other oxidants (Fig. 26).

Work from Schmink and Krska focused on acylsilanes as acyl anion equivalents in the palladium-catalyzed synthesis of diaryl ketones (Fig. 27) [56]. HTE techniques provided successful reaction conditions, with 1,3,5,7-tetramethyl-6-phenyl-2,4,8-trioxa-6-phosphaadamantane as the optimum ligand used in conjunction with a palladacycle precatalyst, leading to the optimal reaction rates and yields (26 examples).

While optimizing a palladium-catalyzed enantioselective Claisen rearrangement of propargyloxy indoles (Fig. 28), Kozlowski and co-workers [57] discovered that the reaction rate was greatly accelerated from *5 days* to *2 h* using Au(III) and (*R*)-BINAP-SbF_6 providing the product in good yield but with only moderate enantioselectivity. HTE optimization studies did not improve the enantioselectivity, but nevertheless the expedient confirmation of the best conditions via HTE should be noted.

Researchers at Merck & Co. developed a practical synthesis of renin inhibitor MK-1597 featuring an asymmetric hydrogenation [58]. HTE allowed the chemists to carry out >384 combinations

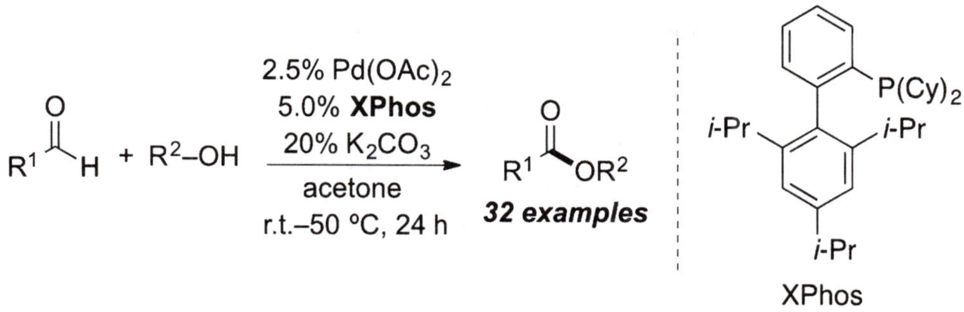

Fig. 26 Palladium-catalyzed oxidative esterification of aldehydes employing acetone as a hydrogen acceptor

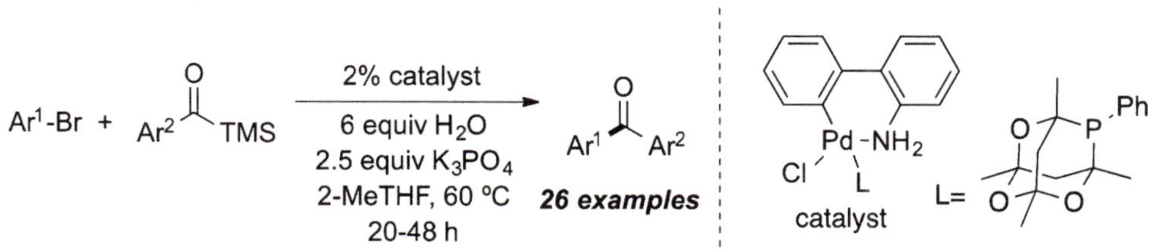

Fig. 27 Palladium-catalyzed synthesis of diaryl ketones using acyl silanes as an acyl anion equivalent

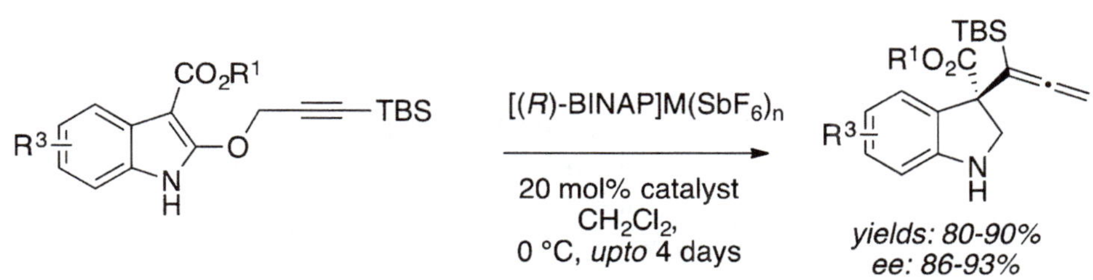

Fig. 28 Enantioselective palladium catalyzed Saucy-Marbet-Claisen rearrangement

of metal precursors, solvents, and chiral ligands, using 500 psi H_2, which led to the identification of one promising catalyst combination of (COD)Ru(Me-allyl)$_2$ and the Josiphos ligand SL-J212-1 affording the product with >90% *ee* though with low yield. Subsequent optimization of this procedure provided 2.3 kg of the key intermediate in 84% isolated yield and 99% *ee* (Fig. 29).

An early example of methodology development using HTE by scientists at Merck & Co. detailed the investigation of conditions that would selectively effect the SMR preferentially on one of two electronically similar aryl chlorides [59] (Fig. 30). Researchers found that the 1,6-naphthyridone dichloride would undergo cross-coupling with the arylboronic acid with high conversion and 92% selectivity when employing one of two ligands, either the tri-*ortho*-anisylphosphine ligand or the bulky NHC, IMes•HCl. These ligands were identified in an HTE screen of 31 ligands providing an early example of one of the most important proof-of-concept results for the HTE approach and subsequently led to a

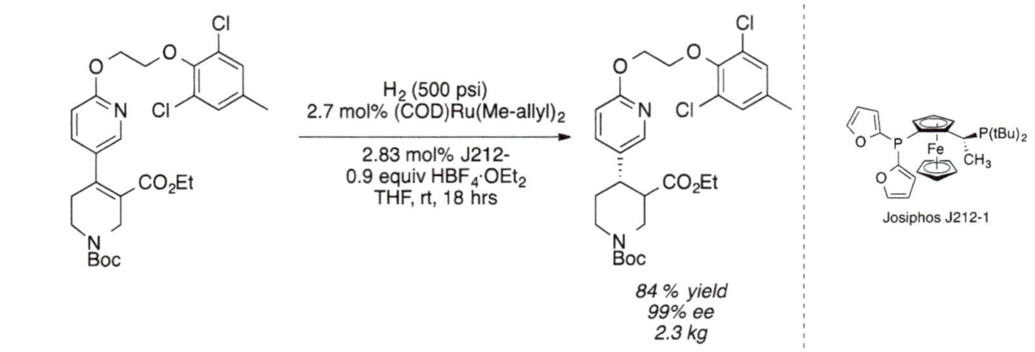

Fig. 29 Asymmetric ruthenium-catalyzed hydrogenation

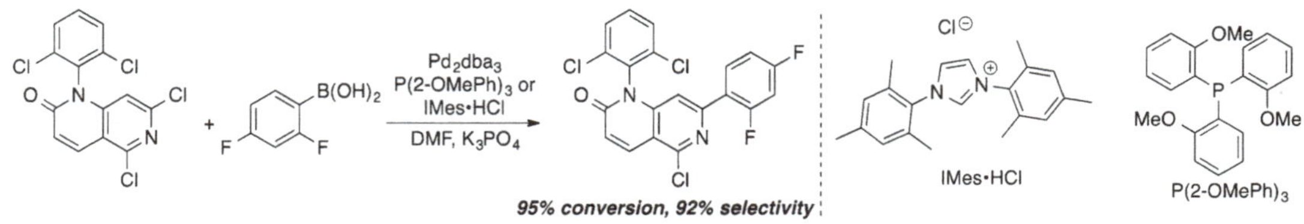

Fig. 30 Palladium-catalyzed regioselective Suzuki-Miyaura cross-coupling reaction

significant push to further develop and optimize this technology by a large cohort of scientists and engineers at Merck & Co.

Recent refinements to the HTE platform have allowed scientists to begin screening photoredox chemistry that relies upon an external light source to initiate a single-electron catalytic cycle with or without a second metal catalyst. In 2014, workers at Merck & Co. reported that they had modified the HTE platform to allow for microscale screening of photoredox chemistry [60]. Here, the researchers reported the methylation, ethylation, and cyclopropylation of unactivated heterocycles via a C-H activation process. Importantly, the researchers focused on elaborated, late-stage druglike molecules with a diverse range of functional groups demonstrating the high tolerance of the developed chemistry to enable the successful functionalization of a range of important chemical motifs (Fig. 31).

In 2015, Molander and co-workers employed a modified HTE platform to allow light from a 26 W compact fluorescent bulb to mediate microscale reactions [61]. In this manner, they developed conditions to cross-couple secondary alkyl potassium trifluoroborates with aryl bromides using a single-electron-mediated alkyl transfer process. An iridium complex was utilized to initiate the single-electron process before a nickel catalyst would mediate the cross-coupling sequence between the alkyl radical and the aryl bromide. After optimization, they scaled the reaction to gram scale and in addition reported the isolation of 28 diverse products (Fig. 32).

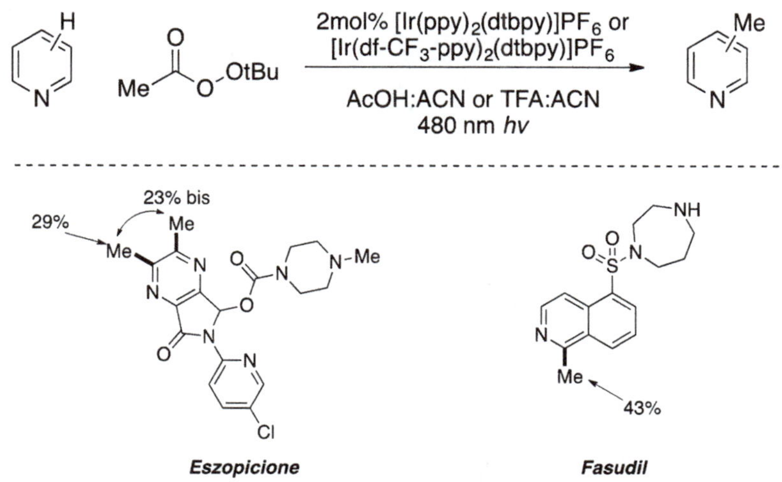

Fig. 31 Late-stage functionalization via photoredox catalysis

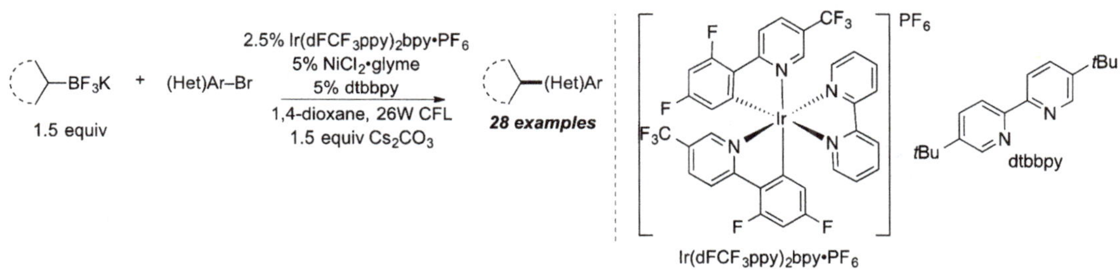

Fig. 32 Dual metal-catalyzed photoredox cross-coupling of 2° alkyl trifluoroborates and aryl bromides

7 Outlook

This brief review illustrates how HTE is ideally suited to both discover and optimize new reaction methodologies. Importantly, working on the microscale, reactions can be optimized using a material-sparing approach with regard to reagents and catalysts while creating minimal waste for disposal. The simplicity of the screening tools allows even the inexperienced user to quickly set up an array of conditions to supplement the existing literature or to develop/optimize novel chemistry. The authors believe that the next generation of scientist will view this approach as an indispensable tool in the synthetic chemist's toolbox.

8 Conclusion

In conclusion this review highlights the "green" nature of microscale, and now nanoscale reaction optimization, which is becoming a crucial aspect of chemistry development. With further stringent regulations always on the horizon, reducing reagent and solvent usage, especially when scouting out new areas of chemistry, can only be achieved by scaling down the screening efforts. HTE is currently and for the foreseeable future will continue to be a cutting-edge technology able to achieve this.

References

1. Gordon EM, Kerwin JF (eds) (1998) Combinatorial chemistry and molecular diversity in drug discovery. Wiley-Liss
2. Drauz K, Gröger H, May O (eds) (2012) Enzyme catalysis in organic synthesis, 3rd edn. Wiley-VCH
3. Gordon EM, Barrett RW, Dower WJ, Fodor SPA, Gallop MA (1994) Applications of combinatorial technologies to drug discovery. 1. Background and peptide combinatorial libraries. J Med Chem 37:1385
4. http://www.hte-company.com/en
5. http://www.chemspeed.com
6. catalysis.avantium.com. Accessed September 14, 2021
7. http://www.biotage.com
8. Kudo N, Perseghini M, Fu GC (2006) A versatile method for Suzuki cross-coupling reactions of nitrogen heterocycles. Angew Chem Int Ed 45:1282–1284
9. Billingsley KL, Barder TE, Buchwald SL (2007) Palladium-catalyzed borylation of aryl chlorides: scope, applications, and computational studies. Angew Chem Int Ed 46:5359
10. Buitrago AB, Regalado EL, Pereira T, Shevlin M, Bateman K, Campeau L-C, Schneeweis J, Berritt S, Shi Z-C, Nantermet P, Liu Y, Helmy R, Welch CJ, Vachal P, Davies IW, Cernak T, Dreher SD (2015) Nanomole-scale high-throughput chemistry for the synthesis of complex molecules. Science 347:49
11. Collins KD, Gensch T, Glorius F (2014) Contemporary screening approaches to reaction discovery and development. Nat Chem 6:859
12. Robbins DW, Hartwig JF (2011) A simple, multidimensional approach to high-throughput discovery of catalytic reactions. Science 333:1423
13. Beeler AB, Su S, Singleton CA, Porco JA Jr (2007) Discovery of chemical reactions through multidimensional screening. J Am Chem Soc 129:1413
14. Kinoshita H, Ingham OJ, Ong WW, Beeler AB, Porco JA Jr (2010) Tandem processes identified from reaction screening: nucleophilic addition to aryl N-phosphinylimines employing La(III)-TFAA activation. J Am Chem Soc 132:6412
15. Goodell JR, McMullen JP, Zaborenko N, Maloney JR, Ho C-X, Jensen KF, Porco JA Jr, Beeler AB (2009) Development of an automated microfluidic reaction platform for multidimensional screening: reaction discovery employing bicyclo[3.2.1]octanoid scaffolds. J Org Chem 74:6169
16. McNally A, Prier CK, MacMillan DWC (2011) Discovery of an α-amino C-H arylation reaction using the strategy of accelerated serendipity. Science 334:1114
17. Anastas P, Eghbali N (2010) Green chemistry: principles and practice. Chem Soc Rev 39:301
18. Hegedus LS (ed) (1999) Transition metals in the synthesis of complex organic molecules, 2nd edn. University Science Books
19. Crawley ML, Trost BM (eds) (2012) Applications of transition metal catalysis in drug discovery and development: an industrial perspective. Wiley
20. Schmink JR, Bellomo A, Berritt S (2013) Scientist-led high-throughput experimentation (HTE) and its utility in academia and industry. Aldrichimica Acta 46(3):71–80
21. Metz AE, Berritt S, Dreher DD, Kozlowski MC (2012) Efficient palladium-catalyzed cross-coupling of highly acidic substrates. Nitroacetates Org Lett 14:760
22. Metz AE, Kozlowski MC (2013) 2-Aryl-2-nitroacetates as central precursors to aryl nitromethanes, α-ketoesters, and α-amino acids. J Org Chem 78:717
23. Walvoord RR, Berritt S, Kozlowski MC (2012) Palladium-catalyzed nitromethylation of aryl halides: an orthogonal formylation equivalent. Org Lett 14:4086
24. Cheng K, Walsh PJ (2013) Arylation of aldehyde homoenolates with aryl bromides. Org Lett 15:2298
25. McGrew GI, Temaismithi J, Carroll PJ, Walsh PJ (2010) Synthesis of polyarylated methanes through cross-coupling of tricarbonylchromium-activated benzyllithiums. Angew Chemie Int Ed 49:5541

26. McGrew GI, Stanciu C, Zhang J, Carroll PJ, Dreher SD, Walsh PJ (2012) Asymmetric cross-coupling of aryl triflates to the benzylic position of benzylamines. Angew Chem Int Ed 51:11510
27. Zhang J, Stanciu C, Wang B, Hussain MM, Da C-S, Carroll PJ, Dreher SD, Walsh PJ (2011) Palladium-catalyzed allylic substitution with (η6-arene-CH2Z)Cr(CO)(3)-based nucleophiles. J Am Chem Soc 133:20552
28. Gossage RA, Jastrzebski JTBH, van Koten G (2005) Hetero-aggregate compounds of aryl and alkyl lithium reagents: a structurally intriguing aspect of organolithium chemistry. Angew Chem Int Ed 44:1448
29. Zhang J, Bellomo A, Creamer AD, Dreher SD, Walsh PJ (2012) Palladium-catalyzed C(sp3)-H arylation of diarylmethanes at room temperature: synthesis of triarylmethanes via deprotonative-cross-coupling processes. J Am Chem Soc 134:13765
30. Ramirez A, Sun X, Collum DB (2006) Lithium diisopropylamide-mediated enolization: catalysis by Hemilabile ligands. J Am Chem Soc 128:10326
31. Remenar JF, Collum DB (1998) Structure and reactivity of lithium diisopropylamide solvated by polyamines: evidence of monomer- and dimer-based dehydrohalogenations. J AmChem Soc 120:4081
32. Remenar JF, Lucht BL, Collum DB (1997) Lithium diisopropylamide solvated by monodentate and bidentate ligands: solution structures and ligand binding constants. J Am Chem Soc 119:5567
33. Bellomo A, Zhang J, Trongsiriwat N, Walsh PJ (2013) Additive effects on palladium-catalyzed deprotonative-cross-coupling processes (DCCP) of sp3 C–H bonds in diarylmethanes. Chem Sci 4:849
34. So CM, Lau CP, Kwong FY (2007) Easily accessible and highly tunable indolyl phosphine ligands for Suzuki-Miyaura coupling of aryl chlorides. Org Lett 9:2795
35. Yucel B, Walsh PJ (2014) Reversed-polarity synthesis of diaryl ketones through palladium-catalyzed direct arylation of 2-Aryl-1,3-dithianes. Adv Synth Catal 356:3659
36. Baker Dockrey SA, Makepeace AK, Schmink JR (2014) Palladium-catalyzed cross-coupling of 2-aryl-1,3-dithianes. Org Lett 16:4730
37. McCabe Dunn JM, Kuethe JT, Orr RK, Tudge MT, Campeau L-C (2014) Development of a palladium-catalyzed α-arylation of cyclopropyl nitriles. Org Lett 16:6314
38. Hussain N, Frensch G, Zhang J, Walsh PJ (2014) Chemo- and regioselective c(sp3)-h arylation of unactivated allylarenes by deprotonative cross-coupling. Angew Chem Int Ed 53:3693
39. Montel S, Jia T, Walsh PJ (2014) Palladium-catalyzed α-arylation of benzylic phosphine oxides. Org Lett 16:130
40. Sandrock DL, Jean-Gérard L, Chen C-Y, Dreher SD, Molander GA (2010) Stereospecific cross-coupling of secondary alkyl β-trifluoroboratoamides. J Am Chem Soc 132:17108
41. Dreher SD, Lim S-E, Sandrock DL, Molander GA (2009) Suzuki–Miyaura cross-coupling reactions of primary alkyltrifluoroborates with aryl chlorides. J Org Chem 74:3626
42. Raushel J, Sandrock DL, Josyula KV, Pakyz D, Molander GA (2011) Reinvestigation of aminomethyltrifluoroborates and their application in Suzuki-Miyaura cross-coupling reactions. J Org Chem 76:2762
43. Molander GA, Sandrock DL (2007) Aminomethylations via cross-coupling of potassium organotrifluoroborates with aryl bromides. Org Lett 9:1597
44. Molander GA, Gormisky PE, Sandrock DL (2008) Scope of aminomethylations via Suzuki–Miyaura cross-coupling of organotrifluoroborates. J Org Chem 73:2052
45. Fleury-Brégeot N, Raushel J, Sandrock DL, Dreher SD, Molander GA (2012) Synthesis and cross-coupling of secondary ammoniomethyltrifluoroborates: rapid and efficient access to secondary arylmethylamines. Chem Eur J 18:9564
46. Schmink JR, Tudge MT (2013) Facile preparation of highly-functionalized, nitrogen-bearing diarylmethanes. Tetrahedron Lett 54:15
47. Molander GA, Argintaru A, Aron I, Dreher SD (2010) Nickel-catalyzed cross-coupling of potassium aryl- and heteroaryltrifluoroborates with unactivated alkyl halides. Org Lett 12:5783
48. Molander GA, Traister KM, Barcellos T (2013) Palladium-catalyzed α-arylation of 2-chloroacetates and 2-chloroacetamides. J Org Chem 78:4123
49. Molander GA, Amani J, Wisniewski SR (2014) Accessing 2-(Hetero)arylmethyl-, -allyl-, and -propargyl-2,1- borazaronaphthalenes: palladium-catalyzed cross-couplings of 2-(chloromethyl)-2,1-borazaronaphthalenes. Org Lett 16:6024
50. Molander GA, Wisniewski SR, Traister KM (2014) Reductive cross-coupling of 3-bromo-2,1-borazaronaphthalenes with alkyl iodides. Org Lett 16:3692

51. Molander GA, Wisniewski SR, Etemadi-Davan E (2014) Suzuki–Miyaura cross-coupling of brominated 2,1-borazaronaphthalenes with potassium alkenyltrifluoroborates. J Org Chem 79:11199
52. Molander GA, Ryu D (2014) Diastereoselective synthesis of vicinally bis(trifluoromethylated) alkylboron compounds through successive insertions of 2,2,2-trifluorodiazoethane. Angew Chem Int Ed 53:14181
53. Preshlock SM, Ghaffari B, Maligres PE, Krska SW, Malezczeka RE Jr, Smith MR III (2013) High-throughput optimization of Ir-catalyzed C-H borylation: a tutorial for practical applications. J Am Chem Soc 135:7572
54. Curto JM, Kozlowski MC (2015) Chemoselective Activation of sp3 vs sp2 C–H Bonds with Pd(II). J Am Chem Soc 137:18
55. Tschaen B, Schmink J, Molander GA (2013) Pd-catalyzed aldehyde to ester conversion: a hydrogen transfer approach. Org Lett 15:500
56. Schmink JR, Krska SW (2011) Reversed-polarity synthesis of diaryl ketones via palladium-catalyzed cross-coupling of acylsilanes. J Am Chem Soc 133:19574
57. Cao T, Linton EC, Deitch J, Berritt S, Kozlowski MC (2012) Copper(II)- and palladium(II)-catalyzed enantioselective claisen rearrangement of allyloxy- and propargyloxy-indoles to quaternary oxindoles and spirocyclic lactones. J Org Chem 77:11034
58. Molinaro C, Shultz S, Roy A, Lau S, Trinh T, Angelaud R, O'Shea PD, Abele S, Cameron M, Corley E, Funel J-A, Steinhuebel D, Weisel M, Krska S (2011) A practical synthesis of renin inhibitor MK-1597 (ACT-178882) via catalytic enantioselective hydrogenation and epimerization of piperidine intermediate. J Org Chem 76:1062
59. Cai C, Chung JYL, McWilliams JC, Sun Y, Shultz CS, Palucki M (2007) From high-throughput catalyst screening to reaction optimization: detailed investigation of regioselective Suzuki coupling of 1,6-naphthyridone dichloride. Org Proc Res Dev 11:328
60. DiRocco DA, Dykstra K, Krska S, Vachal P, Conway DV, Tudge M (2014) Late-stage functionalization of biologically active heterocycles through photoredox catalysis. Angew Chem Int Ed 53:4802
61. Tellis JC, Primer DN, Molander GA (2014) Dual catalysis. Single-electron transmetalation in organoboron cross-coupling by photoredox/nickel dual catalysis. Science 345:6195

Chapter 16

Radiopharmaceutical Discovery with $^{11}CO_2$-Fixation Methods Inspired by Green Chemistry

Benjamin H. Rotstein and Neil Vasdev

Abstract

Efforts to use sustainable and abundant feedstocks for chemical synthesis have produced efficient methods for incorporating carbon dioxide into small organic molecules with high efficiency at ambient temperature and pressure. These methods have inspired radiochemists to develop strategies for using carbon-11 carbon dioxide, which is routinely produced directly in a cyclotron beam target, to prepare radiolabeled compounds possessing functional groups such as ureas, carbamates, oxazolidinones, carboxylic acids, esters, and amides. In turn, novel and clinically relevant radiopharmaceuticals have been developed for targets including fatty acid amide hydrolase, monoamine oxidase B, and glycogen synthase kinase 3β. The versatility of these technologies has also facilitated high-throughput discovery of fluorine-18-labeled radiotracers.

Key words Carbon dioxide, Carbon-11, CO_2-fixation, Radiochemistry, Green chemistry, Positron emission tomography, Radiotracer, Fluorine-18

1 Carbon Dioxide and Practical Radiochemistry

Historically, chemical supply and synthesis have relied heavily on fossil fuel feedstock, as well as some renewable sources, such as corn and sugarcane. This situation persists today, though concerns over depletion of natural resources and the environmental effects of their consumption have stimulated efforts into developing "carbon-neutral" chemical feedstocks that are sustainable and accessible by low-energy processes using nontoxic reagents or catalysts [1]. Carbon dioxide (CO_2), which is readily available as a combustion product, makes up a significant fraction of Earth's atmosphere and is exploited by photosynthesis for growth of plants and algae [2], representing one such feedstock. CO_2 is indeed an economically viable resource for the preparation of polycarbonates and polyurethane materials [3]. In part due to "green chemistry" initiatives, CO_2 has been targeted for use as a C_1-source in the prepara-

Paul F. Richardson (ed.), *Green Chemistry in Drug Discovery: From Academia to Industry*, Methods in Pharmacology and Toxicology, https://doi.org/10.1007/978-1-0716-1579-9_16, © Springer Science+Business Media, LLC, part of Springer Nature 2022

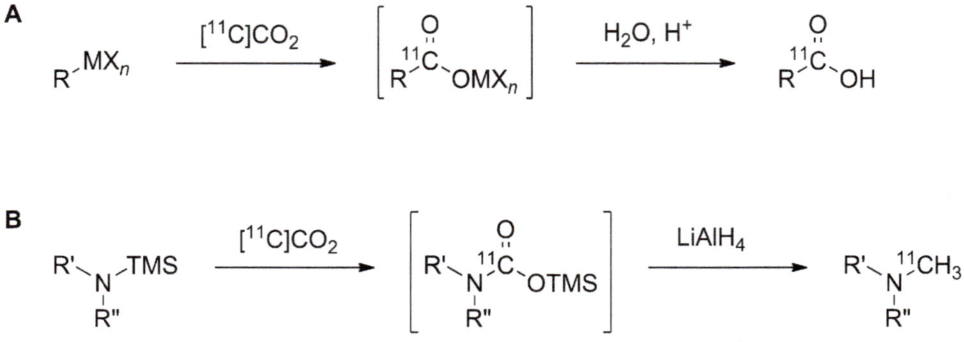

Fig. 1 [^{11}C]CO_2-fixation using organometallics; *M* metal, X_n ligands or counterions

tion of intermediates such as methanol [4–6], urea [2], lactones [7], heterocycles [8–10], biodegradable polymers [11–14], and other fine chemicals [15, 16].

Though CO_2 had been used in chemical synthesis for decades prior to the advent of green chemistry as an ethos, the renewed focus on efficient, mild, and versatile methodologies has had a substantial impact on the utility of this reagent for directly preparing a wide array of complex functional groups and molecules. Similarly, carbon-11 (^{11}C, $t_{1/2} = 20.4$ min)-labeled CO_2 ($^{11}CO_2$) has been applied for the synthesis of radiolabeled carboxylic acids for over 70 years [17], using Grignard reagents, organolithiums, or other organometallics (Fig. 1). Carbon-11-labeled amides, esters, acid chlorides, and ketones could also be prepared in this manner, though substrates are limited to those that can tolerate a highly basic organometallic. Silanamines represent an alternative fixating agent for $^{11}CO_2$ and could be used to generate *O*-silyl carbamates, which could then be reduced to ^{11}C-methylamines using lithium aluminum hydride [18], and these topics have been recently reviewed [19].

In addition to the limitations of chemical reactivity and functional group compatibility, radiosynthesis must contend with practical limitations related to the low abundance, short half-life, and risk of exposure to radioactivity originating from the radionuclide. $^{11}CO_2$ is now routinely produced in medical cyclotrons and used as a no-carrier-added (i.e., [^{11}C]CO_2) reactant with low mass (typically ~100 nmol). [^{11}C]CO_2 is the limiting reagent when used in chemical transformations and therefore demands an efficient method for trapping it in the reaction medium. [^{11}C]CO_2 is produced by proton bombardment of nitrogen gas in the presence of 1% O_2 by the $^{14}N(p,\alpha)^{11}C$ nuclear reaction. Residual oxygen and by-product nitrous oxide species must either be tolerated by the reaction or efficiently removed from the gas mixture [20–22]. Due to the short half-life of carbon-11, its nature as a gaseous reagent at convenient temperatures for chemical reactions, and associated radioactivity, automated or semiautomated apparatus are most frequently used for handling [23]. Apparatus must be well designed and constructed to mitigate leaks, control gas flow rates, and facilitate relatively rapid radiosyntheses of less than an hour, including purification and reformulation.

For the abovementioned reasons, chemical transformations that are of great use for conventional CO_2-fixation may be impractical for radiochemists. Since generating high pressures of $[^{11}C]CO_2$ is not possible due to low abundance, reactions that cannot be performed at or near ambient pressure are unlikely to succeed. An additional concern is related to molar activity, the concentration of radioactivity relative to mass of the product. Since highly basic organometallics, such as Grignard reagents and organolithiums, are susceptible to react with atmospheric levels of CO_2, it may be crucial to store these radiotracer precursors in carefully controlled environments that prevent formation of unwanted nonradioactive products. In practice, the organometallic precursors are usually freshly prepared immediately before use. Furthermore, since these reagents often produce insoluble by-products upon quenching, extra considerations must be made to ensure robust liquid handling of crude reaction mixtures to facilitate purification.

Reactions developed for CO_2-fixation that are designed on the principles of green chemistry have a high probability of being amenable for radiochemistry. Foremost, reactions that are operationally simple and do not require high temperatures or pressure are more easily adapted for automation and as such likely to be compatible with $[^{11}C]CO_2$. Green chemistry promotes reactions that incorporate a greater proportion of reactants into the final product, minimize excess solvent and separation reagents, and obviate the need for protecting groups and derivatives. Similarly, direct incorporation of $[^{11}C]CO_2$ into reactants without the need for multistep syntheses and separations saves valuable time, is easier for automation, and often translates into a simpler purification of the final product. Finally, since the foremost application of radiochemistry with ^{11}C is in vivo imaging with positron emission tomography (PET), there exists a strong preference for avoiding toxic or hazardous solvents or substances that could potentially contaminate the final injectable product.

Radiochemical reactions with ^{11}C are seldom purposefully designed with environmental impact in mind. The scale of these transformations with short-lived isotopes is such that environmental impact is minimal, compared to conventional industrial processes, and the high energy demands of radioisotope production are beyond the control of radiochemists. Still, a green chemistry approach to process design can be highly compatible with radiosynthesis process design [24, 25]. As will be detailed herein, chemical methodologies spurred on by an interest in green chemistry have had a transformative impact on PET radiochemistry.

2 CO_2 as a Feedstock for Small Molecule Carbamates, Isocyanates, and Ureas

Functional groups with high oxidation state carbon centers can be readily prepared from highly oxidized precursors, such as phosgene [26] and isocyanates [27, 28]. The search for less toxic and more environmentally benign alternatives to prepare these functional groups naturally included readily available carbon dioxide, activated using strained rings [29–31], metal complexes [32–36], or organic catalysts [10, 37, 38]. The solubility of CO_2 in organic solvents and in the presence of various bases and additives was also studied [39]. Critically, nitrogenous bases such as amidines [40, 41] and guanidines [42], as well as alkali carbonates [43], were found to efficiently trap CO_2 in solution. Each of these bases facilitates transamidation to amines, though the exact nature of the trapped species in solution remains the subject of some debate [44–47]. Initially, crystallogaphic and spectroscopic studies pointed to bicarbonate salts in solution, which would be formed by hydrolysis of carbamic intermediates by adventitious water (Fig. 2) [41, 42, 48, 49]. More recent crystallographic studies have characterized the carbamate zwitterion formed using a bicyclic guanidine [50]. This species was later demonstrated to be viable as a precursor for intermolecular formal CO_2 insertion into O–H and N–H bonds [51].

Within the context of synthetic utility, replacing phosgene with CO_2 for the preparation of carbamates was shown to be feasible using amines and organic or inorganic bases to trap CO_2 in solution [34, 38]. Following addition of alkyl halides, carbamates were formed though often accompanied by significant fractions of *N*-alkylated species [52]. High selectivity for carbamates could be achieved with pentaalkylguanidines, such as *N*-cyclohexyl-*N'*,*N'*,

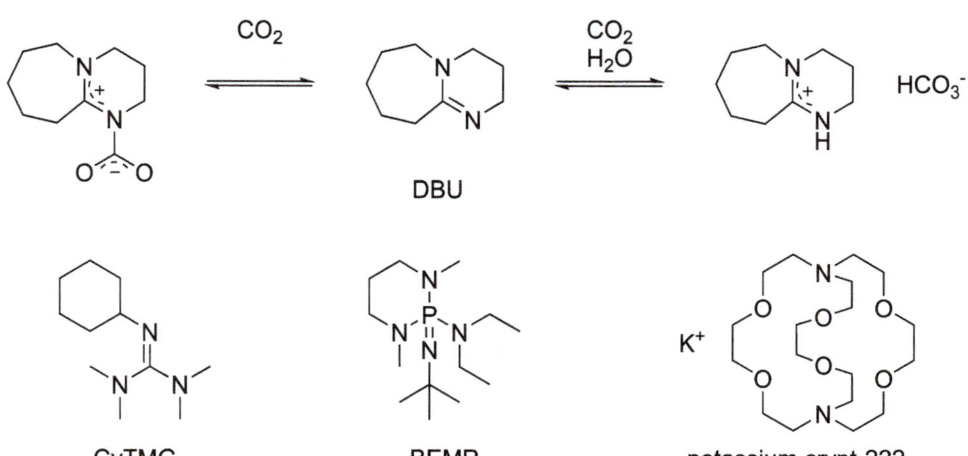

Fig. 2 Modes of coordination for organic base trapping CO_2 in solution and a selection of alternative bases useful for CO_2-fixation

N'',*N''*-tetramethylguanidine (CyTMG), and to a lesser extent amidines, such as 1,8-diazabicyclo[5.4.0]undec-7-ene (DBU) [53, 54]. This selectivity was attributed to the highly polarizable and sterically hindered cation formed, which produced a more nucleophilic counterion, as well as the strong basicity of guanidines, which drives forward the equilibrium of amine and CO_2 toward carbamate anion. Using guanidine bases, alkyl chlorides were eligible alkylating agents, whereas weaker bases, such as potassium cryptand complexes, mandated the use of unhindered or activated alkyl bromides, iodides, or tosylates. Surprisingly, the pressure of CO_2 has an inverse relationship to the rate of reaction with unhindered primary amines. This is apparently due to double CO_2 insertion in N–H bonds, forming the less reactive *bis*-carbamate.

CO_2 was also also used to replace phosgene in the preparation of isocyanates [55] and carbamoyl chlorides [56]. Carbamate anions were prepared similarly to the methods described above, using primary or secondary amines with organic bases in acetonitrile under CO_2 (1 atm) at room temperature or below. A phosphorus-based dehydrating reagent, such as phosphorus oxychloride ($POCl_3$), was then added to form the isocyanate, which could be isolated in generally high yields. Strong organic bases, such as pentaalkylguanidines and phosphazenes, or higher pressures of CO_2 effectively inhibited formation of symmetrical ureas by preventing equilibrium release of amines. Despite the well-documented formation of ionic carbamates by reaction of amines with CO_2, a practical methodology to exploit these intermediates required controlling the nucleophilicity in this way to favor *O*-alkylation or dehydration over *N*-alkylation or *N*-acylation [57].

3 $^{11}CO_2$-Fixation for Radiolabeling Carbonyl Groups

Carbon-11 is the most versatile PET radioisotope, due to the ubiquity of carbon nuclei in biological molecules of interest and the appeal of in vivo studies using direct isotopologues for noninvasive dynamic imaging [58]. Cyclotron-produced ^{11}C is readily available in high molar activity as $[^{11}C]CO_2$ and along with ^{18}F represents the most commonly used isotopes for PET radiotracer development. Dozens of reagents have been developed for incorporation of ^{11}C into radiotracers [59], yet the most common strategy remains methylation using $[^{11}C]$methyl iodide or $[^{11}C]$ methyl triflate [60], themselves produced from $[^{11}C]CO_2$. While ^{11}C-methylation is a powerful strategy that has been employed to prepare a great many important PET radiopharmaceuticals, it is capable only of generating a single type of label—the low oxidation state methyl group. Alternative strategies that would facilitate radiolabeling higher oxidation state functional groups, such as carboxylic acids, carbamates, ureas, and heterocycles, have been avidly pursued using reagents such as [^{11}C]formaldehyde, [^{11}C]carbon monoxide, ammonium [^{11}C]cyanide, [^{11}C]phosgene—which are all prepared from [^{11}C]CO_2—and our focus, direct use of [^{11}C] CO_2 [19].

3.1 ^{11}C-Ureas

Owing to the low mass and concentration of [^{11}C]CO_2 in cyclotron target and delivery gas, efficient trapping of carbon dioxide in the reaction mixture is crucial to obtaining high radiochemical yields of ^{11}C-labeled products relative to starting radioactivity. In early examples of [^{11}C]CO_2-fixation, reactions were carried out using highly reactive organometallic substrates (i.e., Grignards and organolithiums) that react directly with [^{11}C]CO_2 to form C–^{11}C bonds. Chakraborty et al. reported the first synthesis of [^{11}C]urea using lithium hexamethyldisilazide (LHMDS) to trap [^{11}C]CO_2 in a THF solution, followed by aqueous hydrolysis with ammonium chloride [61]. Unfortunately, this method is only applicable to the preparation of simple [^{11}C]urea and derivatives such as [^{11}C-*carbonyl*]uracil by multistep syntheses.

In order to prepare more structurally complex ^{11}C-ureas, organic base-mediated [^{11}C]CO_2-fixation was pursued, inspired by the mild industrial-focused processes discussed above [57]. In its first iteration, triethylamine in dichloromethane was used to facilitate ^{11}C-carboxylation of aniline and aliphatic amines, followed by dehydration with $POCl_3$ [62]. Under these conditions, the ^{11}C-*carbonyl*-isocyanate reacted immediately with the excess amine, to form symmetrical products including ^{11}C-*carbonyl*-ureas and ^{11}C-*carbonyl*-carbodiimides (Fig. 3). While these results provide evidence for the desired reactivity using [^{11}C]CO_2, they also suggest that achieving selectivity for unsymmetrical ureas could be a practical challenge in this context. Whereas in nonradioactive synthesis one could reasonably expect to saturate precursor amines using excess CO_2, analogous stoichiometry using no-carrier-added [^{11}C]CO_2 would demand very precise measurements and high dilution.

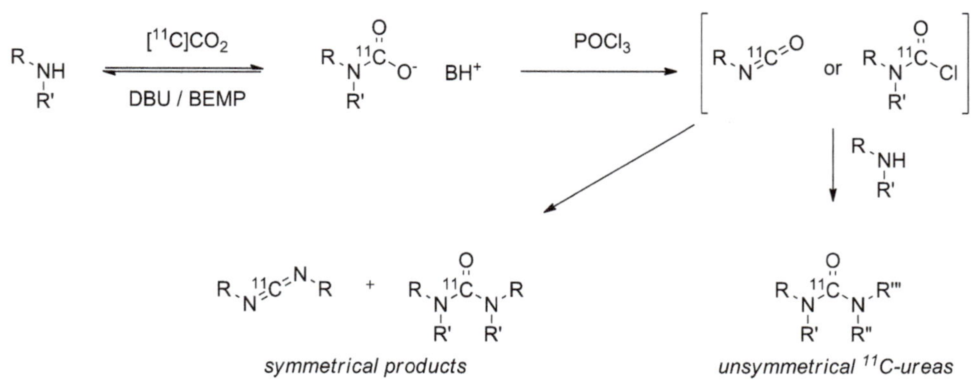

Fig. 3 Preparation of ^{11}C-ureas by [^{11}C]CO_2-fixation and dehydration

To overcome these challenges, radiochemists first sought to identify organic bases that could efficiently trap [^{11}C]CO_2 at both ambient temperature and pressure and at practical delivery flow rates. Hooker et al. demonstrated that DBU was effective for this purpose [63], and Wilson et al. later found that 2-*tert*-butylimino-2-diethylamino-1,3-dimethyl-perhydro-1,3,2-diazaphosphorine (BEMP) was an even stronger [^{11}C]CO_2-trapping agent [64]. With BEMP, selective radiosynthesis of unsymmetrical ^{11}C-*carbonyl*-ureas was finally achieved (Fig. 3). Formation of symmetrical products could be suppressed by using excess $POCl_3$, but this strategy also stipulated using an even larger excess of the second amine nucleophile for attack on the ^{11}C-isocyanate. Consequently, a high concentration reaction mixture is formed under these conditions, which presents a purification challenge. An alternative solution was to reduce the concentration of the initial amine, and fortunately this was found to be compatible with both fast reaction times ($\leq$2 min) and high yields [65]. The ideal substrates for isocyanate formation were found to be aliphatic primary amines. Cyclic secondary amines were also well tolerated, especially less hindered examples. These presumably form ^{11}C-carbamoyl chloride intermediates rather than ^{11}C-isocyanates. Initially, the scope of amine nucleophiles that could be used for ^{11}C-urea formation was limited primarily to dimethylamine, though a variety of aryl amines have now been used for this purpose in the context of radiotracer synthesis (*see* Subheading 4) [65].

An alternative approach to isocyanates used iminophosphorane precursors, prepared from azides or primary amines, for condensation with carbon dioxide [66]. Phenyltriphenylphosphinimine was deployed to prepare [^{11}C-*carbonyl*]phenylisocyanate and a variety of unsymmetrical ^{11}C-ureas from aliphatic and aromatic amines [67]. In this case, no other base was necessary to trap [^{11}C]CO_2 in solution; rather the reaction vessel containing a THF solution was cooled to −60 °C during gas delivery and heated to 60 °C to

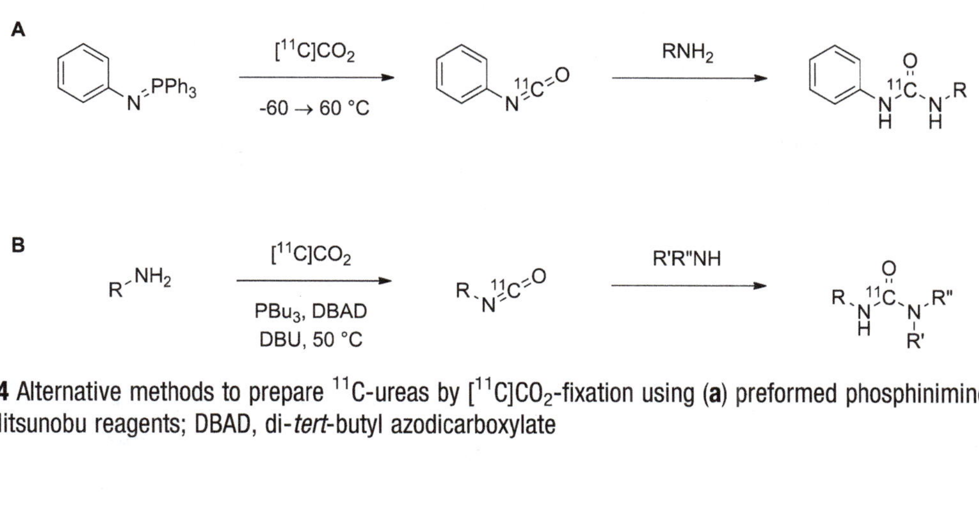

Fig. 4 Alternative methods to prepare ^{11}C-ureas by [^{11}C]CO_2-fixation using (**a**) preformed phosphinimines or (**b**) Mitsunobu reagents; DBAD, di-*tert*-butyl azodicarboxylate

complete the reaction (Fig. 4a). Given the high trapping efficiency in the absence of base, it seems likely that [^{11}C]CO_2 forms a complex with either the iminophosphorane or amine precursor.

A related approach took advantage of Mitsunobu reagents (phosphines and azo compounds) to convert ionic carbamates into isocyanates [68, 69]. In the presence of an additional amine, the isocyanate could be transformed in situ to a urea [70]. Under radiolabeling conditions, the reaction temperature needed careful optimization (50 °C) to prevent [^{11}C]CO_2 release while still promoting the reaction (Fig. 4b) [71]. Aliphatic and aromatic amines could be drawn on for ^{11}C-isocyanate formation, though secondary aliphatic amines were necessary for ^{11}C-urea formation to achieve selectivity for unsymmetrical products [72].

3.2 ^{11}C-Carbamates

Carbamates, like ureas, are attractive functional groups for drug and radiotracer design due to their stability in vivo, their role as a linker of ligand fragments, and for drug-target interactions through the carbamate itself [73]. While these were previously radiolabeled using [^{11}C]phosgene and [^{11}C]carbon monoxide [74], methods using [^{11}C]CO_2 are especially convenient as they may obviate the need for intermediate redox manipulation. Unlike ^{11}C-ureas, *O*-alkyl carbamates may be prepared without intermediate dehydration to form ^{11}C-isocyanates. Simply bubbling [^{11}C]CO_2 into a solution of DBU, amine, alkyl chloride, and DMF, followed by heating to 75 °C, produced high yields of *O*-alkyl [^{11}C]carbamates (Fig. 5) [63]. Primary and secondary alkyl amines worked best, while anilines proved a challenge. Benzylic and allylic chlorides were more selective electrophiles than their bromide congeners, while more hindered and less activated electrophiles generally presented a greater challenge.

This strategy was also applied to methyl [^{11}C]carbamates using methylating agents such as dimethylsulfate (DMS), methyl iodide (CH_3I), or methyl tosylate (CH_3OTs) [64]. For this application the more basic BEMP was used for [^{11}C]CO_2 trapping, and primary or secondary amines, including electron-rich anilines, performed well under the standard conditions. Electron-neutral and

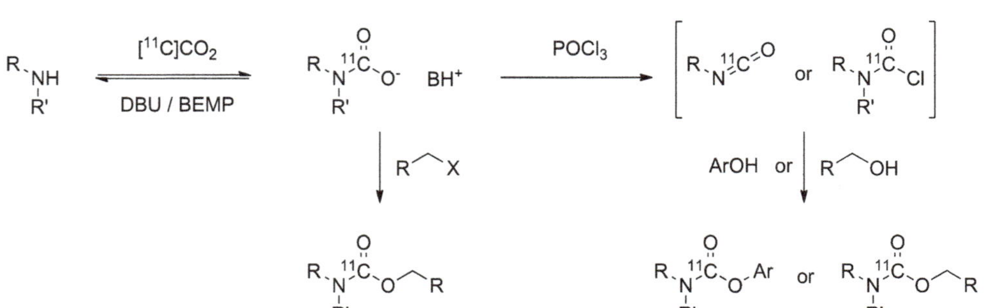

Fig. 5 Two strategies for the preparation of ^{11}C-carbamates by [^{11}C]CO_2-fixation

electron-deficient anilines could also give useful yields at higher concentrations. Reactions generally proceeded very quickly and at room temperature. Excess methylating agent or addition of the methylating agent to the reaction mixture prior to addition of the amine led to low yields of the desired compounds, suggesting the intermediacy of the carbamate ion.

In order to prepare those *O*-substituted ^{11}C-carbamates not readily accessible by alkylation of carbamate ions, alcohols and phenols can be used to quench ^{11}C-isocyanates, prepared as described above [65]. Again, the stoichiometry of reagents must be balanced to prevent formation of symmetrical products. Initially, methanol and phenols were used for quenching the ^{11}C-isocyanate, but more recently hindered and electron-deficient alcohols, such as ethanol, isopropanol, *tert*-butanol, and hexafluoroisopropanol [75], have been successfully incorporated into ^{11}C-carbamates with the high basicity of BEMP likely assisting in these transformations.

Oxazolidinone heterocycles have been radiolabeled with ^{11}C using [^{11}C]phosgene [76, 77]. Although there have been several improvements on the preparation of this reagent, technical challenges relating to the maintenance and reliability of synthesis instruments have limited its widespread use [74]. Using the same amino alcohol ^{11}C-phosgenation precursors for [^{11}C]CO_2-fixation, oxazolidinones are formed after dehydration with $POCl_3$ [78]. Optimization of the fixation base, dehydrating agent, and reaction concentrations allows this method to be used at ambient temperature for a rapid synthesis of ^{11}C-oxazolidinones directly from cyclotron-produced [^{11}C]CO_2.

3.3 ^{11}C-Carboxylic Acids

^{11}C-Carboxylic acids have been consistently produced from [^{11}C]CO_2, using organometallic precursors, such as methylmagnesium bromide for [^{11}C]acetate and *n*-pentadecylmagnesium bromide for [^{11}C]palmitate [79, 80]. This strategy is, however, limited to ^{11}C-carboxylic acids for which a stable organometallic precursor can be prepared. In the interest of expanding the utility of CO_2 as a feedstock for bulk and specialty chemicals, metal and organocatalytic approaches for CO_2-fixation to generate carboxylic acids and their derivatives have been keenly pursued [10, 81]. In particular, the use of pre-functionalized organoboron and organozinc reagents, which have superior stability and functional group tolerance when compared to Grignards and organolithiums, has garnered much attention [82–84]. Boronic esters have been shown to be amenable for [^{11}C]CO_2-fixation using a copper catalyst [85], overcoming the significant challenge of much lower concentrations of [^{11}C]CO_2 available in the reaction mixture compared to CO_2 under nonradioactive conditions (Fig. 6). As such, significant optimization was required, including shifting away from alkoxide bases in favor of *N*,*N*,*N′*,*N′*-tetramethylethylenediamine (TMEDA),

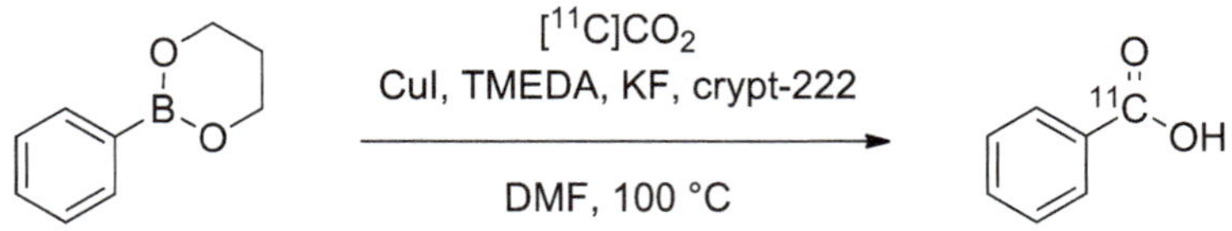

Fig. 6 Copper-mediated [^{11}C]CO_2-fixation for ^{11}C-carboxylation of boronic esters

which acts both as a trapping agent for [^{11}C]CO_2 and a ligand for the copper catalyst. A soluble fluoride additive was also found to significantly improve radiochemical yields. A variety of functional groups were tolerated under the optimized conditions, while the ^{11}C-carboxylic acids could be further transformed to yield ^{11}C-*carbonyl*-esters and ^{11}C-*carbonyl*-amides with high specific activities in the isolated products [85].

4 ^{11}C-*Carbonyl* Radiotracer Development

Radiochemical methodologies using PET radionuclides are inevitably measured by their practical utility for radiotracer development or radiopharmaceutical preparation. Though novel methods may be interesting in their own right and may serve to spur the development of further synthetic tools that are both interesting and practical, the most clinically significant developments are those that can be translated to deliver new biological probes for PET imaging. The methods discussed above have had significant impact on PET radiotracer development and clinical research and offer appealing strategies for radiotracer development that is not limited to the most well-established radiochemical reactions. The radiotracers discussed below are compiled in Table 1.

4.1 Fatty Acid Amide Hydrolase (FAAH) Radiotracers

Fatty acid amide hydrolase (FAAH) is an integral membrane enzyme that regulates signaling at cannabinoid receptors such as CB_1 and CB_2. The main CB_1 neurotransmitter is anandamide, which is a fatty acid amide. It is produced on-demand in the postsynaptic neuron and participates in retrograde signaling by binding CB_1 at the presynaptic neuron to stimulate calcium and potassium ion channels [112]. Since its structure precludes intracellular storage, anandamide levels and the tone of the cannabinoid system are regulated by FAAH.

An irreversible FAAH inhibitor, PF-04457845, has advanced to clinical trials and generated interest for development of FAAH PET radiotracers [86]. A derivative containing an [^{18}F]fluoroethyl group, [^{18}F]PF-9811, was developed [90], followed by the direct isotopologue, [^{11}C-*carbonyl*]PF-04457845, prepared by [^{11}C]CO_2-fixation [89]. Since this ^{11}C-urea is prepared from a highly nucleophilic cyclic secondary amine and the relatively poorly nucleophilic 3-aminopyridazine, the latter needed to be present in

Table 1
Radiotracers and radiopharmaceuticals developed using [^{11}C]CO_2-fixation

Radiotracer	Target	Compound class/synthesis strategy	Development history
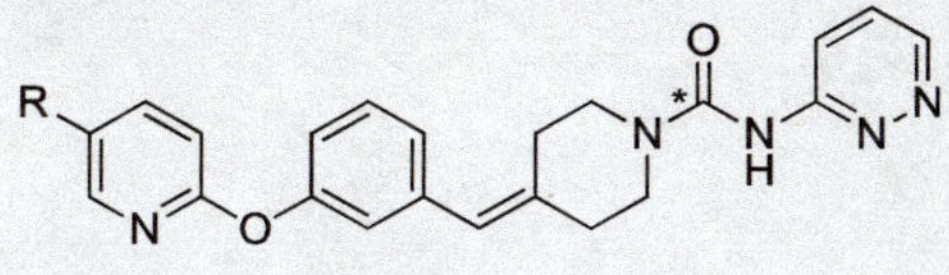 **[^{11}C]PF-04457845**: R = CF_3, * = ^{11}C **[^{18}F]PF-9811**: R = $OCH_2CH_2{}^{18}F$ [^{11}C]PF-04457845: R = CF_3, * = ^{11}C [^{18}F]PF-9811: R = $OCH_2CH_2{}^{18}F$	Fatty acid amide hydrolase (FAAH)	^{11}C-urea/[^{11}C]CO_2-fixation via dehydration	Discovery [86, 87] Therapeutic clinical trial [88] ^{11}C labeling, biodistribution [89] ^{18}F derivative, biodistribution [90]
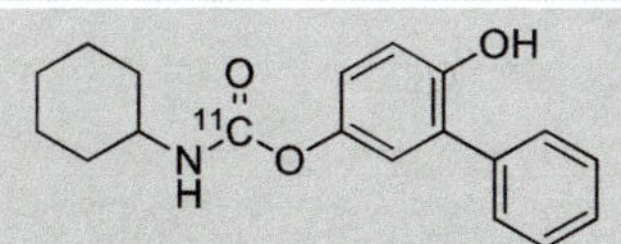 **[^{11}C]CURB (URB694) (not shown: URB597)** [^{11}C]CURB (URB694) (not shown: URB597)	FAAH	^{11}C-carbamate/[^{11}C]CO_2-fixation via dehydration	Discovery [91] ^{11}C labeling [65] Preclinical [92] Clinical [93, 94] Library radiosynthesis [95]
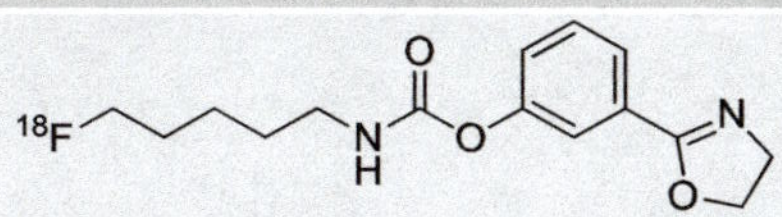 **[^{18}F]DOPP** [^{18}F]DOPP	FAAH	^{18}F-alkyl carbamate	Discovery, preclinical, and kinetic modeling [96, 97]

(continued)

Table 1 (continued)

Radiotracer	Target	Compound class/synthesis strategy	Development history
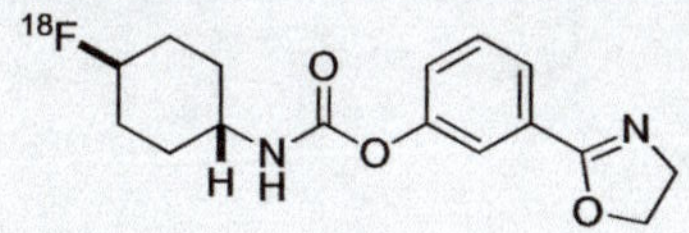 **[^{18}F]FCHC** [^{18}F]FCHC	FAAH	^{18}F-alkyl carbamate	Discovery, preclinical [98]
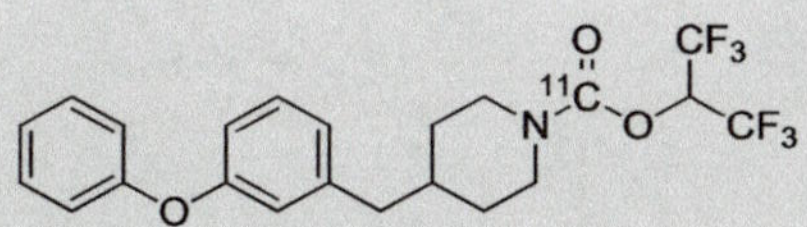 **[^{11}C]JW642** [^{11}C]JW642	Monoacylglycerol lipase (MAGL)	^{11}C-urea or ^{11}C-carbamate/[^{11}C]CO_2-fixation via dehydration	Radiolabeling, preclinical [75]
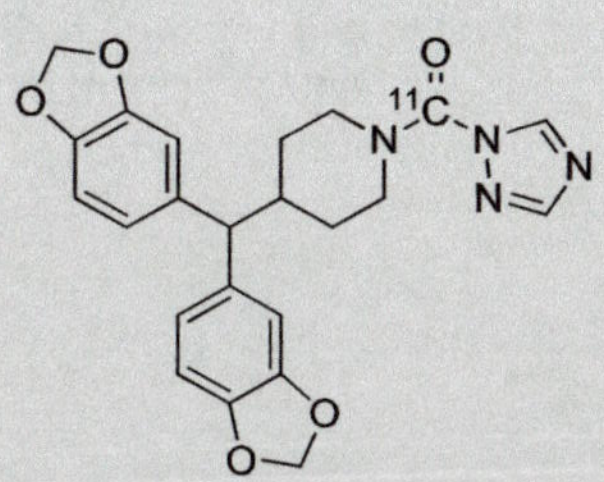 **[^{11}C]JJKK-0048** (not shown: **[^{11}C]KML29**, **[^{11}C]ML30**) [^{11}C]JJKK-0048 (not shown: [^{11}C]KML29, [^{11}C]ML30)			

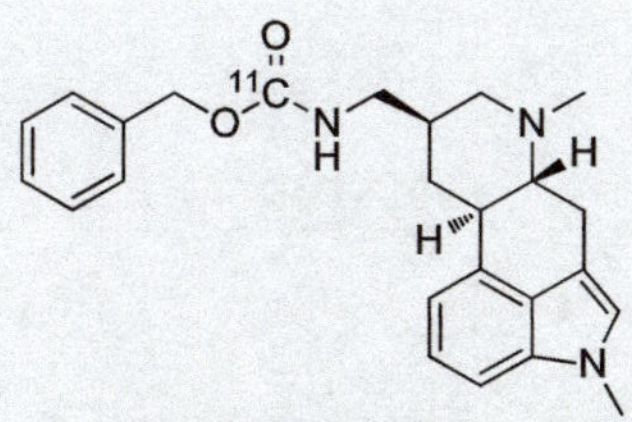 **[^{11}C]metergoline** [^{11}C]metergoline	Serotonin receptors (5HTR)	^{11}C-carbamate/[^{11}C]CO_2-fixation via ^{11}C-carbamate alkylation	Synthesis [63] Preclinical [99]
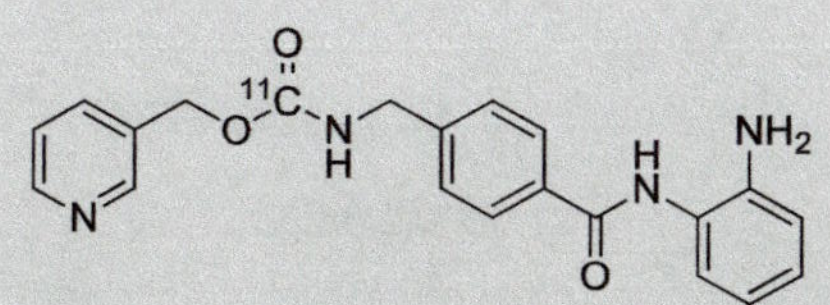 **[^{11}C]MS-275** [^{11}C]MS-275	Histone deacetylases (HDACs)	^{11}C-carbamate/[^{11}C]CO_2-fixation via ^{11}C-carbamate alkylation	Synthesis, preclinical [100]
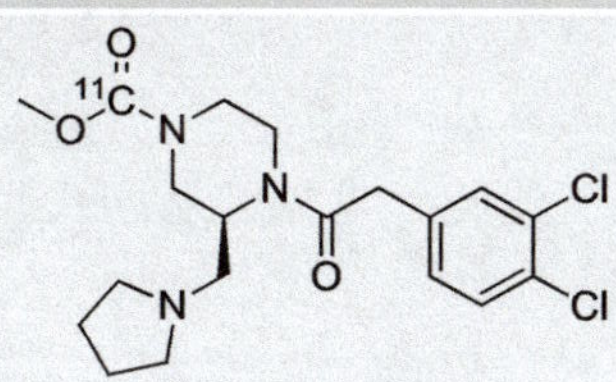 **[^{11}C]GR103545** [^{11}C]GR103545	κ-opioid receptor (agonist)	^{11}C-carbamate/[^{11}C]CO_2-fixation via ^{11}C-carbamate alkylation	Preclinical [101, 102] Alternative synthesis [103, 104] [^{11}C]CO_2-fixation [64] Clinical [105]

(continued)

Table 1 (continued)

Radiotracer	Target	Compound class/synthesis strategy	Development history
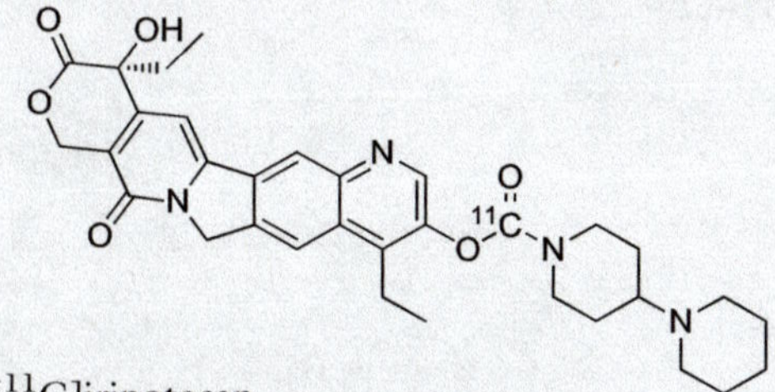 [^{11}C]irinotecan	Topoisomerase 1	^{11}C-carbamate/[^{11}C]CO_2-fixation via dehydration	Synthesis, preclinical [106]
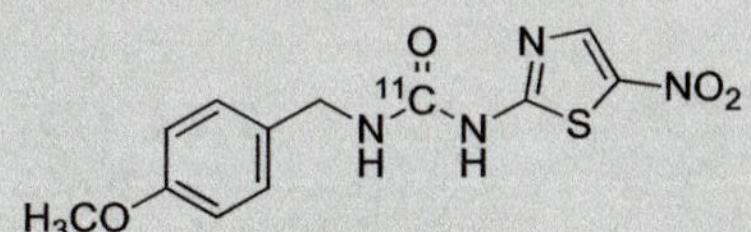 **[^{11}C]irinotecan** **[^{11}C]AR-A014418** [^{11}C]AR-A014418	Glycogen synthase kinase-3β (GSK-3β)	^{11}C-urea/[^{11}C]CO_2-fixation via dehydration	Synthesis, preclinical [107] [^{11}C]CO_2-fixation [108]
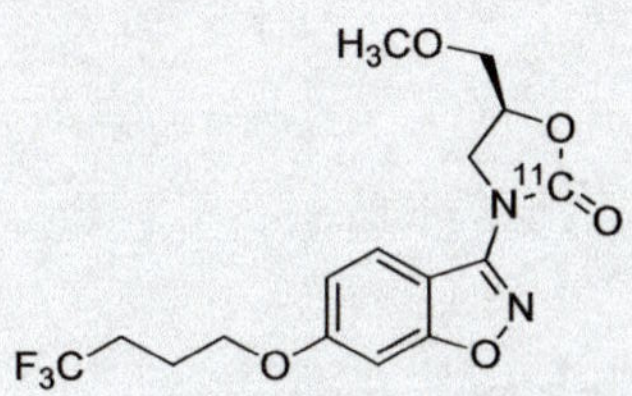 **[^{11}C]SL25.1188** [^{11}C]SL25.1188	Monoamine oxidase B (MAO-B)	^{11}C-oxazolidinone/[^{11}C]CO_2-fixation via dehydration	Synthesis [77] Preclinical [109] [^{11}C]CO_2-fixation [78] Clinical [110]

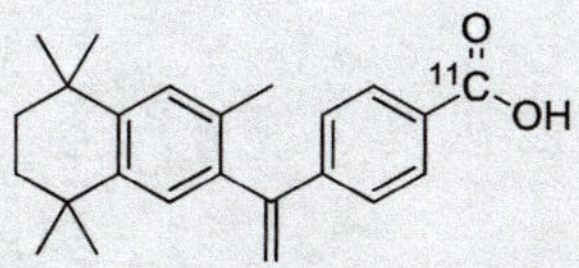 **[11C]bexarotene** [11C]bexarotene	Retinoid X receptors (RXRs)	^{11}C-carboxylic acid/Cu-mediated [^{11}C] CO_2-fixation	Synthesis, preclinical [111]

20-fold excess and could be present prior to carbamate dehydration with $POCl_3$. The radiotracer was isolated in sufficient yield to conduct preclinical studies in rat, which demonstrated high brain uptake and distribution consistent with that of the target. Furthermore, uptake was irreversible and could not be extracted ex vivo from brain homogenates. Translation of this radiotracer for clinical research studies is underway.

[^{11}C]CURB is a FAAH PET radiotracer designed in analogy to the selective FAAH inhibitor and drug candidate URB597 [113]. [^{11}C]CURB is an *O*-aryl carbamate that is prepared by [^{11}C]CO_2-fixation through an intermediate ^{11}C-isocyanate. Since an unsymmetrical dihydroquinone is used to form the ^{11}C-carbamate, a mixture of regioisomers is formed, of which the minor one is [^{11}C]CURB and is isolated by HPLC in 8% non-decay-corrected radiochemical yield (RCY), with high molar activity (92.5 GBq/μmol) 27 min after end of bombardment (EOB) [65]. [^{11}C]CURB was first evaluated in rats and showed high brain penetration and target selectivity for FAAH [92]. This radiotracer has now been translated for human use and appears to be useful for regional quantification of FAAH activity in the brain [93, 94].

Monoacylglycerol lipase (MAGL) plays a similar role to FAAH and is responsible for metabolism of the endocannabinoid 2-arachidonoylglycerol. Using well-established protocols for [^{11}C]CO_2-fixation, a series of ^{11}C-carbamate and ^{11}C-urea inhibitors of MAGL were radiolabeled and evaluated ex vivo in mice [75]. The radiochemistry here is especially notable due to the diversity of nucleophiles used to quench ^{11}C-carbamoyl chlorides, which includes the highly acidic alcohol 1,1,1,3,3,3-hexafluoroisopropanol, as well as benzotriazole and 1,2,4-triazole. Unfortunately these radiolabeled compounds failed to show high brain uptake, possibly due to high lipophilicity and MAGL concentration in the blood.

4.2 Other ^{11}C-Carbamates and ^{11}C-Ureas for Preclinical and Clinical PET Imaging

Among the first ^{11}C-*carbonyl*-carbamates prepared by [^{11}C]CO_2-fixaton was [^{11}C]metergoline, a serotonin receptor antagonist [63]. An ergoline alkaloid served as the amino precursor for ^{11}C-carbamylation using [^{11}C]CO_2 and benzyl chloride in the presence of DBU. Using the PET tracer for imaging in baboons, the pharmacokinetics of the drug could be characterized in the brain and peripheral organs [99].

The histone deacetylase (HDAC) inhibitor MS-275 (entinostat) was also radiolabeled by [^{11}C]CO_2-fixation. MS-275 had been reported to alter region-specific histone acetylation in rodent brains and was therefore considered for development as a CNS-targeting HDAC inhibitor therapeutic. In nonhuman primates, however, [^{11}C]MS-275 showed low blood-brain barrier penetration, which was not altered by pre-treatment with large doses of nonradioactive drug [100]. This suggests that the central nervous system (CNS) effects of MS-275 may be limited to rodents, as the drug may not enter brain tissue of higher species.

[^{11}C]GR103545 is a PET radiotracer for κ-opioid receptors undergoing translation for human imaging studies [101, 102]. A methyl carbamate, [^{11}C]GR103545, was originally radiolabeled using phosgene and [^{11}C]CH_3OH to prepare [^{11}C-*methyl*]methyl chloroformate and later using [^{11}C]CH_3I for methylation of the carbamate anion [103]. [^{11}C]CO_2-fixation presented an operationally simple alternative, as the precursor amine could be loaded into a reactor with BEMP and DMF, followed sequentially by [^{11}C]CO_2 and a solution of dimethyl sulfate in DMF. In this way [^{11}C-*carbonyl*]GR103545 could be prepared at room temperature in high yield (13% RCY) in 23 minutes after synthesis and purification [64].

In order to radiolabel irinotecan, an anticancer therapeutic and carbamate analogue of camptothecin, both [^{11}C]CO_2-fixation and [^{11}C]phosgene were evaluated [106]. Specific activities and time of synthesis were comparable, but the yield of [^{11}C]irinotecan using [^{11}C]CO_2-fixation was nearly twice that of the ^{11}C-phosgenation method. Since the ^{11}C-carbamate possesses an *O*-aryl substituent, the ^{11}C-isocyanate needed to be prepared, and [^{11}C]CO_2-fixation offered much greater synthetic accessibility and convenience than did the preparation of [^{11}C]phosgene. The biodistribution of [^{11}C] irinotecan was subsequently evaluated in mice.

To probe the abundance of glycogen synthase kinase 3β (GSK-3β) in the brain, [^{11}C-*methoxy*]AR-A014418, a selective, urea-based inhibitor, was developed [107]. The [^{11}C]CO_2-fixation approach to this tracer was later developed as an alternative synthesis of [^{11}C-*carbonyl*]AR-A014418 [108]. Since AR-A014418 was found to be insufficiently potent for PET radiotracer development, this method for radiolabeling with [^{11}C]CO_2 could be used to expeditiously prepare more promising GSK-3β targeting analogues using the same urea scaffold without requiring the presence of methyl ethers.

4.3 Monoamine Oxidase B Radiotracers

Monoamine oxidase B (MAO-B) is the principle enzyme for metabolizing endogenous amine neurotransmitters such as dopamine, serotonin, and phenethylamines and is a metabolic barrier to protect neurons from exogenous amines. Irreversible radiotracers, such as [^{11}C]L-deprenyl-D_2, can be used to measure MAO-B activity in addiction research, but is challenging to quantify due to its irreversible uptake and the release of [^{11}C](*R*)-methamphetamine and [^{11}C](*R*)-amphetamine by-products [114]. A structurally dissimilar reversible radiotracer, [^{11}C]SL25.1188, was developed using [^{11}C]phosgene for radiolabeling [77]. In a bid to improve the accessibility and the isolated yield of [^{11}C]SL25.1188, [^{11}C]CO_2-fixation was successfully used to replace [^{11}C]phosgene in the radiosynthesis [78]. As a direct consequence of this improvement, [^{11}C]SL25.1188 has been translated for human PET imaging in clinical research studies [110]. This is an example of a methodological development improving access to a desirable radiotracer, which was otherwise challenging to produce at sites without specialized equipment [115].

4.4 Retinoid X Receptor Radiotracers

Bexarotene is a synthetic retinoid agonist, FDA-approved for treatment of T-cell lymphoma. In recent years, bexarotene has been in demand off-label to treat small cell lung cancer and Alzheimer's disease. The Alzheimer's indication can be traced to a report that demonstrates that transgenic mouse models of Alzheimer's disease, when treated with bexarotene, display soluble and plaque amyloid clearance, as well as cognitive improvements [116]. To evaluate the potential for bexarotene to enter the brain of higher species and engage retinoid X receptors (RXRs), the carboxylic acid was radiolabeled using copper(I)-mediated [^{11}C]CO_2-fixation [111]. Significant optimization of the reaction conditions and reagents was conducted to adapt this reaction for preparation of this substrate, resulting in a 15% non-decay-corrected RCY after 32 min in high purity and adequate specific activity at time of injection. PET neuroimaging was conducted in nonhuman primate and demonstrated reasonable brain uptake reflective of RXR density. Further studies with [^{11}C]bexarotene are underway to identify the in vivo specificity and selectivity for RXRs and determine the utility of this tracer in retinoid drug development for neurodegeneration.

5 $^{11}CO_2$-Fixation for Fluorine-18 Radiotracer Discovery

During the course of radiotracer development efforts, ^{11}C is often selected as an appealing isotope for radiolabeling small molecule tracer candidates. This is due to the ubiquity of carbon atoms in organic small molecules, as well as the convenience of radiolabeling common functional groups with a high probability of success and often without the need for resource-intensive preparation of radiolabeling precursors. Since ^{11}C-methylation conditions tend to be robustly applicable to a variety of tracer scaffolds, functional groups such as methyl ethers, methylamines, and methyl esters may be incorporated by design into radiotracer candidates to facilitate convenient radiolabeling and preliminary biodistribution studies. A drawback of this approach is that it places constraints on the structural diversity available by mandating a methyl-substituted functional group and the absence of additional functional groups that are incompatible with ^{11}C-methylation. [^{11}C]CO_2-fixation, in its recent incarnations [19], represents a complementary approach for high-throughput radiolabeling without the need for specialized precursors beyond amines and alcohols. While this strategy too may be accompanied by certain constraints on structural diversity, it can be appreciated that an alternative set of limitations still makes available a much wider scope of radiotracer candidates than were previously offered. Furthermore, $[^{11}C]CO_2$-fixation for the preparation of ^{11}C-carbamates and ^{11}C-ureas is readily amenable to combinatorial radiosynthesis approaches, in as much as a set of amines and a second set of alcohols, phenols, or amines can be combined to produce a library of more structurally complex radiolabeled compounds. One such approach, and its utility in the design of fluorine-18 (^{18}F)-labeled FAAH PET radiopharmaceuticals, is discussed below.

While ^{11}C is an appealing isotope for radiotracer development, there are several reasons why a ^{18}F-radiotracer may be preferred for PET. The longer half-life of ^{18}F ($t_{1/2} = 109.7$ min) allows for longer synthesis and imaging times, if necessary, and is also often a better match for the pharmacokinetics of the radiotracer. In addition, the longer half-life means that radionuclide production and tracer synthesis need not be occur on-site with imaging facilities; radiotracers can be transported across cities or regions which also more easily facilitates multicenter trials. Due to the lower mean positron energy (^{18}F, 0.25 MeV; ^{11}C, 0.39 MeV), PET images acquired with ^{18}F are of slightly higher resolution. Incorporation of ^{18}F into small molecules can be a significant challenge depending on the position of the radiolabel, and several recent methods for efficient ^{18}F-fluorination of radiotracers will be highlighted in this section.

5.1 Radiotracer Library Development

After the successful development of $[^{11}C]$CURB for FAAH PET (*see* Subheading 4.1), demand grew for an ^{18}F-labeled analogue. The structure of $[^{11}C]$CURB (URB694) does not contain a fluorine atom, so a direct isotopologue was not an option. Since a new radiotracer would need to be developed, this was treated as an opportunity to determine how structural analogues would influence pharmacokinetics in vivo, to develop an optimal ^{18}F-radiotracer for FAAH. Based on the structures of URB597 and the closely related URB694, a synthetic array was constructed from commercially available amines and bicyclic phenols [95]. To begin, nonradioactive carbamates were prepared using phosgene for in vitro FAAH assays and physicochemical evaluation. From these data, eight FAAH inhibitors were radiolabeled with carbon-11 using a standard $[^{11}C]CO_2$-fixation protocol to rapidly generate the ^{11}C-*carbonyl*-carbamate radiotracers.

The method is robust enough that these compounds could be evaluated by ex vivo biodistribution to determine structure-activity relationships for radiotracer optimization. It was found that a linear *N*-alkyl group leads to increased potency over cyclic isomers and the presence of a dihydrooxazole leads to faster and greater brain uptake with less nonspecific binding. These insights were then applied to design the fluorine-18 carbamate radiotracers [^{18}F]DOPP [96] and [^{18}F]FCHC [98]. Both show greater brain uptake in vivo and sensitivity to FAAH activity in vitro than [^{11}C]CURB. [^{18}F]DOPP in particular shows promise for human translation. To better define the metabolism and distribution of this radiotracer in vivo prior to human studies, [^{18}F]DOPP was evaluated in non-human primates at baseline and after pretreatment with a FAAH inhibitor, URB597 [97]. [^{18}F]DOPP shows good brain uptake in baboon and regional distribution similar to that of [^{11}C]CURB in humans. Kinetic parameters for quantification of [^{18}F]DOPP distribution were evaluated and compared favorably with the human [^{11}C]CURB data, suggesting greater sensitivity for FAAH activity in vivo. [^{18}F]DOPP is expected to enter human research studies in the near future.

5.2 Modern Radiofluorination Strategies

After identification of an optimal PET tracer structure using ^{11}C or ^{18}F, it is imperative to develop a radiosynthesis that is robust, automated, and reproducible to facilitate routine PET imaging at both the development and other sites where the compound may be in demand. For example, both [^{18}F]DOPP and [^{18}F]FCHC require multistep syntheses using carefully buffered reaction mixtures due to the low stability of carbamates to radiofluorination conditions [96, 98]. An automated method has been developed to prepare [^{18}F]DOPP, but this still presents a greater challenge than the synthesis of, for example, [^{11}C]CURB, which can be performed in one pot [92]. Modern methods for [^{11}C]CO_2-fixation described above have the advantages of efficiency in terms of yields and number of manipulations, utility for a wide range of compounds, and not requiring much specialized equipment [19]. Here we will briefly discuss recent advances in radiofluorination that share these properties to facilitate access to a greater number of radiotracers. More in-depth reviews have also been recently published [117, 118].

In general, aliphatic radiofluorination can be carried out using nucleophilic [^{18}F]fluoride complexes and primary or secondary alkyl sulfonate precursors [74]. Occasionally, as in the synthesis of the fluorohydrins [^{18}F]FMISO and [^{18}F]FETNIM, the highly basic conditions for radiofluorination necessitate protecting groups or give rise to radioactive by-products that must be carefully separated from the product radiotracer. A cobalt–salen catalyst has been developed for epoxide-ring opening with [^{18}F]fluoride under milder conditions, such that in some examples protecting groups are obviated [119]. Additionally, high levels of enantiomeric excess could be achieved using racemic epoxides in some cases, which may be useful in cases where a single enantiomer precursor is not available. Benzylic C–H [120] and decarboxylative [121] radiofluorination have been developed using manganese–salen catalysts. These reactions appear to operate by formation of a stabilized organic

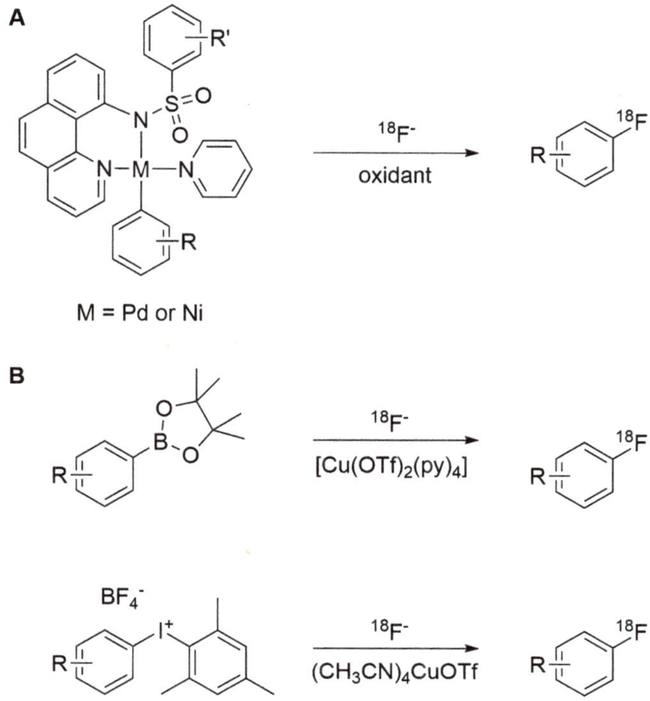

Fig. 7 (**a**) Transition metal-based precursors and (**b**) transition metal-mediated reactions for "late-stage" aryl radiofluorination

radical that undergoes radiofluorination by single-electron transfer. Although benzylic fluorides are likely to have a high propensity for defluorination in vivo, this method has also been demonstrated for the radiofluorination of alternative sites and holds promise for substrate-directed "late-stage" radiofluorination of complex molecules.

The most common method for radiofluorination of arenes is by nucleophilic aromatic substitution of activated arenes [122]. Consequently, methods for "late-stage" fluorination of deactivated arenes have been identified as an important goal for radiochemists. Transition metal-mediated radiofluorination from palladium [123] and nickel [124] precursor complexes appears to be a powerful strategy and aroused much interest in this field (Fig. 7a). The nickel-based method can tolerate small volumes of aqueous fluoride, and these methods have both been used to prepare radiotracers [125, 126], though challenges associated with preparation of the precursors and conducting radiofluorination on preparative scales may limit their widespread appeal.

Copper-mediated methods for radiofluorination of diaryliodonium salts have also been reported [127]. This method appears to be very general and powerful for radiofluorination of electron-rich rings, and in some cases high molar activity products are accessible

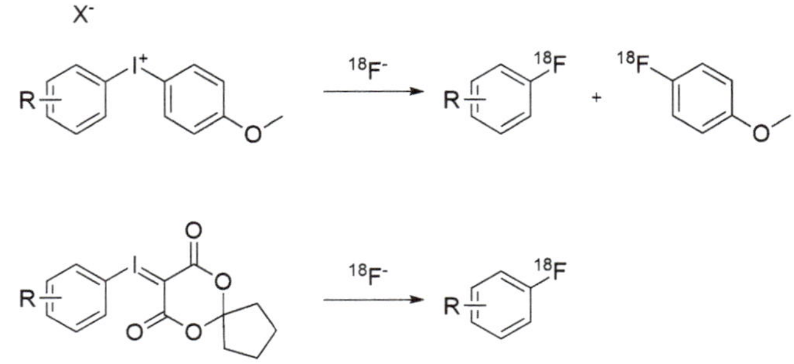

Fig. 8 Hypervalent iodonium based precursors for aryl radiofluorination

despite the presence of tetrafluoroborate counterions. A similar strategy was reported using copper-mediated radiofluorination of arylboronic esters in the presence of atmospheric oxygen [128] (Fig. 7b). This too appears to be a very promising method and was used to prepare [^{18}F]FDOPA in 5% non-decay-corrected RCY.

Hypervalent iodonium precursors offer an alternative mechanistic approach to radiofluorination from S_NAr [129]. Diaryliodonium salts were applied first for radiofluorination, with much attention given to developing aryl auxiliaries that would induce high levels of regioselectivity in radiofluorination, since reductive elimination of an aryl C–^{18}F bond can occur on two positions of the precursor (Fig. 8). These precursors have proven to be useful for the preparation of a number of radiotracers, though continue to be plagued in many cases by low regioselectivity, harsh conditions for radiofluorination, and difficulties in preparation and purification of the required precursors. Iodonium ylides have emerged as alternatives to diaryliodonium salts for radiofluorination, particularly in as much as they offer much greater regioselectivity [130–132]. These precursors are also readily prepared from aryl iodides by oxidation and condensation with dioxodione auxiliaries, do not contain a counterion, and can be purified by recrystallization or flash chromatography. Spirocyclic auxiliaries for iodonium ylides have been identified as a privileged scaffold for iodonium ylide radiofluorination precursors and lead to greater radiochemical yields of labeled products (Fig. 8). This method has been used to prepare challenging radiotracers such as 5-[^{18}F]fluorouracil and [^{18}F]FPEB in much greater RCYs than were previously obtainable, and production of the latter radiopharmaceutical has now been validated by this method for use in human PET imaging studies [133]. It has also been applied for convenient preparation of useful and novel ^{18}F-bioconjugation reagents [134, 135]. This method appears to be promising for the development of improved radiosyntheses of many challenging radiopharmaceuticals.

6 Conclusion

While radiochemists are unlikely to consider their own work in the context of "green" chemistry, the principles of the latter have had a profound influence on methodologies used for development and preparation of radiopharmaceuticals. As an almost certainly unexpected result of green chemistry initiatives to supplant toxic and wasteful phosgene as a reagent for the preparation of bulk and specialty chemicals, $[^{11}C]CO_2$-fixation has undergone a resurgence as novel, versatile, and mild methodologies using $[^{11}C]CO_2$ have emerged. Moreover, this has directly led to development of novel radiopharmaceuticals that are powerful agents for clinical psychiatric and neuroscience research. Library radiosynthesis is readily achievable by new methods in $[^{11}C]CO_2$-fixation and can be used to prepare limitless radiotracer candidates in the forms of carbamates and ureas from simple amines and alcohols. Green chemistry principles may indeed continue to play a role in developing other areas of PET radiochemistry. Ethanol is emerging as an alternative and highly practical solvent for PET radiochemistry [24]. Creative solutions for synthetic challenges, such as those posed by green chemistry, can and already have made positive contributions to superficially unrelated fields.

References

1. Mikkelsen M, Jørgensen M, Krebs FC (2010) The teraton challenge. A review of fixation and transformation of carbon dioxide. Energy Environ Sci 3:43–81. https://doi.org/10.1039/B912904A
2. Aresta M (2010) Carbon dioxide as chemical feedstock, 1st edn. Wiley-VCH
3. Langanke J, Wolf A, Hofmann J, Böhm K, Subhani MA, Müller TE, Leitner W, Gürtler C (2014) Carbon dioxide (CO_2) as sustainable feedstock for polyurethane production. Green Chem 16:1865–1870. https://doi.org/10.1039/C3GC41788C
4. Albo J, Alvarez-Guerra M, Castaño P, Irabien A (2015) Towards the electrochemical conversion of carbon dioxide into methanol. Green Chem 17:2304–2324. https://doi.org/10.1039/C4GC02453B
5. Behrens M (2014) Heterogeneous catalysis of CO_2 conversion to methanol on copper surfaces. Angew Chem Int Ed 53:12022–12024. https://doi.org/10.1002/anie.201409282
6. Li Y, Junge K, Beller M (2013) Improving the efficiency of the hydrogenation of carbonates and carbon dioxide to methanol. ChemCatChem 5:1072–1074. https://doi.org/10.1002/cctc.201300013
7. Huang K, Sun C-L, Shi Z-J (2011) Transition-metal-catalyzed C–C bond formation through the fixation of carbon dioxide. Chem Soc Rev 40:2435. https://doi.org/10.1039/c0cs00129e
8. North M, Pasquale R, Young C (2010) Synthesis of cyclic carbonates from epoxides and CO_2. Green Chem 12:1514–1539. https://doi.org/10.1039/C0GC00065E
9. Kielland N, Whiteoak CJ, Kleij AW (2013) Stereoselective synthesis with carbon dioxide. Adv Synth Catal 355:2115–2138. https://doi.org/10.1002/adsc.201300422
10. Fiorani G, Guo W, Kleij AW (2015) Sustainable conversion of carbon dioxide: the advent of organocatalysis. Green Chem 17:1375–1389. https://doi.org/10.1039/C4GC01959H
11. Darensbourg DJ (2007) Making plastics from carbon dioxide: Salen metal complexes as catalysts for the production of polycarbonates from epoxides and CO_2. Chem Rev 107:2388–2410. https://doi.org/10.1021/cr068363q
12. Kember MR, Buchard A, Williams CK (2011) Catalysts for CO_2/epoxide copolymerisation.

Chem Commun 47:141–163. https://doi.org/10.1039/C0CC02207A

13. Klaus S, Lehenmeier MW, Anderson CE, Rieger B (2011) Recent advances in CO_2/epoxide copolymerization—new strategies and cooperative mechanisms. Coord Chem Rev 255:1460–1479. https://doi.org/10.1016/j.ccr.2010.12.002
14. Lu X-B, Ren W-M, Wu G-P (2012) CO_2 copolymers from epoxides: catalyst activity, product selectivity, and stereochemistry control. Acc Chem Res 45:1721–1735. https://doi.org/10.1021/ar300035z
15. Sakakura T, Choi J-C, Yasuda H (2007) Transformation of carbon dioxide. Chem Rev 107:2365–2387. https://doi.org/10.1021/cr068357u
16. Tsuji Y, Fujihara T (2012) Carbon dioxide as a carbon source in organic transformation: carbon–carbon bond forming reactions by transition-metal catalysts. Chem Commun 48:9956. https://doi.org/10.1039/c2cc33848c
17. Buchanan JM, Hastings AB, Nesbett FB (1943) The role of carboxyl-labeled acetic, propionic, and butyric acids in liver glycogen formation. J Biol Chem 150:413–425
18. Ram S, Ehrenkaufer RL (1988) Synthesis of ^{11}C-radiopharmaceuticals using direct fixation of [^{11}C]carbon dioxide and [^{11}C]carbon monoxide. Int J Rad Appl Instrum B 15:345–355
19. Rotstein BH, Liang SH, Holland JP, Collier TL, Hooker JM, Wilson AA, Vasdev N (2013) $^{11}CO_2$ fixation: a renaissance in PET radiochemistry. Chem Commun 49:5621–5629. https://doi.org/10.1039/c3cc42236d
20. Tewson TJ, Banks W, Franceschini M, Hoffpauir J (1989) A trap for the removal of nitrogen oxides from carbon-11 carbon dioxide. Int J Rad Appl Instrum A 40:765–768
21. Mock BH, Vavrek MT, Mulholland GK (1995) Solid-phase reversible trap for [^{11}C] carbon dioxide using carbon molecular sieves. Nucl Med Biol 22:667–670
22. Kim D, Alexoff DL, Schueller M, Babst B, Ferrieri R, Fowler JS, Schlyer DJ (2014) The design and performance of a portable handheld $^{11}CO_2$ delivery system. Appl Radiat Isot 94:338–343. https://doi.org/10.1016/j.apradiso.2014.09.008
23. Alexoff DL (2002) Automation for the synthesis and application of PET radiopharmaceuticals. In: Welch MJ, Redvanly CS (eds) Handbook of radiopharmaceuticals. Wiley, pp 283–305
24. Shao X, Fawaz MV, Jang K, Scott PJH (2014) Ethanolic carbon-11 chemistry: the introduction of green radiochemistry. Appl Radiat Isot 89:125–129. https://doi.org/10.1016/j.apradiso.2014.01.033
25. Stewart MN, Hockley BG, Scott PJH (2015) Green approaches to late-stage fluorination: radiosyntheses of ^{18}F-labelled radiopharmaceuticals in ethanol and water. Chem Commun 51:14805–14808. https://doi.org/10.1039/C5CC05919D
26. Babad H, Zeiler AG (1973) Chemistry of phosgene. Chem Rev 73:75–91. https://doi.org/10.1021/cr60281a005
27. Ozaki S (1972) Recent advances in isocyanate chemistry. Chem Rev 72:457–496. https://doi.org/10.1021/cr60279a002
28. Delebecq E, Pascault J-P, Boutevin B, Ganachaud F (2013) On the versatility of urethane/urea bonds: reversibility, blocked isocyanate, and non-isocyanate polyurethane. Chem Rev 113:80–118. https://doi.org/10.1021/cr300195n
29. Soga K, Hosoda S, Nakamura H, Ikeda S (1976) A new synthetic route to 2-oxazolidones. J Chem Soc Chem Commun:617. https://doi.org/10.1039/c39760000617
30. Saito N, Hatakeda K, Ito S, Asano T, Toda T (1986) Formation of Bis(2-oxazolidinone) derivatives by reactions of 2-methoxy-3,3-dimethyl-2-phenyloxirane or α-bromoisobutyrophenone with carbon dioxide and aliphatic α,ω-diamines. Bull Chem Soc Jpn 59:1629–1631. https://doi.org/10.1246/bcsj.59.1629
31. Decortes A, Castilla AM, Kleij AW (2010) Salen-complex-mediated formation of cyclic carbonates by cycloaddition of CO2 to epoxides. Angew Chem Int Ed 49:9822–9837. https://doi.org/10.1002/anie.201002087
32. Kojima F, Aida T, Inoue S (1986) Fixation and activation of carbon dioxide on aluminum porphyrin. Catalytic formation of a carbamic ester from carbon dioxide, amine, and epoxide. J Am Chem Soc 108:391–395. https://doi.org/10.1021/ja00263a008
33. Yoshida Y, Ishii S, Kawato A, Yamashita T, Yano M, Inoue S (1988) Novel syntheses of arylcarbamic esters from carbon dioxide and aromatic amine via a zinc carbamate. Bull Chem Soc Jpn 61:2913–2916. https://doi.org/10.1246/bcsj.61.2913
34. Belforte A, Calderazzo F (1989) Formation of alkylurethanes from carbon dioxide by regioselective *O*-alkylation of alkali-metal *N*,

N-diethylcarbamates in the presence of complexing agents. J Chem Soc Dalton Trans:1007–1009. https://doi.org/10.1039/DT9890001007

35. Mahe R, Sasaki Y, Bruneau C, Dixneuf PH (1989) Catalytic synthesis of vinyl carbamates from carbon dioxide and alkynes with ruthenium complexes. J Org Chem 54:1518–1523. https://doi.org/10.1021/jo00268a008
36. Bruneau C, Dixneuf PH (1992) Catalytic incorporation of CO_2 into organic substrates: synthesis of unsaturated carbamates, carbonates and ureas. J Mol Catal 74:97–107. https://doi.org/10.1016/0304-5102(92)80227-8
37. Fournier J, Bruneau C, Dixneuf PH (1989) Phosphine catalysed synthesis of unsaturated cyclic carbonates from carbon dioxide and propargylic alcohols. Tetrahedron Lett 30:3981–3982. https://doi.org/10.1016/S0040-4039(00)99300-6
38. Aresta M, Quaranta E (1992) Role of the macrocyclic polyether in the synthesis of N-alkylcarbamate esters from primary amines, CO_2 and alkyl halides in the presence of crown-ethers. Tetrahedron 48:1515–1530. https://doi.org/10.1016/S0040-4020(01)92239-2
39. Chang CJ (1992) The solubility of carbon dioxide in organic solvents at elevated pressures. Fluid Phase Equilib 74:235–242
40. Yoshida M, Komatsuzaki Y, Ihara M (2008) Synthesis of 5-vinylideneoxazolidin-2-ones by DBU-mediated CO_2-fixation reaction of 4-(Benzylamino)-2-butynyl carbonates and benzoates. Org Lett 10:2083–2086. https://doi.org/10.1021/ol800663v
41. Pérez ER, Santos RHA, Gambardella MTP, de Macedo LGM, Rodrigues-Filho UP, Launay J-C, Franco DW (2004) Activation of carbon dioxide by bicyclic amidines. J Org Chem 69:8005–8011. https://doi.org/10.1021/jo049243q
42. Pereira FS, deAzevedo ER, da Silva EF, Bonagamba TJ, da Silva Agostíni DL, Magalhães A, Job AE, Pérez González ER (2008) Study of the carbon dioxide chemical fixation—activation by guanidines. Tetrahedron 64:10097–10106. https://doi.org/10.1016/j.tet.2008.08.008
43. Salvatore RN, Shin SI, Nagle AS, Jung KW (2001) Efficient carbamate synthesis via a three-component coupling of an amine, CO_2, and alkyl halides in the presence of Cs_2CO_3 and tetrabutylammonium iodide. J Org Chem 66:1035–1037. https://doi.org/10.1021/jo001140u
44. Ma J, Zhang X, Zhao N, Al-Arifi ASN, Aouak T, Al-Othman ZA, Xiao F, Wei W, Sun Y (2010) Theoretical study of TBD-catalyzed carboxylation of propylene glycol with CO_2. J Mol Catal Chem 315:76–81. https://doi.org/10.1016/j.molcata.2009.09.003
45. Edwards PR, Hiscock JR, Gale PA, Light ME (2010) Carbamate complexation by urea-based receptors: studies in solution and the solid state. Org Biomol Chem 8:100–106. https://doi.org/10.1039/B917140A
46. Endo T, Nagai D, Monma T, Yamaguchi H, Ochiai B (2004) A novel construction of a reversible fixation–release system of carbon dioxide by amidines and their polymers. Macromolecules 37:2007–2009. https://doi.org/10.1021/ma0305479
47. Murphy LJ, Robertson KN, Kemp RA, Tuononen HM, Clyburne JAC (2015) Structurally simple complexes of CO_2. Chem Commun 51:3942–3956. https://doi.org/10.1039/C4CC08510H
48. Pereira FS, Lincon da Silva Agostini D, do Espírito Santo RD, de Azevedo ER, Bonagamba TJ, Job AE, ERP G (2011) A comparative solid state ^{13}C NMR and thermal study of CO_2 capture by amidines PMDBD and DBN. Green Chem 13:2146. https://doi.org/10.1039/c1gc15457e
49. Heldebrant DJ, Jessop PG, Thomas CA, Eckert CA, Liotta CL (2005) The reaction of 1,8-diazabicyclo[5.4.0]undec-7-ene (DBU) with carbon dioxide. J Org Chem 70:5335–5338. https://doi.org/10.1021/jo0503759
50. Villiers C, Dognon J-P, Pollet R, Thuéry P, Ephritikhine M (2010) An isolated CO_2 adduct of a nitrogen base: crystal and electronic structures. Angew Chem Int Ed 49:3465–3468. https://doi.org/10.1002/anie.201001035
51. Das Neves Gomes C, Jacquet O, Villiers C, Thuéry P, Ephritikhine M, Cantat T (2012) A diagonal approach to chemical recycling of carbon dioxide: organocatalytic transformation for the reductive functionalization of CO_2. Angew Chem Int Ed 51:187–190. https://doi.org/10.1002/anie.201105516
52. Yoshida Y, Ishii S, Watanabe M, Yamashita T (1989) Novel synthesis of carbamate ester from carbon dioxide, amines, and alkyl halides. Bull Chem Soc Jpn 62:1534–1538. https://doi.org/10.1246/bcsj.62.1534
53. McGhee WD, Pan Y, Riley DP (1994) Highly selective generation of urethanes from amines, carbon dioxide and alkyl chlorides. J Chem

Soc Chem Commun 6:699. https://doi.org/10.1039/c39940000699

54. McGhee W, Riley D, Christ K, Pan Y, Parnas B (1995) Carbon dioxide as a phosgene replacement: synthesis and mechanistic studies of urethanes from amines, CO_2, and alkyl chlorides. J Org Chem 60:2820–2830
55. Waldman TE, McGhee WD (1994) Isocyanates from primary amines and carbon dioxide: "dehydration" of carbamate anions. J Chem Soc Chem Commun:957. https://doi.org/10.1039/c39940000957
56. McGhee WD, Pan Y, Talley JJ (1994) Conversion of amines to carbamoyl chlorides using carbon dioxide as a phosgene replacement. Tetrahedron Lett 35:839–842
57. Riley D, McGhee WD, Waldman T (1994) Generation of urethanes and isocyanates from amines and carbon dioxide. In: Anastas PT, Farris CA (eds) Benign by design. American Chemical Society, pp 122–132
58. Ametamey SM, Honer M, Schubiger PA (2008) Molecular imaging with PET. Chem Rev 108:1501–1516. https://doi.org/10.1021/cr0782426
59. Ferrieri RA (2002) Production and application of synthetic precursors labeled with carbon-11 and fluorine-18. In: Welch MJ, Redvanly CS (eds) Handbook of radiopharmaceuticals. Wiley, pp 229–282
60. Bolton R (2001) Isotopic methylation. J Label Compd Radiopharm 44:701–736. https://doi.org/10.1002/jlcr.497
61. Chakraborty PK, Mangner TJ, Chugani HT (1997) The synthesis of no-carrier-added [^{11}C]urea from [^{11}C]carbon dioxide and application to [^{11}C]uracil synthesis. Appl Radiat Isot 48:619–621
62. Schirbel A, Holschbach MH, Coenen HH (1999) N.C.A. [^{11}C]CO_2 as a safe substitute for phosgene in the carbonylation of primary amines. J Label Compd Radiopharm 42:537–551. https://doi.org/10.1002/(SICI)1099-1344(199906)42:6<537::AID-JLCR215>3.0.CO;2-3
63. Hooker JM, Reibel AT, Hill SM, Schueller MJ, Fowler JS (2009) One-pot, direct incorporation of [^{11}C]CO_2 into carbamates. Angew Chem Int Ed 48:3482–3485. https://doi.org/10.1002/anie.200900112
64. Wilson AA, Garcia A, Houle S, Vasdev N (2010) Direct fixation of [^{11}C]-CO_2 by amines: formation of [^{11}C-*carbonyl*]-methylcarbamates. Org Biomol Chem 8:428–432. https://doi.org/10.1039/b916419g
65. Wilson AA, Garcia A, Houle S, Sadovski O, Vasdev N (2011) Synthesis and application of isocyanates radiolabeled with carbon-11. Chem Eur J 17:259–264. https://doi.org/10.1002/chem.201002345
66. Molina P, Alajarin M, Arques A (1982) Convenient improved syntheses of isocyanates or isothiocyanates from amines. Synthesis 1982:596–597. https://doi.org/10.1055/s-1982-29877
67. van Tilburg EW, Windhorst AD, van der Mey M, Herscheid JDM (2006) One-pot synthesis of [^{11}C]ureas via triphenylphosphinimines. J Label Compd Radiopharm 49:321–330. https://doi.org/10.1002/jlcr.1052
68. Horvath MJ, Saylik D, Elmes PS, Jackson WR, Lovel CG, Moody K (1999) A Mitsunobu-based procedure for the preparation of alkyl and hindered aryl isocyanates from primary amines and carbon dioxide under mild conditions. Tetrahedron Lett 40:363–366
69. Saylik D, Horvath MJ, Elmes PS, Jackson WR, Lovel CG, Moody K (1999) Preparation of isocyanates from primary amines and carbon dioxide using mitsunobu chemistry. J Org Chem 64:3940–3946. https://doi.org/10.1021/jo982362j
70. Peterson SL, Stucka SM, Dinsmore CJ (2010) Parallel synthesis of ureas and carbamates from amines and CO_2 under mild conditions. Org Lett 12:1340–1343. https://doi.org/10.1021/ol100259j
71. Haji Dheere AK, Yusuf N, Gee A (2013) Rapid and efficient synthesis of [^{11}C]ureas via the incorporation of [^{11}C]CO_2 into aliphatic and aromatic amines. Chem Commun 49:8193–8195. https://doi.org/10.1039/c3cc44046j
72. Haji Dheere AK, Bongarzone S, Taddei C, Yan R, Gee AD (2015) Synthesis of ^{11}C-labelled symmetrical ureas via the rapid incorporation of [^{11}C]CO_2 into aliphatic and aromatic amines. Synlett 26:2257–2260. https://doi.org/10.1055/s-0034-1381055
73. Ghosh AK, Brindisi M (2015) Organic carbamates in drug design and medicinal chemistry. J Med Chem 58:2895–2940. https://doi.org/10.1021/jm501371s
74. Miller PW, Long NJ, Vilar R, Gee AD (2008) Synthesis of ^{11}C, ^{18}F, ^{15}O, and ^{13}N radiolabels for positron emission tomography. Angew Chem Int Ed 47:8998–9033. https://doi.org/10.1002/anie.200800222
75. Hicks JW, Parkes J, Tong J, Houle S, Vasdev N, Wilson AA (2014) Radiosynthesis and ex vivo evaluation of [^{11}C-carbonyl]carbamate- and urea-based monoacylglycerol

lipase inhibitors. Nucl Med Biol 41:688–694. https://doi.org/10.1016/j.nucmedbio.2014.05.001

76. Dollé F, Bramoullé Y, Hinnen F, Demphel S, George P, Bottlaender M (2003) Efficient synthesis of [^{11}C]befloxatone, a selective radioligand for the *in vivo* imaging of MAO-A density using PET. J Label Compd Radiopharm 46:783–792. https://doi.org/10.1002/jlcr.718

77. Bramoullé Y, Puech F, Saba W, Valette H, Bottlaender M, George P, Dollé F (2008) Radiosynthesis of (*S*)-5-methoxymethyl-3-[6-(4,4,4-trifluorobutoxy)benzo[*d*]isoxazol-3-yl]oxazolidin-2-[^{11}C]one ([^{11}C] SL25.1188), a novel radioligand for imaging monoamine oxidase-B with PET. J Label Compd Radiopharm 51:153–158. https://doi.org/10.1002/jlcr.1492

78. Vasdev N, Sadovski O, Garcia A, Dollé F, Meyer JH, Houle S, Wilson AA (2011) Radiosynthesis of [^{11}C]SL25.1188 via [^{11}C] CO_2 fixation for imaging monoamine oxidase B. J Label Compd Radiopharm 54:678–680. https://doi.org/10.1002/jlcr.1908

79. Pike VW, Eakins MN, Allan RM, Selwyn AP (1981) Preparation of carbon-11 labelled acetate and palmitic acid for the study of myocardial metabolism by emission-computerised axial tomography. J Radioanal Nucl Chem 64:291–297

80. Tang X, Tang G, Nie D (2013) Fully automated synthesis of ^{11}C-acetate as tumor PET tracer by simple modified solid-phase extraction purification. Appl Radiat Isot 82:81–86. https://doi.org/10.1016/j.apradiso.2013.07.012

81. Appel AM, Bercaw JE, Bocarsly AB, Dobbek H, DuBois DL, Dupuis M, Ferry JG, Fujita E, Hille R, Kenis PJA, Kerfeld CA, Morris RH, Peden CHF, Portis AR, Ragsdale SW, Rauchfuss TB, Reek JNH, Seefeldt LC, Thauer RK, Waldrop GL (2013) Frontiers, opportunities, and challenges in biochemical and chemical catalysis of CO_2 fixation. Chem Rev 113:6621–6658. https://doi.org/10.1021/cr300463y

82. Takaya J, Tadami S, Ukai K, Iwasawa N (2008) Copper(I)-catalyzed carboxylation of aryl- and alkenylboronic esters. Org Lett 10:2697–2700. https://doi.org/10.1021/ol800829q

83. Kobayashi K, Kondo Y (2009) Transition-metal-free carboxylation of organozinc reagents using CO_2 in DMF solvent. Org Lett 11:2035–2037. https://doi.org/10.1021/ol900528h

84. Ohishi T, Zhang L, Nishiura M, Hou Z (2011) Carboxylation of alkylboranes by N-heterocyclic carbene copper catalysts: synthesis of carboxylic acids from terminal alkenes and carbon dioxide. Angew Chem Int Ed 50:8114–8117. https://doi.org/10.1002/anie.201101769

85. Riss PJ, Lu S, Telu S, Aigbirhio FI, Pike VW (2012) Cu^{I}-catalyzed ^{11}C carboxylation of boronic acid esters: a rapid and convenient entry to ^{11}C-labeled carboxylic acids, esters, and amides. Angew Chem Int Ed 51:2698–2702. https://doi.org/10.1002/anie.201107263

86. Johnson DS, Stiff C, Lazerwith SE, Kesten SR, Fay LK, Morris M, Beidler D, Liimatta MB, Smith SE, Dudley DT, Sadagopan N, Bhattachar SN, Kesten SJ, Nomanbhoy TK, Cravatt BF, Ahn K (2011) Discovery of PF-04457845: a highly potent, orally bioavailable, and selective urea FAAH inhibitor. ACS Med Chem Lett 2:91–96. https://doi.org/10.1021/ml100190t

87. Ahn K, Smith SE, Liimatta MB, Beidler D, Sadagopan N, Dudley DT, Young T, Wren P, Zhang Y, Swaney S, Van Becelaere K, Blankman JL, Nomura DK, Bhattachar SN, Stiff C, Nomanbhoy TK, Weerapana E, Johnson DS, Cravatt BF (2011) Mechanistic and pharmacological characterization of PF-04457845: a highly potent and selective fatty acid amide hydrolase inhibitor that reduces inflammatory and noninflammatory pain. J Pharmacol Exp Ther 338:114–124. https://doi.org/10.1124/jpet.111.180257

88. Huggins JP, Smart TS, Langman S, Taylor L, Young T (2012) An efficient randomised, placebo-controlled clinical trial with the irreversible fatty acid amide hydrolase-1 inhibitor PF-04457845, which modulates endocannabinoids but fails to induce effective analgesia in patients with pain due to osteoarthritis of the knee. Pain 153:1837–1846. https://doi.org/10.1016/j.pain.2012.04.020

89. Hicks JW, Parkes J, Sadovski O, Tong J, Houle S, Vasdev N, Wilson AA (2013) Synthesis and preclinical evaluation of [^{11}C-*carbonyl*]PF-04457845 for neuroimaging of fatty acid amide hydrolase. Nucl Med Biol 40:740–746. https://doi.org/10.1016/j.nucmedbio.2013.04.008

90. Skaddan MB, Zhang L, Johnson DS, Zhu A, Zasadny KR, Coelho RV, Kuszpit K, Currier G, Fan K-H, Beck EM, Chen L, Drozda SE, Balan G, Niphakis M, Cravatt BF, Ahn K, Bocan T, Villalobos A (2012) The synthesis and in vivo evaluation of [^{18}F]

PF-9811: a novel PET ligand for imaging brain fatty acid amide hydrolase (FAAH). Nucl Med Biol 39:1058–1067. https://doi.org/10.1016/j.nucmedbio.2012.03.011

91. Clapper JR, Vacondio F, King AR, Duranti A, Tontini A, Silva C, Sanchini S, Tarzia G, Mor M, Piomelli D (2009) A second generation of carbamate-based fatty acid amide hydrolase inhibitors with improved activity in vivo. ChemMedChem 4:1505–1513. https://doi.org/10.1002/cmdc.200900210
92. Wilson AA, Garcia A, Parkes J, Houle S, Tong J, Vasdev N (2011) [^{11}C]CURB: evaluation of a novel radiotracer for imaging fatty acid amide hydrolase by positron emission tomography. Nucl Med Biol 38:247–253. https://doi.org/10.1016/j.nucmedbio.2010.08.001
93. Rusjan PM, Wilson AA, Mizrahi R, Boileau I, Chavez SE, Lobaugh NJ, Kish SJ, Houle S, Tong J (2012) Mapping human brain fatty acid amide hydrolase activity with PET. J Cereb Blood Flow Metab 33:407–414. https://doi.org/10.1038/jcbfm.2012.180
94. Boileau I, Bloomfield PM, Rusjan P, Mizrahi R, Mufti A, Vitcu I, Kish SJ, Houle S, Wilson AA, Tong J (2014) Whole-body radiation dosimetry of ^{11}C-carbonyl-URB694: a PET tracer for fatty acid amide hydrolase. J Nucl Med 55:1993–1997. https://doi.org/10.2967/jnumed.114.146464
95. Wilson AA, Hicks JW, Sadovski O, Parkes J, Tong J, Houle S, Fowler CJ, Vasdev N (2013) Radiosynthesis and evaluation of [^{11}C-*carbonyl*]-labeled carbamates as fatty acid amide hydrolase radiotracers for positron emission tomography. J Med Chem 56:201–209. https://doi.org/10.1021/jm301492y
96. Sadovski O, Hicks JW, Parkes J, Raymond R, Nobrega J, Houle S, Cipriano M, Fowler CJ, Vasdev N, Wilson AA (2013) Development and characterization of a promising fluorine-18 labelled radiopharmaceutical for in vivo imaging of fatty acid amide hydrolase. Bioorg Med Chem 21:4351–4357. https://doi.org/10.1016/j.bmc.2013.04.077
97. Rotstein BH, Wey H-Y, Shoup TM, Wilson AA, Liang SH, Hooker JM, Vasdev N (2014) PET imaging of fatty acid amide hydrolase with [^{18}F]DOPP in nonhuman primates. Mol Pharm 11:3832–3838. https://doi.org/10.1021/mp500316h
98. Shoup TM, Bonab AA, Wilson AA, Vasdev N (2015) Synthesis and preclinical evaluation of [^{18}F]FCHC for neuroimaging of fatty acid amide hydrolase. Mol Imaging Biol 17:257–263. https://doi.org/10.1007/s11307-014-0789-1
99. Hooker JM, Kim SW, Reibel AT, Alexoff D, Xu Y, Shea C (2010) Evaluation of [^{11}C] metergoline as a PET radiotracer for 5HTR in nonhuman primates. Bioorg Med Chem 18:7739–7745. https://doi.org/10.1016/j.bmc.2010.04.039
100. Hooker JM, Kim SW, Alexoff D, Xu Y, Shea C, Reid A, Volkow N, Fowler JS (2010) Histone deacetylase inhibitor MS-275 exhibits poor brain penetration: pharmacokinetic studies of [^{11}C]MS-275 using positron emission tomography. ACS Chem Neurosci 1:65–73. https://doi.org/10.1021/cn9000268
101. Talbot PS, Narendran R, Butelman ER, Huang Y, Ngo K, Slifstein M, Martinez D, Laruelle M, Hwang D-R (2005) ^{11}C-GR103545, a radiotracer for imaging κ-opioid receptors in vivo with PET: synthesis and evaluation in baboons. J Nucl Med 46:484–494
102. Tomasi G, Nabulsi N, Zheng M-Q, Weinzimmer D, Ropchan J, Blumberg L, Brown-Proctor C, Ding Y-S, Carson RE, Huang Y (2013) Determination of in vivo *B*max and *K*d for ^{11}C-GR103545, an agonist PET tracer for κ-opioid receptors: a study in nonhuman primates. J Nucl Med 54:600–608. https://doi.org/10.2967/jnumed.112.112672
103. Schoultz BW, Arstad E, Marton J, Willoch F, Drzezga A, Wester H-J, Henriksen G (2008) A new method for radiosynthesis of ^{11}C-labeled carbamate groups and its application for a highly efficient synthesis of the kappa-opioid receptor tracer [^{11}C] GR103545. Open Med Chem J 2:72–74. https://doi.org/10.2174/1874104500802010072
104. Nabulsi NB, Zheng M-Q, Ropchan J, Labaree D, Ding Y-S, Blumberg L, Huang Y (2011) [^{11}C]GR103545: novel one-pot radiosynthesis with high specific activity. Nucl Med Biol 38:215–221. https://doi.org/10.1016/j.nucmedbio.2010.08.014
105. Naganawa M, Jacobsen LK, Zheng M-Q, Lin S-F, Banerjee A, Byon W, Weinzimmer D, Tomasi G, Nabulsi N, Grimwood S, Badura LL, Carson RE, McCarthy TJ, Huang Y (2014) Evaluation of the agonist PET radioligand [^{11}C]GR103545 to image kappa opioid receptor in humans: kinetic model selection, test–retest reproducibility and receptor occupancy by the antagonist PF-04455242. NeuroImage 99:69–79. https://doi.org/10.1016/j.neuroimage.2014.05.033
106. Kawamura K, Hashimoto H, Ogawa M, Yui J, Wakizaka H, Yamasaki T, Hatori A, Xie L, Kumata K, Fujinaga M, Zhang M-R (2013)

Synthesis, metabolite analysis, and in vivo evaluation of [^{11}C]irinotecan as a novel positron emission tomography (PET) probe. Nucl Med Biol 40:651–657. https://doi.org/10.1016/j.nucmedbio.2013.03.004

107. Vasdev N, Garcia A, Stableford WT, Young AB, Meyer JH, Houle S, Wilson AA (2005) Synthesis and ex vivo evaluation of carbon-11 labelled *N*-(4-methoxybenzyl)-*N'*-(5-nitro-1,3-thiazol-2-yl)urea ([^{11}C]AR-A014418): a radiolabelled glycogen synthase kinase-3β specific inhibitor for PET studies. Bioorg Med Chem Lett 15:5270–5273. https://doi.org/10.1016/j.bmcl.2005.08.037

108. Hicks JW, Wilson AA, Rubie EA, Woodgett JR, Houle S, Vasdev N (2012) Towards the preparation of radiolabeled 1-aryl-3-benzyl ureas: radiosynthesis of [^{11}C-*carbonyl*]AR-A014418 by [^{11}C]CO_2 fixation. Bioorg Med Chem Lett 22:2099–2101. https://doi.org/10.1016/j.bmcl.2011.12.139

109. Saba W, Valette H, Peyronneau M-A, Bramoullé Y, Coulon C, Curet O, George P, Dollé F, Bottlaender M (2010) [^{11}C] SL25.1188, a new reversible radioligand to study the monoamine oxidase type B with PET: preclinical characterisation in nonhuman primate. Synapse 64:61–69. https://doi.org/10.1002/syn.20703

110. Rusjan PM, Wilson AA, Miler L, Fan I, Mizrahi R, Houle S, Vasdev N, Meyer JH (2014) Kinetic modeling of the monoamine oxidase B radioligand [^{11}C]SL25.1188 in human brain with high-resolution positron emission tomography. J Cereb Blood Flow Metab 34:883–889. https://doi.org/10.1038/jcbfm.2014.34

111. Rotstein BH, Hooker JM, Woo J, Collier TL, Brady TJ, Liang SH, Vasdev N (2014) Synthesis of [^{11}C]Bexarotene by Cu-mediated [^{11}C]Carbon dioxide fixation and preliminary PET imaging. ACS Med Chem Lett 5:668–672. https://doi.org/10.1021/ml500065q

112. Blankman JL, Cravatt BF (2013) Chemical probes of endocannabinoid metabolism. Pharmacol Rev 65:849–871. https://doi.org/10.1124/pr.112.006387

113. Kathuria S, Gaetani S, Fegley D, Valiño F, Duranti A, Tontini A, Mor M, Tarzia G, Rana GL, Calignano A, Giustino A, Tattoli M, Palmery M, Cuomo V, Piomelli D (2003) Modulation of anxiety through blockade of anandamide hydrolysis. Nat Med 9:76–81. https://doi.org/10.1038/nm803

114. Fowler JS, Logan J, Volkow ND, Wang G-J, MacGregor RR, Ding Y-S (2002) Monoamine oxidase: radiotracer development and human studies. Methods 27:263–277

115. Hicks JW, Sadovski O, Parkes J, Houle S, Hay BA, Carter RL, Wilson AA, Vasdev N (2015) Radiosynthesis and ex vivo evaluation of [^{18}F]-(*S*)-3-(6-(3-fluoropropoxy)benzo[*d*] isoxazol-3-yl)-5-(methoxymethyl)-oxazolidin-2-one for imaging MAO-B with PET. Bioorg Med Chem Lett 25:288–291. https://doi.org/10.1016/j.bmcl.2014.11.048

116. Cramer PE, Cirrito JR, Wesson DW, Lee CYD, Karlo JC, Zinn AE, Casali BT, Restivo JL, Goebel WD, James MJ, Brunden KR, Wilson DA, Landreth GE (2012) ApoE-directed therapeutics rapidly clear β-amyloid and reverse deficits in AD mouse models. Science 335:1503–1506. https://doi.org/10.1126/science.1217697

117. Brooks AF, Topczewski JJ, Ichiishi N, Sanford MS, Scott PJH (2014) Late-stage [^{18}F] fluorination: new solutions to old problems. Chem Sci 5:4545–4553. https://doi.org/10.1039/C4SC02099E

118. Campbell MG, Ritter T (2015) Modern carbon–fluorine bond forming reactions for aryl fluoride synthesis. Chem Rev 115:612–633. https://doi.org/10.1021/cr500366b

119. Graham TJA, Lambert RF, Ploessl K, Kung HF, Doyle AG (2014) Enantioselective radiosynthesis of positron emission tomography (PET) tracers containing [^{18}F]Fluorohydrins. J Am Chem Soc 136:5291–5294. https://doi.org/10.1021/ja5025645

120. Huang X, Liu W, Ren H, Neelamegam R, Hooker JM, Groves JT (2014) Late stage benzylic C–H fluorination with [^{18}F]Fluoride for PET imaging. J Am Chem Soc 136:6842–6845. https://doi.org/10.1021/ja5039819

121. Huang X, Liu W, Hooker JM, Groves JT (2015) Targeted fluorination with the fluoride ion by manganese-catalyzed decarboxylation. Angew Chem Int Ed 54:5241–5245. https://doi.org/10.1002/anie.201500399

122. Tredwell M, Gouverneur V (2012) ^{18}F labeling of Arenes. Angew Chem Int Ed 51:11426–11437. https://doi.org/10.1002/anie.201204687

123. Lee E, Kamlet AS, Powers DC, Neumann CN, Boursalian GB, Furuya T, Choi DC, Hooker JM, Ritter T (2011) A fluoride-derived electrophilic late-stage fluorination reagent for PET imaging. Science 334:639–642. https://doi.org/10.1126/science.1212625

124. Lee E, Hooker JM, Ritter T (2012) Nickel-mediated oxidative fluorination for PET with aqueous [^{18}F]Fluoride. J Am Chem Soc 134:17456–17458. https://doi.org/10.1021/ja3084797
125. Kamlet AS, Neumann CN, Lee E, Carlin SM, Moseley CK, Stephenson N, Hooker JM, Ritter T (2013) Application of palladium-mediated ^{18}F-Fluorination to PET radiotracer development: overcoming hurdles to translation. PLoS One 8:e59187. https://doi.org/10.1371/journal.pone.0059187
126. Ren H, Wey H-Y, Strebl M, Neelamegam R, Ritter T, Hooker JM (2014) Synthesis and imaging validation of [^{18}F]MDL100907 enabled by Ni-mediated fluorination. ACS Chem Neurosci 5:611–615. https://doi.org/10.1021/cn500078e
127. Ichiishi N, Brooks AF, Topczewski JJ, Rodnick ME, Sanford MS, Scott PJH (2014) Copper-catalyzed [^{18}F]Fluorination of (Mesityl)(aryl)iodonium salts. Org Lett 16:3224–3227. https://doi.org/10.1021/ol501243g
128. Tredwell M, Preshlock SM, Taylor NJ, Gruber S, Huiban M, Passchier J, Mercier J, Génicot C, Gouverneur V (2014) A general copper-mediated nucleophilic ^{18}F fluorination of Arenes. Angew Chem 126:7885–7889. https://doi.org/10.1002/ange.201404436
129. Yusubov MS, Svitich DY, Larkina MS, Zhdankin VV (2013) Applications of iodonium salts and iodonium ylides as precursors for nucleophilic fluorination in positron emission tomography. ARKIVOC 1:364–395
130. Satyamurthy N, Barrio J (2010) No-carrier-added nucleophilic [F-18] fluorination of aromatic compounds WO2010117435 (A2), October 14, 2010
131. Cardinale J, Ermert J, Humpert S, Coenen HH (2014) Iodonium ylides for one-step, no-carrier-added radiofluorination of electron rich arenes, exemplified with 4-(([^{18}F]fluorophenoxy)-phenylmethyl)piperidine NET and SERT ligands. RSC Adv 4:17293. https://doi.org/10.1039/c4ra00674g
132. Rotstein BH, Stephenson NA, Vasdev N, Liang SH (2014) Spirocyclic hypervalent iodine(III)-mediated radiofluorination of non-activated and hindered aromatics. Nat Commun 5:4365. https://doi.org/10.1038/ncomms5365
133. Stephenson NA, Holland JP, Kassenbrock A, Yokell DL, Livni E, Liang SH, Vasdev N (2015) Iodonium ylide-mediated radiofluorination of ^{18}F-FPEB and validation for human use. J Nucl Med 56:489–492. https://doi.org/10.2967/jnumed.114.151332
134. Wang L, Jacobson O, Avdic D, Rotstein BH, Weiss ID, Collier L, Chen X, Vasdev N, Liang SH (2015) *Ortho*-Stabilized ^{18}F-azido click agents and their application in PET imaging with single-stranded DNA aptamers. Angew Chem Int Ed 54(43):12777–12781. https://doi.org/10.1002/anie.201505927
135. Jacobson O, Weiss I, Wang L, Wang Z, Yang X, Dewhurst A, Ma Y, Zhu G, Niu G, Kiesewetter DO, Vasdev N, Liang S, Chen X (2015) ^{18}F-labeled single-stranded DNA aptamer for PET imaging of protein tyrosine kinase-7 expression. J Nucl Med 56(11):1780–1785. https://doi.org/10.2967/jnumed.115.160960

Further Reading[1]

Part I: Greener Approaches to Classical Transformations

Green Syntheses of Common Heterocycles

Ali S, Al-Rashida M, Younus HA, Moin ST, Hameed A (2020) Piperidinium-based deep eutectic solvents: efficient and sustainable eco-friendly medium for one-pot n-heterocycles synthesis. Chem Select 5(41):12697–12703

Brandao P, Pineiro M, Pinho e Melo TMVD (2019) Flow chemistry: towards a more sustainable heterocyclic synthesis. Eur J Org Chem 2019 (43):7188–7217

Cabrele C, Reiser O (2016) The modern face of synthetic heterocyclic chemistry. J Org Chem 81(21):10109–10125

Campos JF, Berteina-Raboin S (2020) Greener synthesis of nitrogen-containing heterocycles in water, PEG and bio-based solvents. Catalysis 10(4):429. https://doi.org/10.3390/catal10040429

Emayavatamban B, Sen M, Sundararaju B (2017) Iron-catalyzed sustainable synthesis of pyrrole. Org Lett 19(1):6–9

Gupta SS, Kumari S, Kumar I, Sharma U (2020) Eco-friendly and sustainable synthetic approaches to biologically significant fused N-heterocycles. Chem Heterocyclic Cmpds 56(4):433–444

Kaur N (2018) Green synthesis of three- to five-membered *O*-heterocycles using ionic liquids. Synth Commun 48(13):1588–1613

Kerru N, Gummidi L, Maddila S, Jonnalagadda SB (2021) A review of recent advances in the green synthesis of azole- and pyran-based fused heterocycles using MCRs and sustainable catalysts. Curr Org Chem 25(1):4–39

M'hamed MO (2015) Ball milling for heterocyclic compounds synthesis in green chemistry: a review. Synth Commun 45(22):2511–2528

Maji M, Panja D, Borthakur I, Kundu S (2021) Recent advancement on sustainable synthesis of N-heterocycles following acceptorless dehydrogenative coupling protocol using alcohols. Organic Chem Front. https://doi.org/10.1039/D0QO01577F

[1] The book contains a collection of chapters highlighting the core concepts and development of several emerging topics with the unifying theme being that, with their implementation, the practitioner can make more judicious choices in terms of the environmental friendliness of their chemistry. The editors are extremely grateful to all the authors for their contributions, though also realize that each of the subjects covered herein represents highly active research areas with innovative breakthroughs continuing to be made. This chapter contains selected additional references on each topic that have appeared during the editing process of this book through to early 2021. In none of the cases are these not intended to be a comprehensive list (this would be impossible particularly when one considers there were almost 1200 publications in 2020 on C-H activation alone) though instead are selected reviews within the each field on a specific topic, or an article which the editor considers to be of particular interest or importance. One interesting point to note when reviewing this collection of references is the number of cases in which there is synergy between the overall topics considered within the book. For example, "cross-couplings carried out in sustainable media," "biocatalysis in flow," and "the combination of transition metal- and photo-catalyzed C-H activations" represent several examples to illustrate this, and highlight the creative thinking of researchers in progressing drug discovery to a more sustainable future.

Paul F. Richardson (ed.), *Green Chemistry in Drug Discovery: From Academia to Industry*, Methods in Pharmacology and Toxicology, https://doi.org/10.1007/978-1-0716-1579-9, © Springer Science+Business Media, LLC, part of Springer Nature 2022

Sayantika B, Ghosh T (2021) Gold-catalyzed carboxylative cyclization reactions: recent advances. Asian J Org Chem. https://doi.org/10.1002/ajoc.202000694

Tundo P, Musolino M, Arico F (2019) Dialkyl carbonates in the green synthesis of heterocycles. Front Chem. https://doi.org/10.3389/fchem.2019.00300

Vachan BS, Karuppasamy M, Vinoth P, Vivek Kumar S, Perumal S, Sridharan V, Menendez CJ (2020) Proline and its derivatives as organocatalysts for multi-component reactions in aqueous media: synergic pathways to the green synthesis of heterocycles. Adv Synth Catal 362(1):87–110

Xia M, Lambu MR, Tatina MB, Judeh ZMA (2021) A practical synthesis of densely functionalized pyrroles via a three-component cascade reaction between carbohydrates, oxoacetonitriles and ammonium acetate. J Org Chem 86(1):837–849

Yan L-J, Wang J-L, Xu D, Burgess KS, Zhu A-F, Rao Y-Y, Chen X-B, Wang Y-C (2018) Chemically sustainable and green one-pot multicomponent synthesis of highly functionalized polycyclic N-fused pyrrolidine heterocycles. Chem Select 3(2):662–665

Zhao F, Masci D, Tomarelli E, Castagnolo D (2020) Biocatalytic and chemoenzymatic approaches for the synthesis of heterocycles. Synthesis 52(20):2948–2961

Amide Bond Formation by Hydrogen Borrowing

Allen CL, Williams JMJ (2011) Metal-catalyzed approaches to amide bond formation. Chem Soc Rev 40:3405–3415

Corma A, Navas J, Sabater MJ (2018) Advances in one-pot synthesis through borrowing hydrogen catalysis. Chem Rev 118(4):1410–1459

Gunanathan C, Milstein D (2013) Applications of acceptorless dehydrogenation and related transformations in chemical synthesis. Science 341:249

Reed-Berendt BG, Polidano K, Morrill LC (2019) Recent advances in homogeneous borrowing hydrogen catalysis using earth-abundant first row transition metals. Org Biomol Chem 17:1595–1607

Sabatini MT, Boulton LT, Sneddon HF, Sheppard TD (2019) A green chemistry perspective on catalytic amide bond formation. Nat Catal 2:10–17

Todorovic M, Perrin DM (2019) Recent developments in catalytic amide bond formation. Peptide Sci 112(6):e24210

Wong CM, McBurney RT, Binding SC, Peterson MB, Goncales VR, Gooding JJ, Messerle BA (2017) Iridium (III) homo- and heterogeneous catalysed hydrogen borrowing C-N bond formation. Green Chem 19:3142–3151

Mitsunobu Reaction

Beddoe RH, Andrews KG, Magné V, Cuthbertson JD, Saska J, Shannon-Little AL, Shanahan SE, Sneddon HF, Denton RM (2019) Redox-neutral organocatalytic Mitsunobu reactions. Science 365(6456):910–914

Beddoe RH, Sneddon HF, Denton RM (2018) The catalytic Mitsunobu reaction: a critical analysis of the current state-of-the-art. Org Biomol Chem 16:7774–7781

Buonomo JA, Aldrich CC (2018) Mitsunobu reactions catalytic in phosphine and a fully catalytic system. Angew Chem Int Ed 54(44):13041–13044

Fletcher S (2015) The Mitsunobu reaction in the 21st century. Org Chem Front 2:739–752

Hirose D, Gazvoda M, Košmrlj J, Taniguchi T (2016a) Advances and mechanistic insight on the catalytic Mitsunobu reaction using recyclable azo reagents. Chem Sci 7(8):5148–5149

Hirose D, Gazvoda M, Košmrlj J, Taniguchi T (2016b) The "Fully Catalytic System" in Mitsunobu reaction has not been realized yet. Org Lett 18(16):4036–4039

Hirose D, Gazvoda M, Košmrlj J, Taniguchi T (2018) Systematic evaluation of 2-arylcarboxylates and 2-arylcarboxamides as Mitsunobu reagents. J Org Chem 83(8):4712–4729

Panday SK (2019) Advances in the Mitsunobu reaction: an excellent organic protocol with versatile applications. Mini-Rev Organic Chem 16 (2):127–140

Pokluda A, Kohout M, Chudoba J, Krupička M, Cibulka R (2019) Nitrosobenzene: reagent for the Mitsunobu esterification reaction. ACS Omega 4 (3):5012–5018

Zhang C, Amar Y, Cao L, Lapkin AA (2020) Solvent selection for Mitsunobu reaction driven by an active learning surrogate model. Org Process Res Dev 24(12):2864–2973

Zou Y, Wong JJ, Houk KN (2020) Computational exploration of a Redox-Neutral organocatalytic Mitsunobu reaction. J Am Chem Soc 142 (38):16403–16408

Alcohol Activation for Nucleophilic Displacement

An J, Denton RM, Lambert TH, Nasca ED (2014) The development of catalytic nucleophilic substitution reactions: challenges, progress and future directions. Org Biomol Chem 12:2993–3003

Denton RM, An J, Adeniran B, Blake A, Lewis W, Poulton A (2011) Catalytic phosphorus(V)-mediated nucleophilic substitution reactions: development of a catalytic appel reaction. J Org Chem 76(16):6749–6767

Dryzhakoz M, Richmond E, Moran J (2016) Recent advances in direct catalytic dehydrative substitution of alcohols. Synthesis 48(7):935–959

Epifanov M, Mo JY, Dubois R, Yu H, Sammis GM (2021) One-pot deoxygenation and substitution of alcohols mediated by sulfuryl fluoride. J Org Chem 86(5):3768–3777

Estopiñá-Durán S, Taylor JE (2021) Brønsted acid-catalysed dehydrative substitution reactions of alcohols. Chem Eur J 27:106–120

Huy PH (2020) Lewis base catalysis promoted nucleophilic substitutions—recent advances and future directions. Eur J Org Chem 2020(1):10–27

Huy PH, Filbrich I (2018) A general catalytic method for highly cost- and atom-efficient nucleophilic substitutions. Chem Eur J 24(29):7410–7416

Huy PH, Hauch T, Filbrich I (2016) Lewis based catalyzed nucleophilic substitutions of alcohols. Synlett 27(19):2631–2636

Huy PH, Motsch S, Kappler SM (2016) Formamides as Lewis base catalysts in SN reactions—efficient transformations of alcohols into chlorides, amines and ethers. Angew Chem Int Ed 55(34):10145–10149

Kohlmeyer C, Schaefer A, Huy PH, Hilt G (2020) Formamide-catalyzed nucleophilic substitutions: mechanistic insight and rationalization of catalytic activity. ACS Catal 10(19):11567–11577

Motsch S, Schuetz C, Huy PH (2018) Systematic evaluation of sulfoxides as catalysts in nucleophilic substitutions of alcohols. Eur J Org Chem 2018 (33):4541–4547

Stach T, Draeger J, Huy PH (2018) Nucleophilic substitution of alcohols in high levels of catalytic efficiency. Org Lett 20(10):2980–2983

Zoller B, Stach T, Huy PH (2020) Synthesis of haloalkanes via Lewis base catalysis enables the activation of alcohols by means of chloroformates as phosgene substitutes. ChemCatChem 12(22):5637–5643

Friedel-Crafts Chemistry

Fang Z, He W, Tu T, Lv N, Qiu C, Li X, Zhu N, Wan L, Guo K (2018) An efficient and green pathway for continuous Friedel-Crafts acylation over α-Fe_2O_3 and $CaCO_3$ nanoparticles prepared in the microreactors. Chem Eng J 331:443–449

Heravi MM, Zadsirjan V, Saedi P, Momeni T (2018) Applications of Friedel-Crafts reactions in total synthesis of natural products. RSC Adv 8 (70):40061–40163

Kokel A, Schafer C, Torok B (2019) Organic synthesis using environmentally benign acid catalysis. Curr Org Synth 16(4):615–649

La Manna P, Soriente A, De Rosa M, Buonerba A, Talotta C, Gaeta C, Neri P (2019) Green, mild and efficient Friedel-Crafts benzylation of scarcely reactive arenes and heteroarenes under on-water conditions. ChemSusChem 12(8):1673–1683

Muthukumar A, Sekar G (2018) Friedel-Crafts hydroxyalkylation of indoles with α-Keto amides using reusable K_3PO_4/nBu_4NBr catalytic system in water. J Org Chem 83(16):8827–8839

Nandi KK (2018) Sustainable catalyst for Friedel-Crafts acylation. Catal Green Chem Eng 1(2):149–153

Pozhydaiev V, Power M, Gandon V, Moran J, Leboeuf D (2020) Exploiting hexafluoroisopropanol (HFIP) in Lewis and Brønsted acid-catalyzed reactions. Chem Commun 56(78):11548–11564

Raviola C, Protti S, Ravelli D, Fagnoni M (2019) Photogenerated acyl/ alkoxycarbonyl radicals for sustainable synthesis. Green Chem 21 (4):748–764

Sadiq Z, Iqbal M, Hussain EA, Naz S (2018) Friedel-Crafts reactions in aqueous media and their synthetic applications. J Mol Liq 255:26–42

Sunke R, Nallapati SB, Kumar JS, Shiva Kumar K, Pal M (2017) Use of $AlCl_3$ in Friedel-Crafts arylation type reactions and beyond: an overview of the development of unique methodologies leading to N-heteroarenes. Org Biomol Chem 15(19):4042–4057

Tan L, Rahman A (2017) From technical efficiency to economic efficiency: development of aza- Friedel-Crafts reaction using phosphoric acid immobilized in glycerol as a sustainable approach. Sustainability 9(7):1176/ 1–1176/10

Tengg M, Stecher H, Offner L, Plasch K, Anderi F, Weber H, Schwab H, Mandana G (2016) Methyltransfereases: green catalysts for Friedel-Crafts alkylations. ChemCatChem 8(7):1354–1360

Tran PH, Nguyen TT-D, Tu T-AT, Le TN (2020) Magnetically recoverable γ-Fe_2O_3 nanoparticles as a highly active catalyst for Friedel-Crafts benzoylation reaction under ultrasound irradiation. Arab J Chem 13(1):290–297

Troschke E, Graetz S, Luebken T, Borchardt L (2017) Mechanochemical Friedel-Crafts alkylation—a sustainable pathway towards porous organic polymers. Angew Chem Int Ed 56(24):6859–6863

Zhao Z, Tian X, Tang P, Ren Y-L, Zhao S, Zheng X, Cheng X (2021) Autocatalytic Friedel-Crafts acylation of arenes without additional catalyst and additive. Chem Select 6(8):1772–1776

Zhao Z, Wang X (2015) Supported phosphotungstic acid catalyst on modified carbon for Friedel-Crafts alkenylation of diverse aromatics to their corresponding α-arylstyrenes. Appl Catal A General 503:103–110

Zhou B, Yang J, Li M, Gu Y (2011) Gluconic acid aqueous solution as a sustainable and recyclable promoting medium for organic reactions. Green Chem 13(8):2204–2211

Ionic Liquids

Dai C, Zhang J, Huang C, Lei Z (2017) Ionic liquids in selective oxidation: catalysts and solvent. Chem Rev 117(10):6929–6983

Deetlefs M, Fanselow M, Seddon KR (2016) Ionic liquids: the view from Mount improbable. RSC Adv 6:4280–4288

Dong K, Liu X, Dong H, Zhang X, Zhang S (2017) Multiscale studies on ionic liquids. Chem Rev 117(10):6636–6695

Egorova KS, Gordeev EG, Ananikov VP (2017) Biological activity of ionic liquids and their application in pharmaceutics and medicine. Chem Rev 117 (10):7132–7189

Gomes JS, Silva SS, Reis RL (2019) Biocompatible ionic liquids: fundamental behaviours and applications. Chem Soc Rev 48:4317–4335

Hajipour AR, Rafiee F (2015) Recent progress in ionic liquids and their applications in organic synthesis. Org Prep Proc Int 47:249–308

Karimi B, Tavakolian M, Akbari M, Mansouri F (2018) Ionic liquids in asymmetric synthesis: an overall view from reaction media to supported ionic liquid catalysis. ChemCatChem 10(15):3173–3205

Kaur N (2017) Ionic liquids: promising but challenging solvents for the synthesis of N-heterocycles. Mini-Rev Organic Chem 14(1):3–23

Kaur N (2018) Perspectives of ionic liquids applications for the synthesis of five- and six-membered *O,N*-heterocycles. Synthetic Commun 48 (5):473–495

Li J, Yang S, Wu W, Jiang H (2018) Recent advances in Pd-catalyzed cross-coupling reaction in ionic liquids. Eur J Org Chem 2018(11):1284–1306

Potdar MK, Kelso GF, Schwarz L, Zhang C, Hearn MTW (2015) Recent developments in chemical synthesis with biocatalysts in ionic liquids. Molecules 20(9):16788–16816

Qiao Y, Ma W, Theyssen N, Chen C, Hou Z (2017) Temperature-responsive ionic liquids: fundamental behaviors and catalytic applications. Chem Rev 117(10):6881–6928

Singh SK, Savoy AW (2020) Ionic liquids synthesis and applications: an overview. J Mol Liq 297:1120238. https://doi.org/10.1016/j.molliq.2019.112038)

Stevens JC, Shi J (2019) Biocatalysis in ionic liquids for lignin valorization: opportunities and recent developments. Biotechnol Adv 37(8):107418

Vanda H, Dai Y, Wilson EG, Verpoorte R, Choi YH (2018) Green solvents from ionic liquids and deep eutectic solvents to natural deep eutectic solvents. C R Chim 21(6):628–638

Ventura SPM, eSilva FA, Quental MV, Mondal D, Freire MG, Coutinho JAP (2017) Ionic-liquid-mediated extraction and separation processes for bioactive compounds: past, present and future trends. Chem Rev 117 (10):6984–7052

Wang B, Qin L, Mu T, Xue Z, Gao G (2017) Are ionic liquids chemically stable? Chem Rev 117(10):7113–7131

Welton T (2018) Ionic liquids: a brief history. Biophys Rev 10:691–706

Part II: Synthetic Strategy

Designing Cascade Reactions for Efficiency

Baccalini A, Faita G, Zanoni G, Maiti D (2020) Transition metal promoted cascade heterocycle synthesis through C-H functionalization. Chem Eur J 26(44):9749–9783

Bellotti P, Koy M, Gutheil C, Heuvel S, Glorius F (2021) Three-component three-bond forming cascade *via* palladium photoredox catalysis. Chem Sci 12:1810–1817

Biemolt J, Ruijter E (2018) Advances in palladium-catalyzed cascade cyclizations. Adv Synth Catal 360(20):3821–3871

Chauhan P, Mahajan S, Enders D (2017) Achieving molecular complexity via stereoselective multiple domino reactions promoted by a secondary amine organocatalyst. Acc Chem Res 50(11):2809–2821

Ciulla MG, Zimmermann S, Kumar K (2018) Cascade reaction based synthetic strategies targeting biologically intriguing indole polycycles. Org Biomol Chem 17:413–431

Elinson MN, Vereshchagin AN, Ryzhkov FV (2016) Catalysis of cascade and multicomponent reactions of carbonyl compounds and C-H acids by electricity. Chem Rec 16(4):1950–1964

Faisca Phillips AM, Pomberiro AJL, Kopylovich MN (2017) Recent advances in cascade reactions initiated by alcohol oxidation. ChemCatChem 9 (2):217–246

Fang X, Zhang J-W, Wang C-J (2016) Asymmetric cycloaddition and cascade addition-cyclization reactions. Topics Organometal Chem 58:183–205

Gajalakshmi S, Valentina P, Maheshwaran A (2017) Comprehensive review on current developments of tandem reaction based organic synthesis. Int J Pharm Sci Rev Res 45(2):165–169

Gandhi S, Sivadas V, Baire B (2021) Thiourea-tertiary amine promoted cascade catalysis: a tool for complexity generation. Eur J Org Chem 2021 (2):220–234

Gu Y, Tan C, Gong J, Yang Z (2018) Diversity-oriented synthesis of natural products via gold-catalyzed cascade reactions. Synlett 29(12):1552–1571

Huang H-M, Garduno-Castro MH, Morrill C, Procter DJ (2019) Catalytic cascade reactions by radical relay. Chem Soc Rev 48(17):4626–4638

Liao J, Yang X, Ouyang L, Lai Y, Huang J, Luo R (2021) Recent advances in cascade radical cyclization of radical acceptors for the synthesis of carbo- and heterocycles. Org Chem Front. https://doi.org/10.1039/D0QO01453B

Lu L-Q, Chen J-R, Xiao W-J (2012) Development of cascade reactions for the concise construction of diverse heterocyclic architectures. Acc Chem Res 45(8):1278–1293

Nairoukh Z, Cormier M, Marek I (2017) Merging C-H and C-C bond cleavage in organic synthesis. Nat Rev Chem 1:0035. https://doi.org/10.1038/s41570-017-0035

Nicolaou KC, Chen JS (2009) The art of total synthesis through cascade reactions. Chem Soc Rev 38:2993–3009

Ohno H, Inuke S (2018) Recent progress in palladium-catalyzed cascade cyclizations for natural product synthesis. Synthesis 50(4):700–710

Ping Y, Li Y, Zhu J, Kong W (2019) Construction of quaternary stereocenters by palladium-catalyzed carbopalladation-initiated cascade reactions. Angew Chem Int Ed 58(6):1562–1573

Plesniak MP, Huang H-M, Procter DJ (2017) Radical cascade reactions triggered by single electron transfer. Nat Rev Chem 1:0077. https://doi.org/10.1038/s41570-017-0077

Reyes E, Uria U, Carrillo L, Vicario JL (2017) Enantioselective cascade reactions under N-heterocyclic carbene catalysis. Synthesis 49(3):451–471

Schmidt S, Castiglione K, Kourist R (2018) Overcoming the incompatibility challenge in chemoenzymatic and multi-catalytic cascade reactions. Chem Eur J 24(8):1755–1768

Song X-R, Yang R, Xiao Q (2021) Recent advances in the synthesis of heterocyclics via cascade cyclization of propargylic alcohols. Adv Synth Catal 363 (4):852–876

Sugimoto K, Matsuya Y (2017) Recent applications in gold-catalyzed cascade reactions in total synthesis of natural product. Tetrahedron Lett 58 (47):4420–4426

Tian L, Luo Y-C, Hu X-Q, Xu P-F (2016) Recent developments in the synthesis of chiral compounds with quaternary centers by organocatalytic cascade reactions. Asian J Org Chem 5(5):580–607

Vavsari VF, Balalaie S (2017) Cascade reaction in the synthesis of heterocyclic natural products. Curr Org Chem 21(15):1393–1426

Wang C, Chen F, Qian P, Cheng J (2021) Recent advances in the Rh-catalyzed cascade arene C-H bond activation/annulation toward diverse heterocyclic compounds. Org Biomol Chem 19(8):1705–1721

Wang Y, Lu H, Xu P-F (2015) Asymmetric catalytic cascade reactions for constructing diverse scaffolds and complex molecules. Acc Chem Res 48 (7):1832–1844

Wang Y, Zhao H (2016) Tandem reactions combining biocatalysts and chemical catalysts for asymmetric synthesis. Catalysts 6(12):194. https://doi.org/10.3390/catal6120194

Zhang B, Studer A (2015) Recent advances in the synthesis of nitrogen heterocycles via radical cascade reactions using isonitriles as radical acceptors. Chem Soc Rev 44(11):3505–3521

Multicomponent Synthesis

Ahmadi T, Ziarani GM, Gholamzadeh P, Mollabagher H (2017) Recent advances in asymmetric multicomponent reactions (AMCRs). Tetrahedron Asymm 28(5):708–724

Collet JW, Roose TR, Weijers B, Maes BUW, Ruijter E, Orru RVA (2020) Recent advances in palladium-catalyzed isocyanide insertions. Molecules 25 (21):4906. https://doi.org/10.3390/molecules25214906)

Ghashghaei O, Seghetti F, Lavilla R (2019) Selectivity in multiple multicomponent reactions: types and synthetic applications. Beilstein J Org Chem 15:521–534

Graebin CS, Ribeiro FV, Rogério KA, Kümmerle AE (2019) Multicomponent reactions for the synthesis of bioactive compounds: a review. Curr Org Synth 16(6):855–899

Ibarra IA, Islas-Jácome A, González-Zamora E (2018) Synthesis of polyheterocycles *via* multicomponent reactions. Org Biomol Chem 16:1402–1418

Insuasty D, Castillo J, Becerra D, Rojas H, Abonia R (2020) Synthesis of biologically active molecules through multicomponent reactions. Molecules 25(3):505. https://doi.org/10.3390/molecules25030505

Melo LR, Silva WA (2016) Ionic liquid in multicomponent reactions: a brief review. Curr Green Chem 3(2):120–132

Neochoritis CG, Zhao T, Dömling A (2019) Tetrazoles via multicomponent reactions. Chem Rev 119(3):1970–2042

Paprocki D, Madej A, Koszelewski D, Brodzka A, Ostaszewski R (2018) Multicomponent reactions accelerated by aqueous micelles. Front Chem. https://doi.org/10.3389/fchem.2018.00502

Ramos LM, Rodrigues MO, Neto BAD (2019) Mechanistic knowledge and noncovalent interactions as the key features for enantioselective catalyzed multicomponent reactions: a critical review. Org Biomol Chem 17:7260–7269

Rocha RO, Rodrigues MO, Neto BAD (2020) Review on the Ugi multicomponent reaction mechanism and the use of fluorescent derivatives as functional chromophores. ACS Omega 5(2):972–979

Rodrigues MO, Eberlin MN, Neto BAD (2021) How and why to investigate multicomponent reactions mechanisms? A critical review. Chem Record. https://doi.org/10.1002/tcr.20200165

Santra S (2019) Baker's yeast catalyzed multicomponent reactions: a new hope. Chem Select 4(43):12630–12637

Serafini M, Murgia I, Giustiniano M, Pirali T, Tron GC (2021) The 115 years old multicomponent bargellini reaction: perspectives and new applications. Molecules 26(3):558. https://doi.org/10.3390/molecules26040558)

Váradi A, Palmer TC, Dardashti RN, Majumdar S (2016) Isocyanide-based multicomponent reactions for the synthesis of heterocycles. Molecules 21 (1):19. https://doi.org/10.3390/molecules21010019)

Wan J-P, Gan L, Liu Y (2017) Transition metal-catalyzed C-H bond functionalization in multicomponent reactions: a tool toward molecular diversity. Org Biomol Chem 15:9031–9043

Younus HA, Al-Rashida M, Hameed A, Uroos A, Salar U, Rana S, Khan KM (2021) Multicomponent reactions (MCR) in medicinal chemistry: a patent review (2010-2020). Expert Opin Ther Patents 31(3):267–289

Zarganes-Tzitzikas T, Chandgude AL, Dömling A (2015) Multicomponent reactions, union of MCRs and beyond. Chem Rec 15(5):981–996

Zhi S, Ma Z, Zhang W (2019) Consecutive multicomponent reactions for the synthesis of complex molecules. Org Biomol Chem 17:7632–7650

C-H Activation Catalyzed with Transition Metals

Basu D, Kumar S, Sudhir VS, Bandichhor R (2018) Transition metal catalyzed C-H activation for the synthesis of medicinally relevant molecules: a review. J Chem Sci 130:71. https://doi.org/10.1007/s12039-018-1468-6)

Chen J, Lv S, Tian S (2019) Electrochemical transition-metal-catalyzed C-H bond functionalization: electricity as clean surrogates of chemical oxidants. ChemSusChem 12(1):115–132

Chen S, Ranjan P, Voskressensky LG, Van der Eycken EV, Sharma UP (2020) Recent Developments in transition-metal catalyzed direct C-H alkenylation, alkylation and alkynylation of azoles. Molecules 25:4970. https://doi.org/10.3390/molecules25214970

Chen Z, Wang B, Zhang J, Yu W, Liu Z, Zhang Y (2015) Transition metal-catalyzed C-H bond functionalizations by the use of diverse directing groups. Org Chem Front 2:1107–1295

Da Silva Júnior EN, Jardim GAM, Gomes RS, Liang Y-F, Ackermann L (2018) Weakly-coordinating *N*-oxide and carbonyl groups for metal-catalyzed C-H activation: the case for A-ring functionalization. Chem Commun 54:7398–7411

Das R, Kapur M (2018) Transition-metal-catalyzed C-H functionalization reactions of π-deficient heterocycles. Asian J Org Chem 7(7):1217–1235

Dong Z, Ren Z, Thompson SJ, Xu Y, Dong G (2017) Transition-metal-catalyzed C-H alkylation using alkenes. Chem Rev 117(13):9333–9403

Gallego D, Baquero EA (2018) Recent advances on mechanistic studies on C-H activation catalyzed by base metals. Open Chem 16:1. https://doi.org/10.1515/chem-2018-0102

Gandeepan P, Cheng C-H (2015) Transition-metal-catalyzed π-bond-assisted C-H bond functionalization: an emerging trend in organic synthesis. Chem Asian J 10(4):824–838

Gandeepan P, Cheng C-H (2016) Advancements in the synthesis and applications of cationic *N*-heterocycles through transition metal-catalyzed C-H activation. Chem Asian J 11(4):448–460

Gandeepan P, Kaplaneris N, Santoro S, Vaccaro L, Ackermann L (2019) Biomass-derived solvents for sustainable transition metal-catalyzed C-H activation. ACS Sustain Chem Eng 7(9):8023–8040

Gandeepan P, Müller T, Zell D, Cera G, Warratz S, Ackermann L (2019) 3D transition metals for C-H activation. Chem Rev 119(4):2192–2452

Gensch T, Hopkinson MN, Glorius F, Wencel-Delord J (2016) Mild metal-catalyzed C-H activation: examples and concepts. Chem Soc Rev 45:2900–2936

Guillemard L, Wencel-Delord J (2020) When metal-catalyzed C-H functionalization meets visible-light photocatalysis. Beilstein J Org Chem 16:1754–1804

Gulías M, Mascareñas JL (2016) Metal-catalyzed annulations through activation and cleavage of C-H bonds. Angew Chem Int Ed 55 (37):11000–11019

Guo T, Huang F, Yu L, Yu Z (2015) Indole synthesis through transition metal-catalyzed C-H activation. Tetrahedron Lett 56(2):296–302

Huang H, Ji X, Wu W, Jiang H (2015) Transition metal-catalyzed C-H functionalization of *N*-oxyenamine internal oxidants. Chem Soc Rev 44:1155–1171

Huang Z, Lim HN, Mo F, Young MC, Dong G (2015) Transition metal-catalyzed ketone-directed or mediated C-H functionalization. Chem Soc Rev 44:7764–7786

Kalepu J, Pilarski LT (2019) Weinreb amides as directing groups for transition metal-catalyzed C-H functionalizations. Molecules 24(5):830. https://doi.org/10.3390/molecules24050830

Landaeta VR, Rodriguez-Lugo RE (2017) Insights on the C-H bond activation by transition metal complexes from groups 8-10 bearing (P-N) chelates. J Mol Catal A Chem 426(B):316–325

Leitch JA, Bhonoah Y, Frost CG (2017) Beyond C2 and C3: transition-metal-catalyzed C-H functionalization of indole. ACS Catal 7(9):5618–5627

Li Y, Xu S (2018) Transition-metal-catalyzed C-H functionalization for construction of quaternary carbon centers. Chem Eur J 24(61):16218–16245

Liao G, Zhang T, Lin Z-K, Shi B-F (2020) Transition metal-catalyzed enantioselective C-H functionalization via chiral transient directing group strategies. Angew Chem Int Ed 59(45):19773–19786

Liu B, Hu F, Shi B-F (2015) Recent advances on ester synthesis via transition-metal catalyzed C-H functionalization. ACS Catal 5(3):1863–1881

Liu B, Yang Y, Li P, Wang F, Li X (2021) Recent advances in transition metal-catalyzed olefinic C-H functionalization. Org Chem Front 8:1085–1101

Ma C, Fang P, Mei T-A (2018) Recent advances in C-H functionalization using electrochemical transition metal catalysis. ACS Catal 8 (8):7179–7189

Maraswami M, Loh T-P (2019) Transition-metal-catalyzed alkenyl sp^2 C-H activation: a short account. Synthesis 51:1049–1062

Mo J, Wang L, Liu Y, Cui X (2015) Transition-metal-catalyzed direct C-H functionalization under external-oxidant-free conditions. Synthesis 47 (4):439–459

Moghimi S, Mahdavi M, Shafiee A, Foroumadi A (2016) Transition-metal-catalyzed acyloxylation: activation of $C(sp^2)$-H and $C(sp^3)$-H bonds. Eur J Org Chem 2016(20):3282–3299

Niu B, Yang K, Lawrence B, Ge H (2019) Transition ligand-enabled transition metal-catalyzed C-H functionalization. ChemSusChem 12 (13):2955–2969

Oeschger R, Su B, Yu I, Ehinger C, Romero E, He S, Hartwig J (2020) Diverse functionalization of strong alkyl C-H bonds by undirected borylation. Science 368(6492):736–741

Ping L, Chung DS, Bouffard J, Lee S-g (2017) Transition metal-catalyzed site- and regio-divergent C-H bond functionalization. Chem Soc Rev 46:4299–4328

Qiu G, Wu J (2015) Transition metal-catalyzed direct remote C-H functionalization of alkyl groups *via* $C(sp^3)$-H bond activation. Org Chem Front 2:169–178

Roudesly F, Oble J, Poli G (2017) Metal-catalyzed C-H activation/functionalization: the fundamentals. J Mol Catal A Chem 426(B):275–296

Sandtorv AH (2015) Transition metal-catalyzed C-H activation of indoles. Adv Synth Catal 357(11):2403–2435

Santhoshkumar R, Cheng C-H (2019) Reaching green: heterocycle synthesis by transition metal-catalyzed C-H functionalization in sustainable medium. Chem Eur J 25(40):9366–9384

Shi H, Lu Y, Weng J, Bay KL, Chen X, Tanaka K, Verma P, Houk KN, Yu J-Q (2020) Differentiation and functionalization of remote C-H bonds in adjacent positions. Nat Chem 12:399–404

Song B, Xu B (2017) Metal-catalyzed C-H functionalization involving isocyanides. Chem Soc Rev 46:1103–1123

Wei Y, Hu P, Zhang M, Su W (2017) Metal-catalyzed decarboxylative C-H functionalization. Chem Rev 117(13):8864–8907

Xiang Y, Wang C, Ding Q, Peng Y (2019) Diazo compounds: versatile synthons for the synthesis of nitrogen heterocycles *via* transition metal-catalyzed cascade C-H activation/carbene insertion/annulation reactions. Adv Synth Catal 361(5):919–944

Xu X, Luo J (2019) Transition metal-catalyzed directing-group-assisted C-H activation of phenols. ChemSusChem 12(20):4601–4616

Yang J (2015) Transition metal-catalyzed *meta*-C-H functionalization of aromatic compounds. Org Biomol Chem 13:1930–1941

Yang T, Kong C, Yang S, Yang Z, Yang S, Ehara M (2019) Reaction mechanism, norbornene and ligand effects, and origins of *meta*-selectivity of Pd/norbornene-catalyzed C-H activation. Chem Sci 11:113–125

Yoshikai N (2015) $Cp^{*}Co^{III}$-catalyzed C-H activation of (Hetero)arenes: expanding the scope of base-metal-catalyzed C-H functionalizations. ChemCatChem 7(5):732–734

Zhao Q, Poisson T, Pannecoucke X, Besset T (2017) The transient directing group strategy: a new trend in transition-metal-catalyzed C-H bond functionalization. Synthesis 49(21):4808–4826

Photocatalytic C–H Activation

Aukland MH, Šiaučiulis M, West A, Perry GJP, Procter DJ (2020) Metal-free photoredox-catalysed C-H/C-H coupling of arenes enabled by interrupted Pummerer activation. Nat Catal 3:163–169

Chalotra N, Sultan S, Shah BA (2020) Recent advances in photoredox methods for ketone synthesis. Asian J Org Chem 9(6):863–881

Crisenza GEM, Melchiorre P (2020) Chemistry glows green with photoredox catalysis. Nat Commun 11:803. https://doi.org/10.1038/s1467-019-13887-8

Dwivedi V, Kalsi D, Sundararaju B (2019) Electrochemical-/photoredox aspects of transition metal-catalyzed directed C-H bond activation. ChemCatChem 11(21):5160–5187

Ener ME, Darcy JW, Menges FS, Mayer JM (2020) Base-directed photoredox activation of C-H bonds by PCET. J Org Chem 85(11):7175–7180

Fabry DC, Rueping M (2016) Merging visible light photoredox catalysis with metal catalyzed C-H activations: on the role of oxygen and superoxide ions as oxidants. Acc Chem Res 49(9):1969–1979

Garrido-Castro AF, Maestro MC, Alemán J (2020) α-functionalization of imines via visible light photoredox catalysis. Catalysts 10(5):562. https://doi.org/10.3390/catal10050562

Ghosh I, Marzo L, Das A, Shaikh R, König B (2016) Visible light mediated photoredox catalytic arylation reactions. Acc Chem Res 49(8):1566–1577

Guo K, Zhang Z, Li A, Li Y, Huang J, Yang Z (2020) Photoredox-catalyzed isomerization of highly substituted allylic alcohols by C-H bond activation. Angew Chem Int Ed 59(28):11660–11668

Jia Z, Yang Q, Zhang L, Luo S (2019) Photoredox mediated acceptorless dehydrogenative coupling of saturated *N*-heterocycles. ACS Catal 9(4):3589–3594

Kalsi D, Barsu N, Chakrabarti S, Dahiya P, Rueping M, Sundararaju B (2019) C-H and N-H bond annulation of aryl amides with unactivated olefins by merging cobalt (III) and photoredox catalysis. Chem Commun 55:11626–11629

Kalsi D, Dutta S, Barsu N, Rueping M, Sundararaju (2018) Room-temperature C-H functionalization by merging cobalt and photoredox catalysis. ACS Catal 8(9):8115–8120

Kim HJ, Fabry DC, Mader S, Rueping M (2019) Photoredox/rhodium catalysis in C-H activation for the synthesis of nitrogen containing heterocycles. Org Chem Front 6:2319–2323

Lee BJ, DeGlopper KS, Yoon TP (2020) Site-selective alkoxylation of benzylic C-H bonds by photoredox catalysis. Angew Chem Int Ed 59(1):197–202

Lima F, Kabeshov MA, Tran DN, Battilocchio C, Sedelmeier J, Sedelmeier G, Schenkel B, Ley SV (2016) Visible light activation of boronic esters enables efficient photoredox $C(sp^2)$-$C(sp^3)$ cross-couplings in flow. Angew Chem Int Ed 55(45):14085–14089

Plutschack MB, Correia CA, Seeberger PH, Gilmore K (2015) Organic photoredox chemistry in flow. Top Organomet Chem 57:43–76

Reischauer S, Pieber B (2021) Emerging concepts in photocatalytic organic synthesis. iScience 24(3):102209. https://doi.org/10.1016/j.isci.2020.102209

Romero NA, Margrey KA, Tay NE, Nicewicz DA (2015) Site-selective arene C-H amination via photoredox catalysis. Science 349(6254):1326–1330

Seo H, Katcher MH, Jamison TF (2017) Photoredox activation of carbon dioxide for amino acid synthesis in continuous flow. Nat Chem 9:453–456

Sharma U, Gemoets HPL, Schroeder FS (2017) Merger of visible-light photoredox catalysis and C-H activation for the room temperature C-2 acylation of indoles in batch and flow. ACS Catal 7(6):3818–3823

Shen T, Lambert TH (2021) Electrophotocatalytic diamination of vicinal C-H bonds. Science 371(6529):620–626

Verschueren RH, De Borggraeve WM (2019) Electrochemistry and photoredox catalysis: a comparative evaluation in organic synthesis. Molecules 24(11):2122. https://doi.org/10.3390/molecules24112122)

Xiang M, Meng Q-Y, Li J-X, Zheng Y-W, Ye C, Li Z-J, Chen B, Tung C-H, Wu L-Z (2015) Activation of C-H bonds through oxidant-free photoredox catalysis: cross-coupling hydrogen-evolution transformation of isochromans and β-keto esters. Chem Eur J 21(50):18080–18084

Yin Y, Zhao X, Qiao B, Jiang Z (2020) Cooperative photoredox and chiral hydrogen-bonding catalysis. Org Chem Front 7:1283–1296

In Situ Protecting Groups for Chemoselective Transformations

Ágoston K, Streicher H, Fügedi P (2016) Orthogonal protecting group strategies in carbohydrate chemistry. Tetrahedron Asymmetry 27(16):707–728

De Figueiredo RM, Suppo J-S, Midrier C, Campagne J-M (2017) Sequential one-pot synthesis of dipeptides through the transient formation of CDI-*N*-protected α-aminoesters. Adv Synth Catal 359(11):1963–1968

Emmert MH, He CQ, Shah AA, Felten S (2021) Lewis acid-mediated, mild C-H aminoalkylation of azoles via three component coupling. Chem Sci. https://doi.org/10.1039/D0SC06868C

Fang J, Zeng J, Sun J, Zhang S, Xiao X, Lu Z, Meng L, Wan Q (2019) Total synthesis of resin glycosides Murucoidins IV and V. Org Lett 21(16):6213–6216

Fernandes RA, Kumar P, Choudhary P (2021) Evolution of strategies in protecting-group-free synthesis of natural products: a recent update. Eur J Org Chem 2021(5):711–740

Henschke JP, Wu P-Y, Lin C-W, Chen S-F, Chiang P-C, Hsiao C-N (2015) β-selective C-arylation of silyl protected 1,6-anhydroglucose with arylalanes: the synthesis of SGLT2 inhibitors. J Org Chem 80(4):2295–2309

Kang SK, Lee SW, Woo D, Sim J, Suh Y-G (2017) Practical and efficient synthesis of gefitinib through selective O-alkylation: a novel concept for a transient protection group. Synth Comm 47(21):1990–1998

Klán P, Šolomek T, Bochet CG, Blanc A, Givens R, Rubina M, Popik V, Kostikov A, Wirz J (2013) Photoremovable protecting groups in chemistry and biology: reaction mechanisms and efficacy. Chem Rev 113:119–191

Knauer S, Koch N, Uth C, Meusinger R, Avrutina O, Kolmar H (2020) Sustainable peptide synthesis enabled by a transient protecting group. Angew Chem Int Ed 59(31):12984–12990

Mong K-K T, Cheng K-C, Lu I-C, Pan C-W, Wang Y-F, Shen L-C (2020) Cascade in situ phosphorylation and one-pot glycosylation for rapid synthesis of heptose-containing oligosaccharides. J Org Chem 85 (24):16060–16071

Mu Y, Nguyen TT, Koh M-J, Schrock RR, Hoveyda AH (2019) *E*- and *Z*-, di- and tri-substituted alkenyl nitriles through catalytic cross-metathesis. Nat Chem 11(5):478–487

Ohta R, Fujioka H (2017) Chemoselective reduction and alkylation of carbonyl functions using phosphonium salts as *in situ* protecting groups. Chem Pharm Bull 65(1):10–18

Panzella L, DellaGreca M, Longobardo L (2018) A facile preparation of hydroxycinnamyl alcohols with simultaneous protection of phenol groups as carbonate. Chem Select 3(38):10637–10640

Peng F, Humphrey GR, Maloney KM, Lehnherr D, Weisel M, Levesque F, Naber JR, Brunskill APJ, Larpent P, Zhang S-W, Lee AY, Arvary RA, Lee CH, Bishara D, Narsimhan K, Sirota E, Whittington M (2020) Development of a green and sustainable manufacturing process for Gefapixant citrate (MK-7264) part 2: development of a robust process for phenol synthesis. Org Process Res Dev 24(11):2453–2461

Rocheleau S, Pottel S, Huckić I, Moitessier N (2017) Highly regioselective monoacylation of unprotected glucopyranoside using transient directing-protecting groups. Eur J Org Chem 2017(3):646–656

Sánchez A, Calderón E, Vazquez A (2013) Using the 9-BBN group as a transient protective group for the functionalization of reactive chains of α-amino acids. Synthesis 45(10):1364–1372

Seo H, Bedard AC, Chen RP, Hicklin RW, Alabugin A, Jamison TF (2018) Selective N-monomethylation of primary amines with dimethyl carbonate in continuous flow. Tetrahedron 74(25):3124–3128

Shi Z, Fan J, Kronenthal DR, Mudryk BM (2018) Development of a practical synthesis of a farnesyltransferase inhibitor. Org Process Res Dev 22 (11):1534–1540

Tay J-H, Dorokhov V, Wang S, Nagorny P (2019) Regioselective single pot C3-glycosylation of strophanthidol using methylboronic acid as a transient protecting group. J Antibiot 72:437–448

Traboni S, Bedini E, Vessella G, Iadonisi A (2020) Solvent-free approaches in carbohydrate synthetic chemistry: role of catalysis in reactivity and selectivity. Catalysts 10:1142. https://doi.org/10.3390/catal10101142)

Yan X, Harutyunyan SR (2019) Catalytic enantioselective addition of organometallics to unprotected carboxylic acids. Nat Commun 10:3402. https://doi.org/10.1038/s41467-019-11345-z

Yang T, Zhu F, Walczak MA (2018) Stereoselective oxidative glycosylation of anomeric nucleophiles with alcohols and carboxylic acids. Nat Commun 9:3650. https://doi.org/10.1038/s41467-018-06016-4

Zheng Y, Wang J, Xia P-J, Zhao Q-I, Xiao J-A, Xiang H-Y, Chen X-Q, Yang H (2017) Organocatalytic asymmetric allylic alkylation of Morita-Baylis-Hillman carbonates with diethyl 2-aminomalonate assisted by in situ protection. J Org Chem 82(23):12202–12208

Part III: Enabling Technologies

Biocatalysis

Abdelraheem EMM, Busch H, Hanefeld U, Tonin F (2019) Biocatalysis explained: from pharmaceutical to bulk chemical production. React Chem Eng 4:1878–1894

Adams JP, Brown MJB, Diaz-Rodriguez A, Lloyd RC, Roiban G-D (2019) Biocatalysis: a pharma perspective. Adv Synth Catal 361(11):2421–2432

Bergquist PL, Siddiqui S, Sunna A (2020) Cell-free biocatalysis for the production of platform chemicals. Front Energy Res. https://doi.org/10.3389/fenrg.2020.00193

Chakrabarty S, Romero EO, Pyser JB, Yazarians JA, Narayan ARH (2021) Chemoenzymatic total synthesis of natural products. Acc Chem Res. https://doi.org/10.1021/acs.accounts.0c00810

Devine PN, Howard RM, Kumar R, Thompson MP, Truppo MD, Turner NJ (2018) Extending the application of biocatalysis to meet the challenges of drug development. Nat Chem 2:409–421

Foley AM, Maguire AR (2019) The impact of recent developments in technologies which enable the increased use of biocatalysts. Eur J Org Chem 23:3713–3734

Fryszkowska A, Devine PN (2020) Biocatalysis in drug discovery and development. Curr Opin Chem Biol 55:151–160

Goodwin NC, Morrison JP, Fuerst DE, Hadi T (2019) Biocatalysis in medicinal chemistry: challenges to access and drives for adoption. ACS Med Chem Lett 10(10):1363–1366

Guarjardo N, Dominguez de Maria P (2019) Continuous biocatalysis in environmentally-friendly media: a triple synergy for future sustainable processes. ChemCatChem 11(14):3128–3137

Hinzmann A, Adebar N, Betke T, Leppin M, Groeger H (2019) Biotransformations in pure organic medium: organic-solvent-labile enzymes in the batch and flow synthesis of nitriles. Eur J Org Chem 41:6911–6916

Kelly SA, Mix S, Moody TS, Gilmore BF (2020) Transaminases for industrial biocatalysis: novel enzyme discovery. Appl Microbiol Biotech 104(11):4781–4794

Porter JL, Rusli RA, Ollis DL (2016) Directed evolution of enzymes for industrial biocatalysis. Chem Biochem 17(3):197–203

Ramsden JI, Cosgrove SC, Turner NJ (2020) Role of biocatalysis as naturally occurring catalysts in fragment-based drug discovery. Chem Sci 11(41):11104–11112

Sandoval BA, Hyster TK (2020) Emerging strategies for expanding the toolbox of enzymes in biocatalysis. Curr Opin Chem Biol 55:45–51

Schmermund L, Jurkas V, Oezgen FF, Barone GD, Buechsenschuetz HC, Winkler CK, Schmidt S, Kourist R, Kroutil W (2019) Photo-biocatalysis: biotransformations in the presence of light. ACS Catal 9(5):4115–4144

Schwarz J, Rosenthal K, Snajdrobav R, Kittelman M, Luetz S (2020) The development of biocatalysis as a tool for drug discovery. Chimia 74(5):368–377

Sheldon RA, Brady D (2019) Broadening the scope of biocatalysis in sustainable organic synthesis. ChemSusChem 12(13):2859–2881

Sheldon RA, Woodley JM (2018) Role of biocatalysis in sustainable chemistry. Chem Rev 118(2):801–838

Tufvesson P, Nordblad M, Kruhne U, Schurmann M, Vogel A, Wohlgemuth R, Woodley JM (2015) Economic considerations for selecting and amine donor in biocatalytic transamination. Org Process Res Dev 19(6):652–660

Winkler CK, Schrittwieser JH, Kroutil W (2021) Power of biocatalysis for organic synthesis. ACS Cent Sci 7(1):55–71

Woodley JM (2019a) Accelerating the implementation of biocatalysis in industry. Appl Microbiol Biotech 103(12):4733–4739

Woodley JM (2019b) Reaction engineering for the industrial implementation of biocatalysis. Top Catal 62(17–20):1202–1207

Zhang X, King-Smith E, Dong L-B, Yang L-C, Rudolf JD, Shen B, Renata H (2020) Divergent synthesis of complex diterpenes through a hybrid oxidative approach. Science 369(6505):799–806

Zhu Y, Chen Q, Shao L, Jia Y, Zhang X (2020) Microfluidic immobilized enzyme reactors for continuous biocatalysis. React Chem Eng 5(1):9–32

New Directions in Coupling Chemistry

Barth EL, Davis RM, Beromi MM, Walden AG, Balcells D, Brudvig GW, Dardir AH, Hazari N, Lant HMC, Mercado BQ, Peczak IL (2019) Bis (dialkylphosphino)ferrocene-ligated nickel (II) precatalysts for Suzuki-Miyaura reactions of aryl carbamates. Organometallics 38(17):3377–3387

Biswas S, Qu B, Desrosiers J-N, Choi Y, Haddad N, Lee NK, Song JJ, Senanayake CH (2020) Nickel-catalyzed cross-electrophile reductive couplings of neopentyl bromides with aryl bromides. J Org Chem 85 (12):8214–8220

Bisz E, Kardela M, Szostak M (2019) Ligand effect on iron-catalyzed cross-coupling reactions: evaluation of amides as O-coordinating ligands. ChemCatChem 11(23):5733–5737

Bisz E, Szostak M (2018) 2-Methyltetrahydrofuran: a green solvent for iron-catalyzed cross-coupling reactions. ChemSusChem 11(8):1290–1294

Campeau L-C, Hazari N (2019) Cross-coupling and related reactions: connecting past success to the development of new reactions for the future. Organometallics 38(1):3–35

Charboneau DJ, Barth EL, Hazari N, Uehling MR, Zultanski SL (2020) A widely applicable dual catalytic system for cross-electrophile coupling enabled by mechanistic studies. ACS Catal 10(21):12642–12656

Chen X, Lin J, Wang B, Tian X (2020) Nickel-catalyzed mizoroki-keck/amination cascade reactions of *o*-dihaloarenes with allylamines: synthesis of indoles. Org Lett 22(19):7704–7708

Cheng L-J, Mankad NP (2020) C-C and C-X coupling reactions of unactivated alkyl electrophiles using copper catalysis. Chem Soc Rev 49:8036–8064

Dorel R, Grugel CP, Haydl AM (2019) The Buchwald-Hartwig amination after 25 years. Angew Chem Int Ed 58(48):17118–17129

Gerleve C, Studer A (2020) Transition-metal-free oxidative cross-coupling of tetraarylborates to biaryls using organic oxidants. Angew Chem Int Ed 59 (36):15468–15473

Greaves ME, Ronson TO, Lloyd-Jones GC, Maseras F, Sproules S, Nelson DJ (2020) Unexpected nickel complex speciation unlocks alternative pathways for the reactions of alkyl halides with dppf-Nickel (0). ACS Catal 10 (18):10717–10725

Haas D, Hammann JM, Greiner R, Knochel P (2016) Recent developments in Negishi cross-coupling reactions. ACS Catal 6(3):1540–1552

He Z, Song F, Sun H, Huang Y (2018) Transition-metal-free Suzuki-type cross-coupling reaction of benzyl halides and boronic acids via 1,2-metalate shift. J Am Chem Soc 140(7):2693–2699

Kim S, Goldfogel MJ, Gilbert MM, Weix DJ (2020) Nickel-catalyzed cross-electrophile coupling of aryl chlorides with primary alkyl chlorides. J Am Chem Soc 142(22):9902–9907

Lavoie CM, Stradiotto M (2018) Bisphosphines: a prominent ancillary ligand class for application in nickel-catalyzed C-N cross-coupling. ACS Catal 8 (8):7228–7250

Li G, Nieves-Quinones Y, Zhang H, Liang Q, Su S, Liu Q, Kozlowski MC, Jia T (2020) Transition-metal-free formal cross-coupling of aryl methyl sulfoxides and alcohols via nucleophilic activation of C-S bond. Nat Commun 11:2890. https://doi.org/10.1038/s41467-020-16713-8

McGuire RT, Simon CM, Yadav AA, Ferguson MJ, Stradiotto M (2020) Nickel-catalyzed cross-coupling of sulfonamides with (Hetero)aryl chlorides. Angew Chem Int Ed 59(23):8952–8956

Merchant RR, Lopez JA (2020) A general $C(sp^3)$–$C(sp^3)$ cross-coupling of benzyl sulfonylhydrazones with alkyl boronic acids. Org Lett 22 (6):2271–2275

Nebra N (2020) High-valent Ni(III) and Ni(IV) species relevant to C-C and C-heteroatom cross-coupling reactions: state of the art. Molecules 25 (5):1141. https://doi.org/10.3390/molecules25051141

Paul A, Smith MD, Vannucci AK (2017) Photoredox-assisted reductive cross-coupling: mechanistic insight into catalytic aryl-alkyl cross-couplings. J Org Chem 82(4):1996–2003

Poremba KE, Dibrell SE, Reisman SE (2020) Nickel-catalyzed enantioselective reductive cross-coupling reactions. ACS Catal 10(15):8237–8246

Sakai HA, Liu W, Le C, MacMillan DWC (2020) Cross-electrophile coupling of unactivated alkyl chlorides. J Am Chem Soc 142(27):11691–11697

Sestelo JP, Sarandeses LA (2020) Advances in cross-coupling reactions. Molecules 25(19):4500. https://doi.org/10.3390/molecules25194500

Shi W-J, Zhao H-W, Wang Y, Cao Z-C, Zhang L-S, Yu D-G, Shi Z-J (2016) Nickel- or iron-catalyzed cross-coupling of aryl carbamates with arylsilanes. Adv Synth Catal 358(15):2410–2416

Sun X, Rocha MVJ, Hamlin TA, Poater J, Bickelhaupt FM (2019) Understanding the differences between iron and palladium in cross-coupling reactions. Phys Chem Chem Phys 21:9651–9664

Teng Q, Wu W, Duong HA, Huynh HV (2018) Ring-expanded N-heterocyclic carbenes as ligands in iron-catalysed cross-coupling reactions of arylmagnesium reagents with aryl chlorides. Chem Commun 54:6044–6047

Truesdell BL, Hamby TB, Sevov CS (2020) General $C(sp^2)$-$C(sp^3)$ cross-electrophile coupling reactions enabled by overcharge protection of homogeneous electrocatalysts. J Am Chem Soc 142(12):5884–5893

Wei X-J, Abdiaj I, Sambiagio C, Li C, Zysman-Colman E, Alcázar J, Noël T (2019) Visible-light-promoted iron-catalyzed $C(sp^2)$-$C(sp^3)$ Kumada cross-coupling in flow. Angew Chem Int Ed 58(37):13030–13034

Wei J, Liang H, Ni C, Sheng R, Hu J (2019) Transition-metal-free desulfinative cross-coupling of heteroaryl sulfinates with Grignard reagents. Org Lett 21(4):937–940

Wu W, Teng Q, Chua Y-Y, Huynh HV, Duong HA (2017) Iron-catalyzed cross-coupling reactions of arylmagnesium reagents with aryl chlorides and tosylates: influence of ligand structural parameters and identification of a general N-heterocyclic Carbene ligand. Organometallics 36 (12):2293–2297

Xue W, Shishido R, Oestreich M (2018) Bench-stable stock solutions of silicon Grignard reagents: application to iron- and cobalt-catalyzed radical C(sp^3)-Si cross-coupling reactions. Angew Chem Int Ed 57(37):12141–12145

Zeng H, Dou Q, Li C-J (2019) Photoinduced transition-metal-free cross-coupling of aryl halides with H-phosphonates. Org Lett 21(5):1301–1305

Zheng Y-L, Newman SG (2020) Cross-coupling reactions with esters, aldehydes and alcohols. Chem Commun 57:2591–2604

Flow Chemistry: Advantages and Strategic Uptake

Baumann M, Baxendale IR (2015) The syntheses of active pharmaceutical ingredients (APIs) using continuous flow chemistry. Beilstein J Org Chem 11:1194–1219

Baumann M, Moody TS, Smyth M, Wharry S (2020) A perspective on continuous flow chemistry in the pharmaceutical industry. Org Process Res Dev 24(10):1802–1813

Bogdan AR, Dombrowski AW (2019) Emerging trends in flow chemistry and applications to the pharmaceutical industry. J Med Chem 62 (14):6422–6468

Britton J, Majumdar S, Weiss GA (2018) Continuous flow biocatalysis. Chem Soc Rev 47:5891–5918

Britton J, Raston CL (2017) Multi-step continuous-flow synthesis. Chem Soc Rev 46:1250–1271

Brzozowski M, O'Brien M, Ley SV, Polyzos A (2015) Flow chemistry: intelligent processing of gas-liquid transformations using a tube-in-tube reactor. Acc Chem Res 48(2):349–362

Coley CW, Thomas DA III, Lummiss JAM, Jaworski JN, Breen CP, Schultz V, Hart T, Fishman JS, Rogers L, Gao H, Hicklin RW, Plehiers PP, Byington J, Piotti JS, Green WJ, Hart AJ, Jamison TF, Jensen KF (2019) A robotic platform for flow synthesis of organic compounds informed by AI planning. Science 365(6453). https://doi.org/10.1126/science.aax.1566

Cossar PJ, Hizartzidis L, Simone MI, McCluskey A, Gordon CP (2015) The expanding utility of continuous flow hydrogenation. Org Biomol Chem 13:7119–7130

Dallinger D, Kappe CO (2017) Why flow means green—evaluating the merits of continuous processing in the context of sustainability. Curr Opin Green Sustainable Chem 7:6–12

Fabry DC, Sugiono E, Rueping M (2016) Online monitoring and analysis for autonomous continuous flow self-optimizing reactor systems. React Chem Eng 1:129–133

Fanelli F, Parisi G, Degenarro L, Luisi R (2017) Contribution of microreactor technology and flow chemistry to the development of green and sustainable synthesis. Beilstein J Org Chem 13:520–542

Gemoets HPL, Su Y, Shang M, Hessel V, Luque R, Noël T (2016) Liquid phase oxidation chemistry in continuous-flow microreactors. Chem Soc Rev 45:83–117

Gérardy R, Emmanuel N, Toupy T, Kassin V-E, Tshinalonza NN, Schmitz M, Monbaliu J-CM (2018) Continuous flow organic chemistry: success and pitfalls at the interface with current societal challenges. Eur J Org Chem 2018(20–21):2301–2351

Gioiello A, Piccinno A, Lozza AM, Cerra B (2020) Medicinal chemistry in the Era of machines and automation: recent advances in continuous flow technology. J Med Chem 63:6624–6647

Guidi M, Seeberger PH, Gilmore K (2020) How to approach flow chemistry. Chem Soc Rev 49:8910–8932

Harper KC, Moschetta EG, Bordawekar SV, Wittenberger SJ (2019) A laser driven flow chemistry platform for scaling photochemical reactions with visible light. ACS Cent Sci 5(1):109–115

Hartman RL, McMullen JP, Jensen KF (2011) Deciding whether to go with the flow: evaluating the merits of flow reactors for synthesis. Angew Chem Int Ed 50(33):7502–7519

Hone CA, Roberge DM, Kappe CO (2017) The use of molecular oxygen in pharmaceutical manufacturing: is flow the way to go? ChemSusChem 10 (1):32–41

Jensen KF (2017) Flow chemistry—microreaction technology comes of age. AIChE 63(3):858–869

Lummiss JAM, Morse PD, Beingessner RL, Jamison TF (2017) Towards more efficient, greener syntheses through flow chemistry. Chem Res 17 (7):667–680

McMullen JP, Jensen KF (2010) Integrated microreactors for reaction automation: new approaches to reaction development. Ann Rev Anal Chem 3:19–42

Movsisyan M, Delbeke EIP, Berton JKET, Battilochhio C, Ley SV, Stevens CV (2016) Taming hazardous chemistry by continuous flow technology. Chem Soc Rev 45:4892–4928

Müller STR, Wirth T (2015) Diazo compounds in continuous-flow technology. ChemSusChem 8(2):245–250

Noël T, Buchwald SL (2011) Cross-coupling in flow. Chem Soc Rev 40:5010–5029

Noël T, Su Y, Hessel V (2015) Beyond organometallic flow chemistry: the principles behind the use of continuous-flow reactors for synthesis. Topics Organometal Chem 57:1–41

Pastre JC, Browne DL, Ley SV (2013) Flow chemistry syntheses of natural products. Chem Soc Rev 42:8849–8869

Pletcher D, Green RA, Brown RCD (2018) Flow electrolysis cells for the synthetic organic chemistry laboratory. Chem Rev 118(9):4573–4591

Plutschack MB, Pieber B, Gilmore K, Seeberger PH (2017) The Hitchhiker's guide to flow chemistry. Chem Rev 117(18):11796–11893

Politano F, Oksdath-Mansilla G (2018) Light on the horizon: current research and future perspectives in flow photochemistry. Org Process Res Dev 22 (9):1045–1062

Porta R, Benaglia M, Puglisi A (2016) Flow chemistry: recent developments in the synthesis of pharmaceutical products. Org Process Res Dev 20 (1):20–25

Puglisi A, Rossi A (2021) Stereoselective organocatalysis and flow chemistry. Phys Sci Rev. https://doi.org/10.1515/psr-2018-0099

Rogers L, Jensen KF (2019) Continuous manufacturing—the Green chemistry promise? Green Chem 21:3481–3498

Rossetti I, Comppagnoni M (2016) Chemical reaction engineering, process design and scale-up issues at the frontier of synthesis: flow chemistry. Chem Eng J 296:56–70

Sambiagio C, Noél T (2020) Flow photochemistry: shine some light on those tubes! Trends Chem 2(2):92–106

Sans V, Cronin L (2016) Towards dial-a-molecule by integrating continuous flow, analytics and self-optimisation. Chem Soc Rev 45:2032–2043

Schweidtmann AM, Clayton AD, Holmes N, Bradford E, Bourne RA, Lapkin AA (2018) Machine learning meets continuous flow chemistry: automated optimization towards the Pareto front of multiple objectives. Chem Eng J 352:277–282

Thompson MP, Peñafiel I, Cosgrove SC, Turner NJ (2019) Biocatalysis using immobilized enzymes in continuous flow for the synthesis of fine chemicals. Org Process Res Dev 23(1):9–18

Tran DN, Battilochhio C, Lou S-B, Hawkins JM, Ley SV (2015) Flow chemistry as a discovery tool to access sp^2-sp^3 cross-coupling reactions *via* diazo compounds. Chem Sci 6:1120–1125

Webb D, Jamison TF (2010) Continuous flow multi-step organic synthesis. Chem Sci 1:675–680

Wegner J, Ceylan S, Kirschning A (2011) Ten key issues in modern flow chemistry. Chem Commun 47:4583–4592

Wegner J, Ceylan S, Kirschning A (2012) Flow chemistry—a key enabling technology for (multistep) organic synthesis. Adv Synth Catal 354 (1):17–57

Reaction Optimization: Solvents and Reagents Choice

Blincoe WD, Lin S, Dreher SD, Sheng H (2020) Practical guide on MALDI-TOF MS method development for high throughput profiling of pharmaceutically relevant, small molecule chemical reactions. Tetrahedron 76 (36):131434

Buitrago Santanilla A, Regalado EL, Pereira T, Shevlin M, Bateman K, Campeau L-C, Schneeweis J, Berritt S, Shi Z-C, Natermet P, Liu Y, Helmy R, Welch CJ, Vachal P, Davies IW, Cernak T, Dreher SD (2015) Nanomole-scale high-throughput chemistry for the synthesis of complex molecules. Science 347:49–53

Grainger R, Whibley S (2021) A perspective on the analytical challenges encountered in high-throughput experimentation. Org Process Res Dev 25(3):354–364. https://doi.org/10.1021/acs.oprd.0c00463

Jaman Z, Logsdon DM, Szilágyi B, Sobreira TJP, Avramova L, Cooks RG, Thompson DH (2019) High throughput and continuous flow evaluation of nucleophilic aromatic substitution reactions. ACS Comb Sci 22 (4):184–196

Krska SW, DiRocco DA, Dreher SD, Shevlin M (2017) The evolution of chemical high-throughput experimentation to address challenging problems in pharmaceutical synthesis. Acc Chem Res 50(12):2976–2985

Loren BP, Ewan HS, Avramova L, Ferreira CR, Sobreira TJP, Yammine K, Liao H, Cooks G, Thompson DH (2019) High throughput experimentation using DESI-MS to guide continuous-flow synthesis. Sci Rep 9:14745

McCullough K, Williams T, Mingle K, Jamshidi P, Lauterbach J (2020) High-throughput experimentation meets artificial intelligence: a new pathway to catalyst discovery. Phys Chem Chem Phys 22:11174–11196

Mennen SM, Alhambra C, Allen CL, Barberis M, Berritt S, Brandt TA, Campbell AD, Castañón J, Cherney AH, Christensen M, Damon DB, de Diego E, Garcia-Cerrada S, Garcia-Losada P, Haro R, Janey J, Leitch DC, Li L, Liu F, Lobben PC, MacMillan DWC, Magano J, McInturff E, Monfette S, Post RJ, Schultz D, Sitter BJ, Stevens JM, Strambeanu II, Twilton J, Wang K, Zajac MA (2019) The evolution of high-throughput experimentation in pharmaceutical development and perspectives on the future. Org Process Res Dev 23(6):1213–1242

Perera D, Tucker JW, Brahmbhatt S, Helal CJ, Chong A, Farrell W, Richardson P, Sach NW (2018) A platform for automated nanomole-scale reaction screening and micromole-scale synthesis in flow. Science 359 (6374):429–434

Sawicki JW, Bogdan AR, Searle PA, Talaty N, Djuric SW (2019) Rapid analytical characterization of high-throughput chemistry screens utilizing desorption electrospray ionization mass spectrometry. React Chem Eng 4 (9):1589–1594

Shevlin M (2017) Practical high-throughput experimentation for chemists. ACS Med Chem Lett 8(6):601–607

Welch CJ (2019) High throughput analysis enables high throughput experimentation in pharmaceutical process research. React Chem Eng 4:1895–1911

Radiopharmaceutical Discovery with $^{11}CO_2$-Fixation Methods Inspired by Green Chemistry

Ahamed M, Verbeek J, Funke U, Lecina J, Verbruggen A, Bormans G (2016) Recent progress in metal catalyzed direct carboxylation of aryl halides and pseudo halides employing CO_2: opportunities for ^{11}C radiochemistry. ChemCatChem 8(24):3692–3700

Artz J, Müller TE, Thenert K, Kleinekorte J, Meys R, Sternberg A, Bardow A, Leitner W (2018) Sustainable conversion of carbon dioxide: an integrated review of catalysis and life cycle assessment. Chem Rev 118(2):434–504

Chen Z, Chen J, Mast N, Rong J, Deng X, Shao T, Fu H, Yu Q, Sun J, Shao Y, Josephson L, Collier TL, Pikuleva I, Liang SH (2020) Synthesis and pharmacokinetic study of a $[^{11}C]CO_2$-labeled cholesterol 24-hydroxylase inhibitor using "in-loop" $[^{11}C]CO_2$ fixation method. Bioorg Med Chem Lett 30(9):127068

Dahl K, Collier TL, Cheng R, Zhang X, Sadovski O, Liang SH, Vasdev N (2018) "In-loop" $[^{11}C]CO_2$ fixation: prototype and proof of concept. J Label Comp Radiopharm 61(3):252–262

Dahl K, Halldin C, Schou M (2017) New methodologies for the preparation of carbon-11 labeled radiopharmaceuticals. Clinc Transl Imaging 5:275–289

Deng X, Rong J, Wang L, Vasdev N, Zhang L, Josephson L, Liang SH (2019) Chemistry for positron emission tomography: recent advances in ^{11}C-, ^{18}F-, ^{13}N-, and ^{15}O-labeling reactions. Angew Chem Int Ed 58(9):2580–2605

Dornan MH, Petrenyov D, Simard J-M, Boudjemeline M, Mititelu R, DaSilva JN, Belanger AP (2020) Synthesis of a ^{11}C-isotopologues of the B-Raf-selective inhibitor encorafenib using in-loop $[^{11}C]CO_2$ fixation. ACS Omega 5(33):20960–20966

Downey J, Bongarzone S, Hader S, Gee AD (2018) In-loop flow $[^{11}C]CO_2$ fixation and radiosynthesis of N,N'-$[^{11}C]$dibenzylurea. J Label Comp Radiopharm 61(3):263–271

Fukumura T, Mori W, Ogawa M, Fujinaga M, Zhang M-R (2021) The $[^{11}C]$ phosgene: synthesis and application for development of PET radiotracers. Nucl Med Biol 92:138–148

Gaudeau M, Zhang M, Tatoulian M, Lescot C, Ognier S (2020) Fast carbonylation reaction from CO_2 using plasma gas/liquid microreactors for radiolabeling applications. React Chem Eng 5:1981–1991

Gunnar A (2015) Development of carbon-11 labelled PET tracers-radiochemical and technological challenges in a historical perspective. J Label Comp Radiopharm 58(3):65–72

Hicks JW, Parkes J, Tong J, Houle S, Vasdev N, Wilson AA (2014) Radiosynthesis and ex vivo evaluation of $[^{11}C$-carbonyl]carbamate- and urea-based monoacylglycerol lipase inhibitors. Nucl Med Biol 41(8):688–694

Ismailani US, Munch M, Mair BA, Rotstein BH (2020) Interrupted aza-Wittig reactions using iminophosphoranes to synthesize 11C-carbonyls. ChemRxiv. https://doi.org/10.26434/chemrxiv.13506702.v1. Accessed 16 Mar 2021

Jakobsson JE, Lu S, Telu S, Pike VW (2020) $[^{11}C]$Carbonyl difluoride—a new and highly efficient $[^{11}C]$Carbonyl group transfer agent. Angew Chem Int Ed 59(18):7256–7260

Kong D, Munch M, Qiqige Q, Cooze C, Rotstein BH, Lundgren R (2020) Fast carbon isotope exchange of carboxylic acids enabled by organic photoredox catalysis. ChemRxiv. https://doi.org/10.26434/chemrxiv.13111505.v1. Accessed 16 Mar 2021

Li L, Shao X, Cole EL, Ohnmacht SA, Ferrari V, Hong YT, Williamson DJ, Fryer TD, Quesada CA, Sherman P, Riss PJ, Scott PJH, Aigbirhio FI (2015) Synthesis and initial in vivo studies of [^{11}C]SB-216763: the first radiolabeled brain penetrative inhibitor of GSK-3. ACS Med Chem Lett 6 (5):548–552

Liang SH, Vasdev N (2015) Total radiosynthesis: thinking outside "the Box". Aust J Chem 68(9):1319–1328

Liger F, Eijsbouts T, Cadarossanesaib F, Tourvieille C, Le Bars D, Billard T (2015) Direct [^{11}C]Methylation of amines from [^{11}C]CO_2 for the synthesis of PET radiotracers. Eur J Org Chem 2015(29):6434–6438

Mossine AV, Brooks AF, Jackson IM, Quesada CA, Sherman P, Cole EL, Donnelly DJ, Scott PJH, Shao X (2016) Synthesis of diverse ^{11}C-labeled PET radiotracers via direct incorporation of [^{11}C]CO_2. Bioconjug Chem 27(5):1382–1389

Munch M, Ismailani U, Karunakaran G, Mair B, Rotstein B (2020) Iminophosphoranes as radiolabeling precursors using [^{11}C]CO_2-fixation. J Nucl Med 61(supplement 1):1103

Pekosak A, Bulc JZ, Korat S, Schuit RC, Kooijam E, Vos R, Rongen M, Verlaan M, Takkenkamp K, Beaino W, Poot AJ, Windhorst AD (2018) Synthesis and preclinical evaluation of the first carbon-11 labeled PET tracers targeting substance P1-7. Mol Pharm 15(11):4872–4883

Pichler V, Berroteran-Infante N, Philippe C, Vraka C, Klebermass E-M, Balber T, Pfaff S, Nics L, Mitterhauser M, Wadsak W (2018) An overview of PET radiochemistry, part 1: the covalent labels ^{18}F, ^{11}C, and ^{13}N. J Nucl Med 59(9):1350–1354

Qu W, Hu B, Babich JW, Waterhouse N, Dooley M, Ponnala S, Urgile J (2020) A general ^{11}C-labeling approach enabled by fluoride-mediated desilylation of organosilanes. Nat Commun 11:1736. https://doi.org/10.1038/s41467-020-15556-7

Rotstein BH, Liang SH, Placzek MS, Hooker JM, Gee AD, Dolle F, Wilson AA, Vasdev N (2016) ^{11}CO bonds made easily for positron emission tomography radiopharmaceuticals. Chem Soc Rev 45(17):4708–4726

Yang L, Scott PJH, Shao X (2017) [11C] Carbon dioxide: starting point for labeling PET radiopharmaceuticals. Carbon dioxide chemistry, capture and oil recovery. IntechOpen. https://doi.org/10.5772/intechopen.72313. Accessed 16 Mar 2021

Index

A

Paul F. Richardson (ed.), *Green Chemistry in Drug Discovery: From Academia to Industry*, Methods in Pharmacology and Toxicology, https://doi.org/10.1007/978-1-0716-1579-9, © Springer Science+Business Media, LLC, part of Springer Nature 2021

D

Index

I

P

U

V

W

Z